Chaos Synchronization and Cryptography for Secure Communications:

Applications for Encryption

Santo Banerjee
Politecnico di Torino, Italy

A volume in the Advances in Information Security, Privacy, and Ethics (AISPE) Book Series

Director of Editorial Content:	Kristin Klinger
Director of Book Publications:	Julia Mosemann
Acquisitions Editor:	Lindsay Johnston
Development Editor:	Christine Bufton
Publishing Assistant:	Casey Conapitski
Typesetter:	Deanna Jo Zombro
Production Editor:	Jamie Snavely
Cover Design:	Lisa Tosheff

Published in the United States of America by
Information Science Reference (an imprint of IGI Global)
701 E. Chocolate Avenue
Hershey PA 17033
Tel: 717-533-8845
Fax: 717-533-8661
E-mail: cust@igi-global.com
Web site: http://www.igi-global.com

Library of Congress Cataloging-in-Publication Data

Chaos synchronization and cryptography for secure communications : applications for encryption / Santo Banerjee, editor.
p. cm.
Includes bibliographical references and index. Summary: "This book explores the combination of ordinary and time delayed systems and their applications in cryptographic encoding"--Provided by publisher. ISBN 978-1-61520-737-4 (hardcover) -- ISBN 978-1-61520-738-1 (ebook) 1. Telecommunication--Security measures. 2. Chaotic behavior in systems. 3. Cryptography. 4. Data encryption (Computer science) 5. Synchronization.
I. Banerjee, Santo, 1976-
TK5102.94.C523 2010
621.382--dc22
2010015993

This book is published in the IGI Global book series Advances in Information Security, Privacy, and Ethics (AISPE) Book Series (ISSN: 1948-9730; eISSN: 1948-9749)

British Cataloguing in Publication Data
A Cataloguing in Publication record for this book is available from the British Library.

Advances in Information Security, Privacy, and Ethics (AISPE) Book Series

ISSN: 1948-9730
EISSN: 1948-9749

Mission

In the digital age, when everything from municipal power grids to individual mobile telephone locations is all available in electronic form, the implications and protection of this data has never been more important and controversial. As digital technologies become more pervasive in everyday life and the Internet is utilized in ever increasing ways by both private and public entities, the need for more research on securing, regulating, and understanding these areas is growing.

The **Advances in Information Security, Privacy, & Ethics (AISPE) Book Series** is the source for this research, as the series provides only the most cutting-edge research on how information is utilized in the digital age.

Coverage

- Access Control
- Device Fingerprinting
- Global Privacy Concerns
- Information Security Standards
- Network Security Services
- Privacy-Enhancing Technologies
- Risk Management
- Security Information Management
- Technoethics
- Tracking Cookies

IGI Global is currently accepting manuscripts for publication within this series. To submit a proposal for a volume in this series, please contact our Acquisition Editors at Acquisitions@igi-global.com or visit: http://www.igi-global.com/publish/.

The Advances in Information Security, Privacy, and Ethics (AISPE) Book Series (ISSN 1948-9730) is published by IGI Global, 701 E. Chocolate Avenue, Hershey, PA 17033-1240, USA, www.igi-global.com. This series is composed of titles available for purchase individually; each title is edited to be contextually exclusive from any other title within the series. For pricing and ordering information please visit http://www.igi-global.com/book-series/advances-information-security-privacy-ethics/37157. Postmaster: Send all address changes to above address.

Titles in this Series

For a list of additional titles in this series, please visit: www.igi-global.com

Theory and Practice of Cryptography Solutions for Secure Information Systems
Atilla Elçi (Aksaray University, Turkey) Josef Pieprzyk (Macquarie University, Australia) Alexander G. Chefranov (Eastern Mediterranean University, North Cyprus) Mehmet A. Orgun (Macquarie University, Australia) Huaxiong Wang (Nanyang Technological University, Singapore) and Rajan Shankaran (Macquarie University, Australia)
Information Science Reference • copyright 2013 • 351pp • H/C (ISBN: 9781466640306) • US $195.00 (our price)

IT Security Governance Innovations Theory and Research
Daniel Mellado (Spanish Tax Agency, Spain) Luis Enrique Sánchez (University of Castilla-La Mancha, Spain) Eduardo Fernández-Medina (University of Castilla – La Mancha, Spain) and Mario Piattini (University of Castilla - La Mancha, Spain)
Information Science Reference • copyright 2013 • 390pp • H/C (ISBN: 9781466620834) • US $195.00 (our price)

Threats, Countermeasures, and Advances in Applied Information Security
Manish Gupta (State University of New York at Buffalo, USA) John Walp (M&T Bank Corporation, USA) and Raj Sharman (State University of New York, USA)
Information Science Reference • copyright 2012 • 319pp • H/C (ISBN: 9781466609785) • US $195.00 (our price)

Investigating Cyber Law and Cyber Ethics Issues, Impacts and Practices
Alfreda Dudley (Towson University, USA) James Braman (Towson University, USA) and Giovanni Vincenti (Towson University, USA)
Information Science Reference • copyright 2012 • 342pp • H/C (ISBN: 9781613501320) • US $195.00 (our price)

Information Assurance and Security Ethics in Complex Systems Interdisciplinary Perspectives
Melissa Jane Dark (Purdue University, USA)
Information Science Reference • copyright 2011 • 306pp • H/C (ISBN: 9781616922450) • US $180.00 (our price)

Chaos Synchronization and Cryptography for Secure Communications Applications for Encryption
Santo Banerjee (Politecnico di Torino, Italy)
Information Science Reference • copyright 2011 • 596pp • H/C (ISBN: 9781615207374) • US $180.00 (our price)

Technoethics and the Evolving Knowledge Society Ethical Issues in Technological Design, Research, Development, and Innovation
Rocci Luppicini (University of Ottawa, Canada)
Information Science Reference • copyright 2010 • 322pp • H/C (ISBN: 9781605669526) • US $180.00 (our price)

www.igi-global.com

701 E. Chocolate Ave., Hershey, PA 17033
Order online at www.igi-global.com or call 717-533-8845 x100
To place a standing order for titles released in this series, contact: cust@igi-global.com
Mon-Fri 8:00 am - 5:00 pm (est) or fax 24 hours a day 717-533-8661

Editorial Advisory Board

List of Reviewers

Table of Contents

Foreword xviii

Preface xx

Acknowledgment xxv

Section 1
Chaotic Dynamical Systems, Theory and Experiments

Chapter 1
Dynamical Systems and Their Chaotic Properties 1
Konstantinos Chlouverakis, National and Kapodistrian University of Athens, Greece

Chapter 2
Chaotic Dynamical Systems Associated with Tilings of R^N 19
Lionel Rosier, Institut Elie Cartan de Nancy, France

Chapter 3
Identification and State Observation of Uncertain Chaotic Systems Using Projectional Differential Neural Networks 42
Alejandro García, National Polytechnic Institute of Mexico, Mexico
Isaac Chairez, National Polytechnic Institute of Mexico, Mexico
Alexander Poznyak, National Polytechnic Institute of Mexico, Mexico

Chapter 4
Simple Chaotic Electronic Circuits 68
M.P. Hanias, Technological and Educational Institute of Chalkis, Greece
G. S. Tombras, National and Kapodistrian University of Athens, Greece

Chapter 5
Experimental Evidences of Shil'nikov Chaos and Mixed-Mode Oscillation in Chua Circuit 91
Syamal Kumar Dana, Indian Institute of Chemical Biology, India
Satyabrata Chakraborty, Indian Institute of Chemical Biology, India

Section 2
Synchronization of Chaotic Systems

Chapter 6
Synchronization of Chaotic Oscillators 105
J. M. González-Miranda, Universidad de Barcelona, Spain

Chapter 7
Synchronization in Integer and Fractional Order Chaotic Systems 127
Ahmed E. Matouk, Mansoura University, Egypt & Hail University, Saudi Arabia

Chapter 8
Chaos Synchronization 152
Hassan Salarieh, Sharif University of Technology, Iran
Mohammad Shahrokhi, Sharif University of Technology, Iran

Chapter 9
Chaotic Gyros Synchronization 183
Mehdi Roopaei, Islamic Azad University, Iran
Mansoor J. Zolghadri, Shiraz University, Iran
Bijan Ranjbar Sahraei, Shiraz University, Iran
Seyyed Hossein Mousavi, Shiraz University, Iran
Hassan Adloo, Shiraz University, Iran
Behnam Zare, Tarbiat Modarres University, Iran
Tsung-Chih Lin, Feng Chia University, Taiwan

Chapter 10
Importance of Chaos Synchronization on Technology and Science 210
Ricardo Aguilar-López, CINVESTAV-IPN, Mexico
Ricardo Femat, IPICYT, México
Rafael Martínez-Guerra, CINVESTAV-IPN, Mexico

Chapter 11
Synchronization of Oscillators 247
Jean B. Chabi Orou, Université d'Abomey-Calavi, Benin

Chapter 12
Synchronization of Uncertain Neural Networks with H_∞ Performance and Mixed Time-Delays 261
Hamid Reza Karimi, University of Agder, Norway

Chapter 13
Adaptive Synchronization in Unknown Stochastic Chaotic Neural Networks with Mixed Time-Varying Delays 289
Jian-an Fang, Donghua University, China
Yang Tang, Donghua University, China

Chapter 14
Type-2 Fuzzy Sliding Mode Synchronization 314
Tsung-Chih Lin, Feng-Chia University, Taiwan
Ming-Che Chen, Feng-Chia University, Taiwan
Mehdi Roopaei, Islamic Azad University, Iran

Section 3
Cryptographic Applications

Chapter 15
Secure Transmission of Analog Information Using Chaos.. 337
A.S. Dmitriev, Inst. of Radio Eng. & Electr. of Russian Academy of Sciences, Russia
E.V. Efremova, Inst. of Radio Eng. & Electr. of Russian Academy of Sciences, Russia
L.V. Kuzmin, Inst. of Radio Eng. & Electr. of Russian Academy of Sciences, Russia
A.N. Miliou, Aristotle University of Thessaloniki, Greece
A.I. Panas, Inst. of Radio Eng. & Electr. of Russian Academy of Sciences, Russia
S.O. Starkov, Inst. of Radio Eng. & Electr. of Russian Academy of Sciences, Russia

Chapter 16
Control-Theoretical Concepts in the Design of Symmetric Cryptosystems..................................... 361
Gilles Millérioux, University Henri Poincaré of Nancy, France & Research Center on Automatic Control of Nancy (CRAN), France
José María Amigó, University Miguel Hernandez of Elche (UMH), Spain & Centro de Investigación Operativa, Spain

Chapter 17
Unmasking Optical Chaotic Cryptosystems Based on Delayed Optoelectronic Feedback............... 386
Silvia Ortín, Instituto de Física de Cantabria (CSIC-Universidad de Cantabria), Spain
Luis Pesquera, Instituto de Física de Cantabria (CSIC-Universidad de Cantabria), Spain

Chapter 18
Encryption of Analog and Digital Signals through Synchronized Chaotic Systems.......................... 415
Kehui Sun, Central South University, China

Chapter 19
Digital Information Transmission Using Discrete Chaotic Signal 439
A.N. Anagnostopoulos, Aristotle University of Thessaloniki, Greece
A.N. Miliou, Aristotle University of Thessaloniki, Greece
S.G. Stavrinides, Aristotle University of Thessaloniki, Greece
A. S. Dmitriev, Inst. of Radio Eng. & Electr. of Russian Academy of Sciences, Russia
E. V. Efremova, Inst. of Radio Eng. & Electr. of Russian Academy of Sciences, Russia

Chapter 20
Mathematical Treatment for Constructing a Countermeasure Against the One-Time Pad Attack on the Baptista Type Cryptosystem 463
M.R.K. Ariffin, Universiti Putra Malaysia, Malaysia
M.S.M.Noorani, Universiti Kebangsaan Malaysia, Malaysia

Chapter 21
Chaos Synchronization with Genetic Engineering Algorithm for Secure Communications 476
Sumona Mukhopadhyay, Army Institute of Management, India
Mala Mitra, Camellia School of Engineering, India
Santo Banerjee, Politecnico di Torino, Italy & Techfab, Italy

Compilation of References 510

About the Contributors 554

Index 563

Detailed Table of Contents

Foreword xviii

Preface xx

Acknowledgment xxv

Section 1
Chaotic Dynamical Systems, Theory and Experiments

Chapter 1
Dynamical Systems and Their Chaotic Properties 1
Konstantinos Chlouverakis, National and Kapodistrian University of Athens, Greece

The chapter presented by Konstantinos Chlouverakis aims to bring together intellectuals to understand chaotic systems and their properties by describing the fundamental concepts of continuous-time chaotic dynamical systems thereby opening the gateway into the cryptic world of secure communications and chaotic dynamics.

Chapter 2
Chaotic Dynamical Systems Associated with Tilings of R^N 19
Lionel Rosier, Institut Elie Cartan de Nancy, France

Lionel Rosier presents a mechanism for chaos synchronization which is applied in masking information. For this, the author has chosen a class of discrete dynamical systems defined on the homogeneous space associated with a regular tiling of R^N because any dynamical system in this class is chaotic in the sense of Devaney, and that it admits at least one positive Lyapunov exponent. A common example of such a system is the N−dimensional torus T^N.

Chapter 3
Identification and State Observation of Uncertain Chaotic Systems Using
Projectional Differential Neural Networks 42
Alejandro García, National Polytechnic Institute of Mexico, Mexico
Isaac Chairez, National Polytechnic Institute of Mexico, Mexico
Alexander Poznyak, National Polytechnic Institute of Mexico, Mexico

Alejandro García, Isaac Chairez and Alexander Poznyak presents an application of the DNN structures to identify chaotic complex systems as well as to carry out its state observation. They have applied Lyapunov stability theory for the convergente of the chaotic complex system. The effectiveness of the novel DNN topology is shown by a classical numerical example of Chua's circuit.

Chapter 4
Simple Chaotic Electronic Circuits..68
M.P. Hanias, Technological and Educational Institute of Chalkis, Greece
G. S. Tombras, National and Kapodistrian University of Athens, Greece

M.P. Hanias and G. S. Tombras considers that simple chaotic electronics circuits in the form of diode resonator circuits, Resistor-Inductor-LED optoelectronic chaotic circuits and Single Transistor chaotic circuits can be used as transmitters and receivers for chaotic cryptosystems. These circuits are flexible to the changes in their circuit parameters and so help in investigating the influence of such changes in complexity of the generated strange attractors. The authors have performed a time series analysis based on Grassberger and Procaccia's method. The invariant parameters such as correlation, and minimum embedding dimension are respectively calculated along with the Kolmogorov entropy. The chapter also examines the RLT circuits when they are in a critical state.

Chapter 5
Experimental Evidences of Shil'nikov Chaos and Mixed-Mode Oscillation in Chua Circuit............91
Syamal Kumar Dana, Indian Institute of Chemical Biology, India
Satyabrata Chakraborty, Indian Institute of Chemical Biology, India

Syamal Kumar Dana and Satyabrata Chakraborty in their chapter have provided the evidence of Shil'nikov type homoclinic chaos and mixed mode oscillations in asymmetry-induced Chua's oscillator through their experiment. In the homoclinic bifurcations, the asymmetry has a vital role. They have subjected the circuit to a forced DC voltage to realize the asymmetry. Their observations reveal transition from large amplitude to homoclinic chaos through a sequence of mixed-mode oscillations interspersed by chaotic states by tuning a control parameter.

Section 2
Synchronization of Chaotic Systems

Chapter 6
Synchronization of Chaotic Oscillators 105
J. M. González-Miranda, Universidad de Barcelona, Spain

J. M. González-Miranda presents a review about the varied forms which the mutually coupled or unidirectionally driven dynamic chaotic oscillators display on synchronization. This will enlighten the readers on the key role which physics and mathematics play in laying the foundation useful for reaping the benefits of chaos synchronization in telecommunications and cryptography.

Chapter 7
Synchronization in Integer and Fractional Order Chaotic Systems 127
Ahmed E. Matouk, Mansoura University, Egypt & Hail University, Saudi Arabia

Ahmed E. Matouk's chapter concentrates on the application of some of the basic synchronization methods to integer and fractional order chaotic systems. Lyapunov stabily theory are also used to control the synchronized error. Numerical examples well support the analytical results.

Chapter 8
Chaos Synchronization 152
Hassan Salarieh, Sharif University of Technology, Iran
Mohammad Shahrokhi, Sharif University of Technology, Iran

Chaos synchronization plays a vital role in various message encryption methods for which Hassan Salarieh and Mohammad Shahrokhi necessitates an overview of the concepts of chaotic system so as to implement its properties in secure communication. The chapter begins by defining complete, lag, phase and generalized synchronization approaches followed by the application of control theory for synchronization of different chaotic systems. They have presented some synchronization algorithms based on different control techniques. They have shown ways to modify the controlling methods in order to cope with parameter uncertainties and measurement noise. The performance of the discussed methods are supported by simulation of several chaotic systems.

Chapter 9
Chaotic Gyros Synchronization 183
Mehdi Roopaei, Islamic Azad University, Iran
Mansoor J. Zolghadri, Shiraz University, Iran
Bijan Ranjbar Sahraei, Shiraz University, Iran
Seyyed Hossein Mousavi, Shiraz University, Iran
Hassan Adloo, Shiraz University, Iran
Behnam Zare, Tarbiat Modarres University, Iran
Tsung-Chih Lin, Feng Chia University, Taiwan

In this chapter, M. Roopaei, M. J. Zolghadri, B. S. Ranjbar, S. H. Mousavi, H. Adloo, B. Zare and T.C. Lin have presented three methods for synchronizing of two chaotic gyros in the presence of uncertainties, external disturbances and dead-zone nonlinearity. In the first method, there is dead-zone nonlinearity in the control input, which limits the performance of accurate control methods. The effects of this nonlinearity have been attenuated using a fuzzy parameters adaptation integrated with sliding mode control method. For the second method, they have proposed a robust adaptive fuzzy sliding mode control scheme to overcome the synchronization problem for a class of unknown nonlinear chaotic gyro. The third method considers two different gyro systems. For this method they also have proposed a fuzzy controller to eliminate chattering phenomena during the reaching phase of sliding mode control. All the mentioned methods are simulated and the results illustrate the effectiveness of the proposed methods.

Chapter 10

Importance of Chaos Synchronization on Technology and Science 210
Ricardo Aguilar-López, CINVESTAV-IPN, Mexico
Ricardo Femat, IPICYT, México
Rafael Martínez-Guerra, CINVESTAV-IPN, Mexico

In this chapter, Ricardo Aguilar-López, Ricardo Femat and Rafael Martínez-Guerra highlight the importance of chaos synchronization on technology and science. The chapter introduces their attempt in three sections. The first sections deals with the topic of synchronized state in the sense of identical synchronization. This is realized with the robust nonlinear observer design, considering corrupted measurements and model uncertainties, coupling uncertainty estimators with nonlinear state observers. In the second part of their chapter, the authors have addressed applications to chaos communications primarily an application of chaos theory which is aimed to provide security in the transmission of information performed through telecommunications technologies mainly dealing with the transmitter. A message is added on to a chaotic signal and then, the message is masked in the chaotic signal. When the chaotic signal carries such a message it is called the chaotic carrier. All this is done via control theory and is a particular case of chaos synchronization. In the last section they have discussed an application to synchronization of biological systems. They have investigated the intercellular communication of Ca^{2+} waves and the mechanism in which they coordinate a multicellular response to a local event. Recently, it has been observed in a variety of systems that calcium signals can also propagate from one cell to another and thereby serve as a means of intercellular communication. Their attempt of introducing the feedback control laws in some biological systems is influenced by the behavior of biophysical mechanisms of cellular dynamics. The authors also present links between feedback control schemes, with an external input, and intracellular calcium functions for coordination and control.

Chapter 11

Synchronization of Oscillators.. 247
Jean B. Chabi Orou, Université d'Abomey-Calavi, Benin

Jean B. Chabi Orou describes the synchronization of oscillators in a lucid manner which can facilitate to get a grasp of the main concepts before delving into the different kinds of synchronization configurations, Chaotic synchronization and also addressed the issue of stability of the synchronization. They have briefly mentioned the influence of noise on the synchronization process.

Chapter 12

Synchronization of Uncertain Neural Networks with H_∞ Performance and Mixed Time-Delays...... 261
Hamid Reza Karimi, University of Agder, Norway

Hamid Reza Karimi introduces an exponential synchronization method for a class of uncertain master and slave neural networks with mixed time-delays. The mixed delays comprise different neutral, discrete and distributed time-delays. In order to design a delayed state-feedback control as a synchronization law in terms of linear matrix inequalities under less restrictive conditions, some delay-dependent sufficient conditions have been established by an appropriate discretized Lyapunov-Krasovskii functional and some free weighting matrices. Irrespective of their initial state, this controller will guarantee the exponential synchronization of the two coupled master and slave neural networks. The author has demonstrated the effectiveness of the established synchronization laws by suitable numerical simulations.

Chapter 13
Adaptive Synchronization in Unknown Stochastic Chaotic Neural Networks with Mixed Time-Varying Delays 289
Jian-an Fang, Donghua University, China
Yang Tang, Donghua University, China

Jian-an Fang and Yang Tang investigate the problem of synchronization and parameter identification for a class of chaotic neural networks. The author identifies the importance of the inherent features of high security of neural networks and their potential in applications such as pattern recognition, image processing to name a few. These features have aroused interest in synchronization of chaotic neural network (CNN). They have carried out their work with stochastic perturbation via state and output coupling, which involve both the discrete and distributed time-varying delays. They have derived several sufficient conditions in order to perform synchronization of stochastic chaotic neural networks with the help of adaptive feedback techniques. They claim to estimate all the connection weight matrices when the lag synchronization and complete synchronization is achieved in mean square at the same time. They have supported the effectiveness of their proposed method with suitable simulation.

Chapter 14
Type-2 Fuzzy Sliding Mode Synchronization 314
Tsung-Chih Lin, Feng-Chia University, Taiwan
Ming-Che Chen, Feng-Chia University, Taiwan
Mehdi Roopaei, Islamic Azad University, Iran

Tsung-Chih Lin, Ming-Che Chen, Mehdi Roopaei presents an adaptive interval type-2 fuzzy neural network (FNN) controller to synchronize chaotic systems with training data corrupted by noise or rule uncertainties involving external disturbances. The scheme is applied for the synchronization of non-identical chaotic systems. The asymptotic stability is also studied by the Lyapunov stability theorem. The simulation results prove the effectiveness of the scheme.

Section 3
Cryptographic Applications

Chapter 15
Secure Transmission of Analog Information Using Chaos........ 337
A.S. Dmitriev, Inst. of Radio Eng. & Electr. of Russian Academy of Sciences, Russia
E.V. Efremova, Inst. of Radio Eng. & Electr. of Russian Academy of Sciences, Russia
L.V. Kuzmin, Inst. of Radio Eng. & Electr. of Russian Academy of Sciences, Russia
A.N. Miliou, Aristotle University of Thessaloniki, Greece
A.I. Panas, Inst. of Radio Eng. & Electr. of Russian Academy of Sciences, Russia
S.O. Starkov, Inst. of Radio Eng. & Electr. of Russian Academy of Sciences, Russia

A.S. Dmitriev, E.V. Efremova, L.V. Kuzmin, A.N. Miliou, A.I. Panas, S.O. Starkov present an experimental study of a practical realization of a complex analog signal transmission system using dynamic chaos. They attempt to use the synchronized chaotic system for secure wireless communications in RF band. An analysis of the restrictions and problems connected with the quality of synchronization of the transmitter and the receiver of the wireless communication systems has also been demonstrated. In wireless transmission various perturbing factors degrade its quality and the primary reason for this can be attributed to the chaotic response desynchronization associated with the phenomenon of "on-off" intermittency. Their investigation reveals that the quality of transmission can be improvised on increasing the level of information signal fed to the transmitter. On the other hand, for secure communication the information signal level must be reduced. Hence, they claim that if a compromise be made on these contradictory requirements then an improvement of the quality of the synchronous chaotic response in the receiver can be achieved.

Chapter 16
Control-Theoretical Concepts in the Design of Symmetric Cryptosystems 361
Gilles Millérioux, University Henri Poincaré of Nancy, France & Research Center on Automatic Control of Nancy (CRAN), France
José María Amigó, University Miguel Hernandez of Elche (UMH), Spain & Centro de Investigación Operativa, Spain

Gilles Millérioux and José Maria Amigó in this chapter focus on the fact that message-embedded chaotic ciphers and conventional self-synchronizing stream ciphers are equivalent under the so-called flatness condition. The flatness condition is borrowed from control theory. The authors claim that this kind of chaotic cipher may be an interesting alternative for the design of Self-Synchronizing Stream Ciphers and suggest new approaches in the design of self-synchronizing stream ciphers. This chapter focusses on digital encryption and hence on discrete-time dynamical systems(maps). The authors highlight the state-of-the-art of the structures involved in chaotic cryptographic schemes in comparison to conventional ciphers especially with symmetric ciphers.The initial sections are dedicated on the background of cryptography and its different modes with significance to stream ciphers with explanation of ways of deriving some permutation or substitution ciphers from chaotic dynamical systems for cryptographic purposes, followed by reviews of popular synchronization-based cryptosystems and outlining important issues to be addressed.

Chapter 17
Unmasking Optical Chaotic Cryptosystems Based on Delayed Optoelectronic Feedback 386
Silvia Ortín, Instituto de Física de Cantabria (CSIC-Universidad de Cantabria), Spain
Luis Pesquera, Instituto de Física de Cantabria (CSIC-Universidad de Cantabria), Spain

Silvia Ortín and Luis Pesquera have presented an analysis of the security of optical chaotic communication systems. Their result shows that such chaotic cryptosystems based on feedback with several fixed time delays are vulnerable to security hazards. To achieve this they have developed a model based on a new type of neural network known as modular neural network. This model reconstructs the nonlinear dynamics of the transmitter from experimental time series in the single-delay case and also from numerical simulations in single and two-delay cases. To support their experiment they have generated the chaotic carrier by a laser diode when it is subjected to delayed optoelectronic feedback. The authors have demonstrated that even if the chaotic attractor manifests itself in huge dynamics, the model's complexity does not increase (remains unaffected ?) on increasing the time delay. Whereas, the reconstruction of nonlinear dynamics is more difficult when the feedback strength is increased. The developed model is utilized as an unauthorized receiver which recovers the message.

Chapter 18

Encryption of Analog and Digital Signals through Synchronized Chaotic Systems........................ 415
Kehui Sun, Central South University, China

Kehui Sun has analyzed the basic principles of chaos based communication tools in cryptography in analogue and digital medium mainly Chaos Masking, Chaos Shift Keying, Chaos Modulation, and Chaos Spreading Spectrum. The author has also analyzed their modifications and performances of chaotic sequence supported by appropriate simulation tools and designed an effective chaotic sequence generator. This has been used in their proposed two encryption schemes namely chaotic sequence encryption and chaotic data stream encryption.

Chapter 19

Digital Information Transmission Using Discrete Chaotic Signal 439
A.N. Anagnostopoulos, Aristotle University of Thessaloniki, Greece
A.N. Miliou, Aristotle University of Thessaloniki, Greece
S.G. Stavrinides, Aristotle University of Thessaloniki, Greece
A. S. Dmitriev, Inst. of Radio Eng. & Electr. of Russian Academy of Sciences, Russia
E. V. Efremova, Inst. of Radio Eng. & Electr. of Russian Academy of Sciences, Russia

A.N. Anagnostopoulos, A.N. Miliou, S.G. Stavrinides, A.S. Dmitriev and E.V. Efremova have studied a digital information transmission system using discrete chaotic signal over cable and proposed a robust,secure approach of transmission with an enhanced encoding scheme to inject further security. This has been designed keeping in mind the simplicity factor of implementing a chaotic system for the purpose. To realize the aim, the authors have used a the non-autonomous 2nd order non-linear oscillator system presented in (Tamaševičious, Čenys, Mycolaitis, & Namajunas, 1998) supported with results on synchronization. The authors also analyze the effect of noise (internal or external) on the synchronization of the drive-response system (unidirectional coupling between two identical systems). They have also taken into account the practical issue pertaining to the mismatch between the parameters of the transmitter and the receiver resulting due to different operating conditions of the otherwise identical transmitter and receiver circuits.

Chapter 20

Mathematical Treatment for Constructing a Countermeasure Against the One-Time Pad
Attack on the Baptista Type Cryptosystem.. 463
M.R.K. Ariffin, Universiti Putra Malaysia, Malaysia
M.S.M.Noorani, Universiti Kebangsaan Malaysia, Malaysia

M.R.K. Ariffin and M.S.M. Noorani present a notion for preventing a one time pad attack on the chaotic cryptosystem proposed by M. S. Baptista in 1998. They have discussed the weakness of this cryptosystem to be its vulnerability to one-time pad attack (a type of chosen plaintext attack) together with the non-uniform distribution of ciphertexts. Hence, the authors have given a mathematical approach with an example to show the Baptista type cryptosystem can be prevented against the one-time pad attack.

Chapter 21

Chaos Synchronization with Genetic Engineering Algorithm for Secure Communications 476

Sumona Mukhopadhyay, Army Institute of Management, India
Mala Mitra, Camellia School of Engineering, India
Santo Banerjee, Politecnico di Torino, Italy & Techfab, Italy

Sumona Mukhopadhyaym, Mala Mitra and Santo Banerjee have proposed a method of digital cryptography inspired from Genetic Algorithm (GA) and synchronization of chaotic delayed system. The chapter introduces a brief idea about the concept of Evolutionary Algorithm (EA) and demonstrates how the potential of dynamical system such as chaos and EA can be utilized in a reliable, efficient and computational cheaper method for secure communication. GA is a subclass of Evolutionary algorithm and as such is governed by the rules of organic evolution. In GA the selection mechanism and both transformation operators-crossover and mutation are probabilistic. The parameters and the keys are secure since the synchronized dynamical system does not necessitate the transmission of keys over the communication channel. The random sequence obtained from chaotic generator further transforms it into a powerful stochastic method of searching the solution space in varied directions for an optimal solution escaping points of local optima. But randomicity can sometimes destabilize the system and there is no guarantee that it yields an improved solution. The authors substituted the random and probabilistic selection operator of GA with problem specific operator to design the cryptosystem to control such random behavior otherwise it would lead to a solution which is uncorrelated with the original message and may also lead to loss of information. The way selection has been modified leads to two versions of the proposed genetic engineering algorithm for cryptography. Simulation results demonstrates that both the flavors of the proposed cryptography successfully recover the message and a comparison with cryptography with $(\mu/\rho,\lambda)$ Evolutionary Strategy's selection scheme shows a computational edge of their proposed work.

Compilation of References 510

About the Contributors 554

Index 563

Foreword

In the past decades, the advent of electronic computers and, later, of the network of such computers, presently known as the World Wide Web, has exponentially increased the amount of data exchanged among the most diverse people, areas and organizations, all over the world. This has brought about many advantages to the academic institutions, to the military structures to the financial institutions, to banks and corporations of all sizes, to the public administrations, governments etc. A detailed list of all those interested in the exchange of information over the internet is presently impossible. One simply observes that practically everybody in the industrialized world is directly or indirectly interested in that, while larger and larger fractions of the populations from the emerging economies are rapidly becoming interested in that. In particular, sensitive information and information protected by privacy laws constitute a large fraction of the total information being transmitted by all forms of telecommunication equipments. Therefore, the development of secure communication has become impelling, in order to prevent an-authorized people from intercepting and embezzling information meant to others. The most obvious way of achieving this goal is to encrypt the data that is to be transmitted. Among the most promising encrypting techniques, we find those based on chaotic synchronized dynamics, which constitute the subject of the present timely collection of research chapters. Chaos is one type of deterministic, hence in principle fully predictable, dynamic behaviour, which, however, turns out to be as unpredictable as a stochastic process, in practice. This is related to the impossibility of knowing with infinite precision the initial state of the system at hand. Also, a chaotic signal enjoys the characteristics of e.g. aperiodicity.

Interacting chaotic oscillators are of interest in many areas of physics, biology, and engineering. For instance, one challenging problem, faced by the current biological sciences, concerns our understanding of the emergence of collective coherent behaviours, from groups of interacting functional units, separately displaying complicated behaviours. In particular, it is remarkable that chaotic systems can be synchronized letting them communicate only a part of the information concerning their state. When this phenomenon was discovered, it became immediately clear that it could have been used to create keys for cryptography using the unsent state spaces. Indeed, by using the proper keys, or interactions, the sender may be synchronized with the receiver, and only part of the data needs to be transmitted, for the whole message to be delivered. This significantly reduces the possibility that the message be understood by an-authorized people. As perfect, 100% secure, ways of encrypting a message may not be realizable, even chaotic synchronization needs to be tested and further developed, while remaining one of the most promising tools for secure communication in the years to come. This makes the present collection of chapters especially timely and useful.

Lamberto Rondoni
Politecnico di Torino, Italy

***Lamberto Rondoni** received his "Laurea" in nuclear engineering in 1986, from University of Bologna. He received a Masters in Physics and a Masters in Mathematics in 1990 from Virginia Tech, where he earned his PhD in Mathematical physics, in 1991. He was awarded a postdoctoral position at Virginia Tech, in academic year 1991-1992, before accepting a Research Associate position at University of New South Wales. From 1995 to 1999 he held a Ricercatore Universitario positon, at Politecnico di Torino, where he is a Professore Associato, and leads a nonequilibrium statistical mechanics research group. His research activity concerns the Boltzmann equation and transport of particles; applications of the theory of stochastic processes to chemical kinetics and pattern formation; applications of dynamical systems theory to understand nonequilibrium phenomena, including fluctuations in nanoscale as well as macroscopic objects, like gravitational waves detectors. He has written about 80 papers, several popularization articles and two books.*

Preface

The Book

This book is an intersection of two major subject area, chaos synchronization and cryptography.

Chaos theory has a huge applicability in different branches of Science,Engineering and Management. It is highly important in physical, biological, social Sciences for its unique capability to model the natural systems which are nonlinear. In the field of engineering,chaos is very useful for control theory and applications. Genetic engineering algorithms, Swarm Intelligence, different global optimization techniques are useful in engineering and also in management for economic research and stock predictions.

Synchronization is a special phenomenon of chaotic systems. Two system can synchronize when they are both chaotic in nature. This phenomenon has wide applications in neural network and cryptography.

The content of the book is recent developments of the subject area.

Organization of the Book

The book is divided in three sections. The first section contains five chapters describing the dynamics of different chaotic systems with theory, simulations and experiments. Section 2 is based on nine chapters on synchronization of chaotic systems. The phenomenon have investigated with chaotic oscillators, integer and fraction order systems and delayed neural networks. Section 3 contains seven chapters based on cryptographic applications of chaotic systems.

Chapter 1: *"Dynamical Systems and their Chaotic Properties"* Konstantinos Chlouverakis presents an analytical observation on ordinary and time delayed dynamical systems and there chaotic properties. The non linearity have quantified with phase space diagrams, lyapunov exponents,bifurcation diagrams. The stability analysis is also computed and a coupled system is also studied for cryptographic encoding.

Chapter 2: *"Chaotic Dynamical Systems Associated with Tilings of R^N"* Lionel Rosier presents a mathematical mechanism for chaos synchronization that can be applied to encryption. A class of discrete dynamical systems is defined on the homogeneous space associated with a regular tiling of R^N because any dynamical system in this class is chaotic in the sense of Devaney, and that it admits at least one positive Lyapunov exponent. A common example of such a system is the N−dimensional torus T^N

Chapter 3: *"Identification and State Observation of Uncertain Chaotic Systems Using Projectional Differential Neural Networks"* Alejandro García, Isaac Chairez and Alexander Poznyak made an approximation of dynamical systems by artificial neural networks and studied the analysis of projectional differential neural network. The effectiveness of the novel DNN topology is shown by a classical numerical example of Chua's circuit.

Chapter 4: *"Simple Chaotic Electronic Circuits"* M.P. Hanias and G.S.Tombras considers that simple chaotic electronics circuits in the form of diode resonator circuits, Resistor-Inductor-LED optoelectronic chaotic circuits and Single Transistor chaotic circuits can be used as transmitters and receivers for chaotic cryptosystems. These circuits are flexible to the changes in their circuit parameters and so help in investigating the influence of such changes in complexity of the generated strange attractors. The authors have performed a time series analysis based on Grassberger and Procaccia's method. The invariant parameters such as correlation, and minimum embedding dimension are respectively calculated along with the Kolmogorov entropy. The chapter also examines the RLT circuits when they are in a critical state.

Chapter 5 : *"Experimental Evidences of Shil'nikov Chaos and Mixed-Mode Oscillation in Chua Circuit"* Syamal Kumar Dana and Satyabrata Chakraborty in their chapter have provided the evidence of Shil'nikov type homoclinic chaos in asymmetry-induced Chua's oscillator. In the experiment, different time scales are artificially created in a double scroll Chua attractor by inducing asymmetry in the system by external DC forcing. One of the double scroll attractors shrinks in size and creates an additional time scales in the overall dynamics that plays a crucial role in the origin of homoclinic chaos, bursting and MMOs.

Chapter 6: *"Synchronization of Chaotic Oscillators"* J. M. González-Miranda presents a review about the varied forms which the mutually coupled or unidirectionally driven dynamic chaotic oscillators display on synchronization. This will enlighten the readers on the key role which physics and mathematics play in laying the foundation useful for reaping the benefits of chaos synchronization in telecommunications and cryptography.

Chapter 7: *"Synchronization in Integer and Fractional Order Chaotic Systems"* Ahmed E. Matouk's, chapter concentrates on the synchronization techniques on integer and fractional order chaotic systems. Lyapunov stabily theory are also used to control the synchronized error. Numerical examples well support the analytical results.

Chapter 8: *"Chaos Synchronization"* Hassan Salarieh and Mohammad Shahrokhi investigates an overview of the synchronization phenomenon in secure communication. The chapter begins by defining complete, lag, phase and generalized synchronization approaches followed by the application of control theory for synchronization of different chaotic systems. They have presented some synchronization algorithms based on different control techniques. They have shown ways to modify the controlling methods in order to cope with parameter uncertainties and measurement noise. The performance of the discussed methods are supported by simulation of several chaotic systems.

Chapter 9: *"Chaotic Gyros Synchronization"* In this Chapter, M. Roopaei, M. J. Zolghadri, B. S. Ranjbar, S. H. Mousavi, H. Adloo, B. Zare and T.C. Lin have presented three methods for synchronizing of two chaotic gyros in the presence of uncertainties, external disturbances and dead-zone nonlinearity. In the first method, there is dead-zone nonlinearity in the control input, which limits the performance of accurate control methods. The effects of this nonlinearity have been attenuated using a fuzzy parameters adaptation integrated with sliding mode control method. For the second method, they have proposed a robust adaptive fuzzy sliding mode control scheme to overcome the synchronization problem for a class of unknown nonlinear chaotic gyro. The third method considers two different gyro systems. For this method they also have proposed a fuzzy controller to eliminate chattering phenomena during the reaching phase of sliding mode control. All the mentioned methods are simulated and the results illustrate the effectiveness of the proposed methods.

Chapter 10: *"Importance of Chaos Synchronization on Technology and Science"* In this chapter, Ricardo Aguilar-López, Ricardo Femat and Rafael Martínez-Guerra highlight the importance of chaos synchronization on technology and science. The chapter introduces their attempt in three sections. The first sections deals with the topic of synchronized state in the sense of identical synchronization. This is realized with the robust nonlinear observer design, considering corrupted measurements and model uncertainties, coupling uncertainty estimators with nonlinear state observers. In the second part of their chapter, the authors have addressed applications to chaos communications primarily an application of chaos theory which is aimed to provide security in the transmission of information performed through telecommunications technologies mainly dealing with the transmitter. A message is added on to a chaotic signal and then, the message is masked in the chaotic signal. When the chaotic signal carries such a message it is called the chaotic carrier. All this is done via control theory and is a particular case of chaos synchronization. In the last section they have discussed an application to synchronization of biological systems. Their attempt of introducing the feedback control laws in some biological systems is influenced by the behavior of biophysical mechanisms of cellular dynamics. The authors also present links between feedback control schemes, with an external input, and intracellular calcium functions for coordination and control.

Chapter 11: *"Synchronization of Oscillators"* Jean B. Chabi Orou describes the synchronization of oscillators in a lucid manner which can facilitate to get a grasp of the main concepts before delving into the different kinds of synchronization configurations, Chaotic synchronization and also addressed the issue of stability of the synchronization. They have briefly mentioned the influence of noise on the synchronization process.

Chapter 12: *"Synchronization of Uncertain Neural Networks with H_∞ Performance and Mixed Time-Delays"* Hamid Reza Karimi, introduces an exponential synchronization method for a class of uncertain master and slave neural networks with mixed time-delays. The mixed delays comprise different neutral, discrete and distributed time-delays. In order to design a delayed state-feedback control as a synchronization law in terms of linear matrix inequalities under less restrictive conditions, some delay-dependent sufficient conditions have been established by an appropriate discretized Lyapunov-Krasovskii functional and some free weighting matrices. Irrespective of their initial state, this controller will guarantee the exponential synchronization of the two coupled master and slave neural networks. The author has demonstrated the effectiveness of the established synchronization laws by suitable numerical simulations.

Chapter 13: *"Adaptive Synchronization in Unknown Stochastic Chaotic Neural Networks with Mixed Time-Varying Delays"* Jian-an Fang andYang Tang investigate the problem of synchronization and parameter identification for a class of chaotic neural networks. The author identifies the importance of the inherent features of high security of neural networks and their potential in applications such as pattern recognition, image processing to name a few. These features have aroused interest in synchronization of chaotic neural network (CNN). They have carried out their work with stochastic perturbation via state and output coupling, which involve both the discrete and distributed time-varying delays. They have derived several sufficient conditions in order to perform synchronization of stochastic chaotic neural networks with the help of adaptive feedback techniques. They claim to estimate all the connection weight matrices when the lag synchronization and complete synchronization is achieved in mean square at the same time. They have supported the effectiveness of their proposed method with suitable simulation.

Chapter 14: *"Type-2 Fuzzy Sliding Mode Synchronization"* Tsung-Chih Lin, Ming-Che Chen, Mehdi Roopaei presents an adaptive interval type-2 fuzzy neural network (FNN) controller to synchronize chaotic systems with training data corrupted by noise or rule uncertainties involving external disturbances. The scheme is applied for the synchronization of non-identical chaotic systems. The asymptotic stability is also studied by the Lyapunov stability theorem. The simulation results prove the effectiveness of the scheme.

Chapter 15: *"Secure Transmission of Analog Information using Chaos"* A.S. Dmitriev, E.V. Efremova, L.V. Kuzmin, A.N. Miliou, A.I. Panas, S.O. Starkov present an experimental study of a practical realization of a complex analog signal transmission system using dynamic chaos. They attempt to use the synchronized chaotic system for secure wireless communications in RF band. An analysis of the restrictions and problems connected with the quality of synchronization of the transmitter and the receiver of the wireless communication systems has also been demonstrated. In wireless transmission various perturbing factors degrade its quality and the primary reason for this can be attributed to the chaotic response desynchronization associated with the phenomenon of "on-off" intermittency. Their investigation reveals that the quality of transmission can be improvised on increasing the level of information signal fed to the transmitter. On the other hand, for secure communication the information signal level must be reduced. Hence, they claim that if a compromise be made on these contradictory requirements then an improvement of the quality of the synchronous chaotic response in the receiver can be achieved.

Chapter 16: *"Control-Theoretical Concepts in the Design of Symmetric Cryptosystems"* Gilles Millérioux and José Maria Amigó in this chapter focus on the fact that message-embedded chaotic ciphers and conventional self-synchronizing stream ciphers are equivalent under the so-called flatness condition. The flatness condition is borrowed from control theory. The authors claim that this kind of chaotic cipher may be an interesting alternative for the design of Self-Synchronizing Stream Ciphers and suggest new approaches in the design of self-synchronizing stream ciphers. This chapter focusses on digital encryption and hence on discrete-time dynamical systems(maps). The authors highlight the state-of-the-art of the structures involved in chaotic cryptographic schemes in comparison to conventional ciphers especially with symmetric ciphers.The initial sections are dedicated on the background of cryptography and its different modes with significance to stream ciphers with explanation of ways of deriving some permutation or substitution ciphers from chaotic dynamical systems for cryptographic purposes, followed by reviews of popular synchronization-based cryptosystems and outlining important issues to be addressed.

Chapter 17: *"Unmasking Optical Chaotic Cryptosystems Based on Delayed Optoelectronic Feedback"* Silvia Ortín, Luis Pesquera presents an analysis of the security of optical chaotic communication systems. Their result shows that such chaotic cryptosystems based on feedback with several fixed time delays are vulnerable to security hazards. To achieve this they have developed a model based on a new type of neural network known as modular neural network. This model reconstructs the nonlinear dynamics of the transmitter from experimental time series in the single-delay case and also from numerical simulations in single and two-delay cases. To support their experiment they have generated the chaotic carrier by a laser diode when it is subjected to delayed optoelectronic feedback. The authors have demonstrated that even if the chaotic attractor manifests itself in huge dynamics, the model's complexity does not increase (remains unaffected ?) on increasing the time delay. Whereas, the reconstruction of nonlinear dynamics is more difficult when the feedback strength is increased. The developed model is utilized as an unauthorized receiver which recovers the message.

Chapter 18: *"Encryption of Analog and Digital Signals through Synchronized Chaotic Systems"* Kehui Sun has analyzed the basic principles of chaos based communication tools in cryptography in analogue and digital medium mainly Chaos Masking, Chaos Shift Keying, Chaos Modulation, and Chaos Spreading Spectrum. The author has also analyzed their modifications and performances of chaotic sequence supported by appropriate simulation tools and designed an effective chaotic sequence generator. This has been used in their proposed two encryption schemes namely chaotic sequence encryption and chaotic data stream encryption.

Chapter 19: *"Digital Information Transmission using Discrete Chaotic Signal"* A.N. Anagnostopoulos, A.N. Miliou, S.G. Stavrinides, A.S. Dmitriev and E.V. Efremova have studied a digital information transmission system using discrete chaotic signal over cable and proposed a robust,secure approach of transmission with an enhanced encoding scheme to inject further security. This has been designed keeping in mind the simplicity factor of implementing a chaotic system for the purpose. To realize the aim, the authors have used a the non-autonomous 2nd order non-linear oscillator system presented in (Tamaševičious, Čenys, Mycolaitis, & Namajunas, 1998) supported with results on synchronization. The authors also analyze the effect of noise (internal or external) on the synchronization of the drive-response system (unidirectional coupling between two identical systems). They have also taken into account the practical issue pertaining to the mismatch between the parameters of the transmitter and the receiver resulting due to different operating conditions of the otherwise identical transmitter and receiver circuits.

Chapter 20: *"Mathematical Treatment for Constructing a Countermeasure Against the One Time Pad Attack on the Baptista Type Cryptosystem."* M.R.K. Ariffin and M.S.M.Noorani present a notion for preventing a one time pad attack on the chaotic cryptosystem proposed by M.S. Baptista in 1998. They have discussed the weakness of this cryptosystem to be its vulnerability to one-time pad attack (a type of chosen plaintext attack) together with the non-uniform distribution of ciphertexts. Hence, the authors have given a mathematical approach with an example to show the Baptista type cryptosystem can be prevented against the one-time pad attack

Chapter 21: *"Chaos Synchronization with Genetic Engineering Algorithm for Secure Communications."* Sumona Mukhopadhyay Mala Mitra and Santo Banerjee have proposed a method of digital cryptography inspired from Genetic Algorithm(GA) and synchronization of chaotic delayed system. The chapter introduces a brief idea about the concept of Evolutionary Algorithm(EA) and demonstrates how the potential of dynamical system such as chaos and EA can be utilized in a reliable, efficient and computational cheaper method for secure communication. GA is a subclass of Evolutionary algorithm and as such is governed by the rules of organic evolution. In GA the selection mechanism and both transformation operators-crossover and mutation are probabilistic. The parameters and the keys are secure since the synchronized dynamical system does not necessitate the transmission of keys over the communication channel . The random sequence obtained from chaotic generator further transforms it into a powerful stochastic method of searching the solution space in varied directions for an optimal solution escaping points of local optima. But randomicity can sometimes destabilize the system and there is no guarantee that it yields an improved solution. The authors substituted the random and probabilistic selection operator of GA with problem specific operator to design the cryptosystem to control such random behavior otherwise it would lead to a solution which is uncorrelated with the original message and may also lead to loss of information. The way selection has been modified leads to two versions of the proposed genetic engineering algorithm for cryptography. Simulation results demonstrates that both the flavors of the proposed cryptography successfully recover the message and a comparison with cryptography with ($\mu/\rho,\lambda$) Evolutionary Strategy's selection scheme shows a computational edge of their proposed work.

Intended Audience

This book can be used as a reference for the scientific and industrial research. The different chapters can provide knowledge of the subject areas, as well as the developments in the field of theory and applications. The science and engineering students will be benefited to learn the subject by some review chapters, also they have the scope to continue some developments with the existing and new research work described here.

Acknowledgment

I wish to express my heartfelt appreciation to all those who have contributed to this body of this work both explicitly and implicitly. I express my gratitude to all my colleagues from universities and professional Institutions in India and abroad for extending their support and encouragement.

This book is a representation of my learning over a period of time. I am deeply indebted to innumerable people for their support and stimulus for shaping the knowledge.

Let me first start by recalling my co-authors for the book, who will inhly responded to my inquisition. It is beyond my capacity to adequately articulate to value their contribution to learning and shaping my thinking in an important way.

My students who inspired me by their thought provoking questions enriched my intellectual depth.

Thanks are also due to all the reviewers who benefited me through their observations.

My heartfelt gratitude to my publisher for their keen interest , and astute effort in bringing out this book. Without their support this book would not have been possible.

Finally My special thanks to my family for their tremendous support and for motivating me to keep learning through the journey of my life.

I would sincerely like to appreciate and acknowledge the comments and suggestions from the readers for this book.

Santo Banerjee
Politecnico di Torino, Italy

Section 1
Chaotic Dynamical Systems, Theory and Experiments

Chapter 1
Dynamical Systems and Their Chaotic Properties

Konstantinos Chlouverakis
National and Kapodistrian University of Athens, Greece

ABSTRACT

This chapter describes the fundamental concepts of continuous-time chaotic dynamical systems so as to make the book self-contained and readable by fairly unfamiliar people with using simple chaotic systems and understanding their chaotic properties, but also by experts who desire to refine their knowledge regarding chaos theory. It will also define the sense in which a chaotic system is suitable for real-world applications such as encryption.

1. INTRODUCTION

In this chapter the basic characteristics of several systems of different types will be described theoretically and examined numerically. The types of dynamical systems are limited to ordinary and delay differential equations and the examples given are the three famous pendulum, Ikeda and Mackey-Glass equations. The chaos quantifiers used in chaos theory namely the attractor dimension and the metric entropy together with the Lyapunov exponents are described with a brief text regarding the way of their numerical evaluation. Furthermore, linearized stability analysis of a delay equation is derived in detail and lastly a system of coupled delay equations that describes a semiconductor laser subjected to optical feedback is presented and investigated since it is widely used for chaos encryption. The purpose of this chapter is to familiarize the reader with the basic principles of chaos and how the latter may be treated numerically.

DOI: 10.4018/978-1-61520-737-4.ch001

1.1 Dynamical Systems

A dynamical system is one whose state changes in time. If the changes are determined by specific rules, rather than being random, the system is said to be deterministic, and in this chapter only deterministic systems will be considered. The changes can occur at discrete time steps or continuously. This chapter will be concerned only with continuous-time, deterministic, dynamical systems, since they best approximate the real world. Ordinary and time-delayed differential equations will be described together with prototypical elegant famous examples together with their chaotic properties. The most obvious examples of dynamical systems are those that involve something moving through space like a planet orbiting the Sun or a pendulum swinging back and forth. But dynamical systems can also be more abstract, such as money flowing though the economy, information propagating across the Internet, or disease moving though the population. We should mention that discrete-time systems have already been extensively explored, in part because they are more computationally tractable and therefore will not be investigated and mentioned herein.

Calculation of the motion of astronomical bodies is one of earliest problems solved by scientists, and the calculations are some the most precise in all of science, allowing, for example, the prediction of eclipses many years in advance both in time and space. Contrast this strong predictability with the difficulty of predicting even a few days in advance whether the sky will be clear enough to observe that eclipse using extremely detailed models and vast computational resources. Therein lays the difference between regular and chaotic dynamics. It is therefore prudent to pose the question: *How one can measure how complex a system is*? The answer lies on the chaos quantifiers, namely the metric entropy, the attractor dimension and generally the Lyapunov exponents. These measures will be defined next and will be calculated for the examples of the chaotic systems. But first we will describe some basic concepts from dynamical systems theory.

1.2 Chaos and Strange Attractors: Background Material from Dynamical Systems Theory

A basic example of an ordinary differential equation (or else termed as *flow*) is the pendulum (Matthews, 2005) that consists of a mass m on the end of massless, rod of length L. In the pendulum, one time-dependent variable is the angle x (in radians) that the pendulum makes from the vertical. At each instant, there is a force equal to $\sim\sin(x)$ pulling the mass back to its equilibrium position at $x = 0$. Newton's second law leads to the following system of motion equations:

$$\begin{aligned} \frac{dx}{dt} &= u \\ \frac{du}{dt} &= -\sin\left(x\right) \end{aligned} \tag{1.1}$$

where u is the angular velocity. In some sense this system has one degree of freedom that spans its motion in two different directions x and u and there is a unique direction and amplitude of the motion given by the vector whose components are dx/dt and dy/dt. The simultaneous motion of all the points in this space resembles a flowing fluid, and hence systems such as Eq. (1.1) are usually called *flows*.

If one tries to integrate the above equations, non-chaotic solutions will occur. Therefore there should be a sufficient criterion that would guarantee the minimal qualifications for a system to be capable to result in chaotic solutions. Chaos requires at least three dimensions so that the clockwise trajectories followed by small particles moving with the flow for various initial conditions can cross by passing behind one another. This notion has been formalized in the Poincarè-Bendixson theorem (Hirsch *et al*, 2005).

1.2.1 Dissipation

In the previous example, there was no friction, and therefore mechanical energy is conserved and the pendulum swings forever. Such systems are said to be *conservative* since they conserve energy. These are also called *Hamiltonian systems* and if one calculates the trace of the Jacobian matrix, the result will always be zero. Furthermore, the area occupied by a cluster of initial conditions remains constant in time (Lichtenberg and Lieberman, 1992). The flow is also time-reversible, since changing the sign of t corresponds to the transformation of one variable (i.e. u), which flips the figure vertically and preserves the shape of the trajectories. Strictly conservative systems are rare in nature, with the standard examples being the motion of astronomical bodies in the near vacuum of outer space and the non-relativistic motion of individual charged particles in magnetic fields.

For the pendulum, it is more realistic to include a friction term in the equations of motion, due primarily to air resistance. Therefore, we may transform the system to the following:

$$\begin{aligned} \frac{dx}{dt} &= u \\ \frac{du}{dt} &= -\sin\left(x\right) - bu \end{aligned} \qquad (1.2)$$

where b is a measure of the friction and is assumed to have a constant value as the motion occurs, but that can be changed to produce different types of motion. More generally, such a term is called *damping*, and it can be a nonlinear function of u and can also depend on x. Systems with damping are said to be *dissipative*, in contrast to the conservative system previously discussed. Dissipative systems do not conserve mechanical energy and are not time-reversible. Their clockwise trajectories followed by small particles moving with the flow for various initial conditions are not contours of constant energy. They are described by a compressible flow as shown in Figure 1 for a case with a small damping of $b = 0.05$ and an initial condition of $(x_0, u_0) = (0, 0.05)$.

In this case, the pendulum rotates several times and then swings back and forth with a decreasing amplitude, approaching ever closer to the stable equilibrium point at $(x_0, u_0) = (0, 0)$. This point acts as an attractor for all initial conditions, since they are drawn to it in the limit of $t \rightarrow \infty$, and it is called a *sink* for obvious reasons.

1.2.2 Limit Cycles

Consider a situation in which the damping is negative near the origin – in contrast to (1.2) – but then becomes positive when the trajectory gets too far from the origin. In such a case, the trajectory is drawn to the region where the two effects offset. Such a system is the following:

Figure 1. Phase plane of (1.2) with b=0.05 and initial conditions $(x_0, u_0) = (0, 0.05)$

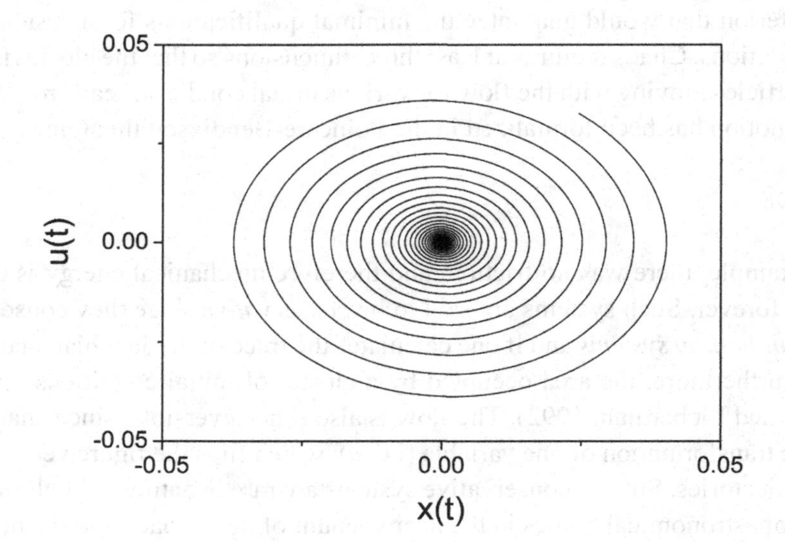

$$
\begin{aligned}
\frac{dx}{dt} &= u \\
\frac{du}{dt} &= -\sin(x) - (1 - x^2)\,u
\end{aligned}
\qquad (1.3)
$$

The attractor that is created in Figure 2 depicts a *limit cycle* for various different initial conditions.

Actually a limit cycle is a closed trajectory in phase space having the property that at least one other trajectory spirals into it as time approaches plus/minus infinity. In dynamical systems it is considered as a periodic state. In the case of (1.3) the anti-damping that occurs near the origin plays the role of an

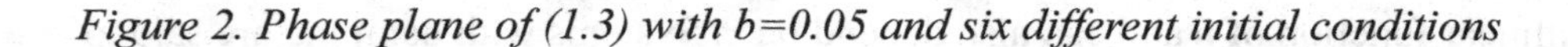

Figure 2. Phase plane of (1.3) with b=0.05 and six different initial conditions

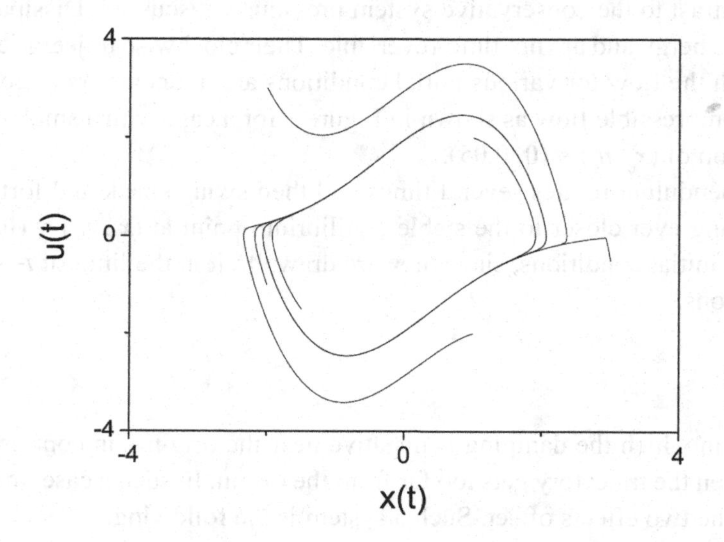

external force of energy and a positive feedback. Limit cycles can occur and in flows that the damping term is a constant like in the famous Lorenz system (Lorenz, 1963) and furthermore in delay differential equations (Farmer, 1982; Vicente *et al*, 2005). Such examples will be examined in the next sessions.

1.2.3 Strange Attractors and Chaos

In the example of (1.3) chaos in not possible to occur. The reason for the latter has to do with the streamlines (trajectories followed by small particles moving with the flow for various initial conditions) cannot cross by passing one another. This is the basic concept of the *Poincarè-Bendixson* theorem (Hirsch *et al.*, 2004). Therefore, for a flow (or else a system of ordinary differential equations) 3 dimensions are needed in order for the system to be able to result in chaotic solutions. Additionally, at least one nonlinearity is needed for the aforementioned scope. Until some years ago, the Lorenz system was supposed to be the simplest chaotic flow, but Sprott (Sprott, 1997A) derived a prototypical elegant minimal system that beat that record. As an example herein for a chaotic ODE we present the simple damped pendulum but with a sinusoidal forcing given by:

$$\begin{aligned} \frac{dx}{dt} &= u \\ \frac{du}{dt} &= -\sin\left(x\right) - bu + \sin\left(\Omega t\right) \end{aligned} \tag{1.4}$$

This is actually a 3-dimensional ODE since a variable z can be added that includes the Ωt term i.e. dz/dt = Ω and sin(Ωt) in (1.4) to be transformed into sin(z). With this mathematical formalism the system is now *autonomous* since on the right-hand side does not include the time t. Otherwise, (1.4) is a *non-autonomous* system. Autonomous systems are more commonly studied and are important and more elegant since there is a unique direction of the flow at almost every point in their state space independent of when the point is visited. Famous autonomous systems are the Lorenz, the Rössler (Rössler, 1976) and many systems derived by Sprott (1997A) etc. In Figure 3 the attractor is shown for b=0.05, Ω=0.8 and initial conditions (x_0, u_0, z_0) = (0, 1, 0) and in contrast to Figure 2 the lines are never repeated. The trajectory winds around forever, never repeating, on an object called a *strange attractor*, which in this case has a dimension greater than 2.0 and is usually not an integer. Usually such strange attractors are signature of chaos but not an absolute requirement. The calculation for the Lyapunov exponent for example is a definite measure of chaos as it will be demonstrated next.

1.2.4 Lyapunov Exponents, Metric Entropy and Attractor Dimension

Chaos can often be identified with some confidence by observing the strange attractor as stated above. However, it is useful to have a quantitative and more objective measure of it, and for that purpose the *Lyapunov exponent* is a perfect candidate. The main defining feature of chaos is the sensitive dependence on initial conditions. Two nearby initial conditions on the attractor separate by a distance that grows exponentially in time when averaged along the trajectory. The Lyapunov exponent is the average rate of growth of this distance, with a positive value signifying sensitive dependence (chaos), a zero value signifying periodicity (or quasiperiodicity), and a negative value signifying a stable equilibrium.

Figure 3. Phase plane of (1.4) with b=0.05, Ω=0.8 and initial conditions $(x_0, u_0, z_0) = (0, 1, 0)$

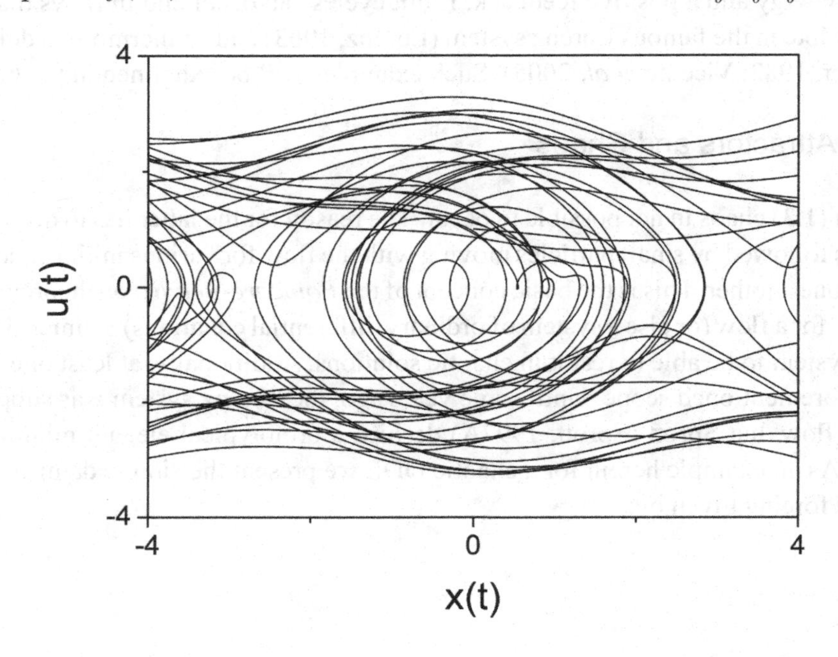

Calculation of the Lyapunov exponent is conceptually simple since one only needs to follow two initially nearby trajectories and fit the logarithm (base-e throughout this book except as noted) of their separation to a linear function of time. The slope of the fit is the Lyapunov exponent. Generally, there are a number of practical difficulties regarding this kind calculation but one has simply to choose carefully the initial seperation of the two orbits (in the order of 10^{-6}-10^{-9}) and average for a long time in order for the trajectory to reach the attractor. It is prudent furthermore to avoid the numerical transients and therefore to choose carefully the initial conditions. For further details see (Kaplan and York, 1979; Sprott 2003).

The aforementioned Lyapunov exponent (LE) is actually the largest Lyapunov exponent out of many. In fact there are as many LEs as the space variables. For the example of (1.4) the number of LEs is 3 since we have the variables *x*, *y* and *z*. Fortunately, the largest one is the only one that is required to identify chaos, since if it is positive, the system exhibits sensitive dependence on initial conditions independent of the values of the others, and if it is zero or negative, none of the others can be positive either. The LEs are sorted from the largest to the smallest with $\lambda_1 \geq \lambda_2 \geq, \ldots, \geq \lambda_N$ for an *N*-dimensional system. For further details see (Kaplan and York, 1979). For a flow, or any differential equation one (at least) exponent should be equal except for fixed points where it is negative. The sum of all the LEs should be equal to the trace of the Jacobian matrix. Therefore for a 3-dimensional flow like (1.4) or the Lorenz system (where the trace is constant), if one calculates λ_1 and knows the former, then the smallest LE may be derived easily. For 3-dimensional flows the combinations of the LEs are given below in Table 1.

The spectrum of Lyapunov exponents contains more information about the dynamics than does the largest exponent by itself. Considering that each exponent can be negative, zero, or positive, that their sum cannot be positive for a bounded system, there are five possible combinations for a 3-dimensional state space as given in Table 1. In higher-dimensional state spaces, other cases are possible including attracting 3-toruses (0, 0, 0, -, …) and hyperchaos (+, +, 0, -, …). Surely, one can calculate even thousands of positive LEs such as in delay differential equations resulting in hyperchaotic systems that almost mimic

Table 1.

λ_1	λ_2	λ_3	Attractor	Dimension	Dynamics
-	-	-	Equilibrium point	0	Static
0	-	-	Limit Cycle	1	Periodic
0	0	-	Attracting 2-torus	2	Quasiperiodic
0	0	0	Invariant torus	1 or 2	Quasiperiodic
+	0	-	Strange	2 to 3	Chaotic

pure randomness. See for example (Chlouverakis and Sprott, 2007), whereas for a coupled system of ODEs the scaling law of the dimension and the number of positive LEs is given according to the number of variables *N*. Noise, is known to have an infinite number of positive LEs and infinite dimension but this is beyond the scope of this chapter.

The spectrum of Lyapunov exponents allows one to calculate the dimension of the attractor. It is reasonable that when all the exponents are negative, corresponding to contraction in all directions, then the attractor must be a point with a dimension of zero. Think of an ellipsoid all of whose axes are shrinking. When λ_1 is zero and all the others are negative, the only non-contracting direction is parallel to the flow, and hence the attractor must be a limit cycle with a dimension of one. When the largest two exponents are zero and all the rest are negative, the trajectory can move freely in one direction perpendicular to the flow, but contracts in the other perpendicular directions, in which case the attractor is a 2-torus, a surface with a dimension of two. The number of leading zero exponents represents the number of directions that the trajectory can fill without contracting or expanding and hence is the dimension of the attractor. If λ_1 is positive and λ_2 is zero, then there is a surface defined by these two perpendicular directions in which a cluster of initial conditions expands without limit. Thus the attractor can have a dimension no less than two and we have a strange chaotic attractor. The sum of the 3 exponents (considering a 3-dimensional flow herein) must be non-positive. Then the volume of the cluster of initial conditions must contract without limit, and so the attractor must have a dimension less than three. This is the reason that in Table 1 the chaotic attractor has a dimension between 2 and 3. It is prudent then to assign a non-integer value between two and three in such a way that the cluster of initial conditions neither expands nor contracts in this fractional dimension. Kaplan and Yorke (1979) proposed a linear interpolation given by:

$$D_{KY} = 2 + \frac{\lambda_1}{|\lambda_3|} \tag{1.5}$$

and the dimension is called *Kaplan-Yorke dimension* or else *Lyapunov dimension*. Ofcourse this dimension can be expanded to any *N*-dimensional flow, DDE or PDE according to:

$$D_{KY} = j + \frac{\lambda_1 + \ldots + \lambda_j}{|\lambda_{j+1}|} \tag{1.6}$$

whereas j is the largest integer that gives $\lambda_1 + \ldots + \lambda_j \geq 0$.

It is concluded therefore that by knowing the whole spectrum of the Lyapunov exponents one can easily calculate the dimension of the dynamical system according to (1.6). Besides that valuable information, with knowing the spectrum we can also calculate the *metric entropy* h_{KS} of the system which by Pesin's identity (Pesin, 1977) is equal simply to the sum of all positive Lyapunov exponents and it quantifies the *chaoticity* of the system. Therefore the dimension is a quantifier for the complexity and the entropy for the chaoticity of a dynamical system. For 3-dimensional systems, the entropy coincides with the largest LE. The entropy can be viewd also as a prediction horizon. Its unit has values of inverse time and its reciprocal gives the time where after that no prediction may take place. Therefore, the larger the entropy, the more chaotic and less predictable a system is. Systems with large values of entropy and dimension are usually time-delayed ones and will be studied next.

1.3 Delay Differential Equations

Up to now we saw some typical examples of ordinary differential equations (ODEs) that are widely used in chaos theory. Usually, these examples result in low-complexity behavior because of the limited degrees of freedom (number of variables). Surely, coupled ODEs can result in whatever values of dimension and metric entropy are desired (Liu et al, 2003) but this is beyond the scope of this chapter. A more elegant example of high-dimensional systems is the delay differential equation (DDE). A delay differential equation is actually an infinite-dimensional dynamical system represented by finitely many discrete variables that advance in small discrete time steps. It has been found that such systems exhibit a wide variety of dynamical complexity, from chaos and hyperchaos (Farmer, 1982) to complex bifurcation phenomena (Luzyanina et al., 2001). Many minimal and prototypical DDEs (Sprott, 2007) and furthermore complicated coupled DDEs have been widely investigated both numerically (Chlouverakis et al., 2007; Vicente et al., 2005) and experimentally (Argyris et al., 2008; Chlouverakis et al., 2008). Such systems may describe various phenomena in nature, like in chemistry (Epstein, 1990), biology (Mackey and Glass, 1977), optical (Vicente et al., 2005) and optoelectronic (Bogris et al., 2007; Chembo et al., 2008; Ikeda, 1979) configurations.

A general form of a DDE is given below:

$$\frac{dx(t)}{dt} = f\left[t, x(t), f(x-\tau)\right] \tag{1.7}$$

where $x(t\text{-}\tau)$ represents the trajectory of the solution in the past for the time interval $(-\tau, 0)$ whereas the delay time τ is a constant. Because of the latter, a DDE can be considered to evolve in an infinite-dimensional space since the aforementioned time interval may be divided to infinite variables. Delay differential equations differ from ordinary differential equations in that the derivative at any time depends on the solution at prior times. The two most famous examples of such DDEs are the Mackey and Glass (1977) in (1.8) and the Ikeda (1979) in (1.9):

$$\frac{dx(t)}{dt} = \frac{ax(t)}{1 + x(t-\kappa)^{10}} - bx(t) \tag{1.8}$$

$$\frac{dx(t)}{dt} = \mu \sin\left[x(t-\tau) - x_0\right] - x(t) \tag{1.9}$$

Herein we will concentrate on the Ikeda DDE given in (1.9). Specifically we will focus on the impact of the parameters μ and τ on the complexity of the resulted attractors. Both the delay terms in the above equations are inside the nonlinearities. Therefore, such systems are expected to be highly-complex. In such DDEs of the form of (1.7) where the delay term is included in the nonlinearity, the two quantifiers of Lyapunov dimension and metric entropy follow the scaling laws below, where *B* and *C* are positive constants:

$$D_{KY} = B\mu\tau \qquad \text{(1.10a)}$$

$$h_{KS} = {}^{C\mu}\!/_{\tau} \qquad \text{(1.10a)}$$

Equations (1.10) conclude that as the bifurcation parameter μ and the delay τ increase, the Lyapunov dimension increases linearly. It is calculated that as these two parameters increase, the number of positive LEs increase as well. Though, the magnitude of the increased positive LEs is decreasing, leading therefore to a saturation of the metric entropy to a constant value. Therefore for the entropy, as τ increases for constant μ the entropy remains the same, whereas as μ increases for constant τ the entropy increases linearly. The latter can be given mathematically in (1.10b). The effects of these two parameters are depicted in Figure 4 and 5 with plotting the phase portraits of $x(t)$ and $x(t-\tau)$. Therefore, for encryption applications where the transmitter is a chaos generator that is described by such a DDE, high values of entropy and dimension are needed and one can easily achieve this with (1.10).

In Figure 4 it is clearly noticed that with increasing the time-delay, the topology of the attractor remains the same. Surely the dimension is known to increase as given in (1.10a). On the contrary, in Figure 5 the topology does not remain the same as the bifurcation parameter μ increases. In that case, both the dimension and the entropy increase linearly and the attractor resembles a very noisy one. Just for the record, the dimension of the cases (μ, τ)=(3,100) and (μ, τ)=(5,20) is greater than 100 and may be easily increased to any desired value.

1.3.1 Lyapunov Exponents in Delay Differential Equations

The Lyapunov exponents' spectrum is rather difficult to be calculated for a DDE. For typical nonlinear functions *f*(.) in (1.7), analytic solutions are impossible especially in the case where the dynamics are chaotic, and hence numerical methods must be used. Such methods entail approximating the continuous evolution of $x(t)$ by an iterative mapping at discrete (but small) time steps. There is no unique way to do this, but perhaps the simplest is the Euler method:

$$x(t+h) = x(t) + hf\left[x(t), x(t-Nh)\right] \tag{1.11a}$$

Figure 4. Phase planes of the Ikeda DDE in (1.9) with μ=3, x_0=0 and the time-delay τ varying

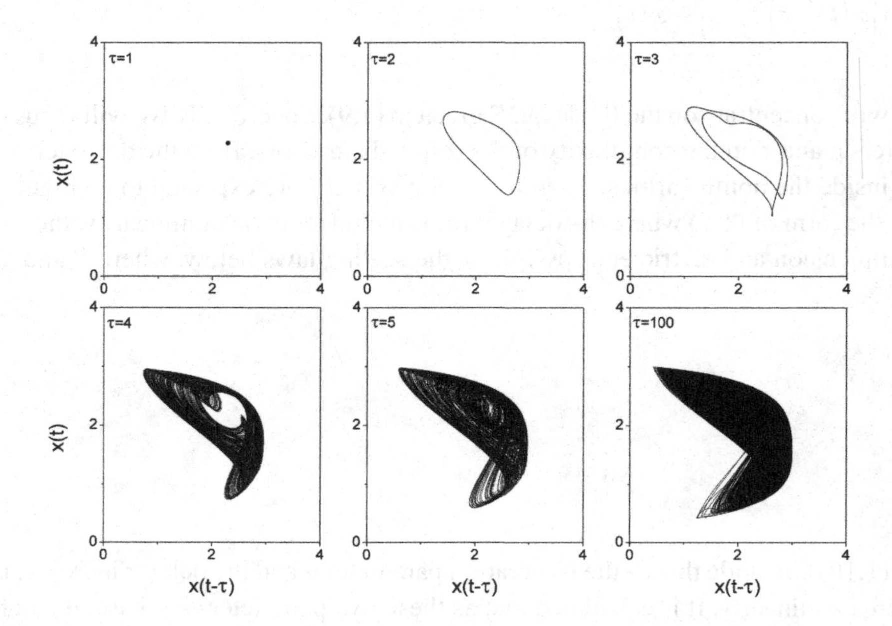

Figure 5. Phase planes of the Ikeda DDE in (1.9) with τ=5, x_0=0 and the μ parameter varying. Note the difference in the axes as μ increases.

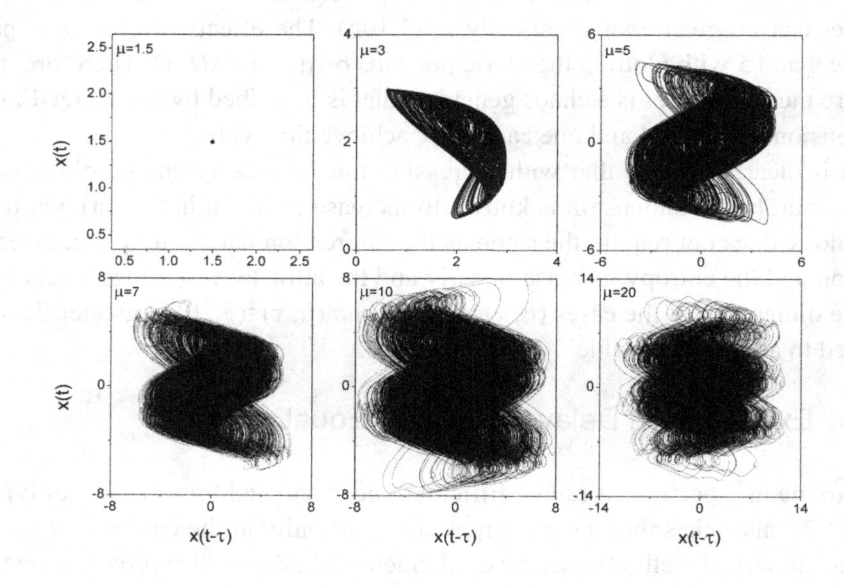

where h is the (small) time step and $N = \tau/h$ is one minus the dimension of the iterated map, which can be equivalently expressed in the form:

$$
\begin{aligned}
x_0(t+h) &= x_0(t) + \tau f\left[x_0(t), x_N(t)\right] \big/ N \\
x_i(t+h) &= x_{i-1}(t), \quad 1 \le i \le N
\end{aligned}
\tag{1.11b}
$$

This map now is easy to be integrated taking in mind that N should always be small. With increasing both τ and μ in the Ikeda DDE, the dimension increases and therefore N should be decreasing as well in order not to get spurious solutions of the attractor. Hence, if one tries to integrate a DDE then the results should be verified next with decreasing the time-step h.

The spectrum of the LEs then is calculated with applying the Wolf algorithm (Wolf et al., 1985). One of the most used and effective numerical techniques to calculate the Lyapunov spectrum for a smooth dynamical system relies on periodic *Gram-Schmidt* orthonormalization of the Lyapunov vectors to avoid a misalignment of all the vectors along the direction of maximal expansion (Shimada and Nagashima, 1979). The Lyapunov exponents describe the behavior of vectors in the tangent space of the phase space and are defined from the Jacobian matrix. Generally, the latter describes how a small change at the initial point X_0 propagates to the final point of the trajectory in the N–dimensional phase space. The limit of the square root of the Jacobian multiplied with its transpose matrix for $t\to\infty$ results in a new matrix $L(X_0)$. Then the spectrum of the LEs is defined as the logarithm of the eigenvalues of $L(X_0)$ (Ruelle, 1979).

1.3.2 Linearized Stability Analysis

As with other types of equations, we derive a lot of insight from a stability analysis of the equilibria. An equilibrium point is a point in the state space for which $x(t) = x^*$ is a solution for all t. Thus, for a DDE of the form (1.7), equilibrium points satisfy:

$$f(x^*, x^*, ..., x^*) = 0 \qquad (1.12)$$

When we work out the stability of the equilibrium points of ODEs, we assume that the system has been displaced by a small distance in phase space from the equilibrium. With ODEs, the phase space is a finite-dimensional coordinate space. We proceed similarly with DDEs, except that the phase space is now infinite-dimensional, so that we have to consider displacements from equilibrium in this space. In other words, our displacements are time-dependent functions $\delta x(t)$ persisting over an interval of at least t_{max}, the longest delay (Mackey and Glass, 1977).

In this section we will give an example of stability analysis of the scaled dimensionless famous Mackey-Glass DDE in (1.13):

$$y' = 1 - \frac{ayy_\tau}{1 + y_\tau^{\ n}} \qquad (1.13)$$

Note that we changed the notation of $x(t-\tau)$ to x_τ and the dot represents the derivative with respect to the dimensionless time now. The steady state satisfies:

$$\frac{\left(y^*\right)^{n+1}}{1 + \left(y^*\right)^n} = \frac{1}{a} \qquad (1.14a)$$

Note that the function on the left-hand side is zero when $y^* = 0$ and that:

$$\frac{d}{dy^*}\left[\frac{\left(y^*\right)^{n+1}}{1+\left(y^*\right)^n}\right] = \frac{\left(y^*\right)^n\left[n+1+\left(y^*\right)^n\right]}{\left[1+\left(y^*\right)^n\right]^2} > 0 \tag{1.14b}$$

Since the derivative is strictly positive and tends to a constant at large n, this function passes through every positive value and so there is a unique, positive steady state for each value of n and α. We cannot determine the value of this steady state without knowing these two parameters. Hopefully that won't be too much of a problem. The (one-dimensional) Jacobians now are:

$$\begin{aligned} J_0 &= -\left.\frac{a y_\tau^{\ n}}{1+y_\tau^{\ n}}\right|_{y_\tau=y} = -\frac{a\left(y^*\right)^n}{1+\left(y^*\right)^n} \\ J_\tau &= -ay\left.\frac{n y_\tau^{\ n-1}}{\left(1+y_\tau^{\ n}\right)^2}\right|_{y=y^*, y_\tau=y^*} = -\frac{an\left(y^*\right)^n}{1+\left(y^*\right)^n} \end{aligned} \tag{1.14c}$$

The characteristic equation now becomes:

$$J_0 + e^{-\lambda\tau} J_\tau - \lambda = 0 \tag{1.14d}$$

We set: $\lambda=\mu+i\nu$ and separate the real and imaginary parts of this equation. We get therefore:

$$\begin{aligned} J_0 + e^{-\mu\tau} J_\tau \cos\left(\nu\tau\right) - \mu = 0 \\ e^{-\mu\tau} J_\tau \sin\left(\nu\tau\right) + \nu = 0 \end{aligned} \tag{1.14e}$$

Negative values of μ are clearly possible in contrast to positive ones. Suppose that the delay x is very small. Then, $\sin(\nu\tau)\to 0$, so that the second equation in (1.14e) implies $\nu\approx 0$. Equation (1.14e) then becomes $\mu\approx J_0+J_\tau$. Since the two Jacobians are negative, then μ is negative. If we know that the real part of the characteristic value can be negative, the standard trick to deal with these problems is to go looking for the bifurcation directly. In other words, set $\mu = 0$ and see if we can eliminate ν to get a sensible bifurcation condition. Setting $\mu = 0$, (1.14e) becomes:

$$\begin{aligned} J_\tau \cos\left(\nu\tau\right) = -J_0 \\ J_\tau \sin\left(\nu\tau\right) = -\nu \end{aligned} \tag{1.15}$$

We cannot satisfy the first of these equations if $n = 0$ since J_0 and J_τ have the same sign. Therefore, a pair of complex eigenvalues will cross the imaginary axis together, so we are looking for an Andronov-Hopf bifurcation. We then square these two equations and add them:

$$J_0^{\ 2} + \nu^2 = J_\tau^{\ 2} \tag{1.16}$$

This equation has real solutions if, and only if:

$$\left|J_0\right| \le \left|J_\tau\right|$$

Using (1.4c), this inequality becomes:

$$n \ge 1 + \left(y^*\right)^n \tag{1.17}$$

We see immediately that $n > 1$ is a necessary, but certainly not sufficient, condition for the bifurcation to occur. The question we now face is whether this inequality can be satisfied, and if so whether we need to seek for additional conditions. Inequality (1.17) can be satisfied provided y^* isn't too large. If the latter is very small, then the right-hand side of the inequality is actually a decreasing function of n, and n can eventually be made sufficiently large to satisfy the inequality. If on the other hand y^* is very large, the right hand-side of (1.17) will increase too fast. The boundary case is obtained when both the values of these functions and their tangents are equal:

$$n = 1 + \left(y^*\right)^n$$
$$\left(y^*\right)^n \ln\left(y^*\right) = 1$$

The system above can be easily numerically. The solution results in: $n = 4.5911$ and $y^*=1.3211$ and hence we must have $y^*<1.3211$ in order for the bifurcation to be possible.

In this section we derived some stability analysis for a simple DDE. Of course this procedure can be numerically programmed for any system of DDEs and especially when these are more complicated than the one studied here.

1.3.3 Coupled Delay Differential Equations: Lasers with Optical Feedback

In the previous section we demonstrated some dynamical properties of two famous one-dimensional DDEs. Herein we will concentrate on a system of coupled DDEs that is used to model semiconductor lasers subjected to optical feedback. Semiconductor lasers are very sensitive to external optical light. Even small external reflections and perturbations may provide a sufficient cause that can lead to an unstable operating behavior. This is the dominant reason why almost all types of commercial semiconductor lasers that apply to the standard telecommunications systems are provided with an optical isolation stage that eliminates – or at least minimizes – optical perturbations by the external environment. However, in applications where the raise of instabilities is of great importance, the isolation stage is omitted and the semiconductor lasers are driven to unstable operation. The optical feedback system is a phase-sensitive delayed-feedback system for which all three known routes, namely, period-doubling, quasi-periodicity, and route to chaos through intermittency can be easily found with varying suitably the key parameters.

Semiconductor lasers with applied optical feedback are very interesting configurations not only from the viewpoint of fundamental physics for nonlinear chaotic systems, but also for their potential for applications. The instability and dynamics of semiconductor lasers with optical feedback are studied by the nonlinear laser rate equations for the field amplitude, the phase, and the carrier density given below in their complex form (Vicente et al., 2005):

$$
\begin{aligned}
\dot{E}(t) &= \frac{(1+i\pm)}{2}\left[G(t)-\frac{1}{t_{ph}}\right]E(t) \; + {}^{o}E(t-\ddot{A})e^{-i\Omega\tau} \\
\dot{N}(t) &= \frac{I}{e}-\frac{1}{t_{n}}N(t)-G(t)\left|E(t)\right|^{2} \\
G(t) &= \frac{g\left[N(t)-N_{0}\right]}{1+s\left|E(t)\right|^{2}}
\end{aligned}
\tag{1.18}
$$

where α =5 is the linewidth enhancement factor, $g = 1.5 \times 10^{-8} ps^{-1}$ is the gain parameter, $s = 5 \times 10^{-7}$ is the gain saturation coefficient, $\gamma^{-1} = 2\ ps$ is the photon lifetime, $\gamma_e^{-1} = 2\ ns$ is the carrier lifetime, $N_0 = 1.5 \times 10^{8}$ is the carrier number at transparency, $e = 1.602 \times 10^{-19}$ C is the electron charge, Ω is the frequency of the free running laser, κ is the feedback coefficient and τ is the external cavity round-trip. In this section we will not give more details on the underlying significance of the parameters included in (1.18). The field equation dE(t)/dt is a complex one and therefore (1.13) consists of two DDEs and one ODE for the carriers dN(t)/dt. One should not that the delay term in (1.13) $\kappa E(t-\tau)$ is not included inside a nonlinearity in contrast to the Ikeda and Mackey-Glass DDEs in (1.8) and (1.9). This plays a crucial role in the development of complex dynamics since we will see next that such a system does not follow specific scaling laws for the attractor dimension and entropy.

In the numerical investigation we will fix the material parameters that appear in (1.18) and vary the control ones κ, I, and the phase $C_p=\Omega\tau$. At the beginning we will fix the length L in to L=1cm that corresponds to τ=200ps.

In Figure 6 we calculate the dimension and entropy of (1.18) as the feedback strength varies. This parameter is equivalent to the bifurcation parameter μ in the Ikeda DDE in (1.9) since it is in front of the delay term. Similarly to the Ikeda DDE, the two chaos quantifiers increase as the parameter increases and this is something to be expected. Nevertheless, there are other parameters in (1.13) that influence the dynamics. In Figure 7 we present complexity calculations as in Figure 6 but with the injection current changing that appears in the ODE of (1.18) for different values of κ.

In Figure 7 we note a maximum of the metric entropy and a monotonic increase of the Lyapunov dimension as the injection current increases for a moderate value of the feedback strength. The entropy maximum is attributed to the gain saturation coefficient included in the model and denoted with s in the gain term $G(t)$ in (1.18). Due to this parameter, the increase of bias current drives the laser into the saturation regime, where the chaotic fluctuations are unavoidably suppressed.

This deterioration of h_{KS} with increasing injection current may be a drawback for encryption applications since high unpredictability is required. In Figure 7 additionally we plot the chaos quantifiers with an enhanced feedback strength κ=25ns^{-1}. It is noted that the maximum of the metric entropy has disappeared and that h_{KS} increases monotonically with the injection current I as the Lyapunov dimension does also. Hence it is concluded that for short-cavity lasers, high values of the feedback strength κ and the injection current I are needed in order to preserve the system under enhanced chaotic dynamics and to

Figure 6. Lyapunov dimension D_{KY} and metric entropy h_{KS} versus feedback strength κ for injection current $I=1.5I_{TH}$

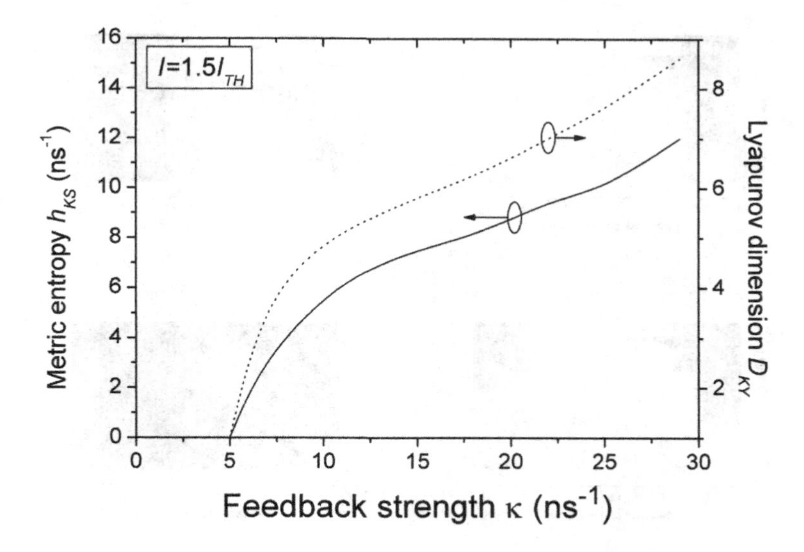

Figure 7. Lyapunov dimension D_{KY} and metric entropy h_{KS} versus injection current I for feedback strength κ=15ns^{-1} (left) and κ=25ns^{-1} (right).

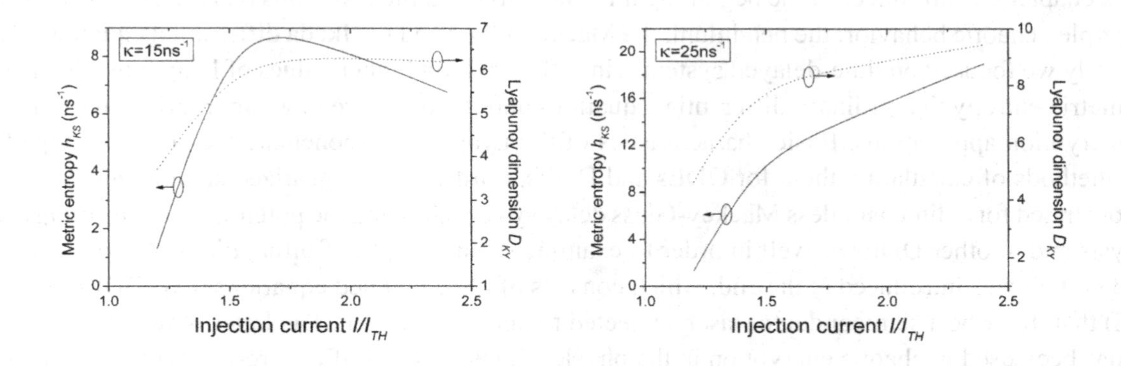

avoid windows of periodicity that would cancel the system's initial aim for cryptographic applications.

Such windows of periodicity appear in the bifurcation diagrams of Figure 8 where the feedback phase varies for various different values of feedback strength and injection current.

From Figure 8 it is clear that in order to transit the system to fully developed chaos and avoiding periodicity pockets, high values of injection current and feedback strength are required. Hence, from the above it is concluded that in order to avoid the phase dependence that may annihilate the desired chaotic carriers, it is a necessity to carefully tune the feedback phase during an experiment, or more preferably to operate the laser in high injection currents $I>2I_{TH}$ and high feedback strengths κ>15ns^{-1} for these specific integrated devices. Generally, DDEs with short delay times are known to result in significant dependence on parameters besides the damping ones like x_0 in (1.9).

Figure 8. Bifurcation diagrams as the feedback phase varies C_p for various values of κ and I.

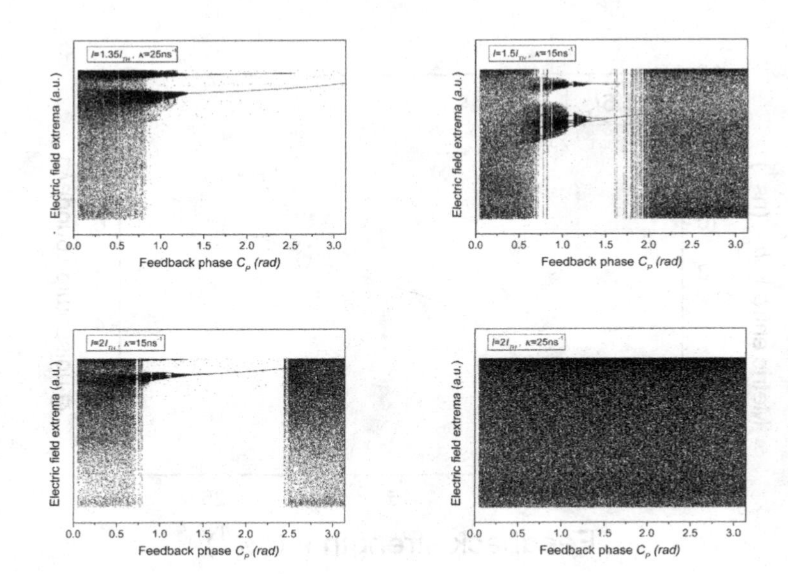

1.4 CONCLUSION

In this chapter we introduced at the beginning three famous dynamical systems that are capable to result in complex chaotic behavior: the pendulum, the Mackey-Glass and the Ikeda differential equations. Deliberately we focused on time-delayed systems since they exhibit higher values of Lyapunov dimension and metric entropy that ordinary differential equations and therefore are more appropriate and suitable for encryption applications. Basic characteristics of the Lyapunov exponents were presented together with methods of calculating them for ODEs and DDEs. Furthermore, linearized stability analysis was demonstrated for a dimensionless Mackey-Glass delay system showing the potential of performing such analysis and to other DDEs as well in order to examine specific types of bifurcations. A more complicated system was introduced at the end, which consists of three coupled equations (two DDEs and one ODE) that describe a semiconductor laser subjected to optical feedback. The latter is an optical system that has been used in chaotic encryption in the physical layer with significant results with using a commercial fibre-link (Argyris et al., 2005).

REFERENCES

Argyris, A., Hamacher, M., Chlouverakis, K. E., Bogris, A., & Syvridis, D. (2008). Photonic integrated device for chaos applications in communications. *Physical Review Letters*, *100*, 194101. doi:10.1103/PhysRevLett.100.194101

Argyris, A., Syvridis, D., Larger, L., Annovazzi-Lodi, V., Colet, P., & Fischer, I. (2005). Chaos-based communications at high bit rates using commercial fibre-optic links. *Nature*, *437*, 343–346. doi:10.1038/nature04275

Bogris, A., Argyris, A., & Syvridis, D. (2007). Analysis of the optical amplifier noise effect on electro-optically generated hyperchaos. *IEEE Journal of Quantum Electronics*, *47*(7), 552–559. doi:10.1109/JQE.2007.898843

Chembo, Y. K., Larger, L., & Colet, P. (2008). Nonlinear dynamics and spectral Stability of optoelectronic microwave oscillators. *IEEE Journal of Quantum Electronics*, *44*(9), 858–866. doi:10.1109/JQE.2008.925121

Chlouverakis, K. E., Argyris, A., Bogris, A., & Syvridis, D. (2008). Complexity and synchronization in chaotic fiber-optic systems. *Physica D. Nonlinear Phenomena*, *237*, 568–572. doi:10.1016/j.physd.2007.09.023

Chlouverakis, K. E., Mikroulis, S., Stamataki, I., & Syvridis, D. (2007). Chaotic dynamics of semiconductor microring lasers. *Optics Letters*, *32*, 2912–2914. doi:10.1364/OL.32.002912

Chlouverakis, K. E., & Sprott, J. C. (2007). Hyperlabyrinth chaos: From chaotic walks to spatiotemporal chaos. *Chaos (Woodbury, N.Y.)*, *17*, 023110. doi:10.1063/1.2721237

Epstein, I. R. (1990). Differential delay equations in chemical kinetics: Some simple linear model systems. *The Journal of Chemical Physics*, *92*, 1702–1712. doi:10.1063/1.458052

Farmer, J. D. (1982). Chaotic attractors of an infinite-dimensional dynamical system. *Physica D. Nonlinear Phenomena*, *4*(3), 366–393. doi:10.1016/0167-2789(82)90042-2

Hirsch, M. W., Smale, S., & Devaney, R. L. (2004). *Differential Equations, Dynamical Systems and an Introduction to Chaos* (2nd ed.). San Diego, CA: Academic Press/Elsevier.

Ikeda, K. (1979). Multiple-valued stationary state and its instability of the transmitted light by a ring cavity system. *Optics Communications*, *30*, 257–261. doi:10.1016/0030-4018(79)90090-7

Kaplan, J., & Yorke, J. (1979). Chaotic behavior of multidimensional difference equations . In Peitgen, H. O., & Walther, H. O. (Eds.), *Functional Differential Equations and Approximation of Fixed Points* (pp. 204–227). New York: Springer. doi:10.1007/BFb0064319

Lichtenberg, A. J., & Lieberman, M. A. (1992). *Regular and Chaotic Dynamics* (2nd ed.). New York: Springer-Verlag.

Liu, Z., Lai, Y.-C., & Matías, M. A. (2003). Universal scaling of Lyapunov exponents in coupled chaotic oscillators. *Physical Review E: Statistical, Nonlinear, and Soft Matter Physics*, *67*, 045203. doi:10.1103/PhysRevE.67.045203

Lorenz, E. N. (1963). Deterministic nonperiodic flow. *Journal of the Atmospheric Sciences*, *20*, 130–141. doi:10.1175/1520-0469(1963)020<0130:DNF>2.0.CO;2

Luzyanina, T., Engelborghs, K., & Roose, D. (2001). Numerical bifurcation analysis of differential equations with state-dependent delay. *International Journal of Bifurcation and Chaos in Applied Sciences and Engineering*, *11*(3), 737–753. doi:10.1142/S0218127401002407

Mackey, M., & Glass, L. (1977). Oscillation and chaos in physiological control systems. *Science*, *197*, 287–289. doi:10.1126/science.267326

Matthews, M. R., Stinner, A., & Gauld, C. F. (2005). *The Pendulum: Scientific, Historical, Philosophical and Educational Perspectives*. New York: Springer.

Pesin Ya, B. (1977). Lyapunov characteristic exponents and smooth ergodic theory. *Russian Mathematical Surveys, 32*(4), 55–114. doi:10.1070/RM1977v032n04ABEH001639

Rössler, O. E. (1976). An equation for continuous chaos. *Physics Letters. [Part A], 57*(5), 397–398. doi:10.1016/0375-9601(76)90101-8

Ruelle, D. (1979). Ergodic theory of differentiable dynamic systems. *Publications Mathématiques de L'IHÉS, 50*, 27–58.

Shimada, I., & Nagashima, T. (1979). A numerical approach to ergodic problem of dissipative dynamical systems. *Progress of Theoretical Physics, 61*, 1605–1616. doi:10.1143/PTP.61.1605

Sprott, J. C. (1997). Simplest dissipative chaotic flow. *Physics Letters. [Part A], 228*, 271–274. doi:10.1016/S0375-9601(97)00088-1

Sprott, J. C. (1997). Some simple chaotic jerk functions. *American Journal of Physics, 65*, 537–543. doi:10.1119/1.18585

Sprott, J. C. (2003). *Chaos and Time-Series Analysis*. Oxford, UK: Oxford University Press.

Sprott, J. C. (2007). A simple chaotic delay differential equation. *Physics Letters. [Part A], 366*, 397–402. doi:10.1016/j.physleta.2007.01.083

Vicente, R., Dauden, J., Colet, P., & Toral, R. (2005). Analysis and characterization of the hyperchaos generated by a semiconductor laser subject to a delayed feedback loop. *IEEE Journal of Quantum Electronics, 41*(4), 541–548. doi:10.1109/JQE.2005.843606

Wolf, A., Swift, J. B., Swinney, H. L., & Vastano, J. A. (1985). Determining Lyapunov exponents from a time series. *Physica D. Nonlinear Phenomena, 16*(3), 285–317. doi:10.1016/0167-2789(85)90011-9

Chapter 2
Chaotic Dynamical Systems Associated with Tilings of R^N

Lionel Rosier
Institut Elie Cartan de Nancy, France

ABSTRACT

In this chapter, we consider a class of discrete dynamical systems defined on the homogeneous space associated with a regular tiling of R^N, whose most familiar example is provided by the N–dimensional torus T^N. It is proved that any dynamical system in this class is chaotic in the sense of Devaney, and that it admits at least one positive Lyapunov exponent. Next, a chaos-synchronization mechanism is introduced and used for masking information in a communication setup.

1 INTRODUCTION

Chaos synchronization has exhibited an increasing interest in the last decade since the pioneering works reported in (Pecora & Carroll, 1990; Pecora & Carroll, 1991), and it has been advocated as a powerful tool in secure communication (Blekhman et al, 2002; Kolumban et al,1998; Nijmeijer, 1997; Kennedy & Ogorzalek, 1997).

Chaotic systems are indeed characterized by a great sensitivity to the initial conditions and a spreading out of the trajectories, two properties which are very close to the Shannon requirements of confusion and diffusion (Massey, 1992).

There are basically two approaches when using chaotic dynamical systems for secure communications purposes. The first one amounts to numerically computing a great number of iterations of a discrete chaotic system, in using e.g. the message as initial data (see (Schmitz, 2001) and the references therein). The second one amounts to hiding a message in a chaotic dynamics. Only a part of the state vector (the "output") is conveyed through the public channel. Next, a synchronization mechanism is designed to retrieve the message at the receiver part (see (Rosier et al, 2006) and the references therein).

DOI: 10.4018/978-1-61520-737-4.ch002

In both approaches, the first difficulty is to "build" a chaotic system appropriate for encryption purposes. In this context, the corresponding chaotic signals must have no patterning, a broadband power spectrum and an auto-correlation function that quickly drops to zero. In (Pecora et al, 1997), a mean for synthesizing volume-preserving or volume-expanding maps is provided. For such systems, there are several directions of expansion (stretching), while the discrete trajectories are folded back into a confined region of the phase space. Expansion can be carried out by unstable linear mappings with at least one positive Lyapunov exponent. Folding can be carried out with modulo functions through shift operations, or with triangular, trigonometric functions through reflection operations. Fully stretching piecewise affine Markov maps have also attracted interest because such maps are expanding in all directions and they have uniform invariant probability densities (see (Hasler et al, 1996; Rovatti & Setti,1998))

Besides, we observe that the word "chaotic" has not the same meaning everywhere, and that the chaotic behavior of a system is often demonstrated only by numerical evidences. The first aim of this chapter is to provide a rigorous analysis, based on the definition given by Devaney (Devaney, 2003), of the chaotic behavior of a large class of affine dynamical systems defined on the homogeneous space associated with a regular tiling of R^N. Classical piecewise affine chaotic transformations, as the *tent map*, belong to that class. The dimension N may be arbitrarily large in the theory developed below, but, for obvious reasons, most of the examples given here will be related to regular tilings of the plane ($N = 2$). The study of the subclass of (time-invariant or switched) affine systems on T^N, the N−dimensional torus, is done in (Rosier et al, 2004 ; Rosier et al, 2006). The folding for this subclass is carried out with modulo maps, which, from a geometric point of view, amounts to "fold back" R^N to $[0,1)^N$ by means of translations by vectors in Z^N. Those translations are replaced here by all the isometries of some crystallographic group for an arbitrary regular tiling of R^N. Notice also that the fundamental domain used in the numerical implementation may be chosen with some degree of freedom. It may be a hypercube (as $[0,1)^N$ for T^N), or a polyhedron, or a more complicated bounded, connected set in R^N.

For ease of implementation and duplication, a cryptographic scheme must involve a map for which the parameters identification is expected to be a difficult task, while computational requirements for masking and unmasking information are not too heavy. The second aim of this chapter is to show that all these requirements are fulfilled for the class of dynamical systems considered here. The way of extracting the masked information is provided through an observer-based synchronization mechanism with a finite-time stabilization property.

Let us now describe the content of the chapter. Section 2 is devoted to the mathematical analysis of the chaotic properties of the following discrete dynamical system

$$x_{k+1} = Ax_k + B \pmod G \tag{1}$$

where $A \in Z^{N\times N}$, $B \in R^N$, and (mod G) means roughly that x_{k+1} is the point in the fundamental domain T derived from $Ax_k + B$ by some transformation g in the group G. (1) may be viewed as a "realization" in $T \subset R^N$ of an abstract dynamical system on the homogeneous space R^N/G of classes modulo G. The torus T^N corresponds to the simplest case when G is the group of all the translations of vectors $u \in Z^N$ and the fundamental domain is $T = [0,1)^N$. Note that most of the examples encountered in the literature are given only for the torus T^N with $N = 1$ and $|A| \geq 2$, or for $N = 2$ and $\det A = 1$ (see e.g. (Katok & Hasselblatt, 1995)). We give here a sufficient condition for (1) to be chaotic in the sense of Devaney for any given regular tiling of R^N ($N \geq 1$), and we investigate the Lyapunov exponents of (1) and the equirepartition of the trajectories of (1).

Finally, a masking/unmasking technique based on a dynamical embedding is proposed in Section 3. The results in this chapter have been announced in (Rosier, 2010).

2 CHAOTIC DYNAMICAL SYSTEMS AND REGULAR TILINGS OF R^N

2.1 Chaotic Dynamical System

Let (M,d) denote a compact metric space, and let $f : M \rightarrow M$ be a continuous map. The following definition of a chaotic system is due to Devaney (Devaney, 2003).

Definition 1. The discrete dynamical system (Σ) $x_{k+1} = f(x_K)$ is said to be chaotic if the following conditions are fulfilled:

- (C1) (Sensitive dependence on initial conditions) There exists a number $\varepsilon>0$ such that for any $x_0 \in M$ and any $\delta>0$, there exists a point $y_0 \in M$ with $d(x_0,y_0) <\delta$ and an integer $k \geq 0$ such that $d(x_k,y_k) \geq \varepsilon$;
- (C2) (One-sided topological transitivity) There exists some $x_0 \in M$ with $(x_k)_{k\geq 0}$ dense in M;
- (C3) (Density of periodic points) The set $D = \{x_0 \in M;\ \exists\ k > 0,\ x_k = x_0\}$ is dense in M.

Recall (Walters, 1982, Thm 5.9), (Vesentin, 1999, Thm 1.2.2) that when f is onto (i.e., f(M) = M), the one-sided topological transitivity is equivalent to the condition:

(C2') For any pair of nonempty open sets U,V in M, there exists an integer $k \geq 0$ such that

$f^k(U) \cap V \neq \theta\ (\Leftrightarrow U \cap f^k(V) \neq \theta)$,

θ denoting the empty set.

2.2 Regular Tiling of R^N

An isometry g of R^N is a map from R^N into R^N such that $||g(X)-g(Y)|| = ||X-Y||$ for all $X,Y\in R^N$. Let G be a group of isometries of R^N such that for any point $X\in R^N$ the orbit of X under the action of G, namely the set

$G\,.\,X = \{g(X);\ g \in G\}$,

is closed and discrete. Let $P \subset R^N$ be a compact, connected set with a nonempty interior. Following (Berger, 2009), we shall say that the pair (G,P) constitutes a regular tiling of R^N if the two following conditions are fulfilled:

$$\bigcup_{g\in G} g(p) = R^N \qquad (2)$$

$$\forall g,h \in G\ (g(P^0) \cap h(P^0) \neq \theta \Rightarrow g=h) \tag{3}$$

Recall that P^0 stands for the interior of P, that is

$$P^0 = \{ x \in P;\ \exists \varepsilon > 0,\ B(x, \varepsilon) \subset P \}.$$

The set $P \subset R^N$ is termed a fundamental tile, and the group G a crystallographic group. An example of a regular tiling of R^2 with a triangular fundamental tile is represented in Figure 1. Note that a point $X \in R^N$ may in general be obtained in several ways as the transformation of a point in P by an isometry in G. We introduce a set *T* (not to be confused with the torus T), called a fundamental domain, with $P^0 \subset T \subset P$ and such that

$$\bigcup_{g \in G} g(T) = R^N \tag{4}$$

$$\forall X, X' \in T,\ \forall g \in G\ (X' = g(X) \Rightarrow X' = X). \tag{5}$$

Introducing the equivalence relation in R^N

$$X \sim Y \Leftrightarrow \exists g \in G,\ Y = g(X),$$

we denote by $x = \overline{X}$ the class of X for ~; i.e., $x = \{ g(X);\ g \in G\} = G\ .\ X$. When several group are considered at some time, we denote by $\overline{X}^G$ the class of X modulo G. Finally, we introduce the homogeneous space of cosets $H = (R^N/G) = \{x = \overline{X}\ ;\ X \in R^N\}$, and define on it the following metric

Figure 1. A regular tiling of R^2 with a triangular fundamental tile

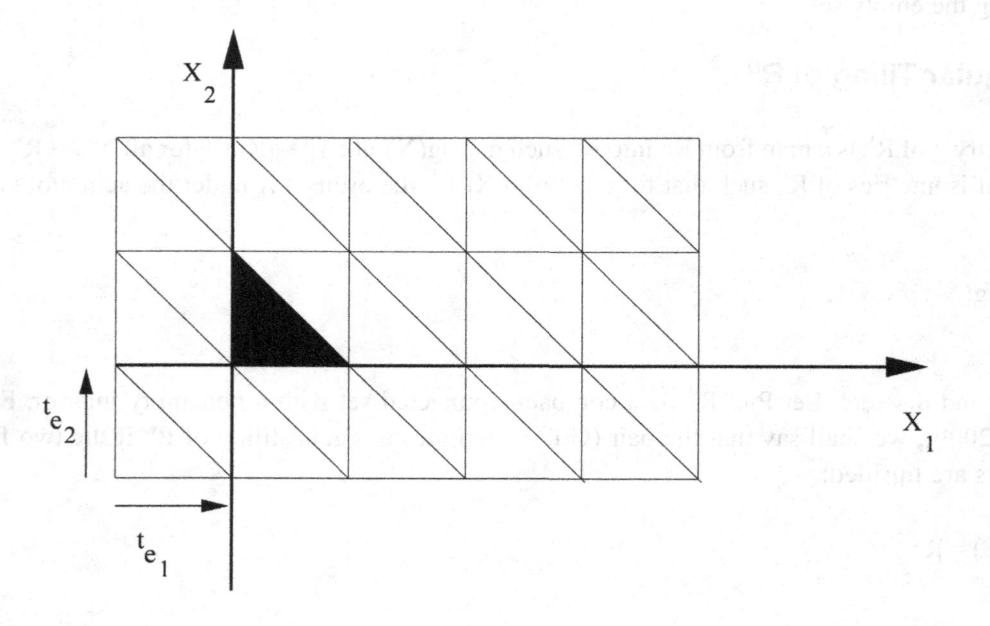

$$d(\bar{X},\bar{Y}) = \inf_{g\in G} \|Y - g(X)\|$$

The natural covering mapping $\pi : R^N \rightarrow H$, defined by $\pi(X) = \bar{X}$, satisfies

$$d(\pi(X),\pi(Y)) \leq \|X - Y\|,$$

hence it is continuous. It follows that $H = \pi(P)$ is a compact metric space. On the other hand, the restriction of π to T is a bijection from T onto H. We may therefore define the projection $\varpi: R^N \rightarrow T$ by $\varpi(X) = (\pi|T)^{-1}\pi(X)$. Note that ϖ is in general not continuous when T is equipped with the topology induced from R^N, while it is continuous when T is endowed with the topology inherited from H.

The simplest example of a regular tiling of R^N is provided by the group of translations by vectors with integral coordinates (which is isomorphic to the lattice subgroup)

$$G = \{ t_u ; u \in Z^N\} \sim Z^N, \quad (6)$$

where $tu(X) = X+u$. In such a situation, a fundamental tile (resp. domain) is given by $P = [0,1]^N$ (resp. $T = [0,1)^N$), and the homogeneous space H is the standard N−dimensional torus T^N. A classification (up to isomorphism) of the crystallographic groups of R^N has been done for a long time for $N \leq 3$. There are 17 such groups in R^2, and 230 groups in R^3, see (Berger, 2009; Burckhardt, 1966).

2.3 Affine Transformation

We aim to define "simple" chaotic dynamical systems on M = H by using affine transformations. Assume given a matrix $A \in Z^{N\times N}$ and a point $B \in R^N$. The following hypotheses will be used at several places in the chapter.

(H1): $\forall X,X' \in R^n \ (X \sim X' \Rightarrow AX + B \sim AX' + B)$

i.e. $X' = g(X)$ for some $g \in G$ implies $AX' + B = g'(AX + B)$ for some $g' \in G$;

(H2): There exist a subgroup $G' \subset G$ of translations and a finite collection of isometries $(gi)^k_{i=1}$ in G such that:

1. G is spanned as a group by the isometries in $G' \cup (gi)^k_{i=1}$;
2. $G' = \{tu \ ; u=\sum^N_{i=1} y_i u_i, y=(y_i)^N_{i=1} \in Z^N \}$ for some basis $(u_i)^N_{i=1}$ of R^N;
3. Setting $P' := \cup_{1\leq i\leq k} gi(P)$ we have that (G',P') is a regular tiling of R^N. We denote by T' a fundamental domain for (G',P')

(H1) is a compatibility condition needed to define a dynamical system on H. If G is given by (6), then (H1) holds for any $A \in Z^{N\times N}$ and any $B \in R^N$. However, if

$$G = \{tu; u =\sum^N_{i=1} y_i u_i, y=(y_i)^N_{i=1} \in Z^N \} \quad (7)$$

for some basis $(u_i)^N_{i=1}$ of R^N, then (H1) holds if and only if

$$U^{-1} AU \in Z^{N \times N} \tag{8}$$

where U is the N ×N matrix with ui as ith column for $1 \leq i \leq N$.

(H2) allows to decompose the projection ϖ onto T into a projection onto T', a fundamental domain for the regular tiling (G',P') of R^N involving only translations, followed by a projection from T' onto T .

Example 1. Let G =< t1, t2, r > and G' =< t1, t2 >, where t1(X) = X + (1,−1), t2(X) = X + (1, 1), and r(X1,X2) = (−X2,X1). Pick k = 4 and (g1, g2, g3, g4) = (r, r^2, r^3, id). Take as fundamental tiles P = {X = (X1,X2); $1 \leq X1 \leq 2$, $0 \leq X2 \leq 2 - X1$} (solid line) and P' = P ∪ r(P) ∪ r^2(P) ∪ r^3(P) (broken line) (see Figure 2).

Assume that (H1) holds. Then we may define

$$A\overline{X} + B = \overline{(AX + B)}$$

for any X ∈ R^N. Thus we may consider the dynamical system (∑A,B) on H defined by

$$(\sum A,B)\ xk+1 = f(x_k) := Ax_k + B,\ x_0 \in H \tag{9}$$

The map f is called an affine transformation of H.

Example 2. Let N = 1, and let G =< t, s > be the group spanned by the translation t(X) = X+2 and the symmetry s(X) = 2 − X. Set P = [0,1]. Then (G, P) constitutes a regular tiling of R. Note that P is also a

Figure 2. A regular tiling of R^2 with a triangular fundamental tile

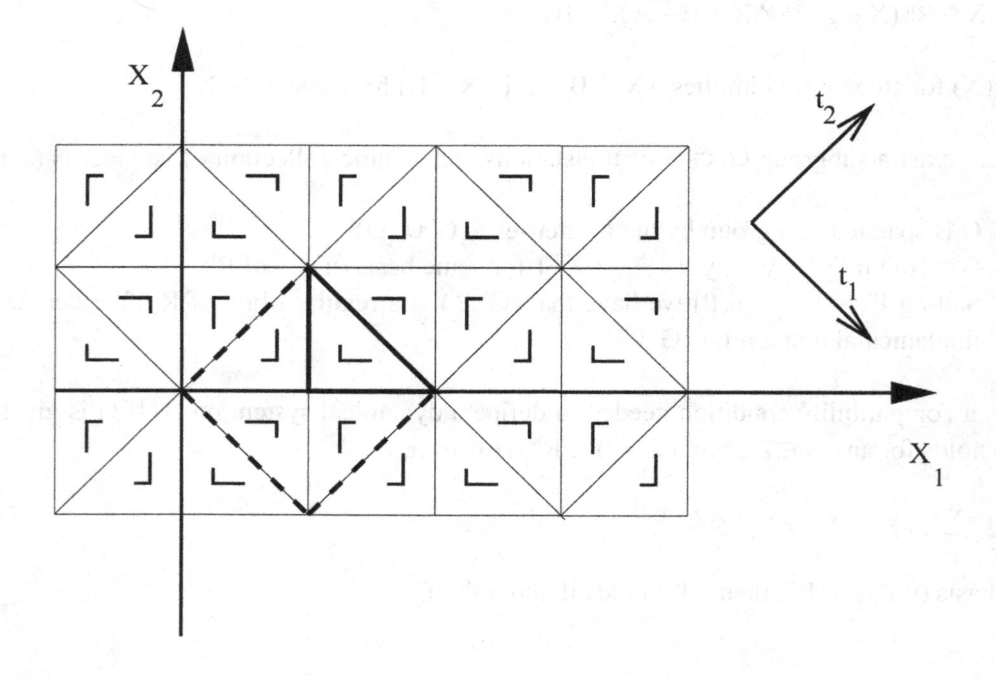

fundamental domain. Pick $(A,B) = (2,0) \in R^2$. (H1) and (H2) are satisfied with $G' = \{tu;\ u \in 2Z\}$, $k = 2$, $g_1 = s$ and $g_2 = s^2 = id$. Let us write the realization of (9) in P. Obviously, $AX \in P$ for $0 \leq X < 1/2$, while $s(AX) = 2(1 - X) \in P$ for $1/2 \leq X \leq 1$. Viewed in $P = [0,1]$, the dynamics reads then

$$x_{k+1} = h(x_k) \tag{10}$$

where h is the familiar tent map (see Figure 3).

$$h(x) = 2x \text{ if } 0 \leq x < \frac{1}{2},$$

$$h(x) = 2(1 - x) \text{ if } \frac{1}{2} \leq x \leq 1.$$

It follows from Theorem 2 (see below) that (10) is chaotic on [0,1].

When $H = T^N$ and $B = 0$, f is nothing else than an endomorphism of the topological group $(T^N,+)$, and f is onto (resp., an isomorphism) if and only if $\det A \neq 0$ (resp., $\det A = \pm 1$) (see (Walters, 1982 Thm 0.15)). Let sp(A) denote the spectrum of the matrix A, that is the set of the eigenvalues of A. A root of unity is any complex number of the form $\lambda = \exp(2\pi i t)$, with $t \in Q$. To see whether a dynamical system $(\sum A,B)$ is chaotic, we need the following key result (Walters, 1982 Thm1.11).

Proposition 1. Let $f(x) = Ax+b$ ($b \in T^N$, $A \in Z^{N\times N}$ with $\det A \neq 0$) be an affine transformation of T^N. Then the following conditions are equivalent:

(i) $(\sum A,b)$ is one-sided topologically transitive;
(ii) (a) A has no proper roots of unity (i.e., other than 1) as eigenvalues, and
 (b) $(A - I)T^N + Zb$ is dense in T^N;

Figure 3. A: Action of s and t; B: the tent map

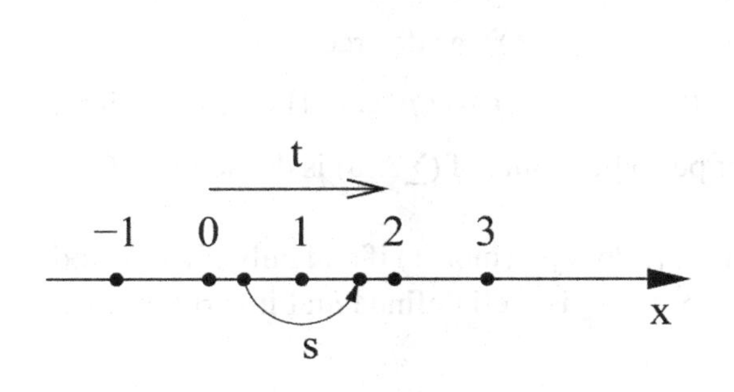

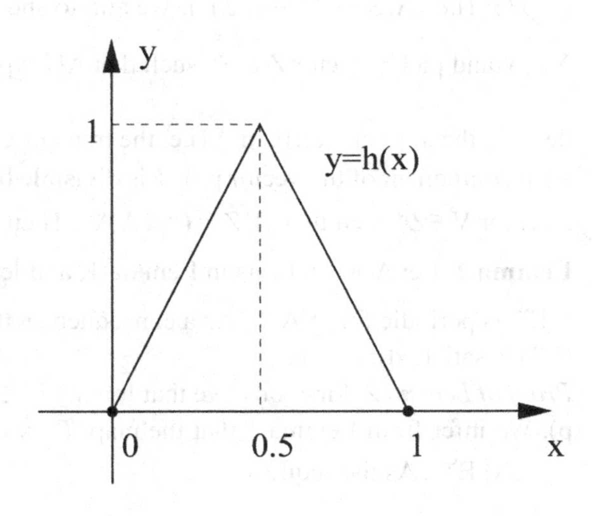

(iii) f is ergodic; that is, f is measure-preserving (i.e. for any Borel set $E \subset T^N$, $m(f^{-1}(E)) = m(E)$, where m denotes the Lebesgue measure on T^N), and the only Borel sets $E \subset T^N$ for which $f^{-1}(E) = E$ satisfy $m(E) = 0$ or $m(E) = 1$.

Notice that (ii) reduces to "A has no roots of unity as eigenvalues" when b = 0. Indeed, it may be seen that $(A - I)T^N$ is dense in T^N if and only if $(A - I)$ is invertible.

2.4 Endomorphism of T^N

The first result in this chapter, which comes from (Rosier et al, 2006), provides a necessary and sufficient condition for $\sum A,0$ to be chaotic in T^N.

Theorem 1. Let $A \in Z^{N\times N}$. Then $(\sum A,0)$ is chaotic in T^N if, and only if, det $A \neq 0$ and A has no roots of unity as eigenvalues.

Proof. Assume first that $(\sum A,0)$ is chaotic. We first claim that A is nonsingular. Indeed, if det A = 0, then the map f defined in (9) is not onto (Walters, 1982 Thm 0.15), i.e. $AT^N \neq T^N$. As AT^N is compact (hence equal to its closure), it is not dense in T^N, hence we cannot find some state $x_0 \in T^N$ such that the sequence $(x_k) = (A^k x_0)$ is dense in T^N, which contradicts (C2). Thus det $A \neq 0$. On the other hand, since $(\sum A,0)$ is one-sided topologically transitive, the matrix A has no roots of unity as eigenvalues by virtue of Proposition 1.

Conversely, assume that det $A \neq 0$ and that A has no roots of unity as eigenvalues. As (C1) is a consequence of (C2) and (C3) (see (Banks et al, 1992), (Vesentini, 1999 Thm 1.3.1)), we only have to establish the later properties. (C2) follows from Proposition 1. To prove (C3) we need to prove two lemmas.

Lemma 1. Let $A \in Z^{N\times N}$ be such that det $A \neq 0$, and pick any $p \in N^*$ with (p,det A) = 1 (i.e. p and det A are relatively prime). Then the map $T : x \in (Z/pZ)^N \to Ax \in (Z/pZ)^N$ is invertible.

Proof of Lemma 1. First, observe that the map T is well-defined. Indeed, if $X, Y \in Z^N$ fulfill $X - Y \in (pZ)^N$, then $AX - AY \in (pZ)^N$ so that AX and AY belong to the same coset in $(Z/pZ)^N = Z^N/(pZ)^N$. As $(Z/pZ)^N$ is a finite set, we only have to prove that T is one-to-one. Let $X, Y \in Z^N$ be such that AX = AY in $(Z/pZ)^N$ (i.e., $A(X - Y) \in (pZ)^N$). We aim to show that X = Y in $(Z/pZ)^N$ (i.e., $X - Y \in (pZ)^N$). Set U = X − Y and pick a vector $Z \in Z^N$ such that AU = pZ. It follows that $U = \left(\frac{p}{\det A}\right) \tilde{A}\, Z$, where $\tilde{A} \in Z^{N\times N}$ denotes the adjoint matrix of A (i.e. the transpose of the matrix formed by the cofactors). Since $U \in Z^N$, each component of the vector $p\tilde{A}\, Z$ is divisible by det A. Since (p, det A) = 1, we infer the existence of a vector $V \in Z^N$ such that $\tilde{A}\, Z = (\det A)V$. Then $X - Y = U = pV \in (pZ)^N$, as desired.

Lemma 2. Let A and p be as in Lemma 1, and let $Ep := \{\bar{0}, \overline{1/p}, \ldots, \overline{(p-1)/p}\} \subset T$. Then each point $x \in E^N_p$ is periodic for $(\sum A,0)$. As a consequence, the set of periodic points of $(\sum A,0)$ is dense in T^N (i.e., (C3) is satisfied).

Proof of Lemma 2. First, observe that for any $i,j \in \{0,\ldots,p-1\}$, $i/p \equiv j/p \pmod 1$ if and only if $i \equiv j \pmod p$. We infer from Lemma 1 that the map $\tilde{T} : x \in E^N_p \to Ax \in E^N_p$ is well defined and invertible. Pick any $x \in E^N_p$. As the sequence

$(\tilde{T}^k x)_{k\geq 1}$ takes its values in the (finite) set E^N_p, there exist two numbers $k2 > k1 \geq 1$ such that $\tilde{T}^{k_1}x = \tilde{T}^{k_2}x$. $\tilde{T}$ being invertible, we conclude that $A^{k1-k2}x = x$ (i.e., x is a periodic point). Finally, the set $E = \cup\{E^N_p\ ;\ p \geq 1, (p,\det A) = 1\}$ is clearly dense in T^N (take for p any large prime number), and all its points are periodic. This completes the proof of Lemma 2 and of Theorem 1.

For an affine transformation, we obtain a result similar toTheorem1 when$1\notin sp(A)$.

Corollary 1. Let $A \in Z^{N\times N}$ and $b \in T^N$. Assume that 1 is not an eigenvalue of A. Then $(\sum A,b)$ is chaotic in T^N if, and only if, $\det A \neq 0$ and A has no roots of unity as eigenvalues.
Proof. Pick any $B \in R^N$ with $\overline{B} = b$. As $1 \notin sp(A)$, we may perform the change of variables

$$x = r - \overline{(A-I)^{-1}B} \tag{11}$$

which transforms (9) into

$$r_{k+1} = Ar_k, \tag{12}$$

$$r_0 = x_0 + \overline{(A-I)^{-1}B}$$

Clearly, the conditions (C2) and (C3) are fulfilled for $(\sum A,b)$ if, and only if, they are fulfilled for (12). Therefore, the result is a direct consequence of Theorem 1.

Corollary 2. Let G be defined by (7) for some basis $(ui)^N_{i=1}$ of R^N. Let $A \in Z^{N\times N}$ and $B \in R^N$. Assume that (8) holds and that 1 is not an eigenvalue of A. Then $(\sum A,B)$ is chaotic in $H = R^N/G$ if, and only if, $\det A \neq 0$ and A has no roots of unity as eigenvalues.
Proof. From Corollary 1, we know that the dynamical system on T^N

$$z_{k+1} = \tilde{f}(z_k) := U^{-1}AUz_k + U^{-1}B \tag{13}$$

is chaotic if, and only if, $\det A \neq 0$ and A has no roots of unity as eigenvalues. To prove that the dynamical system on $H = R^N/G$

$$x_{k+1} = f(x_k) := Ax_k + B \tag{14}$$

is chaotic under the same conditions, it is sufficient to prove that the maps $f: H\rightarrow H$ and $\tilde{f}: T^N \rightarrow T^N$ are topologically conjugate; i.e., there exists a homeomorphism $h: H \rightarrow T^N$ such that $h\circ f = \tilde{f}\circ h$. Define h by $h(\overline{X}) = \overline{Z}$ where $Z = U^{-1}X$, $\overline{X} = G.X$ is the class of X in H and $\overline{Z}$ is the class of Z in T^N. Note first that h is well defined and continuous. Indeed, if $X' = X+UK$ with $K \in Z^N$, then $Z' = U^{-1}X' = U^{-1}X + K = Z + K$, so that h is well defined. On the other hand, the map $X\in R^N \rightarrow U^{-1}X\in T^N$ is clearly continuous. Obviously, h is invertible with $h^{-1}(\overline{Z}) = \overline{X}$ for $X = UZ$. h is therefore a homeomorphism from H onto T^N. Let us check now that $h \circ f = \tilde{f} \circ h$. Pick any $X \in R^N$. Then

$$h \circ f(\bar{X}^G) = h(\overline{AX + B}^G) = \overline{U^{-1}(AX+B)}^{TN} = \tilde{f}(\overline{U^{-1}X}^{TN}) = \tilde{f} \circ h(\bar{X}^G)$$

and the result follows.

We are in a position to state and prove the main result of this chapter.

Theorem 2. Let (G, P) be a regular tiling of R^N, and let $(A,B) \in Z^{N\times N} \times R^N$ be such that both the assumptions (H1) and (H2) are fulfilled. Assume in addition that det $A \neq 0$ and that A has no roots of unity as eigenvalues. Then the discrete dynamical system in R^N/G

$$x_{k+1} = Ax_k + B \tag{15}$$

is chaotic.

Proof. Pick any fundamental domain *T* for (G,P), and let G' and *T'* be as in (H2). In addition to (15), we shall consider the discrete dynamical system in R^N/G'

$$z_{k+1} = Azk + B. \tag{16}$$

For any given $X0 \in R^N$, let $x0 = \bar{X}_0^G$ and $z0 = \bar{X}_0^{G'}$. Clearly, if $X \sim X'$ (mod G'), then $X \sim X'$ (mod G). Therefore, one can define a map $p : R^N/G' \to R^N/G$ by $p(\bar{X}^{G'}) = \bar{X}^G$. p is continuous and onto. We need two claims.

CLAIM 1. $xk = p(z_k)$ for all k.

Indeed, this is true for k = 0, and if for some $k \geq 0$, $x_k = p(z_k)$ (i.e. for some $Xk \in R^N$, $x_k = \bar{X}_k^G$ and $z_k = \bar{X}_k^{G'}$), then we have that

$$x_{k+1} = \overline{AX_k + B}^G = p(\overline{AX_k + B}^{G'}) = p(z_{k+1})$$

which completes the proof of Claim 1.

CLAIM 2. The image by p of any dense set in R^N/G' is a dense set in R^N/G.

Let $A \subset R^N/G'$ be a given dense set. Pick any $X \in R^N$ and any $\varepsilon > 0$. Since A is dense in R^N/G', there exists $Y \in R^N$ such that $\bar{Y}^{G'} \in A$ and $d(\bar{X}^{G'}, \bar{Y}^{G'}) = \inf_{g\in G'} \|Y - g(X)\| < \varepsilon$. It follows that $d(\bar{X}^G, \bar{Y}^G) = \inf_{g\in G} \|Y - g(X)\| < \varepsilon$ for $G' \subset G$. Since $\bar{Y}^G = p(\bar{Y}^{G'}) \in p(A)$ and the pair (X,ε) was arbitrary, this demonstrates that p(A) is dense in R^N/G. Claim 2 is proved.

Let us complete the proof of Theorem 2. To prove that (15) is chaotic, it is sufficient (see (Banks et al, 1992)) to check that the conditions (C2) and (C3) are fulfilled. We know from Corollary 2 that (16) is chaotic. We may therefore pick $X0 \in R^N$ so that, setting $z0 = \bar{X}_0^{G'}$, the sequence $\{z_k\}k \geq 0$ defined by

(16) is dense in R^N/G'. By Claim 1 and Claim 2, the sequence {xk} defined by (15) and $x0 = \bar{X}_0^{\,G}$ is dense in R^N/G; that is, (C2) is fulfilled for (15). On the other hand, the set of periodic points for (16) is dense in R^N/G', since (C3) is fulfilled for (16). By Claim 1, any periodic point z_0 for (16) gives rise to a periodic point $x_0 = p(z_0)$ for (15). By Claim 2, the set of periodic points for (15) is dense in R^N/G ; i.e., (C3) is fulfilled for (15). The proof of Theorem 2 is complete.

Example 3.

1. Let G =< te1 , t2e2, s> where te1(X) = X + (1, 0), t2e2(X) = X + (0,2), s(X1,X2) =(X1,−X2), and P = [0,1]×[0,1]. Pick G' =< te1 , te2 >, k= 2, (g1,g2)=(s,id) (see Figure 4). Finally, pick $A = \begin{pmatrix} -2 & 0 \\ 0 & 3 \end{pmatrix}$ and B = (0.5,−3.2). Note that [A,S] :=AS − SA= 0, where $s=\begin{pmatrix} 1 & 0 \\ 0 & -1 \end{pmatrix}$ is the matrix corresponding to the symmetry s. Then (H1) and (H2) are satisfied, sp (A) = {−2, 3}, and by Theorem 2 the dynamical system (9) is chaotic in $H = R^2/G$.
2. Let G =< t2e1 , t2e2 , s1, s2 > where t2e1(X) = X + (2, 0), t2e2(X) = X+(0,2), s1(X1,X2)=(−X1,X2), s2(X1,X2) = (X1,−X2) = −s1(X1,X2), and P = [0,1] ×[0,1]. Pick G' =<t2e1 , t2e2 >, k = 4, (g1, g2, g3, g4) = (s1, s2, s2 ∘ s1, id). (see Figure 5). Finally, pick $A = \begin{pmatrix} 0 & -3 \\ 4 & 0 \end{pmatrix}$ and B = (−0.2, 1.7). Note that AS = −SA, where S is as above. Then (H1) and (H2) are satisfied, sp (A) = $\{\pm 2i\sqrt{3}\}$, and by Theorem 2 the dynamical system (9) is chaotic in $H = R^2/G$.

2.5 Lyapunov Exponents

Let M denote a compact differentiable manifold endowed with a Riemann metric $< u,v>m$, and let f : M → M be a map of class C^1. The following definition is borrowed from (Mané, 1987).

Definition 2. A point x ∈ M is said to be a regular point of f if there exist numbers $\lambda 1(x) > \lambda 2(x) > \ldots \lambda m(x)$ and a decomposition

$$T_x M = E1(x) \oplus \ldots \oplus Em(x)$$

of the tangent space $T_x M$ of M at x such that

$$\lim_{k \to +\infty} \frac{1}{k} \ln \| (Dx\ f^k)u(x) \| = \lambda j(x)$$

for all 0 ≠ u ∈ Ej(x) and every 1 ≤ j ≤ m. ($\|v\|^2 :=<v,v>x\ \forall v \in T_x M$.) The numbers λj(x) and the spaces Ej(x) are termed the Lyapunov exponents and the eigenspaces of f at the regular point x.

Assume now that the group G is such that each isometry g ∈ G has no fixed point, i.e. g(X) ≠ X for all X ∈ R^N. Then $H = R^N/G$ is a smooth flat Riemannian manifold. Before investigating the Lyapunov

Figure 4. G = < te1 , t2e2, s >

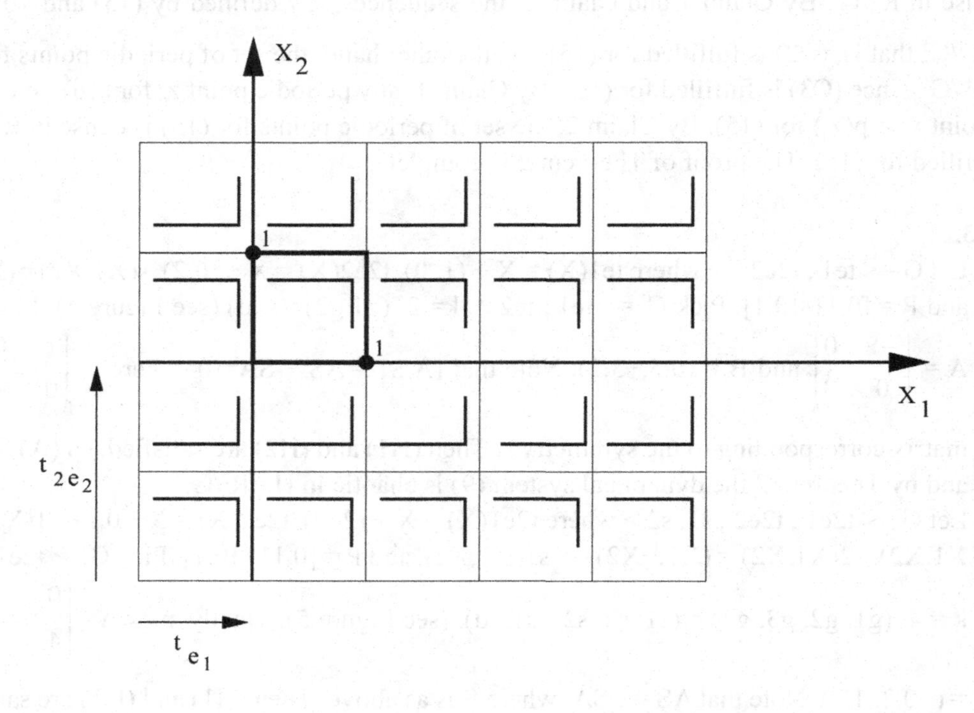

Figure 5. G = < t2e1 , t2e2 , s1, s2 >

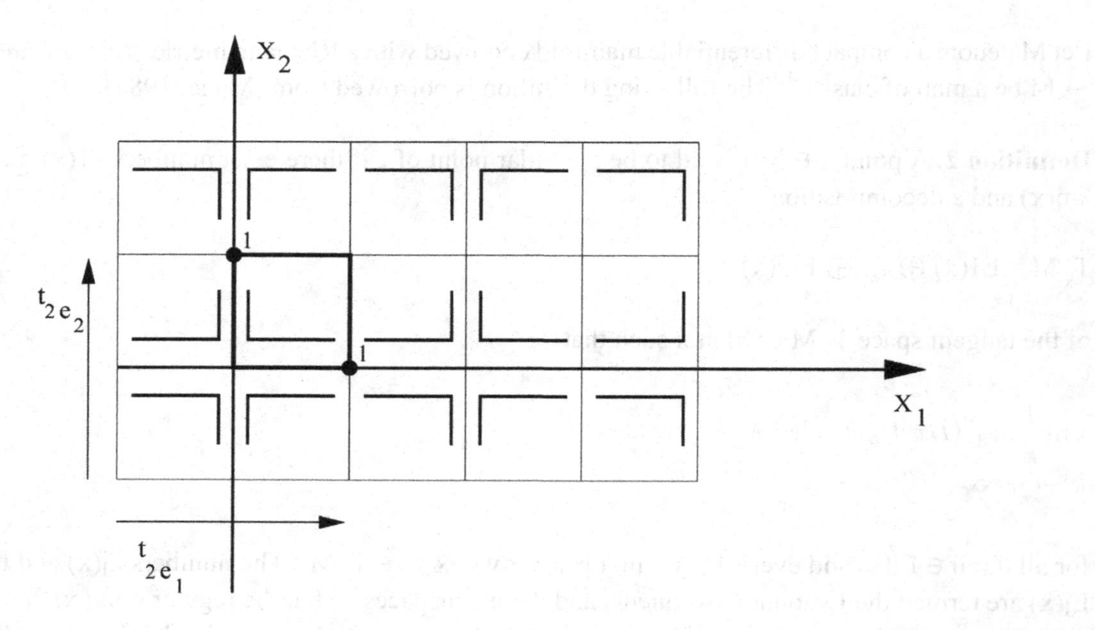

exponents of an affine transformation on H, let us give a few examples.

Example 4.

1. $H = T^N$, and more generally, $H = R^N/G$ where G is as in (7);
2. $H = R^N/G$ for $G = \langle t2e1, te2, te1 \circ s \rangle$ where (e1,e2) is the canonical basis of R^2 and $s(X1,X2) = (X1,-X2)$ (see Figure 6). H is then the Klein bottle. The torus T^2 and the Klein bottle H are the only smooth manifolds obtained in dimension 2. In dimension 3, there are 6 smooth manifolds (see (Wolf, 1984 Section 3.5.5 p. 117)).

Consider now an affine transformation $f(\bar{X}) = \overline{AX + B}$ of H, the pair (A,B) fulfilling (H1). Assume also that $\det A \neq 0$. Then for any $k \geq 1$,

$$f^k(\bar{X}) = \overline{A^k X + A^{k-1}B + \cdots + AB + B}$$

Pick a point $X \in P^0$ such that

$$A^k X + A^{k-1} B + \ldots + AB + B \in \cup_{g \in G}\, g(P^0)$$

(note that such a property holds for almost every $X \in R^N$), and an isometry $g \in G$ such that $g(A^k X + A^{k-1} B + \ldots + AB + B) \in P^0$.

For $\|U\|$ sufficiently small, we also have that

$$g(A^k (X + U) + A^{k-1} B + \ldots + AB + B) \in P^0$$

Figure 6. The regular tiling of R^2 associated with the Klein bottle

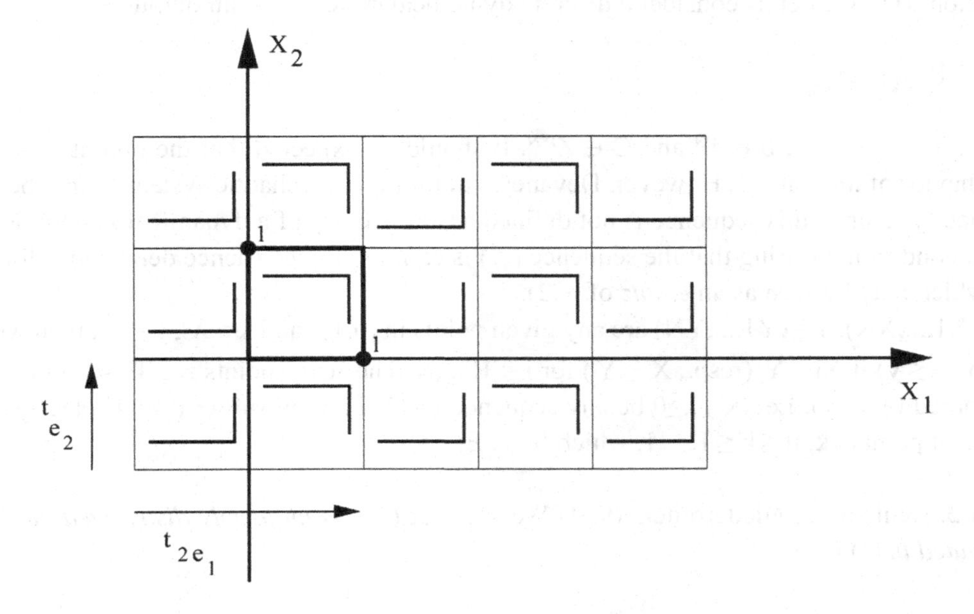

Therefore $(D_{\bar{x}} f^k)\bar{U} = \overline{GA^kU}$, where $G = Dg \in R^{N\times N}$. Since G is an orthogonal matrix, we have that $\|\overline{GA^KU}\| = \|A^kU\|$. Let $\mu 1 > \mu 2 > \ldots \mu m > 0$ denote the absolute values of the eigenvalues of A, and let Ei(x) be the direct sum of the generalized eigenspaces (see (Greub, 1975)) associated with the eigenvalues whose absolute value is μi, for each $i \leq m$. Then, using the Jordan decomposition of A, we easily see that for any $U \in Ej \setminus \{0\}$

$$\lim_{k \to +\infty} \frac{1}{k} \ln \| A^kU \| = \ln \mu j$$

Observe now that if $\sigma(A)$ does not intersect the circle $\{z \in C; |z| = 1\}$, then A has at least one eigenvalue λ with $|\lambda| > 1$ (since the product of all the eigenvalues of A is $\det A \in Z \setminus \{0\}$), hence f admits at least one positive Lyapunov exponent. Therefore, we have proved the following

Proposition 2. Let (G, P) be a regular tiling of R^N such that any isometry $g \in G$ has no fixed point. Let $(A,B) \in R^{N\times N} \times R^N$ be such that (H1) is satisfied, $\det A \neq 0$ and each eigenvalue λ of A satisfies $|\lambda| \neq 1$, and let $f : H = R^N/G \to H$ be defined by $f(x) = Ax + B$. Then almost every point $x \in H$ is regular for f, with Lyapunov exponents $\ln \mu 1 > \ldots > \ln \mu m$, where $\mu 1 > \ldots > \mu m$ are the absolute values of the eigenvalues of A. Furthermore, $\ln \mu 1 > 0$.

Notice that the existence of (at least) one positive Lyapunov exponent is often considered as a characteristic property of a chaotic motion (Van Wyk & Steeb, 1997). That property quantifies the sensitive dependence on initial conditions.

2.6 Equidistribution

In this section, $H = T^N$. Let us consider a discrete dynamical system with an output

$$x_{k+1} = Ax_k + B,\ y_k = Cx_k \tag{17}$$

where $x0 \in T^N$, $A \in Z^{N\times N}$, $b \in T^N$ and $C \in Z^{1\times N}$. It should be expected that the output y_k inherits the chaotic behavior of the state x_k. However, Devaney's definition of a chaotic system cannot be tested on the sequence (y_k), since this sequence is not defined as a trajectory of a dynamical system. Rather, we may give a condition ensuring that the sequence (y_k) is equidistributed (hence dense) in T for a.e. x_0, a property which may be seen as an *ersatz* of (C2).

If $X = (X1,...,XN)$, $Y= (Y1,...,YN)$ are any given points in $[0,1)^N$ and $x= \bar{X}$, $y = \bar{Y}$, then we say that $x < y$ (resp., $x \leq y$) if $Xi < Y_i$ (resp., $X_i \leq Y_i$) for $i = 1, ...,N$. The set of points $z \in T^N$ such that $x \leq z < y$ will be denoted by [x,y). Let (x_k) $k\geq 0$ be any sequence in T^N. For any subset E of T^N, let $S_k(E)$ denote the number of points xk, $0 \leq k \leq K - 1$, which lie in E.

Definition 3. (Kuipers & Niederreiter, 1974) We say that (xk) is *uniformly distributed modulo 1* (or *equidistributed in* T^N) if

$$\lim_{K \to \infty} \frac{S\,k([x,y))}{K} = m([x,y)) = \prod_{i=1}^{N} (Yi-Xi)$$

for all intervals $[x,y) \subset T^N$.

The following result is very useful to decide whether a sequence is equidistributed or not.

Proposition 3. **(Weyl criterion** (Kuipers & Niederreiter, 1974; Rauzy, 1976)) The sequence $(x_k)_{k\geq 0}$ is equidistributed in T^N if, and only if, for every lattice point $p \in Z^N$, $p \neq 0$

$$\frac{1}{K} \sum_{0\leq k<K} e^{2i\pi p.xk} \to 0 \text{ as } K \to +\infty$$

The next result shows that under the same assumptions as in Corollary 1 the sequences (x_k) and (y_k) are respectively equidistributed in T^N and T for a.e. initial state $x0 \in T^N$.

Theorem 3. Let $A \in Z^{N\times N}$, $b \in T^N$ and $C \in Z^{1\times N} \setminus \{0\}$. Assume that det A $\neq 0$ and that A has no roots of unity as eigenvalues (hence $\sum A,b$ is chaotic). Then for a.e. $x_0 \in T^N$ the sequence (x_k) (defined in (17)) is equidistributed in T^N, and the sequence $(y_k) = (Cx_k)$ is equidistributed in T.

Proof: By virtue of Theorem 1, the map $f(x) = Ax + b$ is ergodic on T^N. It follows then from Birkhoff Ergodic Theorem (see e.g. (Walters, 1982 Thm 1.14)) that for any $h \in L^1(T^N, dm)$ and for a.e. $x_0 \in T^N$

$$\frac{1}{K} \sum_{0\leq k<K} h(f^k(x_0)) \to \int_T^N h(y)\, dm(y) \text{ as } K \to +\infty.$$

Therefore, for every lattice point $p \in Z^N$, $p \neq 0$, and for a.e. $x0 \in T^N$

$$\frac{1}{K} \sum_{0\leq k<K} e^{2i\pi p.\, f^k(x0)} \to \int_T^N e^{2i\pi p.y}\, dm(y) = 0 \text{ as } K \to +\infty.$$

As $Z^N \setminus \{0\}$ is countable, the same property holds for a.e. $x_0 \in T^N$ and all $p \in Z^N \setminus \{0\}$. Therefore, we infer from Weyl criterion that the sequence $(x_k) = (f^k(x_0))$ is equidistributed for a.e. $x_0 \in T^N$. Pick any $x_0 \in T^N$ such that (x_k) is equidistributed, and let us show that the output sequence $(y_k) = (Cx_k)$ is also equidistributed provided that $C = (C1,\ldots,CN)$ is different from $(0,\ldots,0)$. Indeed, for any $p \in Z \setminus \{0\}$

$$\frac{1}{K} \sum_{0\leq k<K} e^{2i\pi p yk} = \frac{1}{K} \sum_{0\leq k<K} e^{2i\pi (pc) xk} \to 0 \text{ as } K \to +\infty,$$

hence the equidistribution of (y_k) follows again by Weyl criterion.

Remark 1. For a regular tiling (G,P) of R^N, even if the sequence (x_k) is equidistributed in H, the output (y_k) fails in general to be equidistributed in T. This is clear when one considers a regular tiling of R^2 with the triangle $P = \{X = (X1,X2);\ X1 \geq 0,\ X2 \geq 0,\ X1 + X2 \leq 1\}$ as fundamental tile, and $C = (1\ 0)$.

3 SYNCHRONIZATION AND INFORMATION RECOVERING

The aim of this section is to suggest a chaos-based encryption scheme involving affine transformations on the homogeneous space H associated with some regular tiling of R^N. We shall provide conditions which guarantee a synchronization with a finite-time stability of the error despite the inherent nonlinearity of the chaotic systems under study.

3.1 Encryption Setup

Assume given a regular tiling (G,P) of R^N and a pair $(A,B) \in R^{N\times N}\times R^N$ fulfilling the assumptions of Theorem 2. For the sake of simplicity, assume further that $R^N/G'=T^N$, so that $T' = [0,1)^N$. Let $\varpi: R^N \to T$ and $\varpi': R^N \to T'$ denote the projections on the fundamental domains of (G,P) and (G',P'), respectively. Set for $k\in \mathbf{N}$ and $X \in R^N$

$$(\Sigma_{A,B,M,C})\ \varpi_k(X) = \varpi'(X) \text{ if } k\notin(N+1)\,\mathbf{N}; \tag{18}$$

$$\varpi_k(X) = \varpi(X) \text{ if } k\in(N+1)\,\mathbf{N}.$$

At each discrete time k, a symbol $m_k \in R$ (the *plaintext*) of a sequence $(m_k)_{k\geq 0}$ is encrypted by a (nonlinear) encrypting function e which "mixes" m_k and X_k and produces a *ciphertext* uk = e(Xk,mk). We also assume given a decrypting function d such that $m_k = d(X_k,u_k)$ for each k. Next, the ciphertext u_k is embedded in the dynamics (9). We shall consider the following encryption

$$(\Sigma A,B,M,C)\ Xk+1 = \varpi_k\{A(X_k+Mu_k)+B\},\ Y_k = C(X_k+Mu_k) \tag{19}$$

which corresponds to an embedding of the ciphertext in both the dynamics and the output. In (19), $A \in Z^{N\times N}$, $M \in Z^{N\times 1}$, and $C \in Z^{1\times N}$ are given matrices, and $B\in R^N$. $Y_k \in R$ is the output conveyed to the receiver through the channel.

From the definition of the decrypting function d, it is clear that to retrieve m_k at the decryption side we need to recover the pair (X_k,u_k), which in turn calls for reproducing a chaotic sequence ($\hat{X}_k$) synchronized with (X_k) (i.e., such that $\hat{X}_k - X_k \to 0$). To this end, we propose a mechanism based on some suitable unknown input observers, inspired from the ones given in (Millerioux & Daafouz, 2003; Millerioux & Daafouz, 2004; Rosier et al, 2004; Rosier et al, 2006).

We stress that the gain matrices have to be Z-valued here.

For the encryption considered here, the decryption involves the following observer-like structure

$$(\Sigma A,B,M,C)'\ \hat{X}_{k+1} = \varpi_k\{A\hat{X}_k + L(Y_k - \hat{Y}_k) + B\}\ \ \hat{Y}_k = C\hat{X}_k \tag{20}$$

where $L \in Z^{N\times 1}$, $\hat{X}_k \in R^N$ and $\hat{Y}_k \in R$ ($\hat{X}_0$ being an arbitrary point in R^N). Let $\bar{X}$ denote the class of X modulo G', i.e. in T^N. Set $e_k = \bar{X}_k - \bar{\hat{X}}_k$ for all $k \geq 0$. Noticing that for all $X \in R^N$

$$\overline{\varpi k(X)} = \overline{\varpi'(X)} = \bar{X} \text{ for } 1 \leq k \leq N,$$

we obtain by subtracting (20) from (19) that the error dynamics reads

$$e_{k+1} = (A - LC)e_k + \overline{(A - LC)Mu_k}, \text{ for } 1 \le k \le N. \tag{21}$$

Before proceeding to the design of the observers, we give a few definitions and a preliminary result.

3.2 Definitions and preliminary results

Definition 4. A pair (A^b,C^b) is said to be in a *companion canonical form* if it takes the form

$$A^b = \begin{pmatrix} -\alpha N-1 & 1 & 0 & \cdots & 0 \\ -\alpha N-2 & 0 & 1 & \cdots & 0 \\ \vdots & \vdots & \vdots & \ddots & \vdots \\ -\alpha 1 & 0 & 0 & \cdots & 1 \\ -\alpha 0 & 0 & 0 & \ldots & 0 \end{pmatrix}, C^b=(1\ 0 \ldots\ldots 0\ 0) \tag{22}$$

It is well known that the characteristic polynomial of A^b reads

$$\chi A^b(\lambda) = \lambda^N + \alpha^{N-1}\lambda^{N-1} + \ldots + \alpha^1\lambda + \alpha^0.$$

Definition 5. Two pairs (A,C) and (A^b,C^b) in $Z^{N\times N} \times Z^{1\times N}$ are said to be *similar over Z* if there exists a matrix $T \in Z^{N\times N}$ with $\det T = \pm 1$ (hence $T^{-1} \in Z^{N\times N}$ too) such that $A = T^{-1} A^b T$, $C = C^b T$.

The following result provides a sufficient condition for an observable pair (A,C) to admit a Z-valued gain matrix L such that $A - LC$ is Hurwitz.

Proposition 4. Let $A \in Z^{N\times N}$ and $C \in Z^{1\times N}$ be two matrices such that (A,C) is similar over Z to a pair $(A^b,C^b) \in Z^{N\times N}\times Z^{1\times N}$ in a companion canonical form. Let us denote by $(-\alpha^{N-1} \ldots -\alpha^0)'$ the first column of A^b. Then there exists a unique matrix $L \in Z^{N\times 1}$ such that the matrix $A - LC$ is Hurwitz (i.e., $sp(A - LC) \subset \{z \in C; |z|<1\}$), namely $L = T^{-1} L^b$ with $L^b = (-\alpha^{N-1} \ldots - \alpha^0)'$. Furthermore, $(A - LC)^N = 0$.
Proof. Write $A = T^{-1} A^b T$, $C = C^b T$, with (A^b,C^b) as in (22) and $T \in Z^{N\times N}$ with $\det T = \pm 1$. For any given matrix $L \in Z^{N\times 1}$, we define the matrix $L^b = (l^{N-1} \ldots l^0)'$ by $L^b = TL$. Then, $A-LC=T^{-1}(A^b-L^bC^b)T$ with

$$A^b - L^bC^b = \begin{pmatrix} -\alpha N & -1 & -l N & -1 & 1 & 0 & \cdots & 0 \\ -\alpha N & -2 & -l N & -2 & 0 & 1 & \cdots & 0 \\ & \vdots & & & \vdots & \vdots & \ddots & \vdots \\ -\alpha 1 & -l1 & & & 0 & 0 & \cdots & 1 \\ -\alpha 0 & -l0 & & & 0 & 0 & \ldots & 0 \end{pmatrix} \tag{23}$$

Its characteristic polynomial reads

$$\chi A^b - L^b C^b(\lambda) = \lambda^N + (\alpha^{N-1} + l^{N-1})\lambda^{N-1} + \ldots + (\alpha^1 + l^1)\lambda + (\alpha^0 + l^0).$$

If L is such that $A - LC$ is Hurwitz, then $A^b - L^b C^b = T(A - LC)T^{-1}$ is Hurwitz too, hence we may write $\chi_{A-LC}(\lambda) = \chi_{A^b - L^b C^b}(\lambda) = \lambda^p \chi(\lambda)$, where $p \in \{0,...,N\}$ and $\chi \in Z[\lambda]$ has its roots$\lambda_1,...,\lambda_{N-p}$ in the set $\{z \in C; 0 < |z| < 1\}$. Assume that $p < N$, and denote by q the constant coefficient ofχ.Then $q \neq 0$ (since $\chi(0) \neq 0$), and $|q| = \prod^{N-p}_{i=1} |\lambda_i| < 1$, which is impossible, since $q \in Z$. Therefore $p = N$ and $l^j = -\alpha^j$ for any $j \in \{0,...,N-1\}$ (hence L^b and L are unique). On the other hand

$$A^b - L^b C^b = \begin{pmatrix} 0 & 1 & 0 & \cdots & 0 \\ 0 & 0 & 1 & \cdots & 0 \\ \vdots & \vdots & \vdots & \ddots & \vdots \\ 0 & 0 & 0 & \cdots & 1 \\ 0 & 0 & 0 & \cdots & 0 \end{pmatrix}$$

For this choice of L, $\chi_{A-LC}(\lambda) = \lambda^N$ and $(A - LC)^N = 0$.

It should be emphasized that the above argument shows that a Z-valued matrix N is Hurwitz if and only if it is nilpotent. In other words, the system $v_{k+1} = Nv_k$ is asymptotically stable if and only if it is finite-time stable.

We are now in a position to state the second main result of this chapter.

Theorem 4. Let (G,P) be a regular tiling of R^N, and let $(A,B) \in Z^{N\times N} \times R^N$ be such that (H1) and (H2) are fulfilled with $R^N/G' = T^N$. Assume given $C \in Z^{1\times N}$ such that (A,C) is similar over Z to a pair (A^b,C^b) in a companion canonical form. Then one can pick two matrices $L \in Z^{N\times 1}$ and $M \in Z^{N\times 1}$ so that $(A - LC)M = 0$ and $CM = 1$. Furthermore

$X_k = \hat{X}_k$ and $u_k = Y_k - \hat{Y}_k$ $\forall k \geq N + 1$.

Proof. Let T, A^b, C^b, L and L^b be as in the proof of Proposition 4. Set $M^b = (1\ 0.....\ 0)''$ and $M = T^{-1} M^b$. Then $(A - LC)M = T^{-1}(A^b - L^b C^b)T.T^{-1} M^b = 0$ by (23), and $CM = C^b T.T^{-1} M^b = 1$. On the other hand, it follows from (21) and the choice of M that

$e_{k+1} = (A - LC)e_k$ $\forall k \in \{1,...,N\}$

hence $e_{N+1} = (A - LC)^N e_1 = 0$. Since X_{N+1} and $\hat{X}_{N+1}$ belong to T' by construction, we have that $\hat{X}_{N+1} = X_{N+1}$. To complete the proof, it is sufficient to prove the following

CLAIM. For any $k \geq 0$, $\hat{X}_k = X_k$ implies $\hat{X}_{k+1} = X_{k+1}$.

Indeed, using the fact that $(A - LC)M = 0$ and $\hat{X}_k = X_k$ we obtain that

$$\begin{aligned} \hat{X}_{k+1} &= \varpi_k(A\hat{X}_k + LC(X_k + Mu_k - \hat{X}_k) + B \\ &= \varpi_k(AX_k + AMu_k + B) \\ &= X_{k+1} \end{aligned}$$

This completes the proof of Theorem 4.

Remark 2.

1. The projection $\varpi_k(x)$ allows to switch between the dynamics (15) and (16) in R/G and R/G', respectively. For a dynamics in T^N only (G' = G), one can replace $\varpi_k(x)$ by $\varpi'(x)$ (the projection onto $[0,1)^N$).
2. The result in Theorem 4 remains true if we take $\varpi_k(x) = \varpi'(x)$ for $k \leq N$ and $\varpi_k(x) = \varpi(x)$ for $k \geq N + 1$. However, the definition of $\varpi_k(x)$ in (18) guarantees that a finite time synchronization occurs even if the output Y_k is not transmitted at some times. Such a property may be useful for the secured transmission of video sequences.
3. The output $Y_k = C(X_k + Mu_k)$ may be replaced by $\tilde{Y}_k = h(Y_k)$, where $h : R \rightarrow R$ is a nonlinear invertible map. This renders the analysis of the dynamics of Y_k much more complicated.
4. In practice, when $H = T^N$, the matrices A, C, L and M may be constructed in the following way. Pick any matrix $\hat{T} = [\hat{T}_{i,j}] \in Z^{N \times N}$ with $\hat{T}_{i,j} = 0$ for $i > j$ and $\hat{T}_{i,i} = 1$ for all i. We set $T = \hat{T}' \hat{T}$. Note that $\det \hat{T} = \det T = 1$. Next, we pick a pair (A^b, C^b) in a companion canonical form so that the roots of χA^b do not belong to the set $\{0\} \cup \{z \in C; |z| = 1\}$. Then A, C, L and M are defined by $A = T^{-1} A^b T$, $C = C^b T$, $L = T^{-1} A^b (C^b)'$, and $M = T^{-1} (C^b)'$.

3.3 Numerical Simulations

This section is borrowed from (Rosier et al, 2004, preprint of IECN). Assume $H = T^3$ and consider the dynamical system $(\Sigma A,b,M,C)$ with

$$A = \begin{pmatrix} -19 & 26 & 7 \\ -51 & 65 & 17 \\ 152 & -184 & -47 \end{pmatrix}, C = (6 - 5 - 1), b = 0.$$

$(\Sigma A,b)$ is chaotic by virtue of Theorem 1, since $\det A = 3$ (hence $\det A \neq 0$) and the eigenvalues of A are -3, -0.4142, 2.4142 (A has no roots of unity as eigenvalues). The pair (A,C) is similar over Z to the pair (A^b, C^b) in companion canonical form, where

$$A^b = \begin{pmatrix} -1 & 1 & 0 \\ 7 & 0 & 1 \\ 3 & 0 & 0 \end{pmatrix}, C^b = (1\ 0\ 0) \text{ and } T = \begin{pmatrix} 6 & -5 & -1 \\ -5 & 10 & 3 \\ -1 & 3 & 1 \end{pmatrix}$$

According to Proposition 4, the unique matrix $L \in Z^{3\times 1}$ such that $A - LC$ is Hurwitz is $L = T^{-1} L^b$, with $L^b = (-1\ 7\ 3)'$. We obtain $L = (-2 - 6\ 19)'$. The corresponding matrix $M \in Z^{3\times 1}$ such that $(A - LC)M = 0$ and $CM = 1$ is $M = (1\ 2\ {-5})'$.

The information to be masked is a flow corresponding to integers ranging from 0 to 255. The data are scaled to give an input u_k ranging from 0 to 1, and are embedded into the chaotic dynamics of $(\Sigma A,b,M,C)$. From a practical point of view, the transmitted signal yk cannot be coded with an infinite accuracy and so it has to be truncated for throughput purpose. The observer $(\Sigma A,b,M,C)'$ is used in order to recover the information. Numerical experiments bring out that the number of digits of the conveyed output can actually be limited without giving rise to recovering errors. The results reported in Figure 7 show a perfect recovering for a number of digits of yk equal to 4 (this is the minimum number required for perfect retrieving). The recovering error reaches zero after 3 steps, a fact which is consistent with above theoretical results on finite time synchronization (N = 3). The figure highlights the fact that even though the state reconstruction may not be perfect (residual errors due to truncations), a perfect information reconstruction is nevertheless achieved.

Remark 3. Actually, for any system ΣA,B,M,C, the numerical computations can be performed in an *exact way*, i.e. without rounding errors, provided that the number of digits is sufficiently large.

3.4 Concluding Remarks

The message-embedding masking technique studied here does not originate from the conventional cryptography (see (Menezes, 1996) for a good survey). Nevertheless, it seems to be highly related to some popular encryption schemes, the so-called *streamciphers* (Millerioux et al, 2005). Therefore, it is desirable that the proposed scheme be robust against both statistical and algebraic attacks. On one hand, the robustness against statistical attacks follows from the chaotic behavior of the output. On the other hand, the security against algebraic attacks rests on the difficulty to identify the parameters of the system. The identification of the parameters is here a hard task for two reasons:

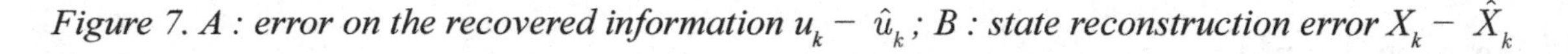

Figure 7. A : error on the recovered information $u_k - \hat{u}_k$; B : state reconstruction error $X_k - \hat{X}_k$

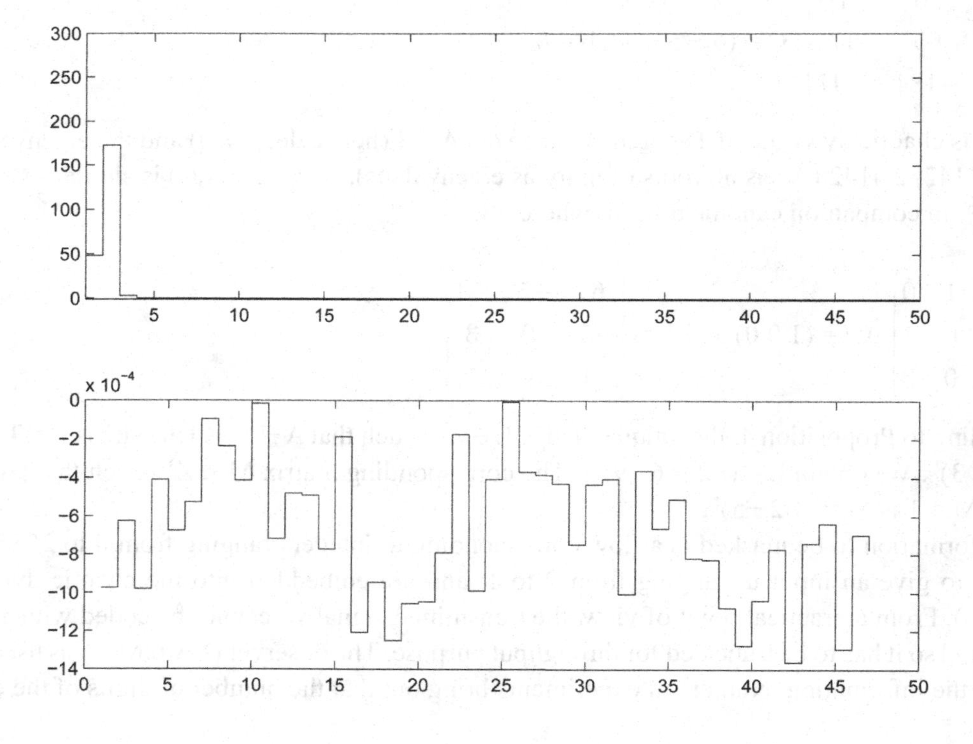

1. The particular structure of the encryption system ($\sum$A,B,M,C), that is the dimension of the matrix A and the tiling of the space used, is assumed to be unknown;
2. The ciphertext uk actually results from a mixing between the plaintext mk and the state X_k ($u_k = e(X_k, m_k)$). This generally results in a nonlinear dynamics ($\sum$A,B,M,C), rendering the parameters hardly identifiable (Ljung & Glad,1994).

A real-time implementation has already been carried out on an experimental platform involving a secured multimedia communication. (For details about the platform, see e.g. (Millerioux et al, 2003)).

REFERENCES

Banks, J., Brooks, J., Cairns, G., Davis, G., & Stacey, P. (1992). On Devaney's definition of chaos. *The American Mathematical Monthly*, *99*(4), 332–334. doi:10.2307/2324899

Berger, M. (2009). *Geometry I*. Berlin: Springer-Verlag.

Blekhman, I. I., Mosekilde, E., & Fradkov, A. L. (Eds.). (2002). *Special Issue on Chaos Synchronization and Control* (*Vol. 58*). Amsterdam: Elsevier.

Blondel, V. D., Sontag, E. D., Vidyasagar, M., & Willems, J. C. (1999). *Open Problems in Mathematical Systems and Control Theory*. Communication and Control Engineering.

Burckhardt, J. J. (1966). *Die Bewegungsgruppen der Kristallographie* (2nd ed.). Basel, Germany: Birkhauser Verlag.

Devaney, R. (2003). *An introduction to chaotic dynamical systems. Studies in nonlinearity*. Boulder, CO: Westview Press.

Greub, W. (1975). *Linear Algebra* (4th ed.). New York: Springer-Verlag.

Hasler, M., Delgado-Restituto, M., & Rodriguez-Vasquez, A. (1996). Markov maps for communications with chaos. In *Proc. of the 1996's Nonlinear Dynamics in Electronic Systems* (*NDES'96*), Sevilla (pp. 161–166).

Katok, A., & Hasselblatt, B. (1995). Introduction to the modern theory of dynamical systems . In *Encyclopedia of Mathematics and its Applications* (*Vol. 54*). Cambridge, UK: Cambridge University Press.

Kennedy, M. P., & Ogorzalek, M. J. (Eds.). (1997). Special Issue. Chaos synchronization and control: theory and applications. *IEEE Trans. Circuits. Syst. I: Fundamental Theo. Appl., 44*(10), 853–1039.

Kolumban, G., Kennedy, M. P., & Chua, L. O. (1998 October). The role of synchronization in digital communications using chaos – Part I: Fundamentals of digital communications. *IEEE Trans. Circuits. Syst. I,* 44, 927–936.

Kuipers, L., & Niederreiter, H. (1974). Uniform distribution of sequences . In *Pure and Applied Mathematics*. New York: Wiley-Interscience.

Ljung, L., & Glad, T. (1994). On global identifiability for arbitrary model parametrizations. *Automatica, 30*, 265–276. doi:10.1016/0005-1098(94)90029-9

Mané, R. (1987). *Ergodic theory and differentiable dynamics*. Berlin: Springer-Verlag.

Massey, J. L. (1992). Contemporary cryptology: an introduction . In Simmons, G. J. (Ed.), *Contemporary Cryptology*. New York: IEEE Press.

Menezes, A. J., Oorschot, P. C., & Vanstone, S. A. (1996 October). *Handbook of Applied Cryptography.* Boca Raton, FL: CRC Press.

Millérioux, G., Bloch, G., Amigo, J. M., Bastos, A., & Anstett, F. (2003). Real-time video communication secured by a chaotic key stream cipher. In *Proc. of IEEE 16th European Conference on Circuits Theory and Design, ECCTD'03*, Krakow, Poland, September 1-4 (pp. 245–248).

Millérioux, G., & Daafouz, J. (2003). An observer-based approach for input independent global chaos synchronization of discrete-time switched systems. *IEEE Transactions on Circuits and Systems. I, Fundamental Theory and Applications, 50*(10), 1270–1279. doi:10.1109/TCSI.2003.816301

Millérioux, G., & Daafouz, J. (2004, July). Input independent chaos synchronization of switched systems. *IEEE Transactions on Automatic Control, 49*(7), 1182–1187. doi:10.1109/TAC.2004.831118

Millérioux, G., Hernandez, A., & Amigo, J. M. (2005 October). Conventional cryptography and message-embedding. In *Proc. of International Symposium on Nonlinear Theory and its Applications, NOLTA'2005*, Bruges.

Nijmeijer, H. (1997). Special Issue. Control of chaos and synchronization. *Systems & Control Letters, 31*, 259–322. doi:10.1016/S0167-6911(97)00042-X

Nijmeijer, H., & Mareels, I. M. Y. (1997, October). An observer looks at synchronization. *IEEE Transactions on Circuits and Systems. I, Fundamental Theory and Applications, 44*, 882–890. doi:10.1109/81.633877

Pecora, L. M., & Carroll, T. L. (1990). Synchronization in chaotic systems. *Physical Review Letters, 64*, 821–824. doi:10.1103/PhysRevLett.64.821

Pecora, L. M., & Carroll, T. L. (1991, August). Driving systems with chaotic signals. *Physical Review A., 44*(8), 2374–2383. doi:10.1103/PhysRevA.44.2374

Pecora, L. M., Carroll, T. L., Johnson, G., & Mar, D. (1997, November). Volume-preserving and volume-expanding synchronized chaotic systems. *Physical Review E: Statistical Physics, Plasmas, Fluids, and Related Interdisciplinary Topics, 56*(5), 5090–5100. doi:10.1103/PhysRevE.56.5090

Rauzy, G. (1976). *Propriétés statistiques de suites arithmétiques. Le Mathématicien, 15*. Paris: Presses Universitaires de France.

Rosier, L. (2010). Chaotic dynamical systems associated with tilings of R^N, arXiv:1002.1125v1 (http://arxiv.org/abs/1002.1125)

Rosier, L., Millérioux, G., & Bloch, G. (2004). Chaos synchronization on the N−torus and cryptography. *Comptes Rendus. Mécanique, 332*(12), 969–972. doi:10.1016/j.crme.2004.09.001

Rosier, L., Millérioux, G., & Bloch, G. (2004). Chaos synchronization for a class of discrete dynamical systems on the N-dimensional torus. *Les prépublications de l'Institut Elie Cartan,* 23.

Rosier, L., Millérioux, G., & Bloch, G. (2006). Chaos synchronization for a class of discrete dynamical systems on the N-dimensional torus. *Systems & Control Letters*, *55*, 223–231. doi:10.1016/j.sysconle.2005.07.003

Rovatti, R., & Setti, G. (1998). On the distribution of synchronization times in coupled uniform piecewise-linear Markov maps. *IEICE Transactions on Fundamentals*, *81*(9), 1769–1776.

Schmitz, R. (2001). Use of chaotic dynamical systems in cryptography. *Journal of the Franklin Institute*, *338*, 429–441. doi:10.1016/S0016-0032(00)00087-9

Van Wyk, M. A., & Steeb, W.-H. (1997). Chaos in electronics . In *Mathematical Modelling: Theory and Applications* (*Vol. 2*). Dordrecht, the Netherlands: Kluwer Academic Publishers.

Vesentini, E. (1999). An introduction to topological dynamics in dimension one. *Sem. Mat. Univ. Politec. Torino*, *55*(4), 303–357.

Walters, P. (1982). *An introduction to ergodic theory*. New York: Springer-Verlag.

Wolf, J. A. (1984). *Spaces of constant curvature* (5th ed.). Wilmington, DE: Publish or Perish, Inc.

Chapter 3

Identification and State Observation of Uncertain Chaotic Systems Using Projectional Differential Neural Networks

Alejandro García
National Polytechnic Institute of Mexico, Mexico

Isaac Chairez
National Polytechnic Institute of Mexico, Mexico

Alexander Poznyak
National Polytechnic Institute of Mexico, Mexico

ABSTRACT

The following chapter tackles the nonparametric identification and the state estimation for uncertain chaotic systems by the dynamic neural network approach. The developed algorithms consider the presence of additive noise in the state, for the case of identification, and in the measurable output, for the state estimation case. Mathematical model of the chaotic system is considered unknown, only the chaotic behavior as well as the maximal and minimal bound for each one of state variables are taking into account in the algorithm. Mathematical analysis and simulation results are presented. Application considering the so-called electronic Chua's circuit is carried out; particularly a scheme of information encryption by the neural network observer with a noisy transmission is showed. Formal mathematical proofs and figures, illustrate the robustness of proposed algorithms mainly in the presence of noises with high magnitude.

DOI: 10.4018/978-1-61520-737-4.ch003

1. INTRODUCTION

Identification, control and synchronization of chaotic systems are gaining great importance and attention in physics and engineering researching topics. Many investigation methods have used modern elegant theories to control chaotic systems, most of them based on the exact chaotic model (differential equations) (Chen & Dong, 1995), (Nijmeijer & Berghuis, 1995), (Gallegos, 1994). If the chaotic system is partly known, for example the differential equation is known but some of the parameters are uncertain, adaptive control methods (Zeng & Singh, 1997) are required.

Alternatively, synchronization of chaotic systems has recently received much attention. Several chaos synchronization schemes have been successfully established (Liao & Huang, 1999), (Jiang, 2002) but their results have been obtained with well-known systems having complete knowledge of system model and full state disposal. Therefore, the synchronization of chaotic systems with parametric uncertainties or different structure is an important issue. Femat proposed an extended form and several schemes have been reported (Femat & Alvarez-Ramírez, 1997), (Femat, Alvarez-Ramírez & Fernández-Anaya, 2000), (Femat & Jauregui-Ortiz, 2001) based on extend state observer theory to solve the problem. On the other hand, (Liao & Tsai, 2000) to solve the chaos synchronization of nonlinear systems specific class with disturbances and unknown parameters by deriving an adaptive observer-based driven system via a scalar transmitted signal.

In all previous control and identification problems, the modeling theory represents the most usual manner to formalize the systems dynamics knowledge. However, in several real chaotically systems, modeling rules may fail generating acceptable reproductions of reality (Meyer, 1992). In those cases, nonparametric identification and estimation (using adaptive methods) can be successfully applied to cover possible model deficiencies. Within the nonparametric identification framework, function approximation techniques play an important role avoiding the necessity of accurate mathematical plant description (Hartmut & Hartmut, 1997).

Nowadays the area of identification has a lot of applications using artificial neural networks (ANN), fuzzy logic and evolutionary algorithms. All these applications born like an alternative to forecast nonlinear dynamics like chaos. However, there exist few theoretical results describing how these algorithms map such dynamics if we compared them with the number of reports and publications in the area. The idea of learning applied to identification and control systems is every day more used and some papers search to reduce the existing gap between the learning and identification theory (Campi & Kumar, 1996), (Ljung, 1996). These models, in many cases, require some kind of previous knowledge about the plant to be identified to define the type of model that it is going to be used. This means that it is possible to find many different combinations of algorithms or functions that represent the system dynamics. However, the problem is that sometimes, can be nonlinear and the identified model does not represent this behavior.

In general, unknown chaotic systems are black box belonging to a given class of nonlinear complex systems. Therefore, a non model-based method is suitable. Among others, NN have become an attractive tool for modeling complex non-linear systems. Their inherent ability to approximate arbitrary continuous functions supports the last statement. NN are particularly powerful for handling large-scale problems. However, NN implementation suffers from lack of efficient constructive approaches, both for choosing network structures and for determining the neuron parameters. It has been proven that artificial NN can approximate a wide range of nonlinear functions to any desired degree of accuracy under specific constrains (Omidvar & Elliot, 1997). It is generally understood that NN training algorithm selection plays an important role for NN applications. In instance, in conventional gradient-descent-type weight adaptation, sensitivity of unknown system is required during on-line training process (Lin, Heang & Wai, 1998). However, it is difficult to acquire sensitivity information for unknown or highly nonlinear dynamics. Besides, local minimum of performance index remains to be the main inconvenient in back propagation algorithm (Haykin, 1994). Radial Basis Function (RBF) networks are often used in order to improve the NN learning efficiency. T. Poggio and F. Girosi (Poggio & Girosi, 1990),(Poggio & Girosi, 2006) analyzed several networks architectures to determine their approximation abilities and pointed out that RBF networks possess the best approximation property. Also, as noted in (Jin & Shi, 1999), (Martell, 200), NN have limited ability to characterize local features, such as discontinuities in curvature, jumps in objective function and others. These local features, which are located in time and/ or frequency, typically embody important information of the system such as aberrant process modes or faults (Bottou & Vapnik, 1992), (Jin & Shi, 1999), has addressed that improved localized modeling can aid both data reduction (compression) and subsequent classification tasks that rely on an accurate representation of local features.

NN are very useful to model nonlinear and chaotic dynamics (Gallegos, 1994) (Nijmeijer & Berghuis, 1995) and there are principally two ways to improve the learning phase for this class of complex system: to process the input and to adapt the architecture of the network to minimize the learning error. Two types of NN can be applied to identify dynamic systems with chaotic trajectories. The static NN connected with a dynamic linear model is used to approximate a chaotic system, but the computing time is very long and some a priori knowledge of chaotic systems is needed. Another approach is identifier-based control using recurrent NN. The recurrent NN can minimize the approximation error of the chaotic behavior (Poznyak, Yu, Ramírez & Sánchez, 1998), however, the number of neurons and the value of their weights are not determined. It is difficult to identify chaotic systems off line via NN, because the dynamic of chaos are very fast. From a practical point of view, the existing results are not satisfactory for the controller design. However, exploiting the fact of being universal approximations, it may straightforwardly to substitute continuous system uncertainties by NN. In general, this class of NN is known as differential neural networks or DNN for short. This class of NN possesses two important characteristics: its adjustable parameters may appear as linear elements in the NN description and they may be modified using differential equations (Haykin, 1994) (Poznyak, Sánchez & Yu, 2001). This approach transforms the original function approximation problem into a nonlinear robust adaptive feedback one. The DNN approach permits to avoid many problems related to global extreme search converting the learning process into a robust feedback design (Poznyak, Yu, Ramírez & Sánchez, 1998). If the mathematical model of a process is incomplete or partially known, the DNN theory provides an effective instrument to attack a wide spectrum of problems such as non-parametric trajectory identification, state estimation, trajectories tracking and etc. (Lewis, Yesildirek & Liu, 1996). Mostly real systems are really difficult to be controlled because lack of information on its internal structure and their current states. In view of DNN continuous structure, more detailed techniques should be applied to solve important questions on the new NN proposal (convergence for example).

Lyapunov's stability theory (specially the so-called controlled Lyapunov theory) has been used within the DNN framework (Poznyak, Sánchez & Yu, 2001) (Rovithakis & Christodoulou, 1994). This is the main tool to prove DNN improvements on the estimation problems or in the control actions design. Even though there exists a general trend to enlarge the nonlinear systems for which the aforementioned researches can be applied, chaotically systems for example. As consequence, there are novel results on stability, convergence to arbitrarily small sets and robustness to modeling imperfections and external perturbations of the closed-loop system.

This chapter addresses the application of the DNN structures to identify chaotic complex systems as well as to carry out its state observation. These approaches using DNN are completely developed, including identifier structure, observer structure, the corresponding convergence schemes and continuous learning algorithms. It is also proved that the identification error converges to a bounded zone by means of a Lyapunov function technique. The effectiveness of the novel DNN topology is shown by a classical numerical example: the Chua's circuit.

2. CHAOTIC SYSTEMS AND ITS APPROXIMATION BY ARTIFICIAL NEURAL NETWORKS

2.1. Class of Systems

From hereafter, it will be considered that the nonlinear continuous-time model given by the following Ordinary Differential Equations (ODE) represents the chaotic system:

$$\begin{aligned} \dot{x}(t) &= f\left(x(t),u(t)\right) + \xi(t), \quad x(0) \text{ is fixed} \\ y(t) &= Cx(t) + \eta(t) \end{aligned} \tag{1}$$

where $\dot{x}(t) = \dfrac{dx(t)}{dt}$, considering $x(t) \in \mathrm{R}^n$ as the state-vector at time $t \geq 0$, $y(t) \in \mathrm{R}^m$ is the corresponding output, available for a designer at any time *t*, the known matrix $C \in \mathrm{R}^{m \times n}$ defines the state-output transformation,, $u(t) \in \mathrm{R}^r$ is a bounded input action $(r \leq n)$ belonging to the following admissible set $U^{adm} := \left\{u(t) \,:\, \left\|u(t)\right\| \leq \Upsilon_u < \infty\right\}$, $\xi(t)$ is the bounded noise vector in the state dynamics, $f\ : \mathrm{R}^{n \times r} \to \mathrm{R}^n$.

2.2 State Constrains

Once we have defined the chaotic system's representation, let us introduce some comments about dynamical systems that will be considered throughout the chapter. For many practical problems designer knows *a priori* that state-vector $x(t)$ always belongs to a given *compact set* $X \subset \mathrm{R}^n$ (even in the presence of noise), i.e.:

$$x(t) \in X, \ \forall t \geq 0 \tag{2}$$

Usually X has a concrete physical sense. For example, in case of electric and electronic circuits we are able to measure the current and voltage in some node of the net, even, we can plot this values over time, it let us observe that these kind of variables stay bounded and belonging at any time to a compact set, this is usual in oscillators circuits and is also true if the circuits present a chaotic behavior. Similar remark can be done for other physical variables such as temperature; chemical concentrations; light intensity and, etcetera.

2.3 Approximation of Dynamic Systems by Artificial Neural Networks

Let us consider the case when the right-hand side $f(x(t),u(t))$ of dynamics (1) is unknown, it means, we do not have a mathematical model for the dynamic chaotic system, but we are able to know the input $u(t)$ and also we can measure each component of the states $x(t)$ at any time. Under this condition, we would like to derive an "artificial" but valid mathematical model for the system, the term "artificial" has been included in order to emphasize that no traditional formulation based on physical laws is considered. In this case, modeling process is carried out taking into account only the available information (input and state measures) in an adaptive algorithm. One possibility is granted by the Artificial Neural Networks theory (Poznyak, Sánchez & Yu, 2001), which establishes that the right-hand side $f(x(t),u(t))$ can be approximated as: $\bar{f}\left(x(t),u(t)\mid W(t)\right)$ where $\bar{f}\in \mathrm{R}^{n}$ defines the approximate mapping depending on the time-varying parameters $W(t)$ which should be adjusted by a concrete "*adaptation law*" suggested by a designer (or derived from stability analysis as in this chapter). It is possible to divide it $\bar{f}\left(x(t),u(t)\mid W(t)\right)$ in two main parts: the first one approximates the linear dynamics by a *Hurwitz* fixed matrix $A\in \mathrm{R}^{n\times n}$ (selected by the designer) and nonlinear part is approximated by variable time parameters $W_{1,2}(t)$ with "sigmoid" multipliers, i.e.:

$$\begin{gathered}\bar{f}\left(x(t),u(t)\mid W_{1,2}(t)\right) := Ax(t)+W_{1}(t)\sigma\left(x(t)\right)+W_{2}(t)\phi\left(x(t)\right)u(t)\\ A\in \mathrm{R}^{n\times n},\ \ W_{1}(t)\in \mathrm{R}^{n\times p},\ \ \sigma\left(\cdot\right)\in \mathrm{R}^{p\times 1}, W_{2}(t)\in \mathrm{R}^{n\times q},\ \ \phi\left(\cdot\right)\in \mathrm{R}^{q\times r}\end{gathered}\tag{3}$$

The activation vector-function $\sigma\left(\cdot\right)$ and the matrix-function $\phi\left(\cdot\right)$ are usually selected as functions with *sigmoid-type components*, i.e.,

$$\begin{gathered}\sigma_{j}\left(x(t)\right) := a_{j}\left(1+b_{j}\exp\left(-\sum_{j=1}^{n}c_{j}x_{j}(t)\right)\right)^{-1}\\ j=\overline{1,n}\end{gathered}\tag{4}$$

And

$$\begin{gathered}\phi_{i,j}\left(x(t)\right) := a_{i,j}\left(1+b_{i,j}\exp\left(-\sum_{s=1}^{n}c_{i,s}x_{s}(t)\right)\right)^{-1}\\ i=\overline{1,q};j=\overline{1,r}\end{gathered}\tag{5}$$

Both set of functions satisfy the following sector conditions

$$\left\|\sigma\left(x(t)\right)-\sigma\left(x'(t)\right)\right\|_{\Lambda_{\sigma}}^{2}\leq L_{\sigma}\left\|x(t)-x'(t)\right\|_{\Lambda_{\sigma}'}^{2}\tag{6}$$

$$\left\|\phi\left(x(t)\right)-\phi\left(x'(t)\right)\right\|_{\Lambda_{\phi}}^{2}\leq L_{\phi}\left\|x(t)-x'(t)\right\|_{\Lambda_{\phi}'}^{2}\tag{7}$$

and bounded on R^n. In (3) the constant parameter A as well as the time-varying parameters $W_{1,2}(t)$ should be properly adjusted to guarantee a good state approximation. It is worth to notice that for any fixed matrices $W_{1,2}(t)=\hat{W}_{1,2}$ the dynamics (1) always could be represented as

$$\begin{aligned}\dot{x}(t) &= Ax(t)+\hat{W}_1\sigma\left(x(t)\right)+\hat{W}_2\phi\left(x(t)\right)u(t)+\tilde{f}(t)+\xi(t)\\ \tilde{f}(t) &:= f\left(x(t),u(t)\right)-\overline{f}\left(x(t),u(t)\mid \hat{W}_{1,2}\right)\end{aligned} \tag{8}$$

Where $\tilde{f}\left(t\right)$ is referred to as the modeling error vector-field called the "*un-modeled dynamics*". It is possible to assume the following upper bound for the un-modeled dynamics $\tilde{f}\left(t\right)$:

$$\begin{aligned}&\left\|\tilde{f}(t)\right\|^2_{\Lambda_f}\leq \tilde{f}_0+\tilde{f}_1\left\|x(t)\right\|^2_{\Lambda^1_{\tilde{f}}}\\ &\tilde{f}_0,\ \tilde{f}_1>0;\ \Lambda_f,\Lambda^1_{\tilde{f}}>0,\ \Lambda_f=\Lambda^?_f,\ \Lambda^1_{\tilde{f}}=\left(\Lambda^1_{\tilde{f}}\right)^T\end{aligned} \tag{9}$$

3 PROJECTIONAL DIFFERENTIAL NEURAL NETWORK IDENTIFIER

The nonparametric identification problem consists in designing a vector-function $\hat{x}(t)\in R^n$ in such a way that it would be "close" to the measurable state-vector $x(t)$. The magnitude of that "closeness" depends on the accepted assumptions concerning the state dynamics as well as the noise effects. Additionally, in such cases where the property (2) is hold by the system, it is desirable that also the identified state $\hat{x}(t)$ belongs to the compact set X at any time, i.e.:

$$\hat{x}(t)\in X,\ \forall t\geq 0 \tag{10}$$

This property will be implemented in the algorithm that will be introduced below. From the nominal representation (3), the common DNN Identifier structure is:

$$\frac{d\hat{x}(t)}{dt}=A\hat{x}(t)+W_1(t)\sigma\left(\hat{x}(t)\right)+W_2(t)\phi\left(\hat{x}(t)\right)u(t) \tag{11}$$

Lets represent (11) in its equivalent integral form:

$$\hat{x}(t)=\hat{x}(t-h(t))+\int_{\tau=t-h(t)}^{t}\left[A\hat{x}(\tau)+W_1(\tau)\sigma\left(\hat{x}(\tau)\right)+W_2(\tau)(\phi\left(x(\tau)\right)u(\tau)\right]d\tau \tag{12}$$

Here $h\left(t\right)\in C^1$ is supposed to be given and non-increasing positive function, that is, $\dot{h}\left(t\right)\leq 0$. In order to hold the condition (10), let us introduce the following projectional DNN structure:

$$\hat{x}(t)=\pi_X\left\{\hat{x}(t-h(t))+\int_{\tau=t-h(t)}^{t}\left[A\hat{x}(\tau)+W_1(\tau)\sigma\left(\hat{x}(\tau)\right)+W_2(\tau)(\phi\left(x(\tau)\right)u(\tau)\right]d\tau\right\} \tag{13}$$

The operator $\pi_X\{\cdot\}$ is the projector to the given convex compact set X satisfying the condition

$$\left\|\pi_X\{x\} - z\right\| \leq \left\|x - z\right\| \tag{14}$$

For any $x \in \mathbf{R}^n$ and any $z \in X$. The operator $\pi_X\{\cdot\}$ may be defined non-uniquely and no restriction about its differentiability is asked for the stability proof. An example of $\pi_X\{\cdot\}$ is given below.

Example 1

$$\pi_x\{x\} = \left[sat\left(x_1\right) \quad sat\left(x_1\right)\ldots sat(x_n)\right] \tag{15}$$

Where, for any $i = 1\ldots n$

$$sat\left(x_i\right) := \begin{cases} \left(x_i\right)^- & x_i \leq \left(x_i\right)^- \\ x_i & \left(x_i\right)^- < x_i < \left(x_i\right)^+ \\ \left(x_i\right)^+ & x_i \geq \left(x_i\right)^+ \end{cases}$$

With $\left(x_i\right)^-, \left(x_i\right)^+ ; \left(x_i\right)^- < \left(x_i\right)^+$ extreme points known a priori for each component.

Remark 1. *As can be seen in (13), projection of the identified state is carried out at each integration time, an also this value is depending on the estimation realized one step before on (t-h(t)) this approach is completely different to the usual one used for some engineers based on the "cut" of undesirable values after the complete identification process as well as introduce a saturation function but without consider it in the stability analysis.*

The weights matrices $W_1(t)$ and $W_2(t)$ supply the adaptive behavior to this class of identifiers if they are adjusted by an adequate manner. We derived (see Appendix) the following nonlinear weight *updating* laws based on the Lyapunov-like stability analysis:

$$\left.\begin{array}{l} \dot{W}_1\left(t\right) = -\left(2k_1\left(t\right)\right)^{-1}\left\{P\Omega_1\left(\hat{x}\left(t\right), \hat{x}\left(t - h\left(t\right)\right)\right)\sigma\left(\hat{x}\left(t\right)\right)^T - \dot{k}_1\left(t\right)\tilde{W}_1\left(t\right)\right\} \\ \quad \Omega_1\left(\hat{x}\left(t\right), \hat{x}\left(t - h\left(t\right)\right)\right) = 2\delta\left(t - h\left(t\right)\right) + \tilde{W}_1\left(t\right)\sigma\left(\hat{x}\left(t\right)\right) \end{array}\right\} \tag{15}$$

$$\left.\begin{array}{l} \dot{W}_2\left(t\right) = -\left(2k_2\left(t\right)\right)^{-1}\left\{P\Omega_2\left(\hat{x}\left(t\right), \hat{x}\left(t - h\left(t\right)\right)\right)\phi\left(\hat{x}\left(t\right)\right)^T u^T\left(t\right) - \dot{k}_2\left(t\right)\tilde{W}_2\left(t\right)\right\} \\ \quad \Omega_2\left(\hat{x}\left(t\right), \hat{x}\left(t - h\left(t\right)\right)\right) = \delta\left(t - h\left(t\right)\right) + \tilde{W}_2\left(t\right)\phi\left(\hat{x}\left(t\right)\right)u\left(t\right) \end{array}\right\} \tag{16}$$

Where:

$$\delta(t) := \hat{x}(t) - x(t) \tag{17}$$

Is the identification error, $\tilde{W}_{1,2}(t) = W_{1,2}(t) - \hat{W}_{1,2}$, $k_{1,2}(t) > 0 \in \mathrm{R}$, $P \in \mathrm{R}^{n \times n}, P = P^T > 0$, the matrix $\boldsymbol{P}$ calculation will be discussed in the corresponding theorem.

3.1 Upper Bound for State Identification Error

3.1.1 Behavior of Weights Dynamics

Here we wish to show that under the adapting weights laws (15) and (16) the weights $W_1(t)$ and $W_2(t)$ are bounded.

Theorem I. *If* $k_{i,t}$ $(i = 1,2)$ in (15) and (16) satisfy

$$\begin{aligned} \dot{k}_{1,t} &\leq -\frac{2\left(k_1(t)\right)^2 \left| tr\left\{\tilde{W}_1^T(t) P\Omega(t)\sigma^T\left(\hat{x}(t)\right)\right\}\right|}{tr\left\{\tilde{W}_1^T(t)\tilde{W}_1(t)\right\} + ck_1(t)\left[k_1(t) - k_{1\min}\right]} \\ \dot{k}_{2,t} &\leq -\frac{2\left(k_2(t)\right)^T \left| tr\left\{\tilde{W}_2(t) P\Phi(t)u^T(t)\phi^T\left(\hat{x}(t)\right)\right\}\right|}{tr\left\{\tilde{W}_2(t)^T \tilde{W}_2(t)\right\} + ck_2(t)\left(k_2(t) - k_{2,\min}\right)} \end{aligned} \tag{18}$$

Then $tr\left\{\tilde{W}_1^T(t)\tilde{W}_1(t)\right\}$ is monotonically non-decreasing function. The complete proof of previous can be consulted in (García, Poznyak A., Chairez & Poznyak T., 2008), an example for the structure of $k(t)$ is given:

Example 2. a) Introduced the following auxiliary function

$$s\left(\tilde{W}_1(t), \delta\left(t - h(t)\right)\right) := \frac{2k_1(t)^2 \left| tr\left\{\tilde{W}_1^T(t) P\Omega\left(\hat{x}(t), \hat{x}\left(t - h(t)\right)\right)\sigma\left(\hat{x}(t)\right)^T\right\}\right|}{tr\left\{\tilde{W}_1^T(t)\tilde{W}_1(t)\right\} + ck_1(t)\left[k_1(t) - k_{1\min}\right]_+^2}$$

Suggesting the representation:

$$k_1(t) = \frac{k_1(0)}{1 + a\exp(bt)} + k_{1\min}$$

$$k_{\min,1} > 0$$

$$\dot{k}_1(t) = -k_1(0)\frac{ba\exp(bt)}{\left[1 + a\exp(bt)\right]^2} < -s\left(\tilde{W}_1(t), \delta\left(t - h(t)\right)\right)$$

Leading to

$$k_1(0)ba\exp(bt) > [1 + a\exp(bt)]s(\tilde{W}_1(t), \delta(t - h(t)))$$
$$k_1(0)ba\exp(bt) > a\exp(bt)s(\tilde{W}_1(t), \delta(t - h(t))) + s(\tilde{W}_1(t), \delta(t - h(t)))$$
$$a\exp(bt)\{k_1(0)b - s(\tilde{W}_1(t), \delta(t - h(t)))\} > s(\tilde{W}_1(t), \delta(t - h(t)))$$

The last inequality is fulfilled if the parameter a is selected as

$$a > \frac{s(\tilde{W}_1(t), \delta(t - h(t)))}{k_1(0)b - s(\tilde{W}_1(t), \delta(t - h(t)))}\exp(-bt)$$

So, the parameter a is dependent on $\tilde{W}_1(t)$ and $\delta(t - h(t))$, i.e.:

$$a(\tilde{W}_1(t), \delta(t - h(t)))$$

b) Analogously, for $\tilde{W}_2(t)$:

$$s(\tilde{W}_2(t), \delta(t - h(t))) := \frac{2k_2(t)^2 \left| tr\{\tilde{W}_2^T(t)P\Omega_2(\hat{x}(t), \hat{x}(t - h(t)))\phi(\hat{x}(t))^T u^T(t)\}\right|}{tr\{\tilde{W}_2^T(t)\tilde{W}_2(t)\} + ck_2(t)[k_2(1) - k_{2\min}]}$$

And

$$k_2(t) = \frac{k_2(0)}{1 + a\exp(bt)} + k_{2\min}$$

Remark 2. *It is worth notice that the learning law (15) and (16) must be realized on-line in parallel with the gain-parameter adaptation procedure (18).*

3.1.2 Main Theorem on the Identification Error Convergence

Hereafter we will assume that

A1) The class of the function $f : \Re^n \to \Re^n$ is Lipchitz continuous in $x \in X$, that is, for all $x, x' \in X$ there exist constants $L_{1,2}$ such that

$$\left\| f(x(t), u(t)) - f(y(t), v(t)) \right\| \le L_1 \left\| x(t) - y(t) \right\| + L_2 \left\| u(t) - v(t) \right\| \tag{19}$$

Where: $x(t)$, $y(t) \in \mathrm{R}^n$; $u(t)$, $v(t) \in \mathrm{R}^m$; $0 \le L_1 < \infty$, $0 \le L_2 < \infty$.

A2) The noise $\xi(t)$ in the system (1) is uniformly (on *t*) bounded such that

$$\left\|\xi(t)\right\|_{\Lambda_\xi}^2 \leq \Upsilon_\xi, \tag{20}$$

Where Λ_ξ is a known "normalizing" non-negative definite matrix which permits to operate with vectors having components of different physical nature (for example, meters, $mole / l$, voltage and etc.), no extra assumption about noise is considered.

Theorem 2 *Under assumptions A1-A2 and if there exist matrices* $\Lambda_i = \Lambda_i^T > 0,\ \Lambda_i \in \mathrm{R}^{n\times n}, i = 1\ldots5,$ and positive parameters ϖ, μ_1, μ_2 and μ_3 such that the following Linear Matrix Inequality

$$\begin{bmatrix} LMI_1 & 0 & 0 & 0 \\ 0 & LMI_2 & 0 & 0 \\ 0 & 0 & LMI_3 & 0 \\ 0 & 0 & 0 & LMI_4 \end{bmatrix} > 0 \tag{21}$$

Where:

$$LMI_1 := \begin{bmatrix} -\Gamma(\delta,\mu_1,\mu_2,\mu_3) & P \\ P & R \end{bmatrix};\ LMI_2 := \begin{bmatrix} \Theta_1 & A^T(K)P \\ PA(K) & \mu_1 P \end{bmatrix}$$

$$LMI_3 := \begin{bmatrix} \Theta_2 & \hat{W}_1^T P \\ P\hat{W}_1 & \mu_2 P \end{bmatrix};\ LMI_4 := \begin{bmatrix} \Theta_3 & \hat{W}_2^T P \\ P\hat{W}_2 & \mu_3 P \end{bmatrix}$$

$tr\left\{\Theta_i\right\} < 1,\ i = 1,2,3$ and

$$\Gamma\left(\delta,\mu_1,\mu_2,\mu_3\right) = A^T P + PA + PR^{-1}P + Q\left(\mu_1,\mu_2,\mu_3\right)$$
$$R^{-1} = \Lambda_1^{-1} + \hat{W}_2^T \Lambda_2 \hat{W}_2 + \hat{W}_2 \Lambda_3 \hat{W}_2^T + \Lambda_4 + \Lambda_5$$
$$Q\left(\mu_1,\mu_2,\mu_3\right) = \left(\left\|\Lambda_2^{-1}\right\| L_\sigma L_\delta^2 + \left\|\Lambda_3^{-1}\right\| L_\phi \Upsilon_u^2 + \mu_1 + L_\sigma \mu_2 + \mu_3 \Upsilon_u^2 L_\phi\right) I + Q_0$$

Has positive definite solution **P**, then the projectional DNN identifier (13) with the weight's learning laws, given by (15), (16), (18), and with *h*(*t*)satisfying

$$\lim_{t\to\infty} h(t) \to \varepsilon, 0 < \varepsilon << 1 \tag{22}$$

then the "identification error" is ultimate bounded as:

$$\frac{1}{T}\int_{\tau=0}^{T} \delta^T\left(\tau - h\left(t\right)(\tau)\right) Q_0 \delta\left(\tau - h\left(t\right)(\tau)\right) d\tau \leq c \tag{23}$$

Where:

$$c := \left(\|P\|\|\Lambda_f^{-1}\| + \|\Lambda_4^{-1}\|\|\Lambda_f^{-1}\|\right)\left[\tilde{f}_0 + \tilde{f}_1\|\Lambda_f^1\| Diam\left(X\right)^2\right] + \left(\|P\| + \|\Lambda_5^{-1}\|\right)\left[\|\Lambda_\xi^{-1}\|^{\frac{1}{2}}\Upsilon_\xi\right]$$

with:

$$Diam(X) = \sup_{x,z\in X}\|x - z\|$$

The proof of this theorem is detailed in the Appendix.

Remark 3. It is easy to see that in the absence of noise ($\xi_t = 0$) and un-modeled dynamics ($\tilde{f} = 0$), we can choose $\tilde{f}_0, \tilde{f}_1$, and Υ_ξ such that:

$$\frac{1}{T}\int_{\tau=0}^{T} \delta^T\left(\tau - h\left(t\right)(\tau)\right)Q_0\delta\left(\tau - h\left(t\right)(\tau)\right)d\tau \to 0$$

3.2 Numerical Identification Example

3.2.1 Chua's Circuit

Let us to introduce the well-known electronic system called *Chua's circuit*, which displays very rich and typical bifurcation and chaotic phenomena such as double scroll, dual double scroll, double hook, etc.. The circuit itself is quite simple: it consists of only one inductor (*L*), two capacitors (C_1,C_2), one linear resistor (R) and one piecewise-linear resistor (g). Its dynamics can be described by the following set of equations (Guanron & Xianonig, 1993):

$$\begin{aligned} \dot{x}_1(t) &= p\left[-x_1(t) + x_2(t) - f(x_1(t))\right] \\ \dot{x}_2(t) &= x_1(t) - x_2(t) + x_3(t) \\ \dot{x}_3(t) &= -qx_2(t) \end{aligned} \qquad (24)$$

where:

$$f(x_1(t)) = m_0x_1(t) + \tfrac{1}{2}(m_1 - m_0)(|x_1(t) + 1| - |x_1(t) - 1|)$$

$x_1(t)$ and $x_2(t)$ are voltages in capacitors while $x_3(t)$ is the current through inductor. Simulation parameters are: $p = 9$, $q = \frac{29}{7}$,, $m_0 = -\frac{5}{7}$, and $m_1 = -\frac{8}{7}$. with these values Chua's circuit presents the

Figure 1. Chaotic Behavior of Chua's circuit system

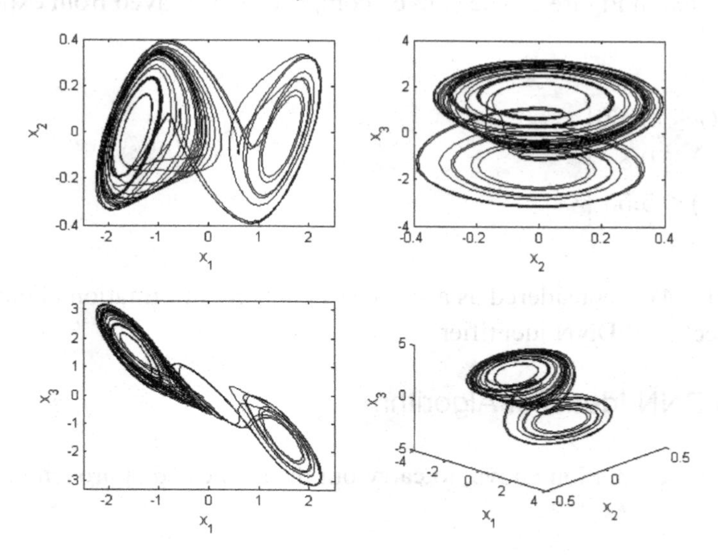

Figure 2. DNN Identifier behavior without projectional operator

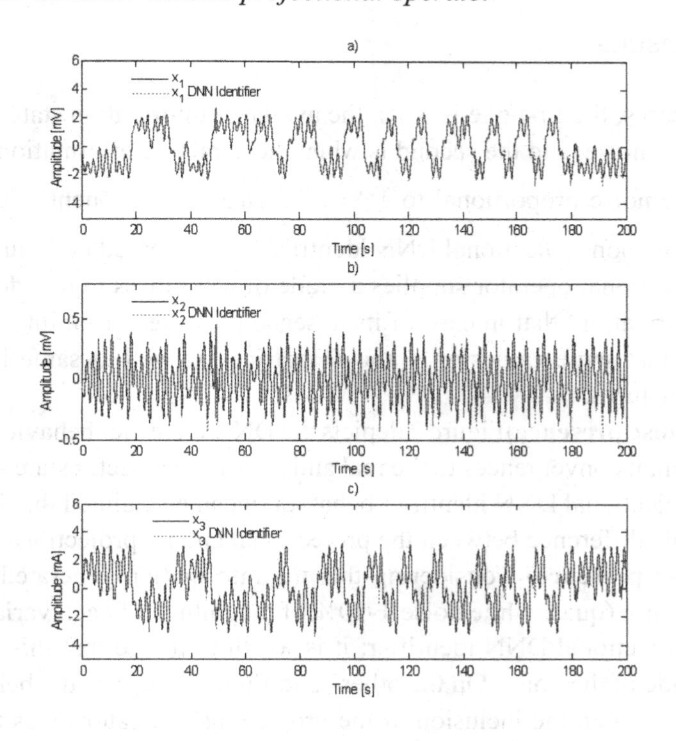

chaotic behavior depicted in Figure 2. The convex compact set X derived from experimental conditions is set as:

$$X := \begin{Bmatrix} -2.16 \leq x_1(t) \leq 3.166 \\ -0.8 \leq x_2(t) \leq 0.8 \\ -5.55 \leq x_3(t) \leq 5.55 \end{Bmatrix}$$

Remark 3. Model (24) is considered as a source of data any information about its structure will be employed in the projectional DNN identifier.

3.2.2 Projectional DNN Identifier Algorithm

As it follows from the presentation above, to carry out the suggested approach one needs to fulfill the following steps:

1. Define the projector.
2. Select Matrices A and $\hat{W}$ (some hints are given by (Chairez, Poznyak A. & Poznyak T., 2006) and (Stepanyan & Hovakimyan, 2007)).
3. Find $\boldsymbol{P}$ as the solution of the *LMI* problem (21), it can be done using the corresponding Matlab Toolbox.
4. Introduce $\boldsymbol{P}$ into the adapting weight law (15), (16) and (18) and realized it on-line.

3.2.3 Simulation Results

We will consider two cases, the first one is when the available information state vector $x(t)$ ($x_1(t)$, $x_2(t)$, $x_3(t)$) is not corrupted by noise, and the second is when the available information is $x(t)+\xi(t)$, where $\xi(t)$ is a "quasi" white noise proportional to 15% of each $x(t)$ component, identification by projectional DNN identifier and non projectional DNN identifier are presented in Figure 5 and 6 respectively. It can be seen that projectional operator supplies a grade of robustness to the identifier structure when it deals with noisy information, that in case of its absence provokes instability after some seconds of simulation, remarking that same values of $\boldsymbol{A}$ and $\boldsymbol{P}$ matrices as well as same initial conditions were considered for both structures.

Case 1. (Without noise presence) Figure 2 depicts the DNN identifier behavior (without projectional operator), as can be seen, it convergences to the real and measurable vector state since the first moment. Figure 3 depicts the Projectional DNN identifier behavior, it can be claimed that in the absence of noise there is not a remarkable difference between the projectional and no projectional operator inclusion.

Case 2. (With noise presence) Considering that the information generated by the model (24) is corrupted by additive noise (quasi white noise +-10% of magnitude of each variable). Figure 4 depicts the behavior of non-projectional DNN identifier; it is worth to notice that this does not convergence because of the magnitude of the noise. On the other hand Figure 5 depicts the behavior for projectional DNN identifier, as can be seen the inclusion of the projectional operator gives to the structure a high grade of robustness, allowing it to identify the chaotic system even for the selected magnitude of noise.

Figure 3. Projectional DNN Identifier behavior

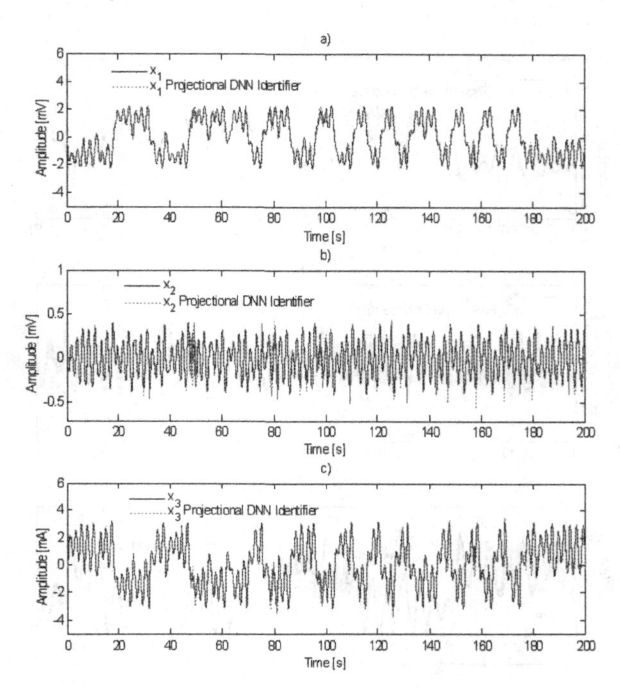

Figure 4. DNN Identifier without projectional operator (additive noise presence in the available information)

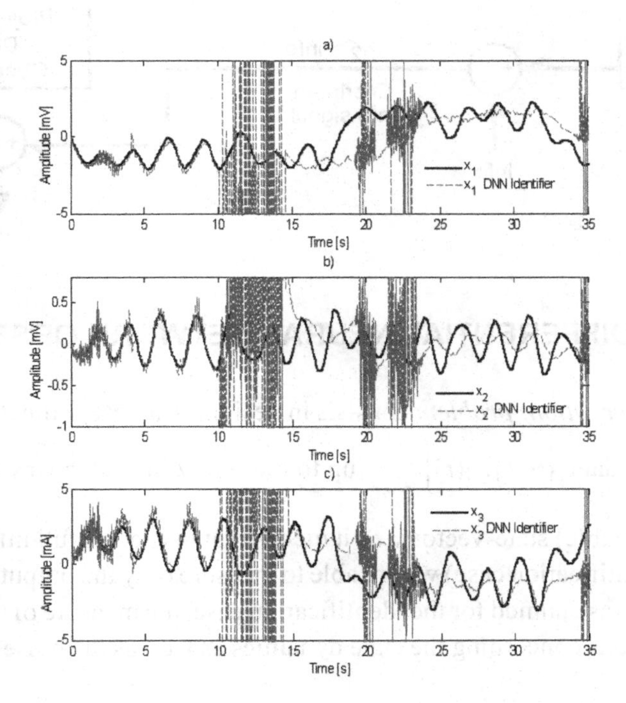

Figure 5. Projectional DNN Identifier (additive noise presence in the available information)

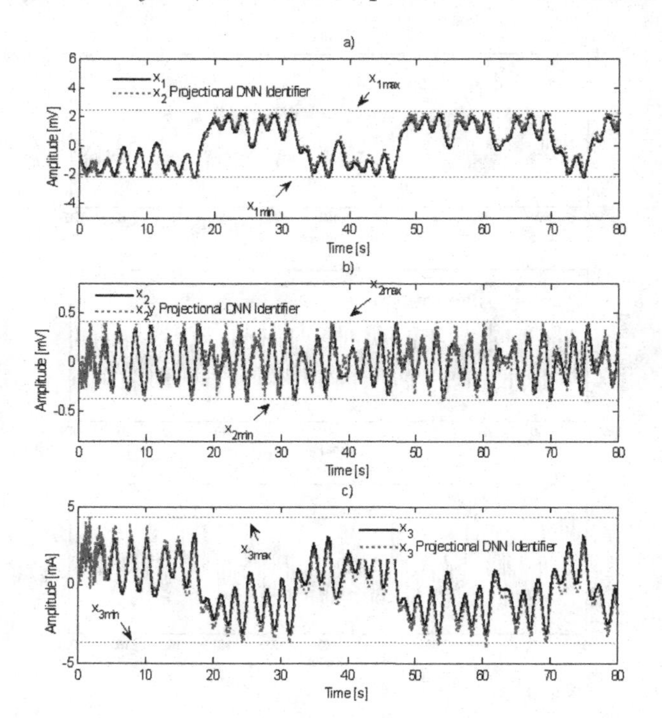

Figure 6. Chaotic synchronization/cryptography based on Projectional DNN Observer

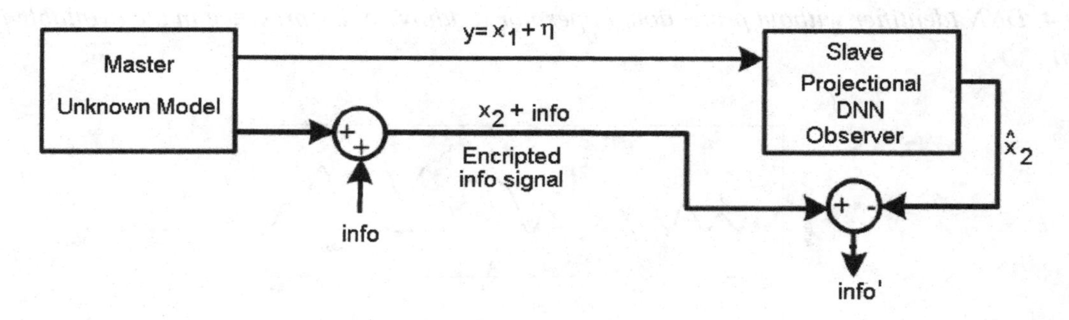

4 PROJECTIONAL DIFFERENTIAL NEURAL NETWORK OBSERVER

The *state estimation (observation) problem* consists in designing a vector-function $\hat{x}(t) \in \mathbf{R}^n$ depending only on the available data $\left\{y\left(t\right), u(t)\right\}_{\tau\in\left[0,t\right]}$ up to the time t in such a way that it would be "close" to its real (but non-measurable) state-vector $x(t)$, it means that instead of full information about state (as was considered in the identification case) we are able to measure only the output, as a linear combination of one or more states. As was claimed for the identification case, the measure of that "closeness" depends on the accepted assumptions concerning the state dynamics as well as the noise effects. Most of observers structures, solve this problem, considering a set of Ordinary Differential Equation structure, usually given by

$$\frac{d}{dt}\hat{x}(t) = F\left(\hat{x}(t), u\left(t\right), y_{\tau\in[0,t]}, t\right),\ \hat{x}_0 \text{ is a fixed vector} \tag{25}$$

Here the mapping $F\ :\ \mathbf{R}^n \times \mathbf{R}^r \times \mathbf{R}^m \times \mathbf{R}^+ \to \mathbf{R}^n$ defines the particular structure of the observer to be implemented. Instead of (25) let us consider the following observer referred hereafter as the *projectional observer*:

$$\hat{x}(t) = \pi_X\left\{\hat{x}(t-h(t)) + \int_{\tau=t-h(t)}^{t} F\left(\hat{x}(\tau), u\left(\tau\right), y_{s\in[0,\tau]}, \tau\right)d\tau\right\},\ \ t > h(0) \tag{26}$$

Here $h\left(t\right) \in C^1$ and the operator $\pi_X\left\{\cdot\right\}$ are as were explained before. If the right-hand side $f(x(t))$ of the dynamics (1) is known then usually the structure F of the observer (25) is selected in the, so-called, Luenberger-type form:

$$F\left(\hat{x}(t), u\left(t\right), y\left(t\right), t\right) = f\left(\hat{x}(t), u(t)\right) + K\left(t\right)\left(y(t) - C\hat{x}(t)\right) \tag{27}$$

So, it repeats the dynamics of the plant and, additionally, contains the correction term, proportional to the output error (see, for example: (Zeits, 1987), (Yaz & Azemi, 1994), (Zak & Walcott, 1990) and (Poznyak, 2004)). The adequate selection of the matrix-gain $K(t)$ provides a good-enough state estimation. Let us now to consider $f\left(x(t), u(t)\right)$ is unknown and use the ANN approximation in expression (26), by this way we derive the projectional DNN observer structure:

$$\begin{gathered}\hat{x}(t) = \pi_X\left\{\hat{x}(t-h(t)) + \int_{\tau=t-h(t)}^{t}\left[A\hat{x}(\tau) + W_1(\tau)\sigma\left(\hat{x}(\tau)\right) + W_2(\tau)(\phi\left(x(\tau)\right)u(\tau) + Ke(\tau)\right]d\tau\right\} \\ e(t) := y(t) - C\hat{x}(t)\end{gathered} \tag{28}$$

Here the weights matrices $W_1(t)$ and $W_2(t)$ supply the adaptive behavior to this class of observers if they are adjusted by an adequate manner. From stability analysis we obtain:

$$\left.\begin{gathered}\dot{W}_1\left(t\right) = -\frac{k_1^{-1}(t)}{2}P\Omega(t)\sigma^T\left(\hat{x}(t)\right) - \dot{k}_1(t)\tilde{W}_1\left(t\right) \\ \Omega(t) := \Pi\tilde{W}(t)\sigma\left(\hat{x}(t)\right) + 2N_{\varpi}C^T e(t-h(t))\ \ \tilde{W}_1(t) := W_1(t) - \hat{W}_1 \\ \Pi = \left(N_{\varpi}\left(\varpi\Lambda_3 + C^T\Lambda_2 C\right)N_{\varpi}P + I\right)\end{gathered}\right\} \tag{29}$$

$$\left.\begin{gathered}\dot{W}_2\left(t\right) = -\frac{k_2^{-1}(t)}{2}P\Phi(t)u^T(\tau)\phi^T\left(\hat{x}\left(\tau\right)\right) - \dot{k}_2(t)\tilde{W}_2\left(t\right) \\ \Phi(t) := \Xi\tilde{W}_2(\tau)(\phi\left(\hat{x}(\tau)\right)u(\tau) + 2N_{\varpi}C^T e(t-h(t))\ \ \tilde{W}_2(t) := W_2(t) - \hat{W}_2 \\ \Xi = \left(N_{\varpi}\left(\varpi\Lambda_7 + C^T\Lambda_6 C\right)N_{\varpi}P + I\right)\end{gathered}\right\} \tag{30}$$

where: $N_{\varpi} = \left(C^T C + \varpi I\right)^{-1}$, $\varpi > 0$. To improve the behavior of this adaptive laws, the matrix $\hat{W}_{1,2}$ can be "provided" by one of the, so-called, *training algorithms* (see, for example, (Chairez, Poznyak A., & Poznyak T., 2006)), either represented by a least square solution or considering some identification structure for some possible set of fictitious values or even an available set of directly measured data of the process.

4.1 Main Theorem on an Upper Bound for the Observation Error

Hereafter we will assume that:

B1) The class of the function $f\ :\Re^n \to \Re^n$ is Lipchitz continuous in $x \in X$, that is, for all $x, x' \in X$ there exist constants $L_{1,2}$ such that

$$\begin{gathered}\left\|f\left(x,u,t\right)-f\left(y,v,t\right)\right\| \le L_1\left\|x-y\right\| + L_2\left\|u-v\right\| \\ \left\|f\left(0,0,t\right)\right\|^2 \le C_1;\quad x,y\in\Re^n;\ u,v\in\Re^m;\ 0\le L_1,L_2<\infty\end{gathered} \tag{31}$$

B2) The pair (*A*,*C*) is observable, that is, there exists the gain matrix $K \in \Re^{n\times m}$ such that matrix

$$\tilde{A}\left(K\right) := A - KC \tag{32}$$

is stable.

B3) The noises ξ_t and η_t in the system (1) are uniformly (on *t*) bounded such that

$$\left\|\xi(t)\right\|^2_{\Lambda_\xi} \le \Upsilon_\xi,\quad \left\|\eta(t)\right\|^2_{\Lambda_\eta} \le \Upsilon_\eta \tag{33}$$

where Λ_ξ and Λ_η are known "normalizing" non-negative definite matrices.

Theorem 3 Under assumptions B1-B3 and if there exist matrices $\Lambda_i = \Lambda_i^T > 0$, $\Lambda_i \in \Re^{n\times n}$, $i = 1\ldots10$, $Q_0 \in \Re^{n\times n}$, $K \in \Re^{n\times m}$ and positive parameters ϖ, μ_1, μ_2 and μ_3 such that the following LMI

$$\begin{bmatrix} \begin{matrix}-\Gamma(K,\varpi,\mu_1,\mu_2) & P \\ P & R\end{matrix} & 0 & 0 & 0 \\ 0 & \begin{matrix}\Theta_1 & \tilde{A}^?\left(K\right)P \\ P\tilde{A}\left(K\right) & \mu_1 P\end{matrix} & 0 & 0 \\ 0 & 0 & \begin{matrix}\Theta_2 & \hat{W}_1^?\left(K\right)P \\ P\hat{W}_1 & \mu_2 P\end{matrix} & 0 \\ 0 & 0 & 0 & \begin{matrix}\Theta_3 & \hat{W}_2^?\left(K\right)P \\ P\hat{W}_2 & \mu_3 P\end{matrix}\end{bmatrix} > 0$$

where $tr\left\{\Theta_i\right\} < 1,\ i = 1,2,3$ and

$$\Gamma(K,\delta,\mu_1,\mu_2) = \left[\tilde{A}^?\left(K\right)P + P\tilde{A}\left(K\right) + Q\left(\delta,\mu_1,\mu_2,\mu_3\right)\right]$$
$$R^{-1} = \Lambda_1^{-1} + \Lambda_9^{-1} + \Lambda_{10}^{-1} + \hat{W}_1\Lambda_5^{-1}\left(\hat{W}_1\right)^? + \hat{W}_2\Lambda_8^{-1}\left(\hat{W}_2\right)^?$$
$$Q\left(\delta,\mu_1,\mu_2,\mu_3\right) = \left[\left\|\Lambda_5\right\|L_\sigma + \left\|\Lambda_8\right\|L_\phi\Upsilon_u^2 + \mu_1 + \mu_2 L_\sigma + \mu_3\Upsilon_u^2 L_\phi\right]I + \varpi\left(\Lambda_3^{-1} + \Lambda_7^{-1}\right) + Q_0$$

has positive definite solution **P** ,then the projectional DNN observer with the weight's learning laws, given by (29),(30),(18), and with $h(t)$ satisfying

$$\lim_{t\to\infty} h(t) \to \varepsilon, 0 < \ \varepsilon << 1$$

Provides the following upper bound for the "averaged estimation" error

$$\overline{\lim_{T\to\infty}}\frac{1}{T}\int_{\tau=0}^{T}\left(\delta^T(\tau - h(\tau))Q_0\delta(\tau - h(\tau)\right)d\tau \leq$$
$$\left\|\Lambda_9\right\|\left(\left(\left\|K\right\|\left\|\Lambda_\eta^{-1}\right\|^{1/2}\Upsilon_\eta + \left\|\Lambda_\xi^{-1}\right\|^{1/2}\Upsilon_\xi\right)\right)^2$$
$$+\left\|\Lambda_{10}\right\|\left\|\Lambda_{\tilde{f}}^{-1}\right\|\left[\tilde{f}_0 + \tilde{f}_1\left\|\Lambda_{\tilde{f}}^1\right\|Diam(x)^2\right] + \left\|K\right\|^2\left\|P\right\|\left\|\Lambda_\eta^{-1}\right\|^{1/2}\Upsilon_\eta$$
$$+\left\|P\right\|\left\|\Lambda_{\tilde{f}}^{-1}\right\|\left[\tilde{f}_0 + \tilde{f}_1\left\|\Lambda_{\tilde{f}}^1\right\|Diam(x)^2\right] + \left\|P\right\|\left\|\Lambda_\xi^{-1}\right\|\Upsilon_\xi + 2\Upsilon_\eta$$

where: $\delta\left(t\right) := \hat{x}\left(t\right) - x\left(t\right)$ and $Diam(x) = \sup_{x,z\in X}\left\|x - z\right\|$. The proof of this theorem can be consulted in (García, Poznyak A., Chairez & Poznyak T., 2008).

Remark 4. It is easy to see that in the absence of noises $(\eta_t = \xi_t = 0)$ and no-modeled dynamics $(\tilde{f} = 0)$, we can choose $\tilde{f}_0, \tilde{f}_1, \Upsilon_\xi$ and Υ_η such that:

$$\overline{\lim_{T\to\infty}}\frac{1}{T}\int_{\tau=0}^{T}\left(\delta^T(\tau - h(\tau))Q_0\delta(\tau - h(\tau)\right)d\tau \to 0$$

4.2 Projectional DNN Observer Algorithm

As it follows from the presentation above, to carry out the suggested approach one needs to fulfill the following steps:

1. Define the projector.
2. Select Matrices A and K.
3. Select $\hat{W}$ (some hints are given by (Chairez, Poznyak A. & Poznyak T.) and (Stepanyan & Hovakimyan, 2007)).

4. Find P as the solution of the LMI problem (34), it can be done using the corresponding Matlab-Toolbox.
5. Introduce P into the adapting weight law (29), (30) and (18) and realized it on-line.

4.2.1 Simulation Results

The same Chua´s system model presented in (24) will be considered as source of data. The first simulated case is the implementation of the Projectional DNN observer based in the available information $y(t) = x_1(t) + \xi_1(t)$, where $\xi_1(t)$ is a "quasi" white noise proportional to 10% of $x_1(t)$ by this information states $x_2(t)$ and $x_3(t)$ are reconstructed. Second simulation considers the case when additionally to the output we can measure $x_2(t)$ and we use this information to the encryption of some desirable info as is shown in the Figure 6, we would like to present this result as an example of possible implementation of Master-Slave synchronization by the Projectional Differential Neural Network Observer.

Case 1. (Projectional DNN Observer) Considering that $x_1(t)$ is the available information and that it is corrupted by additive noise. Figure 7depicts the $\mathbf{x_2}$ and $\mathbf{x_3}$ reconstruction obtained from the Projectional DNN observer it can be claimed that even in the presence of considerable noise the state estimation is carried out with excellent behavior.

Case 2. (Synchronization-Cryptography). Figure 8 depicts the behavior for the approach proposed in Figure 6, a) shows the info selected in this case as random signal, b) presents the cryptography process and c) the recovering process, it is clear that a) and c) have a good coincidence proving the good result obtained. Even under the consideration that the available information is noisy.

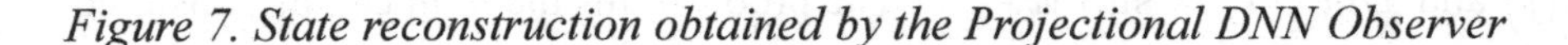

Figure 7. State reconstruction obtained by the Projectional DNN Observer

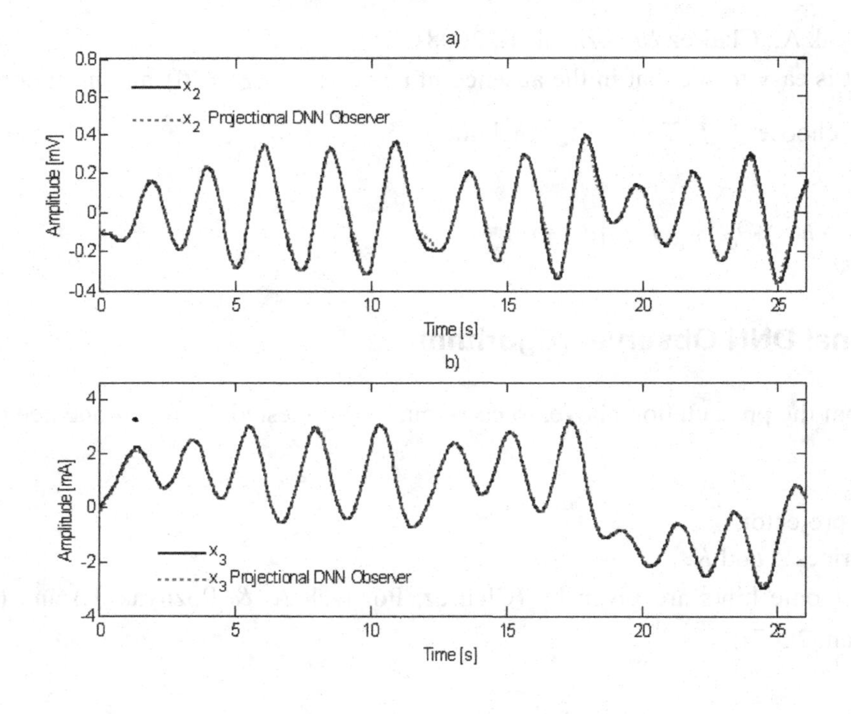

5. CONCLUSION

In this chapter a new type of the non-smooth identifier and an adaptive observer are suggested. The observer was designed to estimate the unknown states for uncertain complex systems affected by external noises. In particular, it was shown that such class of adaptive observers can be successfully applied to reproduce the hesitant information generated by chaotic systems. Actually, they permit to maintain the reproduced state estimates within a-priory given compact set even in presence of input and output noises. Since the obtained estimated trajectories are not smooth, to analyze its stability, a particular Lyapunov-Krasovskii functional candidate is suggested. Using the aforementioned functional, it was possible to prove the practical stability for the estimation error. The upper bound for the estimation error was obtained turning out to be rationally dependent on the lag of the filter and linear with respect to a noise power. In presence of output noise the abovementioned upper bound depends also on the gain matrix K of the filter. Chua circuit mathematical model was used to prove the effectiveness of the suggested approach reproducing complex uncertain dynamics (chaotic systems) for both, identification and state estimation. Encryption by the master-slave configuration was successfully tested considering the presence of high magnitude of noise.

Figure 8. Synchronization – Cryptography information

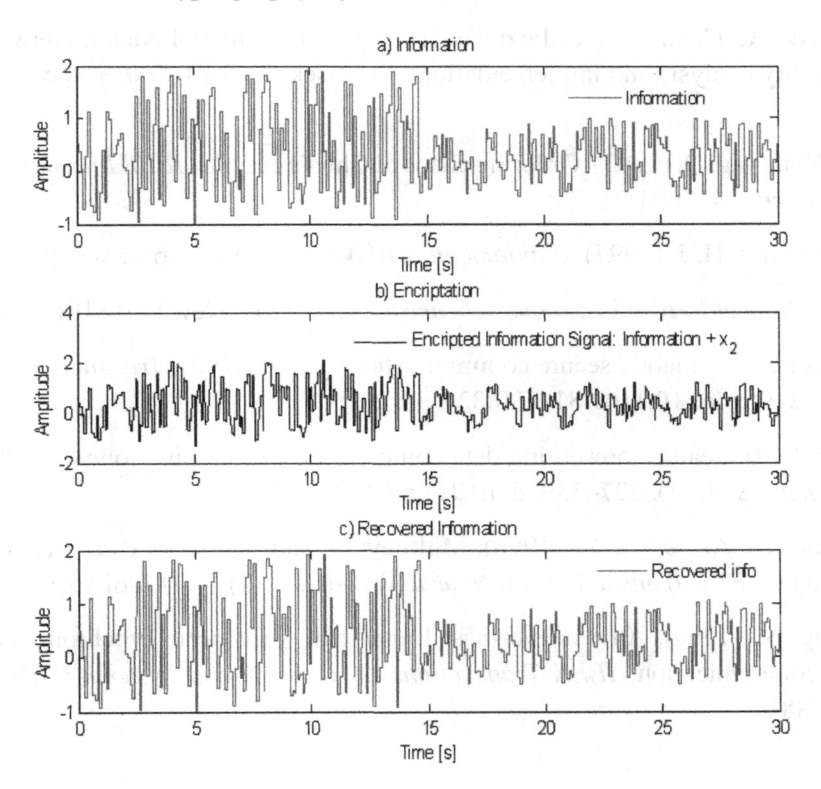

6. REFERENCES

Bottou, L., & Vapnik, V. (1992). Local learning algorithms. *Neural Computation, 4*, 888–900. doi:10.1162/neco.1992.4.6.888

Campi, M., & Kumar, P. R. (1996, December). *Learning dynamical systems in a stationary environment*. Presented at the 35th Conference on Decision and Control, Kobe, Japan.

Chairez, I., Poznayk, A., & Poznayk, T. (2006). New sliding-mode learning law for dynamic neural network observer. *IEEE Transactions on Circuits and Systems, II*, 53, 1338–1342.

Chen, G., & Dong, X. (1995,April-May). *Identification and control of chaotic systems*. Paper presented at the IEEE International Symposium on Circuit and Systems, Seattle, WA.

Femat, R., & Alvarez-Ramírez, J. (1997). Synchronization of a class of strictly different chaotic oscillators. *Physics Letters. [Part A], 236*(12), 307–313. doi:10.1016/S0375-9601(97)00786-X

Femat, R., Alvarez-Ramírez, J., & Fernández-Anaya, G. (2000). Adaptive synchronization of high-order chaotic systems: a feedback with low-order parametrization. *Physica D. Nonlinear Phenomena, 139*(3-4), 231–246. doi:10.1016/S0167-2789(99)00226-2

Femat, R., & Jauregui-Ortiz, R. (2001). A chaos-based communication scheme via robust asymptotic feedback. *IEEE Transactions on Circuits and Systems-I, 48*(10), 1161–1169. doi:10.1109/81.956010

Gallegos, J. A. (1994). Nonlinear regulation of a Lorenz system by feedback linearization techniques. *Dynamics and Control, 4*, 277–298. doi:10.1007/BF01985075

García, A., Poznayk, A., Chairez, I., & Poznyak, T. (2008). Differential Neural Networks Observers: development, stability analysis and implementation . In Husek, P. (Ed.), *System* (pp. 61–82). Structure and Control.

Guanron, Ch., & Xianonig, D. (1993). Ordering chaos of Chua's circuits. *IEEE International Symposium on Circuits and Systems*, 4, 2604-2607.

Hartmut, D., & Hartmut, H. J. (1997). *Fundamentals of Wavelets*. New York: Wiley.

Haykin, S. (1994). *Neural Networks. A comprehensive Foundation*. New York: IEEE Press.

Jiang, Z. (2002). A note on chaotic secure communication systems. *IEEE Transactions on Circuits and Systems-I, 49*(1), 92–96. doi:10.1109/81.974882

Jin, J., & Shi, J. (1999). Feature-preserving data compression of stamping tonnage information using wavelets. *Technometrics, 41*(4), 327–339. doi:10.2307/1271349

Lewis, F. L., Yesildirek, A., & Liu, K. (1996). Multilayer neural-net robot controller with guaranteed tracking performance. *IEEE Transactions on Neural Networks*, 7(2), 1–11. doi:10.1109/72.485674

Liao, T., & Huang, N. (1999). An observer-based approach for chaotic synchronization with application to secure communication. *IEEE Transactions on Circuits and Systems-I, 46*(9), 1144–1150. doi:10.1109/81.788817

Liao, T., & Tsai, S. (2000). Adaptive synchronization of chaotic systems and its application to secure communications. *Chaos, Solitons, and Fractals, 11*(9), 1387–1396. doi:10.1016/S0960-0779(99)00051-X

Lin, F. J., Hwang, W. J., & Wai, R. J. (1998). *Ultrasonic motor servo drive with on-line trained neural network model-following controller* (*Vol. 145*, pp. 105–110). Proc. Inst. Elect. Eng. Electr.

Ljung, L. (1996 December). *Pac-learning and asymptotic system identification theory*. Presented at the 35th Conference on Decision and Control, Kobe, Japan.

Martell, L. (2000). *Wavelet-based data reduction and de-noising procedures.* Unpublished doctoral dissertation, North Carolina State University, Raleigh, USA.

Meyer, Y. (1992). *Wavelets and Applications*. Berlin: Springer-Verlag.

Nijmeijer, H., & Berghuis, H. (1995). On Lyapunov control of the Duffing equation. *IEEE Transactions on Circuits and Systems, 42*, 473–477. doi:10.1109/81.404059

Omidvar, O., & Elliott, D. L. (1997). *Neural Systems for Control.* New York: Academic Publishers.

Poggio, T., & Girosi, F. (1990). Regularization algorithms for learning that are equivalent to multilayer networks. *Science, 247*(27), 978–982. doi:10.1126/science.247.4945.978

Poggio, T., & Girosi, F. (2006). Networks for approximation and learning. In *Proceedings of the IEEE CDC 2006* (Vol. 78, pp. 1481—1497).

Poznyak, A. (2004). Deterministic output noise effects in sliding mode observation . In Sabanovic, A., Fridman, L., & Spurgeon, S. (Eds.), *Variable structure systems: from principles to implementation* (pp. 45–80). London: IET Control Engineering Series.

Poznyak, A. S. (2008). *Advance Mathematical tools for automatic control engineers: Deterministic technique*. London: Elsevier.

Poznyak, A. S., Sanchez, E., & Yu, W. (2001). *Differential Neural Networks for Robust Nonlinear Control (Identification, state Estimation an trajectory Tracking)*. Singapore: World Scientific. doi:10.1142/9789812811295

Poznyak, A. S., Yu, W., Ramirez, H. S., & Sanchez, E. N. (1998). Robust Identification by dynamic neural networks using sliding mode learning. *Applied Mathematics Computer Science, 8*, 101–110.

Rovithakis, G., & Christodoulou, M. (1994). Adaptive control of unknown plants using dynamical neural networks. *IEEE Transactions on Systems, Man, and Cybernetics, 24*, 400–412. doi:10.1109/21.278990

Stepanyan, V., & Hovakimyan, N. (2007). Robust adaptive observer design for uncertain systems with bounded disturbances. *IEEE Transactions on Neural Networks, 18*(5), 1392–1403. doi:10.1109/TNN.2007.895837

Yaz, E., & Azemi, A. (1994). Robust adaptive observers for systems having uncertain functions with unknown bounds. In *Proceedings of American Control Conference*, (pp. 73-74).

Zak, H., & Walcott, B. L. (1990). State observation of nonlinear control systems via the method of Lyapunov . In Zinober, A. S. I. (Ed.), *Deterministic Control of Uncertain Systems* (pp. 333–350). Washington, DC: IEEE Control Engineering Series.

Zeitz, M. (1987). The extended Luenberger observer for nonlinear systems. *Systems & Control Letters*, *9*(28), 149–156. doi:10.1016/0167-6911(87)90021-1

Zeng, Y., & Singh, S. N. (1997). Adaptive control of chaos in Lorenz system. *Dynamics and Control*, *7*, 143–154. doi:10.1023/A:1008275800168

APPENDIX (PROOF OF THEOREM 2)

$$V(t) = \int_{\tau=t-h(t)}^{\tau=t} \left[\left\| \delta(\tau) \right\|_P^2 + k_1(\tau) \mathbf{tr} \left\{ \tilde{W}_1^T(\tau) \tilde{W}_1(\tau) \right\} + k_2(\tau) \mathbf{tr} \left\{ \tilde{W}_2^T(\tau) \tilde{W}_2(\tau) \right\} \right] d\tau$$

$$\begin{aligned} \dot{V}(t) \leq \left\| \delta(t) \right\|_p^2 - \left\| \delta(t-h(t)) \right\|_p^2 + k_1(t) tr \left\{ \tilde{W}_1^T(t) \tilde{W}_1(t) \right\} \\ -k_1(t-h(t)) tr \left\{ \tilde{W}_1^T(t-h(t)) \tilde{W}_1(t-h(t)) \right\} + \\ k_2(t) tr \left\{ \tilde{W}_2^T(t) \tilde{W}_2(t) \right\} - k_2(t-h(t)) tr \left\{ \tilde{W}_2^T(t-h(t)) \tilde{W}_2(t-h(t)) \right\} \end{aligned}$$

$$\left\| a+b \right\|_P^2 = \left\| a \right\|_P^2 + \left\| b \right\|_P^2 + 2(Pa, b)$$

$$\begin{aligned} \tilde{\sigma}(t) &:= \sigma(\hat{x}(t)) - \sigma(x(t)) \\ \tilde{\phi}(t) &:= \phi(\hat{x}(t)) - \phi(x(t)) \end{aligned}$$

$$\begin{aligned} \dot{V}(t) \leq \alpha(t) + \beta(t) + k_1(t) tr \left\{ \tilde{W}_1^T(t) \tilde{W}_1(t) \right\} \\ -k_1(t-h(t)) tr \left\{ \tilde{W}_1^T(t-h(t)) \tilde{W}_1(t-h(t)) \right\} + \\ k_2(t) tr \left\{ \tilde{W}_2^T(t) \tilde{W}_2(t) \right\} - k_2(t-h(t)) tr \left\{ \tilde{W}_2^T(t-h(t)) \tilde{W}_2(t-h(t)) \right\} \end{aligned}$$

$$\begin{aligned} \alpha(t) := \int_{\tau=t-h(t)}^{\tau=t} \left\| A\delta(\tau) \right\|_P^2 d\tau + \int_{\tau=t-h(t)}^{\tau=t} \left\| \tilde{W}_1(\tau) \sigma(\hat{x}(\tau)) \right\|_P^2 d\tau + \\ \int_{\tau=t-h(t)}^{\tau=t} \left\| \hat{W}_1 \tilde{\sigma}(t) \right\|_P^2 d\tau + \int_{\tau=t-h(t)}^{\tau=t} \left\| \tilde{W}_2(\tau) \phi(\hat{x}(\tau)) u(\tau) \right\|_P^2 d\tau \\ \int_{\tau=t-h(t)}^{\tau=t} \left\| \hat{W}_2 \tilde{\phi}(\tau) u(\tau) \right\|_P^2 d\tau + \int_{\tau=t-h(t)}^{\tau=t} \left\| f(\tau) \right\|_P^2 d\tau + \int_{\tau=t-h(t)}^{\tau=t} \left\| \xi(\tau) \right\|_P^2 d\tau \end{aligned}$$

$$\beta(t) := 2\left(P\delta\left(t-h\left(t\right)\right), \int_{\tau=t-h(t)}^{\tau=t} A\delta\left(\tau\right)d\tau\right) + 2\left(P\delta\left(t-h\left(t\right)\right), \int_{\tau=t-h(t)}^{\tau=t} \tilde{W}_1\left(\tau\right)\sigma\left(\hat{x}\left(\tau\right)\right)d\tau\right) +$$
$$2\left(P\delta\left(t-h\left(t\right)\right), \int_{\tau=t-h(t)}^{\tau=t} \hat{W}_1\tilde{\sigma}\left(t\right)d\tau\right) + 2\left(P\delta\left(t-h\left(t\right)\right), \int_{\tau=t-h(t)}^{\tau=t} \tilde{W}_2\left(\tau\right)\phi\left(\hat{x}\left(\tau\right)\right)u\left(\tau\right)d\tau\right) +$$
$$2\left(P\delta\left(t-h\left(t\right)\right), \int_{\tau=t-h(t)}^{\tau=t} \hat{W}_2\tilde{\phi}\left(\tau\right)u\left(\tau\right)d\tau\right) - 2\left(P\delta\left(t-h\left(t\right)\right), \int_{\tau=t-h(t)}^{\tau=t} f\left(\tau\right)d\tau\right) -$$
$$2\left(P\delta\left(t-h\left(t\right)\right), \int_{\tau=t-h(t)}^{\tau=t} \xi\left(\tau\right)d\tau\right)$$

$$XY^T + YX^T \le X\Lambda X^T + Y\Lambda^{-1}Y^T$$

$$\left\|AB\right\| \le \left\|A\right\|\left\|B\right\|, \quad \left\|\int F(\tau)d\tau\right\| \le \int \left\|F(\tau)\right\|d\tau$$

$$\begin{gathered} PA + A^TP + PR^{-1}P + Q(\mu_1,\mu_2,\mu_3) \le 0 \\ R^{-1} = \Lambda_1^{-1} + \hat{W}_1\Lambda_2\hat{W}_1^T + \Lambda_4 + \Lambda_5 \\ Q(\mu_1,\mu_2,\mu_3) = \\ \left(\left\|\Lambda_5\right\|L_\sigma L_\delta^2 + \left\|\Lambda_3^{-1}\right\|L_\phi\Upsilon_u^2 + \mu_1 + L_\sigma\mu_2 + \mu_3\Upsilon_u^2L_\phi\right)I + Q_0 \end{gathered}$$

$$\int_{\tau=t-h(t)}^{\tau=t} 2tr\left\{\tilde{W}_1^T\left(\tau\right)P\delta\left(t-h\left(t\right)\right)\sigma^T\left(\hat{x}\left(\tau\right)\right)\right\}d\tau + \int_{\tau=t-h(t)}^{\tau=t} 2tr\tilde{W}_1^T\left(\tau\right)P\tilde{W}_1\left(\tau\right)\sigma\left(\hat{x}\left(\tau\right)\right)\sigma^T\left(\hat{x}\left(\tau\right)\right)d\tau -$$
$$k_1\left(t\right)tr\left\{\tilde{W}_1^T\left(t\right)\tilde{W}_1\left(t\right)\right\} + k_1\left(t-h\left(t\right)\right)tr\left\{\tilde{W}_1^T(t-h\left(t\right))\tilde{W}_1(t-h\left(t\right))\right\} = 0$$

$$\begin{gathered} \int_{\tau=t-h(t)}^{\tau=t} tr\left\{2\tilde{W}_2^T\left(\tau\right)P\delta\left(t-h\left(t\right)\right)u^T\left(\tau\right)\phi^T\left(\hat{x}\left(\tau\right)\right)\right\}d\tau + \\ \int_{\tau=t-h(t)}^{\tau=t} tr\left\{\tilde{W}_2^T\left(\tau\right)P\tilde{W}_2\left(\tau\right)\phi\left(\hat{x}\left(\tau\right)\right)u\left(\tau\right)u^T\left(\tau\right)\phi^T\left(\hat{x}\left(\tau\right)\right)\right\}d\tau \\ +k_2\left(t\right)tr\left\{\tilde{W}_2^T\left(t\right)\tilde{W}_2\left(t\right)\right\} \\ -k_2\left(t-h\left(t\right)\right)tr\left\{\tilde{W}_2^T(t-h\left(t\right))\tilde{W}_2(t-h\left(t\right))\right\} = 0 \end{gathered}$$

$$\dot{V}(t) \leq ah^3(t) + bh^2(t) + ch(t) - h(t)\delta^T(t - h(t))Q_0\delta(t - h(t))$$

$$\begin{aligned} a &:= \tfrac{1}{3}\left[\left\|\tfrac{\Lambda_1}{h(t)}\right\|\|A\|^2 L_\delta^2 + \left\|\Lambda_2^{-1}\right\| L_\sigma L_\delta^2 + \right. \\ &\left.\left\|\Lambda_3^{-1}\right\| L_\phi \Upsilon_u^2 + L_\delta \mu_1 + L_\delta L_\sigma \mu_2 + \mu_3 \Upsilon_u^2 L_\phi L_\delta\right] \\ b &:= \tfrac{1}{3}\left\|\tfrac{\Lambda_1}{h(t)}\right\|\|A\|^2 L_\delta^2 \\ c &:= \|P\|\left\|\Lambda_\xi^{-1}\right\|\Upsilon_\xi + \left(\|P\| + \left\|\Lambda_4^{-1}\right\|\right)\left\|\Lambda_{\tilde{f}}^{-1}\right\|\left[\tilde{f}_0 + \tilde{f}_1\left\|\Lambda_{\tilde{f}}^1\right\| Diam(x)^2\right] + \\ &\left\|\Lambda_5^{-1}\right\|\left[\left\|\Lambda_\xi^{-1}\right\|^{1/2}\Upsilon_\xi\right] \end{aligned}$$

$$\delta^T(t - h(t))Q_0\delta_{t-h(t)} \leq ah^3(t) + bh(t) + c - \frac{\dot{V}(t)}{h(t)}$$

Integrating both sides from 0 up to T

$$\int_0^T \delta^T(t - h(t))Q_0\delta(t - h(t))d\tau \leq \int_0^T \left[ah^3(\tau) + bh(\tau) + c - \frac{\dot{V}(\tau)}{h(\tau)}\right]d\tau$$

Taking into account the next equallity:

$$-\int_0^T \frac{1}{h(\tau)}\dot{V}(\tau)d\tau = -\int_0^T \frac{1}{h(t)}\frac{dV(\tau)}{d\tau}d\tau = \frac{V(0)}{h(0)} - \frac{V(T)}{h(T)} \leq \frac{V(0)}{h(0)}$$

Implying

$$\int_0^T \delta^T(t - h(t))Q_0\delta(t - h(t))d\tau \leq \int_0^T \left[ah^3(t) + bh(t)\right]d\tau + cT + \frac{V(0)}{h(0)}$$

Dividing over T and taking the upper limit we derive (38).

Chapter 4
Simple Chaotic Electronic Circuits

M. P. Hanias
Technological and Educational Institute of Chalkis, Greece

G. S. Tombras
National and Kapodistrian University of Athens, Greece

ABSTRACT

Simple chaotic electronics circuits as diode resonator circuits, Resistor-Inductor-LED optoelectronic chaotic circuits, and Single Transistor chaotic circuits can be used as transmitters and receivers for chaotic cryptosystems. In these circuits we can change and investigate the influence of various circuit parameters to the complexity of the so generated strange attractors. Time series analysis is performed following Grassberger and Procaccia's method while invariant parameters as correlation, and minimum embedding dimension are respectively calculated. The Kolmogorov entropy is also calculated and the RLT circuits in a critical state are examined.

1. INTRODUCTION

In recent years, a growing number of cryptosystems based on chaos have been proposed. The majority of them are based on parameter modulation of a chaotic oscillator being the transmitter while the receiver is also a chaotic system synchronized by means of an adaptive observer. For example, a Lorenz's attractor is used as the non-linear time-varying system while the encryption process modulates the parameters that affect the shape and complexity of the corresponding strange attractor with the binary encoded plain-text. In general, the security of chaotic cryptosystems is based on the chaotic behaviour of the output of electronic non-linear systems which produce Lorenz type time series. In order to enhance the level of security of the transmission system various novel communication schemes that combine conventional cryptographic methods and synchronization of chaotic systems have been developed. A common feature of these systems is the utilization of the state variables of the chaotic systems (other than the transmitted

DOI: 10.4018/978-1-61520-737-4.ch004

state) as encryption keys and decryption algorithms. In such a system, a low-dimensional Chua's circuit as a chaotic system is used as an independent low-dimensional chaotic system that sends a driving signal to two chaotic slave systems, which are then synchronized and used to generate confidential keys.

A good example of a chaotic communication system is the Tsu-I Chien and Teh–Lu Liao system, the block diagram of which is shown in Figure 1. As it can be readily seen, the complete system comprises four modules, namely the Chaotic Modulator (CM), the Chaotic Secure Transmitter (CST), the Chaotic Secure Receiver (CSR), and the Chaotic Demodulator (CDM).

In the CM, an off-line self-learning process is executed prior to the transmitting period. During this learning process, a novel modulation scheme is applied to establish analog chaotic patterns corresponding to the input bit sets. Subsequently, certain parameters of these chaotic patterns (i.e. the initial peak values) are digitized and stored in the modulator's memory. In the transmitting period, a D/A converter is utilized to retrieve the digitalized parameters corresponding to the input bits from the modulator's memory. These parameters are then used to perturb the chaotic circuit (a Rossler-like system) so that analog chaotic patterns, as opposed to the original input bits, can be transmitted to the CST.

The CST and CSR modules are designed in such a way that a single scalar signal is transmitted across the public channel. By giving certain structural conditions as such of a particular class of chaotic systems (e.g. Chua's circuits), the main transmitter and the nonlinear observer-based main receiver (with an appropriate observer gain) are constructed to synchronize with each other. These two slave systems are driven simultaneously by the transmitted signal and are designed to synchronize and generate confidential keys. Synchronization between the chaotic circuits in the CST and CSR modules is guaranteed through the Lyapunov stability theorem.

The output of the CSR is then supplied to the CDM. The chaotic circuit of this module has the same system parameters as those of the CM, but does not insist on the same initial states as those in the latter, since the demodulator and the modulator can only be synchronized after a short transient period by using the proposed nonlinear observer-based synchronization scheme. In the CDM, a bit detector measures the output of the CDM' s chaotic circuit and matches it to the corresponding bit set, thereby recovering the transmitted input bits. Hence, it becomes clear that control and synchronization of the proposed system

Figure 1. The block diagram of Tsu-I Chien and Teh –Lu Liao proposed communication system

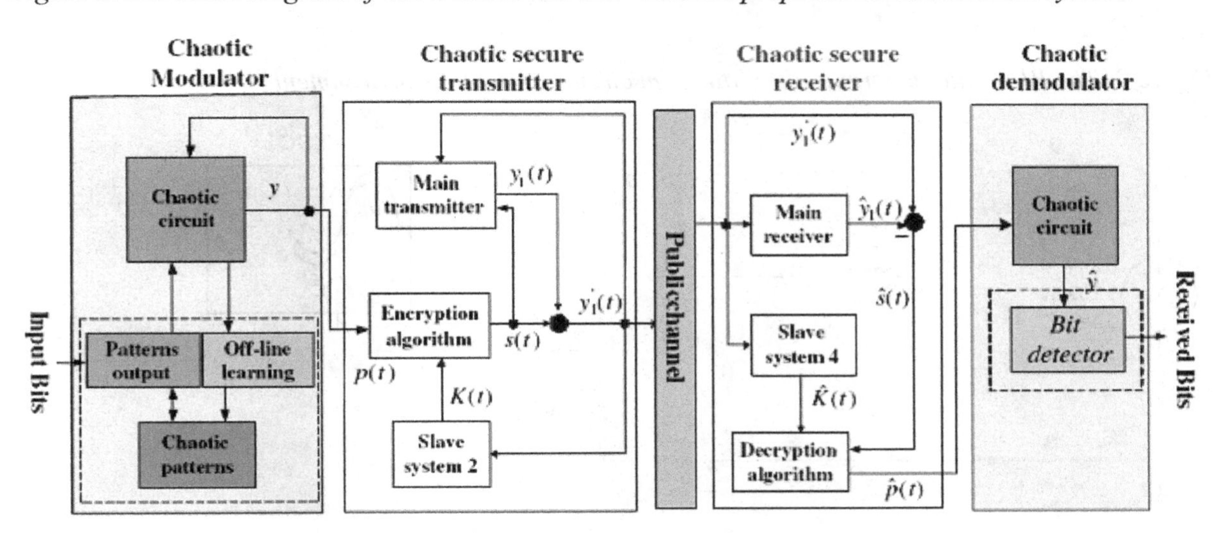

operation is based on analog circuits such as Chua's circuits and Rossler-like systems, i.e. on simple and easily controlled analogue circuits that can readily reveal chaotic operation with low dimensionality.

2. CHAOTIC CIRCUITS

The research direction of dynamical chaos is gradually moving towards practical applications and there is a growing interest for chaotic signal generation sources. In regard to this, various circuits have been proposed, among which active chaotic oscillators are preferably considered due to their relative simplicity and energy efficiency. Such a circuit may be externally triggered, i.e. externally driven to chaotic oscillation and it can typically consist of only one active and a few passive components. In this respect, it is reasonable to expect that elements such as a simple junction diode, a light emitting diode (LED), a bipolar junction transistor (BJT), or a field-effect transistor (FET), will provide for the nonlinear characteristics required towards chaos.

Below we present simple chaotic circuits that can be seen and utilized as chaotic signal generation as well as sources, main transmitters, as well as nonlinear observer-based main receivers for chaos based security communication systems. These circuits are classified as diode resonator circuits, Resistor-Inductor-LED optoelectronic chaotic circuits and Single Transistor chaotic circuits. The operational characteristics of these circuits are derived by simulation using MultiSim a circuit simulation software. Following that, we can change and investigate the influence of various circuit parameters to the complexity of the so generated strange attractors.

2.1 Diode Resonator Circuits

A non autonomous chaotic circuit referred to as the driven RL-diode circuit (*RLD*) is shown in Figure 2.

It consists of a series connection of an ac-voltage source, a linear resistor R_1, a linear inductor L_1 and a diode D_1 type 1N4001GP, that is the only nonlinear circuit element. The state equations describing this circuit are:

Figure 2. The RLD chaotic circuit after MultiSim circuit simulation environment

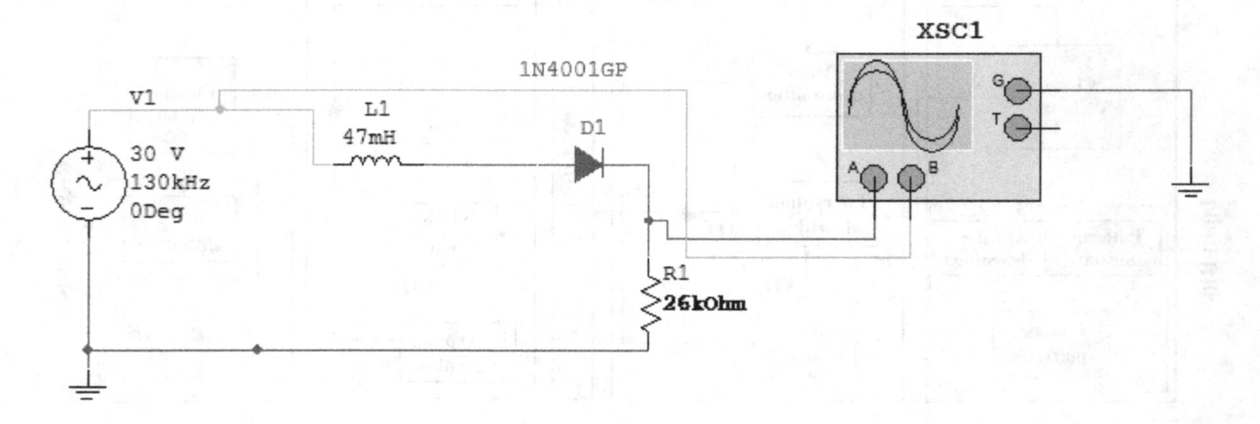

$$V_1 = R_1 i + L_1 \frac{di}{dt} + V_D \tag{2.1.1}$$

$$i = I_s \left(e^{\frac{V_D}{nV_T}} - 1 \right) \tag{2.1.2}$$

Figure 3 Time series V_{R1} (t) (top) Phase portrait of V_1vs. VR1 (bottom)

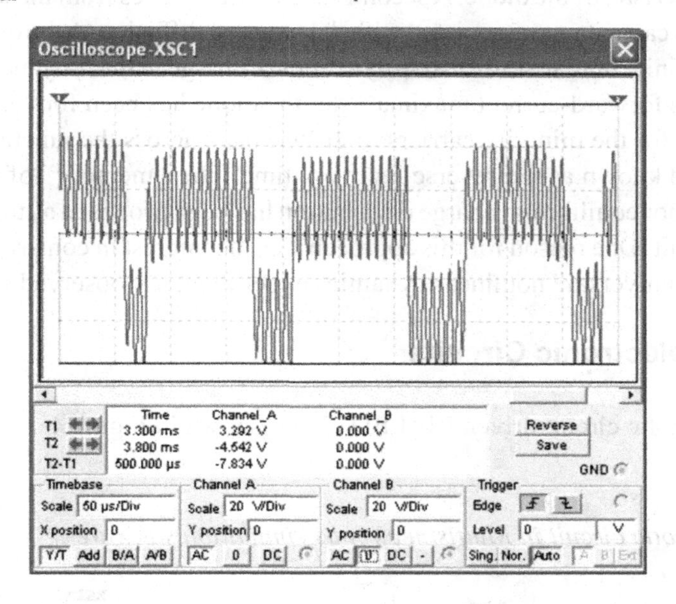

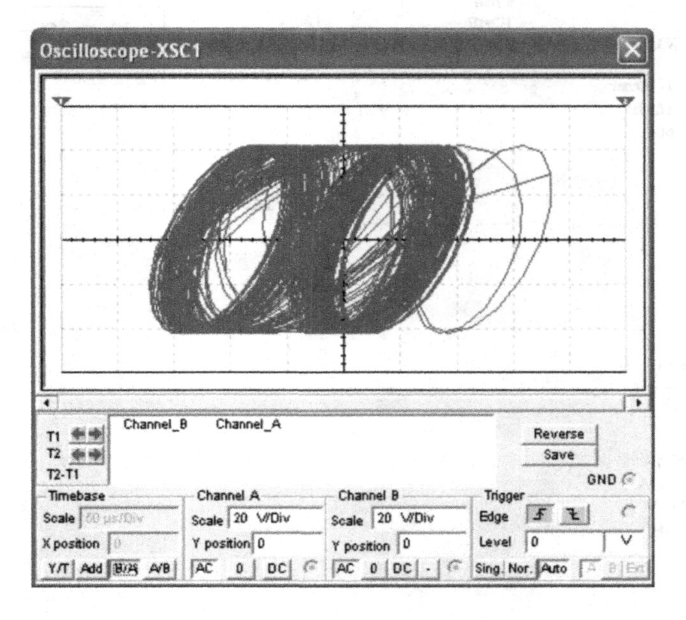

where, V_D is the voltage across the diode, I_S is the diode saturation current, n is a constant which has a value between 1 and 2 depending on the material and the physical structure of the diode, and V_T is the thermal voltage.

An important feature of this circuit is that the current i (or the voltage across the resistor R) can be chaotic although the input voltage V_1 is nonchaotic. The usual procedure is to choose a parameter that strongly affects the system. We found that for V_1=30V RMS and input frequency f=130 KHz, inductance L_1=47mH, the response is a chaotic one. The results of the Multisim simulation are shown in Figure 3 (top). In this figure The RL-diode was implemented and the voltage oscillations across the resistor V_{R1} and its phase portrait V1 vs. V_{R1} are shown in Figure 3 (bottom).

There are three general types of nonlinearity present in the driven RLD circuit. The first is the nonlinear current voltage characteristic of the diode. A second contribution comes from the large and exponentially nonlinear forward-bias capacitance associated with the junction diffusion. A third contribution to nonlinearity comes from the finite time scale diffusive dynamics of charge in the *p-n* junction and the associated "memory" of previous forward-current maxima After the diode has been forward biased and switched off, it takes some time for the minority carriers to diffuse back across the junction, allowing the diode to conduct for a period known as the reverse-recovery time. The "memory" of previous forward-bias currents built into this nonequilibrium charge distribution has been proposed as the main source of chaos in the driven *RLD* circuit. One reason for this approach was the interest in constructing low-dimensional maps to describe the "universal" nonlinear dynamics experimentally observed in the circuit.

2.2 RL-LED Optoelectronic Circuits

A non autonomous chaotic circuit driven RL-LED circuit shown in Figure 4.

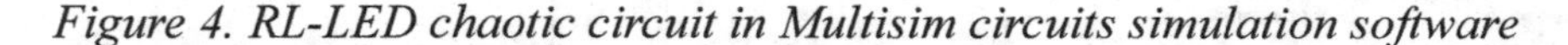

Figure 4. RL-LED chaotic circuit in Multisim circuits simulation software

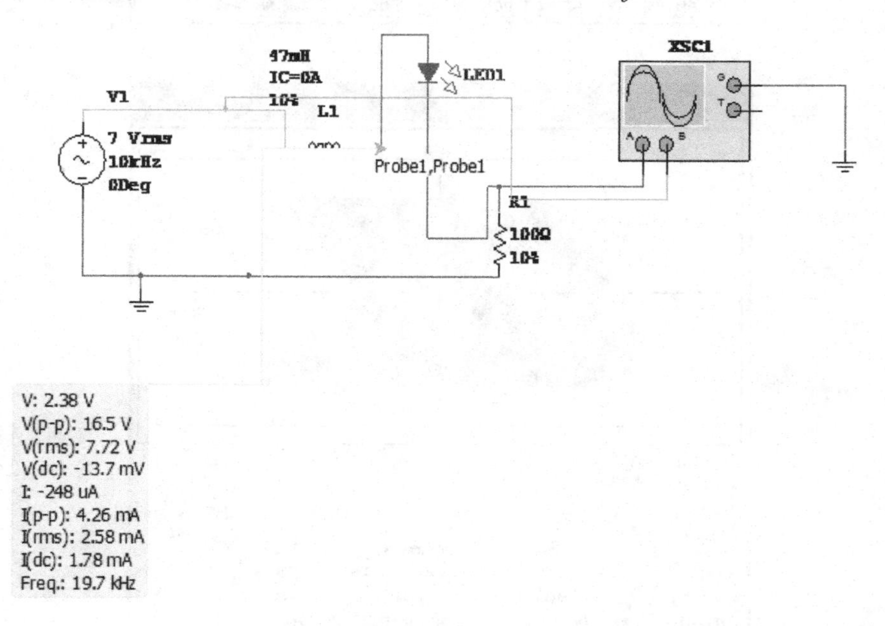

It consists of a series connection of an ac-voltage source, a linear resistor R_1, a linear inductor L_1 and a typical LED. The R_1 is R_1= 100 Ω in series with the LED. The circuit is driven by an input sinusoidal voltage with amplitude V_1 as applied through an inductor L_1=47mH. The simulated circuit operation is monitored by checking the voltage across resistor R_1. Figure 5 shows the simulation obtained chaotic time series of the output signal for input signal amplitude V_{1rms}=7 volts and frequency f=10 KHz.

2.3 Single Transistor Chaotic Circuits

A single BJT chaotic circuit which can be used as an externally driven and controlled chaotic signal generator. The complete circuit is very simple and its software simulated operation demonstrates how chaos can be generated at first and, then, how can it be controlled by varying specific circuit parameters. The complete layout of the considered circuit is shown in Figure 6(a). It consists of a basic common emitter configuration of a BC107BP npn-type BJT with an emitter degeneration resistor R_1=3KΩ and a collector resistor R_2=30Ω in series with the required DC power supply V_2=12 volts. The circuit is driven by an input sinusoidal voltage with amplitude V_1 as applied through an inductor L=75μH directly connected to the transistor base. Clearly, considering the existing base-emitter junction, the input circuit resembles to a typical resistor-inductor-diode (RLD) circuit which is widely known as a simple chaotic circuit. Hence, we call the considered circuit a resistor-inductor-transistor (RLT) circuit and examine its operation for chaos by means of the MultiSim circuit simulation environment, as illustrated in Figure 6(b).

Figure 5. Output chaotic signal V=V (t) across resistor R_1 for the RL-LED circuit of Figure 4

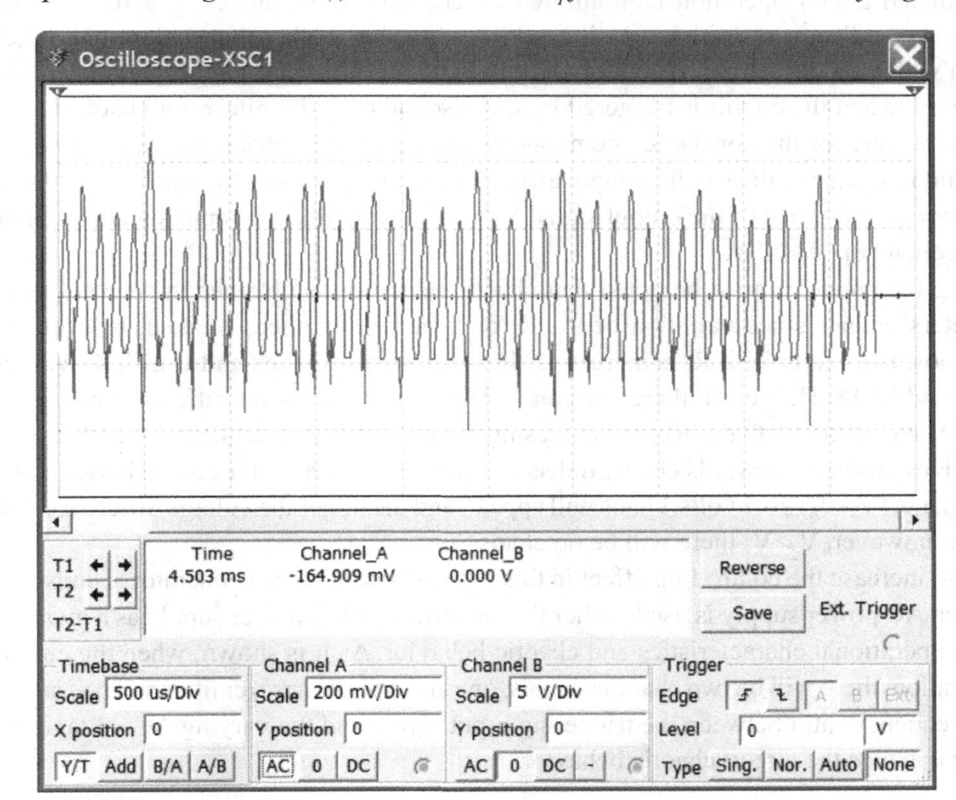

Figure 6. (a) Schematic of the considered transistor circuit, and (b) simulation environment

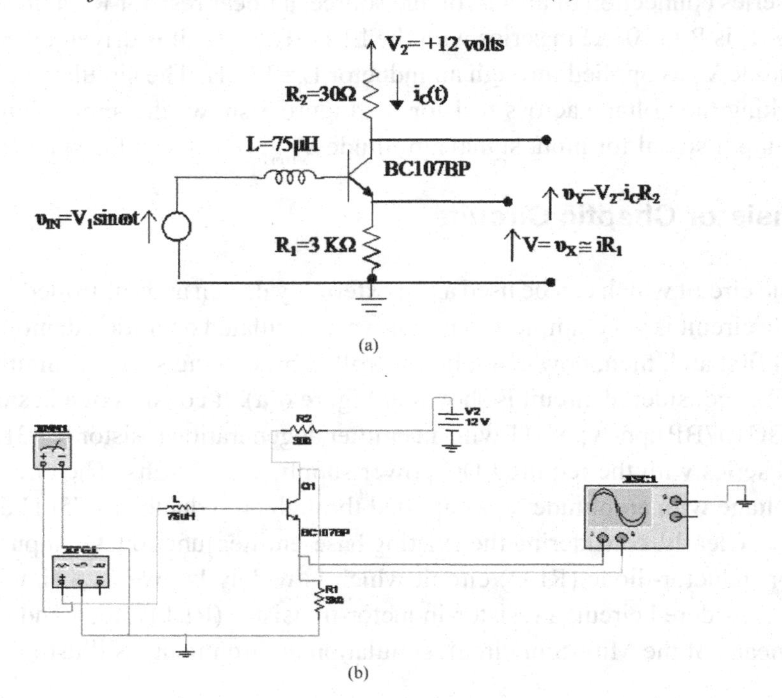

The simulated circuit operation is monitored by checking the voltages υ_X across emitter resistor R_1 and υ_Y across collector resistor R_2. As it can be readily seen both voltages depend on the collector current $i_C(t)$ which, under certain conditions, turns to be an important circuit parameter, since it can become chaotic when the circuit is triggered by a sinusoidal (and definitely not chaotic) input voltage $\upsilon_{IN}=V_1\sin\omega t$. Hence, for the considered component values, it is reasonable to expect that input signal amplitude and frequency values will strongly affect the circuit operation. Figure 7 shows the simulation obtained chaotic time series of the output signal $V(t) = \upsilon_X(t)=i_C(t)R_1$ for input signal amplitude $V_1=13$ volts and frequency $f_{IN}=3$ KHz.

Following the above, it must be noted that chaotic operation is obtained under various conditions and parameters' values since variation of R_1, R_2 or L may strengthen, weaken, or even destroy the achieved chaos. For example, if R_2 is increased, then chaos will weaken and vanish. It will also vanish for L>6mH and L<18μH. We can therefore conclude that it is the instant difference between the input voltage $\upsilon_{IN}=V_1\sin\omega t$ and V_2 that governs the biasing-state of both base-emitter and collector-base junctions and drives current i_C towards chaos. Indeed, as long as the transistor can be biased in the reverse active region, i.e. $(V_1-V_2)\geq 0.7$ volts, chaos will appear, and the larger the voltage difference, the stronger the chaos. If, however, $V_2>V_1$ there will be no chaos.

In order to increase the controlling effect in this type of chaotic signal generator a sinusoidal voltage supply i.e., an AC power supply is used, rather than an ordinary DC power supply as a means to modify the circuit's operational characteristics and chaotic behavior. As it is shown, when the circuit exhibits chaotic operation, there will be two chaotic signals, one at the transistor's emitter and one at its collector, where the frequency ratio between the triggering input signal and the varying AC voltage supply may eliminate or enhance the circuit chaotic behavior.

Figure 7. Output chaotic signal $V=V(t)$ across emitter resistor R_1 for the RLT circuit of Figure 6(b)

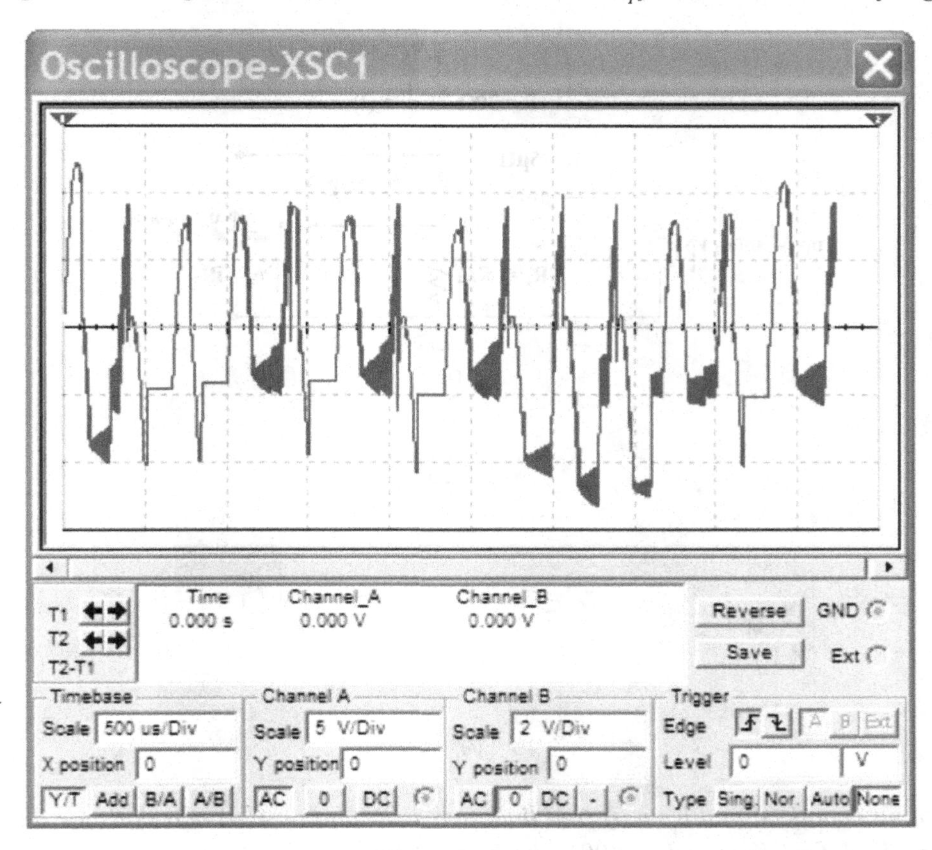

The complete layout of the considered circuit is shown in Figure 8(a). It consists of a basic common emitter configuration of a BC107BP npn-type BJT with an emitter degeneration resistor R_1=3KΩ and a collector resistor R_2=30Ω in series. The circuit is driven by an input sinusoidal voltage with amplitude V_1 as applied through an inductor L=75μH directly to the transistor base and is powered supplied by a sinusoidal voltage υ_2 with amplitude V_2 connected to the transistor collector through R_2. We examine its operation by means of the MultiSim circuit simulation environment, as illustrated in Figure 8 (b), and by monitoring voltage υ_y across the emitter resistor R_1 and voltage υ_x across the collector resistor R_2.

Clearly, both voltages υ_x and υ_y depend on the collector current $i_C(t)$ which, under certain conditions, turns to be an important circuit parameter, since it will become chaotic when the circuit exhibits chaotic operation. For example, the initial RLT circuit exhibits chaotic operation although triggered by a sinusoidal (and definitely not chaotic) input voltage. This chaotic operation has been explained by the chaotic nature of the collector current as a result of the biasing state of both base-emitter and collector-base junctions that could have lead the transistor to operate in its reverse active region. Following that and considering the incorporated RLT circuit as shown in Figure 8, it is reasonable to expect that both the amplitude and frequency values of input signal $\upsilon_1=V_1\sin\omega_1 t$ as well as the supply voltage $\upsilon_2=V_2\sin\omega_2 t$ may strongly affect its operation and, most probably, the presence of chaos. This can be seen in Figure 9, which depicts the obtained chaotic time series of the output signal $\upsilon_y(t)=i_C(t)R_1$ for input signal amplitude V_1=13 volts and frequency f_1=1 KHz and V_2=12 volts and frequency f_2=900 Hz.

Figure 8. (a) The considered RLT circuit, and (b) its simulation environment

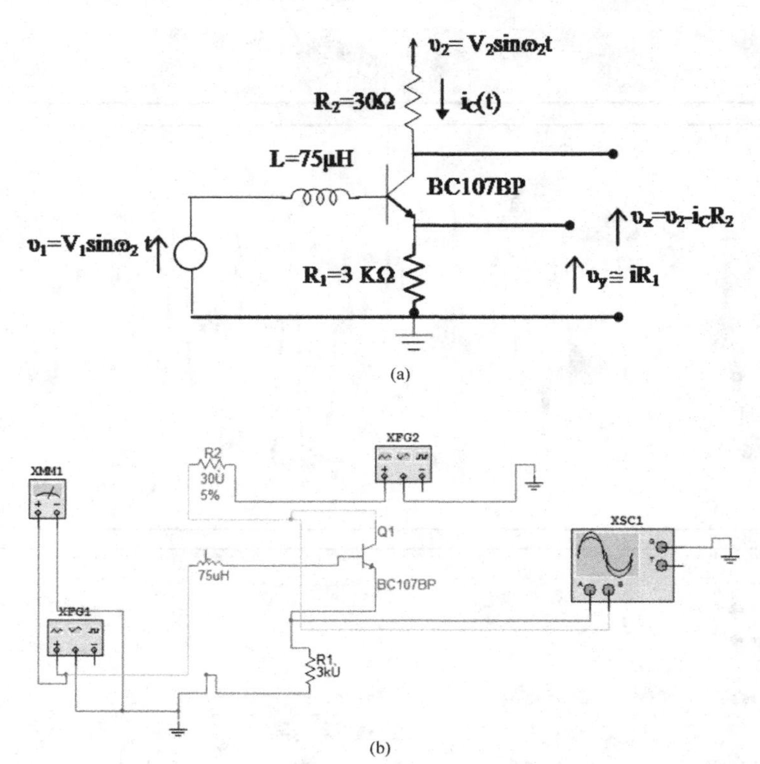

Figure 9. Output chaotic signal $\upsilon_y=\upsilon_y(t)$ (red line) across R_1 and $\upsilon_x=\upsilon_x(t)$ across R_2 for the RLT circuit of Figure 8(b). With f_1=1 KHz and f_2=900Hz both are chaotic.

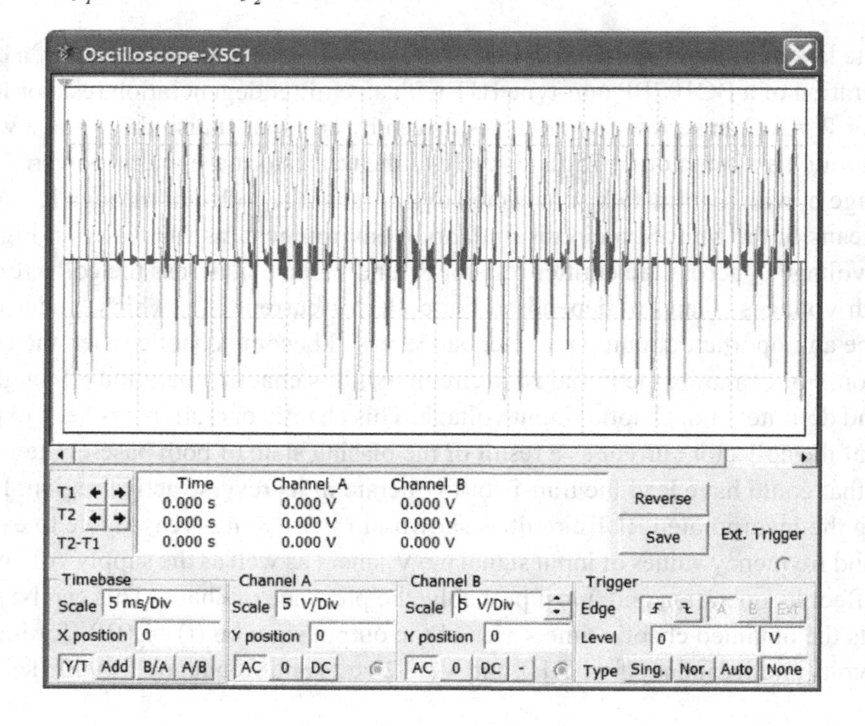

Keeping fixed the values of R_1, R_2 and L we then examine the influence of the AC supply voltage $\upsilon_2=V_2\sin(\omega_2 t)$ relative to its input signal $\upsilon_1=V_1\sin(\omega_1 t)$ towards enhancement or elimination of chaotic state. Indeed, for V_1=13 volts, $f_1=\omega_1/2\pi$=1 KHz, and V_2=12 volts, f_2=900 Hz the time series of voltage υ_y (red line in Figure 9) and the time series of voltage υ_x (green line in Figure 9) will become chaotic. By increasing V_2 to 13 and 14V the situation will not change and both voltages will remain chaotic. However, for V_2=12V and $f_1=f_2$=1KHz both time series $\upsilon_x=\upsilon_x(t)$, $\upsilon_y=\upsilon_y(t)$ will retain their periodic waveform as shown in Figure 10, while for f_1=1KHz and f_2=999Hz the signal $\upsilon_y=\upsilon_y(t)$ (red line) across emitter resistor R_1 will become chaotic despite the fact that the signal $\upsilon_x=\upsilon_x(t)$ across collector resistor R_2 will remain periodic, as shown in Figure 11. Moreover it is interesting to see that as Figure 12 shows, the lower limit of f_2 for which the $\upsilon_x=\upsilon_x(t)$ time series is "weakly" chaotic is f_2=995Hz. Hence we conclude that f_2 strongly affects both the signal across emitter resistor and the signal across collector resistor.

3. CHARACTERIZATION OF STRANGE ATTRACTORS

Finding the invariance parameters, as correlation and embedding dimensions and Kolmogorov entropy, we can monitor and use the corresponding time series and answering the central question of chaos cryptography: whether and under what conditions a chaotic system is unpredictable by probabilistic polyonomian time machines. It is clear that the basic properties characterizing a secure object are randomness increasing and computationally unpredictable. The strength of this randomness and the computationally unpredictable manner it depends on strange attractor properties. The analysis of the obtained chaotic

Figure 10. Output chaotic signal $\upsilon_y=\upsilon_y(t)$ (red or dark grey line) across emitter resistor R_1 and $\upsilon_x=\upsilon_x(t)$ across collector resistor R_2 for the RLT circuit of Figure 8(b). For $f_1=f_2$=1 KHz both are periodic.

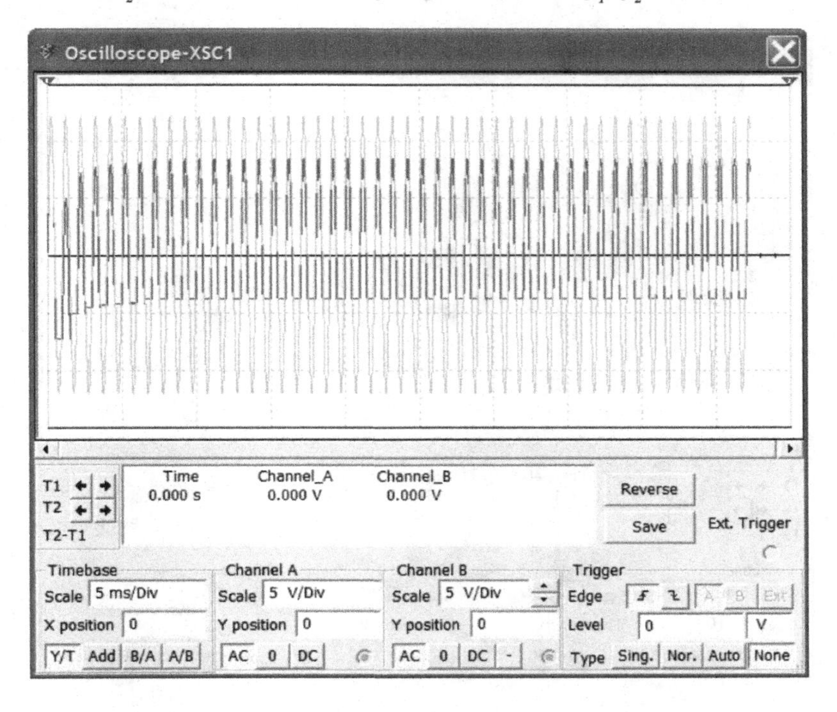

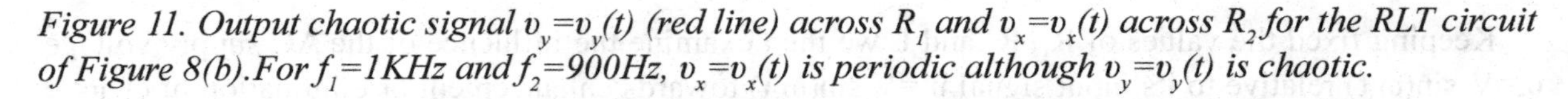

Figure 11. Output chaotic signal $\upsilon_y=\upsilon_y(t)$ (red line) across R_1 and $\upsilon_x=\upsilon_x(t)$ across R_2 for the RLT circuit of Figure 8(b). For f_1=1KHz and f_2=900Hz, $\upsilon_x=\upsilon_x(t)$ is periodic although $\upsilon_y=\upsilon_y(t)$ is chaotic.

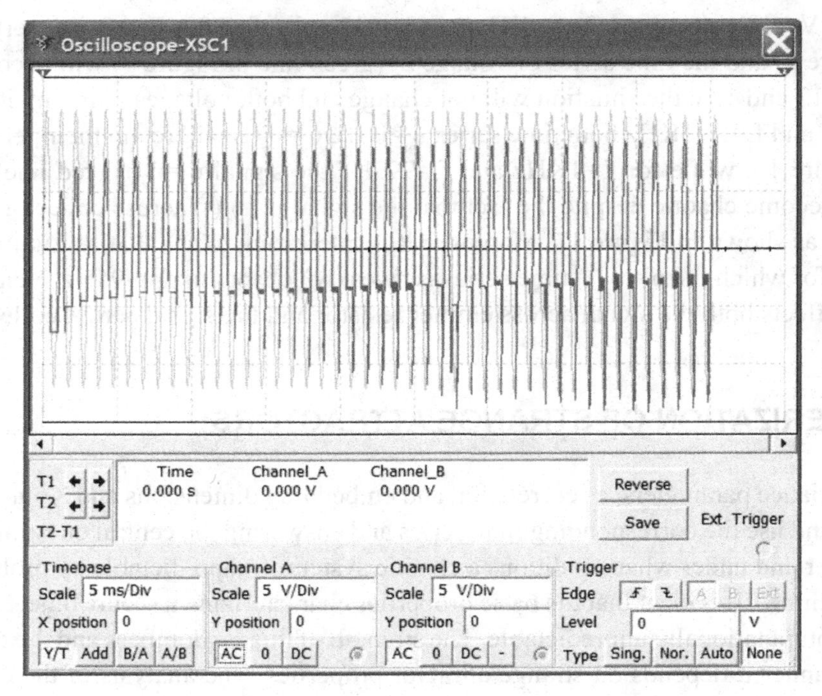

Figure 12. Output chaotic signal $\upsilon_y=\upsilon_y(t)$ (red line) across emitter resistor R_1 and $\upsilon_x=\upsilon_x(t)$ across collector resistor R_2 for the RLT circuit of Figure 8(b). For f_1=1 KHz and f_2=995Hz both are chaotic.

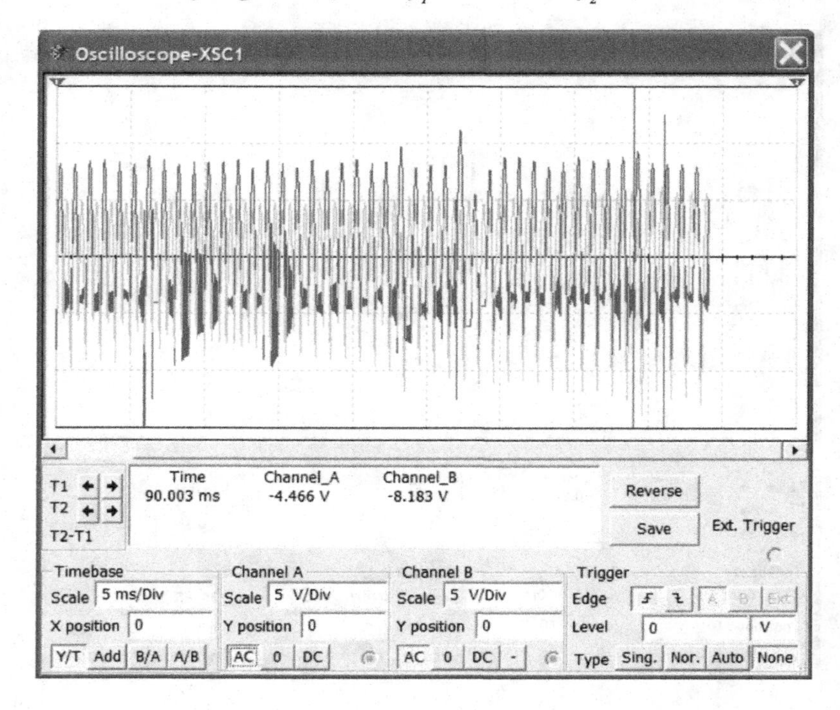

time series is done using the method proposed by Grassberger and Procaccia and successfully applied in similar cases. Moreover, according to Takens theory, the measured time series can be used to reconstruct the original phase space.

At first, we calculate the correlation integral $C(r)$ for the simulated output signal $\upsilon_X(t)=i_C(t)R_1$ for lim $r\text{->}0$ and $N\text{->}\infty$, generally as defined by:

$$C(r) = \frac{1}{N_{pairs}} \sum_{\substack{l=1, \\ j=l+W}}^{N} H\left(r - \left\| \vec{X}_l - \vec{X}_j \right\|\right). \tag{3.1}$$

where N is the number of the corresponding time series points, H is the Heaviside function, and

$$N_{pairs} = \frac{2}{(N-m+1)(N-m+1)} \tag{3.2}$$

with m being the embedding dimension. Clearly, the summation in eq.(3.1) counts the number of pairs $(\vec{X}_l, \vec{X}_j)$ for which the distance, i.e. the Euclidean norm, $\left\| \vec{X}_l - \vec{X}_j \right\|$ is less than r in an m dimensional Euclidean space. In this m dimensional space, each vector $\vec{X}_l$ will be given by:

$$\vec{X}_l = \{ V(t_i), V(t_i+\tau_d), V(t_i+2\tau_d), ..., V[t_i+(m-1)\tau_d] \} \tag{3.3}$$

and represent a point of the m dimensional phase space in which the attractor is embedded each time. In eq. (3.3), τ_d is the time delay determined by the first minimum of the mutual information function.

With eq. (3.1) dividing the considered m dimensional phase space into hypercubes with a linear dimension r, we count all points with mutual distances less than r. Then, it has been proven that if the attractor is a strange one, the correlation integral will be proportional to r^v, where v is a measure of the attractor's dimension called correlation dimension. By definition, the correlation integral $C(r)$ is the limit of correlation sum of eq.(3.1) and is numerically calculated as a function of r from eq.(3.1) for embedding dimensions m=1,...,10.

Following the above and in order to get accurate measurements of the strength of the chaos present in the oscillations of the simulated signals, we introduce the Kolmogorov entropy. Komlogorov entropy quantifies how chaotic a dynamical system is i.e. is zero for a deterministic system that isn't in chaos; it's a positive constant for a chaotic system; and it's infinite for a random process, at least for uniformly distributed data. The method followed so far also leads to an estimate of the Kolmogorov entropy, i.e. the correlation integral $C(r)$ scales with the embedding dimension m, since:

$$C(r) \sim e^{-m\tau_d K_2} \quad 4) \qquad (3.$$

where K_2 is a lower bound to the Kolomogorov entropy.

4. EVALUATION OF INVARIANCE PARAMETERS

The calculation of invariance parameters i.e. the correlation and embedding dimension and the kolmogorov entropy, will give us information about the complexity of corresponding strange attractors. First we calculate the time delay $\tau=i\Delta t$ determined by the first minimum of the time delayed mutual information, I (τ). In case of RL-diode circuit and for time series of Figure 3 (top), with sample rate $\Delta t=4.8\times10^{-7}$ s, the mutual information function exhibits a local minimum at $\tau=6$ time steps as shown at Figure 13(a). In case of RL –Led optoelectronic circuit and for time series of Figure 5, with sample rate $\Delta t=6.25\times10^{-6}$ s, the mutual information function exhibits a local minimum at $\tau=5$ time steps as shown at Figure 13(b). For RLT circuit and for time series of Figure 7 the sample rate $\Delta t=3.33\times10^{-6}$s the mutual information function I (τ_d) exhibits a local minimum at $\tau_d=3$ time steps as shown at Figure 13(c) and, thus, we shall consider $\tau_d=3$ as the optimum delay time.

The next step is to evaluate t correlation integral for m=1 to 10 values of embedding dimension using formula (3.1). Figures 14, 15, 16 (top) depicts the relation between the logarithms of correlation integral $C(r)$ and r for different embedding dimensions m for RL diode, RL-Led and RLT chaotic circuits. As seen in Figures 14, 15, 16 (bottom), the slopes v of the lower linear parts of these log-log curves provide all necessary information for characterizing the attractor. These slopes indicate that for high values of m, v tends to saturate at the non integer value of v=2.11 for RL diode, at the non integer value of v=2.23 for of RL –Led optoelectronic circuit and at the non integer value of v=2.47 for RLT. For all these values of v, the minimum embedding dimension can be $m_{min}=3$ for all chaotic electronic circuits, and thus, the minimum embedding dimension of the attractor for one to one embedding will be equal to 3.

The next step is the calculation of Kolmpgorov entropy. Figure 17 shows the relation between K_2 and the logarithm of r for different embedding dimensions m, indicates that K_2=0.11 bit/s, meaning that there is a steady loose of information at a constant rate given by K_2 for RL diode circuit.

For RL-Led circuit Figure 18 shows the relation between K_2 and the logarithm of r for different embedding dimensions m, indicates that K_2=0.52 bit/s, meaning that there is a steady loose of information at a constant rate given by K_2.

For RL-Led circuit Figure 19 shows the relation between K_2 and the logarithm of r for different embedding dimensions m, indicates that K_2=0.26 bit/s, meaning that there is a steady loose of information at a constant rate given by K_2.

Figure 13. Average Mutual Information I (τ_d) vs. time delay τ_d for RL-diode circuit (a), RL –Led optoelectronic circuit (b), RLT circuit (c)

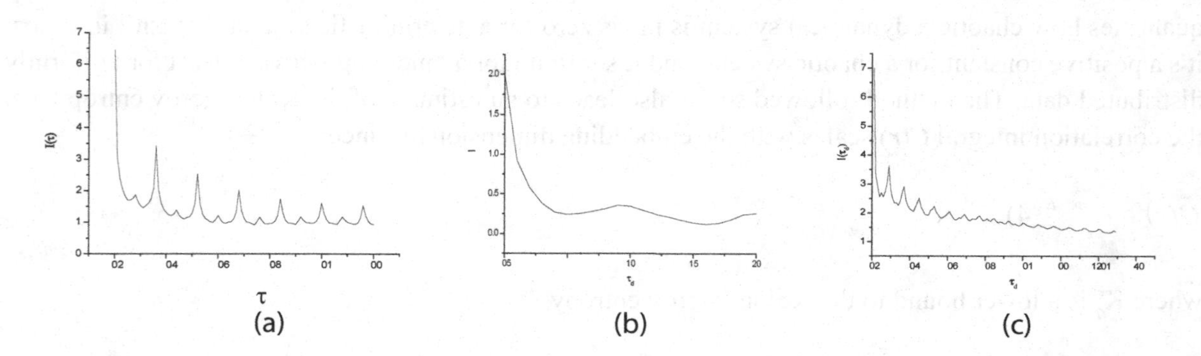

Figure 14. Relation between logC(r) and logr for different embedding dimensions m (top) and the corresponding slopes and scaling region (bottom), for RL diode circuit.

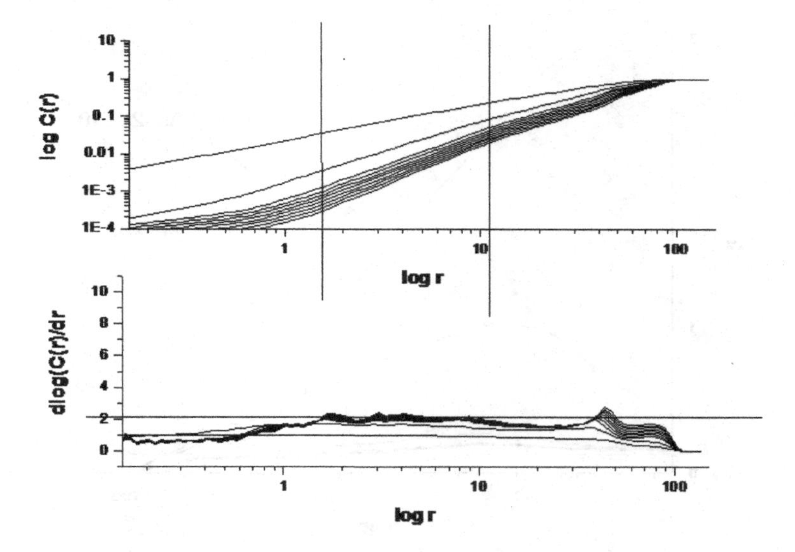

Figure 15. Relation between logC(r) and logr for different embedding dimensions m (top) and the corresponding slopes and scaling region (bottom), for RL-Led circuit.

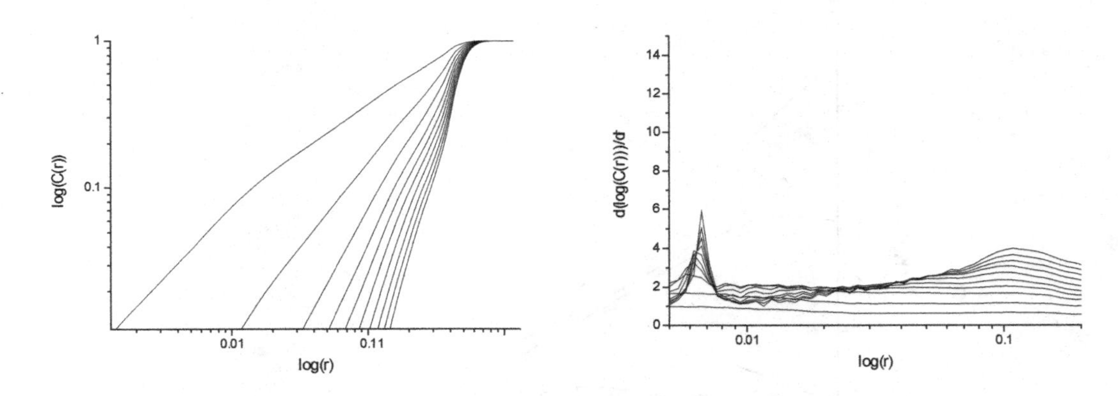

Figure 16. Relation between logC(r) and logr for different embedding dimensions m (top) and the corresponding slopes and scaling region (bottom), for RLT circuit.

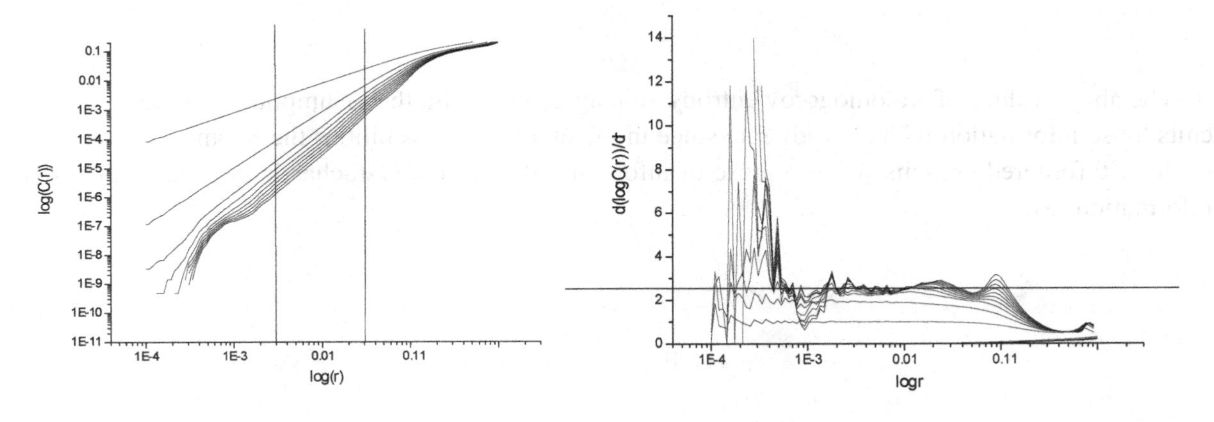

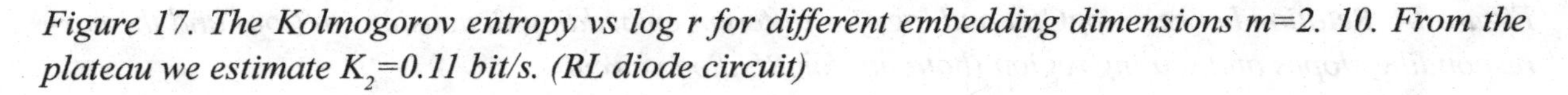

Figure 17. The Kolmogorov entropy vs log r for different embedding dimensions m=2. 10. From the plateau we estimate K_2=0.11 bit/s. (RL diode circuit)

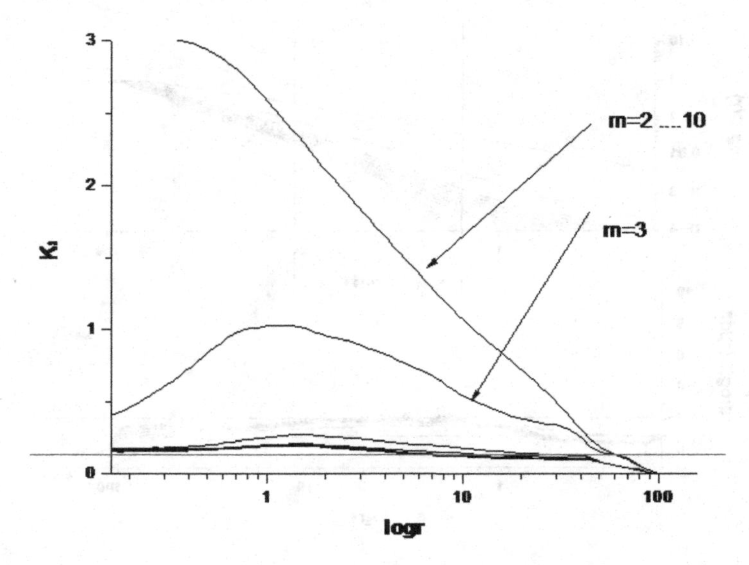

Figure 18. Kolmogorov entropy vs. logr for embedding dimensions m=2,..., 10. (RL-Led circuit)

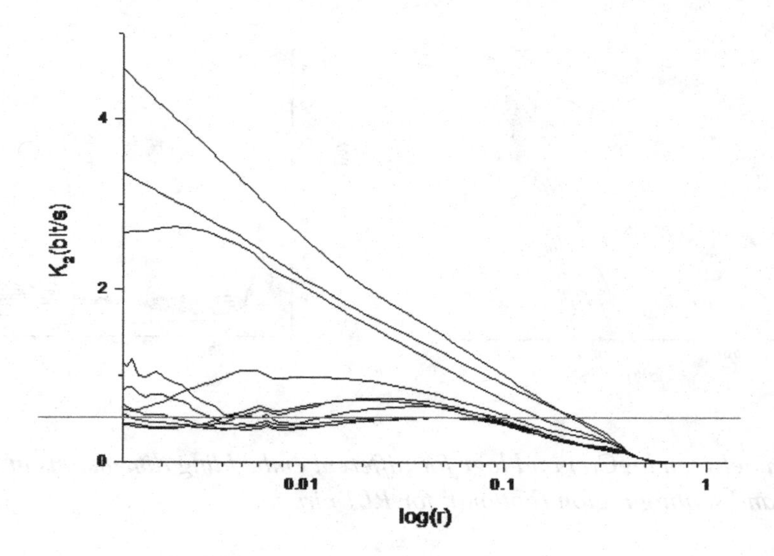

The above values of Kolomogorov entropy is in agreement with the assumption that the above circuits loose information with a steady rate since in the other two possibilities the Kolmogorov entropy is either 0 (ordered systems with no loose of information) or infinite (stochastic systems with a total information lost).

Figure 19. Kolmogorov entropy vs. logr for embedding dimensions m=2,...,10. (RLT circuit)

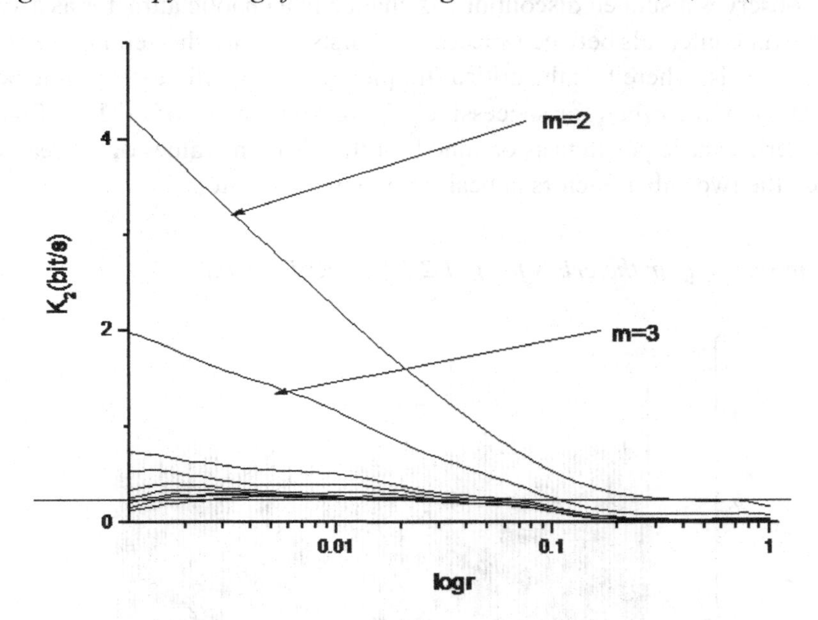

5. MODIFICATION OF STRANGE ATTRACTOR

As mentioned it is important to find simple analogue circuits that work chaotic with low dimensionality that can be controlled easy. For safety reasons it would be useful if easily one can change the properties of corresponding strange attractor changing one circuit parameter. This can be done if we know the route to chaos. As an example if we examine the RLT circuits in a critical state.

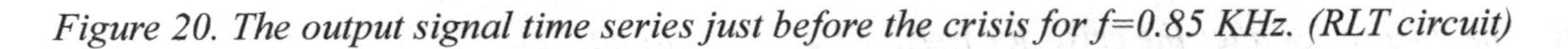

Figure 20. The output signal time series just before the crisis for f=0.85 KHz. (RLT circuit)

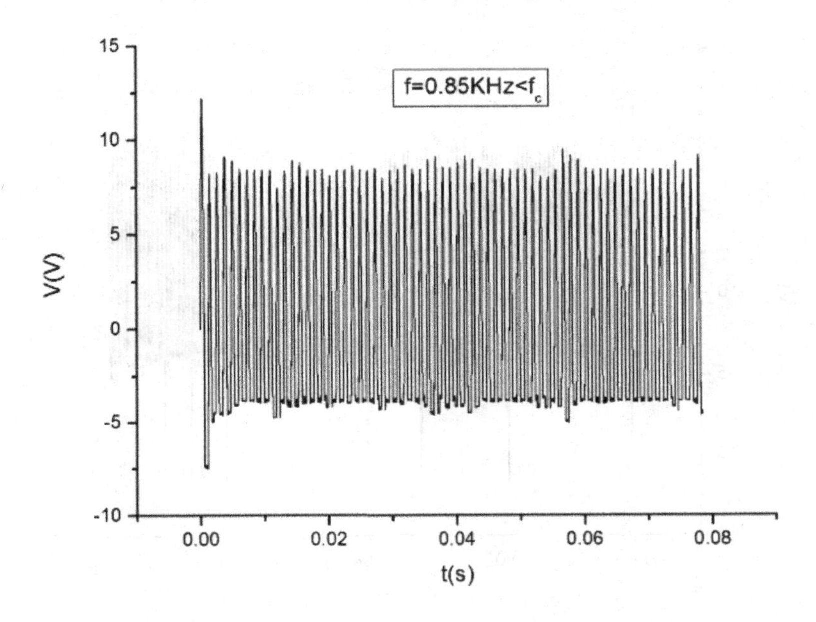

In a crisis, one observes a sudden discontinuous change in a chaotic attractor as a system parameter is varied, e.g. as the time intervals between successive bursts become shorter. Figure 20 shows the time series just before the crisis, where f_c is the critical frequency which will be determine below, while Figures 21, 22, and 23 show the orbits for successively increasing values of f. Then, Figures 24, 25, and 26 depict the time series phase portraits as obtained for the different values of f. Clearly, as f increases, transitions between the two sub attractors appear with increasing rate.

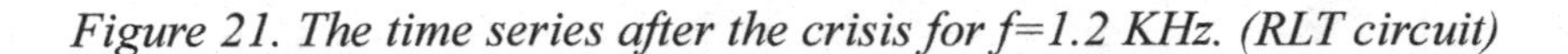

Figure 21. The time series after the crisis for f=1.2 KHz. (RLT circuit)

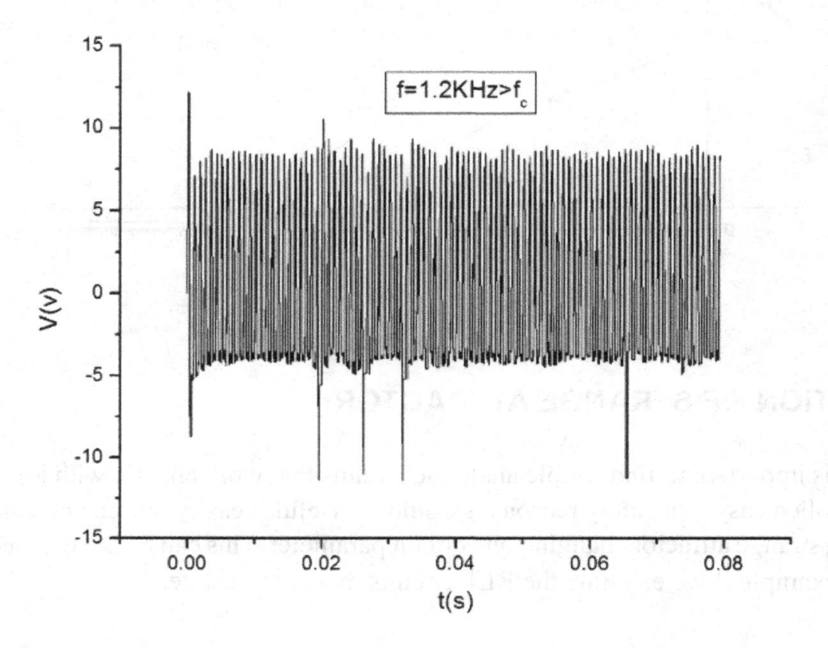

Figure 22. The time series after the crisis for f=1.5 KHz. (RLT circuit)

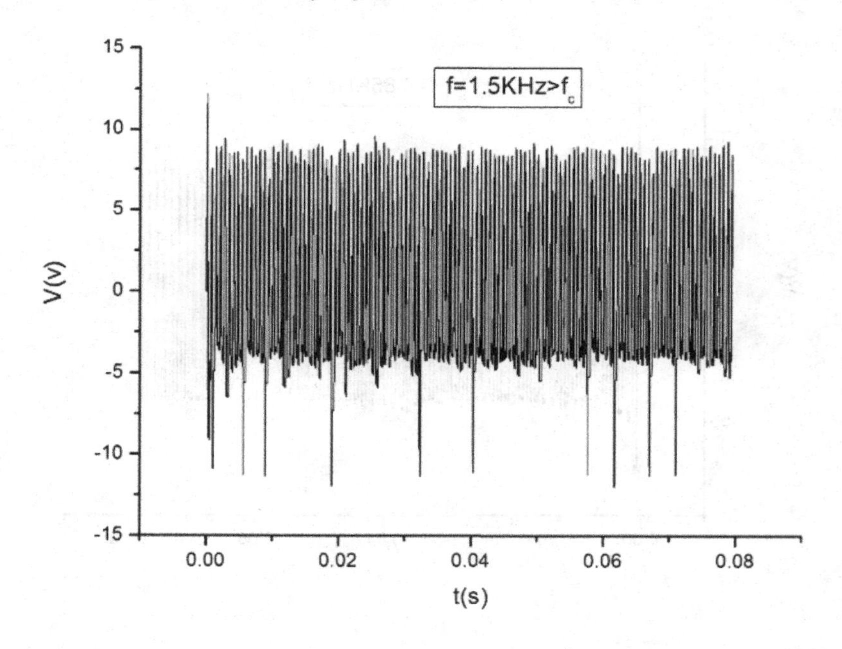

Figure 23. The time series after the crisis for f=2KHz. (RLT circuit)

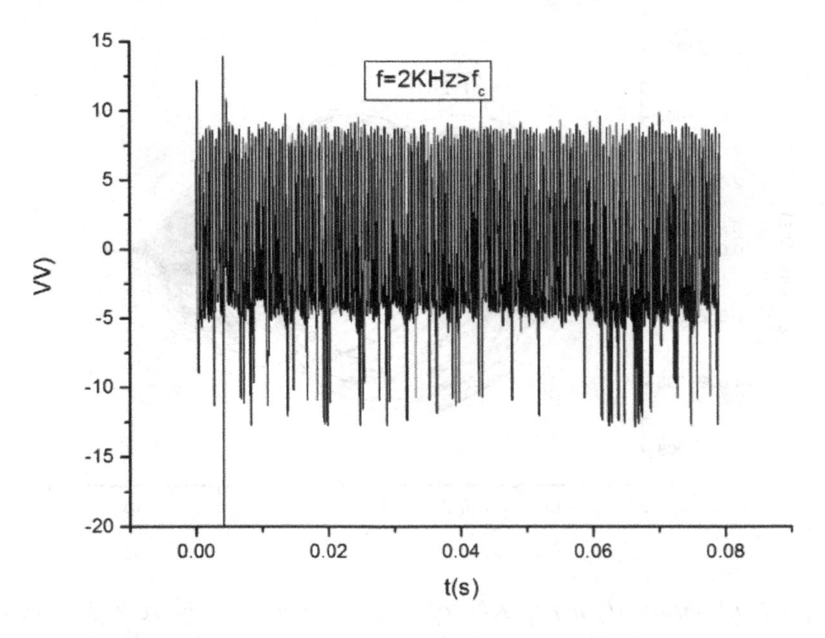

Figure 24. Phase portrait obtained for f=1.5 KHz of the time series of Figure 22. (RLT circuit)

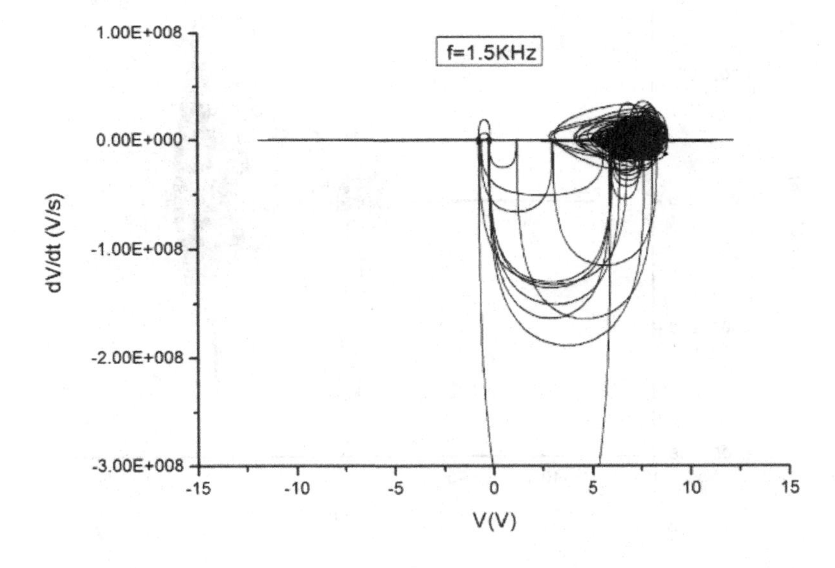

There are three types for changes that attractors can undergo as a system parameter is varied. The first type, the boundary crisis, leads to the sudden destruction of a chaotic attractor. The second type, the interior crisis, leads to the sudden widening of a chaotic attractor. In the third type, the attractor merging crisis, there are two or more attractors merging in order to form a single one. While in the first crisis orbits behave like transients, intermittent bursts or jumps follow the other two crises. This behaviour can be characterized as crisis induced intermittency. A direct consequence of this behaviour is that an orbit spends a lot of time in the region of one of the two attractors. After such a time interval, the orbit

Figure 25. Phase portrait obtained for f=2 KHz of the time series of Figure 23. (RLT circuit)

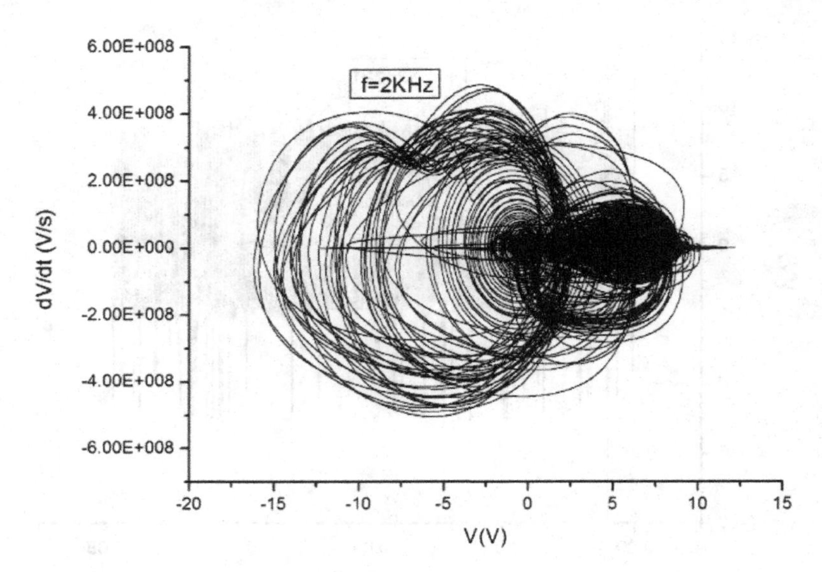

Figure 26. Phase portrait obtained for f=3 KHz of the time series of Figure 7 (RLT circuit)

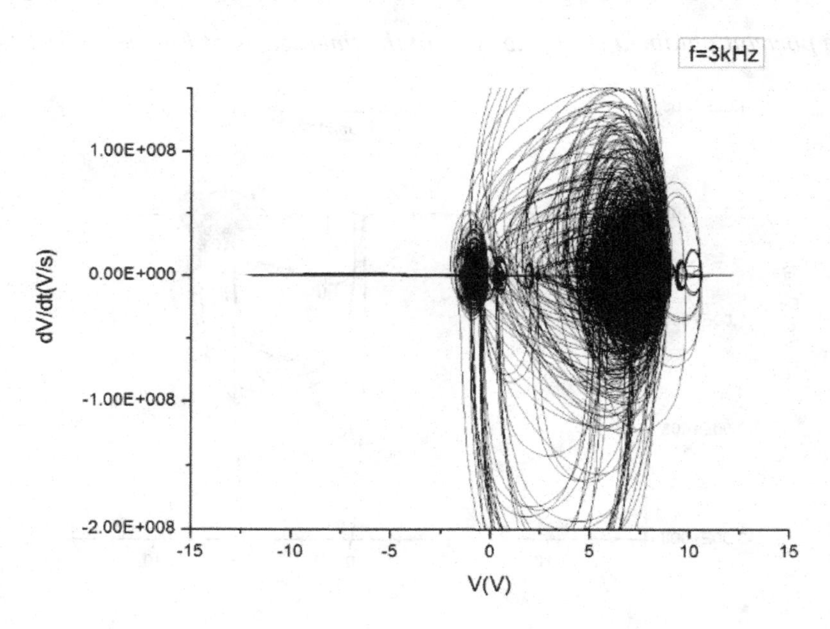

abruptly exits this region and spends a long time interval in the region of the second attractor and so on. Thus, for appropriate values of the control parameter there is one common attractor on which the trajectories switch between different behaviours intermittently, i.e. for finite time stretches the trajectories orbit on the individual attractors. Following that, the plots in Figures 24, 25, and 26 clearly indicate that in our system there is a common attractor which consists of two individual sub attractors with the orbit bursting occasionally between them. Moreover, with f increasing (which is our control parameter), the rate (frequency) of these transitions also increases.

So by this way from same circuit we can have a set of chaotic attractors which can be used for different types of chaotic slave systems, which can be synchronized and used to generate different confidential keys.

6. CONCLUSION

It is easy to product voltage chaotic oscillations with rather simple electronic circuits. The simpler circuit is the RL-diode circuit. With proper values of linear resistor and linear inductor this circuit can product Lorenz type oscillations. We can measure the chaotic nature of these oscillations and find that there is scaling behaviour of the correlation integral and the saturation of correlation dimension with increasing embedding dimensions reflect low dimensionality. The strange attractor that governs the phenomenon has a correlation dimension v=2.11 stretching and folding in a 3 dimension phase space. Thus, the number of degrees of freedom of the whole domain structure is limited and this results in the low value of the correlation dimension. Analogue behaviour exhibits the optoelectronic RL-LED circuit and the RLT circuit. The LED exposes chaotic behaviour even if it works in its operation point. So the proposed circuit can be used to generate chaotic signal, in a light emitting manner, useful in code and decode applications. The strange attractor that governs the chaotic behaviour of LED has a correlation dimension v=2.23 stretching and folding in a 3 dimension phase space as before. The RLT circuit gives us more rich opportunities to modulate the shape and the topological properties of its strange attractor through the type-III intermittency route to chaos. This strange attractor has a correlation dimension v=2.47 and an embedding dimension m=3.

REFERENCES

Aasen, T., Kugiumtzis, D., & Nordahl, S. H. G. (1997). Procedure for estimating the correlation dimension of optokinetic nystagmus signals. *Computers and Biomedical Research, an International Journal, 30*, 95–116. doi:10.1006/cbmr.1997.1441

Aislam, T., & Edwards, J. A. (1996). Secure communications using chaotic digital encoding. *Electronics Letters, 32*, 190–191. doi:10.1049/el:19960107

Alvarez, G., Montoya, F., Romera, M., & Pastor, G. (2004). Breaking parameter modulated chaotic secure communication system. *Chaos, Solitons, and Fractals, 21*, 783–787. doi:10.1016/j.chaos.2003.12.041

Carroll, T. L., & Pecora, L. M. (1991). Synchronizing chaotic circuits. *IEEE Trans Circuits Systems I, 38*, 453–456. doi:10.1109/31.75404

Chen, B., & Wornell, G. W. (1996). Efficient channel coding for analog sources using chaotic systems. In *GLOBECOM '96. Communications: The Key to Global Prosperity* (pp. 131–135).

Chien, T.-I., & Liao, T.-L. (2005). Design of secure digital communication systems using chaotic modulation, cryptography and chaotic synchronization. *Chaos, Solitons, and Fractals, 24*, 241–242.

Chua, L. O., Kocarev, L. J., Eckert, K., & Itoh, M. (1992). Experimental chaos synchronization in Chua's circuit. *International Journal of Bifurcation and Chaos in Applied Sciences and Engineering, 2*, 705–708. doi:10.1142/S0218127492000811

Chua, L. O., & Lin, G. N. (1990). Canonical realization of Chua's circuit family. *IEEE Trans Circuits Systems I*, *37*, 885–902. doi:10.1109/31.55064

Chua, L. O., Wu, C. W., Huang, A., & Zhong, G. Q. (1993). A universal circuit for studying and generating chaos. Part II. Strange attractors. *IEEE Transactions on Circuits and Systems*, *40*, 732–761. doi:10.1109/81.246149

Dedieu, H., Kennedy, M. P., & Hasler, M. (1993). Chaos shift keying: modulation and demodulation of a chaotic carrier using self synchronizing Chua's circuits. *IEEE Trans Circuits Systems II: Analog Digital Signal Process*, *40*, 634–642. doi:10.1109/82.246164

Grassberger, P., & Procaccia, I. (1983). Characterization of Strange Attractors. *Physical Review Letters*, *50*, 346–349. doi:10.1103/PhysRevLett.50.346

Grassberger, P., & Procaccia, I. (1983). Measuring the Strangeness of Strange Attractors. *Physica D. Nonlinear Phenomena*, *9*, 189–208. doi:10.1016/0167-2789(83)90298-1

Grebogi, C., Ott, E., Romeiras, E., & Yorke, J. A. (1987). Critical exponents for crisis-induced intermittency. *Physical Review A.*, *36*, 5365–5380. doi:10.1103/PhysRevA.36.5365

Hanias, M. (2008). Chaotic Behavior of an electrical analogue to the mechanical double pendulum. *Journal of Engineering Science and Technology Review*, *1*, 33–38.

Hanias, M. P., & Anagnostopoulos, J. A. N. (1993). Negative-differential- resistance effects in the $TlGaTe_2$ ternary semiconductor. *Physical Review B: Condensed Matter and Materials Physics*, *47*, 4261–4267. doi:10.1103/PhysRevB.47.4261

Hanias, M. P., Avgerinos, Z., & Tombras, G. S. (2009). Period doubling, Feigenbaum constant and time series prediction in an experimental chaotic RLD circuit. *Chaos, Solitons, and Fractals*, *40*, 1050–1059. doi:10.1016/j.chaos.2007.08.061

Hanias, M. P., Giannaris, G., Spyridakis, A., & Rigas, A. (2006). Time series analysis in chaotic diode resonator circuit. *Chaos, Solitons, and Fractals*, *27*, 569–573. doi:10.1016/j.chaos.2005.03.051

Hanias, M. P., Giannis, L. I., & Tombras, G. S. (2010). Chaotic operation by a single transistor circuit in the reverse active region. *Chaos (Woodbury, N.Y.)*, *20*, 0131051–0131057. doi:10.1063/1.3293133

Hanias, M. P., Kalomiros, J. A., Karakotsou, C., Anagnostopoulos, A. N., & Spyridelis, J. (1994). Quasiperiodic and chaotic self-excited voltage oscillations in $TlInTe_2$. *Physical Review B: Condensed Matter and Materials Physics*, *49*, 16994–16998. doi:10.1103/PhysRevB.49.16994

Hanias, M. P., & Karras, D. A. (2007). Efficient Non Linear Time Series Prediction Using Non Linear Signal Analysis and Neural Networks in Chaotic Diode Resonator Circuits. In *Advances in Data Mining. Theoretical Aspects and Applications* (LNCS 4597, pp. 329-338). Berlin: Springer.

Hanias, M. P., & Karras, D. A. (2007). Improved Multistep Nonlinear Time Series Prediction by applying Deterministic Chaos and Neural Network Techniques in Diode Resonator Circuits. In *IEEE International Symposium on Intelligent Signal Processing, WISP 2007* (pp. 1 - 6).

Hanias, M. P., & Karras, D. A. (2009). On efficient multistep non-linear time series prediction in chaotic diode resonator circuits by optimizing the combination of non-linear time series analysis and neural networks. *Engineering Applications of Artificial Intelligence*, *22*, 32–39. doi:10.1016/j.engappai.2008.04.016

Hanias, M. P., Magafas, L., & Kalomoiros, J. (2008). Non- linear Analysis in RL-LED optoelectronic circuit. *Optoelectronics and advanced materials – Rapid Communications, 2*, 126 – 129.

Hanias, M. P., & Tombras, G. (2009). Time series cross – prediction in a single Transistor Chaotic Circuit. *Chaos, Solitons, and Fractals*, *41*, 1167–1172. doi:10.1016/j.chaos.2008.04.055

Hanias, M. P., & Tombras, G. S. (2009). Time series analysis in single transistor chaotic circuit. *Chaos, Solitons, and Fractals*, *40*, 246–256. doi:10.1016/j.chaos.2007.07.065

He, R., & Vaidya, P. G. (1998). Implementation of chaotic cryptograph with chaotic synchronization. *Physical Review E: Statistical Physics, Plasmas, Fluids, and Related Interdisciplinary Topics*, *57*, 1532–1535. doi:10.1103/PhysRevE.57.1532

Itoh, M., Wu, C. W., & Chua, L. O. (1997). Communication systems via chaotic signals from a reconstruction viewpoint. *International Journal of Bifurcation and Chaos in Applied Sciences and Engineering*, *7*, 275–286. doi:10.1142/S0218127497000194

Kantz, H., & Schreiber, T. (1997). *Nonlinear Time Series Analysis* (2nd ed.). Cambridge, UK: Cambridge University Press.

Kennedy, M. (1994). Chaos in Colpitts oscillator. *IEEE Transaction on Circuits and Systems – I, 41*, 1771-774.

Lonngren, K. E. (1991). Notes to accompany a student laboratory experiment on chaos. *IEEE Transactions on Education*, *34*, 123–128. doi:10.1109/13.79892

Marino, I. P., Lopez, L., Miguez, J., & Sanjuan, M. A. F. (2002). A novel channel coding scheme based on continuous-time chaotic dynamics. In *14th International Conference on Digital Signal Processing* (pp. 1321–1324).

Mykolaitis, G., Tamaševičius, A., & Bumelienė, S. (2004). Experimental demonstration of chaos from the Colpitts oscillator in the VHF and the UHF ranges. *Electronics Letters*, *40*, 91–92. doi:10.1049/el:20040074

Rollins, R. W., & Hunt, E. R. (1982). Exactly Solvable Model of a Physical System Exhibiting Universal Chaotic Behavior. *Physical Review Letters*, *49*, 1295–1298. doi:10.1103/PhysRevLett.49.1295

Schimming, T., & Hasler, M. (2003). Coded modulations based on controlled 1-D and 2-D piecewise linear chaotic maps. In *Proceedings of the 2003 International Symposium ISCAS '03* (Vol. 3, pp. 762–765).

Sobhy, M. I., & Shehata, A.-E. R. (2001). Methods of attacking chaotic encryption and countermeasures. *Proc IEEE Int Conf Acoustics Speech Signal Process (2)*,1001–1004.

Sprott, J. C. (1994). Some simple chaotic flows. *Physical Review E: Statistical Physics, Plasmas, Fluids, and Related Interdisciplinary Topics*, *50*, 647–650. doi:10.1103/PhysRevE.50.R647

Takens F. (1981). Detecting strange attractors in turbulence. *Lecture Notes in Mathematics,* 366-381.

Yang, T. (1996). Recovery of digital signal from chaotic switching. *Int J Circuit Theory Appl.*, *23*, 611–615. doi:10.1002/cta.4490230607

Yang, T., & Chua, L. O. (1997). Impulsive control and synchronization of nonlinear dynamical systems and application to secure communication. *International Journal of Bifurcation and Chaos in Applied Sciences and Engineering*, *7*, 645–664. doi:10.1142/S0218127497000443

Chapter 5
Experimental Evidences of Shil'nikov Chaos and Mixed-Mode Oscillation in Chua Circuit

Syamal Kumar Dana
Indian Institute of Chemical Biology, India

Satyabrata Chakraborty
Indian Institute of Chemical Biology, India

ABSTRACT

Experimental evidences of Shil'nikov type homoclinic chaos and mixed mode oscillations are presented in asymmetry-induced Chua's oscillator. The asymmetry plays a crucial role in the related homoclinic bifurcations. The asymmetry is introduced in the circuit by forcing a DC voltage. The authors observed transition from large amplitude limit cycle to homoclinic chaos via a sequence of mixed-mode oscillations interspersed by chaotic states by tuning a control parameter.

1. INTRODUCTION

Homoclinic orbit is asymptotic to a saddle limit set both forward and backward in time (Wiggins, 1990; Kuznetsov, 1995). In the vicinity of a homoclinic orbit, if a control parameter is tuned, a countable infinity of periodic orbits are observed, which are at the origin of chaos in nonlinear dynamical system. The homoclinic orbit is structurally unstable and not possible to observe either in numerical or physical experiments. However, homoclinic chaos of the *Shilnikov* type has been observed, in the vicinity of the homoclinic orbit, by tuning a suitable control parameter in BZ reaction (Petrov et al, 1992), liquid crystal flow (Peacock & Mullin, 2001), CO_2 laser (Pisarchik et al,2001; Allaria et al, 2001), optothermal bistable device (Herrero et al, 1996) and electronic circuit (Healy et al, 1996). The trajectory of homoclinic chaos is globally stable yet instabilities are bounded to a local domain close to a saddle, which may be a saddle focus or a saddle cycle in 3D system. The local instability has its manifestation in large fluctuations in the return time of spiking oscillation. It is to be noted that homoclinic bifurcation is considered, in recent

DOI: 10.4018/978-1-61520-737-4.ch005

times (Belykh et al, 2000; Izhikevich, 2000), as one of the important mechanisms of emergence of the spiking and bursting behaviors in neurons with inherent fast and slow dynamics.

The *Shil'nikov chaos* (Wiggins, 1990; Kuznetsov, 1995) deals with a saddle focus with real and complex conjugate eigenvalues, (γ, $\sigma \pm j\omega$) in 3D systems. The trajectory of the homoclinic chaos escapes spirally from the saddle focus in 2D eigensapce and re-injects into it along the stable eigendirection for systems with $\gamma<0$, $\sigma>0$ and $|\gamma/\sigma|>1$. A reverse direction of the trajectory of the homolcinic chaos is seen when $\gamma>0$, $\sigma<0$ and $|\gamma/\sigma|>1$. However, it is true, in general, that in period-parameter space, the period of a limit cycle (period-n: n^0, n=1, 2, 3…) increases asymptotically with a control parameter as it approaches the homoclinic orbit or the bifurcation point and in close vicinity of this bifurcation point, instabilities appear yet bounded to a saddle focus which is defined as Shil'nikov chaos. Further studies (Glendinning & Sparrow, 1984) on *Shil'nikov* chaos show that a *Shil'nikov wiggle* (Wiggins, 1990) may appear in period-parameter space if $1/2<|\gamma/\sigma|<1$ for $\gamma<0$ and $\sigma>0$. Under this condition, when a control parameter of a system is tuned from both sides of the homoclinic point, the period of a limit cycle increases in a *wiggle* with alternate sequences of stable and unstable orbits via saddle-node (SN) and period-doubling (PD) bifurcations respectively. The *Shil'nikov wiggle* is beyond the scope of this report. We restrict our discussion here on simpler condition for *Shil'nikov chaos,*$|\gamma/\sigma|>1$, when we find sequences of mixed mode oscillation (MMO) and homoclinic bifurcation.

In many nonlinear dynamical systems such as Chua circuit, the dynamics usually changes with a parameter from stable equilibrium to limit cycle by super-Hopf bifurcation and to chaos via (PD). With further changes in parameter, the system shows period-adding bifurcation when a sequence of periodic windows appear intermediate to chaotic windows in parameter space. The periodic windows are created via SN bifurcation of the chaotic behavior while the periodic states again move to chaotic states via PD. Subsequently, the dynamics follows a reverse PD before moving to period-1 and then to unstable limit cycle via subcritical Hopf bifurcation.

Several numerical studies showed evidences of MMOs as a transition route to the *Shil'nikov* type homoclinic chaos with control parameter in slow-fast systems (Gaspard & Wang, 1987; Koper, 1995;Gory-achev et al, 1997; Rajesh & Ananthakrishna, 2000; Dana et al, 2006). The MMO is seen as a periodic oscillation with alternate appearance of large amplitude oscillations and low amplitude oscillations of different time scales. The sequence of MMOs are actually observed in the intermediate periodic regimes of the bifurcation diagram of the dynamical system, which again exist in isolated bifurcation curves (*isolas*) as elaborated in (Petrov et al, 1992; **Marc**Koper, 1995) and most recently in a jerky flow model (Rajesh & Ananthakrishna, 2000). The MMO as illustrated in Boissonade-De Kepper model (Koper, 1995;Goryachev et al, 1997) arises at a point when a large amplitude limit cycle loses stability via SN bifurcation. It is seen as transition from large amplitude limit cycle to *Farey* sequences of periodic MMO, $1^0 \to \propto^1 \to n^1 \to 1^n \to 1^\propto$ (n is an integer), interspersed by narrow chaotic states when a control parameter is tuned to homoclinicity. The 1^0 denotes the large amplitude limit cycle (period-1) while $\propto^1$ denotes a periodic orbit with transition to small oscillation for once in a long run. The alternate sequences of periodic and chaotic MMOs appear via SN and PD bifurcations respectively. The MMOs are usually denoted by L^s, where L and s are the number of large and small amplitude oscillations respectively. The L remains fixed for a set of selected system parameters. The number of small oscillation (s) is stable in a periodic window but it is highly irregular in chaotic windows. For different sets of selected parameters, MMO of higher L (=2,3,….) may also be observed. It is noteworthy that the width of both the periodic and chaotic windows becomes narrower when the control parameter is tuned to the homoclinic point. As the control parameter is tuned to the bifurcation point, the number of small oscillations increases while the trajectory moves closer and closer to a saddle. The state of homoclinic chaos denoted by $1^{\propto}$ (L=1) is reached when the number of small oscillations s becomes finitely very large but highly irregular, which is reflected as large fluctuations in return time of homoclinic spiking. The homoclinic chaos thus may be seen as infinite number of unstable MMOs embedded into it.

In this paper, we report our experimental observations of the *Shil'nikov* type homoclinic chaos in Chua's circuit from a viewpoint of induced asymmetry. The original Chua's circuit model has inversion symmetry. It has two inherent time scales: one for both the revolving cycles around either of the symmetric equilibrium points and another the large cycle covering the two symmetric scrolls. However, these two time scales cannot help inducing homoclinic bifurcation. A third time scale is introduced in the Chua circuit by inducing asymmetry in the system using external DC forcing. One of the double scroll attractors thereby shrinks in size and creates an additional different time scales in the overall dynamics. In absence of the DC forcing the revolving cycles around both the symmetric equilibrium points has almost nearly equal time scales. In fact, experimental evidences are described here to show that the asymmetry as deliberately induced in a Chua's circuit can be controlled to observe many complex dynamical features, like MMOs and bursting and homoclinic chaos. In real systems, several sources of imperfection always exist, which breaks the inversion symmetry of a model system. These imperfections are usually modeled (Glendinning et al, 2001) by an additional constant term to the normal form of a model flow. Using this concept, we induced asymmetry in a single Chua's circuit by forcing a DC voltage and thereby investigate the role of the asymmetry in the origin of MMOs and homoclinic chaos. We observed sequences of MMOs interspersed by chaotic windows during the intermediate stages of transitions from one large amplitude limit cycle to another large amplitude limit cycle by varying the control parameter. The dynamics follows the sequence of transitions $n^0 \rightarrow \propto^1 \rightarrow n^1 \rightarrow n^m \rightarrow L^s \rightarrow n^p \rightarrow n^1 \rightarrow \propto^1 \rightarrow n^0$ (n, m and p are integers) when a system parameter is controlled keeping the asymmetry parameter at a suitably selected value. We are able to observe, at most, two different scenarios of MMO (L^s: L=1, 2) in the process of transition to homoclinic chaos for two different sets of selected parameters. Evidences of two intermediate bursting regimes ($n^1 \rightarrow n^m$ and $n^p \rightarrow n^1$) are also found for each of the scenarios at the edges of transition from one large amplitude limit cycle to MMO and from MMO to the other large amplitude limit cycle. Many more scenarios with larger L may be observed by appropriate choice of system parameters and asymmetry parameters. However, in experiments, it is extremely difficult to control a parameter very precisely to observe all the details of bifurcation scenarios.

The text of this paper is organized as follows. In the next section, we described the experimental set-up of asymmetric Chua circuit. Evidences of homoclinic chaos and bursting are elaborated in section 3. Details of observed MMO are presented in section 4. The results are summarized with a conclusion in section 5.

2. EXPERIMENTAL SET UP: ASYMMETRIC CHUA'S CIRCUIT

The modified Chua circuit is shown in Figure 1. A DC voltage is forced at the C_1 capacitor node to induce asymmetry in the attractor. The model of the asymmetry-induced Chua's circuit is given by

$$\frac{dV_{C_1}}{dt} = \frac{G}{C_1}(V_{C_2} - V_{C_1}(1+\frac{1}{R_p})) - \frac{1}{C_1}f(V_{C_1}) + \frac{1}{R_C C_1}(V_0 - V_{C_1})$$
$$\frac{dV_{C_2}}{dt} = \frac{1}{C_2}[I_L - G(V_{C_2} - V_{C_1})]$$
$$\frac{dI_L}{dt} = -\frac{1}{L}(V_{C_2} + r_0 I_3) \quad (1)$$

and

$$f(V_{C_1}) = \begin{vmatrix} G_b V_{C_1} + (G_b - G_a)E & \text{if } V_{C_1} < -E \\ G_a V_{C_1} & \text{if } E \le V_{C_1} \le E \\ G_b V_{C_1} + (G_a - G_b)E & \text{if } V_{C_1} > E \end{vmatrix} \quad (2)$$

where $G=1/R_1$ and G_a, G_b are the slopes in the inner and outer regions [22-23] respectively of the piece-wise linear characteristic $f(V_{C1})$.

The slopes G_a and G_b are determined by

$$G_a = (-\frac{1}{R_2} - \frac{1}{R_4}),\ G_b = (\frac{1}{R_3} - \frac{1}{R_4}) \quad (3)$$

The state variables V_{C1}, V_{C2} are the voltages measured at nodes of capacitors C_1 and C_2 respectively and, and I_L is the current through the inductance L respectively. The DC source V_{dc} is connected to the Chua's circuit using a voltage divider network using resistances R_x and R_y, where $V_0=V_{dc}(1+R_x/R_y)$. To facilitate fine control over the strength of asymmetry, a series resistance R_c is connected between the voltage divider and the Chua's circuit. The symmetric Chua's circuit (no DC forcing) shows (Chua et al, 1986; Kennedy, 1993) transition from limit cycle to single scroll chaos via PD and then to alternate period adding and chaotic states via saddle-node (SN) and PD bifurcation respectively, and finally to

Figure 1. Asymmetric Chua's oscillator:±9V supply

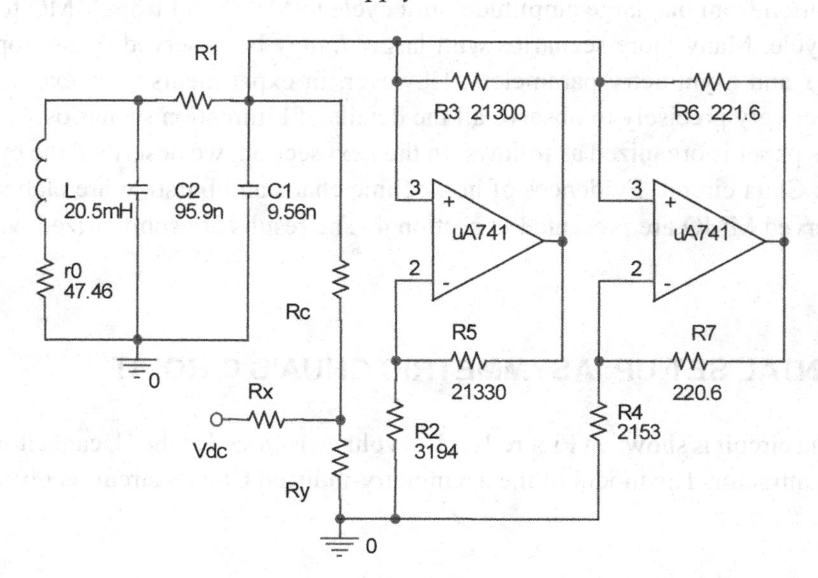

double scroll chaos. The original model has three stable foci, one near the origin with eigenvalues (γ, $-\sigma \pm j\omega$) and other two at inversion symmetric positions with eigenvalues ($-\gamma$, $\sigma \pm j\omega$). The trajectory of double scroll in Chua's circuit revolves most of the time near either of the mirror symmetric equilibrium points and switches irregularly between the two. Thus the system shows two inherent time scales in the dynamics of Chua's double scroll attractor, one due to the single scroll oscillation near either of the saddle foci and other due to a large cycle around both the equilibrium points. However, by inducing asymmetry in the double scroll attractor, one of the scrolls is shrinked in size and thereby one additional time scale is introduced, which characterize the small amplitude oscillation. Effectively, the asymmetry in the double scroll attractor appears as a shift of one of the inversion symmetric saddle foci closer to the origin than the other. However, the basic characteristic of the eigenvalues of all the saddle foci remains unchanged in the asymmetric system.

3. HOMOCLINIC CHAOS AND BURSTING

We force the Chua circuit by an external DC voltage and thereby induce an asymmetry in the system. The DC forcing is controlled by a combination of divider resistances, R_x, R_y and R_C to observe several interesting phenomena such as homolcinic bifurcations, MMOs and bursting. The resistance R_C is only used to control the asymmetry in the attractor, once the resistance R_x and R_y are selected and kept fixed. For appropriate selection of system parameter R_1 and asymmetry parameter R_C, a time series from experiment is shown in Figure 2(a), which shows clear evidences of homoclinic chaos as a train of high amplitude spikes with irregular switching to small amplitude oscillations. The time series of the homoclinic chaos may be seen as a MMO (2^s:L=2) with highly fluctuating number of small amplitude oscillation (s). The number of small amplitude oscillations decides the time interval of the large amplitude spiking, whose successive time intervals are highly uncorrelated as reflected in the time series as well as in the return time of large amplitude oscillations in Figure 2(b). For stable large amplitude oscillations, the return time should show a fixed point along the 45^0 line. For unstable large amplitude oscillations, the return time is highly uncorrelated as it shows large variations as highly scattered points from the stable state. This is considered as an important signature of homoclinic chaos. Here, we have calculated the time interval of the maxima of small amplitude oscillations from the measured time series instead of calculating the time interval of large amplitude oscillations, and then the successive time intervals t(i) is plotted against the previous interval t(i-1) as shown in Figure 2(b). As a result we find more interesting mutivalued structure in the return map. This confirms the complexity of the small amplitude oscillations, which causes large fluctuations in the time interval of the large amplitude oscillations.

It may be noted that the homoclinic trajectory moves closer and closer to the saddle focus as the number of small oscillation increases. For this, a fine tuning of the R_C value is necessary near the bifurcation point so as to maximize the number of small oscillations but with large fluctuations in the interspike intervals. The corresponding 3D trajectories are shown in Figure 3 as obtained both from PSPICE simulation and real experiment. It may be noted that the circuit is first tested in PSPICE simulator before experimenting with real circuit, results of which are quite interestingly matching to each other. The trajectory moves towards the saddle focus at one end along its stable eigendirection as indicated by the arrows in Figure 3. When the trajectory comes in close proximity to this saddle focus, it escapes out spirally. Next it spirals in towards the saddle focus origin before totally escaping away along the unstable eigendirection of this saddle focus. Finally the trajectory takes two global turns around the third saddle

Figure 2. Homoclinic chaos in asymmetry-induced Chua's circuit. (a) Time series of experiment for $R_1=1370.1\Omega$, $R_C=47.25k\Omega$, $R_x=9.33k\Omega$, $R_y=1.73k\Omega$ in lower plot, (b) return time of small amplitude oscillations.

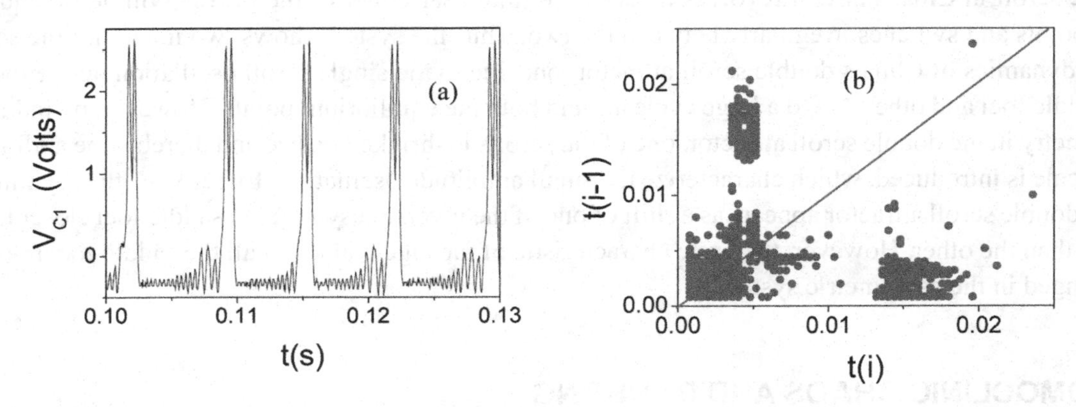

focus at the other end before reinjecting again into the first saddle focus. In reality, the trajectory of the 3D unstable orbit approaches homoclinicity to two different saddle foci, one at the origin and the other at one end of the attractor.

In our first experiment, we set the asymmetry by R_x=9.33kΩ, R_y=1.73kΩ and R_C =47.25kΩ, and then decrease R_1 from 1524Ω. As R_1 is decreased, the circuit dynamics moves from stable focus to limit cycle via supercritical Hopf bifurcation at R_1=1524Ω and a transition to 2-band chaos occurs via a sequence of PD. The period-parameter bifurcation in Figure 4 shows asymptotic increase in period with decrease in R_1 indicating an approach to homoclinicity. A large amplitude limit cycle 2^0 for R_1=1373.3Ω is seen at the highest point (A) of the right bifurcation diagram in Figure 4. This large amplitude limit cycle of period-2 (2^0) is shown in the upper row left of Figure 5. The instability started just beyond this point with further decrease in R_1 and the instabilities continue until R_1=1363.7Ω when it stops with the re-

Figure 3. The 3D trajectory of homoclinic chaos. Left plot from PSPICE simulation using time series of $V_{C1}(t)$ and $V_{C2}(t)$ and $I_L(t)$ along the X-axis, Y-axis and Z-axis respectively for L=2: $R_1=1429.9\Omega$, $R_C=45k\Omega$, $R_x=1580\Omega$, $R_y=507\Omega$. The 3D trajectory in the right plot is reconstructed using experimental time series of $V_{C1}(t)$, $V_{C2}(t)$ and delayed $V_{C1}(t-2)$ along the X-axis, Y-axis and Z-axis respectively. $R_1=1370.1\Omega$, $R_C=47.25k\Omega$, $R_x=9.33k\Omega$, $R_y=1.73k\Omega$.

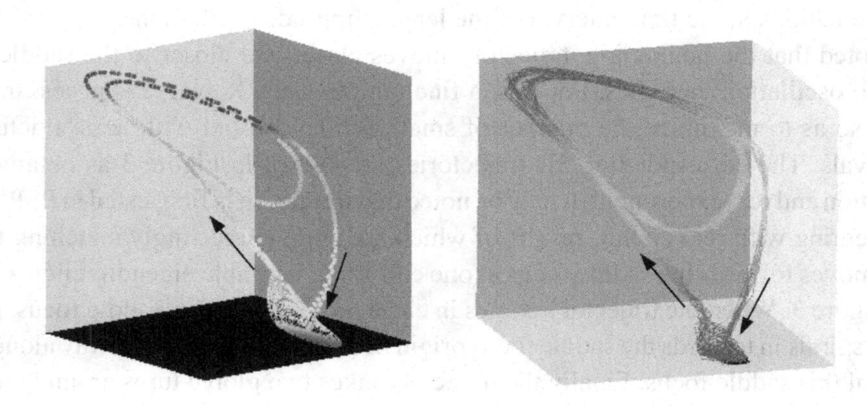

appearance of another large amplitude limit cycle (2^0) shown in the upper row right of Figure 5 which corresponds to the top (A*) of the left bifurcation diagram in Figure 4. The period in lower plot (Figure 4) then decreases with R_1. But the period increases again with an intermediate minimum until another instability region appears at R_1=1352.3Ω. A stable limit cycle (period-2) reappears at R_1=1339Ω, which moves to 1^0 (period-1) limit cycle at R_1=1337Ω via reverse PD and finally loses stability via subcritical Hopf bifurcation for R_1<1337Ω. Each of the local maxima in the period-parameter bifurcation, indicates an approach to homoclincity, however, it is difficult to observe them in a single experiment due to their extreme sensitivity to control parameter. We focus, in the first experiment, on evidences of homoclinicity related to MMO (2^s) in the R_1 interval of (1363.7Ω, 1373.3Ω). In this parameter interval, R_1=[1363.7Ω, 1373.3Ω)], several complex behaviors such as MMO, homoclinic chaos and bursting are observed. The large amplitude limit cycles (2^0) called as spiking at both the edges (A*, A) of the R_1 [1363.7Ω, 1373.3Ω)] interval are shown in the upper row of Figure 5. For small changes in R_1=1364Ω (right: 2^p) and 1373.2Ω (left:2^m) near these edges, two bursting oscillations appear by loss of stability in each of the large amplitude limit cycle as shown in their immediate lower row phase portraits in Figure 5. The corresponding time series of bursting oscillations are shown in Figure 6. The bursting oscillations are seen here as a train of periodic spiking with intermittent transition to small oscillations. The unstable small amplitude limit cycle (2^m) appears as glued (lower row left) to the stable large amplitude limit cycle via SN bifurcation while it is inside the large limit cycle for 2^p. In the process, we also observed a sequence of MMOs interspersed by chaotic states in the parameter interval of R_1=(1363.7Ω, 1373.3Ω) as a route to homoclinic chaos (2^s) at R_1=1370.1Ω already shown in Figures 2-3. However, it becomes very difficult to identify the periodic MMOs, in this experiment, due to their extremely narrow interval in the selected parameter space.

In a second experiment, by appropriate choice of system parameters and control of the asymmetry parameter, we observed a sequence of transitions $1^0 \rightarrow \propto^1 \rightarrow 1^1 \rightarrow 1^m \rightarrow 1^s \rightarrow 1^p \rightarrow 1^1 \rightarrow \propto^1 \rightarrow 1^0$. The large amplitude limit cycle is period-1 (1^0) here. We observed the sequence of transitions as large amplitude limit cycle (1^0), bursting (1^m), homoclinic chaos (1^s), and then again bursting (1^p), large amplitude limit cycle (1^0) with changes in R_1 as shown in Figure 7. The phase portraits of the five different states are shown in Figure 8.

Figure 4. Period parameter bifurcation R_C=47.25kΩ, R_x=9.33kΩ, R_y=1.73kΩ.

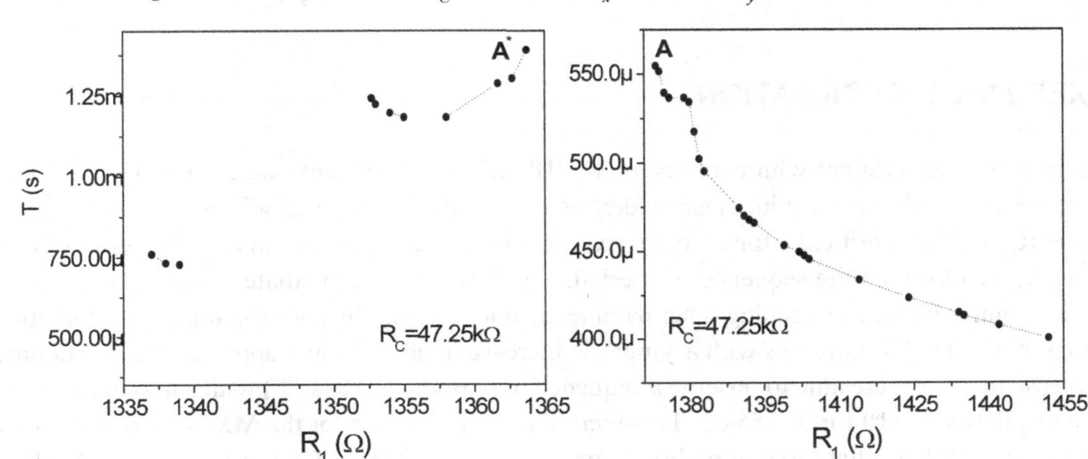

Figure 5. Phase portraits of $V_{C1}(t)$ vs. $V_{C2}(t)$. Upper row: large amplitude limit cycles for R_1=1373.3Ω (left) and for R_1=1363.7Ω (right). Lower row: unstable small amplitude limit cycle glued to the stable large amplitude limit cycle via SN bifurcation for R_1=1373.2Ω (left) and for R_1=1364Ω (right).

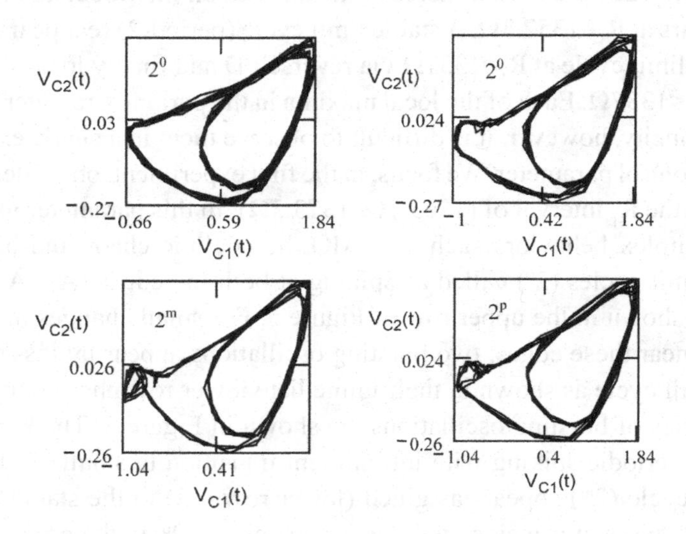

Figure 6. Time series of bursting oscillations for R_1=1364Ω (2^p) in the left plot and for R_1=1373.2Ω (2^m) in right plot. The small oscillations exist as outer circle to the large oscillations in the left plot while they exist inner to the large oscillations in the right plot.

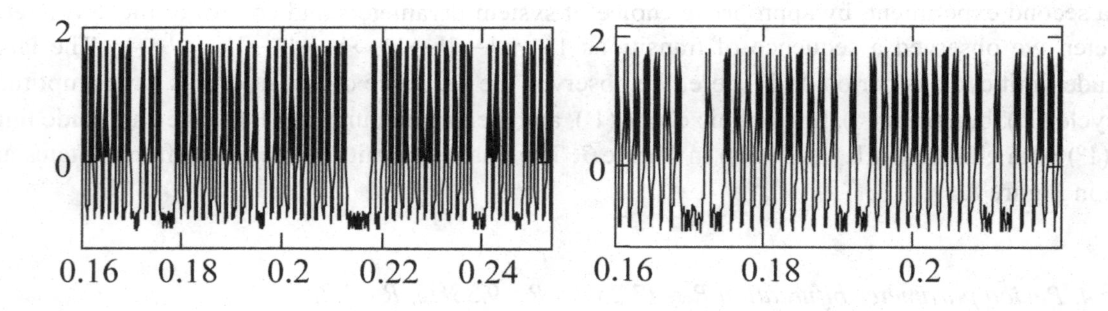

4. MIXED MODE OSCILLATION

We set up a third experiment with parameters, R_x=1858Ω, R_y=333Ω and find the similar sequence of events as described above but with clear evidences of periodic MMO. For selected R_C =47.14kΩ, the parameter R_1=1357Ω is critically tuned to observe homoclinic chaos as seen in the time series of Figure 9. However, we observed the sequences of periodic MMOs with intermediate chaotic states when we use R_C as a control parameter keeping other parameters unchanged. The period-parameter bifurcation in Figure 10 shows devil's staircases with asymptotic increase in period while approaching homoclinicity with decease in R_C. We are able to observe a sequence of periodic MMOs (2^s) with a maximum number of small oscillations s=10 for R_C=55kΩ. The interval in R_C parameter for the MMO becomes narrower with increasing small oscillation (s) as reported earlier (Petrov et al, 1992; Gaspard & Wang, 1987; Koper,

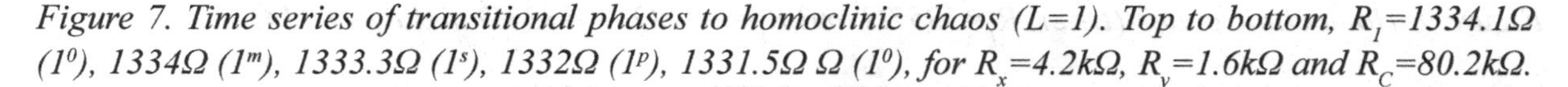

Figure 7. Time series of transitional phases to homoclinic chaos (L=1). Top to bottom, R_l=1334.1Ω (1^0), 1334Ω (1^m), 1333.3Ω (1^s), 1332Ω (1^p), 1331.5Ω Ω (1^0), for R_x=4.2kΩ, R_y=1.6kΩ and R_C=80.2kΩ.

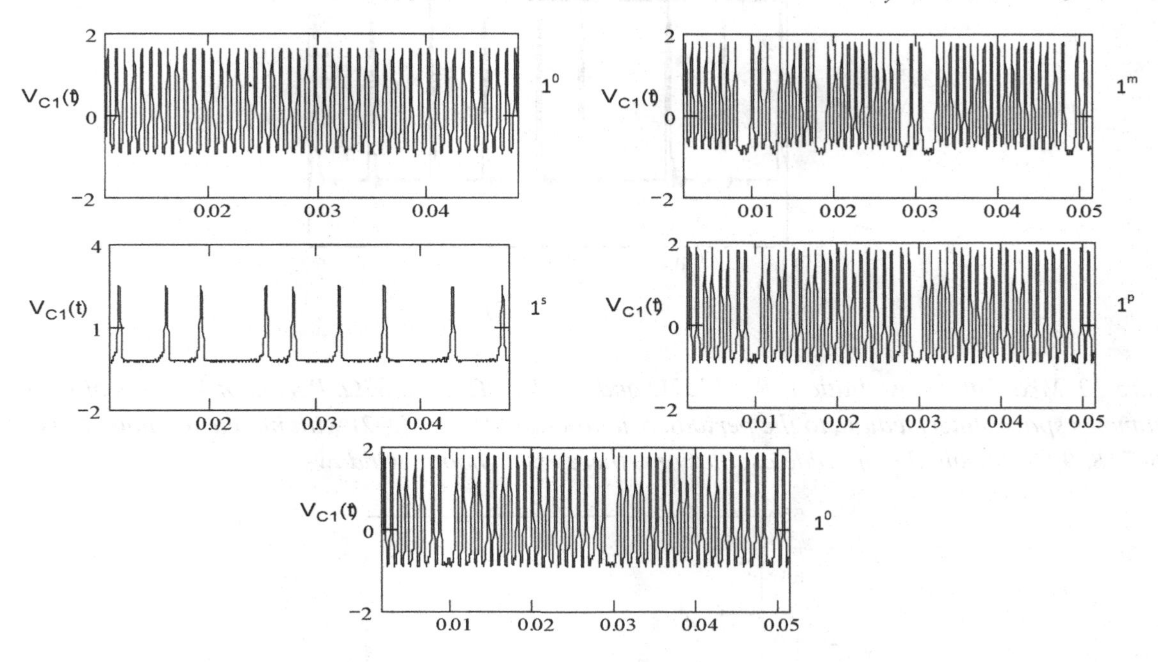

Figure 8. Phase portraits of transitional phases of 1^0 limit cycle: top to bottom, R_l=1334.1Ω (1^0), 1334Ω (1^m), 1333.3Ω (1^s), 1332Ω (1^p), 1331.5Ω (1^0) for R_x=4.2kΩ, R_y=1.6kΩ and R_C=80.2kΩ.

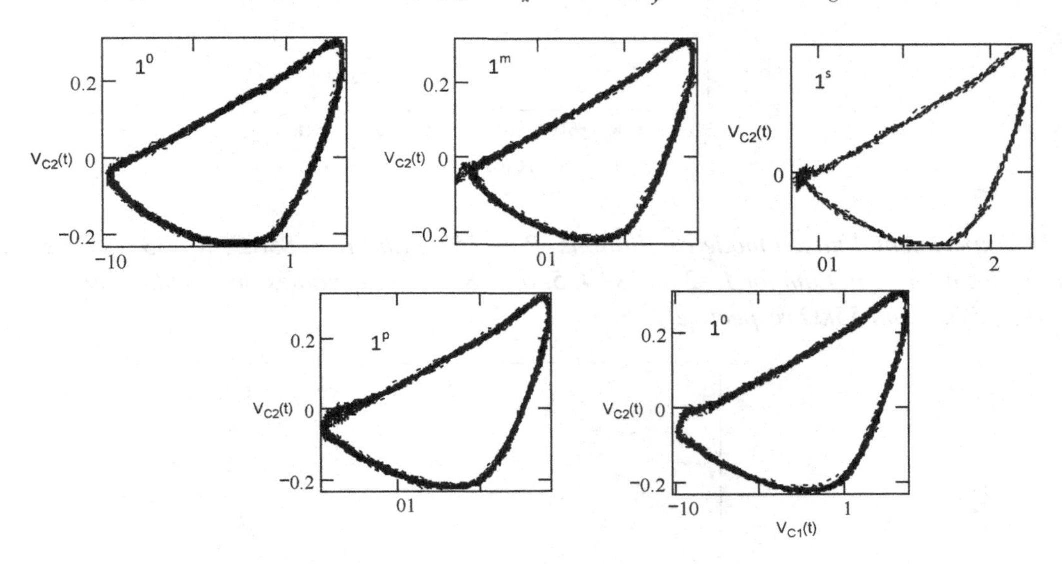

1995; Goryachev et al, 1997; Rajesh, & Ananthakrishna, 2000; Dana & Chakraborty, 2004; Dana et al, 2005; Dana et al, 2007). The intervals between the line plots in Figure 10 indicate the chaotic windows. It is important to note that the time period of the largest (2^{10}) MMO is nearly 5ms for R_C=55kΩ while the time series of homoclinic chaos in Figure 9 for R_C=47.14kΩ show fluctuations in the interspike intervals between 8 ms to 16 ms. It is clearly evident that as R_C is slowly tuned from 85kΩ corresponding

Figure 9. Homoclinic chaos (L=2). R_1=1357Ω and R_x=1858Ω, R_y=333Ω and R_C=47.14kΩ.

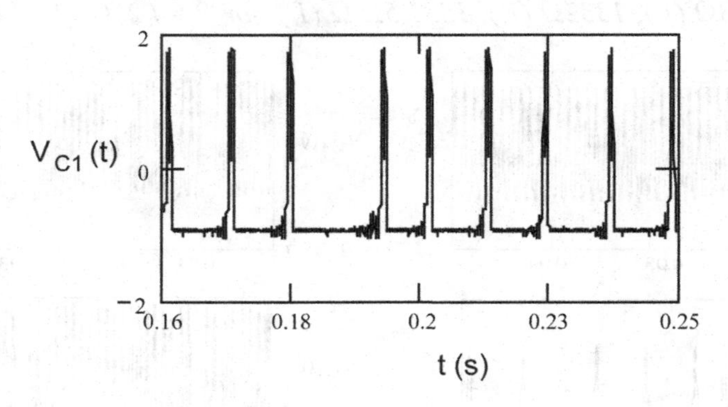

Figure 10. Mixed mode oscillation: R_1=1357Ω and R_x=1858Ω, R_y=333Ω. Period of MMO is plotted in parameter space. Intermediate to the periodic windows of MMOs (L=2) with increasing number (s=4, 5, 6, 7, 8, 9, 10) of small amplitude oscillations, there exist chaotic windows.

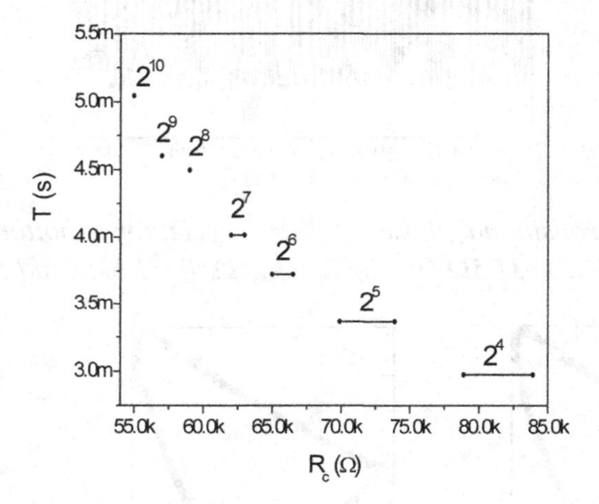

Figure 11. Experimental mixed mode oscillations. R_1=1357Ω and R_x=1858Ω, R_y=333Ω. Mixed mode oscillations from top to bottom for L=2 and s=4, 5, 6, 7, 8, 9, 10. R_C values are 78.9kΩ, 69.9kΩ, 65kΩ, 62kΩ, 59kΩ, 57kΩ and 55kΩ respectively.

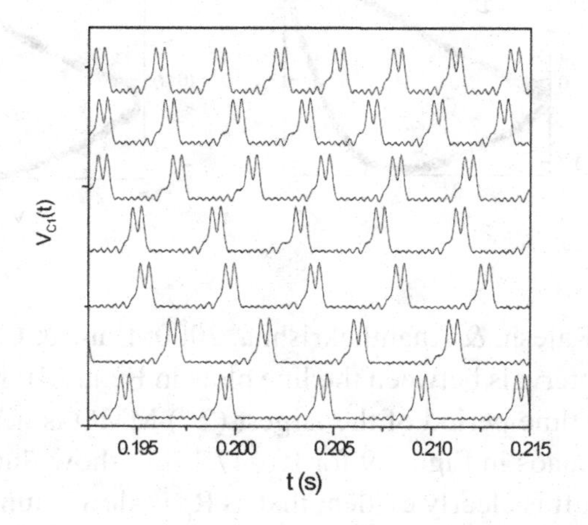

the MMO (*s*=4) to 55kΩ (*s*=10), the interpsike intervals increases by three fold. Thus, by fine-tuning the asymmetry parameter, we are able to reach a close vicinity of the homoclinic orbit with much larger (s>>10) number of small oscillations. The time series of periodic MMOs are shown in Figure 11. The Y-scale is arbitrarily chosen since different time series are either scaled up or down for visual clarity.

5. CONCLUSION

We presented experimental evidences of Shil'nikov type homoclinic chaos in asymmetry-induced Chua's oscillator. The asymmetry plays a crucial role in the bifurcations in the process of transition to homoclinic chaos, when we observed spiking, bursting and MMO. When a system parameter is changed for selected asymmetry, the large amplitude limit cycle (n^0: n=1, 2) moves to MMO through a bursting region ($n^1 \rightarrow n^m$) via SN bifurcation. The system then moves to periodic MMOs with increasing number of small oscillations interspersed by chaotic states. Homoclinic chaos is observed, in this MMO regime, when the asymmetry parameter is fine tuned. The system moves to another large amplitude limit cycle with further change in system parameter. During this later transition, another bursting region ($n^p \rightarrow n^1$) is also observed. The existence of this second bursting regime on the way back to large limit cycle from homoclinic chaos is overlooked in the earlier works. However, the emergence of MMO from large amplitude limit cycle via a bursting regime followed by the transition to homoclincity has been reported earlier in slow-fast system models. On the other hand, in our experiment, different time scales are artificially created in a double scroll Chua attractor by inducing asymmetry in the system by external DC forcing. One of the double scroll attractors shrinks in size and creates an additional time scales in the overall dynamics that plays a crucial role in the origin of homoclinic chaos, bursting and MMOs.

REFERENCES

Allaria, E., Arecchi, F. T., Garbo, A., & Di, ., & Meucci, M. (2001). Synchronization of Homoclinic Chaos. *Physical Review Letters*, *86*(5), 791–794. doi:10.1103/PhysRevLett.86.791

Belykh, V. N., Belykh, I. V., Colding-Jorgensen, M., & Mosekilde, E. (2000). Homoclinic bifurcations leading to bursting oscillations in cell models. *The European Physical Journal E*, *3*(3), 205–219. doi:10.1007/s101890070012

Breakspear, M., Terry, J. R. & Friston, K. J. (2003). Modulation of excitatory synaptic coupling facilitates synchronization and complex dynamics in a biophysical model of neuronal dynamics. *Network: Computation in Neural Computing, 14*, 703.

Chua, L. O., Komuro, M., & Matsumoto, T. (1986). The double scroll family. *IEEE Transactions on Circuits and Systems*, *33*(11), 1072–1118. doi:10.1109/TCS.1986.1085869

Dana, S. K., & Chakraborty, S. (2004). Generation of homoclinic oscillation in the Phase Synchronization Regime in Coupled Chua's oscillators. *International Journal of Bifurcation and Chaos in Applied Sciences and Engineering*, *14*(4), 1375. doi:10.1142/S0218127404009958

Dana, S. K., Chakraborty, S., & Aananthakrishna, G. (2005). Homoclinic bifurcation in Chua's circuit. *PRAMANA - . Journal of Physics*, *64*(3).

Dana, S. K., & Roy, P. K. (2007). Bursting near homoclinic bifurcation in two coupled Chua oscillators. *Int. J. Bifur. Chaos*, *17*(10).

Dana, S. K., Sengupta, D. C., & Hu, C.-K. (2006). Spiking and bursting in Josepshon junction. *IEEE Transactions on Circuits and Wystems. II, Express Briefs*, *50*(10), 1031. doi:10.1109/TCSII.2006.882183

Gaspard, P., & Wang, X.-J. (1987). Homoclinic orbits and mixed-mode oscillations in far from equilibrium systems. *Journal of Statistical Physics*, *48*, 151. doi:10.1007/BF01010405

Glendinning, P., Abshagen, J., & Mullin, T. (2001). Imperfect homoclinic bifurcations. *Phy. Rev. E*, *64*, 036208. doi:10.1103/PhysRevE.64.036208

Glendinning, P., & Sparrow, C. (1984). Local and global behavior near homoclinic orbits. *Journal of Statistical Physics*, *35*, 645. doi:10.1007/BF01010828

Goryachev, A., Strizhak, P., Kapral, R. (1997). Slow manifold structure and the emergence of mixed-mode oscillations. *J. Chem. Phys*.

Healy, J. J., Broomhead, D. S., Cliffe, K. A., Jones, R., & Mullin, T. (1991). The Origins of Chaos in a Modified Van der Pol Oscillator. *Physica D. Nonlinear Phenomena*, *48*, 322. doi:10.1016/0167-2789(91)90091-M

Herrero, R., Pons, R., Farjas, J., Pi, F., & Orriols, G. (1996). Homoclinic dynamics in experimental Shil'nikov attractors. *Phy. Rev. E*, *53*(6), 5627. doi:10.1103/PhysRevE.53.5627

Izhikevich, E. M. (2000). Neural excitability, spiking, and bursting. *International Journal of Bifurcation and Chaos in Applied Sciences and Engineering*, *10*(6), 1171. doi:10.1142/S0218127400000840

Kennedy, M. P. (1993). Three steps to chaso-A Chua's circuit primer. *IEEE Trans. Cir. Systs.*, *40*.

Koper, M. T. M. (1995). Bifurcations of mixed-mode oscillations in a three-variable autonomous van der Pol-Duffing model with a cross-shaped phase diagram. *Physica D. Nonlinear Phenomena*, *80*, 72–94. doi:10.1016/0167-2789(95)90061-6

Kuznetsov, Y. A. (1995). *Elements of Applied Bifurcation Theory*. New York: Springer-Verlag.

Llinás, R. R. (1988). The intrinsic electrophysiological properties of mammalian neurons: insights into central nervous system function. *Science*, *242*, 1654. doi:10.1126/science.3059497

Marino, I. P., Allaria, E., Meucci, R., Boccaletti, S., & Arecchi, F. T. (2003). Information encoding in homoclinic chaotic systems. *Chaos (Woodbury, N.Y.)*, *13*(1), 286–290. doi:10.1063/1.1489115

Peacock, T., & Mullin, T. (2001). Homoclinic bifurcations in a liquid crystal flow. *Journal of Fluid Mechanics*, *432*, 369–386.

Petrov, V., Scott, S. K., & Showalter, K. (1992). Mixed-mode oscillations in chemical systems. *The Journal of Chemical Physics*, *97*(9), 6191–6198. doi:10.1063/1.463727

Pisarchik, A. N., Meucci, R., & Arecchi, F. T. (2001). Theoretical and experimental study of discrete behavior of Shilnikov chaos in a CO2 laser. *The European Physical Journal D*, *13*, 385–391. doi:10.1007/s100530170257

Rajesh, S., & Ananthakrishna, G. (2000). Incomplete approach to homoclinicity in a model with bent-slow manifold geometry. *Physica D. Nonlinear Phenomena*, *140*, 193. doi:10.1016/S0167-2789(99)00241-9

Shilnikov, A. L., & Rulkov, N. F. (2003). Origin of chaos in a two-dimensional map modeling spiking-bursting neural activity. *International Journal of Bifurcation and Chaos in Applied Sciences and Engineering*, *13*(11), 3325. doi:10.1142/S0218127403008521

Wiggins, S. (1990). *Introduction to Applied Nonlinear Dynamical Systems and Chaos*. New York: Springer-Verlag.

Section 2
Synchronization of Chaotic Systems

Chapter 6
Synchronization of Chaotic Oscillators

J. M. González-Miranda
Universidad de Barcelona, Spain

ABSTRACT

In this chapter, the author reviews the variety of forms that the synchronization of the dynamics of mutually coupled or unidirectionally driven chaotic oscillators can display. The aim is to provide a presentation of a background of knowledge on fundamental physics and mathematics that the author expects to be useful in the development of the application of chaos synchronization to telecommunications and cryptography.

INTRODUCTION

One of the most remarkable properties of nonlinear systems is its ability to display chaotic dynamics. The fundamental feature of a system displaying this kind of dynamics is its sensitivity to initial conditions. This means that the dynamics of such system is unpredictable, and consequently irregular and aperiodic. Such features may appear, in principle, undesirable and something to be avoided, at least in engineering applications. However, research on the dynamics of coupled and driven chaotic oscillators, among other issues, has proven these features of nonlinear dynamics to be interesting and potentially useful.

When two chaotic oscillators are mutually coupled, or when one of such oscillators drives another, they may display the phenomenon of chaos synchronization (Pikovsky et al., 2001; González-Miranda, 2004). The study of the synchronization of chaotic oscillators has become a topic on its own within the field of nonlinear dynamics and chaos because of both, its scientific interest as a phenomenon characteristic of nonlinear systems, and its potential applications in many fields. Among them, we have the secure transmission of information in telecommunications, which is based on the use of the unstable and irregular behavior of chaotic systems to conceal the information being transmitted.

DOI: 10.4018/978-1-61520-737-4.ch006

A relevant feature of chaos synchronization is the variety of forms that it can display. In some of them the motion of the synchronized oscillations is strongly correlated, in others this correlation is quite faint, while there are types of synchronization that are somewhere in the middle, and presenting different qualitative features.

Some forms of synchronization may be observed, in principle, on any kind of chaotic systems. Ordered from the strongest to the faintest, being the first and the second equally strong, these are: Identical Synchronization (Fujisaka & Yamada,1983; Pecora & Carroll, 1990), Generalized Synchronization (Rulkov et al., 1995), lag synchronization (Rosenblum et al., 1997), Phase Synchronization (Rosenblum et al., 1996), and Amplitude Envelope Synchronization (González-Miranda, 2002a). A given system is able to display several of these forms of synchronization, is a series of stages of chaotic synchronization, starting from the faintest, as the strength of the coupling is progressively increased.

Moreover, there are other types of synchronization, which are special in the sense they are linked to some particular property, or feature of the chaotic oscillator involved. These include Anticipated Synchronization (Voss, 2000), Marginal Synchronization (González-Miranda, 1996a), and Multistable Synchronization (González-Miranda, 1996b).

The mission of this chapter is to review these different forms of synchronization to provide a systematic presentation of a material that is fundamental for the application of chaos theory to cryptography for secure communications. Each of these forms of synchronization of chaos will be described in qualitative and quantitative form, techniques needed for its observation and measure will be given, as well as conditions for their occurrence.

CHAOTIC OSCILLATORS AND SYNCHRONIZATION LAYOUTS

Deterministic chaos is ubiquitous in nature and has been observed in many fields of physics, chemistry, biology, geology and astrophysics. In particular, it has been observed in systems of interest for telecommunications such as electric circuits and lasers.

An example of electric circuit is the one introduced by Matsumoto et al. (1985) and its many variants (Chua et al., 1993; Bilotta et al., 2007; Gomes & King, 1992). Other interesting chaotic electric circuits exist, some of them reproducing chaotic systems borrowed from other fields. For example the celebrated Lorenz (1963) and Rössler (1976) systems, have been implemented as electric circuits by Cuomo and Oppenheim (1993b) and by Carroll (1995), respectively.

Practically, all kinds of lasers can be prepared to display chaotic behavior (Harrison & Biswas, 1986). In particular, chaotic dynamics has been observed in experiments performed on such a variety of lasers as: CO_2 lasers (Midavaine et al., 1985), semiconductor lasers (Mukai & Otsuka, 1985), Nd:Yttrium Aluminum Garnet lasers (Bracikowski & Roy, 1991) and erbium-doped fiber ring lasers (VanWiggeren & Roy, 1998).

These systems are usually modeled by means of sets of ordinary differential equations. For example, the electric circuit by Matsumoto et al. (1985) which contains a nonlinear resistor, a resistor of resistance R, two capacitors of capacitances C_1 and C_2, and an inductance L, can be described by a system of three non-linear equations on the variables x, y and z, describing the voltages of the capacitors 1 and 2 and the current in the inductance respectively. This reads

$$dx / dt = \alpha\left[y - x - f\left(x\right)\right],$$
$$dy / dt = x - y + z,$$
$$dz / dt = -\beta y,$$

Being $f(x)$ a no-linear function determined by the nonlinear resistor, $\alpha = C_1 / C_2$ and $\beta = C_2 / (LR^2)$ control parameters given by the circuit components.

Another example of chaotic model is given by the real Maxwell-Bloch equations for a resonant single-mode laser (Harrison & Biswas, 1986) which describes the dynamics of the field inside the cavity, *E*, the population inversion, *D*, and the atomic polarization, *P*, by means of

$$dE / dt = -\kappa E + \kappa P,$$
$$dP / dt = \gamma_{\perp} ED - \gamma_{\perp} P,$$
$$dD = \gamma_{\parallel}\left(\text{»} \; +1\right) - \gamma_{\parallel} D - \gamma_{\parallel}\lambda EP,$$

where the bifurcation parameters are the decay rates of the cavity, κ, of the atomic polarization, $\gamma_{\perp}$, of the population inversion, $\gamma_{\parallel}$, and the pumping parameter, λ.

We could pose more examples for electric circuits and laser models allowing mathematical modelization in terms of systems of ordinary differential equations. Therefore, for mathematical representation we will consider here that the systems under consideration are *n*-dimensional flows whose state is described by a vector $x = \left(x_1, x_2, ..., x_n\right) \in \mathbb{R}^n$ whose components are the *n* variables, which describe the state of the system. Its time evolution in continuous time, *t*, is given by the set of *n* ordinary differential equations

$$\frac{dx}{dt} = F\left(x; p\right)$$

with $F\left(x; p\right) = \left[F_1\left(x; p\right), F_2\left(x; p\right), ..., F_n\left(x; p\right)\right]$ a set of *n* nonlinear real functions which describe the system. These depend on the set of *r* real bifurcation parameters $p = \left(p_1, p_2, ..., p_r\right)$ that determine the system dynamics, which we assume to occur in a chaotic or hyperchaotic attractor.

There is a variety of ways to couple two (or more) chaotic oscillators, that have proven to be useful when studying chaos synchronization. They use to be classified in two main groups: (i) drive-response, or master-slave, configuration in which a first oscillator, called the drive, perturbs a second, called the response, but is not perturbed by this second, and (ii) mutual coupling, in which one oscillators perturbs the second and vice versa. These two types of coupling can be performed in a variety of forms, which can be classified in two main groups: replacement of variables, and control techniques.

Let us consider two chaotic oscillators

$$\frac{dx}{dt} = F\left(x;p\right), \frac{dy}{dt} = G\left(y;q\right),$$

Which we identify by the name of its variables, $x = \left(x_1, x_2, ..., x_n\right)$ and $y = \left(y_1, y_2, ..., y_m\right)$. In the replacement of variables coupling scheme the variables of each group are divided in three categories: (i) those that will by injected in the other oscillator, $x^{(d)}$ and, $y^{(d)}$ (ii) those which will be substituted by variables of the other system, $x^{(r)}$ and $y^{(r)}$, and (iii) all other variables, $x^{(u)}$ and $y^{(u)}$. The equations that describe the coupled systems then read

$$\frac{dx}{dt} = F\left(x^{(u)}, x^{(d)}, y^{(d)}; p\right),$$
$$\frac{dy}{dt} = G\left(y^{(u)}, y^{(d)}, x^{(d)}; q\right),$$

where we see that system x evolves under the action of the variables $y^{(d)}$ coming from the other subsystem, and system y evolves under the action of the variables $x^{(d)}$ coming from the other subsystem. The drive-response configuration, being x the drive would read

$$\frac{dx}{dt} = F\left(x^{(u)}, x^{(d)}; p\right),$$
$$\frac{dy}{dt} = G\left(y^{(u)}, x^{(d)}; q\right).$$

This means that the second system is driven by the first because of the injection of the variables $x^{(d)}$.
Two chaotic oscillators, x and y, coupled by means of control techniques can be modeled as

$$\frac{dx}{dt} = F\left(x;p\right) + H_x\left(x,y\right),$$
$$\frac{dy}{dt} = G\left(y;q\right) + H_y\left(x,y\right),$$

with H_x and H_y sets of functions defining the coupling scheme. In the particular case of being $H_x = 0$, we will have the system y driven by the system x. If both $H_x \neq 0$ and $H_y \neq 0$, we have mutual coupling. A particularly simple case of control technique is diffusive coupling; which is defined by

$$-_x\left(x,y\right) = C \cdot \varepsilon \cdot \left(y - x\right),$$
$$-_y\left(x,y\right) = C \cdot \left(1 - \varepsilon\right) \cdot \left(x - y\right),$$

With C a measure of the strength of the coupling, ε a matrix of constants smaller or equal then one, and 1 the unity matrix. For $\varepsilon = 0$, we have the oscillator x as the drive and y as the response.

IDENTICAL AND GENERALIZED SYNCHRONIZATION

The strongest forms of synchronization occur when de values of the observables of one of the coupled oscillators are completely determined by the values of the observables of the other oscillator. In general, we speak about Generalized Synchronization, and define it formally as follows (Rulkov et al., 1995). Given the sets of variables $x = (x_1, x_2, ..., x_n)$ and $y = (y_1, y_2, ..., y_n)$ describing the state of each oscillator, we say that there is synchronization of the oscillator given by *y*(*t*) to the oscillator given by *x*(*t*) when there are sets of initial conditions for each of the oscillators such that when started in these sets we have

$$\lim_{t\to\infty} \left| y(t) - \Phi\left[x(t)\right]\right| = 0,$$

being $\Phi\left[x(t)\right]$ a functional that completely determines the trajectory of one of the oscillators from the trajectory of other as $y(t) = \Phi\left[x(t)\right]$. When the oscillators are mutually synchronized, $\Phi\left[x(t)\right]$ is an invertible functional. An important particular form of Generalized Synchronization is Identical Synchronization (Fujisaka & Yamada, 1983; Pecora & Carroll, 1990), which occurs when that functional is the identity, $\Phi\left[x(t)\right] = x(t)$. Identical Synchronization between two oscillators occurs when, for time large enough, it is $y(t) = x(t)$, within the experimental error bars.

Generalized Synchronization was introduced, after Identical Synchronization, as a form of synchronization to appear when the coupled oscillators are different, assuming that Identical Synchronization is proper of identical oscillators. However, Generalized Synchronization has been observed in several cases of coupled identical oscillators. Multistable Synchronization, which can occur in symmetric chaotic systems (González-Miranda, 1996b), provides examples of this kind, such as the so-called Antiphase Synchronization, which is characteristic of systems that have inversion symmetry. In this case, Generalized Synchronization is characterized by the maxima of the oscillations of one of the oscillators being coincident with the minima of the other. More sophisticated forms of Generalized Synchronization for identical coupled oscillations can be found in the literature (González-Miranda, 2002c).

The detection of Identical Synchronization is straightforward; we just have to compare each of the variables describing the states of the oscillators, $x = (x_1, x_2, ..., x_n)$ and $y = (y_1, y_2, ..., y_n)$. We can do this using parametric plots of the correspondent variables, $y_i = y_i(x_i), i = 1, 2, ..., n$, which should look as straight lines with slope equal to one. Detect Generalized Synchronization requires the use of specific techniques. The auxiliary system approach (Abarbanel et al., 1996; Hramov, & Koronovskii, 2005), generalizes the use of parametric plots. It is to be used in a drive-response configuration, and uses the fact that two identical copies of the response, subject to the same drive, will reach the same state of Generalized Synchronization. Therefore, given an additional independent response system, described by the variables $y' = (y'_1, y'_2, ..., y'_n)$, there is Generalized Synchronization of the response to the drive when $y(t) = y'(t)$. This second system, called the auxiliary system, allows detection of Generalized

Synchronization by means of parametric plots of the auxiliary system versus the response, which should look as if the two systems were in a state of Identical Synchronization.

Other approaches to detect Generalized Synchronization rely on the use of several quantities that measure nonlinear dependence between two time series. These are useful for the two coupling configurations considered: drive-response and mutual coupling. Several approaches have been proposed, we quote some of them. Rulkov et al. (1995) have defined the mutual false nearest neighbors parameter, which is a statistical test of local neighborliness. Schiff et al. (1996) and Breakspear & Terry (2002) used mutual nonlinear prediction techniques, based on the fact that generalized synchronization implies the ability to predict the state one system from the state of the other. Suetani et al. (2006) proposed the use of kernel methods to improve multivariate analysis methods based on the calculation of correlation coefficients. Chen et al. (2007) based their detection of generalized synchronization in clustering analysis.

The stability of the synchronized state has received considerable attention in the literature. Its study is based in a geometrical picture (Fujisaka & Yamada, 1983; Kocarev & Parlitz, 1996; Pecora et al., 1997), where the motion of two coupled chaotic oscillators is observed in a phase space of dimension $d = n + m$. When the synchronized state is achieved, this motion is restricted by the condition $y = \Phi\left[x\right]$, which means that it occurs in a hypersurface of this phase space that is called the synchronization manifold. The subspace orthogonal to it is called the transverse subspace. Stability is then formulated for a perturbation orthogonal to the synchronization manifold applied to a system in the synchronized state. This state is stable when the perturbed system returns exponentially to the synchronization manifold. It is unstable, when the perturbed system escapes exponentially from the synchronization manifold. It is said marginally stable when the perturbed trajectory stays in the neighborhood of the synchronization manifold.

For the case of a drive-response configuration, we formulate these ideas in terms of Lyapunov exponents (Pecora & Carroll, 1990; Kocarev & Parlitz, 1996). If generalized synchronization, and then identical synchronization, has to be asymptotically stable, it must happen that for any trajectory of the drive, $\boldsymbol{x}(t)$, a small perturbations, $\delta y\left(t\right)$, of the synchronized trajectory of the response, $\boldsymbol{y}(t)$, has to result in trajectories such that $\delta y\left(t\right)$ decreases exponentially. Therefore, the linearized equations for $\delta y\left(t\right)$,

$$\frac{d\left(\delta y\right)}{dt} = \left[\frac{dG\left(x,y\right)}{dy}\right]_{\substack{x=x\left(t\right),\\ y=y\left(t\right)}} \delta y$$

must have all its Lyapunov exponents, called conditional Lyapunov exponents, negative.

For a system of mutually coupled oscillators (Fujisaka & Yamada, 1983; Pecora et al., 1997), when studying a perturbation that takes the system state out of the synchronization manifold, we work with the projections of the perturbations onto the synchronization manifold and onto the transverse subspace. The time evolutions of these perturbations are given by two independent equations, one for each projection, that are alike to the above linearized equation. Then, there is asymptotic stable synchronization when perturbations orthogonal to the synchronization manifold vanish exponentially; therefore, the Lyapunov exponents for the equation of the transverse part, called transverse Lyapunov exponents, have to be negative.

In practical cases, when the equations of the system are known, the stability of the synchronized state can be determined by standard methods for the calculation of Lyapunov exponents. Alternatively, we can apply the Lyapunov theory for the stability of solutions of differential equations for that aim (He & Vaidya, 1992). This approach is based on working with error functions, $e(t) = y(t) - \Phi[x(t)]$, instead of perturbations. Then defining appropriate Lyapunov functionals $L = L[e(t)]$ for *e*(*t*), the condition for asymptotically stable synchronization is that *L*(*t*) has to present a negative time derivative. Many applications based on this approach can be found in the literature (see for example: Pyragas, 1998, He & Vaidya, 1999; Shahverdiev, 1999; Jovic et al., 2006; Ghosh et al., 2007; Naghavi & Safavi, 2008; Choon, 2009; Porfiri & Fiorilli, 2010.).

PHASE AND LAG SYNCHRONIZATION

There are other forms of synchronization weaker than Generalized Synchronization, where only one part of the dynamics of the oscillators is synchronized remaining the rest unsynchronized. One of them is Phase Synchronization, which occurs when the phases of the coupled oscillators evolve in synchrony, while its amplitudes are uncorrelated (Stone, 1992; Rosenblum et al., 1996). Other is Lag Synchronization, which occurs when both phases and amplitudes are in synchrony except for a time delay between the time evolutions of the two-coupled oscillators (Rosenblum et al., 1997).

The concept of the phase of a chaotic oscillator is a generalization of the concept of the phase of a periodic oscillator. For chaotic oscillators whose trajectory in phase space rotates around a center along a well-defined direction (i.e., which follow a proper rotation), we can find a plane where the projection of the trajectory follows a proper rotation too (Pikovsky et al., 1997a; Yalçinkaya & Lai, 1997; Pikovsky et al., 2000). The phase is then defined as the polar coordinate of the trajectory, $\varphi(t) \in \mathbb{R}$, by means of the coordinates of the trajectory projection onto that plane, *x*(*t*) and *y*(*t*), as

$$\varphi(t) = \arctan\left[\frac{y(t) - y_0}{x(t) - x_0}\right],$$

Being (x_0, y_0) the center of rotation. When the oscillator dynamics does not follows a proper rotation, techniques have been proposed to decompose the phase dynamics into a superposition of proper rotations (Pikovsky et al., 1997a; Yalçinkaya & Lai, 1997).

The phase dynamics found in most chaotic oscillators can be written as $\varphi(t) = \varphi_0 \pm \Omega t + \xi(t)$, with the plus sign for counter-clockwise rotations and the minus sign for clockwise rotations. In this equation, φ_0 is a constant determined by the initial conditions, $\xi(t)$ a small bounded zero mean chaotic fluctuation, and Ω a positive angular frequency, which gives an overall characterization of the phase dynamics.

In experiments and numerical simulations, Ω can be can be easily obtained from $\varphi(t)$ both, as the slope of a least squares fit to a straight line, and as the absolute value of the time average of $d\varphi/dt$. When only a scalar observable of the system, *s*(*t*), is available a useful approach (Pikovsky et al., 1997a) is to resort to the return map of the system to determine the time intervals between all consecutive maxima or minima, calculate its time average value, $\langle T \rangle$, and then the characteristic frequency $\Omega = 2\pi / \langle T \rangle$.

Phase synchronization may appear when a chaotic oscillator is driven by a weak external periodic force of frequency $\omega_{0,}$ which will be given by the set of *n* ordinary differential equations

$$\frac{dx}{dt} = F(x;p) + Q(t)$$

with $Q(t) = [A_1 \cos(\omega_0 t + \delta_1), A_2 \cos(\omega_0 t + \delta_2), ..., A_n \cos(\omega_0 t + \delta_n)]$ the external force. In this case, depending on the properties of the force applied (frequency, ω_0, amplitudes A_j, and the angles, δ_j), Phase Synchronization may occur as a state where the oscillations stay chaotic, while the phase of the oscillator and the phase of the force (or some of its harmonics) are equal to each other. Phase Synchronization is then defined by means of the function $\psi(t) = \varphi(t) \pm (m / n)\omega_0 t$, with $m, n \in \mathbb{N}$, as the case when there are two numbers, $\varepsilon_1, \varepsilon_2 \in \mathbb{R}$, such that $0 < \varepsilon_2 - \varepsilon_1 < 2\pi$, that verify $\varepsilon_1 < \psi(t) < \varepsilon_2$ for all *t*. From the above ideas on the phase dynamics, it follows that this condition becomes $n\Omega = m\omega_0$ for certain $m, n \in \mathbb{N}$.

We can observe Phase Synchronization in several forms, depending on which of the above approaches to measure the phase is used. If we have the oscillator phase, *φ(t)*, we can compute *ψ(t)* for several appropriate choices of $m, n \in \mathbb{N}$. When there is Phase Synchronization, this function oscillates with an amplitude smaller than 2π. Alternatively, if we resort to return maps to obtain Ω we have to test the condition $n\Omega = m\omega_0$ for appropriate $m, n \in \mathbb{N}$.

The synchronization of the phase has also been studied for non-identical mutually coupled chaotic oscillators (Rosenblum et al. 1996). In this case, Phase Synchronization means that the coupled oscillators dynamics is such that the oscillation amplitudes are uncorrelated, while their phases (or those of their harmonics), stay closer to each other. Given the phases of the two oscillators as functions of time, $\varphi_1(t)$ and $\varphi_2(t)$, the **synchronization of the phase is quantitatively formulated**, by means of the function $\psi(t) = m\varphi_1(t) - n\varphi_2(t)$, with $m, n \in \mathbb{N}$, , as the case when there are two numbers, $\varepsilon_1, \varepsilon_2 \hat{I} R$, such that $0 < \varepsilon_2 - \varepsilon_1 < 2\pi$, that verify $\varepsilon_1 < \psi(t) < \varepsilon_2$ for all *t*. When we only know the oscillator frequencies, Ω_1 and Ω_2, the condition for Phase Synchronization is $m\Omega_1 = n\Omega_2$ for $m, n \in \mathbb{N}$. In this case, Phase Synchronization can be observed in analogous ways as in the case of periodic driving.

The study of Phase Synchronization in Fourier space has proven useful to obtain a qualitative image of the mechanism behind Phase Synchronization. Given a scalar output variable, *s(t)*, the quantity of interest is the one-sided power spectral density

$$P(\omega) = 2\left|\int_0^\infty s(t)\exp(i\omega t)\,dt\right|.$$

Numerical studies performed on several chaotic oscillators for the phase synchronized case when $m = n = 1$ (González-Miranda, 2002a, González-Miranda, 2002b) have show essentially the same mechanism for both externally forced and mutually coupled oscillators. A periodic force of frequency ω_0 applied to a chaotic oscillator modifies $P(\omega)$ in two ways. For one side, new peaks develop for $\omega = \omega_0$, together with the correspondent harmonics; for the other, the spectral peaks that are close to ω_0 shift towards ω_0. The combination of these two mechanisms modifies the structure of $P(\omega)$ as the intensity of the force increases, until the main peak of $P(\omega)$ occurs at ω This coincides with the onset of Phase Synchronization, given by $\Omega = \omega_0$. In mutually coupled chaotic oscillators, the main peak of each oscillator acts on the other as a periodic external force. Phase Synchronization then occurs by a combination of these two effects, which results in the shift of the peaks of each oscillator spectrum to match the position of the corresponding peak of the other. The spectral analysis, then, shows that Phase Synchronization is a process of progressive mutual locking between oscillatory motions in the two systems as the coupling strength increases.

Once Phase Synchronization between two chaotic oscillators has settled down, additional increase of the coupling strength may result in a state of stronger synchronization know as Lag Synchronization (Rosenblum et al., 1997; Zhu & Lai 2001), where the amplitudes of the system variables, $s_1(t)$ and $s_2(t)$, become correlated, although being a time lag, τ, between them. Generalized Time-lagged Synchronization is defined as the case when there is a functional, $\Phi\left[s\left(t\right)\right]$, such that

$$\lim_{t\to\infty}\left|s_1\left(t\right) - \Phi\left[s_2\left(t - \tau\right)\right]\right| = 0,$$

Where τ can be positive or negative, depending on which of the two oscillators is delayed. This form of synchronization is detected by means of the similarity function defined as

$$S\left(T\right) = \sqrt{\frac{\left\langle \left\{s_1\left(t\right) - \Phi\left[s_2\left(t - T\right)\right]\right\}^2 \right\rangle}{\left[\left\langle s_1^2\left(t\right)\right\rangle \left\langle s_2^2\left(t\right)\right\rangle\right]^{1/2}}},$$

where $\left\langle \bullet \right\rangle$ denotes a time average, and $s_1(t)$ and $s_2(t)$ are assumed to have zero mean. $S(T)$ quantifies the degree of synchronization between the two signals as a function of the time shift between them, T. When there is no correlation, $S(T)$ is of the order of one; while, when $s_1(t)$ and $\Phi\left[s_2\left(t - T\right)\right]$ are identical, $S(T)$ becomes null. Therefore, Lag Synchronization with a lag τ is characterized by a minimum of $S(T)$ at $T = \tau$. For coupled oscillators that are nearly identical, we talk of Lag Synchronization. In this case $\Phi[s]$ becomes the identity, and synchronization can also be easily visualized by means of parametric plots of $s_2(t-\tau)$ versus $s_1(t)$.

There is a special form of Lag Synchronization, named Anticipated Synchronization (Voss, 2000), which can be observed when the chaotic oscillators are coupled under a drive-response configuration. This occurs for oscillators described by time-delayed equations (see: Sprott, 2004; González-Miranda, 2004) as a synchronized state in which the actual value of the state variable of the response is the same than the correspondent state variable of the drive is going to take a time τ in the future. Given a time-delayed oscillator with a time delay τ_D, to obtain Anticipated Synchronization using the replacement method, the drive variables, $x^{(d)}(t)$, have to replace to the delayed response variables, $y^{(r)}(t-\tau_D)$. When control techniques are used, the variables to be used in the coupling term are not the actual variables, $x(t)$ and $y(t)$, but the correspondent delayed variables, $x(t-\tau_D)$ and $y(t-\tau_D)$. In these cases, the response anticipation of the state of the drive occurs with a time lag $\tau = \tau_D$.

AMPLITUDE ENVELOPE SYNCHRONIZATION

There is an even weaker form of synchronization named Amplitude Envelope Synchronization. This has been mainly studied when two structurally equal and parametrically different chaotic oscillators are mutually coupled. It develops gradually as the coupling strength increases from zero and it may be observable for coherent chaotic oscillators that are very weakly connected. In this form of synchronization, the phases and amplitudes of the dynamical observables of the oscillators remain uncorrelated; but the amplitudes of the oscillations have envelopes that evolve in synchrony (González-Miranda, 2002a).

Let $s(t)$ be a dynamical observable of a chaotic oscillator, $x = F(x;p)$, in particular it can be one of its variables $x_i(t)$. Its lower envelope is the smoothest curve, $L(t)$, which passes through all the minima of $s(t)$. From a sample of $s(t)$, given in a time interval, $t \in \left[t_I, t_F\right]$ we obtain estimates of $L(t)$ to be used in practice by recording all the relative minima of $s(t)$. Because these will occur at discrete times, $\tau_j \in \left[t_I, t_F\right], j = 1, 2, \ldots, N,$ we have the envelope determined by a discrete function, L_j. The upper envelope of $s(t)$, U_j, is defined and treated in the same way.

Given two coupled oscillators, $\dot{x}^{(1)} = F\left(x^{(1)}; p^{(1)}\right)$ and $\dot{x}^{(2)} = F\left(x^{(2)}; p^{(2)}\right)$, Amplitude Envelope Synchronization is the development of correlated envelopes of the oscillations of the dynamical observables of the two oscillators, $x^{(1)}(t)$ and $x^{(2)}(t)$.

We can see the mechanism for Amplitude Envelope Synchronization by studying the dynamics in Fourier space. This is a mutual excitation of a new low frequency, with its harmonics, in the dynamics of the coupled oscillators. The value of the new frequency is equal to the difference between the dominant frequencies of the oscillators. The oscillation associated to this common frequency appears as the envelope of the chaotic oscillations of each system, and is the part of the dynamics that is synchronized.

We will illustrate these ideas with numerical results for the following two oscillators, each of the type of an electric circuit studied by Gomes & King (1992), that are diffusively coupled oscillators through their $y^{(1)}$ and $y^{(2)}$ variables

$$dx^{(1,2)} / dt = -\alpha^{(1,2)} \cdot \left[\left(x^{(1,2)}\right)^3 - \beta^{(1,2)} x^{(1,2)} - y^{(1,2)}\right],$$

$$dy^{(1,2)} / dt = x^{(1,2)} - y^{(1,2)} + z^{(1,2)} + C \cdot \left(y^{(2,1)} - y^{(1,2)}\right),$$

$$dz^{(1,2)} / dt = -\gamma^{(1,2)} y^{(1,2)},$$

with $\alpha^{(1)} = \alpha^{(2)} = 100$, $\beta^{(1)} = \beta^{(2)} = 0.35$, $\gamma^{(1)} = 610$ and $\gamma^{(2)} = 650$. The constant C measures the strength of the coupling. The development of Amplitude Envelope Synchronization can be seen in Figure 1 and Figure 2, which present plots of time series for $x^{(1)}(t)$ and $x^{(2)}(t)$, their lower envelopes, $L_j^{(1)}$ and $L_k^{(2)}$, and their power spectral densities, $P^{(1)}(\omega)$ and $P^{(2)}(\omega)$, for the independent oscillators, $C = 0$, and for the oscillators under a weak coupling, $C = 0.35$, respectively.

The results in Figure 1, for the independent oscillators, show chaotic time series and lower envelopes that are uncorrelated. The power spectral densities present a marked main peak and harmonics over a broadband noise, which is the characteristic structure of coherent chaotic oscillations. The structure of this noise is different in each oscillator, and the dominant frequencies differ in an amount of the order of $\Delta\omega \approx 0.62$. There is no sign of structure in $P^{(1)}(\omega)$ and $P^{(2)}(\omega)$ for small frequencies, of the order of the difference of frequencies between the main peaks.

In Figure 2, we see a case where Amplitude Envelope Synchronization has developed. This is noticeable by looking to the time series (Figure 2(a)), especially if one pays attention to the lower envelopes (Figure 2(b)). The changes experienced by the power spectral density illustrate the mechanism for the appearance of Amplitude Envelope Synchronization. In Figure 2(c), we see that the main peaks remain nearly the same, which tells us that there is not Phase, nor Generalized Synchronization. The broadband noise like structure persists; therefore, we are still in a regime of chaotic dynamics. The main change occurs at low frequencies where a new structure consisting in a new peak of frequency, much smaller than that of the main peaks, and equal to their difference has developed (Figure 2(d)), together with its harmonics. This common new frequency appears in the time series of the observables as an oscillation of a long period, which is the same for the two oscillators, and can be visualized as a common envelope of the low period oscillations associated to the main peak of each oscillator. This is Amplitude Envelope Synchronization.

Figure 1. (a) Time series for the signal $x^{(1)}(t)$ and $x^{(2)}(t)$, (b) plots of the estimates of the lower envelopes, $L_j^{(1)}$ (full circles) and $L_k^{(2)}$ (hollow circles), with the envelope of $x^{(2)}(t)$ displaced down to ease comparison, and the power spectral densities for $x^{(1)}$ (thick line) and $x^{(2)}$ (thin line) in ranges of frequencies, ω, which (c) include the main peaks and (d) show the region of very small frequencies. All these for a coupling strength, $C = 0$, when the oscillators are uncoupled.

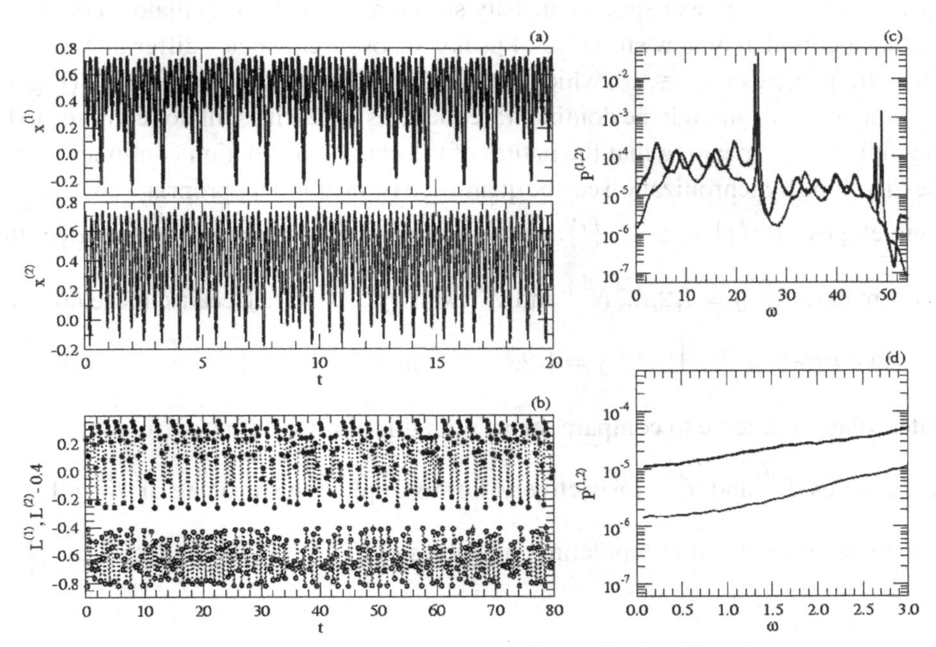

Figure 2. Same than Figure 1 for a coupling strength, C = 0.35, when Amplitude Envelope Synchronization has developed

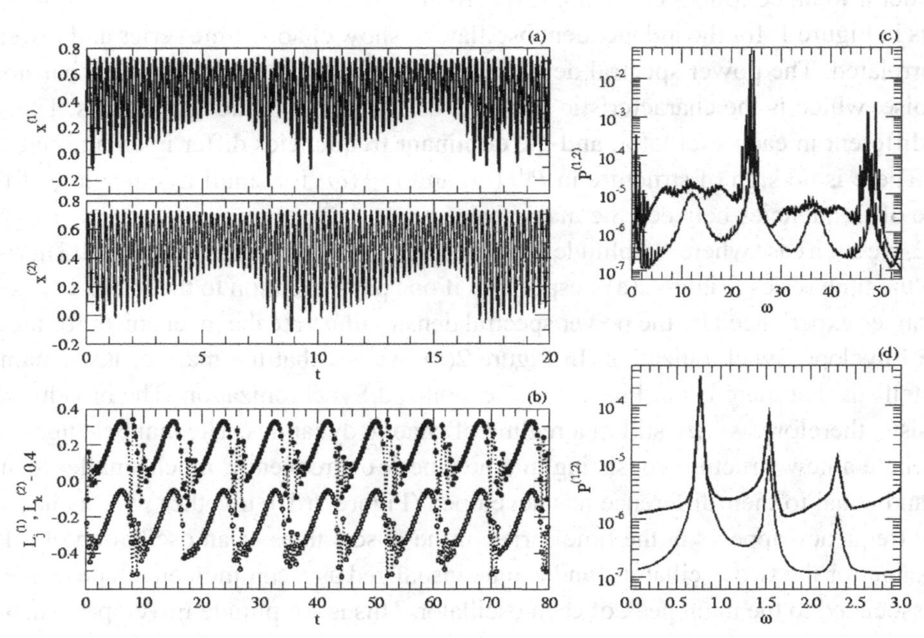

This spectral analysis shows the mechanism for Amplitude Envelope Synchronization. Each oscillator excites a new frequency in the other equal to its main frequency (Figure 2(c)). Being the mean frequencies close, this results in a situation where the power spectral densities of each of the two oscillators has two peaks close to each other: its main peak, with its harmonics, and a secondary one, with its harmonics too, induced by the main peak of the other. Increase in coupling strength results in an enhancement of the induced peaks so that the power spectral density structure of the two oscillators changes drastically in such a way that a completely new structure of peaks, whose frequencies differ in $\Delta\omega$, develops. The smallest of these frequencies is $\omega_0 = \Delta\omega$, which is the main frequency of the envelope (Figure 2(d)). As long as the coupling is small enough the dominant frequencies for each oscillator continue to be the same that in the uncoupled case, meaning that the nature of the chaotic oscillation continues to be the same.

Amplitude Envelope Synchronization can be quantified by means of appropriate correlation functions between the envelopes, $L^{(1)}(t)$ and $L^{(2)}(t)$, of the correspondent variables. In practice, this is made through their estimates, $L_j^{(1)}, j = 1,2,..., N^{(1)}$ and $L_k^{(2)}, k = 1,2,..., N^{(2)}$. Because these do not occur necessarily at the same times, $\tau_j^{(1)} \in [t_I, t_F]\, j = 1,2,..., N^{(1)}$ and $\tau_k^{(2)} \in [t_I, t_F]\, k = 1,2,..., N^{(2)}$, we have to resort to an interpolation scheme to compare them. With this aim we define surrogates, $\tilde{L}_k^{(1)}$ and $\tilde{L}_j^{(2)}$, of the discrete time series $L_j^{(1)}$ and $L_k^{(2)}$, respectively. For example, the surrogate of $L_j^{(1)}$ relative to $L_k^{(2)}$ is the set of numbers resulting from interpolating the values of $L_j^{(1)}$ at the times $\tau_k^{(2)} \in [t_I, t_F]$ through

$$\tilde{L}_k^{(1)} = \frac{L_j^{(1)} \cdot \left(\tau_{j+1}^{(1)} - \tau_k^{(2)}\right) + L_{j+1}^{(1)} \cdot \left(\tau_k^{(2)} - \tau_j^{(1)}\right)}{\tau_{j+1}^{(1)} - \tau_j^{(1)}}, k = 1, 2, ..., N^{(2)},$$

with $\tau_j^{(1)} \leq \tau_k^{(2)} < \tau_{j+1}^{(1)}$. We can quantify the correlation between $L^{(1)}(t)$ and $L^{(2)}(t)$ by an estimate of the correlation between $\tilde{L}_k^{(1)}$ and $L_k^{(2)}$. González-Miranda (2002a) used the Spearman rank-order correlation coefficient, although other estimators could be used.

MARGINAL SYNCHRONIZATION OF CHAOS

A family of especial forms of synchronization, known as Marginal (or uniform) Synchronization of chaos (González-Miranda, 1996a), appears for identical oscillators coupled in a drive response configuration, when they present some spatial symmetry characterized by an invariance under continuous transformations. Depending on the particular symmetry involved, one may observe a variety of synchronization phenomena that have been described by several authors (González-Miranda, 1996a; Matías et al., 1997; Güémez et al., 1997; González-Miranda, 1998b; Maineri & Rehacek, 1999).

The references in the above paragraph assume a coupling scheme of the replacement of variables type, and a particular class of systems whose set of variables $x = \left(x_1, x_2, ..., x_n\right) \in \mathbb{R}^n$ can be decomposed in two subsets, $x = (v,w)$, with $v \in \mathbb{R}^l$, $w \in \mathbb{R}^m$, and $l + m = n$ such that its dynamics is governed by

$$\frac{dv}{dt} = g\left(v, w; p\right),$$
$$\frac{dw}{dt} = h\left(v, w; p\right),$$

with $F = (g, h)$. We say that the flow is continuously symmetric in the subspace $w \in \mathbb{R}^m$ when there is a set of coordinate transformations

$$T_\Pi : \mathbb{R}^m \to \mathbb{R}^m,$$

that transform w in w^*, in such a way that

$$\frac{dw^*}{dt} = h\left(v, w^*; p\right),$$

which means that it leaves the equations of motion invariant. Each of the transformations of the set is defined by numerical values of the parameters, $\Pi = \left(\pi_1, \pi_2, ..., \pi_r\right)$. This set of transformations of coordinates has to verify: (i) the values of the parameters $\pi_i, i = 1,2,\ldots,r,$ change continuously in a given interval, $\pi_i \in (a_i, b_i) \subset \mathbb{R}$ such that $0 \in (a_i, b_i)$, (ii) $T_\Pi(w) \to I$, being I the identity, when $\Pi \to (0,0,\ldots,0)$, and (iii) $\left|T_\Pi(w) - T_{\Pi'}(w)\right| \to 0$ when $\left|\Pi - \Pi'\right| \to (0,0,\ldots,0)$, for all points in the attractor of the chaotic oscillator. This means that T_Π has to be continuous and non-numerable.

We can observe phenomena of Marginal Synchronization of chaos if we use our system as the drive and prepare a response system of the type

$$\frac{dw'}{dt} = h(v, w'),$$

where $w' \in \mathbb{R}^m$ are copies of the ***w*** variables of the drive. The driving variables are the $v \in \mathbb{R}^l$ provided by the drive. In this case, we know that, if the drive is evolving in its chaotic attractor and we set $w'(0) = w(0)$, then $\left|w'(t) - w(t)\right| = 0$ for all $t > 0$. That means that the response evolves in synchrony with the drive. Because of the invariance properties assumed in the equations of motion, if we choose initial conditions $w'(0) = T_\Pi\left[w(0)\right]$ it will happen that $\left|w'(t) - T_\Pi\left[w(t)\right]\right| = 0$ for $t > 0$, which means that the response evolves in a copy of the subsystem ***w***(*t*) that is deformed in a way determined by T_Π.

An interesting, example is the amplitude transformation (González-Miranda, 1996a), $T_\alpha(w) = \alpha w$, that depends on the single parameter $\alpha \in \mathbb{R}$, which multiplies all the coordinates of *w* by a factor of change of scale α. In this case, the response follows a trajectory that reproduces *w*(*t*) but amplified or reduced by a factor α. This particular form of Marginal Synchronization of chaos has been called Projective Synchronization (Maineri & Rehacek, 1999). The Lorenz (1963) model provides a particular example of invariance under an amplitude transformation. It is given by the equations

$$dx / dt = \sigma(y - x),$$
$$dy / dt = x(r - z) - y,$$
$$dz / dt = xy - bz,$$

Being σ, r and b bifurcation parameters. We see that if we define a subsystem whose variables are $w = (x, y) \in \mathbb{R}^2$ the correspondent equations, $dx / dt = \sigma(y - x)$ and $dy / dt = x(r - z)$, are invariant under the transformation $T_\alpha(x, y) = (\alpha x, \alpha y)$.

Projective Synchronization, in the Lorenz model is illustrated in Fig. 3, which has been obtained using the Lorenz equations for the drive and

$$dx'/dt = \sigma(y' - x'),$$
$$dy'/dt = x'(r - z) - y',$$

for the response that is driven by $v = (z) \in \mathbb{R}^1$, and whose variables are $w' = (x', y') \in \mathbb{R}^2$. As an example, in Figure 3(a) and in Figure 3 (b) we present, respectively, the projections onto the x-y plane of a trajectory of the drive and the correspondent trajectory of the response. The initial conditions have been chosen such that $\alpha = 2$; therefore, the trajectory of the response is a copy of the trajectory of the drive amplified by a factor of 2. We demonstrate the synchrony between the two oscillators by means of the parametric plots of $x'(x)$ and $y'(y)$, given in Figures 3 (c,d), which result in straight lines with slopes equal to 2.

We could pose other examples of Marginal Synchronization of chaos different that Projective Synchronization. One case is the Displaced Synchronization (González-Miranda, 1996a; González-Miranda, 1998b), where the response follows a copy of the drive, which is translated from the region of phase space where the attractor exists. Matías et al. (1997) and Güémez et al. (1997) have described others.

Figure 3. Marginal Synchronization of chaos for the Lorenz model: (a) Projection of the drive onto the x-y plane, and (b) plot of the response in the x'-y' plane. (c) Parametric plots of $x^I = x^I(y)$ and $y^I = y^I(y)$, respectively. The parameter values used in this simulation are $\sigma = 10$, $r = 28$ and $b = 8/3$.

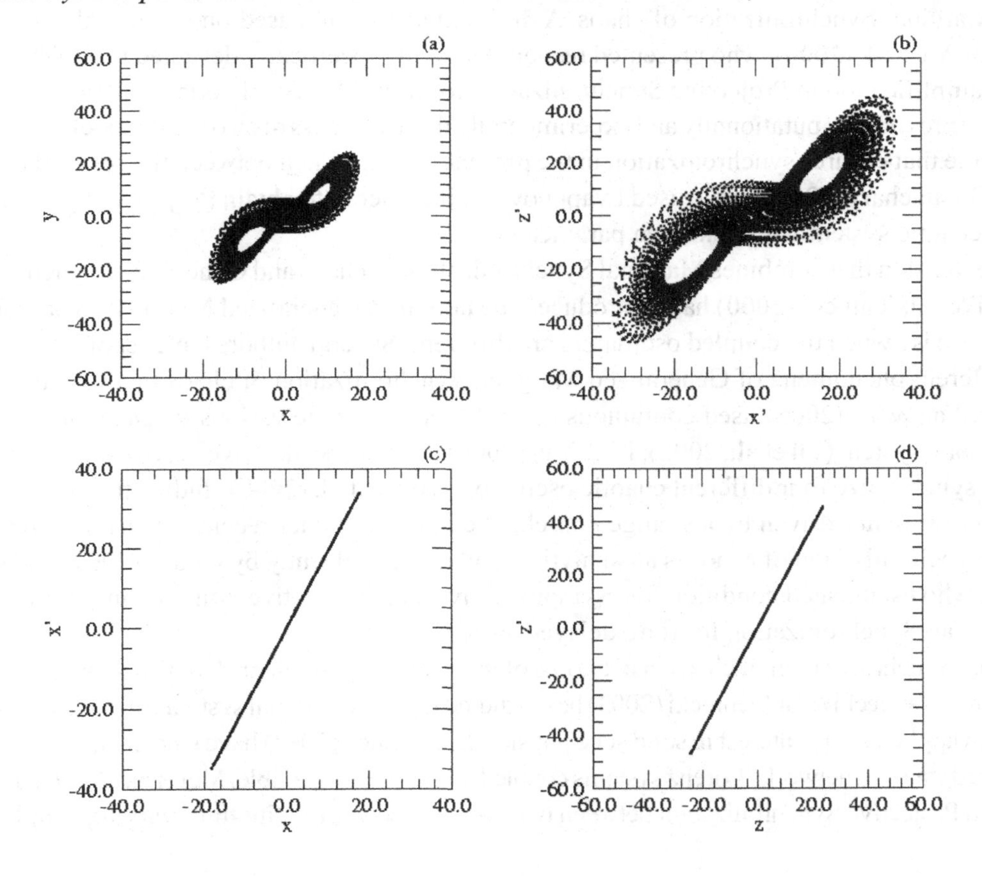

Several authors have studied criteria of occurrence and stability conditions of Marginal Synchronization of chaos (J. M. González-Miranda, 1998a; Krawiecki & Matyjaśkiewicz, 2001; Xu et al., 2002; Xu et al., 2004; Erjaee et al., 2004). Among the results obtained, some show that under a system variable replacement, the particular orbit to be obtained is determined by the initial conditions, and that the existence and continuity of the transformations $w^* = T_\Pi(w)$, implies that the largest conditional Lyapunov exponent for this system has to be null (González-Miranda, 1996a; González-Miranda, 1998a). This means, that close trajectories neither divergence nor convergence. The stability of Marginal Synchronization of chaos is then uniform, also called marginal, in the literature (Szebehely, 1984; Jackson, 1991). Consequently, a perturbation applied to the response evolving in a synchronized state characterized by certain values of the parameters, Π, will send it to another similar synchronized state, characterized by different values of the parameters, with no return to the previous state.

These particular stability properties mean that, under perfect conditions (i.e., a noise free environment for systems with no uncertainty in their parameters), a well-defined synchronization state in time can be obtained, and therefore observed. Under imperfect conditions, the coupled oscillators fluctuate among different states of Marginal Synchronization of chaos; for example, for the case of Projective Synchronization the degree of amplification changes in a series of discrete jumps. Consequently, under noise, or for certain parameter mismatch, a well-defined synchronization state is only observable within a certain time window, whose size varies inversely with the degree of imperfection (González-Miranda, 1998a). To overcome this difficulty the use of control techniques such as continuous control (González-Miranda, 1998b; Shahverdiev, 1999), or the use of tiny control inputs (Xu & Chee, 2002) have been proposed to stabilize Marginal Synchronization of chaos. A different approach, based on stability theory, has been followed by Yu et al. (2006), who presented response systems specifically designed to achieve a desired degree of amplification in Projective Synchronization for four different chaotic oscillators. Grosu et al. (2008) have proved computationally and experimentally the usefulness of an open-loop–closed-loop coupling scheme that ensures synchronization in the presence of mismatch between the coupled oscillators. Ghosh & Bhattacharya (2010) have used Lyapunov stability theory to obtain Projective Synchronization in a hyperchaotic system with unknown parameters.

As an extension that combines Marginal Synchronization of chaos and Generalized Synchronization, Krawiecki & Sukiennicki (2000) have introduced the notion of Generalized Marginal Synchronization, studying the case when the coupled oscillators are different. Several authors have reported the observation of different phenomena of Generalized Marginal Synchronization of chaos in a variety of chaotic oscillators. Yan & Li (2005) used continuous control to synchronize various versions of the so-called unified chaotic system (Lü et al., 2002). Li & Yan (2006) used a cascade drive-response synchronization scheme to synchronize four different chaotic oscillators. Guan et al. (2006) studied the synchronization of a Rössler oscillator driven by a strange non-chaotic attractor. Other recent extension of the ideas of Marginal Synchronization of chaos is an analytical and numerical study by Ghosh (2009) and Ghosh et al. (2010), who established condition for asymptotically stable Projective-Anticipating, Projective, and Projective-Lag Synchronization for time-delayed systems.

Marginal Synchronization of chaos in networks of many chaotic oscillators has also received attention in the literature. Krawiecki & Sukiennicki (2003) have studied high dimensional systems of excited pairs of spin waves, in an application of interest in solid state physics. Liu & Chen (2008) have studied the unbounded and synchronized states of networks Lorenz systems coupled through the y variable. Mei et al. (2009) have studied Generalized Projective Synchronization between two complex networks with time-varying coupling delay.

COMMENTS AND CONCLUSION

For about a quarter of a century, since the theoretical work of Fujisaka & Yamada (1983) on mutually coupled chaotic oscillators, and mainly after the theoretical and experimental work of Pecora & Carrol (1990) on driven chaotic oscillators, synchronization of chaos has been studied intensively. Several different forms of synchronization have been described, as reviewed in this chapter, as well as different forms of coupling have been proven to lead to one or another form of synchronization at least theoretically and computationally. Many experiments have been also performed to demonstrate chaos synchronization in the real world (see, for example, González-Miranda, 2004).

All this fundamental research constitutes nowadays a vast body of knowledge on synchronization of chaos (Pikovsky et al., 2001; González-Miranda, 2004). Although not all is known on the fundamental physics and mathematics of this interesting nonlinear phenomenon, we have a large body of knowledge that is, by far, more than enough to undertake serious applied research focused to make chaos synchronization useful (Gauthier, 1998; Roy, 2005). Among other scientific and technical applications, since the first works on chaos communications by Cuomo & Oppenheim (1993a), an special topic within chaos synchronization has emerged focused to find useful application of the above fundamental results in the engineering of telecommunication, mainly cryptography, which appears promising and fruitful (see for example: Van Wiggeren & Roy, 1998; Argyris et al., 2005).

REFERENCES

Abarbanel, H. D. I., Rulkov, N. F., & Sushchik, M. M. (1996). Generalized synchronization of chaos: The auxiliary system approach. *Physical Review E: Statistical Physics, Plasmas, Fluids, and Related Interdisciplinary Topics*, *53*, 4528–4535. doi:10.1103/PhysRevE.53.4528

Argyris, A., Syvridis, D., Larger, L., Annovazzi-Lodi, V., Colet, P., & Fischer, I. (2005). Chaos-based communications at high bit rates using commercial fibre-optic links. *Nature*, *438*, 343–346. doi:10.1038/nature04275

Bilotta, E., Di Blasi, G., Stranges, F., & Pantano, P. (2007). A Gallery of Chua Attractors, Part VI. *International Journal of Bifurcation and Chaos in Applied Sciences and Engineering*, *17*, 1801–1910. doi:10.1142/S0218127407018105

Bracikowski, C., & Roy, R. (1991). Chaos in a multimode solid-state laser system. *Chaos (Woodbury, N.Y.)*, *1*, 49–64. doi:10.1063/1.165817

Breakspear, M., & Terry, J. R. (2002). Detection and description of non-linear interdependence in normal multichannel human EEG data. *Clinical Neurophysiology*, *113*, 735–753. doi:10.1016/S1388-2457(02)00051-2

Carroll, T. (1995). A simple system for demonstrating regular and synchronized chaos. *American Journal of Physics*, *63*, 377–379. doi:10.1119/1.17923

Chen, S.-S., Chen, L.-F., Wu, Y.-T., Wu, Y.-Z., Lee, P.-L., Yeh, T.-C., & Hsieh, J.-C. (2007). Detection of synchronization between chaotic signals: An adaptive similarity-based approach. *Physical Review E: Statistical, Nonlinear, and Soft Matter Physics*, *76*, 066208. doi:10.1103/PhysRevE.76.066208

Choon, K. A. (2009). A new chaos synchronization method for Duffing oscillator . *IEICE Electronics Express*, *18*, 1355–1360.

Chua, L. O., Wu, C. W., Huang, A., & Zhong, G.-Q. (1993). A universal circuit for studying and generating chaos - Part II. Strange Attractors. *IEEE Transactions on Circuits and Systems*, *CAS-40*, 745–761.

Cuomo, K. M., & Oppenheim, A. V. (1993a). Circuit implementation of synchronized chaos with application to communications. *Physical Review Letters*, *71*, 65–68. doi:10.1103/PhysRevLett.71.65

Cuomo, K. M., & Oppenheim, A. V. (1993b). Synchronization of Lorenz-based chaotic circuits with applications to communications. *IEEE Transactions on Circuits and Systems*, *CAS-40*, 626–633.

Erjaee, G. H., Atabakzade, M. H., & Saha, L. M. (2004). Interesting synchronization-like behavior. *International Journal of Bifurcation and Chaos in Applied Sciences and Engineering*, *14*, 1447–1453. doi:10.1142/S0218127404009934

Fujisaka, H., & Yamada, T. (1983). Stability theory of synchronized motion in coupled-oscillator systems. *Progress of Theoretical Physics*, *69*, 32–47. doi:10.1143/PTP.69.32

Gauthier, D. J. (1998). Chaos Has Come Again. *Science*, *279*, 1156–1157. doi:10.1126/science.279.5354.1156

Ghosh, D. (2009). Generalized projective synchronization in time-delayed systems: Nonlinear observer approach. *Chaos (Woodbury, N.Y.)*, *19*, 013102. doi:10.1063/1.3054711

Ghosh, D., Banerjee, S., & Chowdhury, A. R. (2007). Synchronization between variable time-delayed systems and cryptography. *EPL*, *80*, 30006. doi:10.1209/0295-5075/80/30006

Ghosh, D., Banerjee, S., & Chowdhury, A. R. (2010). Generalized and projective synchronization in modulated time-delayed systems. *Physics Letters. [Part A]*, *374*, 2143–2149. doi:10.1016/j.physleta.2010.03.027

Ghosh, D., & Bhattacharya, S. (2010). (in press). Projective synchronization of new hyperchaotic system with fully unknown parameters. *Nonlinear Dynamics*, *60*.

Gomes, M. G. M., & King, G. P. (1992). Bistable chaos. II. Bifurcation analysis. *Physical Review A.*, *46*, 3100. doi:10.1103/PhysRevA.46.3100

González-Miranda, J. M. (1996a). Chaotic systems with a null conditional Lyapunov exponent under nonlinear driving. *Physical Review E: Statistical Physics, Plasmas, Fluids, and Related Interdisciplinary Topics*, *53*, R5–R8. doi:10.1103/PhysRevE.53.R5

González-Miranda, J. M. (1996b). Synchronization of symmetric chaotic systems. *Physical Review E: Statistical Physics, Plasmas, Fluids, and Related Interdisciplinary Topics*, *53*, 5656–5669. doi:10.1103/PhysRevE.53.5656

González-Miranda, J. M. (1998a). Amplification and displacement of chaotic attractors by means of unidirectional chaotic driving. *Physical Review E: Statistical Physics, Plasmas, Fluids, and Related Interdisciplinary Topics*, *57*, 7321–7324. doi:10.1103/PhysRevE.57.7321

González-Miranda, J. M. (1998b). Using continuous control for amplification and displacement of chaotic signals. *The European Physical Journal B*, *6*, 411–418. doi:10.1007/s100510050568

González-Miranda, J. M. (2002a). Amplitude envelope synchronization in coupled chaotic oscillators. *Physical Review E: Statistical, Nonlinear, and Soft Matter Physics*, *65*, 036232. doi:10.1103/PhysRevE.65.036232

González-Miranda, J. M. (2002b). Phase synchronization and chaos suppression in a set of two coupled nonlinear oscillators. *International Journal of Bifurcation and Chaos in Applied Sciences and Engineering*, *12*, 2105–2112. doi:10.1142/S0218127402005716

González-Miranda, J. M. (2002c). Generalized synchronization in directionally coupled systems with identical individual dynamics. *Physical Review E: Statistical, Nonlinear, and Soft Matter Physics*, *65*, 047202. doi:10.1103/PhysRevE.65.047202

González-Miranda, J. M. (2004). *Synchronization and Control of Chaos*. London, UK: Imperial College Press. doi:10.1142/9781860945229

Grosu, I., Padmanaban, E., Roy, P. K., & Dana, S. K. (2008). Designing Coupling for Synchronization and Amplification of Chaos. *Physical Review Letters*, *100*, 234102. doi:10.1103/PhysRevLett.100.234102

Guan, S., Wan, X., & Lai, C.-H. (2006). Frequency locking by external force from a dynamical system with strange nonchaotic attractor. *Physics Letters. [Part A]*, 298–304. doi:10.1016/j.physleta.2006.01.067

Güémez, J., Martín, C., & Matías, M. A. (1997). Approach to the chaotic synchronized state of some driving methods. *Physical Review E: Statistical Physics, Plasmas, Fluids, and Related Interdisciplinary Topics*, *55*, 122–134. doi:10.1103/PhysRevE.55.124

Harrison, R. G., & Biswas, D. J. (1986). Chaos in Light. *Nature*, *321*, 394–401. doi:10.1038/321394a0

He, R., & Vaidya, P. G. (1992). Analysis and synthesis of synchronous periodic and chaotic systems. *Physical Review A.*, *46*, 7387. doi:10.1103/PhysRevA.46.7387

He, R., & Vaidya, P. G. (1999). Time delayed chaotic systems and their synchronization. *Physical Review E: Statistical Physics, Plasmas, Fluids, and Related Interdisciplinary Topics*, *59*, 4048. doi:10.1103/PhysRevE.59.4048

Hramov, A. E., & Koronovskii, A. A. (2005). Generalized synchronization: A modified system approach. *Physical Review E: Statistical, Nonlinear, and Soft Matter Physics*, *71*, 067201. doi:10.1103/PhysRevE.71.067201

Jackson, E. A. (1991). *Perspectives of Nonlinear Dynamics*. New York: Cambridge University Press.

Jovic, B., Berber, S., & Unsworth, C. P. (2006). A novel mathematical analysis for predicting master–slave synchronization for the simplest quadratic chaotic flow and Ueda chaotic system with application to communications. *Physica D. Nonlinear Phenomena*, *213*, 31. doi:10.1016/j.physd.2005.10.013

Kocarev, L., & Parlitz, U. (1996). Generalized synchronization, predictability, and equivalence of unidirectionally coupled dynamical systems. *Physical Review Letters*, *76*, 1816–1819. doi:10.1103/PhysRevLett.76.1816

Krawiecki, A., & Matyjaśkiewicz, S. (2001). Blowout bifurcation and stability of marginal synchronization of chaos. *Physical Review E: Statistical, Nonlinear, and Soft Matter Physics*, *64*, 036216. doi:10.1103/PhysRevE.64.036216

Krawiecki, A., & Sukiennicki, A. (2000). Generalizations of the concept of marginal synchronization of chaos. *Chaos, Solitons, and Fractals*, *11*, 1445–1458. doi:10.1016/S0960-0779(99)00062-4

Krawiecki, A., & Sukiennicki, A. (2003). Marginal synchronization of spin-wave amplitudes in a model for chaos in parallel pumping. *Physica Status Solidi. B, Basic Research*, *236*, 511–514. doi:10.1002/pssb.200301716

Li, C., & Yan, J. (2006). Generalized projective synchronization of chaos: The cascade synchronization approach. *Chaos, Solitons, and Fractals*, *30*, 140–146. doi:10.1016/j.chaos.2005.08.155

Liu, X., & Chen, T. (2008). Boundedness and synchronization of y-coupled Lorenz systems with or without controllers. *Physica D. Nonlinear Phenomena*, *237*, 630–639. doi:10.1016/j.physd.2007.10.006

Lorenz, E. N. (1963). Deterministic nonperiodic flow. *Journal of the Atmospheric Sciences*, *20*, 130–141. doi:10.1175/1520-0469(1963)020<0130:DNF>2.0.CO;2

Lü, J., Chen, G., Cheng, D., & Celikovsky, S. (2002). Bridge the gap between the Lorenz and the Chen system. *International Journal of Bifurcation and Chaos in Applied Sciences and Engineering*, *12*, 2917–2926. doi:10.1142/S021812740200631X

Maineri, R., & Rehacek, J. (1999). Projective synchronization in three-dimensional chaotic systems. *Physical Review Letters*, *64*, 3042–3045. doi:10.1103/PhysRevLett.82.3042

Matías, M. A., Güémez, J., & Martín, C. (1997). On the behavior of coupled chaotic systems exhibiting marginal synchronization. *Physics Letters. [Part A]*, *226*, 264–268. doi:10.1016/S0375-9601(96)00946-2

Matsumoto, T., Chua, L. O., & Komuro, M. (1985). The Double Scroll. *IEEE Transactions on Circuits and Systems*, *CAS-32*, 798–818.

Mei, S., Chang-Yan, Z., & Li-Xin, T. (2009). Generalized Projective Synchronization between Two Complex Networks with Time-Varying Coupling Delay. *Chinese Physics Letters*, *26*, 010501. doi:10.1088/0256-307X/26/1/010501

Midavaine, T., Dangoisse, D., & Glorieux, P. (1985). Observation of Chaos in a Frequency-Modulated CO_2 Laser. *Physical Review Letters*, *55*, 1989–1992. doi:10.1103/PhysRevLett.55.1989

Mukai, T., & Otsuka, K. (1985). New route to optical chaos: Successive-subharmonic-oscil- lation cascade in a semiconductor laser coupled to an external cavity. *Physical Review Letters*, *55*, 1711–1714. doi:10.1103/PhysRevLett.55.1711

Naghavi, S. V., & Safavi, A. A. (2008). Novel synchronization of discrete-time chaotic systems using neural network observer. *Chaos (Woodbury, N.Y.)*, *18*, 033110. doi:10.1063/1.2959140

Pecora, L. M., & Carroll, T. L. (1990). Synchronization in chaotic systems. *Physical Review Letters*, *64*, 821–824. doi:10.1103/PhysRevLett.64.821

Pecora, L. M., Carroll, T. L., Johnson, G. A., Mar, D. J., & Heagy, J. F. (1997). Fundamentals of synchronization in chaotic systems, concepts, and applications. *Chaos (Woodbury, N.Y.)*, *7*, 520–543. doi:10.1063/1.166278

Pikovsky, A., Rosemblum, M., & Kurths, J. (2001). *Synchronization: A Universal Concept in Nonlinear Sciences*. Cambridge, UK: Cambridge University Press.

Pikovsky, A., Rosenblum, M., & Kurths, J. (2000). Phase synchronization in regular and chaotic systems. *International Journal of Bifurcation and Chaos in Applied Sciences and Engineering*, *10*, 2291–2305.

Pikovsky, A. S., Rosenblum, M. G., Osipov, G. V., & Kurths, J. (1997a). Phase synchronization of chaotic oscillators by external driving. *Physica D. Nonlinear Phenomena*, *104*, 219–238. doi:10.1016/S0167-2789(96)00301-6

Porfiri, M., & Fiorilli, F. (2010). Experiments on node-to-node pinning control of Chua's circuits. *Physica D. Nonlinear Phenomena*, *239*, 454–464. doi:10.1016/j.physd.2010.01.012

Pyragas, K. (1998). Synchronization of coupled time-delay systems: Analytical estimations. *Physical Review E: Statistical Physics, Plasmas, Fluids, and Related Interdisciplinary Topics*, *58*, 3067–3071. doi:10.1103/PhysRevE.58.3067

Rosenblum, M. G., Pikovsky, A. S., & Kurths, J. (1996). Phase Synchronization of Chaotic Oscillators. *Physical Review Letters*, *76*, 1804–1807. doi:10.1103/PhysRevLett.76.1804

Rosenblum, M. G., Pikovsky, A. S., & Kurths, J. (1997). From phase to lag synchronization in coupled chaotic oscillators. *Physical Review Letters*, *78*, 4193–4196. doi:10.1103/PhysRevLett.78.4193

Rössler, O. E. (1976). An equation for continuous chaos. *Physics Letters. [Part A]*, *57*, 397–398. doi:10.1016/0375-9601(76)90101-8

Roy, R. (2005). Communications technology: Chaos down the line. *Nature*, *438*, 298–299. doi:10.1038/438298b

Rulkov, N. F., Sushchijk, M. M., Tsimring, L. S., & Abarbanel, H. D. I. (1995). Generalized synchronization of chaos in directionally coupled chaotic systems. *Physical Review E: Statistical Physics, Plasmas, Fluids, and Related Interdisciplinary Topics*, *51*, 980–994. doi:10.1103/PhysRevE.51.980

Schiff, S. J., So, P., Chang, T., Burke, R. E., & Sauer, T. (1996). Detecting dynamical interdependence and generalized synchrony through mutual prediction in a neural ensemble. *Physical Review E: Statistical Physics, Plasmas, Fluids, and Related Interdisciplinary Topics*, *54*, 6708–6724. doi:10.1103/PhysRevE.54.6708

Shahverdiev, E. M.-O. (1999). Boundedness of dynamical systems and chaos synchronization. *Physical Review E: Statistical Physics, Plasmas, Fluids, and Related Interdisciplinary Topics*, *60*, 3905–3909. doi:10.1103/PhysRevE.60.3905

Sprott, J. C. (2004). *Chaos and time series analysis*. Oxford, UK: Oxford University Press.

Stone, E. F. (1992). Frequency entrainment of a phase coherent attractor. *Physics Letters. [Part A]*, *163*, 367–374. doi:10.1016/0375-9601(92)90841-9

Suetani, H., Iba, Y., & Aihara, K. (2006). Detecting generalized synchronization between chaotic signals: a kernel-based approach. *Journal of Physics. A, Mathematical and General*, *39*, 10723–10742. doi:10.1088/0305-4470/39/34/009

Szebehely, V. (1984). Review of the concept of stability. *Celestial. Mechanical Engineering (New York, N.Y.)*, *34*, 49–64.

Van Wiggeren, G. D., & Roy, R. (1998). Communication with Chaotic Lasers. *Science*, *279*, 1198–1200. doi:10.1126/science.279.5354.1198

Van Wiggeren, G. D., & Roy, R. (1998). Communication with Chaotic Lasers. *Science*, *279*, 1198–1200. doi:10.1126/science.279.5354.1198

Voss, H. U. (2000). Anticipating chaotic synchronization. *Physical Review E: Statistical Physics, Plasmas, Fluids, and Related Interdisciplinary Topics*, *61*, 5115–5119. doi:10.1103/PhysRevE.61.5115

Xu, D., & Chee, C. Y. (2002). Controlling the ultimate state of projective synchronization in chaotic systems of arbitrary dimension. *Physical Review E: Statistical, Nonlinear, and Soft Matter Physics*, *66*, 046218. doi:10.1103/PhysRevE.66.046218

Xu, D., Chee, C. Y., & Li, C. (2004). A necessary condition of projective synchronization in discrete-time systems of arbitrary dimensions. *Chaos, Solitons, and Fractals*, *22*, 175–180. doi:10.1016/j.chaos.2004.01.012

Xu, D., Ong, W. L., & Li, Z. (2002). Criteria for the occurrence of projective synchronization in chaotic systems of arbitrary dimension. *Physics Letters. [Part A]*, *305*, 167–172. doi:10.1016/S0375-9601(02)01445-7

Yalçınkaya, T., & Lai, Y.-C. (1997). Phase Characterization of Chaos. *Physical Review Letters*, *79*, 3885–3888. doi:10.1103/PhysRevLett.79.3885

Yan, J., & Li, C. (2005). Generalized projective synchronization of a unified chaotic system. *Chaos, Solitons, and Fractals*, *26*, 1119–1124. doi:10.1016/j.chaos.2005.02.034

Yu, H., Peng, J., & Liu, Y. (2006). Projective synchronization of unidentical chaotic systems based on stability criterion. *International Journal of Bifurcation and Chaos in Applied Sciences and Engineering*, *16*, 1049–1056. doi:10.1142/S0218127406015301

Zhu, L., & Lai, Y.-C. (2001). Experimental observation of generalized time-lagged chaotic synchronization. *Physical Review E: Statistical, Nonlinear, and Soft Matter Physics*, *64*, 045205. doi:10.1103/PhysRevE.64.045205

Chapter 7
Synchronization in Integer and Fractional Order Chaotic Systems

Ahmed E. Matouk
Mansoura University, Egypt & Hail University, Saudi Arabia

ABSTRACT

In this chapter, the author introduces the basic methods of chaos synchronization in integer order systems, such as Pecora and Carroll method and One-Way coupling technique, applying these synchronization methods to the modified autonomous Duffing-Van der Pol system (MADVP). The conditional Lyapunov exponents (CLEs) are also calculated for the drive and response MADVP systems which match with the analytical results given by Pecora and Carroll method. Based on Lyapunov stability theory, chaos synchronization is achieved for two coupled MADVP systems by finding a suitable Lyapunov function. Moreover, synchronization in fractional order chaotic systems is also introduced. The conditions of Pecora and Carroll method and One-Way coupling method in fractional order systems are also investigated. In addition, chaos synchronization is achieved for two coupled fractional order MADVP systems using One-Way coupling technique. Furthermore, synchronization between two different fractional order chaotic systems is studied; the fractional order Lü system is controlled to be the fractional order Chen system. The analytical conditions for the synchronization of this pair of different fractional order chaotic systems are derived by utilizing the Laplace transform theory. Numerical simulations are carried out to show the effectiveness of all the proposed synchronization techniques.

Synchronization in chaotic systems has particular interest in the past few years; it has many applications, especially in secure communications (Carroll & Pecora, 1991; Chua et al, 1992; Ogorzalek, 1993; Chen, 1997; Bai & Lonngren, 1997).

Fujisaka and Yamada noticed that 'by coupling together oscillators which on their own evolved chaotically, it was possible under certain circumstances to force them to evolve in an identical fashion,'

DOI: 10.4018/978-1-61520-737-4.ch007

(Fujisaka & Yamada, 1983; Yamada & Fujisaka, 1983). They introduced criterion requires that the largest eigenvalue of the Jacobian matrix corresponding to the flow evaluated on the synchronization manifold be negative. Afterwards, He and Vaidya (He & Vaidya, 1992) developed a criterion for chaos synchronization based on the notion of asymptotic stability of dynamical systems. So finding suitable Lyapunov function is one highly useful technique of establishing asymptotic stability of the response subsystem.

In 1990, Louis Pecora and Tom Carroll made an important breakthrough of chaos synchronization (Pecora & Carroll, 1990; Pecora & Carroll, 1991), they showed that there exists the class of chaotic systems for which synchronization can be achieved. Consider a system that can be divided into the drive subsystem (whose largest Lyapunov exponent is positive) and the driven subsystem (with all negative Lyapunov exponents). In this case trajectories from two identical driven subsystems can be synchronized if the same drive system is used. They also indicated that by using chaotic synchronization it might be possible to communicate in a secure manner by using the chaotic signal as a mask with which to hide the message to be sent. If the receiver could synchronize their system to the chaotic signal of the sender then they would be able to remove the mask and the message would be seen. There are so many important applications for synchronization (Pecora et al, 1997), for example the synchronized chaotic systems provide a rich mechanism for signal design and communication applications (Cuomo & Oppenheim, 1993; Kocarev & Parlitz, 1995). It has also potential applications in physical, chemical and biological systems (Uchida et al, 2003; Li et al, 2003; Blasius et al, 1999). Consequently, our prime interest in this chapter is to study and investigate chaos synchronization. After Pecora and Carroll, many effective methods have been presented for synchronizing identical chaotic systems like One-Way coupling method (Lakshmanan & Murali, 1996), adaptive control (Liao & Lin, 1999; Hegazi et al, 2002; Hegazi et al, 2001; Agiza & Matouk, 2006), active control (Bai & Lonngren, 1997), simple global synchronization technique (Jiang et al, 2003), lag synchronization (Taherionl & Lai, 1999), feedback synchronization (Matouk, 2008) and backstepping design approach (Matouk & Agiza, 2008). Chaos synchronization has many types like frequency synchronization (FS), phase synchronization (PS), generalized synchronization (GS), and identical synchronization (IS).

Throughout this chapter we will apply some of the basic synchronization methods to some integer and fractional order chaotic systems.

1. SYNCHRONIZATION IN INTEGER ORDER CHAOTIC SYSTEMS

1.1 Pecora and Carroll Method

Let us consider the system

$$\dot{X} = f[X(t)], \qquad X = (x_1, x_2, \ldots, x_n)^T, \qquad \cdot \equiv \frac{d}{dt} \tag{1}$$

By dividing system (1) into two parts arbitrarily as

$$X = (x_D, x_R)^T,$$

Where x_D are the drive subsystem variables, $x_D = (x_1, ..., x_l)^T$, and x_R are the response subsystem variables, $x_R = (x_1, ..., x_m)^T$. Now, equation (1) is written as

$$\begin{aligned} \dot{x}_D &= g(x_D, x_R), \\ \dot{x}_R &= h(x_D, x_R), \end{aligned} \tag{2}$$

where $g = (f_1(X), ..., f_l(X))^T$, $h = (f_{l+1}(X), ..., f_n(X))^T$, and $l + m = n$.

If we make a copy of the response subsystem called x'_R which driven by the variables x_D (from the original system), it follows that; the system

$$\begin{aligned} \dot{x}_D &= g(x_D, x_R), \\ \dot{x}_R &= h(x_D, x_R), \end{aligned} \tag{3}$$

drives the response system

$$\dot{x}'_R = h(x_D, x'_R). \tag{4}$$

Following the suggestion of Pecora and Carroll, as the variation of the time the x'_R variables will converge asymptotically to the x_R variables and remain with it in step instantaneously. Our goal is to make the response system a stable subsystem in order to obtain synchronization. Suppose we have trajectory $x'_R(t)$ with initial condition $x'_R(0)$ then if a trajectory started at a nearby point $x''_R(t)$ at t = 0, then the conditions needed for synchronization implies that

$$\Delta x'_R(t) = \left| x''_R(t) - x'_R(t) \right| = 0,$$

This leads to

$$\begin{aligned} \Delta \dot{x}'_R(t) &= h(x_D, x''_R) - h(x_D, x'_R) \\ &= D_{x'_R} h(x_D, x'_R) \Delta x'_R + o(x_D, x'_R), \end{aligned} \tag{5}$$

where $D_{x'_R} f$ is the Jacobian of the response vector field with respect to the response variables and $o(x'_R)$ represents the higher order terms. For small $\Delta x'_R$ equation (5) becomes:

$$\Delta \dot{x}'_R(t) = D_{x'_R} h(x_D, x'_R) x'_R. \tag{6}$$

In Ref. (10), Pecora and Carroll used a matrix Z in place of x'_R in Eqn. (6), such that Z(0) is equal to identity matrix. Thus

$$\dot{\Delta} Z = D_{x'_R} h(x_D, x'_R) Z. \tag{7}$$

The solution Z(t) is called the *transfer function* or the *principal matrix solution* (Pecora & Carroll, 1991). Z(t) will determine whether perturbations will grow or shrink in any particular directions. Using the principal matrix solution Z(t) as $t \to \infty$ we can determine the Lyapunov exponents for x'_R subsystem for a particular drive trajectory x_D, which help us to estimate the average convergence rates and their associated directions.

Now, calculating Z(t) for large t from the variational equation (7), the Lyapunov exponents are given by:

$$\lambda_i = \lim_{t \to \infty} \frac{1}{t} \ln \nu_i, \tag{8}$$

where ν_i are the eigenvalues of Z(t), each associated with an eigenvector ξ_i (Pecora & Carroll, 1991).

Definition 1. The Lyapunov exponents for x'_R subsystem (see equation (4)) are called sub Lyapunov exponents or conditional Lyapunov exponents (CLEs).

Now, we have the following theorem:

Theorem 1.(Pecora and Carroll) According to Pecora and Carroll, the coupled chaotic systems can be regarded as drive and response systems which will perfectly synchronize only if CLEs are all negative (Carroll & Pecora, 1991; Pecora & Carroll, 1990).

The above theorem is necessary, but not sufficient condition for synchronization. It says nothing about the set of initial conditions between the drive and response systems.

However, synchronization can be achieved even with positive CLEs (Güémez et al, 1997). **Intermittent synchronization** can occur when CLEs are very small positive or negative values close to zero, while **permanent synchronization** occurs when CLEs take sufficiently large negative values.

Now, suppose that x'_R and x_R are subsystems under perfect synchronization. Hence, if we define $x^*_R = x'_R - x_R$ then $\lim_{t \to \infty} x^*_R(t) = 0$.

Assume that the subsystem is linear, then

$$\begin{aligned} \dot{x}^*_R(t) &= \dot{x}'_R(t) - \dot{x}_R(t) \\ &= h(x_D, x'_R) - h(x_D, x_R) \\ &= J x^*_R, \end{aligned} \tag{9}$$

where J is a (m×m) constant matrix. The real parts of the eigenvalues of the matrix J defined in equation (9), are the conditional Lyapunov exponents (CLEs) we seek.

In the following we are going to apply Pecora and Carroll method to a modified autonomous Duffing-Van der Pol system (Matouk & Agiza, 2008; Matouk, 2005), and known here as MADVP. This system is described by the following equations:

$$\frac{dx}{dt} = -\nu(x^3 - \mu x - y),$$
$$\frac{dy}{dt} = x - \gamma y - z, \tag{10}$$
$$\frac{dz}{dt} = \beta y,$$

The parameters $\beta,\ \nu,\ \gamma$ are all positive real parameters and $\mu \in R$. System (10) exhibits chaotic behaviors at the parameter values $\beta = 200,\ \mu = 0.1,\ \nu = 100$ and $\gamma = 1.6$ (see Figure 1).

1.1.1 Pecora and Carroll Method for MADVP System

The x-drive system is given by equations (10) will be used to drive the following response system:

$$\dot{y}' = x - \gamma y' - z', \tag{11}$$
$$\dot{z}' = \beta y',$$

Figure 1. Phase portrait of system (10), showing the double-band chaos, using the parameter values $\beta = 200,\ \mu = 0.1,\ \nu = 100$ *and* $\gamma = 1.6$.

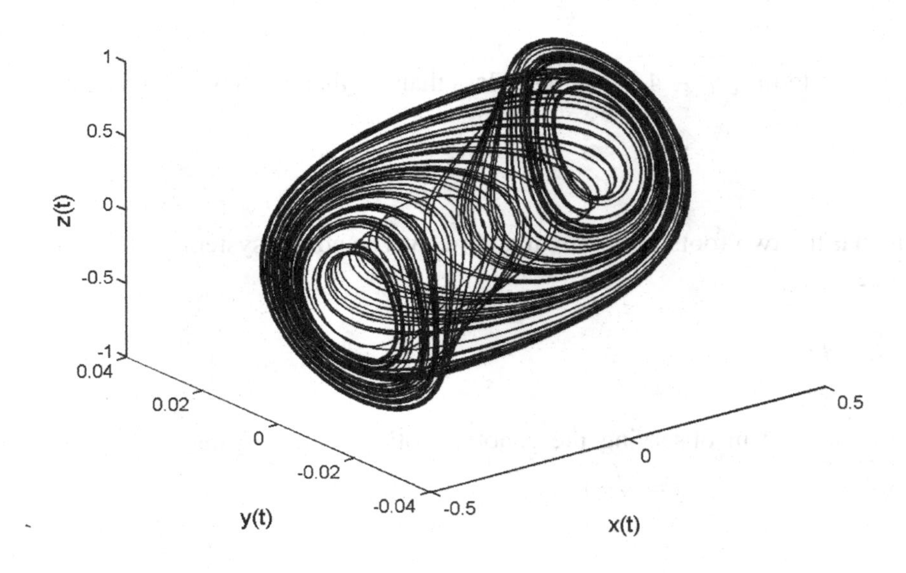

The difference system for $y^* = y - y'$ and $z^* = z - z'$ in matrix form is

$$\begin{bmatrix} \dot{y}^* \\ \dot{z}^* \end{bmatrix} = \begin{bmatrix} -\gamma & -1 \\ \beta & 0 \end{bmatrix} \begin{bmatrix} y^* \\ z^* \end{bmatrix}. \tag{12}$$

Or in vector notation

$$\dot{X}_R^* = JX_R^*,$$

where

$$X_R^* = \begin{bmatrix} y^* \\ z^* \end{bmatrix} \text{ and } J = \begin{bmatrix} -\gamma & -1 \\ \beta & 0 \end{bmatrix}.$$

The real parts of the eigenvalues of the matrix J are the conditional Lyapunov exponents (CLEs). Then the response system synchronizes if all the real parts of the eigenvalues of J are negative.

The characteristic equation of (12) is

$$\lambda^2 + \gamma\lambda + \beta = 0,$$

then the roots are given as follows

$$\lambda_{1,2} = \frac{-\gamma \pm \sqrt{\gamma^2 - 4\beta}}{2}.$$

It is clear that the term $\sqrt{\gamma^2 - 4\beta}$ is always less than γ, then we have two cases;

Case (i) if $\gamma^2 > 4\beta$:

This means that the two roots will be real and negative and the system with x-drive configuration does synchronize.

Case (ii) if $\gamma^2 < 4\beta$:

This is the case used in observing the chaotic motion of this system. The eigenvalues will be $-0.5\gamma \pm 0.5i\sqrt{4\beta - \gamma^2}$, $\qquad i = \sqrt{-1}$.

Hence all eigenvalues have negative real parts, which imply that all solutions tend to zero as t tends to ∞. The conditional Lyapunov exponents (CLEs) are $(-\gamma / 2, -\gamma / 2)$ and as expected, $\lim_{t \to \infty} y^*(t) = 0$ and $\lim_{t \to \infty} z^*(t) = 0$ then the response system with x-drive configuration does synchronize.

1.1.2 Numerical Results

By solving the drive and response systems (10) and (11) numerically and using the above-mentioned parameter values, we find that in spite of the differences in initial conditions of $(y', z') = (0.02, 0.1)$ and $(x, y, z) = (0.1, 0.01, -0.1)$, the response system synchronizes so that for $t \to \infty$, $y - y' \to 0$ and $z - z' \to 0$ (see Figure 2).

Now, we will calculate the conditional Lyapunov exponents (CLEs) numerically. We used a MATLAB code based on the Eckmann and Ruelle QR decomposition technique to calculate the exponents (Eckmann & Ruelle, 1985). Table (1) shows a calculation of the CLEs for various subsystems. From the table we see that, the subsystem (y, z) driven by x is stable subsystem as we expected and the numerical values of its CLEs are close to the analytical values shown above $(-\gamma / 2, -\gamma / 2)$. We also deduce that synchronization can not occur for z-drive configuration because the largest CLE of the (x, y) subsystem is sufficiently positive (see Table 1).

1.2 Method of One-Way Coupling

By One-Way coupling, we mean that the behavior of one full (response) system is dependent on the behavior of another identical (drive) system, but the second one is not influenced by the behavior of the first. In addition, the response system has a different set of initial conditions other than that of the drive system. As time progresses, the two identical chaotic systems can achieve a perfect synchronization among their state variables and maintain it, depending upon the One-Way coupling strength.

Figure 2. (a-b) shows that the synchronization error tends to zero using Pecora and Carroll method

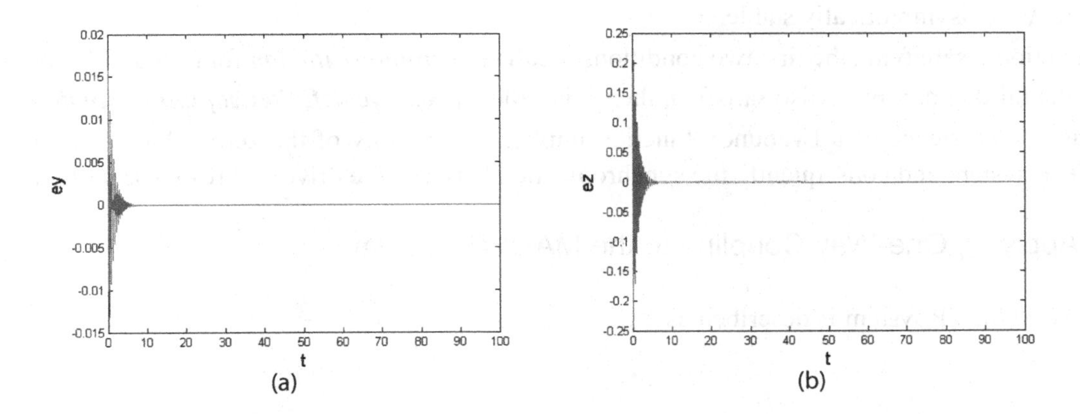

Table 1. Conditional Lyapunov exponents for various drive-response configurations for the MADVP system with the parameter set $\beta = 200$, $\mu = 0.1$, $\nu = 100$, $\gamma = 1.6$.

Drive	Response	CLEs
x	(y, z)	-0.7999, -0.8004
y	(x, z)	0.0021, -12.053
z	(x, y)	+6.0177, -19.4448

In this method, we take two identical chaotic systems $dX / dt = f(X)$, $dX' / dt = f(X')$ where $X = (x_1, ..., x_n)^T$, $X' = (x'_1, ..., x'_n)^T$, $f : R^n \to R^n$ and introduce a coupling term $\delta X = (X - X')$ into the second equation, which leads to the coupled system

$$dX / dt = f(X),$$
$$dX' / dt = f(X') + k(X - X'),$$

where k is a tuning parameter that controls the strength of the feedback into the coupled system. Therefore, it is a kind of linear control. Before we apply this method it is suitable to recall the Lyapunov stability theory which is basically used for this technique.

Theorem 2. (Lyapunov Stability) Let X^* be an equilibrium point for $\dot{X} = F(X)$. Let $V : U \to R$ be a differentiable function defined on an open set U containing X^*. Suppose further that

a) $V(X^*) = 0$ and $V(X) > 0$ if $X \neq X^*$;

b) $\dot{V} \leq 0$ in $U - X^*$.

Then X^* is stable. Furthermore, if V also satisfies

c) $\dot{V} < 0$ in $U - X^*$,

Then X^* is asymptotically stable.

A function V satisfying the first two conditions is called ***Lyapunov function*** for the equilibrium point X^*. If the third condition is also satisfied, then V is called a ***positive definiteLyapunov function.***

Thus, the existence of a Lyapunov function implies the stability of the zero solution of the error dynamical system and consequently the synchronization between the drive and response system.

1.2.1 Applying One-Way Coupling to the MADVP System

The drive MADVP system is described as

$$\begin{aligned}\frac{dx_1}{dt} &= -\nu(x_1^3 - \mu x_1 - y_1),\\ \frac{dy_1}{dt} &= x_1 - \gamma y_1 - z_1,\\ \frac{dz_1}{dt} &= \beta y_1,\end{aligned} \tag{13}$$

and the response system is given as

$$\begin{aligned}\frac{dx_2}{dt} &= -\nu(x_2^3 - \mu x_2 - y_2) - k(x_2 - x_1),\\ \frac{dy_2}{dt} &= x_2 - \gamma y_2 - z_2,\\ \frac{dz_2}{dt} &= \beta y_2,\end{aligned} \tag{14}$$

where $k = \nu\varepsilon$, is a tuning parameter and ε is defined as the One-Way coupling parameter. Define the synchronization errors as

$$e_x = x_1 - x_2,\ e_y = y_1 - y_2,\ e_z = z_1 - z_2,$$

then by subtracting (14) from (13), we obtain:

$$\begin{aligned}\dot{e}_x &= -\nu(c_{x_1,x_2} e_x - \mu e_x - e_y) - k e_x,\\ \dot{e}_y &= e_x - \gamma e_y - e_z,\\ \dot{e}_z &= \beta e_y,\end{aligned} \tag{15}$$

where $c_{x_1,x_2} = x_1^2 + x_1 x_2 + x_2^2 \geq 0$. Now, consider the following Lyapunov function for the error dynamical system (15):

$$V = \frac{\gamma}{2}(\beta e_x^2 + \nu\beta e_y^2 + \nu e_z^2) \geq 0, \tag{16}$$

then

$$\dot{V} = -\nu\gamma\beta c_{x_1,x_2} e_x^2 - \nu\beta(e_x - \gamma e_y)^2 \leq 0,$$

where β, γ, μ and ν are all positive parameters. If we select the coupling parameter ε as; $\varepsilon = (\mu\gamma + 1) / \gamma$, then $\dot{V}$ vanishes identically only at the origin and negative otherwise. i.e., V is a positive definite Lyapunov function, therefore the zero solution of equation (15) is asymptotically stable for this specific choice of the coupling parameter ε. Thus the synchronization is achieved between the drive and response systems (13) and (14).

1.2.2 Numerical Simulation

The drive and response systems (13) and (14) are integrated numerically using the above-mentioned parameter values. The initial conditions are fixed as $x_1(0) = 0.1, y_1(0) = 0.01, z_1(0) = -0.1$ and $x_2(0) = 0.2, y_2(0) = 0.02, z_2(0) = 0.1$. Figure 3, shows that the synchronization errors tend to zero as using the coupling parameter $\varepsilon = (\mu\gamma + 1) / \gamma = 0.725$.

Figure 3. (a-c) The effect of one-way coupling parameter on the error dynamical system (15).

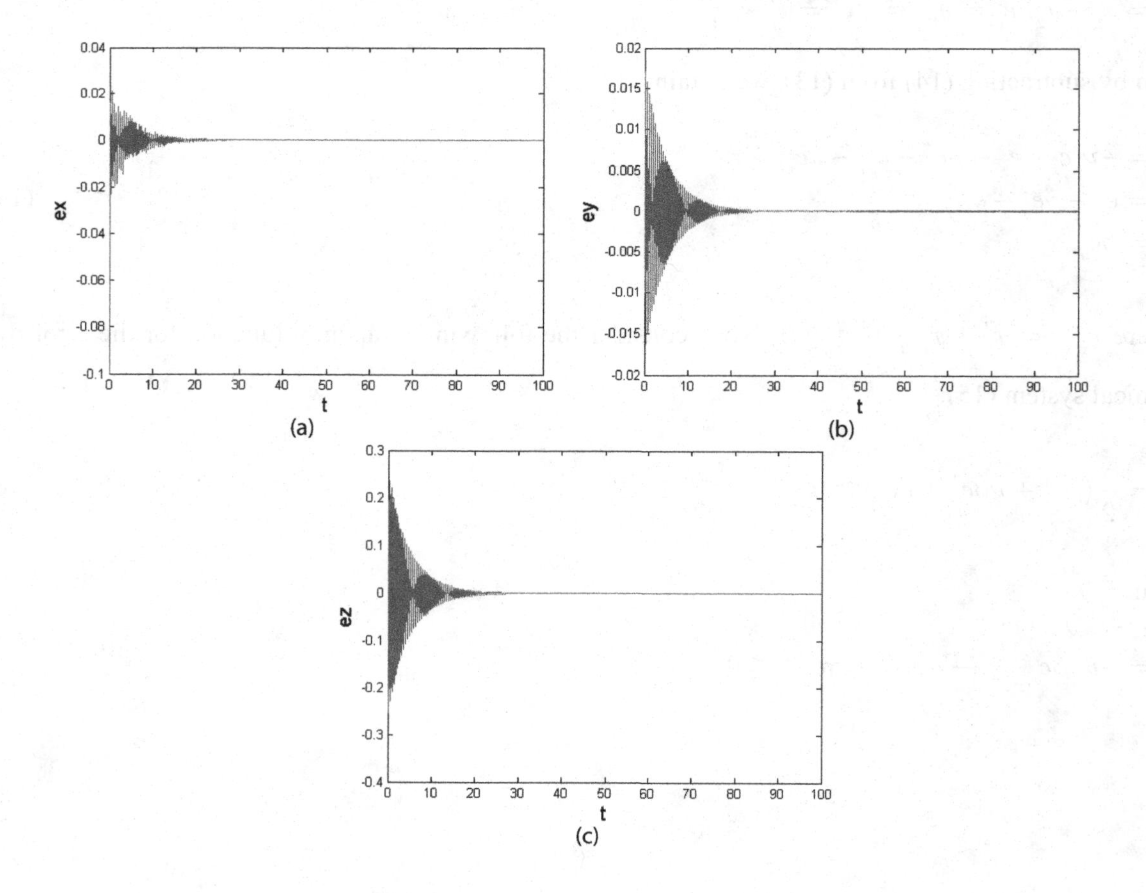

2. SYNCHRONIZATION IN FRACTIONAL ORDER CHAOTIC SYSTEMS

The idea of fractional order has been known since the work of Leibniz and L'Hopital in 1695, where half-order derivative was mentioned (Butzer & Westphal, 2000; Kenneth & Bertram, 1993). There are several definitions of fractional derivatives. One of the most common definitions is the Caputo definition of fractional derivatives (Caputo, 1967):

$$D^{\alpha} f(t) = J^{l-\alpha} f^{(l)}(t), \qquad \alpha > 0, \tag{17}$$

Where $f^{(l)}$ represents the l-order derivative of $f(t)$, l is the minimum integer which is not less than α and J_0^{θ} is the θ-order Riemann-Liouville integral operator and given as

$$J_0^{\theta} u(t) = \frac{1}{\Gamma(\theta)} \int_0^t (t-\tau)^{\theta-1} u(\tau) d\tau, \qquad \theta > 0, \tag{18}$$

where $J_0^{\theta} \equiv J^{\theta}$ is defined on the usual Lebesque space $L_1[a,b]$ for $a \le t \le b$ and $\Gamma(.)$ denotes the gamma function. The operator D^{α} is generally called "the α-order Caputo differential operator". Thus, the fractional derivative of $f(t)$ in the Caputo sense is defined as follows:

$$D^{\alpha} f(t) = J^{l-\alpha} D^{l} f(t) = \frac{1}{\Gamma(l-\alpha)} \int_0^t (t-\tau)^{l-\alpha-1} f^{(l)}(\tau) d\tau. \tag{19}$$

Another definition is the Riemann-Liouville (Podlubny, 1999) which is given by

$$D_*^{\alpha} f(t) = \frac{d^l}{dt^l} J^{l-\alpha} f(t), \tag{20}$$

The question arises now is, "Why do we consider a fractional derivative?" to answer this question we say that; many real dynamical systems are better characterized using a non integer order dynamical model based on fractional calculus or, differentiation or integration of non integer order. Traditional calculus is based on integer order differentiation and integration. The concept of fractional calculus has tremendous potential to change the way we see, model, and control the nature around us. Denying fractional derivatives is like saying that zero, fractional, or irrational numbers do not exist. Recently, fractional calculus has useful applications in physics (Hilfer, 2000), engineering (Sun et al, 1984), mathematical biology (Ahmed & Elgazzar, 2007; El-Sayed et al, 2007) and finance (Chen, 2008). Moreover, it has been found that many systems in interdisciplinary fields can be described by the fractional differential equations, such as viscoelastic systems (Bagley & Calico, 1991; Koeller, 1984), electromagnetic waves (Heaviside, 1971), quantum evolution of complex systems (Kusnezov et al, 1999), psychological and life models (Ahmad & El-Khazali, 2007). Therefore, studying chaos and chaos synchronization in fractional order systems is just a recent focus of interest (Zhou & Li, 2005; Peng, 2007; Wu et al, 2008; Matouk, 2009a; Matouk, 2009b).

An efficient method for solving fractional order differential equations; is the predictor-correctors scheme or more precisely, **PECE** (Predict, Evaluate, Correct, Evaluate) technique which has been investigated in (Diethelm & Ford, 2002; Diethelm et al, 2002), and represents a generalization of the Adams-Bashforth-Moulton algorithm. It is used throughout this section.

To illustrate the **PECE** method; consider the initial value problem:

$$D^{\alpha}x(t) = f(t, x(t)), \ 0 \le t \le T, \ \ x^{(k)}(0) = x_0^{(k)}, \ \ k = 0, 1, ..., l-1. \tag{21}$$

If the function f is continuous, then the initial value problem (21) is equivalent to the nonlinear Volterra integral equation of the second kind (Diethelm & Ford, 2002), which is given as follows:

$$x(t) = \sum_{k=0}^{l-1} \frac{t^k}{k!} x_0^{(k)} + \frac{1}{\Gamma(\alpha)} \int_0^t (t-\xi)^{\alpha-1} f(\xi, x(\xi)) d\xi. \tag{22}$$

By setting $h = T / N$, $t_n = nh$, $n = 0, 1, ..., N \in Z^+$. Then Eqn. (22) can be discretized as follows:

$$x_h(t_{n+1}) = \sum_{k=0}^{l-1} \frac{t_{n+1}^k}{k!} x_0^{(k)} + \frac{h^{\alpha}}{\Gamma(\alpha+2)} f(t_{n+1}, x_h^p(t_{n+1})) + \frac{h^{\alpha}}{\Gamma(\alpha+2)} \sum a_{j,n+1} f(t_j, x_h(t_j)), \tag{23}$$

where

$$a_{j,n+1} = \begin{cases} n^{\alpha+1} - (n-\alpha)(n+1)^{\alpha}, & j = 0, \\ (n-j+2)^{\alpha+1} + (n-j)^{\alpha+1} - 2(n-j+1)^{\alpha+1}, & 1 \le j \le n, \\ 1, & j = n+1, \end{cases}$$

$$x_h^p(t_{n+1}) = \sum_{k=0}^{l-1} \frac{t_{n+1}^k}{k!} x_0^{(k)} + \frac{1}{\Gamma(\alpha)} \sum_{j=0}^{n} b_{j,n+1} f(t_j, x_h(t_j)),$$

$$b_{j,n+1} = \frac{h^{\alpha}}{\alpha}((n+1-j)^{\alpha} - (n-j)^{\alpha}).$$

The error estimate is $\max_{j=0,1,...,N} \left| x(t_j) - x_h(t_j) \right| = O(h^p)$, in which $p = \min(2, 1+\alpha)$.

2.1 Synchronization of Fractional Order Systems via Pecora and Carroll Method

Consider the fractional order dynamical system with the same fractional order

$$\begin{aligned}
\frac{d^\alpha u_1}{dt^\alpha} &= f_1(u_1,\ldots,u_m,u_{m+1},\ldots,u_n),\\
\frac{d^\alpha u_2}{dt^\alpha} &= f_2(u_1,\ldots,u_m,u_{m+1},\ldots,u_n),\\
&\vdots\\
\frac{d^\alpha u_m}{dt^\alpha} &= f_m(u_1,\ldots,u_m,u_{m+1},\ldots,u_n),\\
\frac{d^\alpha u_{m+1}}{dt^\alpha} &= f_{m+1}(u_1,\ldots,u_m,u_{m+1},\ldots,u_n),\\
&\vdots\\
\frac{d^\alpha u_n}{dt^\alpha} &= f_n(u_1,\ldots,u_m,u_{m+1},\ldots,u_n),
\end{aligned} \tag{24}$$

where $f_i : R^n \to R,\ i = 1,2,\ldots,n$. According to the Pecora and Carroll scheme system (24) is decomposed into the following subsystems (the drive and response respectively)

$$\begin{aligned}
\frac{d^\alpha v_1}{dt^\alpha} &= g_1(v_1,\ldots,v_m,w_{m+1},\ldots,w_n),\\
\frac{d^\alpha v_2}{dt^\alpha} &= g_2(v_1,\ldots,v_m,w_{m+1},\ldots,w_n),\\
&\vdots\\
\frac{d^\alpha v_m}{dt^\alpha} &= g_m(v_1,\ldots,v_m,w_{m+1},\ldots,w_n),
\end{aligned} \tag{25}$$

and

$$\begin{aligned}
\frac{d^\alpha w_{m+1}}{dt^\alpha} &= h_{m+1}(v_1,\ldots,v_m,w_{m+1},\ldots,w_n),\\
\frac{d^\alpha w_{m+2}}{dt^\alpha} &= h_{m+2}(v_1,\ldots,v_m,w_{m+1},\ldots,w_n),\\
&\vdots\\
\frac{d^\alpha w_n}{dt^\alpha} &= h_n(v_1,\ldots,v_m,w_{m+1},\ldots,w_n),
\end{aligned} \tag{26}$$

Then we make a copy of the response subsystem w'_R which driven by the variables v_D as follows

$$\begin{aligned}
\frac{d^\alpha w'_{m+1}}{dt^\alpha} &= h_1(v_1,\ldots,v_m,w'_{m+1},\ldots,w'_n),\\
\frac{d^\alpha w'_{m+2}}{dt^\alpha} &= h_2(v_1,\ldots,v_m,w'_{m+1},\ldots,w'_n),\\
&\vdots\\
\frac{d^\alpha w'_n}{dt^\alpha} &= h_m(v_1,\ldots,v_m,w'_{m+1},\ldots,w'_n),
\end{aligned} \tag{27}$$

The idea of Pecora and Carroll is; as $t \to \infty$ the w'_R will converge asymptotically to the w_R variables and remain with it in step instantaneously.

2.2 Synchronization of Fractional Order Systems via One-Way Coupling Method

By considering two identical chaotic fractional order systems $d^\alpha X / dt^\alpha = f(X)$ and $d^\alpha X' / dt^\alpha = f(X')$, where $X = (x_1,\ldots,x_n)^T$, $X' = (x'_1,\ldots,x'_n)^T$ and $f: R^n \to R^n$. By introducing a coupling term $\delta X = (X - X')$, the coupled system is given as follows:

$$\begin{aligned}
\frac{d^\alpha X}{dt^\alpha} &= f(X),\\
\frac{d^\alpha X'}{dt^\alpha} &= f(X') + k(X - X'),
\end{aligned} \tag{28}$$

where k is a tuning parameter that controls the strength of the feedback into the coupled system.

Now, we will apply this method to the two coupled fractional order MADVP systems (Matouk, 2010). The drive system is given as:

$$\begin{aligned}
\frac{d^\alpha x_1}{dt^\alpha} &= -\nu(x_1^3 - \mu x_1 - y_1),\\
\frac{d^\alpha y_1}{dt^\alpha} &= x_1 - \gamma y_1 - z_1,\\
\frac{d^\alpha z_1}{dt^\alpha} &= \beta y_1,
\end{aligned} \tag{29}$$

and the response system is given by:

$$\begin{aligned}\frac{d^\alpha x_2}{dt^\alpha} &= -\nu(x_2^3 - \mu x_2 - y_2) + k(x_1 - x_2),\\ \frac{d^\alpha y_2}{dt^\alpha} &= x_2 - \gamma y_2 - z_2,\\ \frac{d^\alpha z_2}{dt^\alpha} &= \beta y_2,\end{aligned} \tag{30}$$

The drive and response systems (29) and (30) are numerically integrated using the **PECE** method with the parameter values $\beta = 200,\ \mu = 0.1,\ \nu = 100,\ \gamma = 1.6$ and the same fractional order $\alpha = 0.98$. At this set of the parameters and fractional order, the fractional order MADVP system displays chaotic behaviors (see Figure 4a). By selecting $k = \nu\varepsilon = 72.5$, it is easy to see that the states of the drive system (29) approach the states of the response system (30) in an asymptotically way. Thus, chaos synchronization is achieved between the drive and response systems via One-Way coupling (see Figure 4b).

Comparing to the subsection 1.2, we introduce the Lyapunov function in the fractional order case (Zhang & Chen, 2005) as follows; suppose Σ is an open subset of R^n. Consider the autonomous system

$$D^\alpha X = f(X), \tag{31}$$

where $f \in C(\Sigma, R^n),\ \alpha \in (0,1]$.

Assume that $V \in C^1(\Sigma, R),$ then the fractional derivative of order α of the function $V(X)$ along the solution of equation (31) is given as follows:

Figure 4. Shows; (a) 3-dimensional view of the drive system (29); (b) the synchronization errors between the drive and response systems (29) and (30) tend to zero; when using the parameter set $\beta = 200,\ \mu = 0.1,\ \nu = 100,\ \gamma = 1.6,$ *fractional order* $\alpha = 0.98$ *and* $k = 72.5$.

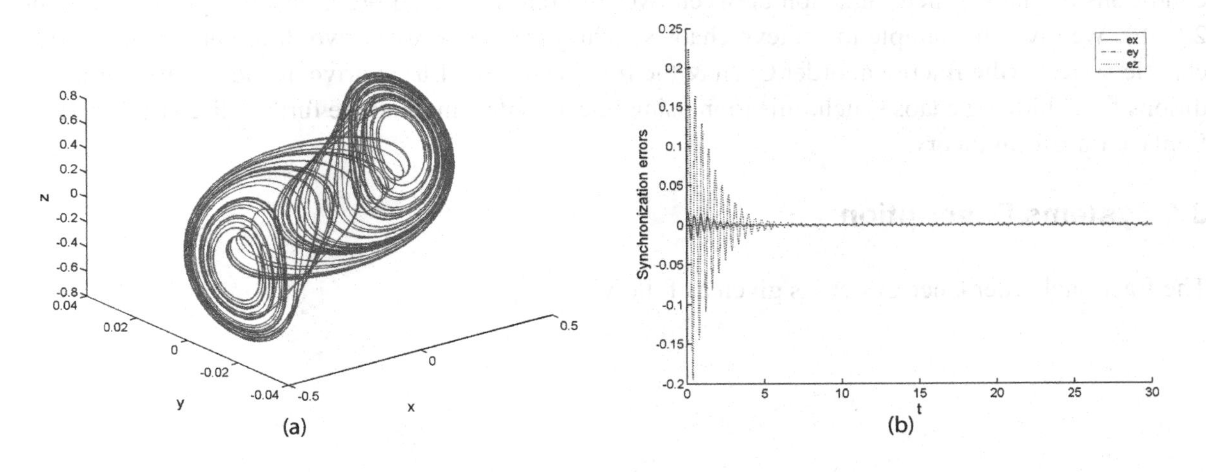

$$D^{\alpha}V = \frac{d^{\alpha}V}{dt^{\alpha}}\bigg|_{(31)} = J^{1-\alpha}DV = J^{1-\alpha}(\frac{dV}{dX}.\frac{dX}{dt}). \tag{32}$$

Definition 2. The function $V(X)$ is said to be a Lyapunov function on a set Σ in R^n relative to equation (31) if $V \in C^1(\Sigma, R)$ and $D^{\alpha}V \leq 0$ on Σ.

Define $\Pi = \left\{X \in clos\Sigma \,\middle|\, D^{\alpha}V = 0\right\}$, and Λ to be the largest set in Π which is invariant with respect to equation (31), where $clos\Sigma$ is the closure of Σ.

Now, we have the following theorem

Theorem 3. Suppose that
(i) V is a Lyapunov function on Σ;

(ii) V is a positive definite function;

(iii) $V(X) \to +\infty$ when $\left\|X\right\| \to +\infty$;

(iv) $\Lambda = \left\{X^c \,\middle|\, f(X^c) = 0\right\}$.

Then the equilibrium point X^c is globally asymptotically stable.

3. SYNCHRONIZATION BETWEEN TWO DIFFERENT FRACTIONAL ORDER CHAOTIC SYSTEMS

In the previous sections, we have used some methods to synchronize two identical chaotic systems. In fact, in many practical cases such as biological systems, laser array and cognitive processes, it is hardly the case that the structure of drive and response systems can be assumed to be identical. Consequently, in these years, more and more applications of chaos synchronization in secure communications make it much more important to synchronize two different chaotic systems. In this section, we investigate the conditions of chaos synchronization between two different fractional order chaotic systems (Matouk, 2009b). We give an example to achieve chaos synchronization between two different fractional order chaotic systems (the fractional order Chen & the fractional order Lü) in drive–response structure. Conditions for achieving chaos synchronization using linear control method are further discussed using the Laplace transform theory.

3.1 Systems Description

The fractional order Chen system is given as follows

Figure 5. Chaotic attractor of the fractional order Chen system with $\alpha = 0.9$ *and* $(a,b,c) = (35,3,28)$

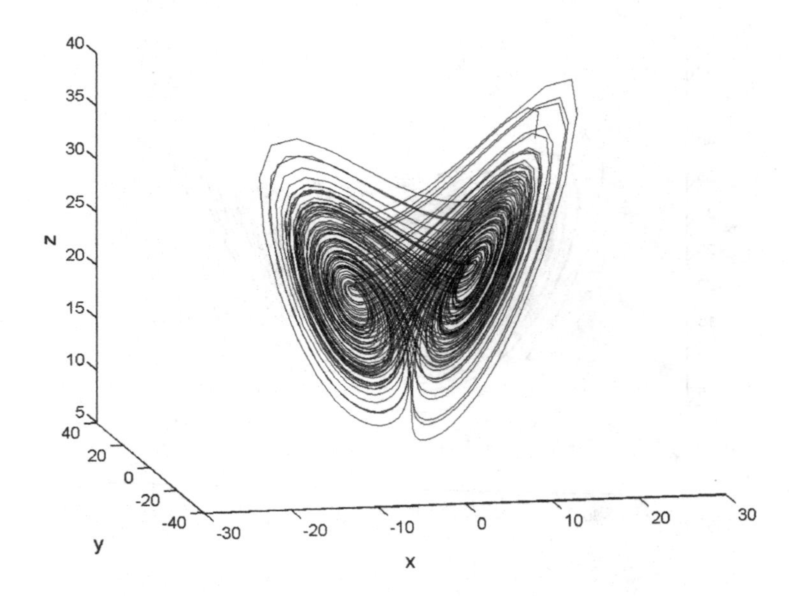

$$\frac{d^\alpha x}{dt^\alpha} = a(y - x),$$
$$\frac{d^\alpha y}{dt^\alpha} = (c - a)x - xz + cy, \tag{33}$$
$$\frac{d^\alpha z}{dt^\alpha} = xy - bz.$$

Here and throughout, $(a,b,c) = (35,3,28)$ where α is the fractional order. In the following we choose $\alpha = 0.9$ at which system (33) exhibits chaotic attractor (see Figure 5).

The fractional order Lü system is given as follows:

$$\frac{d^\alpha x}{dt^\alpha} = r(y - x),$$
$$\frac{d^\alpha y}{dt^\alpha} = -xz + py, \tag{34}$$
$$\frac{d^\alpha z}{dt^\alpha} = xy - qz.$$

Here and throughout, $(r,p,q) = (35,28,3)$. By choosing $\alpha = 0.9$, system (34) has chaotic attractor as shown in figure 6.

Figure 6. Chaotic attractor of the fractional order Lü system with $\alpha = 0.9$ *and* $(r, p, q) = (35, 28, 3)$.

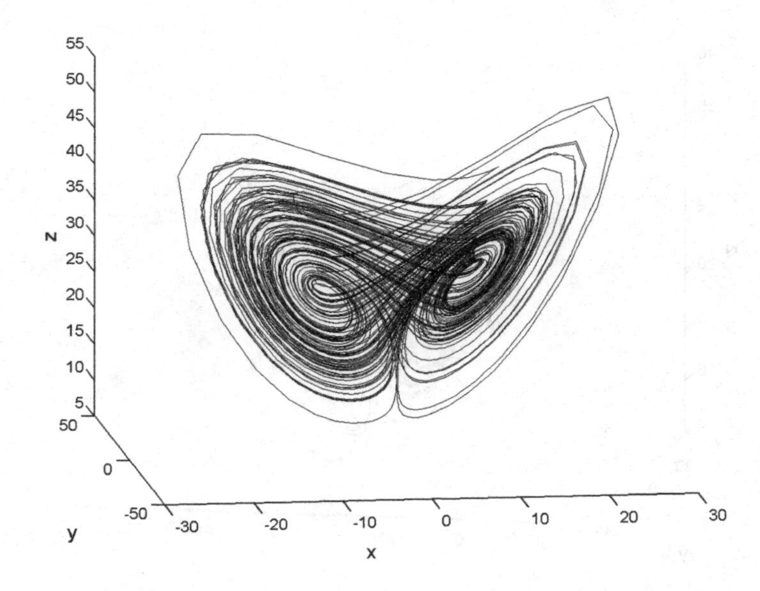

3.2 Synchronization between Two Different Fractional Order Systems

Consider the master-slave (or drive-response) synchronization scheme of two autonomous different fractional order chaotic systems

$$\begin{aligned} \frac{d^{\alpha}X}{dt^{\alpha}} &= f(X), \\ \frac{d^{\alpha}Y}{dt^{\alpha}} &= g(Y) + U(t), \end{aligned} \tag{35}$$

where α is the fractional order, $X \in R^n$, $Y \in R^n$ represent the states of the drive and response systems respectively, $f : R^n \to R^n$, $g : R^n \to R^n$ are the vector fields of the drive and response systems respectively. The aim is to choose a suitable linear control function $U(t) = (u_1, \ldots, u_n)^T$ such that the states of the drive and response systems are synchronized (i.e. $\lim_{t \to \infty} \|X - Y\| = 0$, where $\|.\|$ is the Euclidean norm).

3.2.1 Synchronization between Chen and Lü Fractional Order Systems

Our the goal is to achieve chaos synchronization between the fractional order Chen system and the fractional order Lü system by using the fractional order Chen system to drive the fractional order Lü system. The drive and response systems are given as follow:

$$\begin{aligned}\frac{d^\alpha x_m}{dt^\alpha} &= a(y_m - x_m),\\ \frac{d^\alpha y_m}{dt^\alpha} &= (c-a)x_m - x_m z_m + c y_m,\\ \frac{d^\alpha z_m}{dt^\alpha} &= x_m y_m - b z_m,\end{aligned} \tag{36}$$

and,

$$\begin{aligned}\frac{d^\alpha x_s}{dt^\alpha} &= r(y_s - x_s) + u_1,\\ \frac{d^\alpha y_s}{dt^\alpha} &= -x_s z_s + p y_s + u_2,\\ \frac{d^\alpha z_s}{dt^\alpha} &= x_s y_s - q z_s + u_3,\end{aligned} \tag{37}$$

where u_1, u_2 and u_3 are the linear control functions. Define the error variables as:

$$e_1 = x_s - x_m,\ e_2 = y_s - y_m,\ e_3 = z_s - z_m. \tag{38}$$

By subtracting (36) from (37) and using (38), we obtain

$$\begin{aligned}\frac{d^\alpha e_1}{dt^\alpha} &= r(e_2 - e_1) + (r-a)(y_m - x_m) + u_1,\\ \frac{d^\alpha e_2}{dt^\alpha} &= p e_2 - z_m e_1 - x_m e_3 - e_1 e_3 - (c-a)x_m + (p-c)y_m + u_2,\\ \frac{d^\alpha e_3}{dt^\alpha} &= -q e_3 + y_m e_1 + x_m e_2 + e_1 e_2 - (q-b)z_m + u_3.\end{aligned} \tag{39}$$

Now, by letting

$$\begin{aligned}u_1 &= (a-r)(y_m - x_m),\\ u_2 &= (c-a)x_m + (c-p)y_m - k_1(y_s - y_m),\\ u_3 &= (q-b)z_m - k_2(z_s - z_m),\end{aligned} \tag{40}$$

where $k_1,\ k_2 \geq 0$, then the error system (39) is reduced to

$$\begin{aligned}\frac{d^\alpha e_1}{dt^\alpha} &= r(e_2 - e_1),\\ \frac{d^\alpha e_2}{dt^\alpha} &= (p - k_1)e_2 - z_m e_1 - x_m e_3 - e_1 e_3,\\ \frac{d^\alpha e_3}{dt^\alpha} &= -(q + k_2)e_3 + y_m e_1 + x_m e_2 + e_1 e_2.\end{aligned} \tag{41}$$

By taking the Laplace transform in both sides of (41), letting $E_i(s) = L\{e_i(t)\}$ where $(i = 1,2,3)$, and applying $L\{d^{\alpha}e_i / dt^{\alpha}\} = s^{\alpha}E_i(s) - s^{\alpha-1}e_i(0)$, we obtain

$$\begin{aligned} &s^{\alpha}E_1(s) - s^{\alpha-1}e_1(0) = r(E_2(s) - E_1(s)),\\ &s^{\alpha}E_2(s) - s^{\alpha-1}e_2(0) = (p - k_1)E_2(s) - L\{x_m e_3\} - L\{z_m e_1\} - E_1(s)E_3(s),\\ &s^{\alpha}E_3(s) - s^{\alpha-1}e_3(0) = -(q + k_2)E_3(s) + L\{y_m e_1\} + L\{x_m e_2\} + E_1(s)E_2(s). \end{aligned} \tag{42}$$

Theorem 4. If $E_1(s)$, $E_2(s)$ are bounded and $p - k_1 \neq 0$, then the drive and response systems (36) and (37) will be synchronized under a suitable choice of k_1 and k_2.
Proof. By rewriting equations (42) as follows

$$\begin{aligned} E_1(s) &= \frac{rE_2(s)}{s^{\alpha} + r} + \frac{s^{\alpha-1}e_1(0)}{s^{\alpha} + r},\\ E_2(s) &= -\frac{L\{z_m e_1\}}{s^{\alpha} - p + k_1} - \frac{L\{x_m e_3\}}{s^{\alpha} - p + k_1} - \frac{E_1(s)E_3(s)}{s^{\alpha} - p + k_1} + \frac{s^{\alpha-1}e_2(0)}{s^{\alpha} - p + k_1},\\ E_3(s) &= \frac{L\{y_m e_1\}}{s^{\alpha} + q + k_2} + \frac{L\{x_m e_2\}}{s^{\alpha} + q + k_2} + \frac{E_1(s)E_2(s)}{s^{\alpha} + q + k_2} + \frac{s^{\alpha-1}e_3(0)}{s^{\alpha} + q + k_2}. \end{aligned} \tag{43}$$

Using the final-value theorem of the Laplace transform, it follows that

$$\begin{aligned} &\lim_{t\to\infty} e_1(t) = \lim_{s\to 0^+} sE_1(s) = \lim_{s\to 0^+} sE_2(s) = \lim_{t\to\infty} e_2(t),\\ &\lim_{t\to\infty} e_2(t) = \lim_{s\to 0^+} sE_2(s) = \frac{1}{p - k_1}\lim_{s\to 0^+} sL\{x_m e_3\} + \frac{1}{p - k_1}\lim_{s\to 0^+} sL\{z_m e_1\} + \frac{1}{p - k_1}\lim_{t\to\infty} e_1(t).\lim_{t\to\infty} e_3(t),\\ &\lim_{t\to\infty} e_3(t) = \lim_{s\to 0^+} sE_3(s) = \frac{1}{q + k_2}\lim_{s\to 0^+} sL\{y_m e_1\} + \frac{1}{q + k_2}\lim_{s\to 0^+} sL\{x_m e_2\} + \frac{1}{q + k_2}\lim_{t\to\infty} e_1(t).\lim_{t\to\infty} e_2(t). \end{aligned} \tag{44}$$

Since $E_1(s)$, $E_2(s)$ are bounded and $p - k_1 \neq 0$ then $\lim_{t\to\infty} e_1(t) = \lim_{t\to\infty} e_2(t) = 0$. Now, owing to the attractiveness of the attractors of systems (33) and (34), there exists $\eta > 0$ such that $|x_i(t)| \leq \eta < \infty$, $|y_i(t)| \leq \eta < \infty$ and $|z_i(t)| \leq \eta < \infty$ where i refers to the index of the drive or response variables. Therefore, $\lim_{t\to\infty} e_3(t) = 0$. This implies that

$$\lim_{t\to\infty} e_i(t) = 0, \qquad i = 1,2,3. \tag{45}$$

Consequently, the synchronization between the drive and response systems (36) and (37) is achieved. □

Figure 7. Synchronization errors of the drive system (36) and response system (37) using $k_1 = 20$, $k_2 = 10$ *and fractional orders: (a)* $\alpha = 0.9$, *(b)* $\alpha = 0.95$, *(c)* $\alpha = 0.99$.

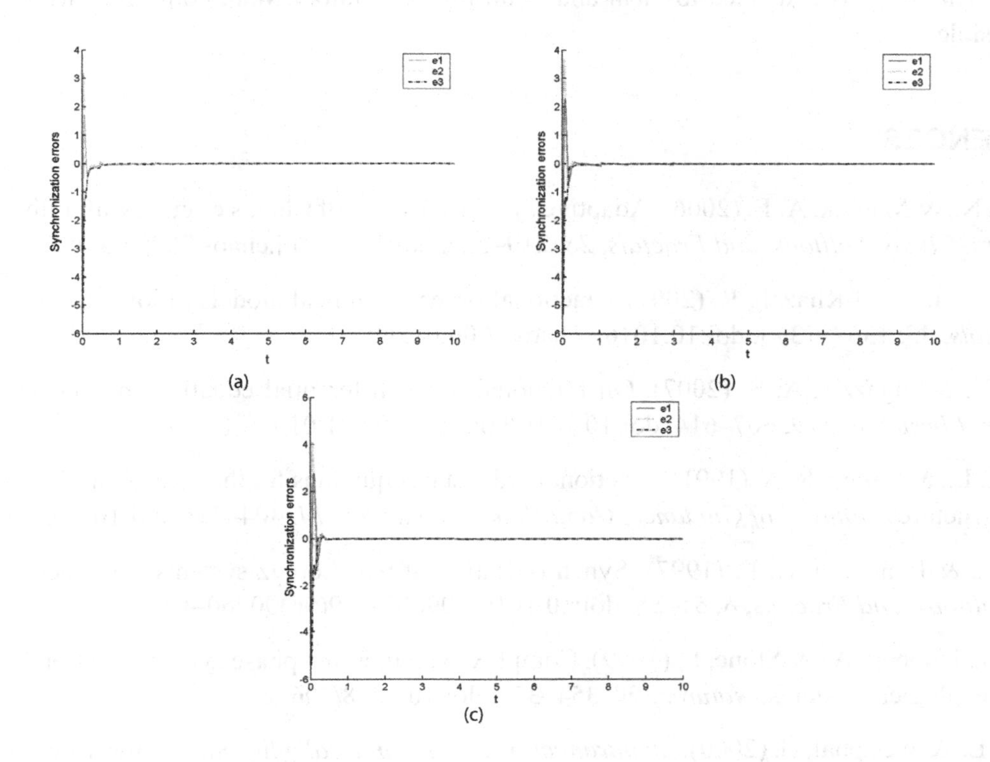

3.2.2 Numerical Results

Based on the **PECE** discretization scheme, the drive and response systems (36) and (37) are integrated numerically with the fractional orders $\alpha = 0.9,\ 0.95,\ 0.99$ and using the initial values $x_m(0) = 15, y_m(0) = 20, z_m(0) = 29$ and $x_s(0) = 10, y_s(0) = 15, z_s(0) = 25$. From Figure 7, it is clear that the synchronization is achieved for all these values of fractional order when $k_1 = 20$ and $k_2 = 10$

ACKNOWLEDGMENT

The results that I present here are the fruit of intensive research activity that was conducted during my tenure at the Preparatory Math Program of the University of Hail, Saudi Arabia, and have been announced in the Preparatory Math Seminar during the academic year 2009 – 2010. This research in progress is sponsored by the University of Hail. I take this opportunity to thank the University and particularly its Dean of the Preparatory Year, Dr. Eid bin Mohia Al-Haisoni, for their generous support for the project.

I am also grateful to Dr. Basil Abdel-Gadir, Director of the UOH Preparatory Math Program for the interest he has shown in our work, his encouragement, and his valuable suggestions. For providing an excellent work environment and for his visionary leadership of our Seminar Series, Dr. Abdel-Gadir has fully merited my deepest respect and gratitude.

I would also like to thank Prof. E. Ahmed, Prof. A.S. Hegazi and Prof. H.N. Agiza for discussion and help. This work is dedicated to them and to all my Preparatory Math colleagues who made this work possible.

REFERENCES

Agiza, H. N., & Matouk, A. E. (2006). Adaptive synchronization of Chua's circuits with fully unknown parameters. *Chaos, Solitons, and Fractals*, *28*, 219–227. doi:10.1016/j.chaos.2005.05.055

Ahmad, W. M., & El-Khazali, R. (2007). Fractional-order dynamical models of love. *Chaos, Solitons, and Fractals*, *33*, 1367–1375. doi:10.1016/j.chaos.2006.01.098

Ahmed, E., & Elgazzar, A. S. (2007). On fractional order differential equations model for nonlocal epidemics. *Physica A*, *379*, 607–614. doi:10.1016/j.physa.2007.01.010

Bagley, R. L., & Calico, R. A. (1991). Fractional order state equations for the control of viscoelastically damped structures. *Journal of Guidance, Control, and Dynamics*, *14*, 304–311. doi:10.2514/3.20641

Bai, E. W., & Lonngren, K. E. (1997). Synchronization of two Lorenz systems using active control. *Chaos, Solitons, and Fractals*, *8*, 51–58. doi:10.1016/S0960-0779(96)00060-4

Blasius, B., Huppert, A., & Stone, L. (1999). Complex dynamics and phase synchronization in spatially extended ecological systems. *Nature*, *399*, 354–359. doi:10.1038/20676

Butzer, P. L., & Westphal, U. (2000). *An introduction to fractional calculus*. Singapore: World Scientific.

Caputo, M. (1967). Linear models of dissipation whose Q is almost frequency independent- II. *Geophysical Journal of the Royal Astronomical Society*, *13*, 529–539.

Carroll, T. L., & Pecora, L. M. (1991). Synchronizing chaotic circuits. *IEEE Transactions on Circuits and Systems*, *1*(38), 453–456. doi:10.1109/31.75404

Chen, G. (1997). *Control and synchronization of chaos*. Houston, TX: University of Houston, Department of Electrical Engineering.

Chen, W. C. (2008). Nonlinear dynamics and chaos in a fractional-order financial system. *Chaos, Solitons, and Fractals*, *36*, 1305–1314. doi:10.1016/j.chaos.2006.07.051

Chua, L. O., Kocarev, L. J., Eckert, K., & Itoh, M. (1992). Experimental chaos synchronization in Chua's circuit. *International Journal of Bifurcation and Chaos in Applied Sciences and Engineering*, *2*, 705–708. doi:10.1142/S0218127492000811

Cuomo, K. M., & Oppenheim, V. (1993). Circuit implementation of synchronized chaos with application to communication. *Physical Review Letters*, *71*, 65–68. doi:10.1103/PhysRevLett.71.65

Diethelm, K., & Ford, N. J. (2002). Analysis of fractional differential equations. *Journal of Mathematical Analysis and Applications*, *265*, 229–248. doi:10.1006/jmaa.2000.7194

Diethelm, K., Ford, N. J., & Freed, A. D. (2002). A predictor-corrector approach for the numerical solution of fractional differential equations. *Nonlinear Dynamics*, *29*, 3–22. doi:10.1023/A:1016592219341

Eckmann, P., & Ruelle, D. (1985). Ergodic theory of chaos and strange attractors. *Reviews of Modern Physics*, *57*, 617–656. doi:10.1103/RevModPhys.57.617

El-Sayed, A. M. A., El-Mesiry, A. E. M., & El-Saka, H. A. A. (2007). On the fractional-order logistic equation. *Applied Mathematics Letters*, *20*, 817–823. doi:10.1016/j.aml.2006.08.013

Fujisaka, H., & Yamada, T. (1983). Stability theory of synchronized motion in coupled oscillator systems. *Progress of Theoretical Physics*, *69*, 32–47. doi:10.1143/PTP.69.32

Güémez, J., Martín, C., & Matías, M. A. (1997). Approach to the chaotic synchronized state of some driving methods. *Physical Review E: Statistical Physics, Plasmas, Fluids, and Related Interdisciplinary Topics*, *55*, 124–134. doi:10.1103/PhysRevE.55.124

He, R., & Vaidya, P. G. (1992). Analysis and synthesis of synchronous periodic and chaotic systems. *Physical Review A.*, *46*, 7387–7392. doi:10.1103/PhysRevA.46.7387

Heaviside, O. (1971). *Electromagnetic theory.* New York: Chelsea.

Hegazi, A. S., Agiza, H. N., & El-Dessoky, M. M. (2001). Synchronization and adaptive synchronization of nuclear spin generator system. *Chaos, Solitons, and Fractals*, *12*, 1091–1099. doi:10.1016/S0960-0779(00)00022-9

Hegazi, A. S., Agiza, H. N., & El-Dessoky, M. M. (2002). Adaptive synchronization for Rössler and Chua's circuit systems. *International Journal of Bifurcation and Chaos in Applied Sciences and Engineering*, *12*, 1579–1597. doi:10.1142/S0218127402005388

Hilfer, R. (Ed.). (2000). *Applications of fractional calculus in physics*. Upper Saddle River, NJ: World Scientific.

Jiang, G. P., Tang, K. S., & Chen, G. (2003). A simple global synchronization criterion for coupled chaotic systems. *Chaos, Solitons, and Fractals*, *15*, 925–935. doi:10.1016/S0960-0779(02)00214-X

Kenneth, S. M., & Bertram, R. (1993). *An introduction to the fractional calculus and fractional differential equations*. New York: Wiley-Interscience.

Kocarev, L., & Parlitz, U. (1995). General approach for chaotic synchronization with application to communication. *Physical Review Letters*, *74*, 5028–5031. doi:10.1103/PhysRevLett.74.5028

Koeller, R. C. (1984). Application of fractional calculus to the theory of viscoelasticity. *Journal of Applied Mechanics*, *51*, 294–298. doi:10.1115/1.3167616

Kusnezov, D., Bulgac, A., & Dang, G. D. (1999). Quantum levy processes and fractional kinetics. *Physical Review Letters*, *82*, 1136–1139. doi:10.1103/PhysRevLett.82.1136

Lakshmanan, M., & Murali, K. (1996). *Chaos in nonlinear oscillators: controlling and synchronization.* Singapore: World Scientific.

Li, Y., Chen, L., Cai, Z., & Zhao, X. (2003). Study on chaos synchronization in the Belousov-Zhabotinsky chemical system. *Chaos, Solitons, and Fractals*, *17*, 699–707. doi:10.1016/S0960-0779(02)00486-1

Liao, T. L., & Lin, S. H. (1999). Adaptive control and synchronization of Lorenz systems. *Journal of the Franklin Institute*, *336*, 925–937. doi:10.1016/S0016-0032(99)00010-1

Matouk, A. E. (2005). *Chaos and synchronization in some nonlinear electronic circuits*. Unpublished master thesis, University of Mansoura, Mansoura.

Matouk, A. E. (2008). Dynamical analysis feedback control and synchronization of Liu dynamical system. *Nonlinear Analysis: TMA*, *69*, 3213–3224. doi:10.1016/j.na.2007.09.029

Matouk, A. E. (2009a). Stability conditions, hyperchaos and control in a novel fractional order hyperchaotic system. *Physics Letters. [Part A]*, *373*, 2166–2173. doi:10.1016/j.physleta.2009.04.032

Matouk, A. E. (2009b). *Chaos synchronization between two different fractional systems of Lorenz family*. Mathematical Problems in Engineering.

Matouk, A. E. (in press). 2010). Chaos, feedback control and synchronization of a fractional-order modified Autonomous Van der Pol-Duffing circuit. *Communications in Nonlinear Science and Numerical Simulation*. doi:.doi:10.1016/j.cnsns.2010.04.027

Matouk, A. E., & Agiza, H. N. (2008). Bifurcations, chaos and synchronization in ADVP circuit with parallel resistor. *Journal of Mathematical Analysis and Applications*, *341*, 259–269. doi:10.1016/j.jmaa.2007.09.067

Ogorzalek, M. J. (1993). Taming chaos-part I: Synchronization. *IEEE Trans. Circ. Sys.*, 1(40, 693-699.

Pecora, L. M., & Carroll, T. L. (1990). Synchronization in chaotic systems. *Physical Review Letters*, *64*, 821–824. doi:10.1103/PhysRevLett.64.821

Pecora, L. M., & Carroll, T. L. (1991). Driving systems with chaotic signals. *Physical Review A.*, *44*, 2374–2383. doi:10.1103/PhysRevA.44.2374

Pecora, L. M., Carroll, T. L., Johnson, G. A., Mar, D. J., & Heagy, J. F. (1997). Fundamentals of synchronization in chaotic systems, concepts, and applications. *Chaos (Woodbury, N.Y.)*, *7*, 520–543. doi:10.1063/1.166278

Peng, G. (2007). Synchronization of fractional order chaotic systems. *Physics Letters. [Part A]*, *363*, 426–432. doi:10.1016/j.physleta.2006.11.053

Podlubny, I. (1999). *Fractional differential equations*. New York: Academic Press.

Sun, H. H., Abdelwahab, A. A., & Onaral, B. (1984). Linear approximation of transfer function with a pole of fractional order. *IEEE Transactions on Automatic Control*, *29*, 441–444. doi:10.1109/TAC.1984.1103551

Taherionl, S., & Lai, Y. C. (1999). Observability of lag synchronization of coupled chaotic oscillators. *Physical Review E: Statistical Physics, Plasmas, Fluids, and Related Interdisciplinary Topics*, *59*, R6247–R6250. doi:10.1103/PhysRevE.59.R6247

Uchida, A., Kinugawa, S., & Yoshimori, S. (2003). Synchronization of chaos in two microchip lasers by using incoherent feedback method. *Chaos, Solitons, and Fractals*, *17*, 363–368. doi:10.1016/S0960-0779(02)00375-2

Wu, X., Li, J., & Chen, G. (2008). Chaos in the fractional order unified system and its synchronization. *Journal of the Franklin Institute*, *345*, 392–401. doi:10.1016/j.jfranklin.2007.11.003

Yamada, T., & Fujisaka, H. (1983). Stability theory of synchronized motion in coupled oscillator systems II. *Progress of Theoretical Physics*, *70*, 1240–1248. doi:10.1143/PTP.70.1240

Zhang, L., & Chen, J. L. G. (2005). Extension of Lyapunov second method by fractional calculus. *Pure and Applied Mathematics*, *21*, 291–294.

Zhou, T., & Li, C. (2005). Synchronization in fractional order differential systems. *Physica D. Nonlinear Phenomena*, *212*, 111–125. doi:10.1016/j.physd.2005.09.012

Chapter 8
Chaos Synchronization

Hassan Salarieh
Sharif University of Technology, Iran

Mohammad Shahrokhi
Sharif University of Technology, Iran

ABSTRACT

Chaos synchronization is the central core of various message encryption methods which are developed based on the properties of chaotic systems. This chapter introduces the concept of chaos synchronization and its application in secure communication. Some standard approaches such as complete, lag, phase and generalized synchronization are defined first. Then application of control theory for synchronization of different chaotic systems is discussed. Some synchronization algorithms based on different control techniques are presented. It is shown that how the controlling methods can be modified in a synchronization framework to cope with parameter uncertainties and measurement noise. Several chaotic systems are simulated and synchronized to show the performance of the reported methods.

INTRODUCTION

Chaotic signals produced by nonlinear systems can be used for encrypting the transmitting messages in secure communications. Synchronization of chaotic systems has an important role in this field of science. Figure 1 shows how chaotic signals and chaos synchronization are utilized for secure communication. The message is a signal that should be transferred in a secure encrypted form. Using two chaotic systems, one in the transmitter side and the other in the receiver side, one can produce a secure link for communication. Indeed the system in the receiver side is synchronized to the one in the transmitter side for signal transmission. The message signal is usually modulated by a high amplitude chaotic signal to provide an encrypted signal for transmission. In the receiver side the other chaotic system which is called response or slave system is synchronized to the chaotic system of transmitter side which is usu-

DOI: 10.4018/978-1-61520-737-4.ch008

ally called drive or master system. Synchronization goal is achieved by coupling two chaotic systems through synchronizing signals which are transferred between two systems. The synchronized chaotic signal is subtracted from modulated signal (demodulation) to extract the message.

There are many techniques for chaotic modulation and synchronization. Synchronization generally occurs when the error between the states or phases of two dynamical systems becomes zero as time approaches to infinity. Synchronization may be achieved normally when a coupling is taken place between two systems. The coupling may be set naturally or intentionally by an error feedback from the modulating system to the demodulating system and vice-versa.

There are two ways for encoding and decoding a data signal, i.e. message, using chaotic systems. The first approach is called *chaos masking* in which the message signal is directly modulated by a chaotic signal (Scholl, and Schuster, 2008). Figure 1a shows a chaos masking scheme. The second approach is to modulate the chaotic system in transmitter side by the message signal and then transmitting the encrypted signal to the receiver. The encrypted signal is used for coupling two chaotic systems and synchronizing them. This method is called the *chaos modulation* or *chaos shift keying* (Scholl, and Schuster, 2008).

Many scientists and engineers have studied the conditions that result in synchronization. Winful and Rahamn (1990) showed that in an array of coupled lasers, identical chaotic signals are produced, and beyond a critical coupling strength, synchronization is substituted by spatiotemporal chaos. Pecora and Carroll (1990) described the conditions necessary for synchronizing a subsystem of a chaotic system with another chaotic system by sending a signal from the chaotic system to the subsystem. Carroll and Pecora (1991) built a simple circuit based on chaotic circuits described by Newcomb et al. (1983, 1986) and showed that a system, consisting of two Lorenz oscillators exhibiting chaos, could achieve synchronization if a portion of the second oscillator is driven by the first one. He and Vaidya (1992) introduced the necessary and sufficient condition for synchronization. Based on obtained results, they designed a high-dimensional chaotic system with nonlinear synchronized subsystems. The possibility of

Figure 1. Schematic diagram of secure data transfer using chaotic systems, (a) chaos masking technique, (b) chaos modulation technique

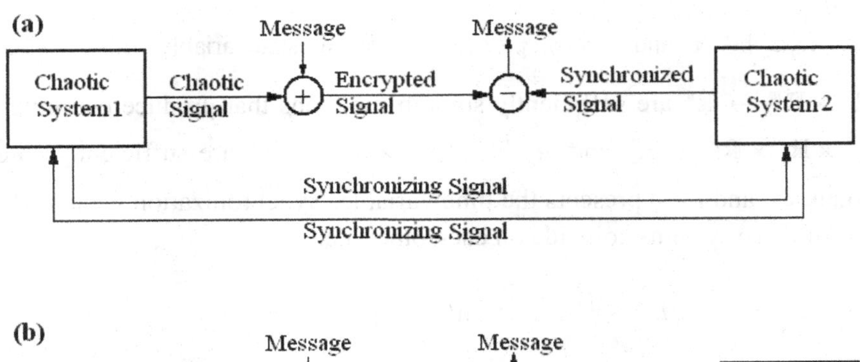

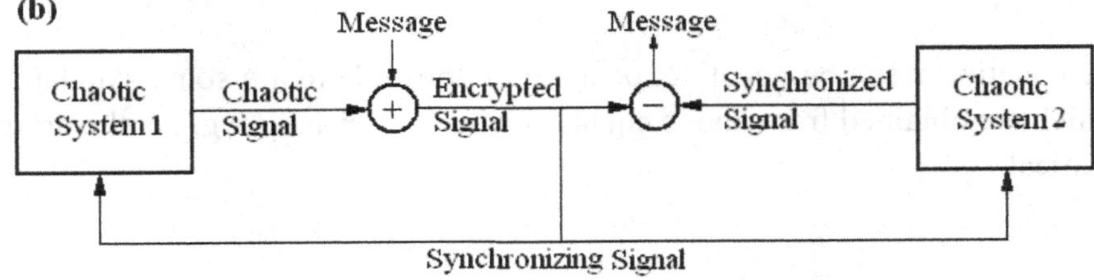

synchronization of systems inherently operating at chaotic modes was analyzed by Ogorzalek (1993), and potential applications of synchronized chaotic systems in signal processing were discussed and analyzed in several examples using the Chua circuit (Ogorzalek, 1993). Cuomo et al. (1993) presented a circuit realization of Lorenz system, and proposed two approaches for secure communication using the Lorenz circuit for both transmitter and receiver. Using a nonlinear digital filter in conjunction with its inverse filter, an encoding and decoding method was proposed for secure digital communication by Frey (1993). Roy and Thornburg (1994) reported an experimental synchronization of two chaotic Nd:YAG lasers.

Coupling between two chaotic systems which leads to synchronization can be modeled as a control action applied to the systems. Kapitaniak (1994) showed that how continuous chaos control can be applied for synchronization between two identical system. Sugawara et al. (1994) showed that two chaotic passive Q-switched lasers can be synchronized by sending the output of one laser to the cavity of the other laser. Heagy et al. (1994) investigated conditions that lead to synchronization in coupled oscillator systems. They presented some general criteria for the stability of synchronized chaotic behavior and tested them experimentally in analog chaotic circuits. Wu and Chua (1995) presented a sufficient condition for chaos synchronization in an array of linearly coupled systems. They stated that "the array will be synchronized if the nonzero eigenvalues of the coupling matrix have real parts that are negative enough". Mirasso et al. (1996) numerically showed that the synchronization of two chaotic semiconductor lasers is possible when a small fraction of output intensity from one is injected into the other.

CHAOS SYNCHRONIZATION

Consider the following chaotic systems:

$$\dot{x} = f\left(t, x, u(t, x, y)\right) \tag{1}$$

$$\dot{x} = f\left(t, x, u(t, x, y)\right) \tag{2}$$

where $x = \left(x_1, x_2, \ldots, x_n\right) \in \mathbb{R}^n$ and $y = \left(y_1, y_2, \ldots, y_m\right) \in \mathbb{R}^m$ arestatevariables, $u : \mathbb{R}^+ \times \mathbb{R}^n \times \mathbb{R}^m \to \mathbb{R}^p$ and $v : \mathbb{R}^+ \times \mathbb{R}^n \times \mathbb{R}^m \to \mathbb{R}^q$ are sufficiently smooth functions that produce couplings between two systems; $f : \mathbb{R}^+ \times \mathbb{R}^n \times \mathbb{R}^p \to \mathbb{R}^n$ and $g : \mathbb{R}^+ \times \mathbb{R}^m \times \mathbb{R}^q \to \mathbb{R}^m$ are sufficiently smooth functions with chaotic dynamics; and $t > 0$ presents the time variable. Synchronization occurs when all or some of state variables of these systems coincide on each other, i.e.

$$\lim_{t \to \infty} \left| x_i(t) - y_i(t - \tau) \right| = 0 \quad \exists i,\ i \in \left\{1, 2, \ldots, \min(m, n)\right\} \tag{3}$$

In the above equation τ can be negative, positive or zero. There are some standard forms of synchronization which are obtained from above equation when $m = n$ and $f = g$, i.e. the drive and response systems are identical.

COMPLETE SYNCHRONIZATION (CS):

This kind of synchronization is obtained when

$$\lim_{t\to\infty} \left|x_i(t) - y_i(t)\right| = 0 \quad \forall i,\ i \in \left\{1,2,\ldots,m=n\right\}, \qquad i.e. \quad \tau = 0 \tag{4}$$

As can be seen for this type of synchronization the time delay is zero.

LAG SYNCHRONIZATION (LS):

This kind of synchronization occurs when the time delay is positive:

$$\lim_{t\to\infty} \left|x_i(t) - y_i(t-\tau)\right| = 0 \quad \forall i,\ i \in \left\{1,2,\ldots,m=n\right\}, \quad \tau > 0 \tag{5}$$

ANTICIPATED SYNCHRONIZATION (AS):

When in lag synchronization the time delay is negative we say anticipated synchronization is occurred, i.e.

$$\lim_{t\to\infty} \left|x_i(t) - y_i(t+\tau)\right| = 0 \quad \forall i,\ i \in \left\{1,2,\ldots,m=n\right\}, \quad \tau > 0 \tag{6}$$

In Equation (1) and Equation (2) when $u(t, x, y) = 0$ and $v(t, x, y) \neq 0$ the coupling is called unidirectional, and usually the first dynamics is called master or drive system and the second dynamics is called slave or response system. When $u(t, x, y) \neq 0$ and $v(t, x, y) \neq 0$ we have a bidirectional coupling between two chaotic systems, and each system plays the roles of slave and master for the other system.

GENERALIZED SYNCHRONIZATION (GS)

The concept of synchronization has been generalized in various forms. One of the first ideas for generalization of synchronization was presented by Rulkov et al. (1995). They suggested that instead of states of two identical chaotic systems, states of the response system to be synchronized with a function of states of the drive system, i.e.

$$\lim_{t\to\infty} \left|\phi_i\left(x(t)\right) - y_i(t)\right| = 0 \quad \exists i,\ i \in \left\{1,2,\ldots,m=n\right\}, \quad f = g \tag{7}$$

where $\phi_i(.)$ is a sufficiently smooth function. The above mentioned definition of generalized synchronization is especially helpful when states are not observed directly, and state measurement is performed through a non identity function, $\phi_i(.)$.

Another generalization of chaos synchronization is obtained when $f \neq g$, i.e. dynamics of master and slave systems are different.

For these cases "generalized synchronization" (GS) term has been usually used and therefore the above definitions are modified to generalized complete synchronization (GCS), generalized lag synchronization (GLS) and generalized anticipated synchronization (GAS).

PHASE SYNCHRONIZATION (PS)

In addition to various kinds of synchronization described above, there is a special form of chaos synchronization in which the phases of two chaotic systems are synchronized. This form of synchronization is effectively used in applications of phase locked loops (PLL) in a large number of radio and telecommunication devices, radio- location, etc. (Scholl, 2008).

The phase of a chaotic oscillator is not a simple one as it is defined for a periodic oscillator. Several definitions of phase and frequency of a chaotic oscillator have been proposed (Gonzalez Miranda, 2004). A straightforward definition is based on the fact that if a chaotic attractor follows a proper rotation in the phase space, there exists a plane in which the projection of the attractor follows a proper rotation too. The phase angle can be defined by using a polar coordinate on that plane. For example in the x-y plane, the phase angle is (Gonzalez Miranda, 2004):

$$\phi(t) = \arctan\left[\frac{y(t) - y(0)}{x(t) - x(0)}\right] \tag{8}$$

The phase angle has a dynamics induced by the main chaotic system:

$$\phi(t) = \phi_0 + \Omega t + \xi(t) \tag{9}$$

where Ω is a first order drift constant and $\xi(t)$ is a nonlinear chaotic fluctuation of the phase angle. To obtain Ω, one of the commonly used methods is time averaging of the phase angle (Scholl, 2008), i.e.

$$\Omega = \lim_{T \to \infty} \frac{\phi(T) - \phi(0)}{T} \tag{10}$$

The chaotic systems may be rewritten using the phase angle as an independent variable:

$$\frac{dx}{d\phi} = f\left(\phi, x\right) \tag{11}$$

Now consider the chaotic systems (1) and (2) re-written with respect to theirs phase angles as:

$$\frac{dx}{d\phi} = f\left(\phi, x, u(\phi, x, y)\right) \tag{12}$$

$$\frac{dy}{d\psi} = g\big(\psi, y, v(\psi, x, y)\big) \tag{13}$$

where $\phi = \phi(t)$ and $\psi = \psi(t)$ are phase angles of the drive and the response systems. In phase synchronization, the goal is to synchronize the oscillation phases (ϕ and ψ) using a feedback control of characteristic time scales (CTS) of two (or many) different interacting oscillators. In order to test the existence of PS between two chaotic oscillators, two criteria should be checked:

1. Frequency locking condition: The mean frequencies of both coupled subsystems should be equal.

$\Omega_1 = \Omega_2$ where

$$\Omega_1 = \lim_{T \to \infty} \frac{\phi(T) - \phi(0)}{T}, \qquad \Omega_2 = \lim_{T \to \infty} \frac{\psi(T) - \psi(0)}{T}$$

2. Phase Bound condition: The phase difference should be bounded

$$\left|\phi - \psi\right| \le const.$$

Since the phase angle is a function of systems states, the phase synchronization can be considered as a kind of generalized synchronization. Zheng and Hu (2000) investigated the relation between generalized synchronization and phase synchronization. They claimed that in presence of parameter misfit, the generalized synchronization may be weaker than the phase synchronization and it does not lead to phase synchronization. In addition, Boccaletti et al. (2001) tried to simplify and unify the definition of synchronization for a coupled dynamical system to a more condensed and concrete form.

There are many works in which the phase synchronization and its properties and applications have been studied. Rosenblum et. al (1997) studied transition from phase synchronization to lag synchronization. They showed that in a coupled chaotic oscillator with synchronized phases, when the coupling strength increases, phase synchronization is substituted by lag synchronization.

Phase synchronization may be used for chaos control when an external periodic force is applied to the system. In this case under specific conditions, the irregular pattern of chaotic trajectory is replaced by a synchronized path whose frequency is locked identically to the frequency of the excitation. Pikovsky et al. (1997) defined phase-locking regions for unstable cycles embedded in a chaotic attractor. They described synchronization in terms of these regions. Chaos control with the help of synchronization by an external force was applied to an electromechanical model (Kiss, and Hudson, 2001). Phase synchronization as a controlling tool is studied in a Rossler chaotic system (Rosa et al., 1998). Experimental real-time synchronization of paced chaotic plasma was investigated in (Ticos et al., 2000). Kiss and Hudson (2001) studied chaos suppression of electrodissolution of nickel by sulphuric acid by synchronizing the reaction to a forcing signal.

Zheng et al. (1998) studied the phenomena of phase slip and phase synchronization in a coupled nonlinear oscillators. They presented a bifurcation tree from synchronization to chaos in the coupled systems versus the strength of coupling. In addition to communication and cryptography, phase synchronization has some applications in biology and neuroscience. Shuai and Durand (1999) investigated the phase synchronization of Chaotically-spiking dynamics of Hindmarsh-Rose neurons as a multiple time scale model. DeShazer et al. (2001) presented a Gaussian filtered phase variable to detect phase synchronization in spatially coupled lasers which shows chaotic behavior. Chen et al. (2003) developed a digital secure communication scheme based on phase synchronization. They utilized the phase of drive chaotic system as a signal transmitted to the response system for synchronization.

ERROR FEEDBACK SYNCHRONIZATION

In most cases the synchronization problem can be stated via a control design problem. Coupling between two systems can be considered as a feedback signal that enters from one system to another. There are many works in the field of chaos synchronization where error feedback control is used as a synchronizing signal (Kapitaniak, 1994). Yang (1996) presented a synchronization scheme between transmitter and receiver by using an adaptive control scheme. He "extended chaotic switching to general chaotic parameter modulation" in secure communication. Femat et al. (2000) proposed an adaptive strategy for chaos synchronization that can be applied to non-identical systems. Bai and Lonngren (2000) applied sequential synchronization to two Lorenz systems using an active control scheme.

Control strategies that have been used for chaos synchronization by scientists and engineers can be divided generally into two main categories; adaptive and non-adaptive schemes. Considering the fact that all dynamical systems may become stochastic due to random uncertainties, noise or random forcing signals, control techniques used for chaos synchronization can be classified into stochastic and deterministic approaches. There is another classification for chaos synchronization; single, dual and multi-synchronization. Single synchronization is achieved when one chaotic system is synchronized to another chaotic system. In dual synchronization, two couples of chaotic systems are synchronized. In multi-synchronization which is an extension to dual synchronization, there are several (more than two) chaotic systems in transmitter and receiver sides that should be synchronized.

NON-ADAPTIVE CONTROL METHODS OF SYNCHRONIZATION

Non-adaptive chaos synchronization can be used when system parameters and functionalities are known (Kapitaniak, 1994) or have some bounded uncertainties with known bounds. There are many linear and nonlinear non-adaptive control methods that have been used for chaos synchronization. Lyapunov based nonlinear control, impulsive control, variable structure systems, sliding mode control, delayed feedback control, backstepping technique and etc. have been widely used to synchronize two identical or non-identical chaotic systems. Some of mentioned techniques can be utilized for systems with parameter uncertainties. The robustness of the synchronized systems against modeled or un-modeled uncertainties can be guaranteed. A systematic approach based on using a linear state observer was developed by Liao and Huang (1999) for constructing two synchronized chaotic systems. They showed how the proposed method can be utilized for secure communication. Liu et al. (2002) presented a linear output feedback control for chaos synchronization of a class of chaotic systems. Yang and Chua (1997) presented an impulsive control technique for chaotic communication. Grassi and Mascolo (1997) derived a nonlinear observer for chaos synchronization between two identical systems. In this work the error feedback for chaos synchronization was a scalar signal. Synchronization of delay differential equations that provide hyperchaos has been considered by Mensour and Longtin (1998) using a scalar feedback control. Agiza and Yassen (2001) studied chaos synchronization between two identical Chen and Rossler systems using active control. Li et al. (2001) proposed an impulsive control with time varying intervals for chaos synchronization of Chua's circuits. Synchronizing two chaotic electronic circuits using modern control theory was studied by Bai et al. (2002). Liao and Chen (2003) investigated chaos synchronization of Lur'e systems via delayed feedback control. Backstepping technique via a recursive algorithm was used for chaos synchronization of two Lorenz, Chua and Duffing systems by Tan et al. (2003). Chen and Han (2003) proposed a nonlinear feedback control for synchronizing two Genesio chaotic systems. They utilized Hurwitz stability analysis to design the controller. Yassen (2005) presented an active control technique for chaos synchronization for Lorenz, Lu and Chen systems. Lyapunove stability theory was used to design a nonlinear control for chaos synchronization between two identical and two different chaotic systems by Chen (2005). Cao et al. (2005) presented a delayed feedback strategy for chaos synchronization of Lur'e systems based on Lyapunov stability theory and linear matrix inequalities (LMI). Park (2006) studied chaos synchronization between Genesio and Rossler systems using Lyapunov based nonlinear controller. Ucar et al. (2006) investigated synchronization between unified chaotic systems via active control. Behzad et al. (2008) presented a sliding mode control method coupled by a nonlinear observer to synchronized two different chaotic systems. The chaotic systems had some bounded random uncertainties, and the measurement system was incomplete and perturbed by bounded noise. Chen et al. (2005) proposed an observer based synchronization scheme using a sliding mode control for secure communication. Yang and Shao (2002) and Zhang et al. (2004) utilized a sliding mode control scheme for chaos synchronization in presence of parameter uncertainty. Salarieh and Alasty (2008) proposed a modified sliding mode control technique for synchronizing two gyro dynamics with stochastic variable inputs.

A TYPICAL SYNCHRONIZATION ALGORITHM BASED ON SLIDING MODE CONTROL (ETEMADI ET AL., 2006)

Consider chaotic systems of Equation (1) and Equation (2) and assume that the control action, i.e. the coupling signal, is unidirectional and has an affine form:

$$x^{(n)} = f(t, x) \tag{14}$$

$$y^{(n)} = g(t,y) + u \tag{15}$$

Where $x = (x, x, \ldots, x^{(n-1)})$ and $y = (y, y, \ldots, y^{(n-1)})$ are state vectors and u is the control action. f and g are two uncertain functions whose nominal values are denoted by $\tilde{f}$ and $\tilde{g}$ are assumed to be known. Besides, the upper bounds of $\left|f - \tilde{f}\right|$ and $\left|g - \tilde{g}\right|$ are given by Δf and Δg. Denoting the state error $e = x\text{-}y$, e = (e,e, …,$e^{(n-1)}$), the synchronization error dynamics is obtained as: $e^{(n)} = f(t, y + e) - g(t, y) - u$ (16)

To design a robust control law for chaos synchronization which endures system uncertainties, an asymptotically stable dynamics for the state error is selected as:

$$S = \sum_{i=0}^{n} \alpha_i \int_0^t e^{(i)}(\tau)d\tau = 0, \qquad \alpha_n > 0 \tag{17}$$

where α_i's are chosen such that the roots of the resulted polynomial $\sum_{i=0}^{n} \alpha_i z^i$ have negative real parts. *S* makes a hyper-plane in error state space that is called the sliding surface. If the control strategy cause the error trajectories converge to the sliding surface in a finite time, the state error will converge to zero, and the synchronization goal is achieved. A Lyapunov function is defined as:

$$V = \frac{1}{2}S^2 \tag{18}$$

The control objective is achieved by satisfying the following reaching condition:

$$\frac{d}{dt}V \leq -\eta|S| \tag{19}$$

where η is an arbitrary positive constant.

Theorem 1.

Using the following control action:

$$u = \tilde{f}(t,x) - \tilde{g}(t,y) + \sum_{0}^{n-1} \frac{\alpha_i}{\alpha_n} e^{(i)} + \frac{K}{\alpha_n} sign(S) \tag{20}$$

where

$$K \geq \Delta f = \Delta g = \eta \tag{21}$$

The reaching condition of Equation (19) is satisfied and the chaotic systems (14) and (15) will be synchronized.

Example 1. Synchronizing the Chaotic Lure and Genesio Systems via Sliding Mode Control (Etemadi et al., 2006)

The drive and response systems are set to be Lure and Genesio systems respectively:

$$x^{(3)} = a_1 x + a_2 \dot{x} + a_3 \ddot{x} + 12\,h(x), \tag{22}$$

$$y^{(3)} = b_1 y + b_2 \dot{y} + b_3 \ddot{y} - b_4 y^2 + u \tag{23}$$

where $a_1 = -6.8$, $a_2 = -3.9$, $a_3 = -1$, $b_1 = -6$, $b_2 = -2.92$, $b_3 = -1.2$, $b_4 = -1$ and

$$h\left(x\right) = \begin{cases} kx & if\left|x\right| < \frac{1}{k} \\ sign(x) & otherwise \end{cases} \tag{24}$$

$K = 1.5$. It is assumed that the nominal known values of the system parameters are $\hat{a}_1 = -7.1$, $\hat{a}_2 = -3.7$, $\hat{a}_3 = -0.95$, $\hat{b}_1 = -6.3$, $\hat{b}_2 = -2.8$, $\hat{b}_3 = -1.25$, $\hat{b}_4 = -0.96$, It is assumed that parameter uncertainties are about 5 percents, i.e.

$$\left|a_i - \hat{a}_i\right| \leq \delta_i, \qquad \left|b_i - \hat{b}_i\right| \leq \Delta_i, \qquad i = 1,2,3 \tag{25}$$

where

$$\delta_i \leq 0.05\left|\hat{a}_i\right|, \qquad \Delta_i \leq 0.05\left|\hat{b}_i\right|, \qquad i = 1,2,3 \tag{26}$$

Considering Equation (22) and Equation (23) we have:

$$\begin{aligned} f(t,x) &= a_1x + a_2\dot{x} + a_3\ddot{x} + 12\,h\left(x\right), & g(t,y) &= b_1y + b_2\dot{y} + b_3\ddot{y} - b_4y^2 \\ \hat{f}(t,x) &= \hat{a}_1x + \hat{a}_2\dot{x} + \hat{a}_3\ddot{x} + 12\,h\left(x\right), & \hat{g}(t,y) &= \hat{b}_1y + \hat{b}_2\dot{y} + \hat{b}_3\ddot{y} - \hat{b}_4y^2 \end{aligned} \tag{27}$$

and

$$\begin{aligned} \left|f(t,x) - \hat{f}(t,x)\right| &\leq \left|a_1 - \hat{a}_1\right|\left|x\right| + \left|a_2 - \hat{a}_2\right|\left|\dot{x}\right| + \left|a_3 - \hat{a}_3\right|\left|\ddot{x}\right| \\ \left|g(t,y) - \hat{g}(t,y)\right| &\leq \left|b_1 - \hat{b}_1\right|\left|y\right| + \left|b_2 - \hat{b}_2\right|\left|\dot{y}\right| + \left|b_3 - \hat{b}_3\right|\left|\ddot{y}\right| + \left|b_4 - \hat{b}_4\right|y^2 \end{aligned} \tag{28}$$

Therefore Δf and Δg can be written as:

$$\begin{aligned} \Delta f &= \delta_1\left|x\right| + \delta_2\left|\dot{x}\right| + \delta_3\left|\ddot{x}\right| \\ \Delta g &= \Delta_1\left|y\right| + \Delta_2\left|\dot{y}\right| + \Delta_3\left|\ddot{y}\right| + \Delta_4y^2 \end{aligned} \tag{29}$$

By setting $\eta = 1$ and using $x(0) = x(0) = x(0) = 1$, $y(0) = y(0) = y(0) = 0$ numerical simulation is performed. Figure 2 shows the results.

The master and slave chaotic systems are perturbed by white Gaussian noise generated by a Weiner process:

$$x^{(n)} = f(x,t) = h(x,t)v \tag{30}$$

$$y^{(n)} = g(y,t) = k(y,t)w = b(y,t)u \tag{31}$$

Figure 2. Synchronization errors between Lure and Genesio systems A Typical Synchronization Algorithm Based on Sliding Mode Control for Stochastic Systems (Salarieh, and Alasty, 2008-a)

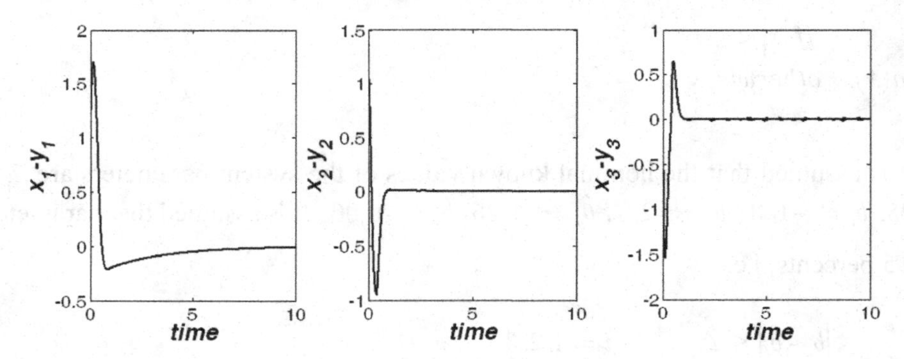

where $x = (x, x, \ldots, x^{(n-1)})$ and $y = (y, y, \ldots, y^{(n-1)})$ are the state vectors, $f, g, h, k, b : \Re^n \times \Re^+ \rightarrow \Re$ are nonlinear and sufficiently smooth functions, v and w are two standard independent Wiener processes and. It is assumed that f, g and b are unknown functions whose nominal known values are denoted by $\hat{f}$, $\hat{g}$ and $\hat{b}$. f, g, b, $\hat{f}$, $\hat{g}$ and $\hat{b}$ satisfy the following conditions:

$$\left|f(x,t) - \hat{f}(x,t)\right| < \Delta f, \quad \left|g(y,t) - \hat{g}(y,t)\right| < \Delta g \tag{32}$$

$$0 < b_m < b(\mathbf{y},t), \hat{b}(\mathbf{y},t) < b_M \tag{33}$$

where F, G, b_m and b_M are known. It is assumed that all the state variables of drive and response systems are available, and functions $h(x,t)$ and $k(y,t)$ are bounded with known bounds M_h and M_k, i.e. $\left|h(x,t)\right| \leq M_h$ and $\left|k(y,t)\right| \leq M_k$. The following two theorems are useful for chaos synchronization of systems (30) and (31).

Theorem 2. (Salarieh, and Alasty, 2008-a):

Define a sliding surface as:

$$S(t) = \left(\frac{d}{dt} + \lambda\right)^{n-1} e(t) = \sum_{m=0}^{n-1} \binom{n-1}{m} e^{(n-1-m)} \lambda^m, \quad \binom{n-1}{m} = \frac{(n-1)!}{m!(n-1-m)!} \tag{34}$$

where $\lambda > 0$ is a positive constant, and a Lyapunov function as

$$V(t) = \frac{1}{2} E[S(t)]^2 \tag{35}$$

where E [.] denotes expected value function. Let the control action be

$$u = -\frac{1}{b_m(y,t)}\left[\hat{g}(y,t) - \hat{f}(x,t) + \sum_{m=1}^{n-1}\binom{n-1}{m}e^{(n-m)}\lambda^m + Ksign(S(t)) + \theta S(t)\right] \tag{36}$$

where $\theta > 0$ is a positive constant and K satisfies the following inequality:

$$K \geq \left\{G + F + (\frac{b_M}{b_m} - 1)\left(\left|\hat{g} - \hat{f}\right| + \left|\sum_{m=1}^{n-1}\binom{n-2}{m}e^{(n-m)}\lambda^m\right|\right)\right\} \tag{37}$$

Using the above control law, the following region in the error phase space is an attracting set for $S(t)$.

$$\Omega = \left\{e \,\middle|\, E\left[S^2\right] < \frac{M_k^2 + M_h^2}{\theta}\right\} \tag{38}$$

It means that:

$$E\left[S^2(t)\right] < \frac{M_k^2 + M_h^2}{\theta} \qquad as \ \ t \to ¥ \tag{39}$$

Theorem 3 (Salarieh, and Alasty, 2008-a):

Let

$$A = \begin{bmatrix} 0 & 1 & 0 & \cdots & 0 \\ 0 & 0 & 1 & & \vdots \\ \vdots & \vdots & \ddots & \ddots & 0 \\ 0 & 0 & \cdots & 0 & 1 \\ -\lambda^{n-1}\binom{n-1}{n-1} & -\lambda^{n-2}\binom{n-1}{n-2} & \cdots & \cdots & -\lambda\binom{n-1}{1} \end{bmatrix}, \quad B = \begin{bmatrix} 0 \\ 0 \\ \vdots \\ 0 \\ 1 \end{bmatrix}, \quad \eta(t) = \begin{bmatrix} e(t) \\ \dot{e}(t) \\ \vdots \\ e^{(n-2)}(t) \\ e^{(n-1)}(t) \end{bmatrix} \tag{40}$$

For any $\lambda > 0$, matrix **A** is Hurwitz and hence the following Lyapunov equation gives a positive definite symmetric matrix **P**:

$$PA = A^T P = -I \tag{41}$$

where **I** is the identity matrix. The control law (36) and (37) results in the following attracting set:

$$\Delta = \left\{\eta \,\middle|\, \left\|\eta(t)\right\| < 2\left\|B^T P\right\|\left(\frac{M_h^2 + M_k^2}{\theta}\right)^{1/2}\right\} \tag{42}$$

Besides

$$E\left[\left(e^{(n-1)}(t)\right)^2\right] \le \left(\sum_{m=1}^{n-1}\binom{n-1}{m}\lambda^m \left\|B^T P\right\| + 1\right)\left(\frac{M_h^2 + M_k^2}{\theta}\right)^{1/2} \tag{43}$$

Theorem 3. Shows That the Tracking Error Can Be Reduced By Increasing The Value Of θ Which Is A Design Parameter

Example 2. Stochastic Chaos Synchronization between Two Nonlinear Gyros (Salarieh, and Alasty, 2008)

The following equations describe the nonlinear dynamics of two gyros with harmonic base excitation. The base excitation is perturbed by the white Gaussian noise. The first and the second equations describe dynamics of drive and response systems respectively.

$$\ddot{x} = -\alpha^2 \frac{(1 - \cos x)^2}{\sin^3 x} - c_1\dot{x} - c_2\dot{x}^3 + \beta \sin x + \left(f \sin \omega t + \mu\, \dot{v}\right)\sin x \tag{44}$$

$$\ddot{y} = -\alpha_s^2 \frac{(1 - \cos y)^2}{\sin^3 y} - c_{1s}\dot{y} - c_{2s}\dot{y}^3 + \beta_s \sin y + (f_s \sin \omega t + \mu_s\, \dot{w})\sin y + (1 + |y|)u \tag{45}$$

where θ is the nutation angle, i.e. the angle which the spin axis of gyro makes with the vertical axis. $\dot{v}$ and $\dot{w}$ are two independent white Gaussian noises; μ and μ_S are two real constants and u is the control action. According to Equations (30) and (31) the functions $f(.)$, $g(.)$, $h(.)$, $k(.)$ and $b(.)$are:

$$\begin{aligned}
f(x,t) &= -\alpha^2 \frac{(1 - \cos x)^2}{\sin^3 x} - c_1\dot{x} - c_2\dot{x}^3 + \beta \sin x + f \sin \omega t \sin x \\
h(x,t) &= \mu \sin x \\
g(y,t) &= -\alpha_s^2 \frac{(1 - \cos y)^2}{\sin^3 y} - c_{1s}\dot{y} - c_{2s}\dot{y}^3 + \beta_s \sin y + f_s \sin \omega t \sin y \\
k(y,t) &= \mu_s \sin y \\
b(y,t) &= 1 + |y|
\end{aligned} \tag{46}$$

where $x = (x_1 x_2) = (x\ \dot{x})$ and $y = (y_1 y_2) = (y\ \dot{y})$. For $\alpha^2 = 100$, $\beta = 1$, $c_1 = 0.5$, $c_2 = 0.05$, $\omega = 2$, $f = 35.5$, $\alpha_s^2 = 94$, $\beta_s = 1.2$, $c_{1s} = 0.45$, $c_{2s} = 0.04$, $f_s = 34$, $u = 0$ and $\mu = \mu_s = 0$ the master and the slave systems show chaotic behaviors.

Due to system uncertainties, the nominal values of parameters which are assumed to be known are denoted by $\hat{\alpha}^2 = 110$, $\hat{\beta} = 1.4$, $\hat{c}_1 = 0.7$, $\hat{c}_2 = 0.06$, $\hat{f} = 37$, $\hat{\alpha}_s^2 = 90$, $\hat{\beta}_s = 0.8$, $\hat{c}_{1s} = 0.34$, $c_{2s} = 0.046$ and $\hat{f}_s = 33$. Thus the nominal functions $\hat{f}(.)$, $\hat{g}(.)$, $\hat{b}(.)$ are obtained as:

$$\hat{f}(x,t) = -\hat{\alpha}^2 \frac{(1 - \cos x)^2}{\sin^3 x} - \hat{c}_1\dot{x} - \hat{c}_2\dot{x}^3 + \hat{\beta}\sin x + \hat{f}\sin\omega t \sin x$$
$$\hat{g}(y,t) = -\hat{\alpha}_s^2 \frac{(1 - \cos y)^2}{\sin^3 y} - \hat{c}_{1s}\dot{y} - \hat{c}_{2s}\dot{y}^3 + \hat{\beta}_s \sin y + \hat{f}_s \sin\omega t \sin y \qquad (47)$$
$$\hat{b}(y,t) = 1 + 0.5\left|y\right|$$

The bound functions $F(.)$, $G(.)$, $b_m(.)$ and $b_M(.)$ may be chosen as:

$$F(x,t) = 20\frac{(1 - \cos x)^2}{\left|\sin^3 x\right|} + 0.5\left|\dot{x}\right| + 0.1\left|\dot{x}^3\right| + 7\left|\sin x\right|$$
$$G(y,t) = 20\frac{(1 - \cos y)^2}{\left|\sin^3 y\right|} + 0.5\left|\dot{y}\right| + 0.1\left|\dot{y}^3\right| + 7\left|\sin y\right| \qquad (48)$$
$$b_m(y,t) = 0.5 + 0.5\left|y\right|, \qquad b_M(y,t) = 2 + \left|y\right|$$

Parameters μ and μ_S are set to 1, and θ and λ are set to 4 and 1 respectively. Simulation results are shown in Figure 3a and 3b. It is seen that the synchronization objective is achieved and mean values of tracking errors ($e_i = x_i - y_i$, $i = 1{,}2$) converge to zero (Figure 3a) and mean square norms of errors (standard deviations, std) converge to a vicinity of zero (Figure 3b).

Synchronization errors time series and time series of master and slave states are shown in Figure 3c.

ADAPTIVE CONTROL METHODS OF SYNCHRONIZATION

Adaptive schemes for chaos synchronization are used when there are some unknown parameters in equations describing the system dynamics. Usually the control law is accompanied by an identification algorithm which is called adaptation law. The adaptive scheme may be direct or indirect. In indirect method an identification algorithm is used to estimate the unknown system parameters, and based on these estimates and selected control strategy, control action is calculated. In direct method controller parameters are estimated directly. Indirect adaptive synchronization between two Lorenz systems based on the Lyapunov stability theorem was developed by Liao (1998). Liao and Tsai (2000) presented an indirect adaptive control for chaos synchronization between two identical chaotic systems. In this work a Lumberger like observer was used to estimate the system states. Wang and Ge (2001) proposed an adaptive backstepping control technique for chaos synchronization. In their work an adaptation law for identification of uncertainties was developed via backstepping method. Bowong and Kakmeni (2004) presented an adaptive scheme for chaos synchronization between two uncertain systems based on backstepping approach. Chen and Lu (2002) proposed a combination of nonlinear control scheme and an identifier for chaos synchronization of nonlinear systems with unknown coefficients. Feki (2003) extended their work by using an output feedback for chaos synchronization between two identical systems. Wang et al. (2004) and Elabassy et al. (2004) presented similar adaptive approaches for chaos synchronization between two Chen and two Lu systems. Park (2005) also used an adaptive scheme for synchronizing two hyperchaotic Chen systems. A novel method for chaos synchronization between two identical systems

Figure 3. (a) Mean values of synchronization errors, (b) Mean square norms of synchronization errors, c) Time series of synchronization errors, the master states (continuous lines) and the slave states (discrete lines) (Salarieh, and Alasty, 2008-a)

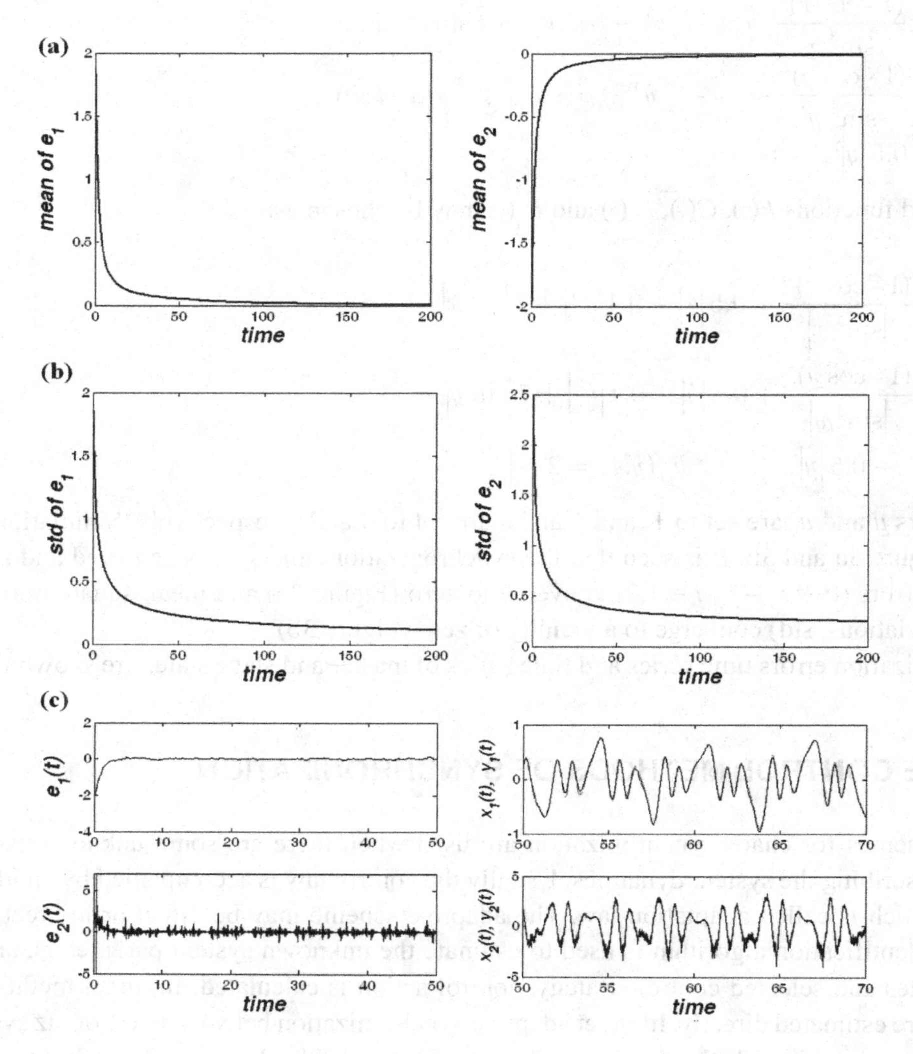

with unknown parameters and external disturbance acting on slave system was proposed by Wang and Su (2004). The presented method is an example of applying the direct adaptive control technique for chaos synchronization. Yau (2004) proposed an adaptive sliding mode control for chaos control of two unified chaotic systems in presence of uncertainties and disturbances. Recently chaos synchronization between two non-identical chaotic systems has attracted engineers and scientists attention. Many adaptive control methods have been developed for chaos synchronization of two different systems (Lu and Cao, 2005 and Zhang et al., 2006). Lu and Cao (2005) studied synchronization of chaotic and hyper-chaotic systems through adaptive control techniques. Their method can be used for non-identical systems as well as identical systems. Yu and Cao (2007) presented an adaptive control algorithm for lag synchronization between two different chaotic systems. Banerjee et al. (2004) proposed an adaptive scheme for phase synchronization between two atmospheric dynamic models. Banerjee and Chowdhury (2009) presented an adaptive control for chaos synchronization between Chen and Lorenz systems and described the application of the proposed method in cryptography. Saha et al. (2004) discussed an adaptive approach for secure communication based on chaos synchronization. Salarieh and Shahrokhi (2008) proposed an adaptive synchronization technique that can be used for chaotic systems with unknown and time varying parameters. They showed that the system parameters can be identified under a weak condition which is usually satisfied for chaotic systems.

A TYPICAL ADAPTIVE ALGORITHM FOR DETERMINISTIC SYSTEMS

Generally, in applying adaptive control techniques for synchronizing chaotic master and slave systems, systems of Equation (1) and Equation (2) have the following forms:

$$x = f(x) + F(x)\,\Theta \tag{49}$$

$$y = g(y) = G(y)\psi + H(y)u \tag{50}$$

Where $x \in R^n$ and $y \in R^n$ are the state vector of systems, $\Theta \in R^m$ and $\psi \in R^l$ are the vector of system parameters which are unknown and may be time varied, $f \in C^1(R^n, R^n)$, $g \in C^1(R^n, R^n)$, $F \in C^1(R^n, R^{n\times m})$ and $G \in C^1(R^n, R^{n\times l})$ are prescribed functions. For simplicity, it is assumed that all states of the master and slave systems are measurable or available. If the states are not completely available one can use an observer for state estimation, as done in (Behzad et al., 2008). To present a general method for finding adaptive synchronization law, define two auxiliary dynamic systems as:

$$\dot{\hat{x}} = f(\hat{x}) + F(\hat{x})\alpha + w_1 \tag{51}$$

$$\dot{\hat{y}} = g(\hat{y}) + G(\hat{y})\beta + H(\hat{y})u + w_2 \tag{52}$$

where α and β are estimates of θ and ψ respectively, and w_1 and w_2 are two control laws that must be designed appropriately.

The following theorem can be used to design an adaptive control scheme for chaos synchronization between two systems of Equation (49) and Equation (50) (Salarieh, and Shahrokhi, 2008).

Theorem 4. (Salarieh, and Shahrokhi, 2008):

Consider the drive-response systems (49) and (50), and suppose $H(y)$ is invertible and assume that θ and ψ are both bounded, randomly time varying with known bounds, i.e.

$$\left\|\theta - \bar{\theta}\right\| < \Theta,\ \left\|\psi - \bar{\psi}\right\| < \Psi \tag{53}$$

where $\bar{\theta}$ and $\bar{\psi}$ are the unknown nominal values of θ and ψ, and Θ and ψ are known constants. The following controller and adaptive laws will synchronize systems (49) and (50),

$$u = H^{-1}(y)[f(x) - g(y) + F(x)\alpha - G(y)\beta + ke_1 + \|F(x)\|\Theta sign(e_1) + \|G(y)\|\Psi sign(e_1)] \tag{54}$$

$$\alpha = PF^T(x)(e_2 + e_1) \tag{55}$$

$$\beta = QG^T(y)(e_3 - e_1) \tag{56}$$

$$w_1 = f(x) - f(\hat{x}) + F(x)\alpha - F(\hat{x})\alpha + ke_2 + \|F(x)\|\Theta sign(e_2) \tag{57}$$

$$w_2 = g(y) - g(\hat{y}) + G(y)\beta - G(\hat{y})\beta + H(y)u - H(\hat{y})u + ke_3 + \|G(y)\|\Psi sign(e_3) \tag{58}$$

where $k > 0$ is chosen arbitrarily, $\mathbf{e}_1 = \mathbf{x} - \mathbf{y}$, $e_2 = x - \hat{x}$ and $e_3 = y - \hat{y}$, and P and Q are two positive definite symmetric matrices. $sign(e_i), i = 1,2,3$ is a vector whose elements are sign function of each elements of e_i, $i = 1,2,3$. $\|...\|$ denotes the Euclidian norm of vectors and norm of matrices, i.e.

$$X = \left[X_1\ X_2\ \cdots\ X_n\right]^T, \quad \|X\| = \left(\sum_{i=1}^{n} X_i^2\right)^{1/2}$$
$$A = \left[A_{ij}\right]_{n\times n}, \quad \|A\| = \left(\sum_{i,j} A_{ij}^2\right)^{1/2} \tag{59}$$

Example 3. Chaos Synchronization between the Lorenz and the Chen Systems (Salarieh, and Shahrokhi, 2008)

As an example the chaotic Lorenz and Chen systems are selected as drive and response dynamics. The Lorenz system is described by

$$\begin{bmatrix}\dot{x}_1\\ \dot{x}_2\\ \dot{x}_3\end{bmatrix} = \begin{bmatrix}0\\ -x_2 - x_1x_3\\ x_1x_2\end{bmatrix} + \begin{bmatrix}x_2 - x_1 & 0 & 0\\ 0 & x_1 & 0\\ 0 & 0 & -x_3\end{bmatrix}\begin{bmatrix}\theta_1\\ \theta_2\\ \theta_3\end{bmatrix} \tag{60}$$

Where $\Theta_1 = 10$, $\Theta_2 = 28$, $\Theta_3 = 8/3$ are positive constant parameters which are assumed to be unknown. The Chen system is given as:

$$\begin{bmatrix}\dot{y}_1\\ \dot{y}_2\\ \dot{y}_3\end{bmatrix} = \begin{bmatrix}0\\ -y_1y_3\\ y_1y_2\end{bmatrix} + \begin{bmatrix}y_2 - y_1 & 0 & 0\\ -y_1 & y_1 + y_2 & 0\\ 0 & 0 & -y_3\end{bmatrix}\begin{bmatrix}\psi_1\\ \psi_2\\ \psi_3\end{bmatrix} + \begin{bmatrix}1 & 0 & 0\\ 0 & 1 & 0\\ 0 & 0 & 1\end{bmatrix}\begin{bmatrix}u_1\\ u_2\\ u_3\end{bmatrix} \tag{61}$$

which has a chaotic attractor when $u_1 = u_2 = u_3 = 0$, $\psi_1 = 35$, $\psi_2 = 28$, and $\psi_3 = 3$.

To apply the control and adaptation laws given in Equations (54) to (58), three different cases are considered. At first it is assumed that $\theta = \overline{\theta} + \delta_1(t)$ and $\psi = \overline{\psi} + \delta_2(t)$ where $\overline{\theta} = [10, 28, 8/3]^T$ and $\overline{\psi} = [35, 28, 3]^T$, and $\delta_i(t), i = 1,2$ are two continuous random variables with zero mean and

Figure 4. (a) Synchronization errors, (b) System parameters and their estimates for $\Theta = \Psi = 3\sqrt{3}$ *(Salarieh, and Shahrokhi, 2008)*

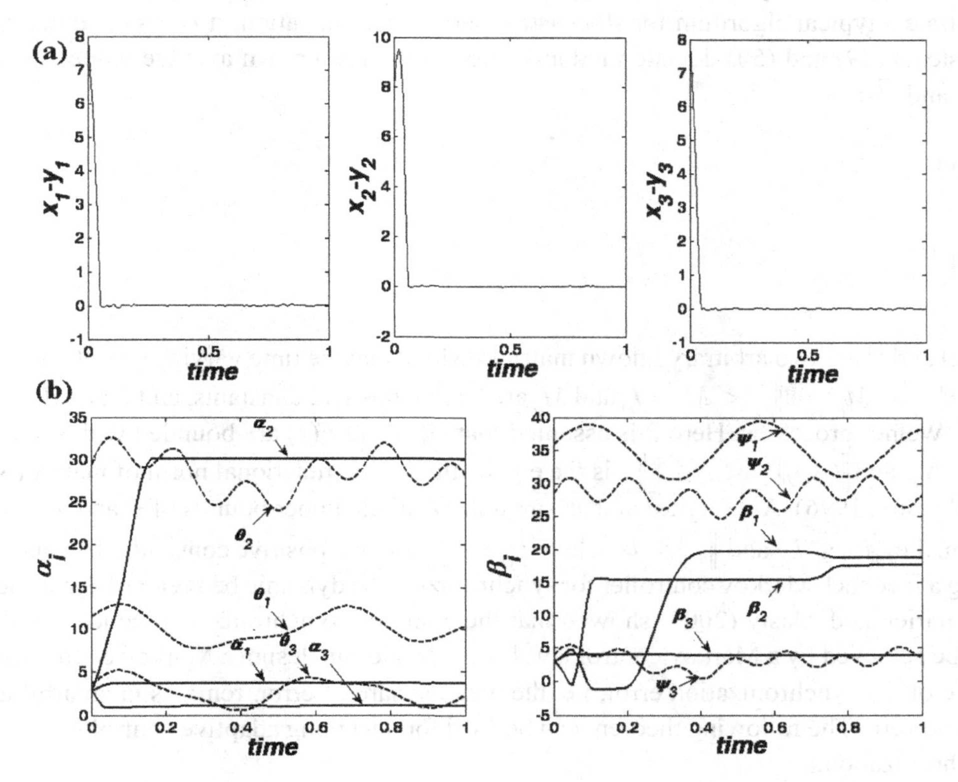

$|\delta_i(t)| < 3,\ i = 1,2$, hence $|\Theta| = |\Psi| = 3\sqrt{3}$. The results of applying the proposed controller are illustrated in Figures 4a and 4b. The initial values are set to $x_i(0) = 10,\ i = 1,2,3,\ y_i(0) = 2,\ i = 1,2,3$, $\alpha_i(0) = \beta_i(0) = 3,\ i = 1,2,3$ and $\hat{x}_i(0) = \hat{y}_i(0) = 0,\ i = 1,2,3$. The parameter *k* is set to 1. As shown in Figure 4b parameter vectors α and β do not converge to θ and ψ but the synchronization objective is completely achieved.

A TYPICAL ADAPTIVE ALGORITHM FOR STOCHASTIC SYSTEMS

All of mentioned works have modeled the chaotic systems in deterministic form, but in real world due to random uncertainties such as stochastic forces acting on physical systems and noisy measurements, a stochastic chaotic dynamics is produced instead of a deterministic one. In this case the deterministic differential equation of system must be substituted by a stochastic differential equation. There are a few works in the field of stochastic chaos regarding control or synchronization (Wu et al., 2005, Wu et al., 2007, and Salarieh, and Alasty, 2009). Stochastic chaos in a Duffing system has been studied where a bounded random process has been added to one of the system parameters, and a controller has been designed to quench the chaos in the system (Wu et al., 2005). Using feedback control, synchronization problem of the stochastic Duffing system described by Wu et al. (2005) has been investigated by Wu et al. (2007).

To illustrate a typical algorithm for stochastic chaos synchronization, it is assumed that the parameters of systems (49) and (50) deviate randomly around their unknown average values which are denoted by $\bar{\theta}$ and $\bar{\psi}$:

$$\theta = \bar{\theta} + \Theta \dot{V} \tag{62}$$

$$\psi = \bar{\psi} + \Psi \dot{W} \tag{63}$$

Where Θ and ψ are two arbitrary known matrices which may be time varying or state dependent and bounded $\|\Theta\|_{eig} < M_{\theta}$, $\|\Psi\|_{eig} < M_{\psi}$. M_{Θ} and M_{ψ} are known positive constants, and V and W are independent vector Weiner processes. Here it is assumed that $G(y)$ and $F(y)$ are bounded with known bounds, $\|F(x)\|_{eig} < N_F$ and $\|G(y)\|_{eig} < N_G$. $\|.\|_{eig}$ is the eigenvalue or the functional norm of matrix (Skogestad, and Postlethwaite, 1996). Although $\bar{\theta}$ and $\bar{\psi}$ are unknown, the upper bounds of $\bar{\theta}$ and $\bar{\psi}$ are assumed to be known, .i.e. $\|\bar{\theta}\| \le L_{\theta}$ and $\|\bar{\psi}\| \le L_{\psi}$ where L_{Θ} and L_{ψ} are two positive constants. The main objective is designing a feedback Markov controller for synchronizing the dynamic behavior of drive and response systems. Salarieh and Alasty (2009) showed that the complete synchronization condition of Equation (4) cannot be achieved by a Markov control law, however one can design a Markov controller by which the variance of the synchronization error, i.e. the mean square of error, remains in an arbitrarily small region around zero. The following theorem can be used for designing adaptive controllers for stochastic chaos synchronization.

Theorem 5. (Salarieh, and Alasty, 2009):

Let **e** = **x**−**y** and $k > 0$, a positive constant, and define:

$$S = \left\{ e\hat{I}\Re^n \middle| E\left[e^T e\right] \le n \frac{N_F^2 M_\theta^2 + N_G^2 M_\psi^2}{k} \right\} \tag{64}$$

Consider the following controller:

$$u = H^{-1}(\mathbf{y})\left[f(\mathbf{x}) - g(\mathbf{y}) + F(\mathbf{x})\alpha - G(\mathbf{y})\beta + 2\left[\lambda_1 \|F(\mathbf{x})\Theta\|_{eig} + \lambda_2 \|G(\mathbf{y})\Psi\|_{eig}\right] sign(\mathbf{x}-\mathbf{y}) + k(\mathbf{x}-\mathbf{y})\right]$$

$$\lambda_i = \begin{cases} 0, & (\mathbf{x}-\mathbf{y}) \notin S \\ 1, & (\mathbf{x}-\mathbf{y}) \in S \end{cases}, \quad i = 1,2 \tag{65}$$

$$\dot{\alpha} = \begin{cases} F^T(x)(x - y), & (x - y) \notin S \quad and \quad \|\alpha\|_2 < \theta_m \\ 0, & (x - \overline{x}) \in S \quad or \quad \|\alpha\|_2 \geq \theta_m \end{cases} \tag{66}$$

$$\dot{\beta} = \begin{cases} -G^T(y)(x - y), & (x - y) \notin S \quad and \quad \|\beta\|_2 < \psi_m \\ 0, & (x - \overline{x}) \in S \quad or \quad \|\beta\|_2 \geq \psi_m \end{cases} \tag{67}$$

Applying the above controller and adaptive laws results in synchronization error that its variance converges into set of S given by Eq. (64), i.e.

$$E\left[(x(t) - y(t))^T(x(t) - y(t))\right] \leq n\frac{N_F^2 M_\psi^2 + N_G^2 M_\theta^2}{k} \quad as \quad t \to \infty \tag{68}$$

Example 4: Stochastic Chaos Synchronization between Two Non-Identical Chua Circuits (Salarieh, and Alasty, 2008-b)

The dynamical equations of the Chua circuit as the master and slave systems are:

$$\begin{aligned} \dot{x}_1 &= p(x_2 - \tfrac{1}{7}(2x_1^3 - x_1)) \\ \dot{x}_2 &= x_1 - x_2 + x_3 \\ \dot{x}_3 &= -qx_2 + rx_1^2 \end{aligned} \tag{69}$$

$$\begin{aligned} \dot{y}_1 &= p_1(y_2 - \tfrac{1}{7}(2y_1^3 - y_1)) + u_1 \\ \dot{y}_2 &= y_1 - y_2 + y_3 + u_2 \\ \dot{y}_3 &= -q_1 y_2 + r_1 y_1^2 + u_3 \end{aligned} \tag{70}$$

The functions defined in Equation (49) and Equation (50) become:

$$f(x) = \begin{bmatrix} 0 \\ x_1 - x_2 + x_3 \\ 0 \end{bmatrix}, \quad F(x) = \begin{bmatrix} x_2 - \tfrac{1}{7}(2x_1^3 - x_1) & 0 & 0 \\ 0 & 0 & 0 \\ 0 & -x_2 & x_1^2 \end{bmatrix}, \quad G(y) = \begin{bmatrix} 1 & 0 & 0 \\ 0 & 1 & 0 \\ 0 & 0 & 1 \end{bmatrix}$$

The first system acts as drive and the second system reacts as response system. Here $\Theta = [p\ q\ r]^T$, $\psi = [p_1 q_1 r_1]^T$, $\theta_i = \overline{\theta}_i + \sum_{j=1}^{3} \Theta_{ij} \dot{V}_j$ and $\psi_i = \overline{\psi}_i + \sum_{j=1}^{3} \Psi_{ij} \dot{W}_j$, $i = 1,2,3$, where V and W are two independent Weiner processes. For $\overline{\theta}_1 = p = 10$, $\overline{\theta}_2 = q = 100/7$, $\overline{\theta}_3 = r = 0.07$, $\overline{\psi}_1 = p_1 = 12$, $\overline{\psi}_2 = q_1 = 135/7$, $\overline{\psi}_3 = r_1 = 0.09$, $\Theta = 0$, $\psi=0$ and $u=0$, both systems show chaotic behaviors. It is assumed that $\Theta_{ii} = 0.3$, $\Theta_{ij} = 0$, $i^1 j$ and $\Psi_{ii} = 0.3$, $\Psi_{ij} = 0$, $i^1 j$ and the adaptive controller given by Equation (65), Equation (66) and Equation (67) is applied for chaos synchronization objective. The controller parameters are set to be $k = 20$, $M_{\Theta} = 0.3$, $M_{\psi} = 0.3$, $L_{\Theta} = 20$ and $L_{\psi} = 20$. The initial conditions of both systems are set to $x_1(0)=1$, $x_2(0)=0.2$, $x_3(0)=-1$, $y_1(0)=y_2(0)=y_3(0)=0$. Figure 5a shows the mean values of synchronization errors, $e_1 = x_1 - y_1$, $e_2 = x_2 - y_2$ and $e_3 = x_3 - y_3$. It is observed that the mean values of error trajectories converge to zero. The standard deviations of the errors, i.e. mean square norms, are shown in Figure 5b. As it is observed they converge into a bounded region around zero which is smaller than that of given by Equation (64).

DUAL AND MULTI-SYNCHRONIZATION OF CHAOS

Recently problem of dual synchronization in chaotic systems is introduced and used experimentally in communication applications. Uchida et al. (2003-a) investigated the dual synchronization of chaos in two pairs of one-way coupled Colpitts electronic oscillators. In dual synchronization problem, there are two slave systems that should be synchronized with two master systems. The state vectors of the master systems are combined linearly to generate a signal that is sent to the slave systems. Problem of multiplexing chaotic signal and synchronizing more than one pair of chaotic systems using only one communication channel is investigated by Tsimring and Sushchik (1996). Liu and Davids (2000) studied the dual synchronization between two one dimensional master and slave chaotic maps by a scalar error signal. Ning et al. (2007) presented a nonlinear method for dual synchronization of two master systems with their corresponding slave systems. In that work the output signal of the master system is a scalar signal constructed by linear combination of master system states. In design of synchronization scheme it is assumed that all states of master and slave systems are available and are used for updating the gain

Figure 5. (a) Mean value of synchronization errors, (b) Mean square norm of synchronization errors (Salarieh, and Alasty, 2008-b)

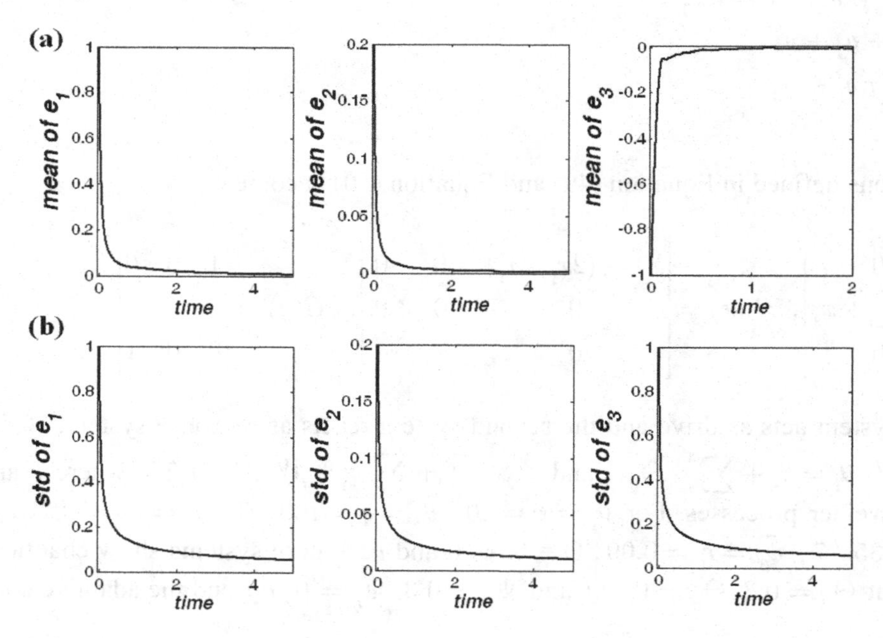

of synchronizing signal. Dual and dual cross synchronization of chaotic external cavity laser diodes have been investigated by Shahverdiev et al. (2003). Experimental dual synchronization of chaos in two pairs of one-way coupled microchip lasers using only one transmission channel have been studied (Uchida et al., 2003-a and Uchida et al., 2003-b). Salarieh and Shahrokhi (2008, 2009) proposed a linear proportional feedback control for dual and multi-synchronization of chaos. For dual synchronization it is assumed that the coupling signal is a scalar output feedback which is used for synchronizing two identical couples of chaotic systems simultaneously (Salarieh, and Shahrokhi, 2008). The proposed technique for dual synchronization was generalized for synchronizing two sets of many chaotic systems simultaneously and called multi-synchronization. Dual and multi-synchronization methods help us to produce more complicated modulation techniques which can encrypt the signals with high degree of security.

A LINEAR CONTROL ALGORITHM FOR MULTI-SYNCHRONIZATION OF CHAOS (SALARIEH, AND SHAHROKHI, 2009)

Multi-synchronization and dual synchronization have similar patterns. Here a typical algorithm for multi-synchronization is explained which can be used for dual synchronization as well (Salarieh, and Shahrokhi, 2009). Figure 6 shows the schematic diagram of multi-synchronization algorithm.

Master dynamics are given by:

$$X^{(i)} = f^{(i)}(t, X^{(i)}),\ i = 1, \ldots, N \tag{71}$$

where $X^{(i)} = \left[X_1^{(i)} \quad L \quad X_n^{(i)}\right]^T$ are the state vectors of the master systems; $f^{(i)} \in C^1(R^+ \times R^n, R^n)$ are known functions. By linear combination of master systems state vectors, a combined chaotic signal is generated as given below:

$$v_m = \sum_{i=1}^{N} A^{(i)} X^{(i)} = C\eta \tag{72}$$

Figure 6. Multi-synchronization scheme (Salarieh, and Shahrokhi, 2009)

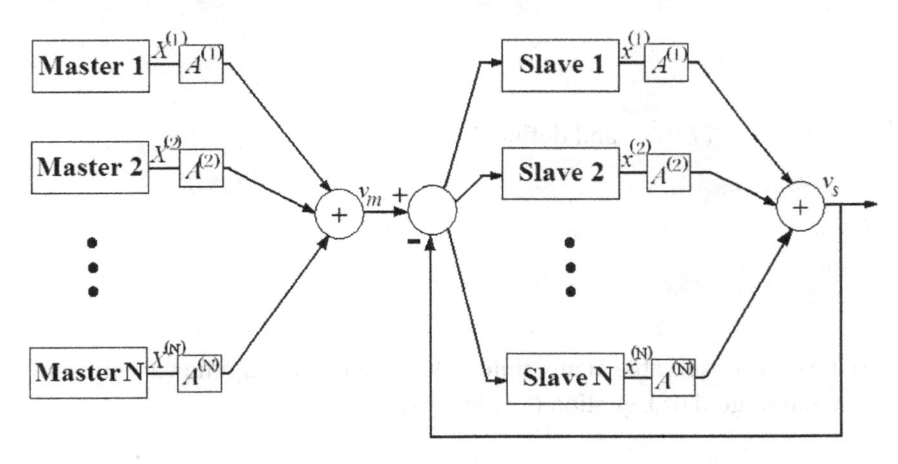

where $A^{(i)}$, i=1,…,N is an $n \times n$ matrix with known constant elements; $C = \begin{bmatrix} A^{(1)} & A^{(2)} & \cdots & A^{(N)} \end{bmatrix}$ and $\eta = \begin{bmatrix} (X^{(1)})^T & (X^{(2)})^T & \cdots & (X^{(N)})^T \end{bmatrix}^T$. The generated signal of (72) is sent to the slave systems which have the same dynamics as the master systems.

The dynamic of ith slave system is given by:

$$x^{(i)} = f^{(i)}(t, x^{(i)}) + u^{(i)}, \; i = 1,2,\ldots, N \tag{73}$$

where $x^{(i)} = \begin{bmatrix} x_1^{(i)} & \cdots & x_n^{(i)} \end{bmatrix}^T$, and $u^{(i)}$ is the manipulated variable, i.e. the control action. The main goal is to synchronize each of the slave system with its corresponding master system, i.e.

$$\lim_{t \to \infty} \left\| x^{(i)}(t) - X^{(i)}(t) \right\| = 0 \tag{74}$$

It should be noted that the only signal available from the master system is the vector v_m defined in Equation (72); and by using this signal, the multi-synchronization objective must be achieved.

Considering the block diagram shown in Figure 6, the error signal for multi-synchronization is:

$$\begin{aligned} e &= v_s - v_m \\ &= C(\xi - \eta) \end{aligned} \tag{75}$$

where $\xi = \begin{bmatrix} (x^{(1)})^T & (x^{(2)})^T & \cdots & (x^{(N)})^T \end{bmatrix}^T$ The proportional state feedback is proposed for chaos synchronization, and the closed loop equation is given below:

$$\dot{x}^{(i)} = f^{(i)}(t, x^{(i)}) + K^{(i)}e, \quad i = 1,2,\ldots,N \Rightarrow \quad \dot{\xi} = \Phi(t,\xi) + Ke \tag{76}$$

where $K = \begin{bmatrix} (K^{(1)})^T & (K^{(2)})^T & \cdots & (K^{(N)})^T \end{bmatrix}^T$ is the proportional feedback gain which is designed to synchronize all chaotic systems simultaneously.

Theorem 6.

Let $\Phi = \begin{bmatrix} (f^1)^T & (f^2)^T & \cdots & (f^N)^T \end{bmatrix}^T$ and define:

$$G(t) = \left. \frac{\partial \Phi(t,\xi)}{\partial \xi} \right|_{\xi = \xi(t)} \tag{77}$$

Assume that $(G(t),C)$ is observable at any time t, then there exists a time varying gain K which guaranties the synchronization goal of Equation (74) locally.

If the observability $(G(t),C)$ is satisfied, one can use different methods to design K; one approach is satisfying the following equation (Salarieh, and Shahrokhi, 2009):

$$\forall t_0 > 0, \ \lim_{T\to\infty} \int_{t_0}^{t_0+T} \mu\left[G(t) + KC\right] dt = -\infty \tag{78}$$

Where $\mu(.)$ is any matrix measure (Vidyasagar, 1993). For example one can use the following matrix measure:

$$\mu_2(M) = \lambda_{\max}\left[\tfrac{1}{2}(M + M^*)\right] \tag{79}$$

Where M^* is transpose-conjugate of M. Setting $M = G(t) + KC^T$, Equation (79) yields:

$$\mu_2\left[G(t) + KC^T\right] = \lambda_{\max}\left[\tfrac{1}{2}\left(G(t) + G^T(t) + KC + C^T K^T\right)\right] \tag{80}$$

To obtain a proper gain K one can select K such that:

$$\lambda_{\max}\left[\tfrac{1}{2}\left(G(t) + G^T(t) + KC + C^T K^T\right)\right] = m \tag{81}$$

where m is a negative real number. Equation (81) provides a controller design criterion (Salarieh, and Shahrokhi, 2009).

Example 5. Dual Synchronization of the Duffing-Van der Pol Dynamical Systems (Salarieh, and Shahrokhi, 2009)

Master and Slave systems are chosen as follows:

Master 1. Duffing system

$$\begin{aligned} \dot{X}_1^{(1)} &= X_2^{(1)} \\ \dot{X}_2^{(1)} &= aX_1^{(1)} + b(X_1^{(1)})^3 + cX_2^{(1)} + f_0 \cos t \end{aligned} \tag{82}$$

Master 2. Van der Pol system

$$\begin{aligned} \dot{X}_1^{(2)} &= X_1^{(2)} - (X_1^{(2)})^3 / 3 - X_2^{(2)} + f_1 \cos t \\ \dot{X}_2^{(2)} &= \gamma(X_1^{(2)} + \alpha - \beta X_2^{(2)}) \end{aligned} \tag{83}$$

The slave systems are similar to the master systems, and their states are denoted by $x_j^{(i)},\ i, j = 1, 2$. The $G(t)$ matrix is obtained as:

$$G(t) = \begin{bmatrix} 0 & 1 & 0 & 0 \\ a + 3b(x_1^{(1)})^2 & c & 0 & 0 \\ 0 & 0 & 1 - (x_1^{(2)})^2 & -1 \\ 0 & 0 & \gamma & -\gamma\beta \end{bmatrix} \tag{84}$$

By setting a=1, b=−1, c=−0.15, f_0=0.3, α = 0.7, β = 0.8, γ = 0.1 and f_1=0.74, both systems show chaotic behaviors. The output signal of the master systems is selected to be:

$$v_m = \begin{bmatrix} X_1^{(1)} + X_1^{(2)} \\ X_2^{(1)} + X_2^{(2)} \end{bmatrix} \tag{85}$$

and the initial conditions are set to $X_1^{(1)}(0) = 3$, $X_2^{(1)}(0) = 3$, $X_1^{(2)}(0) = 3$, $X_2^{(2)}(0) = 3$ and $x_i^{(j)}(0) = 0,\ i = 1,2$. For dual synchronization, the proposed method is applied by setting *m* = -2 in Equation (81). The results are shown in Figure 7, where $e_1 = x_1^{(1)} - X_1^{(1)}$, $e_2 = x_2^{(1)} - X_2^{(1)}$, $e_3 = x_1^{(2)} - X_1^{(2)}$ $e_3 = x_1^{(2)} - X_1^{(2)}$ and $e_4 = x_2^{(2)} - X_2^{(2)}$.

CONCLUSION

In this chapter various methods of chaos synchronization are explained briefly. The concepts of unidirectional and bidirectional chaos synchronization are introduced. Complete, lag, phase and generalized chaos synchronization are defined. Application of control theory to produce coupling signals for chaos synchronization is reviewed and some commonly used control methods for synchronization are described. An adaptive control algorithm and a sliding mode control algorithm are presented for chaos

Figure 7. Dual synchronization results for Duffing-Van der Pol systems with v_m given by (85) (Salarieh, and Shahrokhi, 2009)

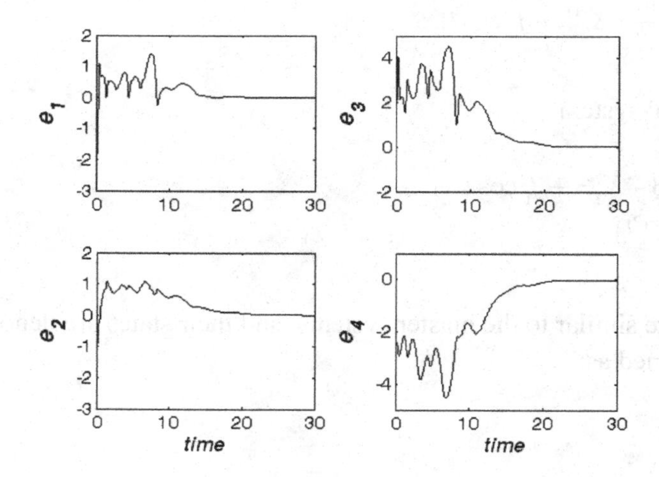

synchronization between two non-identical chaotic systems. In addition, two stochastic synchronization schemes are discussed. Finally concept of dual and multi synchronization is defined and a typical linear control technique is presented for multi-synchronization of chaos.

REFERENCES

Agiza, H. N., & Yassen, M. T. (2001). Synchronization of Rossler and Chen chaotic dynamical systems using active control. *Physics Letters. [Part A]*, *278*(4), 191–197. doi:10.1016/S0375-9601(00)00777-5

Bai, E.-W., & Lonngren, K. E. (2000). Sequential synchronization of two Lorenz systems using active control. *Chaos, Solitons, and Fractals*, *11*(7), 1041–1044. doi:10.1016/S0960-0779(98)00328-2

Bai, E.-W., Lonngren, K. E., & Sprott, J. C. (2002). On the synchronization of a class of electronic circuits that exhibit chaos. *Chaos, Solitons, and Fractals*, *13*(7), 1515–1521. doi:10.1016/S0960-0779(01)00160-6

Banerjee, S., & Chowdhury, A. R. (2009). Lyapunov function, parameter estimation, synchronization and chaotic cryptography. *Communications in Nonlinear Science and Numerical Simulation*, *14*, 2248–2254. doi:10.1016/j.cnsns.2008.06.006

Banerjee, S., Saha, P., & Chowdhury, A. R. (2004). On the application of adaptive control and phase synchronization in non-linear fluid dynamics. *International Journal of Non-linear Mechanics*, *39*, 25–31. doi:10.1016/S0020-7462(02)00125-7

Behzad, M., Salarieh, H., & Alasty, A. (2008). Chaos synchronization in noisy environment using non-linear filtering and sliding mode control. *Chaos, Solitons, and Fractals*, *36*(5), 1295–1304. doi:10.1016/j.chaos.2006.07.058

Boccaletti, S., Pecora, L. M., & Pelaez, A. (2001). Unifying framework for synchronization of coupled dynamical systems. *Physical Review E*, *63*(6), 066219/1-066219/4.

Bowong, S., & Kakmeni, F. M. M. (2004). Synchronization of uncertain chaotic systems via backstepping approach. *Chaos, Solitons, and Fractals*, *21*(4), 999–1011. doi:10.1016/j.chaos.2003.12.084

Cao, J., Li, H. X., & Ho, D. W. C. (2005). Synchronization criteria of Lur'e systems with time-delay feedback control. *Chaos, Solitons, and Fractals*,*23*(4), 1285–1298.

Carroll, T. L., & Pecora, L. M. (1991). Synchronizing chaotic circuits. *IEEE Transactions on Circuits and Systems*, *38*(4), 453–456. doi:10.1109/31.75404

Chen, H.-K. (2005). Global chaos synchronization of new chaotic systems via nonlinear control. *Chaos, Solitons, and Fractals*,*23*(4), 1245–1251.

Chen, J. Y., Wong, K. W., Cheng, L. M., & Shuai, J. W. (2003). A secure communication scheme based on the phase synchronization of chaotic systems. *Chaos (Woodbury, N.Y.)*, *13*(2), 508–514. doi:10.1063/1.1564934

Chen, M., & Han, Z. (2003). Controlling and synchronizing chaotic Genesio system via nonlinear feedback control. *Chaos, Solitons, and Fractals*,*17*(4), 709–716. doi:10.1016/S0960-0779(02)00487-3

Chen, M., Zhou, D., & Shang, Y. (2005). A new observer-based synchronization scheme for private communication. *ChaosSolitons and Fractals, 24*, 1025-1030.

Chen, S., & Lu, J. (2002). Parameters identification and synchronization of chaotic systems based upon adaptive control. *Physics Letters. [Part A], 299*(4), 353–358.

Cuomo, K. M., Oppenheim, A. V., & Strogatz, S. H. (1993). Synchronization of Lorenz-based chaotic circuits with applications to communications. *IEEE Transactions on Circuits and Systems II: Analog and Digital Signal Processing, 40*(10), 626–633. doi:10.1109/82.246163

DeShazer, D. J., Breban, R., Ott, E., & Roy, R. (2001). Detecting phase synchronization in a chaotic laser array. *Physical Review Letters, 87*(4), 441011–441014. doi:10.1103/PhysRevLett.87.044101

Elabbasy, E. M., Agiza, H. N., & El-Dessoky, M. M. (2004). Adaptive synchronization of Lu system with uncertain parameters. *Chaos, Solitons, and Fractals,21*(3), 657–667. doi:10.1016/j.chaos.2003.12.028

Etemadi, S., Alasty, A., & Salarieh, H. (2006). Synchronization of chaotic systems with parameter uncertainties via variable structure control. *Physics Letters. [Part A], 357*(1), 17–21. doi:10.1016/j.physleta.2006.04.101

Feki, M. (2003). An adaptive chaos synchronization scheme applied to secure communication. *Chaos, Solitons, and Fractals,18*(1), 141–149. doi:10.1016/S0960-0779(02)00585-4

Femat, R., Alvarez-Ramírez, J., & Fernández-Anaya, G. (2000). Adaptive synchronization of high-order chaotic systems: A feedback with low-order parametrization. *Physica D. Nonlinear Phenomena, 139*(3-4), 231–246. doi:10.1016/S0167-2789(99)00226-2

Frey, D. R. (1993). Chaotic digital encoding: an approach to secure communication. *IEEE Transactions on Circuits and Systems II: Analog and Digital Signal Processing, 40*(10), 660–666. doi:10.1109/82.246168

Gonzalez Miranda, J. M. (2004). *Synchronization and control of chaos: an introduction for scientists and engineers*. London, UK: Imperial College Press. doi:10.1142/9781860945229

Grassi, G., & Mascolo, S. (1997). Nonlinear observer design to synchronize hyperchaotic systems via a scalar signal. *IEEE Transactions on Circuits and Systems, 44*(10), 1011–1014. doi:10.1109/81.633891

He, R., & Vaidya, P. G. (1992). Analysis and synthesis of synchronous periodic and chaotic systems. *Physical Review A., 46*(12), 7387–7392. doi:10.1103/PhysRevA.46.7387

Heagy, J. F., Carroll, T. L., & Pecora, L. M. (1994). Synchronous chaos in coupled oscillator systems. *Physical Review E: Statistical Physics, Plasmas, Fluids, and Related Interdisciplinary Topics, 50*(3), 1874–1885. doi:10.1103/PhysRevE.50.1874

Kapitaniak, T. (1994). Synchronization of chaos using continuous control. *Physical Review E: Statistical Physics, Plasmas, Fluids, and Related Interdisciplinary Topics, 50*(2), 1642–1644. doi:10.1103/PhysRevE.50.1642

Kiss, I. Z., & Hudson, J. L. (2001). Phase synchronization and suppression of Chaos through intermittency in forcing of an electrochemical oscillator. *Physical Review E: Statistical, Nonlinear, and Soft Matter Physics, 64*(4II), 462151–462158. doi:10.1103/PhysRevE.64.046215

Li, Z. G., Wen, C. Y., Soh, Y. C., & Xie, W. X. (2001). The stabilization and synchronization of Chua's oscillators via impulsive control. *IEEE Transactions on Circuits and Systems. I, Fundamental Theory and Applications*, *48*(11), 1351–1355. doi:10.1109/81.964427

Liao, T.-L. (1998). Adaptive synchronization of two Lorenz systems. *Chaos, Solitons, and Fractals*,*9*(9), 1555–1561. doi:10.1016/S0960-0779(97)00161-6

Liao, T.-L., & Huang, N.-S. (1999). An observer-based approach for chaotic synchronization with applications to secure communications. *IEEE Transactions on Circuits and Systems. I, Fundamental Theory and Applications*, *46*(9), 1144–1150. doi:10.1109/81.788817

Liao, T.-L., & Tsai, S.-H. (2000). Adaptive synchronization of chaotic systems and its application to secure communications. *Chaos, Solitons, and Fractals*,*11*(9), 2387–2396. doi:10.1016/S0960-0779(99)00051-X

Liao, X., & Chen, G. (2003). Chaos synchronization of general Luré systems via time-delay feedback control . *International Journal of Bifurcation and Chaos in Applied Sciences and Engineering*, *13*(1), 207–213. doi:10.1142/S0218127403006455

Liu, F., Ren, Y., Shan, X., & Qiu, Z. (2002). A linear feedback synchronization theorem for a class of chaotic systems. *Chaos, Solitons, and Fractals*,*13*(4), 723–730. doi:10.1016/S0960-0779(01)00011-X

Liu, Y., & Davids, P. (2000). Dual synchronization of chaos . *Physical Review E: Statistical Physics, Plasmas, Fluids, and Related Interdisciplinary Topics*, *61*, 2176–2179. doi:10.1103/PhysRevE.61.R2176

Lu, J., & Cao, J. (2005). Adaptive complete synchronization of two identical or different chaotic (hyperchaotic) systems with fully unknown parameters. *Chaos (Woodbury, N.Y.)*, *15*(4), 043901. doi:10.1063/1.2089207

Mensour, B., & Longtin, A. (1998). Synchronization of delay-differential equations with application to private communication. *Physics Letters A*, *244*(1-3), 59-70.

Mirasso, C. R., & Colet, P., & García-Fernández, P. (1996). Synchronization of chaotic semiconductor lasers: Application to encoded communications. *IEEE Photonics Technology Letters*, *8*(2), 299–301. doi:10.1109/68.484273

Newcomb, R. W., & El-Leithy, N. (1986). Chaos using hysteretic circuits. In *Proceedings - IEEE International Symposium on Circuits and Systems* (pp. 56-61). San Jose, CA: IEEE.

Newcomb, R. W., & Sathyan, S. (1983). RC Op Amp chaos generator. *IEEE Transactions on Circuits and Systems*, *CAS-30*(1), 54–56. doi:10.1109/TCS.1983.1085277

Ning, D., Lu, J., & Han, X. (2007). Dual synchronization based on two different chaotic systems: Lorenz systems and Rossler systems. *Computational & Applied Mathematics*, *206*(2), 1046–1050. doi:10.1016/j.cam.2006.09.007

Ogorzalek, M. J. (1993). Taming Chaos - Part I: synchronization. *IEEE Transactions on Circuits and Systems*, *40*(10), 693–699. doi:10.1109/81.246145

Park, J. H. (2005). Adaptive synchronization of hyperchaotic Chen system with uncertain parameters. *Chaos, Solitons, and Fractals*,*26*(3), 959–964. doi:10.1016/j.chaos.2005.02.002

Park, J. H. (2006). Chaos synchronization between two different chaotic dynamical systems. *Chaos, Solitons, and Fractals,27*(2), 549–554. doi:10.1016/j.chaos.2005.03.049

Pecora, L. M., & Caroll, T. L. (1990). Synchronization in chaotic systems. *Physical Review Letters, 64*, 821–824. doi:10.1103/PhysRevLett.64.821

Pikovsky, A., Zaks, M., Rosenblum, M., Osipov, G., & Kurths, J. (1997). Phase synchronization of chaotic oscillations in terms of periodic orbits. *Chaos (Woodbury, N.Y.), 7*(4), 680–687. doi:10.1063/1.166265

Rosa, E. Jr, Ott, E., & Hess, M. H. (1998). Transition to phase synchronization of chaos. *Physical Review Letters, 80*(8), 1642–1645. doi:10.1103/PhysRevLett.80.1642

Rosenblum, M. G., Pikovsky, A. S., & Kurths, J. (1997). From phase to lag synchronization in coupled chaotic oscillators. *Physical Review Letters, 78*(22), 4193–4196. doi:10.1103/PhysRevLett.78.4193

Roy, R., & Thornburg, J. K. S. (1994). Experimental synchronization of chaotic lasers. *Physical Review Letters, 72*(13), 2009–2012. doi:10.1103/PhysRevLett.72.2009

Rulkov, N. F., Sushchik, M. M., Tsimring, L. S., & Abarbanel, H. D. I. (1995). Generalized synchronization of chaos in directionally coupled chaotic systems. *Physical Review E: Statistical Physics, Plasmas, Fluids, and Related Interdisciplinary Topics, 51*(2), 980–994. doi:10.1103/PhysRevE.51.980

Saha, P., Banerjee, S., & Chowdhury, A. R. (2004). Chaos, signal communication and parameter estimation. *Physics Letters. [Part A], 326*, 133–139. doi:10.1016/j.physleta.2004.04.025

Salarieh, H., & Alasty, A. (2009). Adaptive synchronization of two chaotic systems with stochastic unknown parameters. *Communications in Nonlinear Science and Numerical Simulation, 14*(2), 508–519. doi:10.1016/j.cnsns.2007.09.002

Salarieh, H., & Alasty, A.(2008-a). Chaos synchronization of nonlinear gyros in presence of stochastic excitation via sliding mode control. *Journal of Sound and Vibration, 313*(3-5), 760–771. doi:10.1016/j.jsv.2007.11.058

Salarieh, H., & Alasty, A.(2008-b). Adaptive chaos synchronization in Chua's systems with noisy parameters. *Mathematics and Computers in Simulation, 79*(3), 233–241. doi:10.1016/j.matcom.2007.11.007

Salarieh, H., & Shahrokhi, M. (2008). Adaptive synchronization of two different chaotic systems with time varying unknown parameters. *Chaos, Solitons, and Fractals,37*(1), 125–136. doi:10.1016/j.chaos.2006.08.038

Salarieh, H., & Shahrokhi, M. (2008). Dual synchronization of chaotic systems via time-varying gain proportional feedback. *Chaos, Solitons, and Fractals,38*(5), 1342–1348. doi:10.1016/j.chaos.2008.02.015

Salarieh, H., & Shahrokhi, M. (2009). Multi-synchronization of chaos via linear output feedback strategy. *Computational & Applied Mathematics, 223*(2), 842–852. doi:10.1016/j.cam.2008.03.002

Scholl, E., & Schuster, H. G. (2008). *Handbook of chaos control* (2nd ed.). Weinheim, Germany: Wiley-VCH Verlag.

Shahverdiev, E. M., Sivaprakasam, S., & Shore, K. A. (2003). Dual and dual-cross synchronization in chaotic systems. *Optics Communications, 216*, 179–183. doi:10.1016/S0030-4018(02)02286-1

Shuai, J.-W., & Durand, D. M. (1999). Phase synchronization in two coupled chaotic neurons. *Physics Letters. [Part A], 264*(4), 289–297. doi:10.1016/S0375-9601(99)00816-6

Skogestad, S., & Postlethwaite, I. (1996). *Multivariable Feedback Control Analysis and Design*. Sussex, UK: John Wiley & Sons.

Sugawara, T., Tachikawa, M., Tsukamoto, T., & Shimizu, T. (1994). Observation of synchronization in laser chaos . *Physical Review Letters, 72*(22), 3502–3505. doi:10.1103/PhysRevLett.72.3502

Tan, X., Zhang, J., & Yang, Y. (2003). Synchronizing chaotic systems using backstepping design. *Chaos, Solitons, and Fractals,16*(1), 37–45. doi:10.1016/S0960-0779(02)00153-4

Ticos, C. M., Rosa, E. Jr, & Pardo, W. B. (2000). Walkenstein, J.A., Monti, M., Experimental real-time phase synchronization of a paced chaotic plasma discharge. *Physical Review Letters, 85*(14), 2929–2932. doi:10.1103/PhysRevLett.85.2929

Tsimring, L. S., & Sushchik, M. M. (1996). Multiplexing chaotic signals using synchronization. *Physics Letters. [Part A], 213*, 155–166. doi:10.1016/0375-9601(96)00118-1

Ucar, A., Lonngren, K. E., & Bai, E.-W. (2006). Synchronization of the unified chaotic systems via active control. *Chaos, Solitons, and Fractals,27*(5), 1292–1297. doi:10.1016/j.chaos.2005.04.104

Uchida, A., Kawano, M., & Yoshimori, S.(2003-a). Dual synchronization of chaos in Colpitts electronic oscillators and its applications for communications. *Physical Review E, 68*, 056207. Uchida, A., Kinugawa, S., Matsuura, T., & Yoshimori, S. (2003-b). Dual synchronization of chaos in microchip lasers. *Optics Letters, 28*, 19–21. doi:10.1364/OL.28.000019

Vidyasagar, M. (1993). *Nonlinear Systems Analysis* (2nd ed.). Upper Saddle River, NJ: Prentice-Hall.

Wang, C., & Ge, S. S. (2001). Adaptive synchronization of uncertain chaotic systems via backstepping design. *Chaos, Solitons, and Fractals,12*(7), 1199–1206. doi:10.1016/S0960-0779(00)00089-8

Wang, C. P., & Su, J. P. (2004). A new adaptive variable structure control for chaotic synchronization and secure communication. *Chaos, Solitons, and Fractals,20*(5), 967–977. doi:10.1016/j.chaos.2003.10.026

Wang, Y., Guan, Z. H., & Wen, X. (2004). Adaptive synchronization for Chen chaotic system with fully unknown parameters. *Chaos, Solitons, and Fractals,19*(4), 899–903. doi:10.1016/S0960-0779(03)00256-X

Winful, H., & Rahman, L. (1990). Synchronized chaos and spatiotemporal chaos in arrays of coupled lasers. *Physical Review Letters, 65*(13), 1575–1578. doi:10.1103/PhysRevLett.65.1575

Wu, C., Fang, T., & Rong, H. (2007). Chaos synchronization of two stochastic Duffing oscillators by feedback control. *Chaos, Solitons, and Fractals,32*(3), 1201–1207. doi:10.1016/j.chaos.2005.11.042

Wu, C., Lei, Y., & Fang, T. (2005). Stochastic chaos in a Duffing oscillator and its control. *Chaos, Solitons, and Fractals,27*, 459–469. doi:10.1016/j.chaos.2005.04.035

Wu, C. W., & Chua, L. O. (1995). Synchronization in an array of linearly coupled dynamical systems. *IEEE Transactions on Circuits and Systems. I, Fundamental Theory and Applications, 42*(8), 430–447. doi:10.1109/81.404047

Yang, T. (1996). Secure communication via chaotic parameter modulation. *IEEE Transactions on Circuits and Systems. I, Fundamental Theory and Applications*, *43*(9), 817–819. doi:10.1109/81.536758

Yang, T., & Chua, L. O. (1997). Impulsive stabilization for control and synchronization of chaotic systems: Theory and application to secure communication. *IEEE Transactions on Circuits and Systems*, *44*(10), 976–988. doi:10.1109/81.633887

Yang, T., & Shao, H. H. (2002). Synchronizing chaotic dynamics with uncertainties based on a sliding mode control design. *Physical Review E, 65*(4), 046210/1-046210/7.

Yassen, M. T. (2005). Chaos synchronization between two different chaotic systems using active control. *Chaos, Solitons, and Fractals,23*(1), 131–140. doi:10.1016/j.chaos.2004.03.038

Yau, H. T. (2004). Design of adaptive sliding mode controller for chaos synchronization with uncertainties. *Chaos, Solitons, and Fractals,22*(2), 341–347. doi:10.1016/j.chaos.2004.02.004

Yu, W., & Cao, J. (2007). Adaptive synchronization and lag synchronization of uncertain dynamical system with time delay based on parameter identification. *Physica A, 375*(2), 467–482. doi:10.1016/j.physa.2006.09.020

Zhang, H., Huang, W., Wang, Z., & Chai, T. (2006). Adaptive synchronization between two different chaotic systems with unknown parameters. *Physics Letters. [Part A], 350*(5-6), 363–366. doi:10.1016/j.physleta.2005.10.033

Zhang, H., Ma, X.-K., & Liu, W.-Z. (2004). Synchronization of chaotic systems with parametric uncertainty using active sliding mode control. *Chaos, Solitons, and Fractals,21*(5), 1249–1257. doi:10.1016/j.chaos.2003.12.073

Zheng, Z., & Hu, G. (2000). Generalized synchronization versus phase synchronization. *Physical Review E: Statistical Physics, Plasmas, Fluids, and Related Interdisciplinary Topics*, *62*(6B), 7882–7885. doi:10.1103/PhysRevE.62.7882

Zheng, Z., Hu, G., & Hu, B. (1998). Phase slips and phase synchronization of coupled oscillators. *Physical Review Letters*, *81*(24), 5318–5321. doi:10.1103/PhysRevLett.81.5318

Chapter 9
Chaotic Gyros Synchronization

Mehdi Roopaei
Islamic Azad University, Iran

Mansoor J. Zolghadri
Shiraz University, Iran

Bijan Ranjbar Sahraei
Shiraz University, Iran

Seyyed Hossein Mousavi
Shiraz University, Iran

Hassan Adloo
Shiraz University, Iran

Behnam Zare
Tarbiat Modarres University, Iran

Tsung-Chih Lin
Feng Chia University, Taiwan

ABSTRACT

In this chapter, three methods for synchronizing of two chaotic gyros in the presence of uncertainties, external disturbances and dead-zone nonlinearity are studied. In the first method, there is dead-zone nonlinearity in the control input, which limits the performance of accurate control methods. The effects of this nonlinearity will be attenuated using a fuzzy parameter approximator integrated with sliding mode control method. In order to overcome the synchronization problem for a class of unknown nonlinear chaotic gyros a robust adaptive fuzzy sliding mode control scheme is proposed in the second method. In the last method, two different gyro systems have been considered and a fuzzy controller is proposed to eliminate chattering phenomena during the reaching phase of sliding mode control. Simulation results are also provided to illustrate the effectiveness of the proposed methods.

DOI: 10.4018/978-1-61520-737-4.ch009

INTRODUCTION

Although chaotic systems have deterministic behavior, they are extremely sensitive to initial conditions and difficult to predict. Furthermore, some noises, disturbances and uncertainties always exist in the physical systems that can make system instability.

Synchronization of chaotic systems is one of the most interesting fields of chaos control. Many approaches have been introduced to treat this problem in the few past decades. Synchronization can be defined as two coupled systems conducting coupling evolution in time with given different initial conditions (Li, Chen, Shi & Han, 2003), in other words the purpose of synchronization is to use the output of master system to control the slave system, so that the output of slave system achieves asymptotic synchronization with the output of master system.

Since the pioneering work of Pecora & Carroll (1990), various synchronizations such as feedback control (Wu, Yang & Chua, 1996;Salarieh & Shahrokhi, 2008; Li, Xu & Xiao, 2008); Sliding Mode Control (SMC) (Yan, Hung & Liao, 2006;), backstepping design (Parmananda,1998; Yin, Ren & Shan, 2002), H_∞ control (Slotine, 1991) and adaptive control (Hwang, Hyun, Kim & Park, 2009; Roopaei & Zolghadri Jahromi, 2008; Yassen, 2006; Yassen, 2007) methods have been developed for them.

In mechanical devices, such as positioning tables, overhead crane mechanisms, robot manipulators, gyroscopes, etc. Many accurate control methods are required. For many of them, the performance is limited by friction and dead-zone (Zhou, Shen &Tamura, 2009; Lei, Xu & Zheng, 2005). In particular, precise positioning control of very small displacement is an especially difficult problem for micro positioning devices. Due to lack of precise knowledge about the nonlinearities present in actuators and the fact that their exact parameters (e.g. width of dead-zone) are unknown, these systems present a challenge for the control engineering communities.

A variety of physical principles are utilized for rotation sensing, including mechanical sensing, the Sagnac effect for photons (Stedman, 1997; Andronova & Malykin, 2002), the Josephson effect in super fluid and nuclear spin precession (Woodman, Franks & Richards, 1987). However, mechanical gyroscopes operating in a low gravity environment remain so far unchallenged. (Buchman et al. 2000).

A gyroscope is a device for measuring or maintaining orientation, based on the principles of angular momentum. The device is a spinning wheel or disk whose axle is free to take any orientation (Figure 1). This orientation changes much less in response to a given external torque than it would without the large angular momentum associated with the gyroscope high rate of spin. Since external torque is minimized by mounting the device in gimbals, its orientation remains nearly fixed, regardless of any motion of the platform on which it is mounted. Sensitive gyroscopes are used in many applications, from inertial navigation to studies of the Earth rotation and tests of general relativity (Stedman, 1997).

Recent researchs has specified various types of gyro systems with linear/nonlinear damping characteristics. These systems exhibit a diverse range of dynamic behavior including both sub-harmonic and chaotic Motions (Slotin & Li, 1991; Stedman, 1997; Andronova & Malykin 2002).

The chaotic behavior of gyros was initially introduced by Leipnik and Newton (1981). In (Tong & Mrad, 2001; Ge, H.K Chen & H.H. Chen, 1996) the nonlinear dynamics of a symmetric heavy gyroscope, mounted on a vibrating platform was studied. In these works, a linear damping coefficient was assumed for the gyro system. In (Chen, 2002; Dooren, 2003) it was shown that under base harmonic excitation and a nonlinear damping force; the gyro system exhibit chaotic behavior.

Recently, synchronization of chaotic gyros has been widely investigated by many researchers. Synchronization of two gyros is usually used in areas of secure communications (Chen & Lin, 2003) and attitude control of long-duration spacecrafts (Zhou, Shen & Tamura, 2006).

Figure 1. A mechanical gyroscope

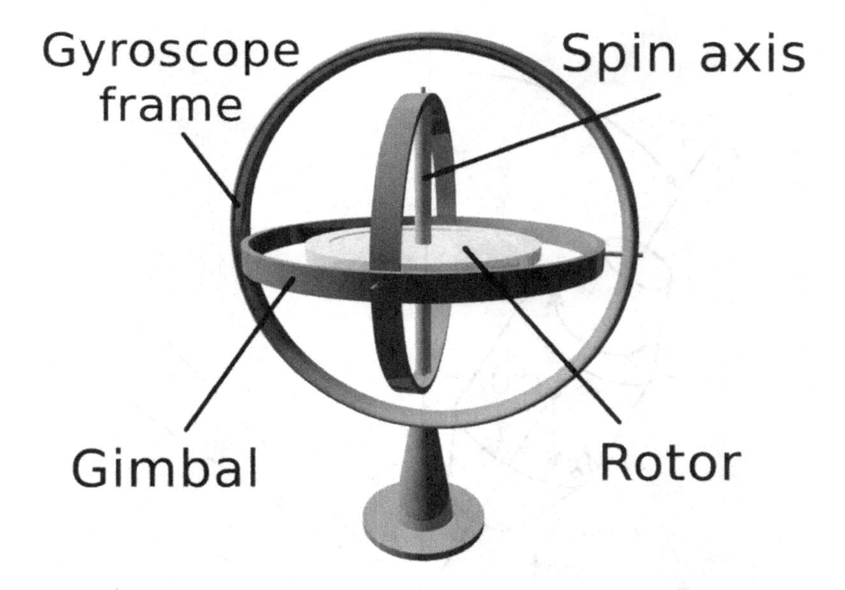

Yan, Hung, Lin and Liao (2007) designed a variable structure control for a chaotic symmetric gyro when linear-plus-cubic damping is considered. Salarieh and Alasty (2008) used a modified sliding mode control, designed a Markov synchronization control law and modeled a stochastic excitation by applying a Gaussian white noise to the deterministic model of gyro. Yau (2008) presented a robust fuzzy sliding mode control (FSMC) scheme for the synchronization of two chaotic nonlinear gyros subject to uncertainties and disturbances.

In this chapter, we will represent three different methods for the synchronization of two gyros that have different structures. Detailed simulations for each method will be presented separately.

The rest of this chapter is organized as follows. Section II presents the system description. In section III some control design methodologies is reviewed. Simulation results show the effectiveness of the mentioned methods in section IV. Finally, conclusions are given in section V.

DESCRIPTION OF NONLINEAR GYRO SYSTEMS

A schematic diagram of a symmetric gyroscope is shown in (Figure 2). The equation governing the motion of such a symmetric gyro mounted on a vibrating base is given by Chen (2002) as

$$\ddot{\theta} + \alpha^2 \frac{(1 - \cos\theta)^2}{\sin^3\theta} - \beta\sin\theta + c_1\dot{\theta} + c_2\dot{\theta}^3 = f\sin\omega t\sin\theta, \tag{1}$$

where θ is the angle between the gyro spin axis and the vertical axis, $\alpha = \frac{I_3 w_z}{I_1}$, $c_1 = \frac{D_1}{I_1}$, $c_2 = \frac{D_2}{I_1}$, $\beta = \frac{M_g l}{I_1}$, $f = \frac{M_g \bar{l}}{I_1}$ such that I_1 is the polar moment of inertia of

Figure 2. A schematic diagram of a symmetric gyroscope

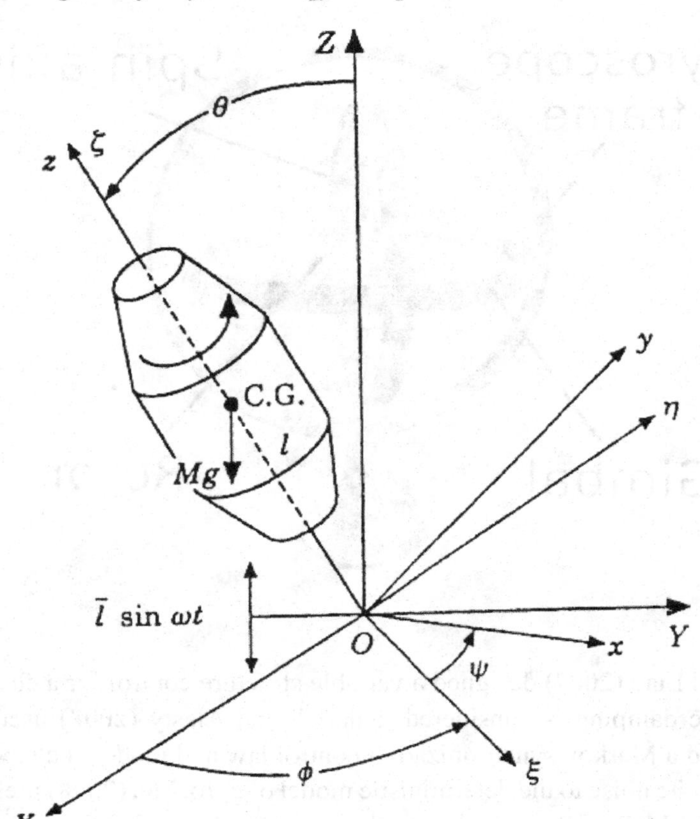

symmetric gyro, M_g the gravity force, $\bar{l}$ the amplitude of the external excitation disturbance, and w the frequency of the external excitation disturbance. D_1 and D_2 are known positive constants, where the term $f\sin(wt)$ represents a parametric excitation, $c_1\Theta$ and $c_2\Theta^3$ are linear and nonlinear damping terms, respectively; and $\alpha^2 \frac{(1-\cos\theta)^2}{\sin^3\theta} - \beta\sin\theta$ is a nonlinear resilience force.

Given the states $x_1 = \Theta$, $x_2 = \Theta$ and $g(\theta) = -\alpha^2 \frac{(1-\cos\theta)^2}{\sin^3\theta}$, this system can be transformed into the nominal form as

$$\begin{cases} \dot{x}_1 = x_2, \\ \dot{x}_2 = g(x_1) - c_1x_2 - c_2x_2^3 + (\beta + f\sin\omega t)\sin(x_1). \end{cases} \tag{2}$$

This gyro system exhibits complex dynamics and has been studied by Chen (2002) for values of f in the range $32 < f < 36$ and constant values of $\alpha^2 = 100$, $\beta = 1$, $c_1 = 0.5$, $c_2 = 0.05$, and $\omega = 2$.

The chaotic behavior of a gyro with the above parameters and $x(0) = [-1,1]$ as the initial states is shown in (Figure 3), (Figure 4), and (Figure 5).

Figure 3. Gyro trajectory State 1

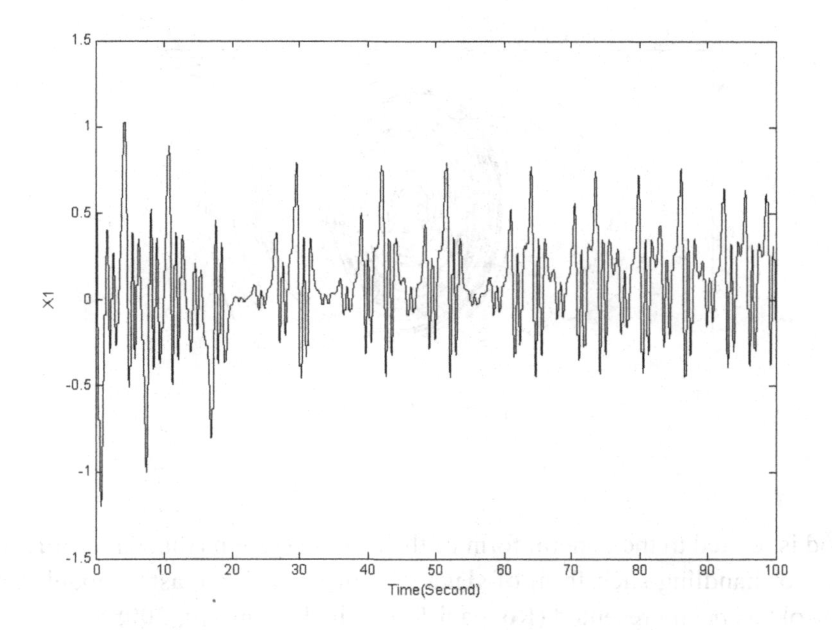

Figure 4. Gyro trajectory State 2

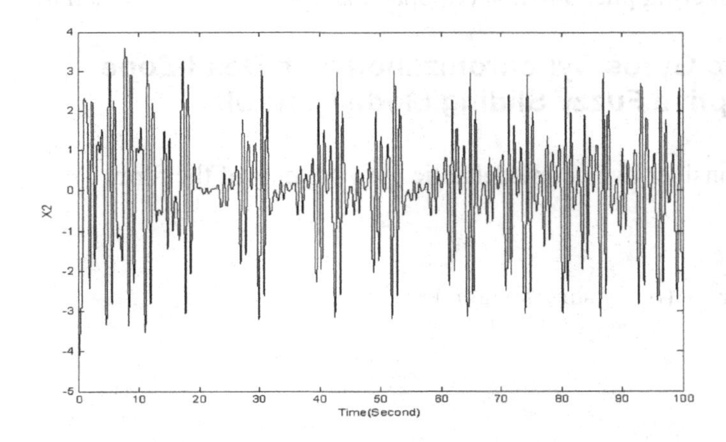

CHAOS SYNCHRONIZATION CONTROL DESIGN

In this section, three different methods are presented. In the first method, two coupled chaotic gyros have been considered such that there is dead-zone nonlinearity limitation in the actuator model and time varying disturbance in the controlled gyro system. For synchronization of two gyros, a sliding mode controller has been proposed in which unknown functions are approximated by the fuzzy logic concept. In addition, the unknown parameters of the fuzzy system are adapted using derived adaptive laws through the time.

Figure 5. Gyro state space trajectory

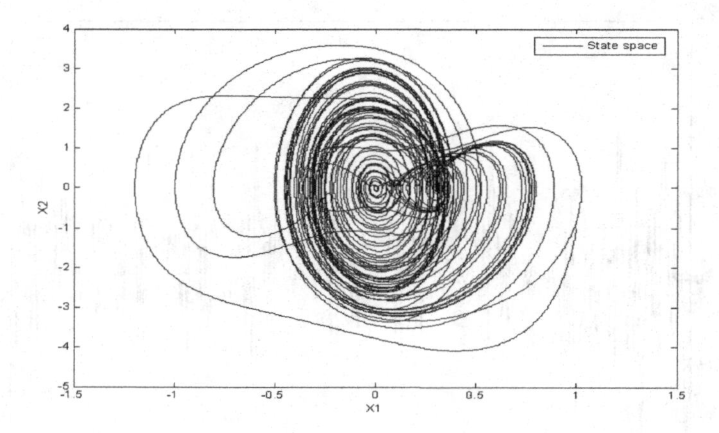

Second method is related to the general form of the chaotic systems that have structured uncertainty and disturbances. For handling such form of slave system to track a master model, a robust adaptive sliding mode control has been presented (Roopaei & Zolghadri Jahromi, 2008).

In the last method, chaotic gyro system with unstructured uncertainty is considered and a sliding mode controller has been engaged. In this proposed method, an adaptive gain fuzzy method is introduced to remove effects of chattering phenomenon (Roopaei, Zolghadri Jahromi & Jafari, 2009).

Method I. Chaotic Gyros Synchronization with Dead-Zone Input Using Adaptive Fuzzy Sliding Mode Control

In this method, we consider two coupledchaotic gyro systems of the form

$$\begin{cases} \dot{x}_1 = x_2, \\ \dot{x}_2 = g(x_1) - c_1 x_2 - c_2 x_2^3 + (\beta + f \sin \omega t) \sin(x_1) \end{cases} \tag{3}$$

and

$$\begin{cases} \dot{y}_1 = y_2, \\ \dot{y}_2 = g(y_1) - c_1 y_2 - c_2 y_2^3 + (\beta + f \sin \omega t) \sin(y_1) + d(t) + \Phi(u), \end{cases} \tag{4}$$

where, $d(t)$is the time-varying disturbance. In general the disturbance is assumed to be bounded (i.e., $\left| d(t) \right| \le \beta$) and $\Phi(u)$: $R \to R$ is the dead-zone input nonlinearity. The dead-zone function $\Phi(\cdot)$ (shown in Figure 6) can be expressed as:

$$\Phi(u) = \begin{cases} m(u - b) & u \geq b \\ 0 & -b < u < b \\ m(u + b) & u \leq -b, \end{cases} \tag{5}$$

where, b is the width of the dead-zone and m is the slope of lines in Figure 6.

In order to investigate the key feature of the dead-zone in control problems, the following assumptions are made:

Assumption 1. The Dead-Zone Output Is Not Available

Assumption 2. The Dead-Zone Parameters *B* And *M* Are Assumed To Be Bounded And The Bounds Of Each Parameter Is Known:

$b \in [b_{\min}, b_{\max}]$ and $m \in [m_{\min}, m_{\max}]$.

Having these assumptions, (5) is represented as

$$\Phi(u) = mu = v(u) \tag{6}$$

where $v(u)$ is

$$v(u) = \begin{cases} -mb & u \geq b \\ -mu & -b < u < b \\ mb & u \leq -b. \end{cases} \tag{7}$$

Figure 6. Dead Zone Nonlinearity

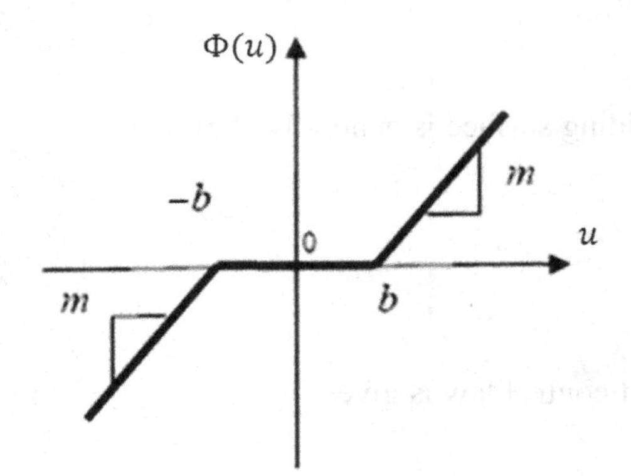

From Assumption 2, *v (u)* is bounded, and satisfies: $|v(u)| \le \rho$, where ρ is the upper bound that can be chosen as: $\rho = mb_{max}$.

By defining $h(x,t) = g(x_1) - c_1 x_2 - c_2 x_2^3 + (\beta + f \sin \omega t)\sin(x_1)$, equation (3) and equation (4) can be rewritten as

$$\begin{cases} \dot{x}_1 = x_2 \\ \dot{x}_2 = h(x,t) \end{cases} \tag{8}$$

and

$$\begin{cases} \dot{y}_1 = y_2, \\ \dot{y}_2 = h(y,t) + d(t) + \Phi(u) \end{cases} \tag{9}$$

where $x = [x_1, x_2]^T$ and $y = [y_1, y_2]^T$.

Assumption 3. The Unknown Function *H(.,.)* Satisfies the Following Condition:

$$|h(x,t)| < H \tag{10}$$

where, *H* is an unknown positive constant. The above assumption is reasonable for the chaotic systems, since chaotic motion is believed to be locally unstable and globally bounded.

If we define the errors in states of the coupled systems as $e_1 = y_1 - x_1$ and $e_2 = y_2 - x_2$, the dynamic equation of these errors can be determined by subtracting the relation (8) from (9). That is

$$\begin{cases} \dot{e}_1 = e_2, \\ \dot{e}_2 = h(y,t) - h(x,t) + d(t) + \Phi(u). \end{cases} \tag{11}$$

In the SMC method, the sliding surface is generally chosen as

$$\begin{cases} s = e_2 + \lambda e_1 \\ \dot{s} = \dot{e}_2 + \lambda \dot{e}_1 = \dot{e}_2 + \lambda e_2. \end{cases} \tag{12}$$

Accordingly, the equivalent control law is given by

$$u_{eq} = \frac{1}{m}\left\{-\lambda e_2 + h(x,t) - h(y,t) - d(t) - v(u)\right\}. \tag{13}$$

In practical, *d*(t) and *v*(*u*) are usually unknown. Therefore, the equivalent control input is modified to

$$u_{eq} = \frac{1}{m}\left\{-\lambda e_2 + h(x,t) - h(y,t)\right\} \tag{14}$$

In presence of uncertainties, the function h is unknown, which makes the equivalent control law (14) generally inapplicable. Thus, we construct two fuzzy systems $\hat{h}(x \mid \theta_x)$ and $\hat{h}(y \mid \theta_y)$ to approximate $h(x, t)$ and $h(y, t)$, respectively.

The fuzzy rule base of $\hat{h}(x \mid \theta_x)$ and $\hat{h}(y \mid \theta_y)$ consists of the following rules:

$$R_{\hat{h}x}^{(m)}: \quad IF\ x_1\ is\ F_{x1}^{m}\ and \cdots and\ x_n\ is\ F_{xn_1}^{m},\ THEN\ \hat{h}(x \mid \theta_x)\ is\ F_{\hat{h}x}^{m} \tag{15a}$$

$$R_{\hat{h}y}^{(m)}: \quad IF\ y_1\ is\ F_{y1}^{m}\ and \cdots and\ y_n\ is\ F_{yn_1}^{m},\ THEN\ \hat{h}(y \mid \theta_y)\ is\ F_{\hat{h}y}^{m} \tag{15b}$$

where, $m = 1,2 \ldots, Q$ (i.e., Q is the total number of fuzzy rules for each fuzzy model), $F_{xi}^{m}(i = 1,\cdots,n)$ and $F_{yi}^{m}(i = 1,\cdots,n)$ are the fuzzy sets assigned to x_i ($i = 1,\ldots,n$) and y_i ($i = 1,\ldots, n$), respectively and $F_{\hat{h}x}^{m}$ and $F_{\hat{h}y}^{m}$ are fuzzy singletons for $\hat{h}(x \mid \theta_x)$ and $\hat{h}(y \mid \theta_y)$.

The outputs of the fuzzy models $\hat{h}(x \mid \theta_x)$ and $\hat{h}(y \mid \theta_y)$ can be expressed as

$$\hat{h}(x \mid \theta_x) = \theta_x^T \xi(x) \tag{16a}$$

$$\hat{h}(y \mid \theta_y) = \theta_y^T \xi(y) \tag{16b}$$

where, $\theta_x = [F_{\hat{h}x}^{1}, F_{\hat{h}x}^{2},\cdots,F_{\hat{h}x}^{Q}]$ and $\theta_y = [F_{\hat{h}y}^{1}, F_{\hat{h}y}^{2},\cdots,F_{\hat{h}y}^{Q}]$ are the adjustable parameter vectors, $\xi(x) = [\xi^1(x), \xi^2(x),\cdots,\xi^Q(x)]^T$ and $\xi(y) = [\xi^1(y), \xi^2(y),\cdots,\xi^Q(y)]^T$ are the vectors of fuzzy basis functions (Kung & Chen, 2005), which are defined as

$$\xi^i(x) = \frac{\prod_{i=1}^{n} F_{xi}^{j}(x_i)}{\sum_{j=1}^{Q}[\prod_{i=1}^{n} F_{xi}^{j}(x_i)]}, \qquad j = 1,2,\cdots,Q, \text{and} \tag{17a}$$

$$\xi^i(y) = \frac{\prod_{i=1}^{n} F_{yi}^{j}(y_i)}{\sum_{j=1}^{Q}[\prod_{i=1}^{n} F_{yi}^{j}(y_i)]}, \qquad j = 1,2,\cdots,Q. \tag{17b}$$

$F_{xi}^{j}(x_i)$, $F_{yi}^{j}(y_i)$ represent the membership degree of x_i in F_{xi}^{j} and y_i in F_{yi}^{j}, respectively.

Therefore, the equivalent control input can be expressed as

$$u_{eq} = \frac{1}{m}\left\{-\lambda e_2 + \hat{h}(x,t) - \hat{h}(y,t)\right\}. \tag{18a}$$

To get rid of chattering phenomenon, the reaching control input is selected as (Yau, 2008)

$$u_r = \frac{k_{fs}}{m} u_{fs}. \tag{18b}$$

and by defining $k_{afs} = \frac{k_{fs}}{m}$. the overall control input can be expressed as

$$u = u_{eq} + u_r = \frac{1}{m}\{-\lambda e_2 + \hat{h}(x,t) - \hat{h}(y,t)\} + k_{afs} u_{fs}. \tag{18c}$$

To derive the adaptive law for adjusting Θ_x and Θ_y, we first define the optimal parameter vectors θ_x^* and θ_y^* as

$$\theta_x^* = \arg \min_{\theta_x Î \Omega_{hx}} [\sup_{x Î U_x} |h(x \mid \theta_x) - h(x,t)|] \text{ and} \tag{19a}$$

$$\theta_y^* = \arg \min_{\theta_y Î \Omega_{hy}} [\sup_{y Î U_y} |h(y \mid \theta_y) - h(y,t)|], \tag{19b}$$

where, Ω_{hx} and Ω_{hy} are defined as $\Omega_{hx} = \{\theta_x Î R^Q \mid \|\theta_x\| \le m_x\}$, $\Omega_{hy} = \{\theta_y \in \Re^Q \mid \|\theta_y\| \le m_y\}$ and m_x and m_y are positive constants. (15)(16)(17)(18)(19)

Define the minimum approximation error as

$$w = [h(y,t) - \hat{h}(y \mid \theta_y^*)] - [h(x,t) - \hat{h}(x \mid \theta_x^*)]. \tag{20}$$

According to (Kung & Chen, 2005) w is bounded; i.e. $w \in L_\infty$.

$$\begin{aligned} |w| &\le |h(y,t) - \hat{h}(y \mid \theta_y^*)| + |h(x,t) - \hat{h}(x \mid \theta_x^*)| \\ &\le |h(y,t)| + |\hat{h}(y \mid \theta_y^*)| + |h(x,t)| + |\hat{h}(x \mid \theta_x^*)| \\ &\le |h(y,t)| + \|\theta_y^{*T}\| \|\xi(y)\| + \|\theta x\| \|\xi(x)\| + |h(x,t)| \\ &\le m_y + H + m_x + H \le \zeta \end{aligned} \tag{21}$$

where ζ is an unknown positive constant.

In order to derive the adaptive law for adjusting Θ_x and Θ_y, we consider the following Lyapunov function

$$V = \frac{1}{2}s^2 + \frac{1}{2\gamma_x}\Phi_x^T\Phi_x + \frac{1}{2\gamma_y}\Phi_y^T\Phi_y + \frac{1}{2\gamma_k}(k_{afs} - \hat{k})^2 \tag{22}$$

Where $\Phi_x = \theta_x - \theta_x^*$, $\Phi_y = \theta_y - \theta_y^*$ and γ_x, γ_y, γ_k are arbitrary positive constants. The time derivative of the (22) is

$$\dot{V} = s\dot{s} + \frac{1}{\gamma_x}\Phi_x^T\dot{\Phi}_x + \frac{1}{\gamma_y}\Phi_y^T\dot{\Phi}_y + \frac{1}{\gamma_k}(k_{afs} - \hat{k})\dot{k}_{afs} \tag{23}$$

$$\begin{aligned}
\dot{V} &= s.[\dot{e}_2 + \lambda\dot{e}_1] + \frac{1}{\gamma_x}\Phi_x^T\dot{\Phi}_x + \frac{1}{\gamma_y}\Phi_y^T\dot{\Phi}_y + \frac{1}{\gamma_k}(k_{afs} - \hat{k})\dot{k}_{afs} \\
&= s.[h(y,t) - h(x,t) + d(t) + \Phi(u) + \lambda\dot{e}_1] + \frac{1}{\gamma_x}\Phi_x^T\dot{\Phi}_x + \frac{1}{\gamma_y}\Phi_y^T\dot{\Phi}_y + \frac{1}{\gamma_k}(k_{afs} - \hat{k})\dot{k}_{afs} \\
&= s.[h(y,t) - \hat{h}(y \mid \theta_y^*) + \hat{h}(y \mid \theta_y^*) - \hat{h}(y \mid \theta_y) - h(x,t) - \hat{h}(x \mid \theta_x^*) + \hat{h}(x \mid \theta_x^*) \\
&+ \hat{h}(x \mid \theta_x) + d(t) + v(u) + k_{afs}u_{fs}] + \frac{1}{\gamma_x}\Phi_x^T\dot{\Phi}_x + \frac{1}{\gamma_y}\Phi_y^T\dot{\Phi}_y + \frac{1}{\gamma_k}(k_{afs} - \hat{k})\dot{k}_{afs} \\
&\le sw - s[\hat{h}(y \mid \theta_y) - \hat{h}(y \mid \theta_y^*)] + s[\hat{h}(x \mid \theta_x) - \hat{h}(x \mid \theta_x^*)] + \beta|s| + \rho|s| - k_{afs}|s| \\
&+ \frac{1}{\gamma_x}\Phi_x^T\dot{\Phi}_x + \frac{1}{\gamma_y}\Phi_y^T\dot{\Phi}_y + \frac{1}{\gamma_k}(k_{afs} - \hat{k})\dot{k}_{afs} \\
&\le \zeta|s| + \beta|s| + \rho|s| - |s|(k_{afs} - \hat{k}) - \hat{k}|s| - (k_{fs} - \beta)|s| + sw + \Phi_y^T(-s\xi(y) + \frac{1}{\gamma_y}\dot{\theta}_y) \\
&\quad + \Phi_x^T(s\xi(x) + \frac{1}{\gamma_x}\dot{\theta}_x) + \frac{1}{\gamma_k}(k_{afs} - \hat{k})\dot{k}_{afs} \\
&\le \zeta|s| + \beta|s| + \rho|s| - |s|(k_{afs} - \hat{k}) - \hat{k}|s| - (k_{fs} - \beta)|s| + sw + \Phi_y^T(-s\xi(y) + \frac{1}{\gamma_y}\dot{\theta}_y) \\
&\quad + \Phi_x^T(s\xi(x) + \frac{1}{\gamma_x}\dot{\theta}_x) + (k_{afs} - \hat{k})(-|s| + \frac{1}{\gamma_k}\dot{k}_{afs}).
\end{aligned} \tag{24}$$

By the above equation, we choose the adaptive laws

$$-s\xi(y) + \frac{1}{\gamma_y}\dot{\theta}_f = 0 \Rightarrow \dot{\theta}_y = \gamma_y s\xi(y), \tag{25a}$$

$$s\xi(x) + \frac{1}{\gamma_x}\dot{\theta}_x = 0 \Rightarrow \dot{\theta}_x = -\gamma_x s\xi(x), \text{ and} \tag{25b}$$

$$-|s| + \frac{1}{\gamma_k}\dot{k}_{afs} = 0 \Rightarrow \dot{k}_{afs} = \gamma_k |s| \tag{25c}$$

It is clear that the scalar $\hat{k}$ can be chosen in such a way that the value of $\zeta + \rho + \beta - \hat{k}$ remains negative (i.e., $\zeta + \rho + \beta - \hat{k} = -\eta$ where $\eta > 0$)

Therefore we obtain (25)

$$\dot{V} \le (\zeta + \rho + \beta - \hat{k})|s| = -\eta|s|. \tag{26}$$

Using Barbalat's lemma (Slotin & Li, 1991), it can be concluded that $s, \dot{s} \in L_\infty$ and s approaches zero as $t \to \infty$. This means that the system is stable and the error asymptotically converges to zero.

Method II. Adaptive Sliding Mode Control Method for Synchronization of Nonlinear Gyros

In this method, we consider the system with more general form than pervious method and some special uncertainties in system will be mentioned. At last, a gyro system controller will be designed using an adaptive sliding mode controller.

Consider a class of n-dimensional chaotic systems with the following state equations. It is assumed that one of the systems acts as the master and the other as the slave.

$$\text{Master System: } \begin{cases} \dot{x}_i = x_{i+1}, & 1 \leq i \leq n-1, \\ \dot{x}_n = g(x,t) \end{cases} \quad x = [x_1, x_2, \cdots, x_n] \in \Re^n \tag{27}$$

$$\text{Slave System: } \begin{cases} \dot{y}_i = y_{i+1}, & 1 \leq i \leq n-1, \\ \dot{y}_n = f(y,t) + F(y,t)\tilde{\theta} + d(t) + u \end{cases} \quad y = [y_1, y_2, L, y_n] \in \Re^n \tag{28}$$

Where $x \in R^n$ and $y \in R^n$ denote the state vectors, $\tilde{\theta} \in \Re^m$ represents the uncertain parameter vector of system (28), $f(y, t)$ and $g(x, t)$ belong to $R_n \to R$ space, $F(y, t)$ is an $1 \times m$ vector, $u \in R$ is the control input and $d(t)$ may come from exoteric perturbation or is the random component added to the system. It is assumed that $|d(t)| \leq \tilde{k} < \infty$ for all t and the value of $\tilde{k} > 0$ is not known in advance.

It is assumed that $g(x, t)$, $f(y, t)$ and $F(y, t)$ satisfy the following necessary condition:

Systems (27) and (28) should have unique solutions in the time interval $[t_0, +\infty)]$, $t_0 > 0$, for any given initial condition $x_0 = x(t_0)$ and $y_0 = y(t_0)$.

For synchronizing systems (27) and (28) a robust adaptive sliding mode control design is used such that:

$$\lim_{t \to \infty} \left\| y(t) - x(t) \right\| \to 0 \tag{29}$$

where $\| . \|$ denotes the Euclidian norm of a vector.

Now, let us define the state errors between the master and slave system as

$$e_1 = y_1 - x_1,\ e_2 = y_2 - x_2 \ldots,\ e_n = y_n - x_n \tag{30a}$$

$$\begin{cases} \dot{e}_1 = e_2, \\ \dot{e}_2 = e_3, \\ \vdots \\ \dot{e}_{n-1} = e_n, \\ \dot{e}_n = f(y,t) + F(y,t)\tilde{\theta} - g(x,t) + d(t) + u(t). \end{cases} \tag{30b}$$

To control and identify the unknown part of (28), the following adaptive controller can be used.

$$u_{ad} = -F(y,t)\hat{\theta} - \hat{k}sign(s) - l\,s, \tag{31}$$

where, $l > 0$ is a constant parameter. The parameters $\hat{\theta} = [\hat{\theta}_1, \hat{\theta}_2, \cdots, \hat{\theta}_m]^T$ and $\hat{k}$ are updated according to the following update laws:

$$\dot{\hat{\theta}}_i = s\gamma_i F_i(y,t)\ (i = 1, \cdots, m), \qquad \dot{\hat{k}} = \gamma_k |s| \tag{32}$$

where, γ_i $(i = 1,\ldots,m) > 0$ and $\gamma_k > 0$ are arbitrary positive constants. In this way, the overall controller is constructed as

$$u = u_{eq} + u_{ad} = -f(y,t) + g(x,t) - \sum_{i=1}^{n-1} c_i e_{i+1} - F(y,t)\hat{\theta} - \hat{k}sign(s) - l\,s \tag{33}$$

Theorem 2.1.

If the controller is selected as (33) then the systems (27) and (28) can be synchronized.

Proof:

Let $\theta = \tilde{\theta} - \hat{\theta}$, $k = \tilde{k} - \hat{k}$ and choose a Lyapunov function a:

$$V = \frac{1}{2}[s^2 + \left\{\sum_{i=1}^{m} \frac{1}{\gamma_i}(\tilde{\theta}_i - \hat{\theta}_i)^2\right\} + \frac{1}{\gamma_k}(\tilde{k} - \hat{k})^2], \tag{34}$$

where, γ_i $(i = 1,\ldots,m)$ and γ_k are arbitrary positive constants. The time derivative of V is

$$\begin{aligned}
\dot{V} &= s\dot{s} - \left\{\sum_{i=1}^{m} \frac{1}{\gamma_i}\left(\tilde{\theta}_i - \hat{\theta}_i\right)\dot{\hat{\theta}}\right\} - \frac{1}{\gamma_k}\left(\tilde{k} - \hat{k}\right)\dot{\hat{k}} \\
&= s[f(y,t) + F(y,t)\tilde{\theta} - g(x,t) + d(t) + u(t) + \sum_{i=1}^{n-1} c_i e_i] - \left\{\sum_{i=1}^{m} \frac{1}{\gamma_i}\left(\tilde{\theta}_i - \hat{\theta}_i\right)\dot{\hat{\theta}}_i\right\} - \frac{1}{\gamma_k}\left(\tilde{k} - \hat{k}\right)\dot{\hat{k}}
\end{aligned} \tag{35}$$

$$\begin{aligned}
&= s[F(y,t)\tilde{\theta} + d(t) + u_{ad}(t)] - \left\{\sum_{i=1}^{m} \frac{1}{\gamma_i}(\tilde{\theta}_i - \hat{\theta}_i)\dot{\hat{\theta}}_i\right\} - \frac{1}{\gamma_k}(\tilde{k} - \hat{k})\dot{\hat{k}} \\
&= s[F(y,t)\tilde{\theta} + d(t) - F(y,t)\hat{\theta} - \hat{k}sign(s) - l\,s] - \left\{\sum_{i=1}^{m} \frac{1}{\gamma_i}(\tilde{\theta}_i - \hat{\theta}_i)\dot{\hat{\theta}}_i\right\} - \frac{1}{\gamma_k}(\tilde{k} - \hat{k})\dot{\hat{k}} \\
&\leq -l\,s^2 + \tilde{k}|s| - \hat{k}|s| + s\sum_{i=1}^{m}[F_i(y,t)(\tilde{\theta}_i - \hat{\theta}_i)] - \left\{\sum_{i=1}^{m} \frac{1}{\gamma_i}(\tilde{\theta}_i - \hat{\theta}_i)\dot{\hat{\theta}}_i\right\} - \frac{1}{\gamma_k}(\tilde{k} - \hat{k})\dot{\hat{k}} \\
&\leq -l\,s^2 + (\tilde{k} - \hat{k})[|s| - \frac{1}{\gamma_k}\dot{\hat{k}}] + s\left\{\sum_{i=1}^{m}[(\tilde{\theta}_i - \hat{\theta}_i)(sF_i(y,t) - \frac{1}{\gamma_i}\dot{\hat{\theta}}_i]\right\}
\end{aligned} \tag{36}$$

By the above equation, we choose the adaptive laws as

$$\left|s\right| - \frac{1}{\gamma_k}\dot{\hat{k}} = 0 \Rightarrow \dot{\hat{k}} = \gamma_k \left|s\right| \text{ and} \tag{37}$$

$$sF_i(y,t) - \frac{1}{\gamma_i}\dot{\hat{\theta}}_i = 0 \Rightarrow \dot{\hat{\theta}}_i = s\gamma_i F_i(y,t)\,. \tag{38}$$

Using adaptive laws (37) and (38) we obtain

$$V \leq - ls^2 \leq 0 \tag{39}$$

Using Barbalat's lemma (Slotin & Li, 1991), it can be concluded that $s, \dot{s} \in L_\infty$. Furthermore s approaches zero as $t \rightarrow \infty$, which shows that the system is stable and the error asymptotically converges to zero.

Method III. Gyros Synchronization using Adaptive Gain Fuzzy Sliding Mode Control

In this method, we consider a class of the following two n-dimensional chaotic systems

$$\begin{cases} \dot{x}_i = x_{i+1}, & 1 \leq i \leq n-1, \\ \dot{x}_n = f(x,t) \end{cases} \quad x = [x_1, x_2, \cdots, x_n] \in \Re^n \text{ and} \tag{40}$$

$$\begin{cases} \dot{y}_i = y_{i+1}, & 1 \leq i \leq n-1, \\ \dot{y}_n = f(y,t) + \Delta f(y) + d(t) + u \end{cases} \quad y = [y_1, y_2, L, y_n] \in \Re^n, \tag{41}$$

where $u \in R$ is the control input, f is a given nonlinear function of x and t, $\Delta f(y)$ is an uncertain term representing the unmodeled dynamics or structural variation of system (41) and $d(t)$ is the disturbance of system (41).

In general, the uncertainty and the disturbance are assumed to be bounded as $\left|\Delta f(y)\right| \leq \alpha$ and $\left| d(t) \right| \leq \beta$.

Moreover, it is assumed that $f(x, t)$ and $f(y, t)$ satisfy all the necessary conditions, such that, systems (40) and (41) having unique solution in the time interval $[t_0, +\infty)$, $t_0 > 0$, for any given initial condition $x_0 = x(t_0)$ and $y_0 = y(t_0)$.

Let us define the state errors between the master and slave system as:

$$e_1 = y_1 - x_1,\ e_2 = y_2 - x_2, \ldots,\ e_n = y_n - x_n \tag{42}$$

$$\begin{cases} \dot{e}_1 = e_2, \\ \dot{e}_2 = e_3, \\ \vdots \\ \dot{e}_{n-1} = e_n, \\ \dot{e}_n = g(x_1, x_2, \cdots, x_n, e_1, e_2, \cdots, e_n) + \Delta f(e + x) + d(t) + u(t) \end{cases} \quad (43)$$

where $g = f(y, t) - f(x, t)$ is a known nonlinear function. The synchronization problem can be viewed as the problem of choosing an appropriate control law $u(t)$ such that the error states e_i $(i = 1, 2\ldots, n)$ in (43) generally converge to zero.

This proposed AGFSMC scheme is shown in (Figure 7). It contains an equivalent control part and a reaching law. The reaching law is

$$u_r = k_{afs} u_{fs} \quad (44)$$

where, k_{afs} is the reaching gain that is achieved by adaptation law and u_{fs} is the output of the *FSMC*, which is determined by the normalized s and $\dot{s}$. The overall control u is chosen as

$u = u_{eq} + u_r = u_{eq} = k_{afs} u_{fs}$ (45) and the adaptation law for k_{afs} is

$$\dot{k}_{afs} = \gamma |s| \quad (46)$$

where, γ is a positive constant. The fuzzy control rules of FSMC provide the mapping of input linguistic variables s and $\dot{s}$ to the output linguistic variable u_{fs} as

$$u_{fs} = FSMC(s, \dot{s}). \quad (47)$$

The membership functions of input variables s and $\dot{s}$, and output variable u_{fs} are shown in (Figure 8) and the fuzzy rule table of FSMC is designed as in (Table 1) (Yau & Chen,2008).

Figure 7. Block diagram of the AGFSMC scheme

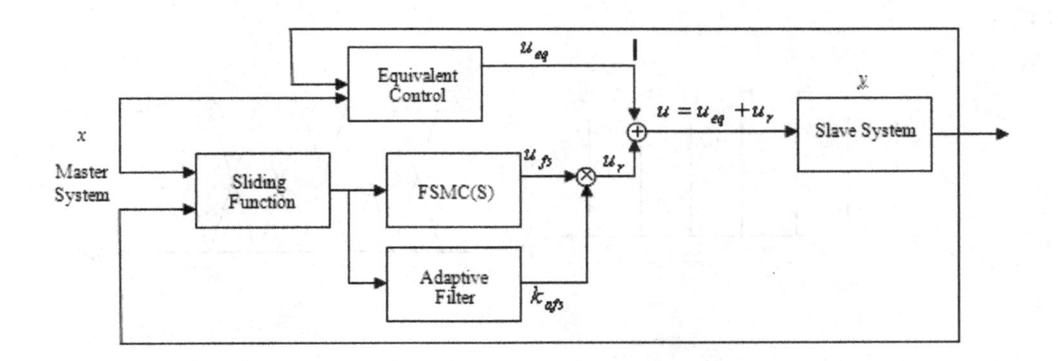

In practical systems, the system uncertainty $\Delta f(x)$ and external disturbance $d\,(t)$ are unknown and the implemented equivalent control input is modified as:

$$u_{eq} = -g(x,e) - \sum_{i=1}^{n-1} c_i e_{i+1} \tag{48}$$

In the following theorem, we provide the proof of stability of the scheme presented in (45) and with the aid of this, it is possible to derive the nonlinear system (40) onto the sliding mode *S=0*. That is, the reaching condition $s(t)\dot{s}(t) < 0$ is guaranteed.

Note that, we can always choose an arbitrary scalar $\hat{k}$ such that ±+² -$\hat{k}$ is negative (i.e., $\alpha + \beta - \hat{k} = -\tau$ where $\tau > 0$) (Roopaei, Karimaghaee & Soleimanifar, 2006).

Theorem 3.1.

Assume that the uncertain nonlinear system (40) is controlled by *u(t)* in (45), where u_{eq} is (48), u_{fs} is (47) and k_{afs} is (46). Then, the error state trajectory converges to the sliding surface $s(t) = 0$.

Proof:

Consider the following Lyapunov function:

Table 1. Rule Base of FSMC

s							u_{fs}	
PB PM PS ZE NS NM NB								
ZE	NS	NM	NB	NB	NB	NB	PB	$\dot{S}$
PS	ZE	NS	NM	NB	NB	NB	PM	
PM	PS	ZE	NS	NM	NB	NB	PS	
PB	PM	PS	ZE	NS	NM	NB	ZE	
PB	PB	PM	PS	ZE	NS	NM	NS	
PB	PB	PB	PM	PS	ZE	NS	NM	
PB	PB	PB	PB	PM	PS	ZE	NB	

Figure 8. Membership functions of the fuzzy sets assigned to (a) output variable (u_{fs}) (b) input variables (s and $\dot{s}$)

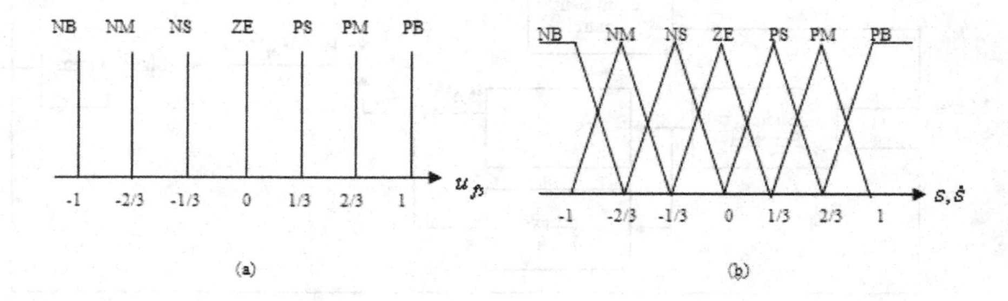

$$V = \frac{1}{2}s^2 + \frac{1}{2\gamma}(\hat{k} - k_{afs})^2 \tag{49}$$

Then, we have

$$\begin{aligned}
\dot{V} &= s\dot{s} - \frac{1}{\gamma}(\hat{k} - k_{afs})\dot{k}_{afs} \\
&= s.[\dot{e}_n + \sum_{i=1}^{n-1} c_i \dot{e}_i] - \frac{1}{\gamma}(\hat{k} - k_{afs})\dot{k}_{afs} \\
&= s.[-g(x,e) + \Delta f(x+e) + d(t) + u_{eq} + k_{afs}u_{fs} + \sum_{i=1}^{n-1} c_i \dot{e}_i] - \frac{1}{\gamma}(\hat{k} - k_{afs})\dot{k}_{afs} \\
&= s.[\Delta f(x+e) + d(t) + k_{afs}u_{fs}] - \frac{1}{\gamma}(\hat{k} - k_{afs})\dot{k}_{afs} \\
&\leq \alpha|s| + \beta|s| - k_{afs}|s| + \hat{k}|s| - \hat{k}|s| - \frac{1}{\gamma}(\hat{k} - k_{afs})\dot{k}_{afs} \\
&\leq (\alpha + \beta - \hat{k})|s| + |s|(\hat{k} - k_{afs}) - \frac{1}{\gamma}(\hat{k} - k_{afs})\dot{k}_{afs} \\
&\leq (\alpha + \beta - \hat{k})|s| + (\hat{k} - k_{afs})(-\frac{1}{\gamma}\dot{k}_{afs} + |s|)
\end{aligned} \tag{50}$$

It is clear that the scalar $\hat{k}$ can be chosen in such a way that the value of $\alpha + \beta - \hat{k}$ remains negative (i.e., $\alpha + \beta - \hat{k} = -\tau$ where $\tau > 0$). Considering that: $-\frac{1}{\gamma}\dot{k}_{afs} + |s| = 0$, the adaptive gain of fuzzy controller is obtained as $\dot{k}_{afs} = \gamma|s|$ and the derivative of the mentioned Lyapunov function will satisfy the following condition.

$$\dot{V} \leq -\tau|s| \tag{51}$$

Using Barbalat's lemma (Slotin & Li, 1991), it can be concluded that s, $\dot{s} \in L_\infty$ and s approaches zero as $t \to \infty$. That is, the system is stable and the error asymptotically converges to zero.

SIMULATION RESULTS

Method I

Consider the system as:

$$\begin{cases} \dot{x}_1 = x_2, \\ \dot{x}_2 = g(x_1) - c_1 x_2 - c_2 x_2^3 + (\beta + f\sin\omega t)\sin(x_1) \end{cases}$$

where $g(x) = -\alpha^2 \frac{(1 - \cos x)^2}{\sin^3 x}$. In experiments, the parameter of the nonlinear gyro systems were specified as follows: $\alpha^2 = 100$, $\beta =$, $c_1 = 0.5$, $c_2 = 0.05$, $\omega = 2$ and $f = 35.5$, which give raise to chaotic state. The initial conditions were defined as: $x_1(0) = 1$, $x_2(0) = -1$, $y_1(0) = 1.6$, $y_2(0) = 0.8$. We assumed that the disturbance term $d(t) = 0.2\cos(\pi t)$, is bounded by $|d(t)| \leq \beta = 0.2$. In the sliding surface design, the value of $\lambda = 6$ was used. To construct the fuzzy logic system (for the equivalent control part), $\hat{h}(x,t)$ and $\hat{h}(y,t)$ were specified as (16-a) and (16-b), respectively. The following Gaussian membership functions were assigned to x_i $(i = 1, 2)$ and y_i $(i = 1, 2)$ over the interval [-3, 3].

$$\mu_{NB}(x) = 1/(1+\exp(5(x+2))) \qquad \mu_{NB}(y) = 1/(1+\exp(5(y+2)))$$
$$\mu_{NM}(x) = \exp(-(x+1.5)^2) \qquad \mu_{NM}(y) = \exp(-(y+1.5)^2)$$
$$\mu_{NS}(x) = \exp(-(x+0.5)^2) \qquad \mu_{NS}(y) = \exp(-(y+0.5)^2)$$
$$\mu_{PS}(x) = \exp(-(x-0.5)^2) \qquad \mu_{PS}(y) = \exp(-(y-0.5)^2)$$
$$\mu_{PM}(x) = \exp(-(x-1.5)^2) \qquad \mu_{PM}(y) = \exp(-(y-1.5)^2)$$
$$\mu_{PB}(x) = 1/(1+\exp(5(x-2))) \qquad \mu_{PB}(y) = 1/(1+\exp(5(y-2)))$$

Using these membership functions, each of the fuzzy systems use thirty six rules to model $\hat{h}(x,t)$ and $\hat{h}(y,t)$. The initial $\Theta_x(0)$ and $\Theta_y(0)$ were chosen randomly in the interval [-2, 2]. The learning rates were specified as: $\gamma_x = \gamma_y = 0.3$. The parameters of the dead-zone were selected as: m=1, m_{min}=0.85, m_{max}=1.25, b_{min}=0.45 *and* b_{max}=0.6.

To build the reaching control part of the proposed controller, u_{fs} was specified as (18-b) and $k_{afs}(0)$ = 0.5.

Simulation results are shown in (Figures 9), (Figure 10), and (Figure 11). (Figure 9) shows the states of the master and slave systems. The control signal is shown in (Figure 10) and the synchronization errors are shown in (Figure 11).

The simulation results show that the proposed method is successful in synchronizing the two chaotic nonlinear gyros when the gyro structures and external disturbances are unknown and we have dead-zone nonlinearity in the control input.

Method II

To verify the effectiveness of this method, compared results with methods proposed in (Yan, Hung & Liao, 2006), conventional SMC method and the scheme presented in (Yau, 2008) is presented.

The external disturbance was set to $d(t) = 0.2\cos(2t)$ and the sliding surface was selected as $s = e_2 = e_1$.

All other parameters and initial values were set to the same values used in (Yan, Hung & Liao, 2006).

We have presented the results of some various methods in (Figure 12) and (Figure 13). In these Figures, we have used M, A, B, C and D to denote the master signal, the method proposed in (Yau, 2008), the proposed method of this section, the method proposed in (Yan, Hung & Liao, 2006) and conventional SMC method, respectively. In this simulations, the values of k_{fs}=25 (normalization factor of the output

Figure 9. Gyros Synchronization Master and Slave States, (a) State 1, (b) State 2

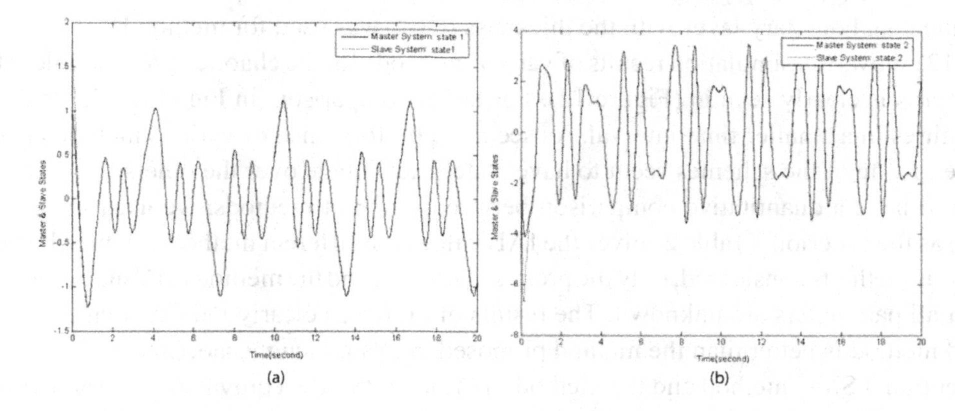

Figure 10. Proposed Controller

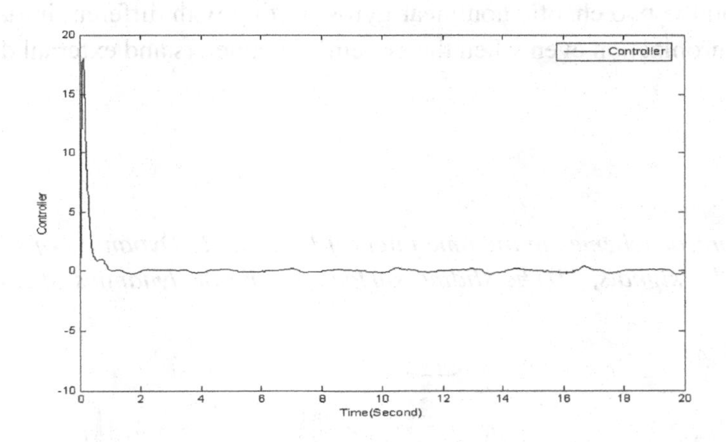

Figure 11. Error Dynamic

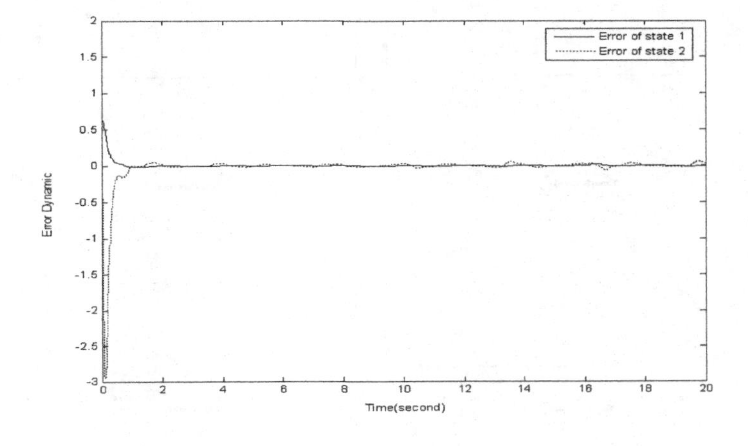

variable) and k_r=12 (reaching gain) were used for methods A and D, respectively. To remove the chattering phenomena, boundary layer with the thickness of 0.5 was used for method D.

(Figure 12) shows the simulation results of various methods for the chaotic gyro example. The chaotic motion of gyros is clearly seen in (Figure 12). For better comparison, in Fig.=ure 13, we have plotted the same figures in a smaller time interval. As seen, the performance of various methods is very close to each other. None of the schemes seem to have a clear advantage over the others.

In order to have a quantitative comparison between different methods, we used Integral Absolute Error (IAE) as the criterion. (Table 2) gives the IAE values for different methods. It must be noticed that among various methods considered, only the proposed method and the method in (Yan, Hung, Liao, 2006) assume that all parameters are unknown. The results of (Table 2) clearly indicate that the performance of proposed method is better than the method proposed in (Yan, Hung, Liao, 2006).

In conventional SMC method and the method in (Yau, 2008), the equivalent control part is assumed to be known. As expected, these methods have better performance in comparison with proposed method of this section. Simulation results show that the trajectory of error dynamics converges to $S=0$ and the synchronization error also converges to zero in the proposed method. Generally, the method of this section performs well and the two chaotic nonlinear gyros starting with different initial values are indeed achieving chaos synchronization even when the system's parameters and external disturbance are fully unknown.

Figure 12. Various control schemes in the time interval [0, 20], (1) Dynamics of state 1, (2) Dynamics of state 2, (3)Controller signals, (4)The sliding surface, (5) Error dynamics of state 1, and (6) Error dynamics of state 2

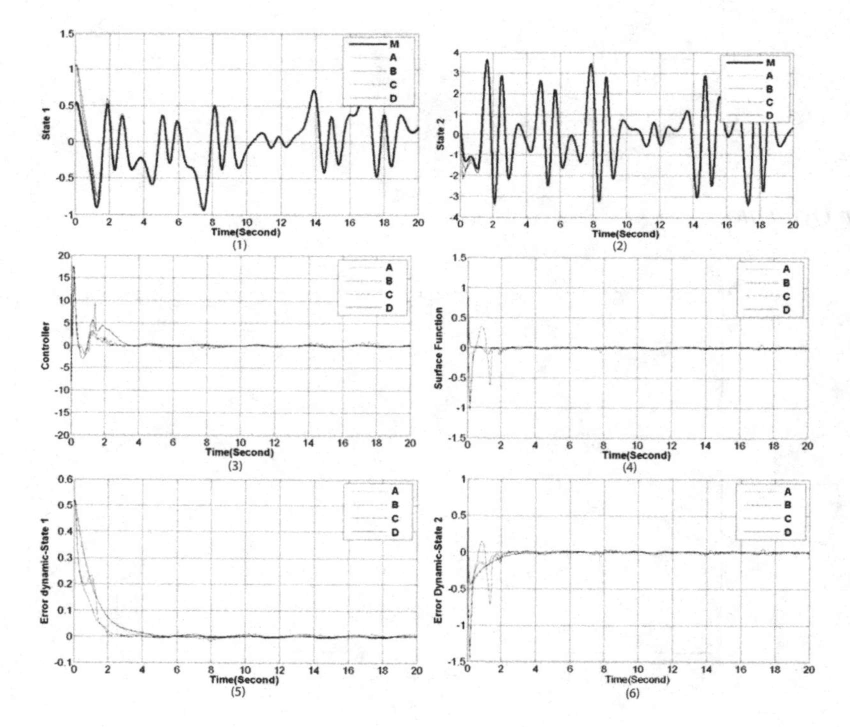

Table 2. Integral Absolute Error (IAE) values for different control methodologies

D	C	B	A	
603.0352	454.5871	371.7151	595.7138	IAE of State 1
599.5184	775.7614	712.1731	605.2504	IAE of State 2

Method III

In simulations, the parameters of the gyro systems were specified as $\alpha^2 = 100$, $\beta = 1$, $c_1 = 0.5$, $c_2 = 0.05$, $\omega = 2$ and $f = 35.5$. Using these values, the gyro system shows chaotic behavior.

Consider two chaotic gyros as (40) and (41). While $f(x,t)$ and $f(y,t)$ are as

$$f(x,t) = g(x_1) - c_1 x_2 - c_2 x_2^3 + (\eta + f \sin \omega t) \sin(x_1)$$
$$f(y,t) = g(y_1) - c_1 y_2 - c_2 y_2^3 + (\eta + f \sin \omega t) \sin(y_1). \quad (52)$$

The initial conditions were specified as $x_1(0) = 1$, $x_2(0) = -1$, $y_1(0) = 1.6$, $y_2(0) = 0.8$. It was assumed that the uncertainty term, i.e. $\Delta f(y_1, y_2) = 0.1 \sin(y_1)$, and the disturbance term, i.e. $d(t) = 0.2 \cos(\pi t)$,

Figure 13. Various control schemes in the time interval [0, 4]; (1) Dynamics of state 1, (2) Dynamics of state 2, (3) Controller signals, (4) The sliding surface, (5) Error dynamics of state 1, and (6) Error dynamics of state 2

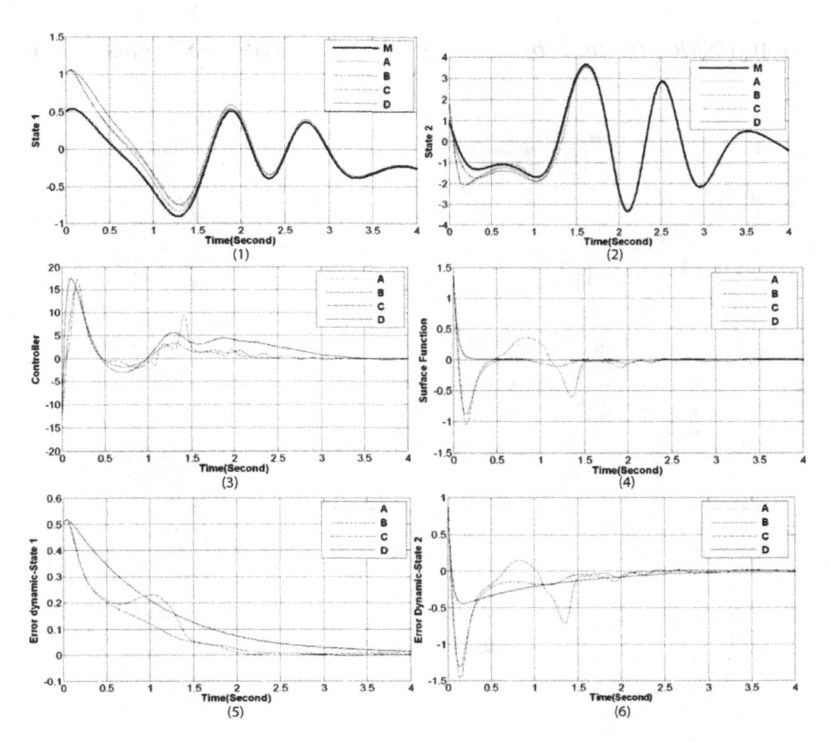

are bounded by $\left|\Delta f(y_1, y_2)\right| \leq \hat{\alpha} = 0.1$ and $\left|d(t)\right| \leq \hat{\beta} = 0.2$, respectively. For the sliding surface, design $s = \dot{e}_2 + 6e_2$ was selected.

In our simulations, k_{fs}=*0.5* (normalization factor of the output variable) and k_r=*0.6* (reaching gain) were used for the method presented in (Yau & Chen, 2008) and the traditional SMC design, respectively. To remove the chattering phenomena, boundary layer with the thickness of *0.5* was used.

Simulation results of the proposed AGFSMC scheme of this section, the FSMC method (Yau & Chen, 2008) and traditional SMC with/without boundary layer is given in Figures 14-17. As seen, the error trajectory converges to *S=0* and the synchronization error also converges to zero, which confirms the theoretical analysis. Overall, this scheme performs well and the two chaotic nonlinear gyros started with different initial values did achieve synchronization in the presence of uncertainty and external disturbance.

In comparison with the FSMC method, this proposed method performs better, which in turn proves that the adaptive gain scheme is effective. We can also see that the chattering phenomenon is avoided in this scheme.

In order to have a quantitative comparison of different methods, we used Integral Absolute Error (IAE) as the criterion. (Table 3) gives the IAE values for different methods. The results of (Table 3) indicate that the performance of this method is better than the method proposed in (Yau & Chen, 2008) and conventional SMC method, however, the conventional SMC method (with boundary layer technique) has better performance rather than this method.

Figure 14. Proposed AGFSMC, (a) and (b) States of Master and Slave Systems (c) controller signal (d) Errors Dynamic

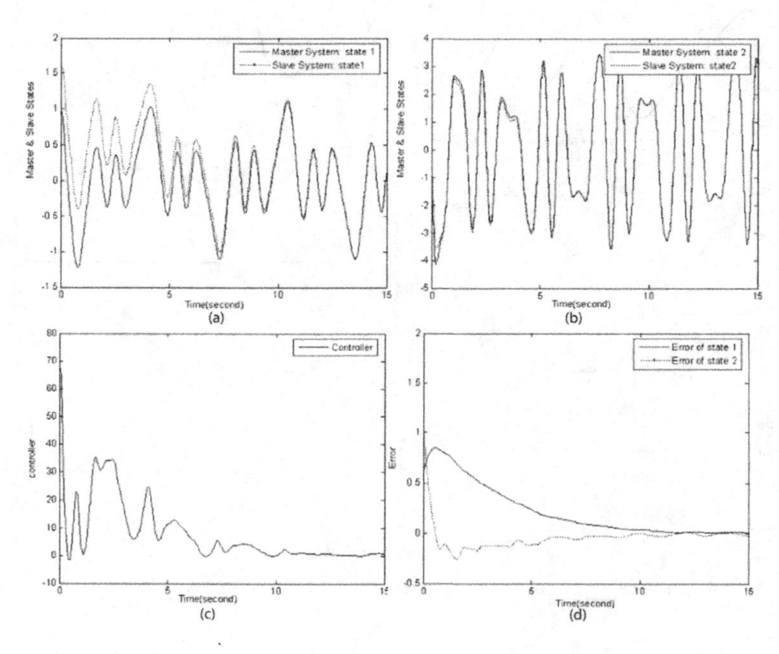

Table 3. Integral Absolute Error (IAE) values for various methods

Traditional SMC with boundary layer technique	Traditional SMC	Proposed method	The method of Yau & Chen(2008)	Method
186.1	484.1	442.1	787.1	IAE

Figure 15. FSMC (Yau & Chen, 2008)(a) and (b) States of Master and Slave Systems (c) controller signal (d) Errors Dynamic

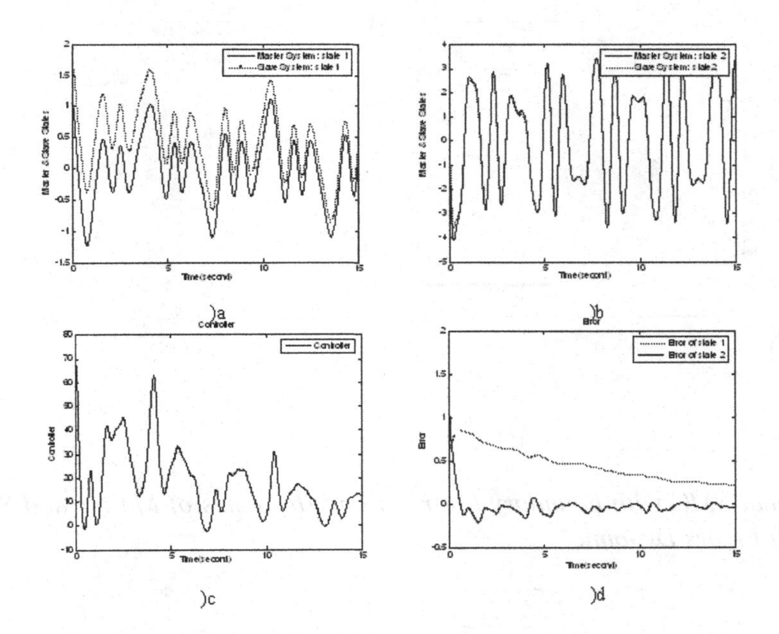

CONCLUSION

This chapter investigated the chaotic behavior of gyro systems and proposed three methods for synchronization of them. In the proposed methods, based on the type of uncertainty and disturbances in gyro systems several new combinations of sliding mode control, fuzzy logic systems and adaptive control have been used. In the first approach, an adaptive fuzzy sliding mode control method was proposed to design a controller for the synchronization of chaotic gyros when uncertainties and disturbances are present and there is dead-zone nonlinearity in the control input. In the second method, for fully unknown systems a robust adaptive SMC has been proposed and the last method used a fuzzy logic approach to reduce effects of the chattering phenomena in the SMC design. Unlike some well-known methods of the SMC, no knowledge on the bound of uncertainty and disturbance was required in these three methods and only some fixed points were required. Finally, the effectiveness of all of these approaches was shown through simulation results.

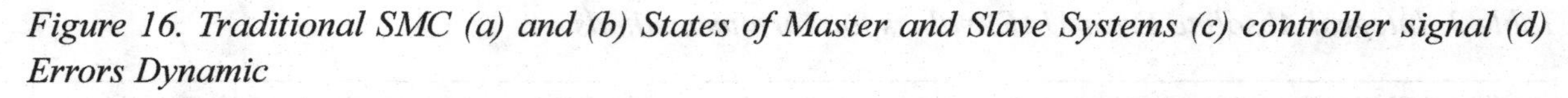

Figure 16. Traditional SMC (a) and (b) States of Master and Slave Systems (c) controller signal (d) Errors Dynamic

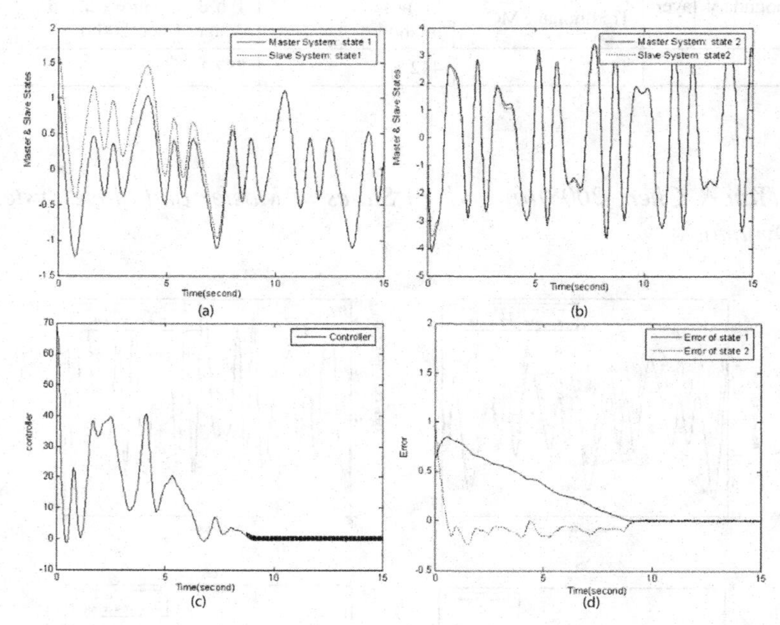

Figure 17. Traditional SMC with boundary layer (a) and (b) States of Master and Slave Systems (c) controller signal (d) Errors Dynamic

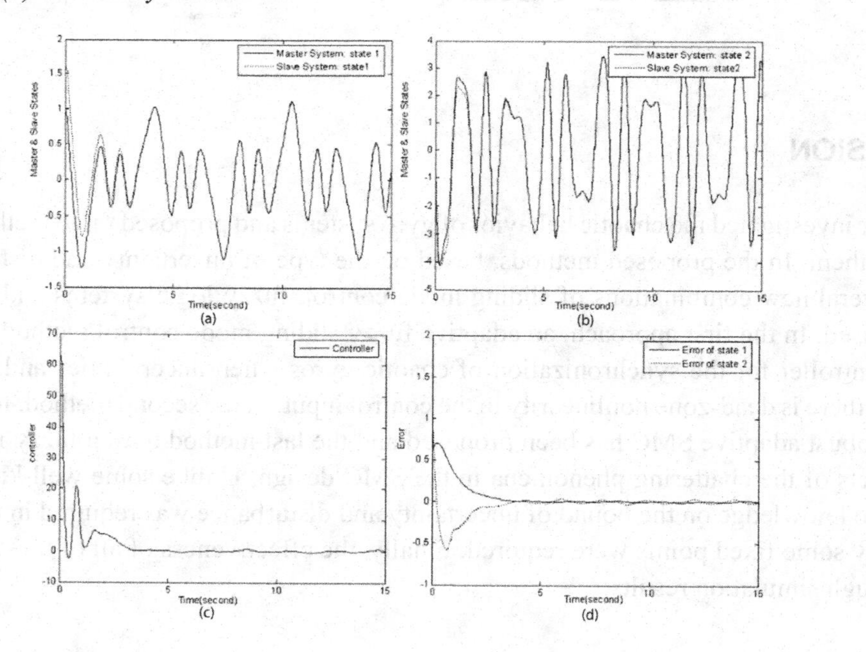

REFERENCES

Andronova, I. A., & Malykin, G. B. (2002). Physical problems of fiber gyroscopy based on the Sagnac effect. *Phys. Usp., 45*, 793. Retrieved from doi: 10.1070/PU2002v045n08ABEH001073

Buchman, S. (2000). Cryogenic gyroscopes for the relativity mission. *Physica B, Condensed Matter, 280*, 497. doi:10.1016/S0921-4526(99)01846-3

Chen, H. K. (2002). Chaos and chaos synchronization of a symmetric gyro with linear-plus-cubic damping. *Journal of Sound and Vibration, 255*(4), 719–740. doi:10.1006/jsvi.2001.4186

Chen, H. K., & Lin, T. N. (2003). Synchronization of chaotic symmetric gyros by one-way coupling conditions. *Proceedings of the Institution of Mechanical Engineers. Part C, Journal of Mechanical Engineering Science, 217*, 331–340. doi:10.1243/095440603762869993

Dooren, R. V. (2003). Comments on Chaos and chaos synchronization of a symmetric gyro with linear-plus-cubic damping. *Journal of Sound and Vibration, 268*, 632–634. doi:10.1016/S0022-460X(03)00343-2

Fradkov, A. L., & Evans, R. J. (2005). Control of chaos: Methods and applications in engineering. *Annual Reviews in Control, 29*, 33–56. doi:10.1016/j.arcontrol.2005.01.001

Ge, Z. M., Chen, H. K., & Chen, H. H. (1996). The regular and chaotic motion of a symmetric heavy gyroscope with harmonic excitation. *Journal of Sound and Vibration, 198*, 131–147. doi:10.1006/jsvi.1996.0561

Hwang, E. J., Hyun, C. H., Kim, E., & Park, M. (2009). Fuzzy model based adaptive synchronization of uncertain chaotic systems: Robust tracking control approach. *Physics Letters. [Part A], 373*, 1935–1939. doi:10.1016/j.physleta.2009.03.057

Kung, C. C., & Chen, T. H. (2005). Observer-based indirect adaptive fuzzy sliding mode control with state variable filters for unknown nonlinear dynamical systems. *Fuzzy Sets and Systems, 155*, 292–308. doi:10.1016/j.fss.2005.04.016

Lei, Y., Xu, W., & Zheng, H. (2005). Synchronization of two chaotic nonlinear gyros using active control. *Physics Letters. [Part A], 343*, 153–158. doi:10.1016/j.physleta.2005.06.020

Leipnik, R. B., & Newton, T. A. (1981). Double strange attractors in rigid body motion. *Physics Letters. [Part A], 86*, 63. doi:10.1016/0375-9601(81)90165-1

Li, X., Xu, W., & Xiao, Y. (2008). Adaptive tracking control of a class of uncertain chaotic systems in the presence of random perturbations. *Journal of Sound and Vibration, 314*, 526–535. doi:10.1016/j.jsv.2008.01.035

Li, Z., Chen, G., Shi, S., & Han, C. (2003). Robust adaptive tracking control for a class of uncertain chaotic systems. *Physics Letters. [Part A], 310*, 40–43. doi:10.1016/S0375-9601(03)00115-4

Parmananda, P. (1998). Recursive proportional-feedback and its use to control chaos in an electrochemical system. *Physics Letters. [Part A]*, 240–255.

Pecora, L. M., & Carroll, T. L. (1990). Synchronization in chaotic systems. *Physical Review Letters*, *64*, 821–824. doi:10.1103/PhysRevLett.64.821

Roopaei, M., Karimaghaee, P., & Soleimanifar, M. (2006). Control of the Chaotic Systems by the Linear State Feedback. *Nonlinear Studies Journal*, *133*, 167–173.

Roopaei, M., Zolghadri Jahromi, M. (2008). Synchronization of a class of chaotic systems with fully unknown parameters using adaptive sliding mode approach. *Chaos*, 18, 043112. Retrieved from doi:10.1063/1.3013601

Roopaei, M., Zolghadri Jahromi, M., & Jafari, S. (2009). Adaptive Gain Fuzzy Sliding Mode Control for the Synchronization of Nonlinear Chaotic Gyros. *Chaos, 19*(1), 013125-013125-9. Retrieved from doi: 10.1063/1.3072786

Salarieh, H., & Alasty, A. (2008). Chaos synchronization of nonlinear gyros in presence of stochastic excitation via sliding mode control. *Journal of Sound and Vibration*, *313*(3-5), 760–771. doi:10.1016/j.jsv.2007.11.058

Salarieh, H., & Shahrokhi, M. (2008). Adaptive synchronization of two different chaotic systems with time varying unknown parameters. *Chaos, Solitons, and Fractals*, *37*, 125–136. doi:10.1016/j.chaos.2006.08.038

Slotine, J. J. E., & Li, W. (1991). *Applied Nonlinear Control*. Englewood Cliffs, NJ: Prentice Hall.

Stedman, G. E. (1997). Ring laser tests of fundamental physics and geophysics. *Reports on Progress in Physics*, *60*, 615–688. doi:10.1088/0034-4885/60/6/001

Tong, X., & Mrad, N. (2001). Chaotic motion of a symmetric gyro subjected to a harmonic base excitation. *Journal of Applied Mechanics—. Transactions of the American Society of Mechanical Engineers*, *68*, 681–684.

Woodman, K. F., Franks, P. W., & Richards, M. D. (1987). The nuclear magnetic resonance gyroscope: a review. *Navigation*, *40*, 366–384. doi:10.1017/S037346330000062X

Wu, C. W., Yang, T., & Chua, L. O. (1996). On adaptive synchronization and control of nonlinear dynamical systems. *International Journal of Bifurcation and Chaos in Applied Sciences and Engineering*, *6*, 455–471. doi:10.1142/S0218127496000187

Yan, J. J., Hung, M. L., & Liao, T. L. (2006). Adaptive sliding mode control for synchronization of chaotic gyros with fully unknown parameters. *Journal of Sound and Vibration*, *298*, 298–306. doi:10.1016/j.jsv.2006.05.017

Yan, J. J., Hung, M. L., Lin, J. S., & Liao, T. L. (2007). Controlling chaos of a chaotic nonlinear gyro using variable structure control. *Mechanical Systems and Signal Processing*, *21*(6), 2515–2522. doi:10.1016/j.ymssp.2006.07.002

Yassen, M. T. (2006). Chaos control of chaotic dynamical systems using backstepping design. *Chaos, Solitons, and Fractals*, *27*, 537–548. doi:10.1016/j.chaos.2005.03.046

Yassen, M. T. (2007). Controlling, synchronization and tracking chaotic Liu system using active backstepping design. *Physics Letters. [Part A]*, *360*, 582–587. doi:10.1016/j.physleta.2006.08.067

Yau, H. T. (2008). Chaos synchronization of two uncertain chaotic nonlinear gyros using fuzzy sliding mode control. *Mechanical Systems and Signal Processing*, *22*, 408–418. doi:10.1016/j.ymssp.2007.08.007

Yin, X., Ren, Y., & Shan, X. (2002). Synchronization of discrete spatiotemporal chaos by using variable structure control. *Chaos, Solitons, and Fractals*, *14*, 1077–1082. doi:10.1016/S0960-0779(02)00048-6

Zhou, D., Shen, T., & Tamura, K. (2006). Adaptive nonlinear synchronization control of twin gyro procession. *ASME Journal of Dynamics System Measurement and Control*, *128*, 592–599. doi:10.1115/1.2232683

Chapter 10
Importance of Chaos Synchronization on Technology and Science

Ricardo Aguilar-López
CINVESTAV-IPN, Mexico

Ricardo Femat
IPICYT, México

Rafael Martínez-Guerra
CINVESTAV-IPN, Mexico

ABSTRACT

In this chapter the authors talk about importance of chaos synchronization on technology and science. This chapter is developed in three sections. In the former the authors tackle the subject related with the so-called the synchronized state in the sense of identical synchronization. It is done via robust nonlinear observer design, considering corrupted measurements and model uncertainties, coupling uncertainty estimators with nonlinear state observers. The second part treats the subject related to the applications to chaos communications, that is to say, an application of chaos theory which is aimed to provide security in the transmission of information performed through telecommunications technologies, speaking roughly at the transmitter, a message is added on to a chaotic signal and then, the message is masked in the chaotic signal. As it carries the information, the chaotic signal is also called chaotic carrier. This is done via control theory and is a particular case of chaos synchronization. Finally, in the latter section the authors talk about application to synchronization of biological systems that is to say, the intercellular Ca^{2+} *waves have been seen like a mechanism by means of which a group of cells can communicate with one another and coordinate a multicellular response to a local event. Recently, it has been observed in a variety of systems that calcium signals can also propagate from one cell to another and thereby serve as a means of intercellular communication. The desire to understand the biophysical mechanisms of cellular dynamics has lead to introduce feedback control laws in some biological systems. Departing from the above ideas, this section explores links between feedback control schemes, with an external input, and intracellular calcium functions for coordination and control.*

DOI: 10.4018/978-1-61520-737-4.ch010

Synchronization of oscillations has been known to scientists since the historical observation of this phenomenon by Huygens in pendulum clocks. With the development of radio and electronics in the 20th century, synchronization occupied a very special place in science and technology. As many phenomena studied by nonlinear dynamics, synchronization was observed and shown to play an important role in many problems of a most diverse nature (physical, ecological, physiological, meteorological, and chemical to name a few). There is hardly a single communication or data storage application that does not rely on synchronization. Original notion and theory of synchronization implies periodicity of oscillators (Pecora, 1990; Pecora, 1998).

The discovery of deterministic chaos introduced a new kind of an oscillating system, a chaotic generator. Chaotic oscillations are found in many dynamical systems of various origins. The behavior of such systems is characterized by instability of bounded trajectories and, as a result, limited predictability along time. Intuitively it would seem that chaos and synchronization are two mutually exclusive terms. Synchronization of chaos is a phenomenon that may occur when two, or more, chaotic oscillators are coupled (namely complex networks), or when a chaotic oscillator drives another chaotic oscillator, namely, master-slave synchronization. Yet it has been shown that synchronization can be observed even in chaotic systems. However, the special features of chaotic systems make it impossible to directly apply the methods developed for synchronization of periodic oscillations. Although synchronization has been observed in chaotic systems, particular properties of chaos provoke synchrony methods of periodic oscillators cannot be applicable on chaos. Even defining the notion of synchronization for chaotic systems is difficult without running into a paradox or controversy.

Clear understanding of chaos synchronization phenomena and dynamical mechanisms behind it open new opportunities both for applications of chaotic signals in engineering and for understanding functionality of neurobiological networks, where irregular (chaotic) dynamics of neurons occurs naturally. This research activity focuses on the developments of theoretical foundations for chaos synchronization based on the experimental studies of synchronization phenomenon in physical, neurobiological and other systems.

From the above, in this book chapter proposal, there are considered several cases for chaos synchronization, in particular:

- **Identical master-slave synchronization.** This is a straightforward form of synchronization that may occur when two identical chaotic oscillators are mutually coupled, or when one of them drives the other. Although synchronization has been observed in chaotic systems, particular properties of chaos provoke synchrony methods of periodic oscillators cannot be applicable on chaos. Consider $(x_1,x_2,,...,x_n)$ and $(x'_1, x'_2,...,x'_n)$ denote the set of dynamical variables that describe the state of the first and second oscillator, respectively, it is said that identical synchronization occurs when there is a set of initial conditions $[x_1(0), x_2(0),...,x_n(0)]$, $[x'_1(0), x'_2(0),...,x'_n(0)]$ such that, denoting the time by t, $|x'_i(t)-x_i((t)|\to 0$, for $i=1,2,...,n$, when $t\to\infty$. That means that for time large enough the dynamics of the two oscillators verifies $x'_i(t)=x_i(t)$, for $i=1,2,...,n$, in a good approximation. This is called the synchronized state in the sense of identical synchronization. It is done vía robust nonlinear observer design, considering corrupted measurements and model uncertainties, coupling uncertanty estimators with nonlinear state observers.
- **Applications to chaos communications** is an application of chaos theory which is aimed to provide security in the transmission of information performed through telecommunications technologies. By secure communications, one has to understand that the contents of the message transmit-

ted are inaccessible to possible eavesdroppers. In chaos communications security (i.e., privacy) is based on the complex dynamic behaviors provided by chaotic systems. Some properties of chaotic dynamics, such as complex behaviour, noise-like dynamics (pseudorandom noise) and spread spectrum, are used to encode data. On the other hand, being chaos a deterministic phenomenon, it is possible to decode data using this determinism. In practice, implementations of chaos communications devices resort to one of two chaotic phenomena: synchronization of chaos, or control of chaos. To implement chaos communications using such properties of chaos, two chaotic oscillators are required as a transmitter (or master) and receiver (or slave). At the transmitter, a message is added on to a chaotic signal and then, the message is masked in the chaotic signal. As it carries the information, the chaotic signal is also called chaotic carrier. It is done via control theory and is a particular case of chaos synchronization.

- **Application to synchronization of biological systems.** The ubiquity of oscillations in biological systems is well established. Oscillations are observed in all types of organisms from the simplest to the most complex. Periods can range from fractions of a second to months or years. From time to time, it has been suggested that many biological oscillations are the result of the breakdown of effective self-regulation. The opposite view is defended here. It is argued that most periodic behavior is not pathological but rather constitutes the normal operation for these systems. They are present because they confer positive functional advantages for the organism. In particular the neuron-science research has developed important tools which allow to basic understanding of biological information procedure in single neurons and neural networks. Medical evidence indicates that the synchronization of large neural networks is much related with the processing of information of the brain. Theoretical studies into nonlinear oscillators' synchronization have focused on the question to what to extend the degree of synchronization can be regulated through of the neuron-coupling degree. Almost everything that we do is controlled by Ca^{2+}, how we move, how our hearts beat and how our brains process information and store memories. To do all of this, Ca^{2+} acts as an intracellular messenger, relaying information within cells to regulate their activity. For example, Ca^{2+} triggers life at fertilization, and controls the development and differentiation of cells into specialized types. It mediates the subsequent activity of these cells and, finally, is invariably involved in cell death. To coordinate all of these functions, Ca^{2+} signals need to be flexible yet precisely regulated. This incredible versatility arises through the use of a Ca^{2+} signaling 'tool kit', whereby the ion can act in the various contexts of space, time and amplitude. It is widely believed that intercellular Ca^{2+} waves are a mechanism by which a group of cells can communicate with one another and coordinate a multicellular response to a local event. Recently, it has been observed in a variety of systems that calcium signals can also propagate from one cell to another and thereby serve as a means of intercellular communication. The desire to understand the biophysical mechanisms of cellular dynamics has lead to introduce feedback control laws in some biological systems. The aims of feedback control laws in biological systems are to cause excitation or suppression of oscillations, entrainment and synchronization, or transitions from chaotic to periodic oscillations and vice versa (Freeman, 2000; Tyson, 2003) using realistic control inputs. Departing from the above ideas, this section explores links between feedback control schemes, with an external input, and intracellular calcium functions for coordination and control.

IDENTICAL MASTER-SLAVE SYNCHRONIZATION

As is well known the study of the synchronization problem for chaotic oscillators has been very important from the nonlinear science point of view, in particular the applications to biology, medicine, cryptography, secure data transmission and so on.

In general, the synchronization research has been focused onto two areas, the first one, related with the employ of *state observers*, where the main applications lies on the synchronization of nonlinear oscillators with the same model structure and order, but different initial conditions and/or parameters (Celikovsky, 2005; Ling, 2004; Martinez-Guerra, 2008; Morgül, 1996; Nijmeijer, 1997a; Shihua, 2004; Solak, 2004).

As is well known observer schemes are widely used for the reconstruction of no measured state dynamics. The only available information is the measured system's output, which represents a function of some current inner states of the system. Usually, the dimension of the vector of output measured signals is smaller than the dimension of the corresponding vector of states; therefore it is necessary to develop estimation techniques, known as observer design dealing with on-line state estimation.

The most successful observation schemes need a nominal model for their implementation, but as is well known the exact knowledge of the nonlinearities of nonlinear plants is a hard task, such that uncertain systems must be tackled. This situation leads to the standard observers not be realizable. Interesting research of observer based chaotic system synchronization have been done, (Mörgul, 1996), applied an observer for nonlinear oscillators which can be transformed in an observability canonical form, reduced order observers have been employed for synchronization purposes and parameters identification in chaotic oscillators (Martinez-Guerra, 2006; Mendoza-Camargo 2004; Aguilar-Ibañez, 2006) with adequate performance, recently in (Martinez-Guerra, 2006) a reduced order observer for observable uncertain chaotic oscillators via algebraic differential approach have been presented. (Alvarez, 2003) presented an observer design which estimate unobservable states of nonlinear plants under the assumption of nominal plant model knowledge, (Aguilar-Lopez, 2002; Aguilar, 2003; Martinez-Guerra, 2004) developed high gain observers for uncertainty estimation in nonlinear systems and integral-type observer for state estimation for partially unknown nonlinear plants, however the problem of state estimation with unobservable uncertainties still remains. Following these ideas, researches were oriented to the observation/estimation problem subjected to bounded nonlinearities or uncertainties. If the plant model is uncertain or incomplete, which is the most common case, the implementation of high gain observers turns out to be adequate. Besides, the designs of new robust observers based on adaptive techniques, such as neural networks, have been proposed (Huaguang, 2007; Pogromsky, 1998; Shihua, 2004).

Another approach related to the construction of observers (basically, asymptotic) for nonlinear processes, is the geometric differential approach (Besançon 1999; Femat, 2001; Isidori, 1995). The main idea is to find some state transformation that represents the original system as a linear equation plus a nonlinear term, which is a function of the system output. However, finding a nonlinear transformation, that places a system of order n into observer form, requires simultaneous integration of n-coupled partial differential equations. Furthermore, this approach needs accurate knowledge of the nonlinear dynamics of the system; hence, turns to be inapplicable if the model for the process includes uncertainties.

However, the problems when the system's states and the uncertain terms are not observable, and the output measured signal is corrupted by noise have not been studied enough and it still remains. From the above, new develops in estimation theory related with both state and uncertainty estimation must be considered.

On the other hand, the use of *control laws* allows toachieve the synchronization between nonlinear oscillators, with different structure and order, where the variable states of the slave system are forced to follow the trajectories of the master system. This approach can be seen as a tracking problem (Bowong, 2004; Femat, 2001; González, 1999; Moukam, 2004; Muraly, 2000) some authors design the controller from the dynamic of the synchronization error, because this approach allows transforming the tracking problem to a regulation problem with the origin (zero) as the corresponding *set point* (Femat, 1999b; Bowong, 2004).

Besides, several control approaches have used neural-network, fuzzy, adaptive, and sliding and other techniques (Aguilar-Lopez, 2008; Huaguang, 2007; Martinez-Guerra, 2008). Other traditional control methods (Ott, 1990) consider introducing an additive feedback controller, to force to the system to reach the desired reference (set point), i.e. $\|x(t)-x_{sp}(t)\|\to 0$ as $t\to\infty$ The above mentioned methodologies are based on the cancellation of the non-linear terms of the chaotic systems in order to impose a desired behavior. Under the above philosophy the non-linear geometric-differential control techniques have been successfully employed (Isidori, 1995; Nijmejer, 1990; Nijmeijer, 1997b).

They correspond to systems that can be fully or partially linearized by a change of coordinates and/or state feedback following differential-geometric concepts (Gonzalez, 1999). Such class of non-linear systems can be linearized by a state feedback control, which cancels all the nonlinearities assuming perfect knowledge of the mathematical model, producing global asymptotic stability (Femat, 2001; Isidori, 1995). A drawback of exact linearization techniques and the others model based controllers is that they rely on exact cancellation of nonlinearities.

In practice, exact knowledge of system dynamics is not possible. A more realistic situation is to know some nominal functions of the corresponding nonlinearities, which are employed in the control design. However, the use of nominal model nonlinearities can lead to performance degradation and even closed-loop instability. In fact, when the systems posses strong nonlinearities, the standard linearizing, generic model, and active controllers, cannot cancel completely such nonlinearities and instabilities can be induced. The worst case is when the knowledge of the nonlinearities is very poor or null, such that, conventional linearizing techniques are inadequate. To avoid the above problems, the geometric approach for the design of nonlinear controllers based on uncertainty observers has been employed; due this kind of techniques are shown satisfactory capabilities for a wide kind of systems (Aguilar, 2002; Aguilar, 2003).

The use of proportional observers coupled with linearizing controllers have been very successful, but the proportional observers have several problems, for example, they are very sensitive for noisy measurements, robustness issues are not completely saved. For this reasons a more sophisticated observers have been designed in order to generate more adequate open-loop and closed-loop performances. PI observers, sliding-mode, numeric, etc. have been developed (Aguilar, 2001; Martinez-Guerra, 2004).

In particular, in this section is proposed an uncertainty proportional-derivative reduced order observer which is able to accelerate the estimation rate and then it is coupled to a linearizing controller to produce a robust Proportional-Integral-Derivative (PID) structure. This methodology is applied to the synchronization of one signal of the slave system, the nonlinear oscillators considered here have different order and structure and the examples proposed are de Duffing-Chen systems and the Van der Pol-Lorenz systems.

OBSERVER BASED SYNCHRONIZATION

Problem Description

Consider the following nonlinear plant with linear output which can be described by Equations 1a and 1b:

$$\dot{X} = f(X,U) = \Im(X) + \ell(X,\mathrm{U}) \tag{1a}$$

$$\mathbf{Y}=g(X, \varpi) = CX + \varpi \tag{1b}$$

Here, $X \in \Re^n$ is the vector of states; $\mathbf{U} \in \Re^q$ is the control input vector; $\Im(\circ):\Re^n \rightarrow \Re^n$ is a nonlinear, partially known vector field; $\ell(\circ):\Re^{n+q} \rightarrow \Re^n$ is a linear vector of arguments; $\varpi \in \Re^m$ is an additive bounded measurement noise and $\mathbf{Y} \in \Re^q$ is the system output. Now, consider that the unknown part of the nonlinear vector field $\Im(\circ):\Re^n \rightarrow \Re^n$ contains observable ($\Im_o$) and unobservable ($\Im_U$) uncertain terms, such that:

$$\Im(\circ) = \begin{cases} \Im_o(\circ) \\ \Im_v(\circ) \end{cases}$$

where $\Im_O$ is a sub-set of observable uncertainties and $\Im_U$ is a sub-set of unobservable uncertainties.

Now, let us consider the following assumptions:

A1. ϖ is a vector function representing an external (may be unknown), bounded perturbation:

$$\left\|\varpi\right\|_\Delta^2 = \wp < \infty, \quad 0 < \Delta = \Delta^T$$

The normalized matrix Δ is introduced to ensure the possibility to deal with components of different physical structure and is assumed as given *a priori*; $\wp$ represents the power of the corresponding perturbation.

A2. All the trajectories $X(\mathrm{t}, X_0), X_0 \in \Re^n$ of the system (1a and 1b) are bounded.

A3. The linear vector field $\ell(X,U)$ is bounded, *i.e.* for any $X \in \Re^n \|\ell(X,U)\| \leq \ell^+ < \infty$.

A4. A nominal model for the unobservable uncertainties is available, *i.e.* $\Im_v = \psi_0(X)$, consequently,

$$\dot{\Im} v = \frac{\partial \psi_0}{\partial X}\frac{\partial X}{\partial t} = \frac{\partial \psi_0}{\partial X}\dot{X}$$

The task is to design an observer to estimate the vector of state variables X, despite of the unknown part of the nonlinear vector $\Im$ (which will be estimated, also); considering that Y and U are measured on-line at each time interval *t*.

Observer Synthesis

In order to avoid the standard proportional observer drawbacks, the following modifications to its structure are proposed:

a. An uncertainty estimator based on *corrective* output term is introduced in the observation methodology, in order to estimate the unknown observable uncertainties of the nonlinear vector I (1a).
b. An uncertainty estimator based on *predictive* term is considered on the observation methodology, in order to estimate the unknown unobservable uncertainties of the nonlinear vector I (1a).

Furthermore, it is well known that classical proportional observers tend to amplify the noise of on-line measurements, which can lead to degradation of the observer performance. In order to save these drawbacks (Aguilar, 2003), the following representation of the system (1a and 1b) is done:

$$\dot{X} = \Im(X) + \ell(X, \mathrm{U}) \tag{2a}$$

$$\dot{X}_{*} = CX + \varpi \tag{2b}$$

$$\dot{\Im}_{0} = \Theta(X) \tag{2c}$$

$$\dot{\Im}_{U} = \frac{\partial \psi_{0}}{\partial X}\frac{\partial X}{\partial t} \tag{2d}$$

$$\mathrm{Y}_{*} = X_{*} \tag{2e}$$

The main issue of this new representation of the system is to eliminate the additive noise from the new system output **Y**$_*$, transforming the original output disturbance into a system disturbance. So the observable and unobservable uncertain terms of the vector I, are considered as a new states, which obey the following assumption:

A7. The dynamics of the uncertain terms, $\aleph_{\mathrm{O}}$ and $\aleph_{\mathrm{U}}$ are bounded, therefore:

$$|\Theta(X)| \le v \text{ for } v > 0 \text{ and } \left|\frac{\partial \psi_{0}}{\partial X}\frac{\partial X}{\partial t}\right| \le \mu \ \ for \ \ \mu > 0$$

Proposition 1: The following dynamic system is an asymptotic-type observer of the system given by Equations (2a to 2d):

$$\dot{\hat{X}} = \hat{\Im} + \ell(\hat{X}, \mathrm{U}) + k_{1}(\mathrm{Y}_{*} - \mathrm{C}\hat{X}_{*}) \tag{3a}$$

$$\dot{\hat{X}}_{*} = \mathrm{C}\hat{X} + k_{2}(Y_{*} - \mathrm{C}\hat{X}_{*}) \tag{3b}$$

$$\dot{\hat{\Im}}_{O} = k_{3}(\mathrm{Y}_{*} - \mathrm{C}\hat{X}_{*}) \tag{3c}$$

$$\dot{\hat{\Im}}_{U} = \left.\frac{\partial \psi_{0}}{\partial X}\right|_{X=\hat{X}} \dot{\hat{X}} \tag{3d}$$

The vector of proportional observer gains, $\mathbf{K} = (k_1 k_2 k_3\ 0)^T$, is proposed as:

$$\mathrm{K} = -N_{\pi}^{-1}\mathrm{C}^T$$
$$N_{\pi} = \left(\frac{1}{\pi^{i+j-1}} N_{ij}\right)_{i,j=1,n+2} \tag{4}$$

The parameter $\pi>0$ determines the desired convergence velocity. Moreover, in order to assure stabilizing properties, N_{π} should be a positive solution of the algebraic Riccati equation:

$$N_{\pi} = \left(R + \frac{\pi}{2}\mathrm{I}\right) + \left(R^T + \frac{\pi}{2}\mathrm{I}\right) N_{\pi} = \mathrm{C}^T\mathrm{C}$$
$$R = \begin{pmatrix} 0 & \mathrm{I}_{n-1,n-1} \\ 0 & 0 \end{pmatrix} \tag{5}$$

Stability Remarks

Sketch of Proof of Proposition 1

In accordance to Equations (2) and (3), the corresponding dynamic error equation, can be written as:

$$\dot{\xi} = \mathrm{A}\xi + \Omega(Z, \mathrm{U}, \varpi) \tag{6}$$

Here:

$$\xi = \begin{bmatrix} \dfrac{\left(X_* - \hat{X}_*\right)}{\left(\pi^{i+j-1}_{i,j=1,n+2}\right)} \\ \dfrac{\left(X - \hat{X}\right)}{\left(\pi^{i+j-1}_{i,j=1,n+2}\right)} \\ \dfrac{\left(\Im_O - \hat{\Im}_O\right)}{\left(\pi^{i+j-1}_{i,j=1,n+2}\right)} \\ \left(\Im_U - \hat{\Im}_U\right) \end{bmatrix}; \mathrm{A} = \begin{bmatrix} 0 & -k_1 & 1 & 0 \\ 1 & -k_2 & 0 & 0 \\ 0 & -k_3 & 0 & 0 \\ 0 & 0 & 0 & 0 \end{bmatrix}; \Omega \begin{bmatrix} \varpi \\ \Delta\ell \\ \Theta \\ \Delta\left(\dfrac{\partial \psi_0}{\partial Z}\bigg|_{X=\hat{X}} \dot{\hat{X}}\right) \end{bmatrix}; |\Omega| \le \begin{bmatrix} \wp \\ \Gamma \\ v \\ \rho \end{bmatrix} = \grave{\mathrm{E}}$$

Consider the following assumptions:

A7. The function $\Delta\ell = \ell(\mathrm{X},\mathrm{U}) - \ell(\hat{\mathrm{X}},\mathrm{U})$ is bounded, i.e. $\left\|\ell(\mathrm{X},\mathrm{U}) - \ell(\hat{\mathrm{X}},\mathrm{U})\right\| \le \Gamma$ for $\Gamma>0$.

A8. There exist two positive constants $j > 0$ and $\lambda > 0$, which satisfy:

$$\left\|\exp\left(\pi^{i+j=1}_{i,j=1,n+2}\mathrm{A}t\right)\xi\right\| \le j\exp\left(-\pi^{i+j=1}_{i,j=1,n+2}\lambda t\right)\left\|\xi\right\|$$

Now, solving the Equation (6), the following expression is obtained:

$$\xi = \exp\left(\pi_{i,j=1,n+2}^{i+j=1} \mathrm{A}t\right) \xi_0 + \int_0^t \exp\left\{\pi_{i,j=1,n+2}^{i+j=1} \mathrm{A}(t-s)\right\} \Omega(s) ds \tag{7}$$

Considering the assumptions *A6, A7* and *A8* and taking norms for both sides of the Equation (7), the following equation is generated:

$$\left\|\xi\right\| \leq j \exp\left(-\pi_{i,j=1,n+2}^{i+j=1} \lambda t\right) \left[\left\|\xi_0\right\| - \frac{j\Psi}{\pi_{i,j=1,n+2}^{i+j=-1} \lambda}\right] + \frac{j\Psi}{\pi_{i,j=1,n+2}^{i+j=-1} \lambda} \tag{8}$$

Taking the limit, when $t \rightarrow \infty$:

$$\left\|\xi\right\| \leq \frac{j\Psi}{\pi_{i,j=1,n+2}^{i+j=-1} \lambda} \tag{9}$$

The above inequality implies that the estimation error can be as small as is desired, if the parameter gain of the observer, π, is chosen large enough.

Application Example: The Chen Oscillator

As signal synchronization example, both two third-order Chen chaotic oscillators are considered as master (driven) and slave systems. The corresponding mathematic model is as follows:

Chen's model:

$$\begin{aligned} \dot{z}_{1,m} &= a(z_{2,m} - z_{1,m}) \\ \dot{z}_{2,m} &= \left(c - a\right) z_{1,m} - c z_{2,m} - \Im_O \\ \dot{z}_{3,m} &= \Im_U - b z_{3,m} \\ s_s &= z_{2,m} + \varpi \end{aligned}$$

where the observable and unobservable uncertainties were considered from the nonlinear terms of the oscillator as:

$\Im_O = z_{1,m} z_{3,m}$ and $\Im_U = z_{1,m} z_{2,m}$

with parameters:

$a = 35$; $b = 3$; $c = 28$ where s_s is the measured signal.

The corresponding observer structure to synchronize the slave system is:

$$\dot{\hat{z}}_{1,s} = a\left(\hat{z}_{2,s} - \hat{z}_{1,s}\right) + k_1\left(s_* - \hat{z}_{4,s}\right)$$
$$\dot{\hat{z}}_{2,s} = \left(c - a\right)\hat{z}_{1,s} - c\hat{z}_{2,s} - \hat{\Im}_O + k_2\left(s_* - \hat{z}_{4,s}\right)$$
$$\dot{\hat{z}}_{3,s} = \hat{\Im}_U - b\hat{z}_{3,s} + k_3\left(s_* - \hat{z}_{4,s}\right)$$
$$\dot{\hat{z}}_4 = \hat{z}_{2,s} + k_4\left(s_* - \hat{z}_{4,s}\right)$$
$$\dot{\Im}_O = k_5\left(s_* - \hat{z}_{4,s}\right)$$
$$\dot{\Im}_U = \left(a\left(\hat{z}_{2,s} - \hat{z}_{1,s}\right) + k_1\left(s_s - \hat{z}_{2,s}\right)\right)\hat{z}_{3,s} + \left(\hat{\Im}_U - b\hat{z}_{3,s} + k_3\left(s_s - \hat{z}_{2,s}\right)\right)\hat{z}_{1,s}$$

where the filtered measured output is $s_* = z_4$.

The master system has as initial condition the vector Z_{m0} = [1 1 1] and the slave system Z_{s0} = [0.5 0.5 0.5]. The synchronization procedure starts at five time units, considering the following observer gain vector K = [100 100 100 100 500].

As can be observed in Figures 1 to 3 from zero to five time units the chaotic signal is unsynchronized, such that the different initial conditions considered, when the synchronization methodology starts the slave signal tends to converge to the master signal without great effort, as can be seen.

Figure 1.

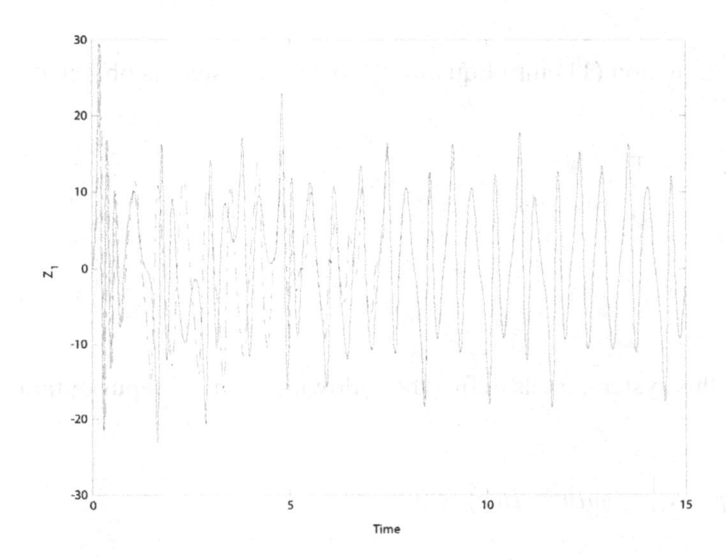

OBSERVER BASED CONTROLLER

Synchronization Methodology Design

Let us to assume that exists a coordinate transformation $(x,v)=\Psi(\aleph)$such that, the system (3) can be locally transformed into canonical form, given by:

$$\begin{aligned} \dot{x}_i &= x_{i+1} \quad i = 1,2,....,r-1 \\ \dot{x}_r &= \mathrm{f}(x,v) + B(x,v)u \\ \dot{v} &= \vartheta(x,v) \\ y &= x_1 \end{aligned} \tag{10}$$

Here, the following definitions were used:

- $B(x,v) = \bar{B}(x) + \Delta B$ where f(x,v) and ΔB are considered model uncertainties related to the non-linear system,
- $\bar{B}(x)$ is the nominal value of the control-input coefficient,
- u is the system input,
- x are the system states,
- y is the slave measured signal.
- r is the relative degree of the system

Now, it is possible to define a convenient change of variables:

$$\eta(x,v)=\mathrm{f}(x,v)+\Delta Bu \tag{11}$$

By substitution of Equation (11) into Equation (10) a new system is obtained:

$$\begin{aligned} \dot{x}_i &= x_{i+1} \quad i = 1,2,...,r-1 \\ \dot{x}_r &= \eta(x,u) + \bar{B}(x)u \\ \dot{v} &= \vartheta(x,v) \\ y &= x_1 \end{aligned} \tag{12}$$

In order to control this system, let us define the following nominal input-output linearizing feedback control.

$$u = \bar{B}^{-1}(x)\left[\tau_g x_r - \eta(x,u)\right] \quad with \quad \bar{B}(x) \neq 0 \tag{13}$$

The controller defined by Equation (13) guarantees exponential stability of non-linear systems with no uncertainties and perfect measurements, *i.e.*, $\Delta B=0$ and $\eta(x,u)$ known, under the assumption that the named inner dynamics $\dot{v} = \vartheta\left(x, v\right)$ *i.e.* the state equations which are not controlled present an stable dynamic behavior (minimum phase system), such that in accordance with the control theory point of view the system is stabilizable.

However, since the uncertainty term, $\eta(x,u)$ is unknown and, moreover, is function of the states, x and the control input, u this ideal control law is not causal and therefore is not realizable.

Nonetheless, there is another way to develop an input-output linearizing controller that is robust against uncertainties. The procedure shown below defines a method to estimate the uncertainty term, $\eta(x,u)$ This approach is based on observer theory, where the uncertainty is only a function of the estimation error. Let us define the following dynamic subsystem under the assumption that system (12) has a relative degree $r = 1$.

$$\dot{x}_r = \eta + \bar{B}u \tag{14}$$

$$\dot{\eta} = \Phi(x, u) \tag{15}$$

The uncertain term, η is considered as a new state and $\Phi(x,u)$ is a non-linear unknown function that describes η's dynamics. We can note that the uncertain term η, is observable, since:

$$\eta = \dot{s} - \bar{B}\, u \tag{16}$$

From the system given by Equations (14, 15), it can be seen that a standard observer structure design, *i.e.* a system copy plus output feedback, is not possible to construct since the term Φ is unknown and considering that variable state X_3 is the system output, let us propose the following uncertainty observer.

$$\dot{\hat{\eta}} = \tau_1(\eta - \hat{\eta}) + \tau_2(\dot{\eta} - \dot{\hat{\eta}}) \tag{17}$$

This uncertainty estimator is a reduced-order observer, which infers the uncertain term; η is obtained from the corresponding state equation. Note that the observer contains proportional and derivative actions; the aim of the derivative term is to improve the speed of the estimation algorithm, because it enhances the anticipatory and stabilizing effects of derivative actions.

Substituting the estimate of the uncertain term in the ideal controller defined by Equation (13), the following non-ideal controller is obtained:

$$u = \bar{B}^{-1}\left[\tau_g x_r - \hat{\eta}\right] \tag{18}$$

From the corresponding state equation $\eta = \dot{x}_r - \bar{B}(x_r)\, u$; substituting into it the Eq. (18) and taking the time derivative of the resulting equation the following expression is obtained:

$$\dot{\eta} = \ddot{x}_r - \tau_g \dot{x}_r + \dot{\hat{\eta}} \tag{19}$$

Introducing this result in Equation (17) yields:

$$\dot{\hat{\eta}} = \tau_2 \ddot{x}_r + (\tau_1 - \tau_2\tau_g) \dot{x}_r - \tau_g \tau_1 x_r \tag{20}$$

Since this controller uses an estimated value of the uncertainty, it cannot cancel the system non-linearities, completely. Practical stability is achieved as long as the uncertainty estimation error is bounded (Aguilar-Lopez, 2002; Martinez-Guerra, 2004). Thus, the system trajectories remain inside a neighborhood close to the defined trajectory of reference.

The final expression for the input-output linearizing controller with uncertainty estimation can be obtained integrating the estimator (Equation 20) and substituting it into the non-ideal controller (Equation 18), to obtain:

$$u = \bar{B}^{-1}(x)\left[(\tau_g - (\tau_1 - \tau_2\tau_g))\, x_r + \tau_g \tau_1 \int_o^t x_r(\theta)d\theta - \tau_2 \dot{x}_r\right] \tag{21}$$

Note that this controller (Equation 21) exhibits PID structure and is equivalent to the linearizing controller based on proportional-derivative uncertainty observer (Equations 17 and 18).

Stability Issues of the Proposed Observer

In order to show the stability properties of the closed-loop system, firstly, a convergence analysis of the uncertainty observer has to be done.

Proposition 1. Let us define η and $\hat{\eta}$ be the uncertainty term and its estimate. The dynamic system $\dot{\hat{\eta}} = \tau_1(\eta - \hat{\eta}) + \tau_2(\dot{\eta} - \dot{\hat{\eta}})$ is an asymptotic-type reduced order observer for the system (14-15).

Proof 1. Let us define the uncertainty estimation error as:

$$\varepsilon = \eta - \hat{\eta} = \Delta\eta \tag{22}$$

Now, the dynamic scalar equation of the estimation error is given by Equation (23), according to Equations (15) and (17) as follows.

$$\dot{\varepsilon} = -\frac{\tau_1}{1+\tau_2}\varepsilon + \frac{\Phi(x,u)}{1+\tau_2} \tag{23}$$

Integrating this last expression, it renders:

$$\varepsilon = \varepsilon_0 \exp\left(-\frac{\tau_1}{1+\tau_2}t\right) + \int_0^t \exp\left(\frac{\tau_1}{1+\tau_2}(t-s)\right)\frac{\Phi(x,u)}{1+\tau_2}ds \tag{24}$$

Now, consider the following assumption:
A1. - $\Phi(x,u)$ is bounded, $\|\Phi\|\leq\Psi$ with $0<\Psi<\infty$
Considering the norms of both sides of Equation (24):

$$\|\varepsilon\| \leq \|\varepsilon_0\| \exp\left(-\frac{\tau_1}{1+\tau_2}t\right) + \int_0^t \exp\left(\frac{\tau_1}{1+\tau_2}(t-s)\right)\frac{\|\Phi(x,u)\|}{1+\tau_2}ds$$

Applying the assumption ***A1*** the following expression is obtained:

$$\|\varepsilon\| \leq \exp\left(-\frac{\tau_1}{1+\tau_2}t\right)\left(\|\varepsilon_0\| - \frac{\Psi}{\tau_1}\right) + \frac{\Psi}{\tau_1} \tag{25}$$

In the limit, when $t\to\infty$:

$$\|\varepsilon\| \leq \frac{\Psi}{\tau_1} \tag{26}$$

It is important to analyze the structure of the Equation (25) in order to obtain some characteristics of the proposed observer. As be desired, in the limit when $t\to\infty$, the estimation error remain around a closed-ball with radius proportional to $\frac{\Psi}{\tau_1}$, which can be made as small as desired by taken τ_1 large enough.

In order to improve the speed to the uncertainty observer convergence to the steady-state estimation error, two actions were taken. The first one is to consider the parameter τ_1 large enough, which is necessary to obtain a small steady-state error, as it was mentioned above. The second one is faced with the influence of the parameter τ_2 related to the derivative action of the uncertainty observer. If $\tau_2 \to -1$, the exponential term of the right side of the equation (25) can be accelerated enough and consequently, the convergence of the uncertainty observer will exhibit better performance.

Note that if the measurements of the system are corrupted by additive noise, *i.e.* $s=x_3+v$, and this noise is considered bounded, $\|v\|\leq\Omega$ a methodology similar to the one used to analyze the estimation error ε can be applied in order to prove that the steady state estimation error becomes $\|\varepsilon\| \leq \frac{\Psi+\Omega}{\tau_1}$. This confirms robustness against noisy measurements.

Now, it is possible to implement a non-ideal controller using the uncertainty estimated,

In order to prove closed-loop stability, it is necessary to analyze the closed-loop equation of the state equation when the non-ideal control law is introduced.

Stability Issues of the Proposed Controller

Consider the state equation of the measured signal of the master system:

$$\dot{x}_m = f_m(x_m) + \ell_m(x_m) \tag{27}$$

With the following assumption:

$$\dot{x}_m < \mathrm{M} < \infty$$

and the closed-loop of the state equation of the slave signal:

$$\dot{x}_s = \Delta\eta - \tau_g\left(x_s - x_m\right) \quad (28)$$

Defining the synchronization error as:

$$\aleph = x_s - x_m$$

The corresponding dynamic is as follows:

$$\begin{aligned} \dot{\aleph} &= \tau_g \aleph + h\left(x_m, x_s, u,\right) \\ where \quad h &= f_m\left(x_m\right) + \ell_m\left(x_m\right) - \Delta\eta \end{aligned} \quad (29)$$

Now, for the above considerations the following assumptions are satisfied. There exist $\tau_g \in \Re$ and $N \in \Re^+$ such that:

A1. $\left\| h\left(x_m, x_s, u\right) \right\| \leq N \quad and \quad \lim_{t \to t_0} \frac{N}{\tau_g} = 0.$

A2. $\lim_{t \to t_0} \left\| \exp\left(-\int_0^t \tau_g \, d\sigma \right) \right\| = 0$, where t_0 is large enough.

Equation (29) can be solved to obtain the error dynamics in the time domain (Equation 22).

$$\aleph = \aleph_0 \exp\left(-\int \tau_g \, dt\right) + \int_0^t \exp\left(-\tau_g\left(t - \sigma\right)\right) h(x_m, x_s, u) \, d\sigma \quad (30)$$

Taking norms of both sides of Equation (30) yields to an inequality (expression 31), which is limited after assumptions **A1** and **A2**.

$$0 \leq \limsup_{t \to t_0} \left\| \xi \right\| \leq \left\| \xi_0 \right\| \limsup_{t \to t_0} \left[\left\| \exp\left(-\int \tau_g \, dt\right) \right\| \right] + \frac{\limsup_{t \to t_0} \left[\int_0^t \left\| \exp\left(\int \tau_g \, dt\right) h(x_m, x_s, u) \, d\sigma \right\| \right]}{\limsup_{t \to t_0} \left\| \exp\left(\int \tau_g \, dt\right) \right\|}$$

(31)

$$0 \leq \limsup_{t \to t_0} \left\| \xi(t) \right\| \leq \frac{\limsup_{t \to t_0} \left[N \int_0^t \left\| \exp \int \tau_g \, d\sigma \right\| \right]}{\limsup_{t \to t_0} \left\| \exp\left(\int \tau_g \, dt \right) \right\|} \tag{32}$$

Expression (32) is an example of the $\frac{\infty}{\infty}$ case of uniform L'Hôpital's rule, which can be applied to solve the undefined quotient (Equation 32). Now, taking the limit when $t \to t_0$ it is possible to obtain a bound for the estimation error (Equation 33).

$$0 \leq \limsup_{t \to t_0} \left\| \aleph(t) \right\| \leq \limsup_{t \to t_0} \frac{N \left\| \exp\left(\int \tau_g \, dt \right) \right\|}{\left\| \exp\left(\int \tau_g \, dt \right) \right\| \left\| \tau_g \right\|} = \limsup_{t \to t_0} \frac{N}{\left\| \tau_g \right\|} \tag{33}$$

and

$$\|\aleph\| \to 0 \tag{34}$$

Results and Discussion

In this section, it is presented the results from the numerical experiments carried off to show the goodness of the proposed synchronization methodology. Both, two cases are considered as application examples; the first one is related with the synchronization of the Van der Pol (master system) – Lorenz (slave system). The mathematical model of both nonlinear oscillators is:

- Van der Pol model:

$$\begin{aligned} \dot{x}_{1,m} &= x_{2,m} \\ \dot{x}_{2,m} &= \mu\left(1 - x_1^2\right) x_2 - \omega_0^2 x_1 - \alpha x_1^3 - \lambda x_1^5 + f_0 \cos \omega t \\ s_m &= x_{2,m} \end{aligned} \tag{35}$$

where

$\mu=0.4$; $\omega_0=0.46$; $\alpha=1$; $\lambda=0.1$; $\omega=0.86$; $f_0=4.5$

With the corresponding initial conditions:

$$\begin{aligned} x_{1,m}(0) &= 0 \\ x_{2,m}(0) &= 0 \end{aligned}$$

- Lorenz model:

$$\begin{aligned}
\dot{x}_{1,s} &= \sigma\left(x_{2,s} - x_{1,s}\right) + u_1 \\
\dot{x}_{2,s} &= \rho x_{1,s} - x_{2,s} - x_{1,s}x_{3,s} \\
\dot{x}_{3,s} &= x_{1,s}x_{2,s} - \beta x_{3,s} \\
s_s &= x_{1,s}
\end{aligned} \tag{36}$$

where: $\sigma = 10$; $\rho = 28$; $\beta = \frac{8}{3}$ and the corresponding initial conditions: $\begin{matrix} x_{1,s}(0) = 1 \\ x_{2,s}(0) = 0 \\ x_{3,s}(0) = -5 \end{matrix}$

The corresponding uncertain term $\eta_1 = \sigma x_{2,s}$ such that $x_{2,m}$ is an unmeasured variable, the control gain was chosen as $\tau_g = 10$, the observer gains were considered as $\tau_1 = 100$ and $\tau_2 = 0.8$. Figure 1 is related with the synchronization procedure, it was turned on at 25 minutes, where the slave signal track the master signal with satisfactory agreement, in accordance with the theory developed.

On the other hand, the systems Duffing-Chen whose mathematical models are the following:

- Duffing model:

$$\begin{aligned}
\dot{z}_{1,m} &= z_{2,m} \\
\dot{z}_{2,m} &= \delta x_{2,m} - \pi_0^2 x_{1,m} - \gamma x_1^3 + K_1 \cos\left(\omega_1 t + \theta_1\right) + K_2 \cos\left(\omega_2 t + \theta_2\right) \\
s_m &= x_{2,m}
\end{aligned} \tag{37}$$

where:

$\delta = -0.2$; $\pi = 0.5$; $\gamma = 1$; $K_1 = 11$; $K_2 = 0$; $\theta_1 = \theta_2 = 0$; $\omega_1 = 1$; $\omega_2 = 4$

Figure 2.

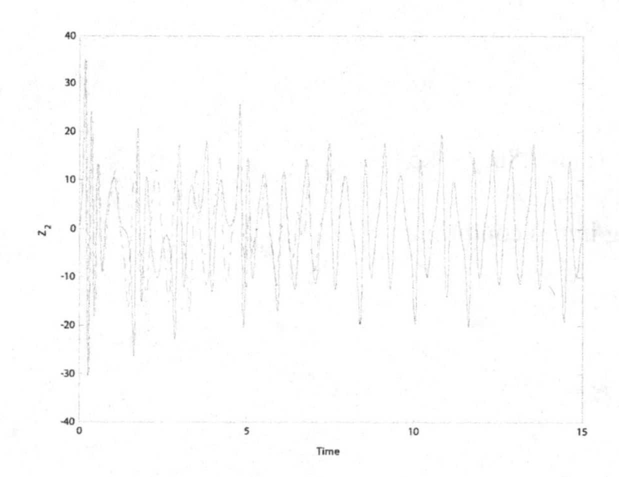

Figure 3.

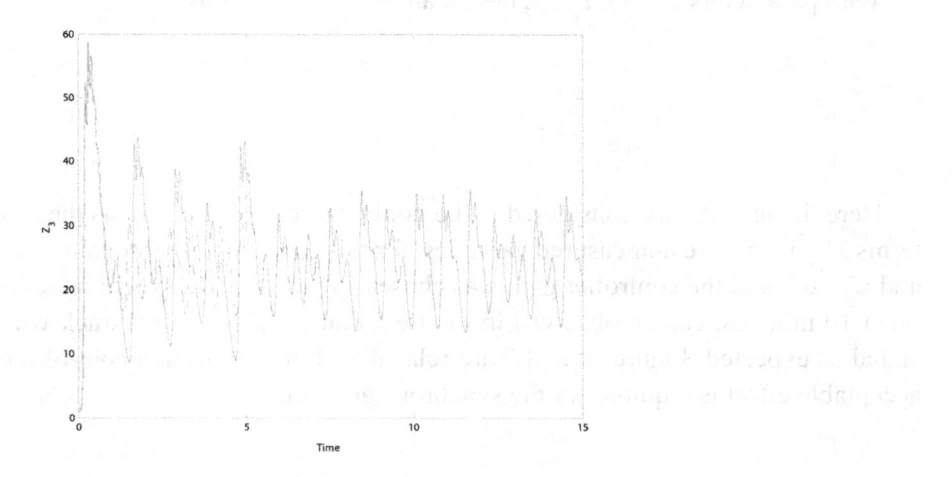

Figure 4.

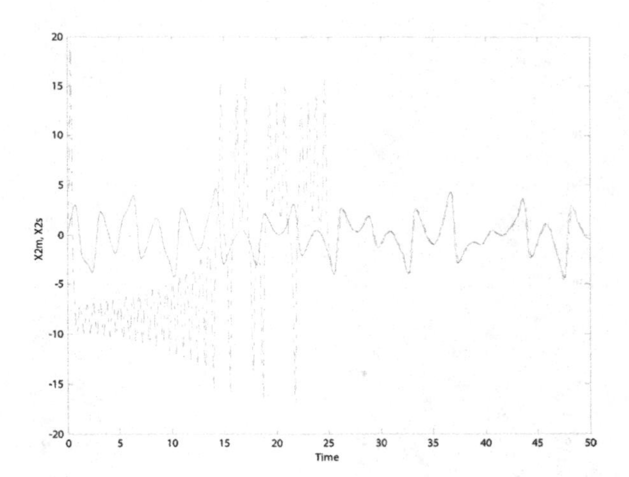

With initial conditions:

$$z_{1,m}\left(0\right)=0$$
$$z_{2,m}\left(0\right)=0$$

- Chen model:

$$\begin{aligned}
\dot{z}_{1,s} &= a\left(z_{2,s}-z_{1,s}\right)\\
\dot{z}_{2,s} &= \left(c-a\right)z_{1,s}-cz_{2,s}-z_{1,s}z_{3,s}+u_2\\
\dot{z}_{3,s} &= z_{1,s}z_{2,s}-b\,z_{3,s}\\
s_s &= z_{2,s}
\end{aligned} \tag{38}$$

With parameters $a = 35$; $b = 3$; $c = 28$ and initial conditions:

$$z_{1,s} = 1$$
$$z_{2,s} = 1$$
$$z_{3,s} = 1$$

Here the uncertainty considered is the nonlinear term $\eta_2 = z_{1,s} z_{3,s}$ as the same in the above case the terms $z_{1,s}$ and $z_{3,s}$ are unmeasured variables, for this case the observer gains were considered as $\tau_1 = 500$ and $\tau_2 = 0.9$ and the controller gain was chosen as $\tau_g = 10$. The synchronization procedure was started on at 10 minutes; can be observed in Figure 5 that the slave signal track without problem the master signal as expected. Figures 6 and 7 are related with the respective control inputs efforts; note that an acceptable effort is required for the synchronization task.

Figure 5.

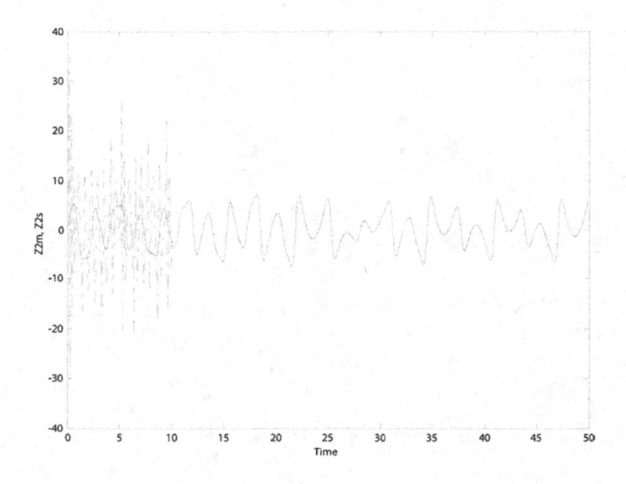

Figure 6.

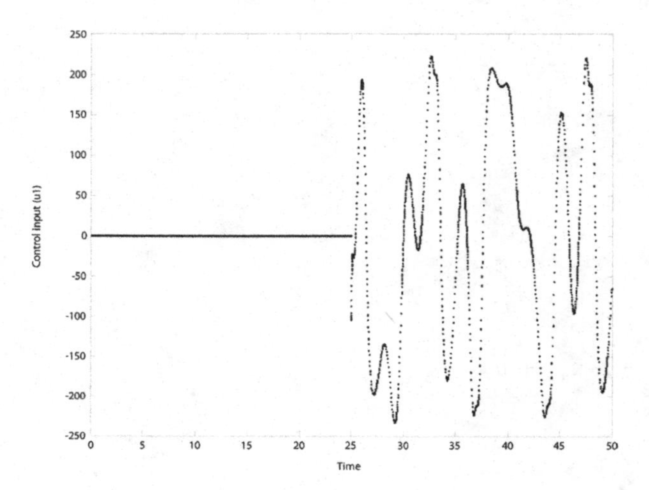

Figure 7.

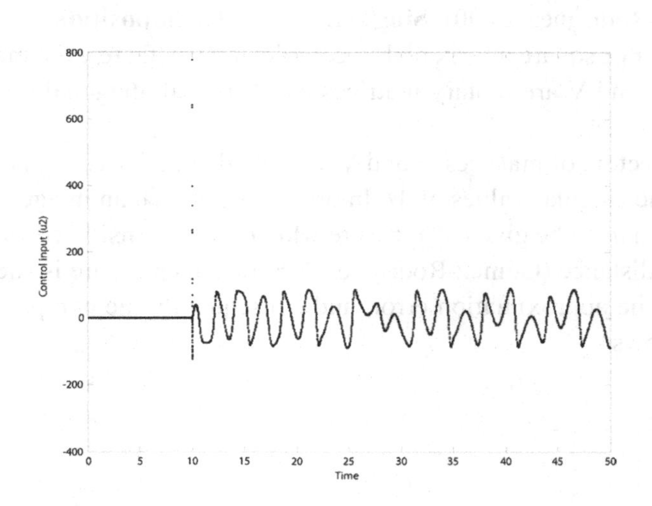

APPLICATIONS TO CHAOS COMMUNICATIONS

Our concern is specifically on the secure image transmission with desired resolution: via master-slave synchronization. Here, a procedure for images compressing and its secure transmission is presented. Our proposal is such that one can choose the desired resolution of the compressed image and to transmit it using a chaotic system as carrier. The recovering of the transmitted image is performed via robust chaos synchronization between two Chua's oscillators (chaos-based communication). The synchronization is achieved by means of a nonlinear feedback in the receiver. The results are illustrated by an example.

Let us start to say image compression is a very important tool in the visualisation issues (e.g., to store a typical gray scale image, size 480x500 and medium resolution, is required a 0.3 Mbytes, approximately). Then a compression strategy is needed for image handling (e.g., in transmission). Another problem is the secure transmission of images, i.e. to protect the privacy of the information. A secure transmission scheme can be based on the synchronization of two chaotic oscillators (Femat, 2001). The image information is injected into the transmitter system and the receiver is forced to track the carrier trajectory in order to recover the information. This problem is solved by a control feedback (Femat, 1999a) which is able to stabilize chaos synchronization. Singular Value Decomposition (SVD) is used to reduce a given image information. It is important to note that the compressed information is the information required to recover an image approximation in the receiver. For instance, a gray scale image 480x500, in terms of SVD, can be transmitted with 0.2 Mbytes under 85% of resolution. The procedure provides a simple manner to determine the error in the reconstruction of the image. As the carrier and receiver have different parameters values as the recovered image information has imperfections.

The Compression Procedure

The SVD has been used to reduce noise effect in images, see for instance (Huang, 1979; Stark, 1987), and references there in. In this paper, the SVD is used to reduce the image information in order to transmit it. Images can be represented in general as a matrix $M \in \mathbb{R}^{n \times m}$. A matrix M can be reconstructed completely by its SVD. To this end, a useful theorem is used.

Theorem 1 (Gómez-Rodriguez, 2000). **Singular Value Decomposition**

Let M be a rectangular or square matrix real or complex matrix there exist matrices U, V and Σ such that M = UΣV, where U and V are unitary matrices and Σ is real, diagonal matrix with non-negative elements.

In this theorem the vectors of matrices U and V are called singular vectors and Σ is a matrix whose diagonal elements are the singular values of M. In order to approach an image, a measure of the goodness of the approximation must be given. A measure which can be considered as a truncation parameter is the so-called relative distance (Gómez-Rodriguez, 2000). This measure is such that one can chose a desired truncation, i.e., the approximation error, and consequently the compression rate. The relative distance is given as follows

$$\rho_d = \frac{\left\| M - M_k \right\|}{\left\| M \right\|} = \frac{\sigma_{k+1}}{\sigma} \tag{39}$$

where σ_{k+1} is the singular value which represents the norm of the truncated part and σ is the norm of the whole matrix and Δ_d. satisfy $0 \leq \rho_d \leq 1$. Then, the compressed image can be approached by

$$M_k = \sum_{i=1}^{k} \sigma_i \left| u_i \right\rangle\left\langle v_i \right| \tag{40}$$

where $k < \min\{n,m\}$. Hence, k determines the truncation error. Note that, this procedure can compress an image which can be decompressed later by the eq (40).

The Secure Transmission Scheme

The secure transmission scheme is based on the synchronization of two chaotic systems. This systems are synchronized by means of a feedback control. Consider the following nonlinear systems

$$\begin{aligned} \dot{x}_m &= F(x_m) + G(x_m)S, \quad & y_m &= h(x_{1,m}) \\ x_s &= F(x_s) + G(x_s)u, \quad & y_s &= h(x_{1,s}) \end{aligned}$$

where x_M, x_S, F, $G \in \mathbb{R}^3$ are smooth vector fields, u, $S \in \mathbb{R}^1$ are the control law and the message (image information) respectively and y_M and y_S are the system outputs. Defining $x_i = x_{i,M} - x_{i,S}$ from the previous systems a dynamical error can be expressed as follows

$$\begin{aligned} \dot{x} &= \Delta F(x;\pi) + G(x_m)S - G(x_s)u \\ y &= h(x) \end{aligned} \tag{41}$$

whereΔF represents the discrepancy between the drive and response vector fields. In order to achieve synchronization, system (41) must be stabilized at origin. Notice that, if all the synchronization states x's tend to zero implies G(x_M)S = G(x_S)u; and finally the field G(x_S) must be chosen as G(x_M). This synchronization is defined as Exact Synchronization. In this application we seek on Complete Practical Synchronization which means G(x_M)S ≅ G(x_S)u, due to the systems are chosen to have different parameters. System (3) can be transformed into the following canonical form, if there exist a coordinate transformation z = Φ(x), (Femat, 2001).

$$\begin{aligned} \dot{z} &= \eta + \beta(z)u \\ \dot{\eta} &= \Gamma(z, v, \eta, u) \\ \dot{v} &= \varsigma(z, v) \\ y &= z \end{aligned} \tag{42}$$

where η is an augmented state which lumps the uncertain terms, β represents the Lie derivative of the output $h(x)$ along the vector field $G(x_S)$, $\Gamma(\cdot)$ represents the dynamic of the lumping state finally is a subsystem that represents the internal dynamics. Note that system (42) can be partially linearized via the control feedback u. In order to achieve synchronization the internal dynamics of (42) must be at least stable. Then a control feedback $u = -(\eta + Kz)/\beta(z)$, where K is a control parameter. Hence, the complete practical synchronization is achieved by means of the control feedback u and the message is recovered. In this case the information recovered is the compressed information for reconstructing the image.

Illustrative Example

In order to illustrate the use of the singular value decomposition for secure communication we consider two Chua's oscillators with different parameters. From the previous section the dynamical error system is given as follows

$$\begin{aligned} \dot{z} &= \Delta f_1 + S - u \\ \dot{v}_1 &= \Delta f_2 \\ \dot{v}_2 &= \Delta f_3 \\ y &= z \end{aligned} \tag{43}$$

where Δf_i are unknown functions and the output system is $y = x_{1,M} - x_{1,S}$. Then system (43) can be rewritten as follows

$$\begin{aligned} \dot{z} &= \eta + S - u \\ \dot{\eta} &= \Delta f_1 \\ y &= z \end{aligned}$$

Where the control low is $u = -(\eta + Kz)$ and the control parameter K = 7.0,which makes besides, the minimum phase condition is demonstrated in (Gomez-Rodriguez, 2000), then with this results one can choose an image to be compressed and transmitted we chose the well-known image (see Figure 8). This image has a high activity, and it affects the amount of information to be considered for transmission.

From image in Figure 10, one can obtain the curve representing the relative distance for every singular value see eq. (39). The error can be calculated by the area under the curve. This measure can be seen as the level of activity of the image, or the entropy. For this image the area is $\xi = 7.5246$. Notice that, in order to have a small error in the compression one has to take into account the percentage of the whole area ξ. Then the criterion is to fix the percentage of the error. In this case, for an error of 15% the error $e = 1.128$, it is needed k = 219 singular values. From eq. (40), to reconstruct the image the information required are k singular values and 2k singular vectors which represent the information to be transmitted. This information is injected into system (43) as the signal S. in order to recover the image, system (43) must be stabilized around origin. Once the systems are synchronized the information can be recovered, Figure 10 shows the approximated image to be transmitted.

Figure 8.

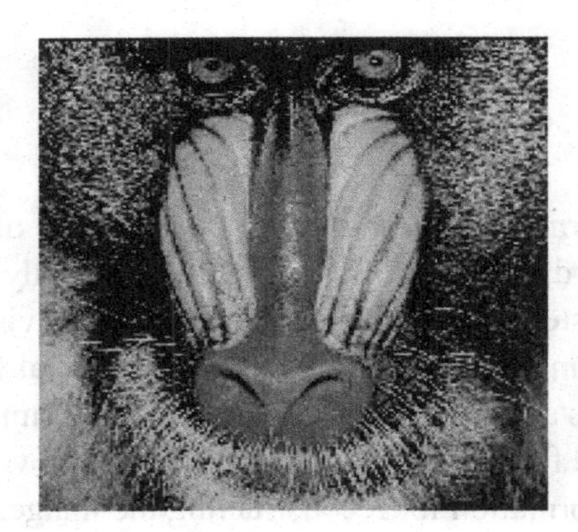

Figure 9.

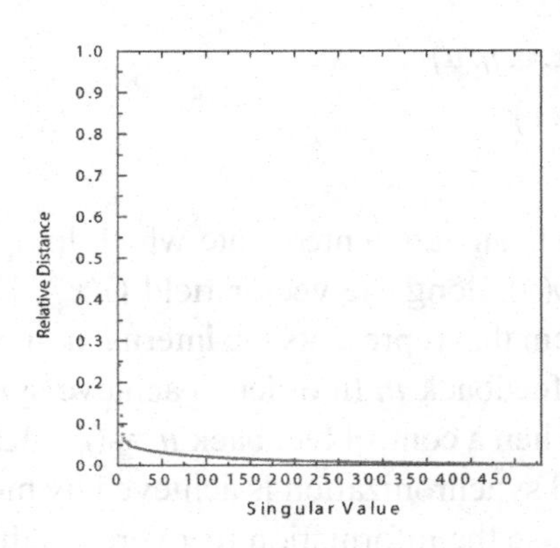

The recovered image, after synchronization is achieved is shown in Figure 11. Notice that the recovered image has error in some areas, in particular where the image have abrupt changes, this error is not in the original image, then it are highlighted by the transmission. Recall that, we are synchronizing two systems with different parameters and initial conditions, *i.e.*, the slave (receiver) trajectories are close to the master system (transmitter), practical synchronization.

Remarks

We presented a systematic procedure for secure transmission of images. The procedure is based on the approximation of an image by its singular value decomposition (SVD). From the whole information a truncation is required to obtain the compressed information. Later, the information recovered is used to reconstruct (decompress) the transmitted image. In this approximation the information required is transmitted by the synchronization of two chaotic systems. A measure of the error is given by the percentage of the truncation in the relative distance graph. The secure transmission system is formed by two Chua's systems, with same parameters and different initial conditions to diminish the error in the reception of the information. Notice that this error is different from that of the error in the approximation of the image.

Figure 10.

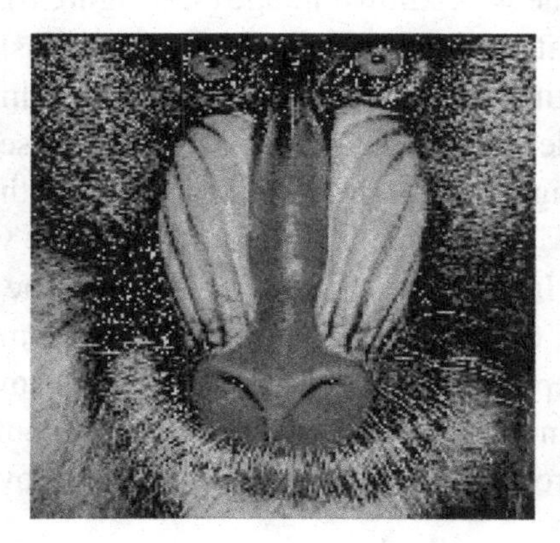

Figure 11.

Figure 12.

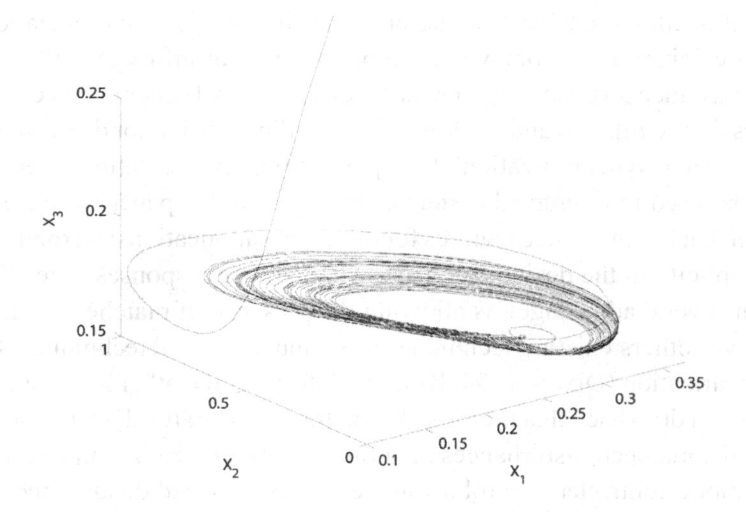

APPLICATION TO SYNCHRONIZATION OF BIOLOGICAL SYSTEMS

The ubiquity of oscillations in biological systems is well established. Oscillations are observed in all types of organisms from the simplest to the most complex. Periods can range from fractions of a second to months or years. From time to time, it has been suggested that many biological oscillations are the result of the breakdown of effective self-regulation. The opposite view is defended here. It is argued that most periodic behavior is not pathological but rather constitutes the normal operation for these systems. They are present because they confer positive functional advantages for the organism. The advantages fall into five general categories: temporal organization, spatial organization, prediction of repetitive events, efficiency and precision of control.

Almost everything that we do is controlled by Ca^{2+}, how we move, how our hearts beat and how our brains process information and store memories. To do all of this, Ca^{2+} acts as an intracellular messenger, relaying information within cells to regulate their activity. For example, Ca^{2+} triggers life at fertilization, and controls the development and differentiation of cells into specialized types. It mediates the subsequent activity of these cells and, finally, is invariably involved in cell death. To coordinate all of these functions, Ca^{2+} signals need to be flexible yet precisely regulated. This incredible versatility arises through the use of a Ca^{2+} signaling 'tool kit', whereby the ion can act in the various contexts of space, time and amplitude. It is widely believed that intercellular Ca^{2+} waves are a mechanism by which a group of cells can communicate with one another and coordinate a multicellular response to a local event. Recently, it has been observed in a variety of systems that calcium signals can also propagate from one cell to another and thereby serve as a means of intercellular communication.

The desire to understand the biophysical mechanisms of cellular dynamics has lead to introduce feedback control laws in some biological systems. The aims of feedback control laws in biological systems are to cause excitation or suppression of oscillations, entrainment and synchronization, or transitions from chaotic to periodic oscillations and vice versa (Falcke, 2003; Green, 1995) using realistic control inputs. Departing from the above ideas, this paper explores links between feedback control schemes, with an external input, and intracellular calcium functions for coordination and control. By manipulation of an external input with a feedback control scheme, we show that (a) chaotic oscillations at the external input can suppress intracellular calcium oscillations, and (b) chaotic synchronization can be achieved between intracellular calcium oscillators with a simple function of influx of Ca^{2+}.

In this work, we introduce a versatile nonlinear feedback control scheme that can be used in the synchronization, suppression/regulation and tracking of the nonlinear behavior displayed by an intracellular Ca^{2+} model. For control and synchronization of simple to complex oscillations there are a lot of control approaches that can be used in biological systems. However, in this paper we introduce a new sliding type control approach that has three nice features for biological applications: (i) robustness against model uncertainties, (ii) simplicity in the design, and (iii) switched type responses. The sliding-mode control schemes have shown several advantages as allowing the presence of matched model uncertainties and convergence speed over others existing techniques as Lyapunov-based techniques, feedback linearization and extended linearization (Dixon, 1990; Kramer, 2005). On the other hand, standard sliding-mode controllers have the main drawback that the closed-loop trajectory, of the designed solution, is not robust even with respect to the matched disturbances on a time interval preceding the sliding motion. Indeed, the classical sliding-mode controllers are robust in the case of matched disturbances only; the designed controller ensures the optimality only after the entrance point into the sliding mode (Freeman, 2000). To try to avoid the above disadvantage a relatively new kind of sliding-mode scheme has been proposed, the high-order sliding-mode technique. This control scheme considers a fractional power of the absolute value of the tracking error, coupled with the sign function; this structure provides several advantages as simplification of the control law, higher accuracy and chattering prevention (Gauthier, 1992; Tyson, 2003). In this paper it is proposed a second order sliding-mode controller coupled with an integral action for the control and synchronization of intracellular calcium dynamics.

Intracellular Calcium Model

The mechanisms underlying the spatial and temporal patterns of the global Ca^{2+} response have been investigated extensively in recent years (Dupont, 2003; Schuster, 2002). The mechanism of Ca^{2+} oscillations and that of associated waves rests on the regulation of Ca^{2+} levels within the cell.

A variety of models for Ca^{2+} oscillations and waves have been proposed (Falcke, 2003; Schuster, 2002). Differing by the degree of detail with which the dynamics and control of the $InsP_3$ receptor are treated, most of these models are based on CICR as the main instability-generating mechanism. We consider the model of Houart et al. (Houart, 1999), which exhibits a diversity of calcium responses, notably steady states, spiking and bursting oscillations, multirhythmic and chaotic regimes.

The model of Houart et al. (Houart,1999) is an extension of the minimal model proposed by Dupont and Goldbeter (Dupont, 1993) to account for the existence of simple Ca^{2+} oscillations in response to extra cellular estimulation. The original model only involves two variables, namely cytosolic and intravascular Ca^{2+} concentrations. The release of Ca^{2+} from the internal stores into the cytosol is activated by IP_3 and cytosolic Ca^{2+}, such as autocatalytic process of IP-sensitive CICR is at the core of the oscillatory mechanism. The extended model incorporates Ca^{2+} pumping into the stores, Ca^{2+} exchange with the external medium, as well as stimulus-activated Ca^{2} entry (Dupont, 1993). Besides simple periodic behavior, this model for cytosolic Ca^{2+} oscillations in no excitable cells shows complex oscillatory phenomena as bursting or chaos.

The model contains three variables, namely the concentrations of free Ca^{2+} in the cytosol (x_1) and in the internal pool (x_2), and the IP_3 concentration (x_3).

$$x_1 = V_{in} - V_2 + V_3 + kfx_2 - kx_3$$
$$x_2 = V_2 - V_3 - kfx_2 \qquad (44)$$
$$x_3 = \beta V_4 - V_5 + ex_3$$

The considered control input $V_{in} = u$

$$x_1 = u - V_2 + V_3 + kfx_2 - kx_3$$
$$x_2 = V_2 - V_3 - kfx_2 \qquad (45)$$
$$x_3 = \beta V_4 - V_5 + ex_3$$

where:

$$V_{in} = V_0 + V_1\beta \; ; \; V_2 = V_{M2}\frac{x_1^2}{K_2^2 + x_1^2};$$

$$V_3 = V_{M3}\frac{x_1^m}{K_{x1}^m + x_1^m}\frac{x_2^2}{K_{x2}^2 + x_2^2}\frac{x_3^4}{K_{x3}^4 + x_3^4}; V_5 = V_{M5}\frac{x_3^p}{K_5^p + x_3^p}\frac{x_1^n}{K_d^n + x_1^n}$$

V_{in} refers to a constant input of Ca^2 from the extra cellular medium and V_1 is the maximum rate of stimulus-induced influx of Ca^{2+} from the extra cellular medium. Parameter β reflects the degree of stimulatio n of the cell by an agonist and thus only varies between 0 and 1. The rates V_2 and V_3 refer, respectively, to pumping of cytosolic Ca^{2+} into the internal stores and to the release of Ca^{2+} from these stores into the cytosol in a process activated by cytosolic calcium (CICR), V_{M2} and V_{M3} denote the maximum values of these rates. Parameters K_2, K_{x2}, K_{x1} and K_{x3} are threshold constants for pumping, release, and activation of release by Ca^{2+} and by IP^3, k_f is a rate constant measuring the passive, linear leak of x_2 into x_1, k relates to the assumed linear transport of cytosolic Ca^{2+} into the extra cellular medium, V_4 is the maximum rate of estimulus-induced synthesis of $InsP_3$. V_5 is the rate of phosphorylation of IP_3 by the 3-kinase, it is characterized by a maximum value V_{M5} and a half-saturation constant K_5 (Houart, 1999).

Controller Design

In this section we develop a nonlinear feedback control scheme for control and synchronization of intracellular calcium oscillations. A difficulty in the models of calcium signaling is the uncertainty about the values of the rate constants, kinetic values and diffusion coefficient. For control design we choose the free calcium (Ca^{2+}) concentration as the measurable dynamical variable (Levant, 1993; Shtessel, 2003), because this is the quantity most frequently measured (for instance, Ca^{2+} can be measured using fluorescent dyes). On the other hand, in the following, we consider an external influx of Ca^{2+} as the control input. Influx of Ca^{2+} from outside the cell is known to affect the frequency of Ca^{2+} oscillations and waves (Schuster, 2002).

Control objective are (i) the regulation or tracking of Ca^{2+} dynamics to a steady state or a desired periodic or chaotic behavior, (ii) the synchronization of the output of a single Ca^{2+} oscillator to an array of Ca^{2+} oscillators by manipulation of the influx of Ca^{2+}, $u = V_{in}$ from the extra cellular medium. Then,

$$\dot{x}_1 = u - V_2 + V_3 + kfx_2 - kx_3 \quad (46)$$
$$y = h(x) = x_1$$

The control input V_{in} is a plausible manipulated variable since it is more readily amenable to experimental manipulation (Schuster, 2002). Similar control input was introduced by Puebla (Puebla, 2005) with a modeling error compensation approach.

Since there exists high uncertainties in the rate constants and kinetic values (Shtessel, 2003), the above equation can be written as:

$$\dot{x}_1 = f(x) + g(x)u \quad (47)$$
$$f(x) = -V_2 + V_3 + kfx_2 - kx_3$$

Now, consider the below nonlinear system representation:

$$\dot{x} = f(x) + g(x)u \quad (48)$$

With measured output

$$y = h(x) \quad (49)$$

In accordance with the geometric-differential theoretical frame the *Lie* derivative of the function $h(x)$ respect to the vector field f is defined as $L_f h(x)$, where $L_f^0 h(x) = h(x)$ *and* $L_f^{k+1} h(x) = L_f\left(L_f^k h(x)\right)$.

The above implies that there exist an invertible diffeomorphism, such that $(\xi,\mu)=\Psi(x)$, therefore, the system (48) can be expressed as:

$$\dot{\xi}_i = \xi_{i+1}\ ; i = 1,2,\ldots,r-1$$
$$\dot{\xi}_r = \eta(\xi,\mu) + \phi(\xi,\mu)u \quad (50)$$
$$\dot{\mu} = \gamma(\xi,\mu)$$
$$y = \xi_1$$

Where $\eta = L_f^r h(x);\ \phi = L_g L_f^{r-1} h(x)$

Now consider the following change of variable (Aguilar, 2003):

$$\xi_{r+1} = \eta + \phi u \quad (51)$$

Therefore:

$$\dot{\xi}_{r+1} = \dot{\eta} + \dot{\phi} u + \phi \dot{u} \quad (52)$$

With this the system (50) is represented as:

$$\begin{aligned}
\dot{\xi}_i &= \xi_{i+1} \;; i = 1,2,\ldots,r-1 \\
\dot{\xi}_r &= \xi_{r+1} \\
\dot{\xi}_{r+1} &= \dot{\eta}(\xi,\mu) + \dot{\phi}(\xi,\mu)u + \phi(\xi,\mu)\dot{u} \\
\dot{\mu} &= \gamma(\xi,\mu) \\
y &= \xi_1
\end{aligned} \tag{53}$$

Where:

$$\dot{\eta} = \frac{\partial \eta}{\partial \xi}\frac{\partial \xi}{\partial t} + \frac{\partial \eta}{\partial \mu}\frac{\partial \mu}{\partial t}$$

$$\dot{\phi} = \frac{\partial \phi}{\partial \xi}\frac{\partial \xi}{\partial t} + \frac{\partial \phi}{\partial \mu}\frac{\partial \mu}{\partial t}$$

Consider the following sliding surface to provide stability to the system (53):

$$s = \xi_{r+1} - \xi_{r+1(0)} + \int \sum_{i=1}^{r+1} a_i \xi_i = 0 \tag{54}$$

The above, yields:

$$\begin{aligned}
\dot{\xi}_i &= \xi_{i+1} \;; i = 1,2,\ldots,r-1 \\
\dot{\xi}_r &= \xi_{r+1} \\
\dot{\xi}_{r+1} &= -\sum_{i=1}^{r+1} a_i \xi_i \\
\dot{\mu} &= \gamma(\xi,\mu) \\
y &= \xi_1
\end{aligned} \tag{55}$$

or

$$\begin{aligned}
\dot{\xi} &= A\xi \\
\dot{\mu} &= \gamma(\xi,\mu) \\
y &= \xi_1
\end{aligned} \tag{56}$$

where $A = \begin{bmatrix} 0 & 1 & \dots & 0 \\ 0 & 0 & 1 & \dots 0 \\ \dots & \dots & \dots & \dots \\ -a_1 & -a_2 & \dots & -a_{r+1} \end{bmatrix}$; Is a Hurwitz matrix with an adequate choosing of the design parameters a_i.

Now, suppose that the vector field η is unknown and the vector field φ is uncertain such that $\phi = \phi_o + \Delta\phi$ where φ_o is a nominal function and $\Delta\varphi$ is the corresponding modeling error, with this an alternative representation of the system (53) considering unknown and uncertain terms is given as follows:

$$\begin{aligned}
&\dot{\xi}_i = \xi_{i+1} \; ; i = 1, 2, \dots, r-1 \\
&\dot{\xi}_r = \xi_{r+1} \\
&\dot{\xi}_{r+1} = \dot{\eta}(\xi, \mu) + \left(\dot{\phi}_o(\xi, \mu) + \Delta\dot{\phi} + \dot{\phi}_0 \right) u + \left(\phi_o(\xi, \mu) + \Delta\phi \right) \dot{u} \\
&\dot{\mu} = \gamma(\xi, \mu) \\
&y = \xi_1
\end{aligned} \tag{57}$$

or

$$\begin{aligned}
&\dot{\xi}_i = \xi_{i+1} \; ; i = 1, 2, \dots, r-1 \\
&\dot{\xi}_r = \xi_{r+1} \\
&\dot{\xi}_{r+1} = \zeta(\xi, \mu) + \phi_o(\xi, \mu) \dot{u} \\
&\dot{\mu} = \gamma(\xi, \mu) \\
&y = \xi_1
\end{aligned} \tag{58}$$

Where: $\zeta(\xi, \mu) = \dot{\eta} + \left(\frac{\partial \phi_o}{\partial \xi} \frac{\partial \xi}{\partial t} + \frac{\partial \phi_o}{\partial \mu} \frac{\partial \mu}{\partial t} + \Delta\dot{\phi} + \dot{\phi}_0 \right) u + \Delta\phi u$ is the whole uncertain term.

Proposing the following reaching law:

$$\dot{s} = g_1 s - g_2 sign(s) |s|^{1/w} \tag{59}$$

where $g_1 \in [0,1)$ and $w \in Z^+$

From equations (59) and (54), the following is obtained:

$$\dot{u} = \phi_o(\xi, \mu)^{-1} \left(g_1 s - g_2 sign(s) |s|^{1/w} - \zeta - \sum_{i=1}^{r+1} a_i \xi_i \right) \tag{60}$$

Substituting the above expression onto equation (58), produce the following closed-loop system:

$$\begin{aligned} \dot{\xi}_i &= \xi_{i+1} \; ; i = 1,2,\dots,r-1 \\ \dot{\xi}_r &= \xi_{r+1} \\ \dot{\xi}_{r+1} &= g_1 s - g_2 sign(s)|s|^{1/w} - \sum_{i=1}^{r+1} a_i \xi_i \\ \dot{\mu} &= \gamma(\xi,\mu) \\ y &= \xi_1 \end{aligned} \tag{61}$$

Now, analyzing the closed-loop stability under ideal conditions, i.e. under the assumption of perfect model knowledge, let us to propose the following Lyapunov function:

$$V = \frac{s^2}{2} \tag{62}$$

Considering its time derivative:

$$\dot{V} = s\left(g_1 s - g_2 sign(s)|s|^{1/w} \le |s|\left(|s| - g_2 |s|^{1/w}\right)\right) < 0 \tag{63}$$

The above, if:

$$g_2 > |s|^{1-1/w} \Rightarrow g_2 > 0 \tag{64}$$

Non Ideal Case

Let us to consider a more real situation, where a perfect model knowledge it is not available, i.e. the term ζ is unknown. Under the above, there exits methodologies to generate the lack of model information, one of these is the called uncertainty estimators, which have been published in the literature, we consider an uncertainty estimator as proposed by (Aguilar, 2003), to infer the unknown term ζ and simultaneously filter the noise of the measured output. The estimation methodology is as follows:

$$\begin{aligned} \dot{\xi}_{r+1} &= \zeta(\xi,\mu) + \phi_o(\xi,\mu)\dot{u} \\ \dot{\xi}_* &= \xi_1 + \vartheta \\ \dot{\varsigma} &= \Theta(\xi,\nu) \\ y_* &= \xi_* \end{aligned} \tag{65}$$

The main issue of this new representation of the system is to eliminate the additive noise from the new system output y_*, transforming the original output disturbance into a system disturbance. So the uncertain term ζ, is considered as a new state and $\Theta(\xi,\nu)$ is a nonlinear unknown function that describes ζ's dynamics, which is assumed bounded.

Proposition 1: The following dynamic system is an asymptotic-type observer of the system (65):

$$\dot{\hat{\xi}}_{r+1} = \hat{\zeta} + \hat{\phi}_o \dot{u} + k_1 \left(\xi_* - \hat{\xi}_* \right)$$
$$\dot{\hat{\xi}}_* = \hat{\xi}_1 + k_2 \left(\xi_* - \hat{\xi}_* \right)$$
$$\dot{\varsigma} = k_3 \left(\xi_* - \hat{\xi}_* \right)$$

The vector of proportional observer gains, $\mathbf{K}=(k_1 k_2 k_3)$, is proposed as:

$$\mathbf{K} = -\mathbf{S}_\theta^{-1}\mathbf{C}^T$$
$$\mathbf{S}_\theta = \left(\frac{1}{\theta^{i+j-1}} S_{ij} \right)_{i,j=1,n+2} \tag{66}$$

The parameter θ>0 determines the desired convergence velocity. Moreover, in order to ensure stabilizing properties, $\mathbf{S}_\theta$ should be a positive solution of the algebraic Riccati equation:

$$\mathbf{S}_\theta \left(\mathbf{E} + \frac{\theta}{2}\mathbf{I} \right) + \left(\mathbf{E}^T + \frac{\theta}{2}\mathbf{I} \right)\mathbf{S}_\theta = \mathbf{C}^T\mathbf{C}$$
$$\mathbf{E} = \begin{pmatrix} 0 & \mathbf{I}_{n-1,n-1} \\ 0 & 0 \end{pmatrix} \tag{67}$$

The convergence properties of the above estimation methodology are given in (Aguilar, 2003). Now, employing the estimation procedure the non-ideal controller is expressed as:

$$\dot{u} = \hat{\phi}_o^{-1} \left(g_1 s - g_2 sign(s)|s|^{1/w} - \hat{\zeta} - \sum_{i=1}^{r+1} a_i \hat{\xi}_i \right) \tag{68}$$

This control law produce practical closed-loop stability such that it cannot cancel completely the nonlinearities, but it forces the trajectories to remains is a close neighborhood around of the set point or the reference trajectory.

Numerical Experiments

A set numerical simulations was carry out to show the open loop behavior of the Ca model, the characteristics of the zero dynamics (for regulation case) and the performance of the proposed control methodology. The initial conditions for the Ca model are taken as $x_{1,0}$=0.2; $x_{2,0}$=0.14; $x_{3,0}$=0.43; the controller's gains were chosen as g_1=0.95; g_2=25. From the process's start-up to 10 minutes, the systems behave at open-loop conditions and the proposed controller acts at 10 minutes. Now, in Figure 13 is shown the close-loop performance of the controlled variable (control output) for the regulation case, it is observed a fast response of the system when the controller acts and force the trajectory to the desired set point (sp = 0.35), Figure 14 is related with the controller effort, a suddenly change of the control input is observed in the first two minutes of the closed-loop regimen, after that the control effort is much more smooth, with this a satisfactory performance can be considered. The uncertainty estimation performance, for the regulation case, can be seen in Figure 15, can be concluded that the estimation procedure is satisfactory; note the performance during the open-loop and closed-loop periods. The behavior of the zero dynamics is observed in Figure 16, where, in accordance with the frame of section 3, the time derivatives of the no controlled variables tend to numerical values around to zero, with this a stable zero dynamic is determinate. The Figure 17 considers the synchronization case, where the proposed controller forces the slave signal to track the master signal, with satisfactory performance, the corresponding control input shows an satisfactory effort as is seen in Figure 18. Finally a tracking case is considered too. The reference signal is chosen as yref = 0.1sin(2 t) + 1, note that the large overshoot showed for the controlled variable when the controller acts, however after 2 minutes de control effort is much more smooth and the tracking signal performance can be also considered as adequate (see Figures 18 and 19).

Figure 13.

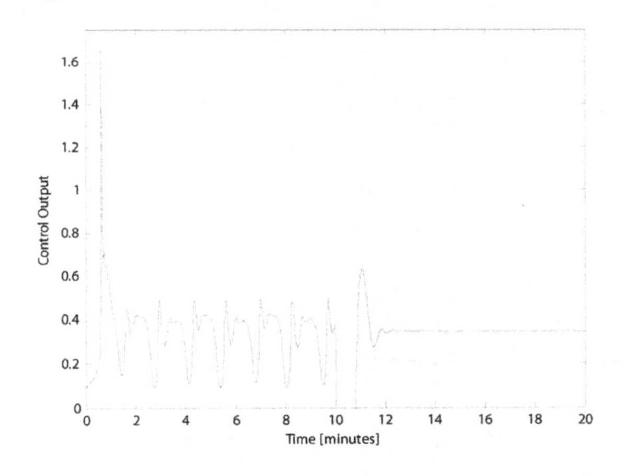

Figure 14.

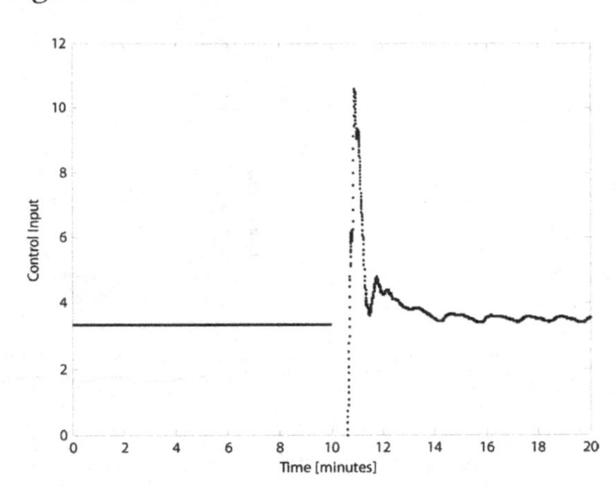

Figure 15.

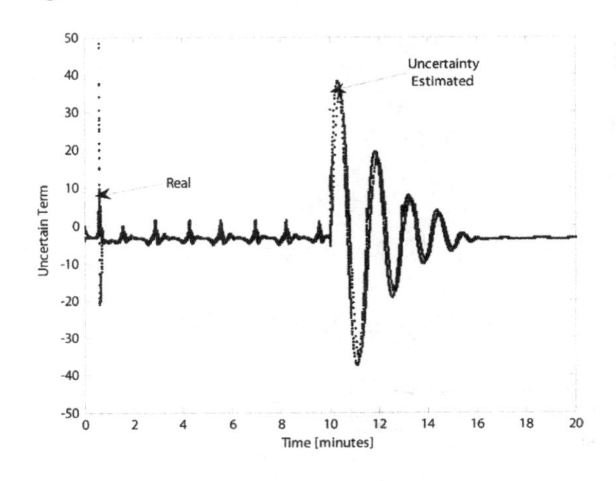

Figure 16.

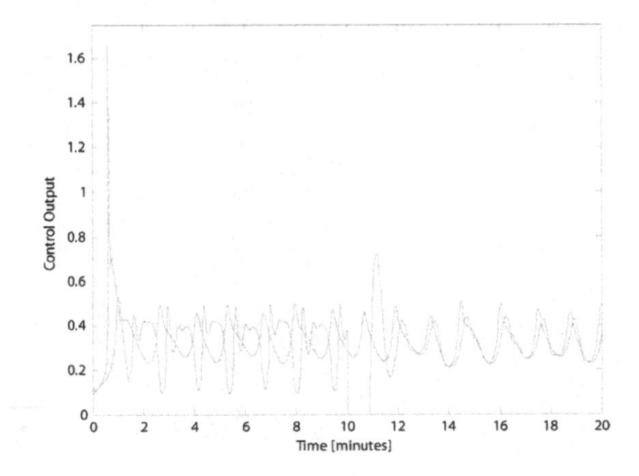

Figure 17.

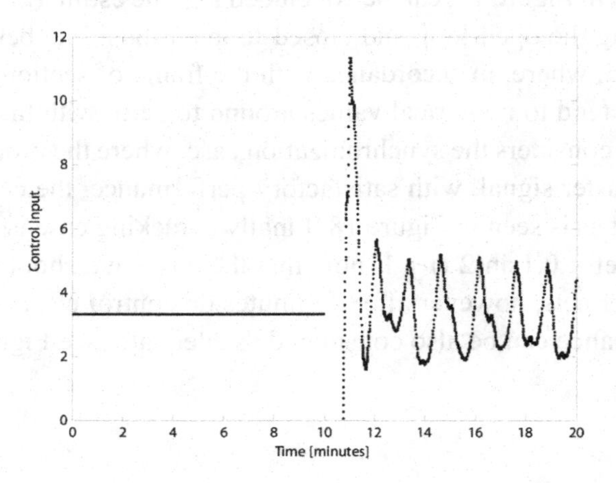

Figure 18.

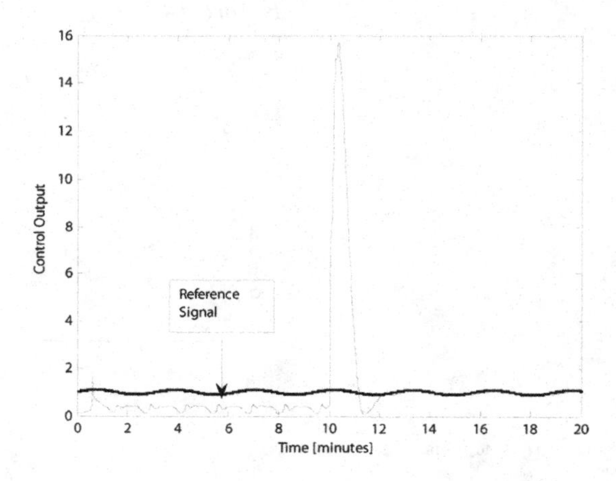

Figure 19.

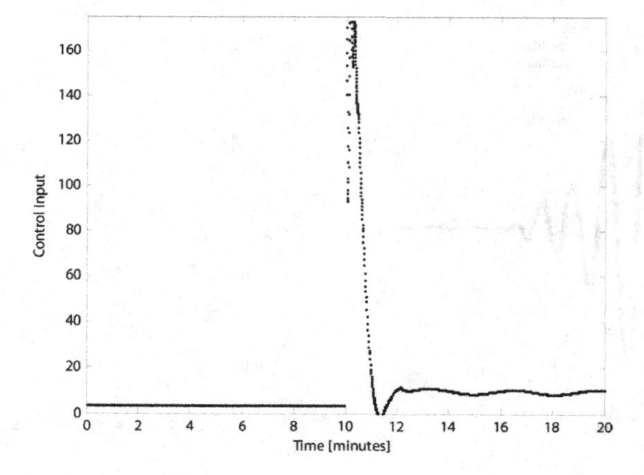

REFERENCES

Aguilar, R., Martínez, R., & Poznyak, A. (2001). PI observers for uncertainty estimation in continuous chemical reactors. In *IEEE Conference on Control Applications/ISIC* (pp. 1037-1041).

Aguilar, R., Martínez-Guerra, R., & Maya-Yescas, R. (2003). State Estimation for Partially Unknown Nonlinear Systems: A Class of Integral High Gain Observers. *IEE Proceedings. Control Theory and Applications*, *150*(3), 240–244. doi:10.1049/ip-cta:20030400

Aguilar, R., Poznyak, A., Martínez-Guerra, R., & Maya-Yescas, R. (2002). Temperature control in catalytic cracking reactors via robust PID controller. *Journal of Process Control*, *12*(6), 695–705. doi:10.1016/S0959-1524(01)00034-8

Aguilar-Ibañez, C., Suarez, C. M. S., Fortunato Flores, A., Martínez-Guerra, R., & Garrido, M. R. (2006). Reconstructing and Identifying the Rossler system by using a high gain observer. *Asian Journal of Control*, *8*(4), 401–407.

Aguilar-López, R., & Alvarez-Ramírez, J. (2002). Sliding-mode control scheme for a class of continuous chemical reactors. *IEE Proceedings. Control Theory and Applications*, *149*(4), 263–268. doi:10.1049/ip-cta:20020558

Aguilar-López, R., & Martínez-Guerra, R. (2008). Synchronization of a coupled Hodgkin-Huxley neurons via high order sliding-mode feedback. *Chaos, Solitons, and Fractals*, *37*, 539–546. doi:10.1016/j.chaos.2006.09.029

Aguilar-López, R., Martínez-Guerra, R., Puebla H., & Hernandez Suarez, R. (2008). High order sliding-mode dynamic control for chaotic intracellular calcium oscillations. *Journal of Nonlinear Analysis-B: Real World Applications*.

Alvarez, J., & Hernandez, H. (2003). Robust estimation of contiunuous nonlinear plants with discrete measurements. *Journal of Process Control*, *13*(1), 69–89. doi:10.1016/S0959-1524(02)00010-0

Besançon, G. (1999). A viewpoint on observability and observer design for nonlinear systems. *Lecture Notes in Control and Information Sciences*, *244*, 3–22. doi:10.1007/BFb0109918

Bowong, S. (2004). Stability analysis for the synchronization of chaotic systems with different order: application to secure communications. *Physics Letters. [Part A]*, *326*, 102–113. doi:10.1016/j.physleta.2004.04.004

Celikovsky, S., & Chen, G. (2005). Secure synchronization of a class of chaotic systems from a nonlinear observer approach. *IEEE Transactions on Automatic Control*, *50*(1), 76–82. doi:10.1109/TAC.2004.841135

Dixon, C. J., Woods, N. M., Cuthbertson, K. S. R., & Cobbold, P. H. (1990). Evidence for two Ca^{2+} -mobilizing purinoreceptors on rat hepatocytes. *The Biochemical Journal*, *269*, 499–502.

Dupont, G., & Goldbeter, A. (1993). A one-pool model for Ca^{+2} oscillations involving Ca^{+2} and inositol 1,4,5-trisphosphate as co-agonists for Ca^{+2} release. *Cell Calcium*, *14*, 311. doi:10.1016/0143-4160(93)90052-8

DuPont, G., Houart, G., & De Koninck, P. (2003). Sensitivity of CaM kinase II to the frequency of Ca^{2+} oscillations: a simple model. *Cell Calcium*, *34*, 485–497. doi:10.1016/S0143-4160(03)00152-0

Falcke, M. (2003). Deterministic and stochastic models of intracellular Ca^{2+} waves. *New Journal of Physics*, *5*, 1–28. doi:10.1088/1367-2630/5/1/396

Femat, R., Alvarez-Ramirez, J., & Castillo-Toledo, B. (1999a). On robust chaos suppression in a class of nondriven oscillators: Application to Chua's circuit. *IEEE Transactions on Circuits and Systems*, *46*(1), 1150.

Femat, R., Jauregui-Ortiz, R., & Solís-Perales, G. (2001). A Chaos-Based Communication Scheme via Robust Asymptotic Feedback. *IEEE Transactions on Circuits and Systems*, *48*(10).

Femat, R., & Solís-Perales, G. (1999b). On the chaos synchronization phenomena. *Physics Letters. [Part A]*, *262*, 50–60. doi:10.1016/S0375-9601(99)00667-2

Freeman, M. (2000). Feedback control of intracellular signaling in developed. *Nature*, *408*, 313–319. doi:10.1038/35042500

Gauthier, J. P., Hammouri, H., & Othman, S. (1992). A simple observer for nonlinear systems with application to bioreactors. *IEEE Transactions on Circuits and Systems*, *37*, 875–880.

Golub, G. H., & Van Loan, C. F. (1989). *Matrix Computating*. Baltimore, MD: The Johns Hopkins University Press.

Gómez-Rodriguez, A. (2000). Entanglement states and the singular value decomposition. *Revista Mexicana de Fisica*, *46*(5).

González, J., Femat, R., Alvarez-Ramírez, J., Aguilar, R., & Barrón, M. A. (1999). A discrete approach to the control and synchronization of a class of chaotic oscillators. *IEEE Transactions on Circuits and Systems . Part I*, *46*(9), 1139–1143.

Green, A. K., Cobbold, P. H., & Dixon, J. C. (1995). Cytosolic free Ca^{2+} oscillations induced by diadenosine 5′,5′′′-P1,P3-triphosphate and diadenosine 5′,5′′′-P1,P4-tetraphosphate in single rat hepatocytes are indistinguishable from those induced by ADP and ATP respectively. *The Biochemical Journal*, *310*, 629–635.

Houart, G., DuPont, G., & Goldbeter, A. (1999). Bursting, chaos and birhythmicity originating from self-modulation of the inositol 1,4,5-trisphosphate signal in a model for intracellular Ca^{+2} oscillations. *Bulletin of Mathematical Biology*, *61*, 507. doi:10.1006/bulm.1999.0095

Huaguang, Z., Huanxin, G., & Zhanshan, W. (2007). Adaptive synchronization of neural networks with different attractors. *Progress in Natural Science*, *17*(6), 687–695. doi:10.1080/10002007088537459

Huang, T. S. (1979). *Topics in Applied Physics: Picture Processing and Digital Filtering*. New York: Springer-Verlag.

Isidori, A. (1995). *Nonlinear Control Theory*. New York: Springer-Verlag.

Kramer, U., Krajnc, B., Pahle, J., Green, A. K., Dixon, C. J., & Marhl, M. (2005). Transition from stochastic to deterministic behavior in calcium oscillations. *Biophysical Journal, 89*, 1603–1611. doi:10.1529/biophysj.104.057216

Levant, A. (1993). Sliding order and sliding accuracy in sliding mode control. *International Journal of Control, 58*, 1247–1263. doi:10.1080/00207179308923053

Ling, L., Xiaogang, W., & Hanping, H. (2004). Estimating system parameters of Chua's circuit from synchronizing signal. *Physics Letters. [Part A], 324*, 36–41. doi:10.1016/j.physleta.2004.02.047

Martínez-Guerra, R., Aguilar, R., & Poznyak, A. (2004). A new robust sliding-mode observer design for monitoring in chemical reactors. *Journal of Dynamic System, Measurement and Control . ASME Journal, 126*(3), 473–478.

Martínez-Guerra, R., Cruz-Victoria, J. C., Gonzalez-Galan, R., & Aguilar-López, R. (2006). A new reduced-order Observer design for the Synchronization of Lorenz Systems. *Chaos, Solitons, and Fractals, 28*(12), 511–517. doi:10.1016/j.chaos.2005.07.011

Martínez-Guerra, R., & Wen, Y. (2008). Chaotic synchronization and secure communication via sliding-mode observer. *International Journal of Bifurcation and Chaos in Applied Sciences and Engineering, 18*(1), 235–243. doi:10.1142/S0218127408020264

Mendoza-Camargo, J., Aguilar-Ibañez, C., Martínez-Guerra, R., & Garrido, R. (2004). On the parameters identification of the Duffing's System by means of a reduced order observer. *Physics Letters. [Part A], 331*(5), 316–324. doi:10.1016/j.physleta.2004.09.005

Morgül, O., & Solak, E. (1996). Observed based synchronization of chaotic systems. *Physical Review E: Statistical Physics, Plasmas, Fluids, and Related Interdisciplinary Topics, 54*(5), 4803–4811. doi:10.1103/PhysRevE.54.4803

Moukam, K. F. M., Bowong, S., Tchawoua, C., & Kaptouom, E. (2004). Chaos control and synchronization of a Φ^6-Van der Pol oscillator. *Physics Letters. [Part A], 322*, 305–323. doi:10.1016/j.physleta.2004.01.016

Muraly, K. (2000). Synchronization based signal transmission with heterogeneous chaotic systems. *International Journal of Bifurcation and Chaos in Applied Sciences and Engineering, 10*(11), 2489–2497. doi:10.1142/S0218127400001729

Nijmeijer, H. (1997a). Control of chaos and synchronization. *Systems & Control Letters, 31*(5), 299–305. doi:10.1016/S0167-6911(97)00042-X

Nijmeijer, H., & Mareels, I. M. Y. (1997b). An observer looks at synchronization. *IEEE Transactions on Circuits and Systems. I, Fundamental Theory and Applications, 44*(10), 882–890. doi:10.1109/81.633877

Nijmeijer, H., & Van der Shaft, A. (1990). *Nonlinear dynamical control systems*. New York: Springer-Verlag.

Ott, E., Grebogi, C., & York, J. A. (1990). Controlling chaos. *Physical Review Letters, 64*(11), 1196–1199. doi:10.1103/PhysRevLett.64.1196

Pecora, L. M., & Caroll, T. L. (1990). Synchronization in Chaotic System. *Physical Review Letters*, *64*(8), 821–824. doi:10.1103/PhysRevLett.64.821

Pecora, L. M., & Caroll, T. L. (1998). Master stability functions for synchronized coupled systems. *Physical Review Letters*, 2109–2112. doi:10.1103/PhysRevLett.80.2109

Pogromsky, A., & Nijmeijer, H. (1998). Observer-based robust synchronization of dynamical systems. *International Journal of Bifurcation and Chaos in Applied Sciences and Engineering*, *8*(11), 2243–2254. doi:10.1142/S0218127498001832

Puebla, H. (2005). Controlling intracellular calcium oscillations. *Journal of Biological System*, *13*, 173. doi:10.1142/S021833900500146X

Schuster, S., Marhl, M., & Hofer, T. (2002). Modeling of simple and complex calcium oscillations: from single-cell responses to intracellular signaling. *European Journal of Biochemistry*, *269*, 1333. doi:10.1046/j.0014-2956.2001.02720.x

Shihua, C., Ja, H., Changping, W., & Jinhu, L. (2004). Adaptive synchronization of uncertain Rössler hyper chaotic system based on parameter identification. *Physics Letters. [Part A]*, *321*, 50–55. doi:10.1016/j.physleta.2003.12.011

Shtessel, Y. B., Shkolnikov, I. A., & Brown, D. J. (2003). An asymptotic second order smooth sliding mode control. *Asian Journal of Control*, *5*(4), 498–504.

Solak, E. (2004). A reduced-order observer for the synchronization of Lorenz Systems. *Physics Letters. [Part A]*, *325*, 276–278. doi:10.1016/j.physleta.2004.04.001

Stark, H. (1987). *Image Recovery: Theory and application*. New York: Academic Press.

Tyson, J. J., Chen, K. C., & Novak, B. (2003). Sniffers, buzzers, toggles and blinkers: dynamics of regulatory and signaling pathways in the cell. *Current Opinion in Cell Biology*, *15*, 221. doi:10.1016/S0955-0674(03)00017-6

Chapter 11
Synchronization of Oscillators

Jean B. Chabi Orou
Université d'Abomey-Calavi, Benin

ABSTRACT

A simple approach is proposed in this chapter to get started on the synchronization of oscillators study. The basics are given in the beginning such that the reader can get quickly familiar with the main concepts which lead to many kinds of synchronization configurations. Chaotic synchronization is next addressed and is followed by the stability of the synchronization issue. Finally, a short introduction of the influence of noise on the synchronization process is mentioned.

INTRODUCTION

Nonlinearity in physics has become a wide topic under investigation in many areas of scientific research such as biology, engineering, social sciences and others. It is a challenge for many researchers to address many issues related to nonlinearity which induces somehow regular and chaotic behaviors in systems governed by non linear differential or partial differential equations. In what follows an attempt is made to give some basics in synchronization of oscillators. Most of the issues have been addressed in a simple and standard manner. Both regular and chaotic dynamical systems have been taken into account such that any beginner in this topic will find the way to derive fundamental quantities and parameters that characterize the system they are dealing with.

DOI: 10.4018/978-1-61520-737-4.ch011

FUNDAMENTALS OF THE SYNCHROIZATION

Kuramoto's Model

One of the most important basic phenomena in sciences is synchronization which was discovered by C. Huyghens (Pecora & Carroll, 1990; Pikovsky, Rosenblum, & Kurths, 2001; Rulkov et al., 1995; Kuramoto, 1984). Synchronization is a phenomenon which means there is an adjustment of the frequencies of periodic self-sustained oscillators through a weak interaction. The weakness of the interaction is to some extent a requirement. The frequencies adjustment is also referred to as phase locking or frequency entrainment.

Many types of synchronizations can be distinguished depending on the dynamic property of the system under investigation. One might have identical synchronization, generalized synchronization and phase synchronization for chaotic and non-chaotic systems (i.e. regular and chaotic systems) Synchronization might also happen in a system of two or many identical or not chaotic systems. It also may be seen as the appearance of a relation relating the phases of interacting systems or between the phase of a system and that of an external force. The synchronization phenomenon is not to be confused with resonance phenomenon and synchronous variation of two variables which does not lead to synchronization. In many areas of science collective synchronization phenomena have been observed. These areas overlaps biology, chemistry, physical and social systems as examples. This phenomenon is mainly observed when two or many oscillators lock on to a common frequency despite differences in the frequency of the individual oscillators. A popular model known as Kuramoto model has been developed and can be found in many textbooks and articles in the literature. Many methods such as the stability analysis can be used to address the synchronization phenomenon. The Kuramoto model consists of N oscillators whose dynamics are governed by the following equations:

$$\dot{\theta}_i = \omega_i + \frac{K}{N}\sum_{j=1}^{N} \sin(\theta_j - \theta_i) \tag{1}$$

where Θ_i is the phase of the *i th* oscillator, ω_i is its natural frequency, K is positive and stands for the coupling gain.

The oscillators are said to synchronize if

$$\dot{\theta}_j - \dot{\theta}_i \rightarrow 0 \text{ as } t \rightarrow \infty \ \forall i, j = 1, \ldots. N \tag{2}$$

The oscillators can also be said to synchronize when $\theta_j - \theta_i$ becomes constant asymptotically as the time goes on. If we represent the oscillators on a circle by points that move with the same angular frequency, then the phase difference which means the angular distance between these points remain constant over the time. Thus, one can define the order parameter r that measures the phase coherence of the oscillators population and takes values between 0 and 1 inclusively as follow

$$re^{i\psi} = \frac{1}{N}\sum_{j=1}^{N} e^{i\theta_j} \tag{3}$$

The order parameter will play a key role in describing how synchronous the system is. If the oscillators synchronize, then the parameter converges to a constant magnitude, $r \leq 1$ after a long time behavior, but if the oscillators add incoherently then the order parameter r remains close to zero. At this point the problem is to characterize the coupling gain K so that the oscillators synchronize.

Kuramoto showed that there exists a critical value K_{rc} for K such that for all $K < K_{rc}$, the oscillators remain unsynchronized, but for $K > K_{rc}$ the incoherent state becomes unstable, the oscillators start synchronizing and eventually $r(t)$ will settle at some $r_\infty(K) < 1$.

There still exist many open problems to address on the influence of the magnitude of the order parameter combined to that of the gain on the behavior of the system.

SYNCHRONIZATION OF PERIODIC OSCILLATIONS

We consider first an autonomous self-sustained oscillator to which we apply a periodic force and next we consider only the dynamics of the phase. The equation of the perturbed phase dynamics in the simple case of a limit cycle oscillator is in the following form

$$\frac{d\varphi}{dt} = \omega_0 + \varepsilon g(\varphi, \omega t) \tag{4}$$

g stands for the coupling function in which the 2π-periodic in the argument depends on the form of the limit cycle and of the forcing g contains fast oscillating and slow varying terms in the vicinity of the resonance ie when $\omega \approx \omega_o$. In that case g as the following form: $g = q(\phi - \omega t)$ the averaging of which gives the phase dynamics equation as follow:

$$\begin{gathered} \frac{d\Delta\varphi}{dt} = -\left(\omega - \omega_0\right) + \varepsilon q(\Delta\varphi) \\ \Delta\varphi = \varphi - \omega t \end{gathered} \tag{5}$$

q is 2π-periodic and the simplest function satisfying this statement is a sine function ie $q\left(.\right) = \sin\left(.\right)$, then we obtain the so-called Adler equation as follow:

$$\frac{d\Delta\varphi}{dt} = -\left(\omega - \omega_0\right) + \varepsilon \sin(\Delta\varphi) \tag{6}$$

The Adler equation can be plotted in $\left(\varepsilon, \left(\omega_0 - \omega\right)\right)$ plane. This plot shows many regions in which $\varepsilon q_{\min} < \omega - \omega_0 < \varepsilon q_{\max}$ and where the Adler equation has a stationary solution which corresponds to phase locking. This means that $\varphi = \omega t + cst$ and the frequency of the oscillator under consideration ie $\Omega = \langle \dot{\varphi} \rangle$ coincides with the forcing frequency ω. That region (there might be several regions) is called synchronization region or Arnold tongue (Figures 1 and 2). There are many synchronization regions when the locking phenomenon is of high order which means when the frequencies of the oscillation fulfill this relation $n\omega \approx m\,\omega_0$ where n and m are integers

Then $\Delta\varphi = m\varphi - n\omega t$ and the phase difference is described by the equation similar to that of Adler

$$\frac{d\Delta\varphi}{dt} = -\left(n\omega - m\omega_0\right) + \varepsilon q(\Delta\varphi) \quad (7)$$

For a synchronous regime of oscillation, $\Omega = \frac{n}{m}\omega$ and the phase locking occurs when

$$m\varphi = n\omega t + cst \quad (8)$$

When using a broad range of frequencies of the forcing, one can see a large family of synchronization regions with a triangular shape touching the frequency axis at the rationals of the natural frequencies $\frac{m}{n}\omega_0$.

Figure 1. Arnold tougue

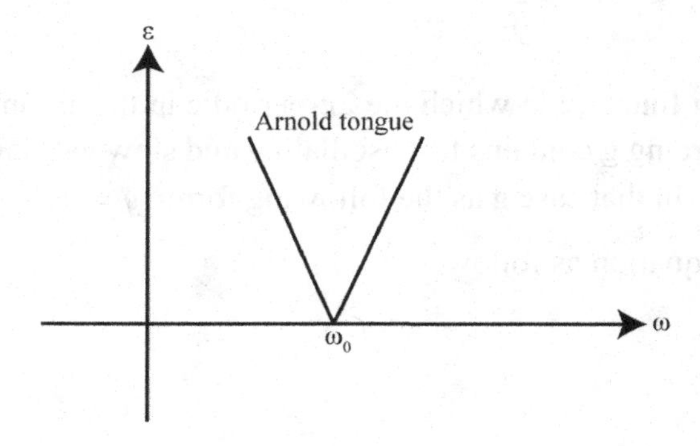

Figure 2. Synchronization region

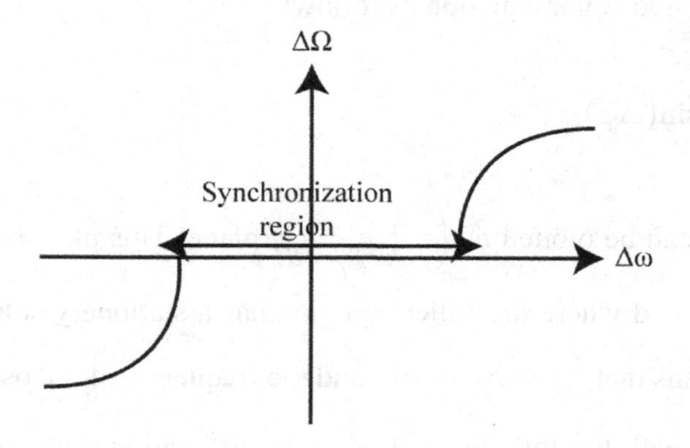

The picture if drawn will show many tongues whose shapes generally are not triangular when the forcing force is not weak but is moderate. In this case, the phase of oscillations does not follow precisely the phase of the forcing force but we must have the following $\left|m\varphi - n\omega t\right| < cst$ and we have the entrainment of the frequencies $\Omega = \dfrac{n}{m}\omega$.

When the amplitude ε of the forcing is maintained fixed and ω varies, different phase locking intervals are observed when the motion is periodic. In between these intervals the motion is quasi-periodic. The Ω - ω curve is called the Devil's staircase. For a strong forcing, the motion is no more close to that in the phase approximation and it may lead to chaos.

SYNCHRONIZATION WITH TWO COUPLED OSCILLATORS

Two coupled oscillators synchronization can be described in the same way as earlier mentioned. We consider the following two equations:

$$\begin{aligned} \frac{d\varphi_1}{dt} &= \omega_1 + \varepsilon g_1(\varphi_1, \varphi_2) \\ \frac{d\varphi_2}{dt} &= \omega_2 + \varepsilon g_2(\varphi_2, \varphi_1) \end{aligned} \tag{9}$$

Once again $\Delta\varphi$ stands for the phase difference such that $\Delta\varphi = \varphi_2 - \varphi_1$ from which the Adler's equation will be derived .We are dealing with two non-identical oscillators. The synchronization here implies that both oscillate with the same frequency or with rationally related frequencies in general such that the common frequency has value between ω_1 and ω_2.

COUPLING CONFIGURATIONS

Contracting System

Let us focus on the specific oscillators with positive-definite diffusion coupling to which we are going to apply the nonlinear contraction theory in order to point out many kind of synchronization.

In short, a dynamical system will be said to be contracting if all solutions of the dynamical equation describing it converge exponentially to a single trajectory regardless of the initial conditions.

One Way Coupling Configuration

We first begin with two identical oscillators coupled in a one-way fashion. This is also called a unidirectional coupled oscillators. This system is governed by this set of equations as follow

$$\dot{x}_1 = h(x_1, t)$$
$$\dot{x}_2 = h(x_2, t) + u(x_1) - u(x_2) \tag{10}$$

where x_1 and x_2 are the state vectors and have values such that x_1 and $x_2 \in R^m$, $h(x_i, t)$ stands for the dynamics of the uncoupled oscillators and $u(x_1) - u(x_2)$ is the coupling force.

Theorem I

If the function $h - u$ is contracting in (10), two systems x_1 and x_2 will reach synchrony exponentially regardless of the initial conditions.

The proof of theorem I can be found in Wang and Slotine (2003).

Here is an example of two identical oscillators. This example consists of two coupled identical Van der Pol oscillators that will be considered with these two equations coupled in a one-way coupling fashion.

$$\ddot{x}_1 + \lambda(x_1^2 - 1)\dot{x}_1 + \omega^2 x_1 = 0$$
$$\ddot{x}_2 + \lambda(x_2^2 - 1)\dot{x}_2 + \omega^2 x_2 = \lambda\kappa\left(\dot{x}_1 - \dot{x}_2\right) \tag{11}$$

where λ, ω and κ are strictly positive constants. Since the system $\ddot{x} + \lambda(x^2 + \kappa - 1)\dot{x} + \omega^2 x = u(t)$ is semi-contracting for $\kappa > 1$, $\mathbf{x}_2$ will synchronize to $\mathbf{x}_1$ asymptotically.

Two Way Coupling Configuration

There are many meanings of synchronization in the literature depending on the context. Here for example we define synchronization of two (or more) oscillators $\mathbf{x}_1$, $\mathbf{x}_2$ as corresponding to a complete match, i.e., $\mathbf{x}_1 = \mathbf{x}_2$. Similarly, we define anti-synchronization as $\mathbf{x}_1 = -\mathbf{x}_2$. These two cases are called respectively in-phase synchronization and anti-phase synchronization in many papers and books.

Synchronization and Anti-Synchronization

Theorem II

Consider two coupled systems. If the dynamics equations verify $\dot{x}_1 - g(x_1, t) = \dot{x}_2 - g(x_2, t)$ where the function g is contracting, then $\mathbf{x}_1$ and $\mathbf{x}_2$ will converge to each other exponentially, regardless of the initial conditions.

Theorem II also can be extended to coupled discrete-time systems, with an appropriate discrete versions of contraction analysis.

The following example based on once again two coupled identical Van der Pol oscillators with these equations describing the coupling

$$\ddot{x}_1 + \lambda(x_1^2 - 1)\dot{x}_1 + \omega^2 x_1 = \lambda\kappa_1\left(\dot{x}_2 - \dot{x}_1\right)$$
$$\ddot{x}_2 + \lambda(x_2^2 - 1)\dot{x}_2 + \omega^2 x_2 = \lambda\kappa_2\left(\dot{x}_1 - \dot{x}_2\right) \tag{12}$$

These two oscillators will reach synchrony asymptotically if $\kappa_1 + \kappa_2 > 1$ for non-zero initial conditions.

A nice example of the anti-synchronization can be found in a model describing two identical biological cells, inert by themselves, but can be excited into oscillations through diffusion interaction across their membranes.

From theorem II a coupled system to describe this model could be

$$\begin{aligned} \dot{x}_1 &= g(x_1, t) + u(x_2, t) - u(x_1, t) \\ \dot{x}_2 &= g(x_2, t) + u(x_1, t) - u(x_2, t) \end{aligned} \tag{13}$$

Next, another theorem can help solve and get the anti-synchronization case.

Theorem III

If the uncoupled dynamics g in (13) is contracting and odd in $\mathbf{x}$, $\mathbf{x}_1 + \mathbf{x}_2$ will converge to zero exponentially regardless of the initial conditions. Moreover, for non-zero initial conditions, $\mathbf{x}_1$ and $\mathbf{x}_2$ will oscillate and reach anti-synchrony if the system $\dot{z} = g(z, t) - 2u(z, t)$ has a stable limit-cycle.

Amplitude Death

The amplitude-death phenomenon also called oscillator-death may occur when the oscillators stop oscillating and stabilize at constant steady states once they are coupled. This means the overall dynamics is contracting and implies that the system tends exponentially to a unique equilibrium.

Identical Oscillators with General Couplings

Let us now assume we have a more general case of coupling with two identical oscillators. The oscillators will be two identical Van der Pol oscillators coupled in a very general way. They will be described by

$$\begin{aligned} \ddot{x}_1 + \lambda(x_1^2 - 1)\dot{x}_1 + \omega^2 x_1 &= \lambda\left(\gamma \dot{x}_2 - \kappa \dot{x}_1\right) \\ \ddot{x}_2 + \lambda(x_2^2 - 1)\dot{x}_2 + \omega^2 x_2 &= \lambda\left(\gamma \dot{x}_1 - \kappa \dot{x}_2\right) \end{aligned} \tag{14}$$

where λ is a positive constant. It can be proved as above that, as long as the condition $|\gamma| > 1 - \kappa$ is satisfied, x_1 converges to x_2 asymptotically for all $\gamma \geq 0$ while x_1 converges to $-x_2$ asymptotically for all $\gamma \leq 0$. Note that if $\gamma = 0$ we get two independent stable subsystems. Both x_1 and x_2 tend to the origin, which can be considered as a continuous connection between $\gamma > 0$ and $\gamma < 0$.

More speculations on the behaviors of this system can be made and be found in Wang and Slotine (2003). The evolution of a dissipative chaotic system in its state-space is characterized both by contraction in some directions and by stretching in other directions. The average rate of contraction and stretching is quantified by the *Lyapunov exponents* of the system. For a chaotic system at least one Lyapunov exponent is positive, while the sum over all exponents is negative because of dissipation. Lyapunov exponents are useful quantities for characterizing chaotic systems because they are topological invariants, i.e. their values are independent of the chosen coordinate system.

CHAOTIC SYNCHRONIZATION

Synchronization of Two Chaotic Systems

A system is said to be chaotic if it is deterministic, has a long-term aperiodic behavior, and shows sensitive dependence on initial conditions on a closed invariant set. One can deal with chaotic oscillators in many dynamical systems of various origines, the behaviour of such systems is characterized by instability as they evolve over time. It has been shown that two chaotic systems could be synchronized by coupling them. This kind of synchronization is of great interest with regards to its applications such as in increasing the power of lasers, in synchronizing the output of electronic circuits, in controlling oscillations in chemical reactions or in encoding electronic messages for secure communications.

Synchronization of chaos is a phenomenon that may occur when two, or more, chaotic oscillators are coupled, or when a chaotic oscillator drives another chaotic oscillator. The main thing to be noticed during this process is the exponential divergence of the trajectories of any chaotic system in the phase space. This type of synchronization of coupled or driven chaotic oscillators is a phenomenon well established experimentally and has been addressed theoretically. When two chaotic oscillators are considered, these include: identical synchronization, generalized synchronization, phase synchronization and others forms.

Identical Synchronization

It is a form of synchronization that occur when two identical chaotic oscillators are mutually coupled under appropriate circumstances, or when one of them drives the other. If $\vec{X} = \vec{X}\left(x_1, \ldots\ldots x_n\right)$ and $\vec{X}' = \vec{X}'\left(x_1', \ldots\ldots x_n'\right)$ are the variables that describe the state of both oscillators respectively, it is said that identical synchronization occurs when there is a set of initial conditions $[x_1(0), x_2(0), \ldots, x_n(0)]$, $[x'_1(0), x'_2(0), \ldots, x'_n(0)]$ such that, $|x'_i(t) - x_i((t)| \to 0$, for $i=1,2,\ldots,n$, when $t \to \infty$ which means $x'_i(t) = x_i(t)$, for $i=1,2,\ldots,n$, when the time is large enough. Here is an example.

We consider two identical and unidirectionally coupled chaotic systems (oscillators) described by the following equations

$$\begin{aligned} \frac{d\vec{X}_1}{dt} &= \vec{F}\left(\vec{X}_1\right) \\ \frac{d\vec{X}_1}{dt} &= \vec{F}\left(\vec{X}_2\right) + \lambda K\left(\vec{X}_2 - \vec{X}_1\right) \end{aligned} \qquad (15)$$

$\bar{X}_1$ is related to the n-dimensional state space variable of the drive system and $\vec{X}2(t)$ goes with the receiver. The flow of the single oscillation is governed by the nonlinear function $\vec{F}$, K is a $\vec{F}$, coupling matrix and finally λ is the scalar coupling strength. The energy in such a system must be dissipative because it models real situation in which dissipation takes place .The description of chaotic and dissipative system will be characterized by contraction (dissipation) and stretching (chaos) in same and other direction accordingly. The contraction and the stretching is related to the Lyapunov exponents of the system .The parameter λ will play a key role in describing the evolution of the systems.

The two systems will be uncoupled if $\lambda = 0$ and there is no synchronization, and if $\lambda \neq 0$, chaos synchronization becomes possible.

For the so called identical synchronization (complete synchronization) $\vec{X}_2(t) = \vec{X}_1(t)$ is a stable solution of the equation. Synchronization implies geometrically that the attraction of the system lies on a n-dimensional hyperplane called the synchronization manifold .

The stability of the synchronous solution will be connected to the Lyapunov function. When the Lyapunov exponents are computed numerically, synchronization occurs when all transverse Lyapunov exponents are negative (Pecora & Carroll, 1990). These Lyapunov exponents also called conditional Lyapunov exponents are the average rate of exponential expansion a contraction in direction transverse to the synchronization manifold.

Generalized Synchronization

This is a type of synchronization that takes place when the coupled chaotic oscillators are different. Here the dynamical variables will be $\vec{X} = \vec{X}\left(x_1, \ldots\ldots x_n\right)$ and $\vec{Y} = \vec{Y}\left(y_1, \ldots\ldots y_n\right)$. Given the dynamical variables $(x_1, x_2, \ldots, x_n)$ and $(y_1, y_2, \ldots, y_m)$ that determine the state of the oscillators, generalized synchronization takes place when $[y_1(t), y_2(t), \ldots, y_m(t)] = \varphi\, [x_1(t), x_2(t), \ldots, x_n(t)]$ where is φ a functional. φ is the identity when the synchronization is the identity.

Phase Synchronization

Phase synchronization is concerned when the coupled oscillators are not identical and their phases evolve in synchrony while the amplitudes remain none synchronized. The key thing here will be to define what is meant by phase in this case. A clear idea will be to imagine that the projection of the trajectories of the oscillator follows a rotation around a centre, then the phase is defined by the angle described by the segment joining the centre of rotation and the projection of the trajectory point onto the plane. A more detailed definition can be found in reference Pikovsky, Rosenblum, and Kurths (2001). Many other forms of chaotic synchronization can be found also in Pikovsky, Rosenblum, and Kurths (2001).

As a standard example, consider two Lorentz chaotic systems in which x represents a signal sent by a first chaotic system to another. The equations governing both completed systems are:

$$
\begin{aligned}
\frac{dx_1}{dt} &= -\delta(y_1 - x_1) \\
\frac{dy_1}{dt} &= -x_1 z_1 + r x_1 - y_1 \\
\frac{dy_2}{dt} &= -x_1 z_2 + r x_1 - y_2 \\
\frac{dz_2}{dt} &= x_1 y_2 - b z_2
\end{aligned}
\qquad (16)
$$

Subscripts (1) and (2) stand for each system label. x_1 is here the variable considered as driving the second system. x_1 has replaced x_2 in the first equation of the second system .From any arbitrary initial conditions, y_2 converges to y_1 and z_2 does the same towards z_1 such that after a certain period of times one can get $y_1 = y_2$, and $z_2 = z_1$ as the system evolves. This kind of synchronization is called identical synchronization. Both systems share the same attractor and a plot of variables x_1, y_1, and y_2 shows that the motion of the system remains in the plane defined by $y_2 = y_1$ and $z_2 = z_1$. This defines a hyperplane in five dimensions called the ***synchronization manifold***. The space orthogonal to the synchronization manifold is called the ***transverse space*** as already mentioned earlier in the identical synchronization.

SYNCHRONIZATION AND STABILITY

The problem of stability is broad and many cases can be thought of. Here we intend to focus on the drive-response of two identical Van der Pol oscillators. The coupling here is dissipative.

Many cases in which the drive is dissipative or the example of a conservative system with a dissipative coupling are addressed in Leung (1999) as follows.

We consider two identical Van Der Pol oscillators described by

$$\begin{aligned} \ddot{x} &= f\left(x, \dot{x}; \beta\right) = -x - \beta\dot{x}\left(x^2 - 1\right) \\ \ddot{u} &= f\left(u, \dot{u}; \beta\right) \end{aligned} \tag{17}$$

The nonlinear term $\beta\dot{x}\left(x^2 - 1\right)$ provides the self sustaining mechanism to the oscillator such that one can observe a perpetual oscillation, β measures the strength of the dissipation. When $\beta = 0$ there is no coupling and the two oscillators are independent and there is no synchronization if there are not launched with the same phase. We suppose now that we are dealing with a one-way drive system classically described by the following equations

$$\begin{aligned} \ddot{x} &= f\left(x, \dot{x}; \beta\right) \\ \ddot{u} &= f\left(u, \dot{u}; \beta\right) + K\left(u - x\right)H\left(t - T_0\right) \end{aligned} \tag{18}$$

$H(x)$ is a Heaviside function, T_0 is the onset time of the driving.The slave oscillator is driven by a nonsinusoidal function of time for $t \geq T_0$ and the master oscillator stays independent over the time. Next we rewrite equation (18) as:

$$\begin{aligned} \dot{x} &= y \\ \dot{u} &= v \\ \dot{y} &= f\left(x, y; \beta\right) \\ \dot{v} &= f\left(u, v; \beta\right) + K\left(u - x\right)H\left(t - T_0\right) \end{aligned} \tag{19}$$

These equations are solved numerically and synchronization is found to be possible with a coupling strength chosen in the domain $0 < k < 1$ ouside this domain the trajectory fails to contract to the original cycle (Van der Pol limit cycle)and synchronization is possible.

One can define a synchronization *Ts* as the time duration which overlaps the onset time to the time when the two externe trajectories (u,v) and (x,y) are within a specific criterion, $\sqrt{(x-u)^2+(y-v)^2}<\delta$ in which δ is a computatinnal accuracy. By defining another variable as $k = |K-Kc|$ where Kc is 0 or 1, one plots *Ts* versus *k*. From many plots over a good range of computational parameters one can come up with a power law relationship such as $T_s \approx |K-K_c|^{-\gamma}$ where $\gamma = 1$.This relationship remainds a behavior of some nonequilibrium transitions.

This has been applied in Germain (2006) to two coupled Van der Pol oscillators as follow:

A self-sustained system possesses a mechanism to damp oscillations that grow too large and a source of energy to pump those that become too small. These kinds of oscillations generally found in many systems in physics, biology, and engineering are marked by the existence of only none stable limit cycle. But under some conditions, some of these systems can exhibit more than one stable limit cycle. In Germain (2006), the Lindsted's perturbation method has been used to analytically establish for this model the amplitudes and frequencies of the limit cycles. The obtained values are then compared to the ones acquired numerically using the fourth order Runge-Kutta algorithm. The one-way and two-way couplings are then used to study the master-slave synchronization. The Whitthaker method and the properties of the Hill equation enable one to analytically investigate the linear stability boundaries of the synchronization process. The addition of other harmonics to the main solution component, the effects of nonlinearities as the areas where initial conditions are chosen are of impact on the stability boundaries. The expressions of the synchronization time are evaluated analytically and numerically.

This study of the stability of the boundary of the synchronization process has a great deal of impacts in implications in biology, biochemistry, as well as in neuroscience.

The critical boundary of the coupling strength has been found in both cases and it has been shown that the synchronization process can be realized more quickly with the two way coupling scheme. The existence of the nonlinear stability and nonlinear instability domains in the synchronization process additionally to the linear ones have been established. The results obtained here can have implications in biology, neuroscience, and biochemistry. Indeed in biology, the master-slave coupling situation is of interest, for example, in the case of displacement from one limit cycle to the other.

This is one-way, *diffusive* coupling, also called negative feedback control.

Some light have been shed on the dynamics of the Van der Pol oscillators with multi-limit cycle (Goldobin & Pikovsky, 2005). The objective now is to considerer two coupled Van der Pol oscillators both in the autonomous and chaotic regimes in order to investigate conditions under which they are synchronized. Since one particular characteristic of this model is the high sensitivity of its phase to the initial conditions, if two Van der Pol oscillators are launched with two different sets of initial conditions, their trajectories will finally circulate either on different limit cycles or on the same limit cycle but with different phases. The purpose of the synchronization is then to phase-lock both oscillators or to call one of the oscillators from its limit cycle to that of the other oscillator.

In the case of a one-way synchronization, the stability boundaries of the synchronization process is studied. All the result can be found in Germain (2006). Oscillatory states have been derived both in the non-autonomous and autonomous cases, using respectively the Lindsted's perturbation method and the harmonic balance method. In the unforced case, the phenomenon of *birhythmicity* has been found. The stability boundaries of the harmonic oscillations have been derived through the Floquet theory and it has been established that the model exhibited Neimark instability. Superharmonic and subharmonic oscillatory states have been investigated and bifurcation diagrams showing transitions from regular to chaotic motion have been drawn. We have also tackled the problem of the stability boundaries and duration time of the synchronization process of two such a model both in the autonomous and chaotic states. In the autonomous state, the analytical investigation through the first component and the full expression of the solution $x(t)$ has used properties of the Hill equation which describes the deviation between the master and the slave oscillators to examine the stability boundaries. The one-way and two way coupling have been considered and it has been shown that the synchronization process can be realized more quickly with the two way coupling (See Figures 3 and 4)

SYNCHRONIZATION AND NOISE

Let us assume an external noise which acts on the dynamics of the limit cycle of the oscillators. It is knows that the Lyapunov exponent becomes negative under influence of small white noise. Large white noise can lead to desynchronization of oscillators, provided they are nonisochronous (Goldobin & Pikovsky, 2005). A good study of the effect of noise can be found in the litterature. Numerical stimulations can be found in Goldobin and Pikovsky (2005). One can find in that reference excellent results to be taken into account for future research.

CONCLUSION

In this chapter, it has been outlined the basics to keep in mind for addressing the synchronization phenomenon. From now, some dynamical processes including autonomous and non-autonomous self oscillating systems by using recent developments in the theory of synchronization can be investigated in a more confortable maner. Most technics to be used can be found in the references given based on examples which are made available.It will be useful to keep gathering in the litterature more issues in the area of the synchronization of dynamical systems. A more general theory based on Riemannian geometry theory might be necessary to have a strong tool later on in studying the synchronization issues.

REFERENCES

Germain, E. K. H. (2006). *Synchronization dynamics of nonlinear self-sustained oscillations with applications in physics, engineering and biology*. PhD thesis, Université d'Abomey-Calavi, Bénin.

Goldobin, D. S., & Pikovsky, A. (2005). Synchronization and desynchronization of self-sustained oscillators by common noise. *Physical Review E: Statistical, Nonlinear, and Soft Matter Physics, 71*, 045201. doi:10.1103/PhysRevE.71.045201

Kuramoto, Y. (1984). *Chemical Oscillations, Waves and Turbulence*. Berlin: Springer-Verlag.

Leung, H. K. (1999, June). Dissipative effects on synchronization of nonchaotic oscillators. *The Chinese Journal of Physiology, 37*(3).

Pecora, L., & Carroll, T. (1990). Synchronization in chaotic systems. *Physical Review Letters*, *64*, 821–824. doi:10.1103/PhysRevLett.64.821

Pikovsky, A., Rosenblum, M., & Kurths, J. (2001). *Synchronization: A Universal Concept in Nonlinear Science*. Cambridge, UK: Cambridge University Press.

Rulkov, N., Sushchik, M., Tsimring, L., & Abarbanel, H. (1995). Generalized synchronization of chaos in directionally coupled chaotic systems. *Physical Review E: Statistical Physics, Plasmas, Fluids, and Related Interdisciplinary Topics*, *51*(2), 980–994. doi:10.1103/PhysRevE.51.980

Wang, W., & Slotine, J.-J. E. (2003). *On partial contraction analysis for coupled nonlinear oscillators*. NSL-030301. Retrieved from http://www.springerlink.com

KEY TERMS AND DEFINITIONS

Chaos: typically refers to a state lacking order or predictability

Synchronization of Oscillators: Process of precisely coordinating or matching two or more oscillators in time

APPENDIX

Figure 3. Synchronization time versus the feedback coupling coefficient with the precision $h = 10^{-2}$ when the master and the slave are on the same limit cycle amplitude A_1

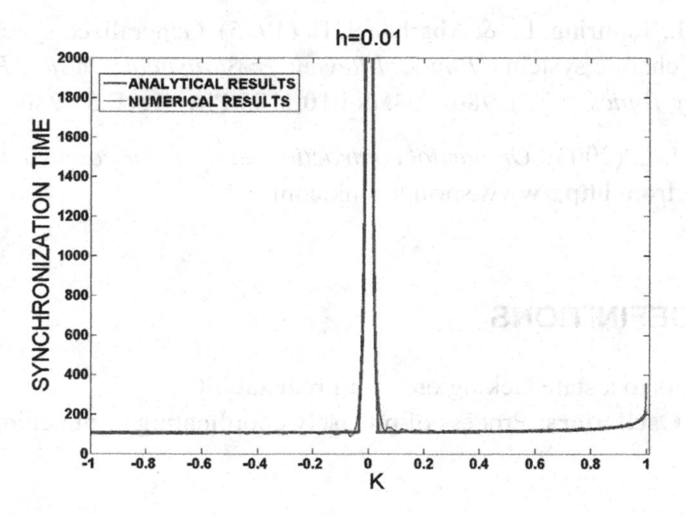

Figure 4. Synchronization time versus the feedback coupling coefficient with the precision $h = 10^{-2}$ when the master and the slave are on two different limit cycles (respectively A_3 et A_1)

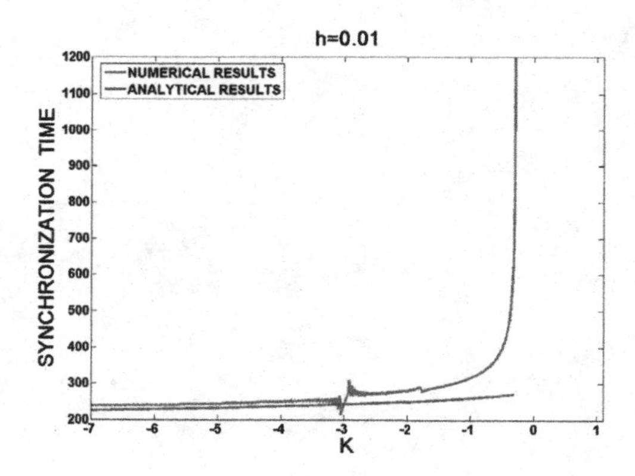

Chapter 12
Synchronization of Uncertain Neural Networks with H_∞ Performance and Mixed Time-Delays

Hamid Reza Karimi
University of Agder, Norway

ABSTRACT

An exponential H_∞ synchronization method is addressed for a class of uncertain master and slave neural networks with mixed time-delays, where the mixed delays comprise different neutral, discrete and distributed time-delays. An appropriate discretized Lyapunov-Krasovskii functional and some free weighting matrices are utilized to establish some delay-dependent sufficient conditions for designing a delayed state-feedback control as a synchronization law in terms of linear matrix inequalities under less restrictive conditions. The controller guarantees the exponential H_∞ synchronization of the two coupled master and slave neural networks regardless of their initial states. Numerical simulations are provided to demonstrate the effectiveness of the established synchronization laws.

INTRODUCTION

In the last few years, synchronization in neural networks (NNs), such as cellular NNs, Hopfield NNs and bi-directional associative memory networks, has received a great deal of interest among scientists from various fields (Chen & Dong,1993 ; Sun et al., 2007; Wang et al., 2008; Cheng et al., 2006; Cao et al., 2007). In order to better understand the dynamical behaviours of different kind of complex networks, an important and interesting phenomenon to investigate is the synchrony of all dynamical nodes. In fact, synchronization is a basic motion in nature that has been studied for a long time, ever since the discovery of Christian Huygens in 1665 on the synchronization of two pendulum clocks. The results of chaos synchronization are utilized in biology, chemistry, secret communication and cryptography, nonlinear oscillation synchronization and some other nonlinear fields. The first idea of synchronizing two identical

DOI: 10.4018/978-1-61520-737-4.ch012

chaotic systems with different initial conditions was introduced by Pecora and Carroll (Pecora & Carroll, 1990), and the method was realized in electronic circuits. The methods for synchronization of the chaotic systems have been widely studied in recent years, and many different methods have been applied theoretically and experimentally to synchronize chaotic systems, such as feedback control (Fradkov & Pogromsky, 1996; Gao et al., 2006; Karimi & Maass, 2009; Wen et al., 2006; Hou et al., 2007; Lu & van Leeuwen, 2006), adaptive control (Liao & Tsai, 2008; Feki, 2003; Wang et al., 2006a; Fradkov & Markov, 1997; Fradkov et al., 2000), backstepping (Park, 2006) and sliding mode control (Yan et al., 2006; García-Valdovinos et al., 2007). Recently, the theory of incremental input-to-state stability to the problem of synchronization in a complex dynamical network of identical nodes, using chaotic nodes as a typical platform was studied in (Cai & Chen, 2006).

On the other hand, in practice, due to the finite switching speed of amplifiers or finite speed of information processing, time delays including delays in the state (discrete delays) or in the derivative of the state (neutral delays) are often encountered in hardware implementation (Hale & Verduyn Lunel, 1993; Fridman, 2006; Gu, 2003; Park, 1999; Gao et al., 2008), which may be a source of oscillation, divergence, and instability in NNs. Another type of time-delays, namely, distributed time-delays, have begun to receive research attention (Wang et al., 2006b; Wang et al., 2006c). The main reason is that, since a NN usually has a spatial nature due to the presence of an amount of parallel pathways of a variety of axon sizes and lengths, continuously distributed delays should be introduced in modelling of the NNs over certain duration of time such that the distant past has less influence compared to the recent behaviour of the state (Wang et al., 2006b). Therefore, the stability problems of NNs with mixed time-delays have gained great research interest (Wang et al., 2007; Cao et al., 2006; Cao & Wang, 2005; Wang & Cao, 2009; Song & Wang, 2008; Liu et al., 2007, Wang et al., 2006d). Recently, both delay-independent and delay-dependent sufficient conditions have been proposed to verify the asymptotical or exponential stability of delayed NNs, see for instance the references (He et al., 2006; Zhang et al., 2005; Wang et al., 2005a; Xu et al., 2005; Xu et al., 2006; Lou et al., 2006; He et al., 2005; Mou et al., 2008a; Mou et al., 2008b; Ho et al., 2006) and references therein. Furthermore, many results have been reported on the stability analysis issue for various NNs with distributed time-delays, such as recurrent NNs (Liang & Cao, 2007; Liang & Cao, 2006), bi-directional associative memory networks (Liang & Cao, 2004), Hopfield NNs (Zhao, 2004a), cellular NNs (Zhao, 2004b). It is noted that both discrete and distributed time delays have been recently considered in the references (Wang et al., 2006b; Wang et al., 2006c), (Song & Wang, 2008) and (Wang et al., 2005a). It can be realized that in Huang et al. (2006), Karimi and Maass (2009) and others (Cao et al. 2008; Yu et al., 2008; Liang et al., 2008; Wang et al., 2005b; Huang et al., 2008) several sufficient conditions in terms of LMIs were presented to solve the synchronization and estimation problems of NNs with time-delays. The authors in (Huang et al., 2008) studied the exponential synchronization problem for a class of chaotic Lur'e systems by using delayed feedback control by employing an integral inequality and introducing several slack variables to reduce the conservatism of the developed synchronization criterion. In (Liang et al., 2008), the problem of synchronization for stochastic discrete-time drive-response networks with time-varying delay was investigated by employing the Lyapunov functional method combined with the stochastic analysis as well as the feedback control technique. The advantage of this approach was that a less conservative condition which depends on the lower and upper bounds of the time-varying delay was obtained. Furthermore, from the published results, it appears that general results pertaining to exponential synchronization of master-slave systems with mixed neutral, discrete and distributed delays and an H_∞ performance criteria are few and restricted, despite its practical importance, mainly due to the mathematical difficulties in dealing with such mixed delays. Hence, it is our intention in this paper to tackle such an important yet challenging problem.

In this paper, we contribute to the further development of an exponential H_∞ synchronization method for a class of uncertain master and slave neural networks (MSNNs) with mixed time-delays, where the mixed delays comprise different neutral, discrete and distributed time-delays. Both the polytopic and the norm-bounded uncertainties are taken into consideration separately. An appropriate discretized Lyapunov-Krasovskii functional (DLKF) is constructed in order to establish some delay-dependent sufficient conditions for designing a delayed state-feedback control as a synchronization law in terms of linear matrix inequalities (LMIs) under less restrictive conditions by introducing some free weighting matrices. Then, the controller is developed based on the available information of the size of the discrete and distributed delays so as to guarantee that the controlled slave system can exponentially synchronize with the master system regardless of their initial states. It is shown that the decay coefficient can be easily calculated by solving the derived delay-dependent conditions. All the developed results are expressed in terms of convex optimization over LMIs and tested on a representative example to demonstrate the feasibility and applicability of the proposed synchronization approach.

The notations used throughout the paper are fairly standard. I_n and 0_n represent, respectively, n by n identity matrix and n by n zero matrix; the superscript 'T' stands for matrix transposition; R^n denotes the n-dimensional Euclidean space; $R^{n\times m}$ is the set of all real m by n matrices. The vector v_i denotes the unit column vector having a '1' element on its ith row and zeros elsewhere. $\|.\|$ refers to the Euclidean vector norm or the induced matrix 2-norm and $diag\{\cdots\}$ represents a block diagonal matrix. $\lambda_{\min}(A)$ and $\lambda_{\max}(A)$ denote, respectively, the smallest and largest eigenvalue of the square matrix A. The operator $sym\{A\}$ denotes $A+A^T$. The notation $P>0$ means that P is real symmetric and positive definite and the symbol * denotes the elements below the main diagonal of a symmetric block matrix. In addition, $L_2[0,\infty)$ is the space of square-integrable vector functions over $[0,\infty)$. Matrices, if the dimensions are not explicitly stated, are assumed to have compatible dimensions for algebraic operations.

PROBLEM DESCRIPTION

In this paper, the problem of characterizing the delay-dependent coupling technique for the synchronization of a class of MSNNs with mixed time-delays is considered. More specifically, consider the master neural network which is described as follows:

$$\begin{cases} \dot{x}(t) = -A\,x(t) + W_1\,f(x(t)) + W_2\,g(x(t-\tau_1)) + W_3\,\dot{x}(t-\tau_2) + W_4 \int\limits_{t-\tau_3}^{t} h(x(s))\,ds + o, \\ x(t) = \varphi(t), \qquad t \in \left[-\bar{\tau},\, 0\right], \\ z_x(t) = C_1\,x(t) + C_2\,x(t-\tau_1) + C_3 \int\limits_{t-\tau_3}^{t} h(x(s))\,ds, \end{cases} \tag{1}$$

with $x(t) = [x_1(t), x_2(t), \cdots, x_n(t)]^n \in \Re^n$ where $x_i(t)$ are the master system's state vector associated with the ith neuron and $z_x(t) \in \Re^s$ is the controlled output of the master network.

$$f(x(t)) = [f_1(x_1(t)), f_2(x_2(t)), \cdots, f_n(x_n(t))]^T,$$

$$g(x(t-\tau_1)) = [g_1(x_1(t-\tau_1)), g_2(x_2(t-\tau_1)), \cdots, g_n(x_n(t-\tau_1))]^T$$

and $h(x(t)) = [h_1(x_1(t)), h_2(x_2(t)), \cdots, h_n(x_n(t))]^T$ denote the activation functions, $A=diag\{a_i\}>0$, the vector $0 = [0_1, 0_2, \ldots, 0_n]^T$ is the constant external input and the constant scalars $o = [o_1, o_2, \cdots, o_n]^T$, for $\tau_i \geq 0$, denote the known neutral, discrete and distributed time-delays, respectively, with $i = 1, 2, 3$. If all $\bar{\tau} := \max\{\tau_1, \tau_2, \tau_3\}$, the network (1) has no time delay. The time-varying vector valued initial function $\phi(t)$ is a continuously differentiable functional.

Now, given the master signal $x(t)=x(t,\phi(t))$, we are to design a feasible coupling technique to realize the synchronization between two identical neural networks with different initial conditions. Actually, the slave neural network is described as follows:

$$\tau_i = 0 \tag{2}$$

with

$$\begin{cases} \dot{y}(t) = -A\,y(t) + W_1\,f(y(t)) + W_2\,g(y(t-\tau_1)) + W_3\,\dot{y}(t-\tau_2) + W_4 \int\limits_{t-\tau_3}^{t} h(y(s))\,ds + E\,w(t) + u(t) + o, \\ y(t) = \phi(t), \qquad t \in \left[-\bar{\tau},\, 0\right], \\ z_y(t) = C_1\,y(t) + C_2\,y(t-\tau_1) + C_3 \int\limits_{t-\tau_3}^{t} h(y(s))\,ds, \end{cases}$$

where $y_i(t)$ are the slave system's state vector associated with the ith neuron; $u(t)\in R^n$ is a coupled term which is considered as the control input; $w(t)\in R^q$ is the disturbance, $z_y(t)\in R^s$ is the controlled output of the slave network and $\varphi(t)$ is a continuously differentiable functional.

Remark 1. The model (1)-(2) can describe a large amount of well-known dynamical systems with time-delays, such as the delayed Logistic model, the chaotic models with time-delays, the artificial neural network model with discrete and distributed time-delays, and the predator-prey model with distributed delays. In real application, these coupled systems can be regarded as interacting dynamical elements in the entire system, such as physical particles, biological neurons, ecological populations, genetic oscillations, and even automatic machines and robots. A feasible coupling design for successful synchronization leads us to fully command the intrinsic mechanism regulating the evolution of real systems, to fabricate emulate systems, and even to remotely control the machines and nodes in networks with large scales [Pecora & Carroll, 1990], [Fradkov et al., 2000], [García-Valdovinos et al., 2007], [Gu et al., 2003], [Yu et al., 2008].

One can define a difference operator $y(t) = [y_1(t), y_2(t), \cdots, y_n(t)]^n \in \Re^n$ such that

$$\nabla : C([-\bar{\tau}, 0], \Re^n) \to \Re^n \tag{3}$$

Definition 1. [Hale & Verduyn Lunel,1993] The difference operator ∇ is said to be stable if the zero solution of the homogeneous difference equation $\nabla x_t = x(t) - W_3\ x(t-\tau_2)$ is uniformly asymptotically stable.

The stability of the difference operator ∇ is necessary for the stability of the MSNNs (1)-(2).

Assumption 1. It follows from [22] that a delay-independent sufficient condition for the asymptotic stability of the MSNNs (1)-(2) is that all the eigenvalues of the matrix W_3 are inside the unit circle, i.e. $\lambda_{\max}(W_3)<1$.

Definition 2. [Huang et al., 2008] The MSNNs (1)-(2) are synchronized globally exponentially if there exist scalars $\alpha>0$ and $M\geq 1$ such that

$$\nabla x_t = 0,\ t \geq 0,\ x_0 = \Psi \in \{\Phi \in C([-\bar{\tau}, 0]) : \nabla\Phi = 0\}$$

where $\left|e(t)\right| \leq M\ e^{-\alpha t}[\left\|\varsigma\right\| + \left\|\dot{\varsigma}\right\|]$, is an initial condition and $e(t) = x(t) - y(t)$ is the synchronization error such α and M are called the exponential decay rate and decay coefficient, respectively.

Let $\varsigma(t) \in C([-\bar{\tau}, 0]; \Re^n)$ The error dynamics between the MSNNs (1)-(2), namely synchronization error network, can be expressed by

$$\hat{e}(t) = e^{\alpha t} e(t). \tag{4}$$

where

$$\begin{cases} \dot{\hat{e}}(t) = -(A - \alpha I)\hat{e}(t) + W_1\ \hat{\psi}_1(\hat{e}(t)) + W_2\, e^{\alpha\tau_1}\ \hat{\psi}_2(\hat{e}(t-\tau_1)) - \alpha\, e^{\alpha\tau_2}\ W_3\, \hat{e}(t-\tau_2) + e^{\alpha\tau_2}\ W_3\, \dot{\hat{e}}(t-\tau_2) \\ \qquad + W_4 \displaystyle\int_{t-\tau_3}^{t} e^{\alpha(t-s)}\ \hat{\psi}_3(\hat{e}(s))\, ds - E\, \hat{w}(t) - \hat{u}(t), \\ \hat{e}(t) = \varphi_e(t;\alpha), \qquad t \in \left[-\bar{\tau},\ 0\right], \\ \hat{z}_e(t) = C_1\, \hat{e}(t) + C_2\, \hat{e}(t-\tau_1) + C_3 \displaystyle\int_{t-\tau_3}^{t} e^{\alpha(t-s)}\ \hat{\psi}_3(\hat{e}(s))\, ds, \end{cases}$$

$\hat{u}(t) = e^{\alpha t} u(t)$, $\hat{w}(t) = e^{\alpha t} w(t)$, $\hat{z}_e(t) = \hat{z}_s(t) - \hat{z}_m(t) = e^{\alpha t}(z_s(t) - z_m(t))$, $\hat{\psi}_1(\hat{e}(t)) = [\hat{\psi}_{11}(\hat{e}_1(t)),$ $\hat{\psi}_{12}(\hat{e}_2(t)), \cdots, \hat{\psi}_{1n}(\hat{e}_n(t))]^T$, and $\hat{\psi}_2(\hat{e}(t-\tau_1)) = [\hat{\psi}_{21}(\hat{e}_1(t-\tau_1)), \hat{\psi}_{22}(\hat{e}_2(t-\tau_1)), \cdots, \hat{\psi}_{2n}(\hat{e}_n(t-\tau_1))]^T$ with $\hat{\psi}_3(\hat{e}(t)) = [\hat{\psi}_{31}(\hat{e}_1(t)), \hat{\psi}_{32}(\hat{e}_2(t)), \cdots, \hat{\psi}_{3n}(\hat{e}_n(t))]^T$ $\hat{\psi}_{1i}(\hat{e}_i(t)) = e^{\alpha t}(f_i(x_i(t)) - f_i(y_i(t)))$, and $\hat{\psi}_{2i}(\hat{e}_i(t)) = e^{\alpha t}(g_i(x_i(t)) - g_i(y_i(t)))$, $\hat{\psi}_{3i}(\hat{e}_i(t)) = e^{\alpha t}(h_i(x_i(t)) - h_i(y_i(t)))$

In this paper, we make the following assumption for the neuron activation functions in (1)-(2), which is more general than the descriptions on the conventional sigmoid activation functions as well as the recently popular Lipschitz-type activation functions.

Assumption 2. [27] The nonlinear functions $f_i(s), g_i(s), h_i(s)$, for any $\varphi_e(t;\alpha) = e^{\alpha t}(\varphi(t) - \phi(t))$. satisfy, respectively, $i = 1, \cdots, n,$, $f_i^- \leq \frac{f_i(s_1) - f_i(s_2)}{s_1 - s_2} \leq f_i^+$, $g_i^- \leq \frac{g_i(s_1) - g_i(s_2)}{s_1 - s_2} \leq g_i^+$ where $h_i^- \leq \frac{h_i(s_1) - h_i(s_2)}{s_1 - s_2} \leq h_i^+$, are some constants.

Remark 2. According to Assumption 2, one can easily check that, for any $f_i^-, f_i^+, g_i^-, g_i^+, h_i^-, h_i^+$ the functions $i = 1, \cdots, n,$ and $\hat{\psi}_{1i}(\hat{e}_i(t)), \hat{\psi}_{2i}(\hat{e}_i(t))$ satisfy, respectively: $\hat{\psi}_{3i}(\hat{e}_i(t))$, $f_i^- \leq \frac{\hat{\psi}_{1i}(\hat{e}_i(t))}{\hat{e}_i(t)} \leq f_i^+$, $g_i^- \leq \frac{\hat{\psi}_{2i}(\hat{e}_i(t))}{\hat{e}_i(t)} \leq g_i^+$

The problem to be addressed in this paper is formulated as follows: given the delayed MSNNs (1)-(2) with a prescribed level of disturbance attenuation $\gamma > 0$, find a driving signal $h_i^- \leq \frac{\hat{\psi}_{3i}(\hat{e}_i(t))}{\hat{e}_i(t)} \leq h_i^+$. of the form

$$\hat{u}(t) \tag{5}$$

where the matrices $\hat{u}(t) = K_1\,\hat{e}(t) + K_2\,\hat{e}(t - \tau_1) + K_3 \int_{t-\tau_3}^{t} \hat{e}(s)\,ds$ are the control gains to be determined such that

1. the synchronization error network (4) is globally exponentially stable;
   under zero initial conditions and for all non-zero $w(t) \in L_2[0,\infty]$, the H_∞ performance measure, i.e., $\{K_i\}_{i=1}^{3}$, satisfies $J_\infty < 0$ (or the induced L_2–norm of the operator form $J_\infty = \int_0^\infty [\hat{z}_e^T(t)\,\hat{z}_e(t) - \gamma^2 \hat{w}^T(t)\,\hat{w}(t)] \quad dt$ to the controlled outputs $\hat{w}(t)$ is less than γ); in this case, the MSNNs (1)-(2) are said to be asymptotically stable with an H_∞ performance measure.

Remark 3. The delay-dependent coupling (5) utilizes the available information of the size of the discrete and distributed delays. However, in many real applications, if the information of the size of the delays is not available for feedback, a memoryless coupling, i.e., $\hat{z}_e(t)$, will be designed to synchronize the master and slave systems. Recently, in [Cao et al., 2008], a global synchronization was given for an array of coupled delayed NNs with linear diffusive hybrid coupling, containing constant, discrete, and distributed-delay coupling. In comparison, our model extends the model structure in [Cao et al., 2008] to a MSNN with a hybrid coupling, containing constant, neutral, discrete, and distributed-delay coupling.

MAIN RESULTS

In this section, we present our new sufficient conditions for the solvability of the problem of the delayed state-feedback control design using the Lyapunov method and an LMI approach.

Theorem 1. Let $\hat{u}(t) = K_1 \hat{e}(t)$, i=1,2, be given for any positive integer N. Under Assumptions 1 and 2, a state feedback controller given in the form (5) exists such that the controlled slave system (2) exponentially synchronizes with the master system (1) with the H_∞ performance level γ >0 and an exponential decay rate α>0, if there exist some scalars $\delta, \sigma_i, \rho_i, \lambda_i$ $h_i = \tau_i / N$ matrices P_2, L_1, L_2, $(i = 1, 2, \cdots, N)$, and positive-definite matrices L_3, Q_i, S_i, H_i, $R_{i,j} = R_{i,j}^T, T_{i,j} = T_{i,j}^T\ (i, j = 0, 1, \cdots, N)$ satisfying the following LMIs

$$P_1, Z_1, Z_2, \bar{U}_1, \bar{U}_2 \tag{6a}$$

$$\begin{bmatrix} P_1 & \tilde{Q} \\ * & \tilde{R} + \tilde{S} \end{bmatrix} > 0, \tag{6b}$$

$$\begin{bmatrix} \bar{U}_1 & -\bar{U}_1 \\ * & S_d \end{bmatrix} > 0, \tag{6c}$$

$$\begin{bmatrix} \bar{U}_2 & -\bar{U}_2 \\ * & H_d \end{bmatrix} > 0, \tag{6d}$$

where

$$\Pi := \begin{bmatrix} \hat{\Xi}_e & -\hat{D}^s & D^a & O^a & -O^s \\ * & -S_d - R_{ds} & 0 & 0 & 0 \\ * & * & -3\bar{U}_1 & 0 & 0 \\ * & * & * & -3\bar{U}_2 & 0 \\ * & * & * & * & -H_d - T_{ds} \end{bmatrix} < 0,$$

$$\hat{\Xi}_e = \begin{bmatrix} \Sigma_{11} & \begin{bmatrix} -L_2 - Q_N + C_1^T C_2 \\ -\delta L_2 \end{bmatrix} & \begin{bmatrix} -\alpha\ P_2^T W_3 e^{\alpha\tau_2} \\ -\alpha\,\delta\ P_2^T W_3 e^{\alpha\tau_2} \end{bmatrix} & \begin{bmatrix} P_2^T W_3 e^{\alpha\tau_2} \\ \delta\ P_2^T W_3 e^{\alpha\tau_2} \end{bmatrix} & \Sigma_{15} \\ * & -S_N + C_2^T C_2 & 0 & 0 & 0 \\ * & * & -H_N & 0 & 0 \\ * & * & * & -U & 0 \\ * & * & * & * & \Sigma_{55} \end{bmatrix},$$

$$\Sigma_{11} = sym(\begin{bmatrix} -P_2^T(A-\alpha I) - L_1 & P_1 - P_2^T \\ -\delta P_2^T(A-\alpha I) - \delta L_1 & -\delta P_2^T \end{bmatrix}) + diag\{sym(Q_0) + S_0 + H_0 - F^+\Lambda_1 F^- - G^+\Lambda_2 G^- - H^+\Lambda_3 H^-, U\},$$

$$\Sigma_{15} = \begin{bmatrix} \begin{bmatrix} P_2^T W_1 + \frac{1}{2}(F^+ + F^-)\Lambda_1 \\ \delta P_2^T W_1 \end{bmatrix} & \begin{bmatrix} \frac{1}{2}(G^+ + G^-)\Lambda_2 \\ 0 \end{bmatrix} & \begin{bmatrix} \frac{1}{2}(H^+ + H^-)\Lambda_3 \\ 0 \end{bmatrix} & \begin{bmatrix} P_2^T W_4 + C_1^T C_3 \\ \delta P_2^T W_4 \end{bmatrix} & \begin{bmatrix} P_2^T W_2 e^{\alpha\tau_1} \\ \delta P_2^T W_2 e^{\alpha\tau_1} \end{bmatrix} & \begin{bmatrix} -P_2^T E \\ -\delta P_2^T E \end{bmatrix} \\ 0 & 0 & 0 & 0 & 0 & 0 \end{bmatrix},$$

$$\Sigma_{55} = diag\{-\Lambda_1, Z_1 - \Lambda_2, \tau_3^2 Z_2 - \Lambda_3, -e^{-2\alpha\tau_3} Z_2 + C_3^T C_3, -Z_1, -\gamma^2 I\},$$

$$R_{ds} = h_1 \begin{bmatrix} R_{0,0} - R_{1,1} & R_{0,1} - R_{1,2} & \cdots & R_{0,N-1} - R_{1,N} \\ R_{1,0} - R_{2,1} & R_{1,1} - R_{2,2} & \cdots & R_{1,N-1} - R_{2,N} \\ \vdots & \vdots & \ddots & \vdots \\ R_{N-1,0} - R_{N,1} & R_{N-1,1} - R_{N,2} & \cdots & R_{N-1,N-1} - R_{N,N} \end{bmatrix},$$

$$T_{ds} = h_2 \begin{bmatrix} T_{0,0} - T_{1,1} & T_{0,1} - T_{1,2} & \cdots & T_{0,N-1} - T_{1,N} \\ T_{1,0} - T_{2,1} & T_{1,1} - T_{2,2} & \cdots & T_{1,N-1} - T_{2,N} \\ \vdots & \vdots & \ddots & \vdots \\ T_{N-1,0} - T_{N,1} & T_{N-1,1} - T_{N,2} & \cdots & T_{N-1,N-1} - T_{N,N} \end{bmatrix},$$

$$\hat{D}^s = h_1 \begin{bmatrix} 2Q_1^a + R_{0,1}^s - L_3 & 2Q_2^a + R_{0,2}^s - L_3 & \cdots & 2Q_N^a + R_{0,N}^s - L_3 \\ Q_1^s - L_3 & Q_2^s - L_3 & \cdots & Q_N^s - L_3 \\ -2R_{N,1}^s & -2R_{N,2}^s & \cdots & -2R_{N,N}^s \end{bmatrix},$$

$$O^s = h_2 \begin{bmatrix} T_{0,1}^s & T_{0,2}^s & \cdots & T_{0,N}^s \\ 0 & 0 & \cdots & 0 \\ -2T_{N,1}^s & -2T_{N,2}^s & \cdots & -2T_{N,N}^s \end{bmatrix},$$

$$D^a = h_1 \begin{bmatrix} R_{0,1}^a & R_{0,2}^a & \cdots & R_{0,N}^a \\ Q_1^a & Q_2^a & \cdots & Q_N^a \\ -R_{N,0}^a & -R_{N,1}^a & \cdots & -R_{N,N-1}^a \end{bmatrix},$$

$$O^a = h_2 \begin{bmatrix} T_{0,1}^a & T_{0,2}^a & \cdots & T_{0,N}^a \\ 0 & 0 & \cdots & 0 \\ -T_{N,0}^a & -T_{N,1}^a & \cdots & -T_{N,N-1}^a \end{bmatrix},$$

$$\tilde{R} = \begin{bmatrix} R_{0,0} & R_{0,1} & \cdots & R_{0,N} \\ R_{1,0} & R_{1,1} & \cdots & R_{1,N} \\ \vdots & \vdots & \ddots & \vdots \\ R_{N,0} & R_{N,1} & \cdots & R_{N,N} \end{bmatrix}, , \ S_d = diag\{S_0 - S_1, S_1 - S_2, \cdots, S_{N-1} - S_N\},$$

$H_d = diag\{H_0 - H_1, H_1 - H_2, \cdots, H_{N-1} - H_N\}$, $\tilde{Q} = [Q_0, Q_1, \cdots, Q_N]$,with

$\tilde{S} = 1/h_1 \ diag\{S_0, S_1, \cdots, S_N\}$, $Q_p^s = (Q_p + Q_{p-1})/2$, $Q_p^a = (Q_p - Q_{p-1})/2$, $R_{p,q}^s = (R_{p,q} + R_{p,q-1})/2$,

$R_{p,q}^a = (R_{p,q} - R_{p,q-1})/2$, $T_{p,q}^s = (T_{p,q} + T_{p,q-1})/2$, $T_{p,q}^a = (T_{p,q} - T_{p,q-1})/2$, $F^+ = diag\{f_1^+, f_2^+, \cdots, f_N^+\}$,

$G^+ = diag\{g_1^+, g_2^+, \cdots, g_N^+\}$, $H^+ = diag\{h_1^+, h_2^+, \cdots, h_N^+\}$, $F^- = diag\{f_1^-, f_2^-, \cdots, f_N^-\}$,

$G^- = diag\{g_1^-, g_2^-, \cdots, g_N^-\}$, $H^- = diag\{h_1^-, h_2^-, \cdots, h_N^-\}$, $\Lambda_1 = diag\{\sigma_1, \sigma_2, \cdots, \sigma_N\}$ and

$\Lambda_2 = diag\{\rho_1, \rho_2, \cdots, \rho_N\}$

The decay coefficient can be calculated by

$\Lambda_3 = diag\{\lambda_1, \lambda_2, \cdots, \lambda_N\}$. with

$$M = \sqrt{\frac{\max\{\Delta_1, \Delta_2\}}{\lambda_{\min}(P_1)}}$$

and
$$\begin{aligned} \Delta_1 = {} & \lambda_{\max}(P_1) + \tau_1 \lambda_{\max}(Q_p)_{p=0}^N + \tau_1 \lambda_{\max}(G^{+T} Z_1 G^+) + \tau_1 \lambda_{\max}(S_p)_{p=0}^N \\ & + \tau_1^2 \lambda_{\max}(R_{p,p})_{p=0}^N + \tau_2 \lambda_{\max}(H_p)_{p=0}^N + \tau_2^2 \lambda_{\max}(T_{p,p})_{p=0}^N + \frac{2}{3} \tau_3^3 \lambda_{\max}(H^{+T} Z_2 H^+) \end{aligned}$$
. Moreover, the controller gains in (5) can be designed as $\Delta_2 = \tau_2 \lambda_{\max}(U)$.

Proof: To prove the theorem, choose a Lyapunov-Krasovskii functional (LKF) candidate as

$$K_i = (P_2^T)^{-1} L_i (i = 1, 2, 3). \tag{7}$$

where

$$V(t) = \sum_{i=1}^{5} V_i(t), \tag{8a}$$

$$V_1(t) = \hat{e}(t)^T P_1 \hat{e}(t) + 2 \hat{e}(t)^T \int_{-\tau_1}^{0} Q(\xi) \, \hat{e}(t + \xi) \ d\xi, \tag{8b}$$

$$V_2(t)=\int_{t-\tau_1}^{t}\hat{\psi}_2(\hat{e}(s))^T Z_1\,\hat{\psi}_2(\hat{e}(s))\;ds+\int_{t-\tau_2}^{t}\dot{\hat{e}}(s)^T U\,\dot{\hat{e}}(s)\;ds, \tag{8c}$$

$$V_3(t)=\int_{-\tau_1}^{0}\hat{e}(t+\xi)^T S(\xi)\hat{e}(t+\xi)\;d\xi+\int_{-\tau_1}^{0}\int_{-\tau_1}^{0}\hat{e}(t+s)^T R(s,\xi)\;\hat{e}(t+\xi)\;ds\;d\xi, \tag{8d}$$

$$V_4(t)=\int_{-\tau_2}^{0}\hat{e}(t+\xi)^T H(\xi)\hat{e}(t+\xi)\;d\xi+\int_{-\tau_2}^{0}\int_{-\tau_2}^{0}\hat{e}(t+s)^T T(s,\xi)\;\hat{e}(t+\xi)\;ds\;d\xi \tag{8e}$$

where

$$V_5(t)=\int_{t-\tau_3}^{t}[\int_{s}^{t}\hat{\psi}_3(\hat{e}(\theta))^T\;d\theta]Z_2[\int_{s}^{t}\hat{\psi}_3(\hat{e}(\theta))\;d\theta]\;\;ds+\int_{0}^{\tau_3}\int_{t-s}^{t}(\theta-t+s)\hat{\psi}_3(\hat{e}(\theta))^T Z_2\,\hat{\psi}_3(\hat{e}(\theta))\;d\theta\;ds.$$

and $H(\xi)=H(\xi)^T$ are continuous matrix functions.

Derivatives of $V_i(t)$, $Q(\xi), R(s,\xi)=R(s,\xi)^T, S(\xi)=S(\xi)^T, T(s,\xi)=T(s,\xi)^T$, are given, respectively, by

$$i=1,\cdots,5 \tag{9a}$$

$$\begin{aligned}\dot{V}_1(t)&=2\dot{\hat{e}}(t)^T[P_1\,\hat{e}(t)+\int_{-\tau_1}^{0}Q(\xi)\,\hat{e}(t+\xi)\;\;d\xi]+2\,\hat{e}(t)^T\int_{-\tau_1}^{0}Q(\xi)\;\dot{\hat{e}}(t+\xi)\;\;d\xi\\&=2\dot{\hat{e}}(t)^T[P_1\,\hat{e}(t)+\int_{-\tau_1}^{0}Q(\xi)\,\hat{e}(t+\xi)\;\;d\xi]+2\,\hat{e}(t)^T\int_{-\tau_1}^{0}Q(\xi)\;\dot{\hat{e}}(t+\xi)\;\;d\xi\end{aligned} \tag{9b}$$

$$\begin{aligned}&\dot{V}_2(t)=\hat{\psi}_2(\hat{e}(t))^T Z_1\,\hat{\psi}_2(\hat{e}(t))+\dot{\hat{e}}(t)^T U\,\dot{\hat{e}}(t)\\&-\hat{\psi}_2(\hat{e}(t-\tau_1))^T Z_1\,\hat{\psi}_2(\hat{e}(t-\tau_1))-\dot{\hat{e}}(t-\tau_2)^T U\,\dot{\hat{e}}(t-\tau_2),\end{aligned} \tag{9c}$$

$$\dot{V}_3(t)=2\int_{-\tau_1}^{0}\dot{\hat{e}}(t+\xi)^T S(\xi)\hat{e}(t+\xi)\;\;d\xi+2\int_{-\tau_1}^{0}\int_{-\tau_1}^{0}\dot{\hat{e}}(t+s)^T R(s,\xi)\;\hat{e}(t+\xi)\;\;ds\;d\xi, \tag{9d}$$

and

$$\dot{V}_4(t)=2\int_{-\tau_2}^{0}\dot{\hat{e}}(t+\xi)^T H(\xi)\hat{e}(t+\xi)\;\;d\xi+2\int_{-\tau_2}^{0}\int_{-\tau_2}^{0}\dot{\hat{e}}(t+s)^T T(s,\xi)\;\hat{e}(t+\xi)\;\;ds\;d\xi$$

$$\begin{aligned}\dot{V}_5(t)=&-[\int_{t-\tau_3}^{t}\hat{\psi}_3(\hat{e}(\theta))^T\;d\theta]Z_2[\int_{t-\tau_3}^{t}\hat{\psi}_3(\hat{e}(\theta))\;\;d\theta]+2\int_{t-\tau_3}^{t}\hat{\psi}_3(\hat{e}(t))^T Z_2[\int_{s}^{t}\hat{\psi}_3(\hat{e}(\theta))\,d\theta]\;\;ds\\&+\int_{0}^{\tau_3}s\hat{\psi}_3(\hat{e}(t))^T Z_2\,\hat{\psi}_3(\hat{e}(t))ds-\int_{0}^{\tau_3}\int_{t-s}^{t}\hat{\psi}_3(\hat{e}(\theta))^T Z_2\,\hat{\psi}_3(\hat{e}(\theta))\;\;d\theta\;ds\end{aligned} \tag{9e}$$

According to Remark 2, we have

$$
\begin{aligned}
&= -[\int_{t-\tau_3}^{t} \hat{\psi}_3(\hat{e}(\theta))^T \, d\theta] Z_2 [\int_{t-\tau_3}^{t} \hat{\psi}_3(\hat{e}(\theta)) \, d\theta] + 2\int_{t-\tau_3}^{t} (\theta - t + \tau_3)\hat{\psi}_3(\hat{e}(t))^T Z_2 \hat{\psi}_3(\hat{e}(\theta)) \, d\theta \\
&\quad + \int_0^{\tau_3} s\hat{\psi}_3(\hat{e}(t))^T Z_2 \hat{\psi}_3(\hat{e}(t))ds - \int_{t-\tau_3}^{t}\int_{t-\theta}^{\tau_3} \hat{\psi}_3(\hat{e}(s))^T Z_2 \hat{\psi}_3(\hat{e}(s)) \, ds\, d\theta \\
&\le \int_{t-\tau_3}^{t} (\theta - t + \tau_3)[\hat{\psi}_3(\hat{e}(t))^T Z_2 \hat{\psi}_3(\hat{e}(t)) + \hat{\psi}_3(\hat{e}(\theta))^T Z_2 \hat{\psi}_3(\hat{e}(\theta))] \, d\theta \\
&\quad - [\int_{t-\tau_3}^{t} \hat{\psi}_3(\hat{e}(\theta))^T \, d\theta] Z_2 [\int_{t-\tau_3}^{t} \hat{\psi}_3(\hat{e}(\theta)) \, d\theta] + \int_0^{\tau_3} s\hat{\psi}_3(\hat{e}(t))^T Z_2 \hat{\psi}_3(\hat{e}(t))\, ds \\
&\quad - \int_{t-\tau_3}^{t} (\theta - t + \tau_3)\hat{\psi}_3(\hat{e}(\theta))^T Z_2 \hat{\psi}_3(\hat{e}(\theta)) \, d\theta \\
&\le \tau_3^2 \hat{\psi}_3(\hat{e}(t))^T Z_2 \hat{\psi}_3(\hat{e}(t)) - [\int_{t-\tau_3}^{t} \hat{\psi}_3(\hat{e}(\theta))^T \, d\theta] Z_2 [\int_{t-\tau_3}^{t} \hat{\psi}_3(\hat{e}(\theta)) \, d\theta] \\
&\le \tau_3^2 \hat{\psi}_3(\hat{e}(t))^T Z_2 \hat{\psi}_3(\hat{e}(t)) - e^{-2\alpha\tau_3}[\int_{t-\tau_3}^{t} e^{\alpha(t-\theta)}\hat{\psi}_3(\hat{e}(\theta))^T \, d\theta] Z_2 [\int_{t-\tau_3}^{t} e^{\alpha(t-\theta)}\hat{\psi}_3(\hat{e}(\theta)) \, d\theta]
\end{aligned}
\tag{10a}
$$

$$-(\hat{\psi}_{1i}(\hat{e}_i(t)) - f_i^+\hat{e}_i(t))^T(\hat{\psi}_{1i}(\hat{e}_i(t)) - f_i^-\hat{e}_i(t)) \ge 0, \tag{10b}$$

$$-(\hat{\psi}_{2i}(\hat{e}_i(t-\tau_1)) - g_i^+\hat{e}_i(t-\tau_1))^T(\hat{\psi}_{2i}(\hat{e}_i(t-\tau_1)) - g_i^-\hat{e}_i(t-\tau_1)) \ge 0, \tag{10c}$$

which are, respectively, equivalent to

$$-(\hat{\psi}_{3i}(\hat{e}_i(t)) - h_i^+\hat{e}_i(t))^T(\hat{\psi}_{3i}(\hat{e}_i(t)) - h_i^-\hat{e}_i(t)) \ge 0, \tag{11a}$$

$$\overleftrightarrow{\psi}_1(t)^T \Delta_{f_i} \overleftrightarrow{\psi}_1(t) \ge 0, \tag{11b}$$

$$\overleftrightarrow{\psi}_2(t-\tau_1)^T \Delta_{g_i} \overleftrightarrow{\psi}_2(t-\tau_1) \ge 0, \tag{11c}$$

where $\overleftrightarrow{\psi}_3(t)^T \Delta_{h_i} \overleftrightarrow{\psi}_3(t) \ge 0$, and

$$\overleftrightarrow{\psi}_i(t) := [\hat{e}(t)^T, \hat{\psi}_i(\hat{e}(t))^T]^T$$

$$\Delta_{f_i} := \begin{bmatrix} -f_i^+ f_i^- v_i v_i^T & \frac{f_i^+ + f_i^-}{2} v_i v_i^T \\ * & -v_i v_i^T \end{bmatrix},$$

$$\Delta_{g_i} := \begin{bmatrix} -g_i^+ g_i^- v_i v_i^T & \frac{g_i^+ + g_i^-}{2} v_i v_i^T \\ * & -v_i v_i^T \end{bmatrix},$$

Moreover, from (4)-(5), the following equation holds for any matrices P_2, P_3 with appropriate dimensions:

$$\Delta_{h_i} := \begin{bmatrix} -h_i^+ h_i^- v_i v_i^T & \frac{h_i^+ + h_i^-}{2} v_i v_i^T \\ * & -v_i v_i^T \end{bmatrix}. \tag{12}$$

Using the obtained derivative terms in (9) and adding the left-hand sides of the equations (11) and (12) into, we obtain the following result for

$$2(\hat{e}(t)^T P_2^T + \dot{\hat{e}}(t)^T P_3^T)(-\dot{\hat{e}}(t) - (A + K_1 - \alpha I)\hat{e}(t) - K_2\,\hat{e}(t-\tau_1) + W_1\,\hat{\psi}_1(\hat{e}(t)) + W_2 e^{\alpha\tau_1}\,\hat{\psi}_2(\hat{e}(t-\tau_1))$$
$$-\alpha\, W_3 e^{\alpha\tau_2}\,\hat{e}(t-\tau_2) + W_3 e^{\alpha\tau_2}\,\dot{\hat{e}}(t-\tau_2) - K_3 \int_{t-\tau_3}^{t} \hat{e}(s)\,ds + W_4 \int_{t-\tau_3}^{t} e^{\alpha(t-s)}\hat{\psi}_3(\hat{e}(s))\,ds - E\,\hat{w}(t)) = 0$$

$$\dot{V}(t), \tag{13}$$

where

$$\dot{V}(t) \le \chi^T(t)\Xi\chi(t) + 2\dot{\hat{e}}(t)^T \int_{-\tau_1}^{0} (Q(\xi) - P_3^T K_3)\,\hat{e}(t+\xi)\;d\xi - 2\hat{e}(t-\tau_1)^T \int_{-\tau_1}^{0} R(-\tau_1,\xi)\,\hat{e}(t+\xi)\;d\xi$$
$$-\int_{-\tau_1}^{0}\int_{-\tau_1}^{0} \hat{e}(t+s)^T\,(\frac{\partial}{\partial s}R(s,\xi) + \frac{\partial}{\partial \xi}R(s,\xi))\,\hat{e}(t+\xi)\;ds\,d\xi - \int_{-\tau_1}^{0} \hat{e}(t+\xi)^T\,\dot{S}(\xi)\hat{e}(t+\xi)\;d\xi$$
$$+ 2\,\hat{e}(t)^T \int_{-\tau_1}^{0} (-P_2^T K_3 - \dot{Q}(\xi) + R(0,\xi))\,\hat{e}(t+\xi)\;d\xi - \int_{-\tau_2}^{0} \hat{e}(t+\xi)^T\,\dot{H}(\xi)\hat{e}(t+\xi)\;d\xi$$
$$-\int_{-\tau_2}^{0}\int_{-\tau_2}^{0} \hat{e}(t+s)^T\,(\frac{\partial}{\partial s}T(s,\xi) + \frac{\partial}{\partial \xi}T(s,\xi))\,\hat{e}(t+\xi)\;ds\,d\xi + \dot{\hat{e}}(t)^T\,U\,\dot{\hat{e}}(t) - \dot{\hat{e}}(t-\tau_2)^T\,U\,\dot{\hat{e}}(t-\tau_2)$$
$$+ \hat{\psi}_2(\hat{e}(t))^T Z_1\,\hat{\psi}_2(\hat{e}(t)) - \hat{\psi}_2(\hat{e}(t-\tau_1))^T Z_1\,\hat{\psi}_2(\hat{e}(t-\tau_1)) + \tau_3^2\,\hat{\psi}_3(\hat{e}(t))^T Z_2\hat{\psi}_3(\hat{e}(t))$$
$$+ 2(\hat{e}(t)^T P_2^T + \dot{\hat{e}}(t)^T P_3^T)(W_1\,\hat{\psi}_1(\hat{e}(t)) + W_2 e^{\alpha\tau_1}\,\hat{\psi}_2(\hat{e}(t-\tau_1)) + W_3\,e^{\alpha\tau_2}\,\dot{\hat{e}}(t-\tau_2)$$
$$+ W_4 \int_{t-\tau_3}^{t} e^{\alpha(t-s)}\hat{\psi}_3(\hat{e}(s))\,ds - E\,\hat{w}(t)) - e^{-2\alpha\tau_3}[\int_{t-\tau_3}^{t} e^{\alpha(t-\theta)}\hat{\psi}_3(\hat{e}(\theta))^T\;d\theta]Z_2[\int_{t-\tau_3}^{t} e^{\alpha(t-\theta)}\hat{\psi}_3(\hat{e}(\theta))\;d\theta]$$
$$+ \sum_{i=1}^{n} \sigma_i \overleftrightarrow{\psi}_1(t)^T \Delta_{f_i} \overleftrightarrow{\psi}_1(t) + \sum_{i=1}^{n} \rho_i \overleftrightarrow{\psi}_2(t)^T \Delta_{g_i} \overleftrightarrow{\psi}_2(t) + \sum_{i=1}^{n} \lambda_i \overleftrightarrow{\psi}_3(t)^T \Delta_{h_i} \overleftrightarrow{\psi}_3(t)$$

and

$$\chi(t) := col\{\hat{e}(t), \dot{\hat{e}}(t), \hat{e}(t-\tau_1), \hat{e}(t-\tau_2)\} \tag{14}$$

with $\Xi = \begin{bmatrix} \hat{\Sigma}_{11} & \begin{bmatrix} P_2^T K_2 - Q(-\tau_1) \\ P_3^T K_2 \end{bmatrix} & \begin{bmatrix} -\alpha e^{\alpha\tau_2} P_2^T W_3 \\ -\alpha e^{\alpha\tau_2} P_3^T W_3 \end{bmatrix} \\ * & -S(-\tau_1) & 0 \\ * & * & -H(-\tau_1) \end{bmatrix}$ where

$$\hat{\Sigma}_{11} = sym(P^T \begin{bmatrix} 0 & I \\ -(A+K_1-\alpha I) & -I \end{bmatrix}) + diag\{sym(Q(0)) + S(0) + H(0), 0\}$$

According to the discretization technique in [Fridman, 2006] and [Gu et al., 2003], the delay intervals $[-\tau_1,0]$ and $[-\tau_2,0]$ are, respectively, divided into N segments $[\theta_p,\theta_{p-1}]$ and $P = \begin{bmatrix} P_1 & 0 \\ P_2 & P_3 \end{bmatrix}$., $[\hat{\theta}_p, \hat{\theta}_{p-1}]$ of equal length (or uniform mesh case), i.e. $h_i = \tau_i/N$, $i=1,2$, where $p = 1,\cdots,N$ and $\theta_p = -p\,h_1$ For instance, this scheme divides the square $\hat{\theta}_p = -p\,h_2$. into $N\times N$ small squares $[\theta_p,\theta_{p-1}] \times [\theta_q,\theta_{q-1}]$ and each small square is further divided into two triangles. It is easily seen using [Gu et al., 2003, Lemma 7.7] that although the LKF candidate for the nonuniform mesh case is no more complicated than the uniform mesh case, it is not the case for the LKF derivative condition. Also, a uniform mesh is not possible for the incommensurate delay case and is not practical in the case of commensurate delays with small common factor. In the sequel, $Q(.), S(.), H(.), R(.,.)$ and $T(.,.)$ are chosen to be piecewise linear. i.e. $[-\tau_1,0]\times[-\tau_1,0]$, $Q(\theta_p + \kappa h_1) = (1-\kappa)Q_p + \kappa Q_{p-1}$, $S(\theta_p + \kappa h_1) = (1-\kappa)S_p + \kappa S_{p-1}$ where

$$H(\hat{\theta}_p + \kappa h_2) = (1-\kappa)H_p + \kappa H_{p-1}$$

and

$$R(\theta_p + \kappa h_1, \theta_q + \beta h_1) = \begin{cases} (1-\kappa)R_{pq} + \beta R_{p-1,q-1} + (\kappa-\beta)R_{p-1,q}, & \kappa \geq \beta \\ (1-\beta)R_{pq} + \kappa R_{p-1,q-1} + (\beta-\kappa)R_{p,q-1}, & \kappa < \beta \end{cases}$$

with

$$T(\hat{\theta}_p + \kappa h_2, \hat{\theta}_q + \beta h_2) = \begin{cases} (1-\kappa)T_{pq} + \beta T_{p-1,q-1} + (\kappa-\beta)T_{p-1,q}, & \kappa \geq \beta \\ (1-\beta)T_{pq} + \kappa T_{p-1,q-1} + (\beta-\kappa)T_{p,q-1}, & \kappa < \beta \end{cases}, \dot{S}(\xi) = h_1^{-1}(S_{p-1} - S_p),$$

$\dot{Q}(\xi) = h_1^{-1}(Q_{p-1} - Q_p)$, $\dot{H}(\xi) = h_2^{-1}(H_{p-1} - H_p)$ and $\frac{\partial}{\partial\xi}R(\xi,\theta) + \frac{\partial}{\partial\theta}R(\xi,\theta) = h_1^{-1}(R_{p-1,q-1} - R_{p,q})$

Thus, one obtains

$$\frac{\partial}{\partial \xi} T(\xi,\theta) + \frac{\partial}{\partial \theta} T(\xi,\theta) = h_2^{-1}(T_{p-1,q-1} - T_{p,q}). \tag{15a}$$

$$\begin{aligned} 2\dot{\hat{e}}(t)^T \int_{-\tau_1}^{0} Q(\xi)\, \hat{e}(t+\xi) \; d\xi &= 2\dot{\hat{e}}(t)^T \sum_{p=1}^{N} h_1 \int_0^1 [(1-\kappa)Q_p + \kappa\, Q_{p-1}]\, \hat{e}(t+\theta_p + \kappa\, h_1)\, d\kappa \\ &= 2\dot{\hat{e}}(t)^T \sum_{p=1}^{N} h_1 \int_0^1 [Q_p^s + (1-2\kappa)Q_p^a)]\, \hat{e}(t+\theta_p + \kappa\, h_1)\, d\kappa \end{aligned} \tag{15b}$$

$$2\hat{e}(t-\tau_1)^T \int_{-\tau_1}^{0} R(-\tau_1,\xi)\, \hat{e}(t+\xi) \; d\xi = 2\hat{e}(t-\tau_1)^T \sum_{p=1}^{N} h_1 \int_0^1 [R_{N,p}^s + (1-2\kappa)R_{N,p-1}^a]\, \hat{e}(t+\theta_p + \kappa\, h_1)\, d\kappa. \tag{15c}$$

$$\int_{-\tau_1}^{0} \hat{e}(t+\xi)^T\, \dot{S}(\xi)\, \hat{e}(t+\xi) \; d\xi = \sum_{p=1}^{N} \int_0^1 \hat{e}(t+\theta_p + \kappa\, h_1)^T (S_{p-1} - S_p)\, \hat{e}(t+\theta_p + \kappa\, h_1)\, d\kappa, \tag{15d}$$

$$\int_{-\tau_2}^{0} \hat{e}(t+\xi)^T\, \dot{H}(\xi)\, \hat{e}(t+\xi) \; d\xi = \sum_{p=1}^{N} \int_0^1 \hat{e}(t+\hat{\theta}_p + \kappa\, h_2)^T (H_{p-1} - H_p)\, \hat{e}(t+\hat{\theta}_p + \kappa\, h_2)\, d\kappa, \tag{15e}$$

and

$$\begin{aligned} &\int_{-\tau_1}^{0}\int_{-\tau_1}^{0} \hat{e}(t+s)^T \left(\frac{\partial}{\partial s} R(s,\xi) + \frac{\partial}{\partial \xi} R(s,\xi)\right) \hat{e}(t+\xi) \; ds\; d\xi \\ &= h_1 \sum_{q=1}^{N}\sum_{p=1}^{N} \int_0^1 \hat{e}(t+\theta_p + \beta\, h_1)^T (R_{p-1,q-1} - R_{p,q})\, \hat{e}(t+\theta_p + \kappa\, h_1)\, d\kappa\; d\beta \end{aligned} \tag{15f}$$

Now, from (4), (13)-(15), one has

$$\begin{aligned} &2\hat{e}(t)^T \int_{-\tau_1}^{0} (-\dot{Q}(\xi) + R(0,\xi))\, \hat{e}(t+\xi) \; d\xi \\ &= 2h_1 \hat{e}(t)^T \sum_{p=1}^{N} \int_0^1 (-Q_{p-1} + Q_p + (1-\kappa)R_{0,p} + \kappa\, R_{0,p-1})\, \hat{e}(t+\theta_p + \kappa\, h_1)\, d\kappa \\ &= 2h_1 \hat{e}(t)^T \sum_{p=1}^{N} \int_0^1 (2Q_p^a + R_{0,p}^s + (1-2\kappa)R_{0,p}^a)\, \hat{e}(t+\theta_p + \kappa\, h_1)\, d\kappa \end{aligned} \tag{16}$$

with

$$\begin{aligned}
&\hat{z}_e(t)^T \hat{z}_e(t) - \gamma^2 \hat{w}(t)^T \hat{w}(t) + \dot{V}(t) \\
&\quad \le \chi_e^T(t)\Xi_e \chi_e(t) - \int_0^1 \varphi_e(\kappa;\alpha)^T S_d \, \varphi_e(\kappa;\alpha)\, d\kappa - \int_0^1 \hat{\varphi}_e(\kappa;\alpha)^T H_d \, \hat{\varphi}_e(\kappa;\alpha)\, d\kappa \\
&\quad\quad + 2\chi_e^T(t)\int_0^1 (D^s + (1-2\kappa)D^a)\, \varphi_e(\kappa;\alpha)\, d\kappa - \int_0^1 \varphi_e(\kappa;\alpha)^T \, d\kappa \, R_{ds} \int_0^1 \varphi_e(\kappa;\alpha)\, d\kappa \\
&\quad\quad + 2\chi_e^T(t)\int_0^1 (O^s + (1-2\kappa)O^a)\, \hat{\varphi}_e(\kappa;\alpha)\, d\kappa - \int_0^1 \hat{\varphi}_e(\kappa;\alpha)^T \, d\kappa \, T_{ds} \int_0^1 \hat{\varphi}_e(\kappa;\alpha)\, d\kappa \ , \\
&\quad \le \chi_e^T(t)(\Xi_e + D^s\tilde{U}_1 D^{sT} + O^s\tilde{U}_2 O^{sT} + \frac{1}{3}(D^a \tilde{U}_1 D^{aT} + O^a \tilde{U}_2 O^{aT}))\chi_e(t) \\
&\quad\quad - \int_0^1 \int_0^1 \varphi_e(\kappa;\alpha)^T R_{ds} \, \varphi_e(s;\alpha)] \, d\kappa \, ds - \int_0^1 \int_0^1 \hat{\varphi}_e(\kappa;\alpha)^T T_{ds} \, \hat{\varphi}_e(s;\alpha)\, d\kappa \, ds \\
&\quad\quad - \int_0^1 \varphi_D(\kappa;\alpha)^T \Theta_1 \, \varphi_D(\kappa;\alpha)\, d\kappa - \int_0^1 \varphi_O(\kappa;\alpha)^T \Theta_2 \, \varphi_O(\kappa;\alpha)\, d\kappa
\end{aligned}$$

$$\chi_e(t) = col\{\chi(t), \dot{\hat{e}}(t-\tau_2), \hat{\psi}_1(\hat{e}(t)), \hat{\psi}_2(\hat{e}(t)), \hat{\psi}_3(\hat{e}(t)), \int_{t-\tau_3}^{t} e^{\alpha(t-\theta)}\hat{\psi}_3(\hat{e}(\theta)) \, d\theta \, , \hat{\psi}_2(\hat{e}(t-\tau_1)), \hat{w}(t)\},$$

$$\chi_D(t) := (D^s + (1-2\kappa)D^a)^T \chi_e(t), \ \chi_O(t) := (O^s + (1-2\kappa)O^a)^T \chi_e(t),$$

$$\varphi_D(\kappa;\alpha) := [\chi_D(t)^T, \varphi_e(\kappa;\alpha)^T]^T, \ \varphi_O(\kappa;\alpha) := [\chi_O(t)^T, \hat{\varphi}_e(\kappa;\alpha)^T]^T,$$

$\varphi_e(\kappa;\alpha) = col\{\hat{e}(t+\theta_1+\kappa h_1), \hat{e}(t+\theta_2+\kappa h_1), \cdots, \hat{e}(t+\theta_N+\kappa h_1)\}$ and

$$\hat{\varphi}_e(\kappa;\alpha) = col\{\hat{e}(t+\hat{\theta}_1+\kappa h_1), \hat{e}(t+\hat{\theta}_2+\kappa h_1), \cdots, \hat{e}(t+\hat{\theta}_N+\kappa h_1)\},$$

$$\Theta_1 := \begin{bmatrix} \tilde{U}_1 & -I \\ * & S_d \end{bmatrix},$$

$$\Theta_2 := \begin{bmatrix} \tilde{U}_2 & -I \\ * & H_d \end{bmatrix},$$

$$\Xi_e = \begin{bmatrix} \tilde{\Sigma}_{11} & \begin{bmatrix} -P_2^T K_2 - Q_N + C_1^T C_2 \\ -P_3^T K_2 \end{bmatrix} & \begin{bmatrix} -\alpha \, P_2^T W_3 e^{\alpha\tau_2} \\ -\alpha \, P_3^T W_3 e^{\alpha\tau_2} \end{bmatrix} & \begin{bmatrix} P_2^T W_3 e^{\alpha\tau_2} \\ P_3^T W_3 e^{\alpha\tau_2} \end{bmatrix} & \tilde{\Sigma}_{15} \\ * & -S_N + C_2^T C_2 & 0 & 0 & 0 \\ * & * & -H_N & 0 & 0 \\ * & * & * & -U & 0 \\ * & * & * & * & \Sigma_{55} \end{bmatrix},$$

$$\tilde{\Sigma}_{11} = \hat{\Sigma}_{11} + diag\{-F^{+}\Lambda_1 F^{-} - G^{+}\Lambda_2 G^{-} - H^{+}\Lambda_3 H^{-}, U\},$$

$$\tilde{\Sigma}_{15} = \begin{bmatrix} \begin{bmatrix} P_2^T W_1 + \frac{1}{2}(F^{+} + F^{-})\Lambda_1 \\ P_3^T W_1 \\ 0 \end{bmatrix} & \begin{bmatrix} \frac{1}{2}(G^{+} + G^{-})\Lambda_2 \\ 0 \\ 0 \end{bmatrix} & \begin{bmatrix} \frac{1}{2}(H^{+} + H^{-})\Lambda_3 \\ 0 \\ 0 \end{bmatrix} & \begin{bmatrix} P_2^T W_4 + C_1^T C_3 \\ P_3^T W_4 \\ 0 \end{bmatrix} & \begin{bmatrix} P_2^T W_2 e^{\alpha\tau_1} \\ P_3^T W_2 e^{\alpha\tau_1} \\ 0 \end{bmatrix} & \begin{bmatrix} -P_2^T E \\ -P_3^T E \\ 0 \end{bmatrix} \end{bmatrix},$$

Let $D^s = h_1 \begin{bmatrix} 2Q_1^a + R_{0,1}^s - P_2^T K_3 & 2Q_2^a + R_{0,2}^s - P_2^T K_3 & \cdots & 2Q_N^a + R_{0,N}^s - P_2^T K_3 \\ Q_1^s - P_3^T K_3 & Q_2^s - P_3^T K_3 & \cdots & Q_N^s - P_3^T K_3 \\ -2R_{N,1}^s & -2R_{N,2}^s & \cdots & -2R_{N,N}^s \end{bmatrix}$. with

$\zeta_i = diag\{\tilde{U}_i, I\}$ Premultiplying $\tilde{U}_i = \bar{U}_i^{-1}$. and postmultiplying ζ_i to the LMIs (6b) and (6c), one obtains ζ_i^T Applying [24, Prop. 5.21] to (16) we conclude that

$$\Theta_i > 0, \; i = 1, 2. \tag{17}$$

where $\hat{z}_e(t)^T \hat{z}_e(t) - \gamma^2 \hat{w}(t)^T \hat{w}(t) + \dot{V}(t) \le \tilde{\chi}_e(t)^T \; \tilde{\Xi}_e \; \tilde{\chi}_e(t)$ and

$$\tilde{\chi}_e(t) = [\chi_e(t)^T, \int_0^1 \varphi_e(\kappa;\alpha)^T \, d\kappa, \int_0^1 \hat{\varphi}_e(\kappa;\alpha)^T \, d\kappa]^T$$

On the other hand, for a prescribed $\gamma > 0$ and under zero initial conditions, J_∞ can be rewritten as

$$\tilde{\Xi}_e = \begin{bmatrix} \Xi_e + \frac{1}{3}(D^a \tilde{U}_1 D^{aT} + O^a \tilde{U}_2 O^{aT}) & -D^s & -O^s \\ * & -S_d - R_{ds} & 0 \\ * & * & -H_d - T_{ds} \end{bmatrix}. \tag{18}$$

and the condition

$$J_\infty \le \int_0^\infty [\hat{z}_e(t)^T \hat{z}_e(t) - \gamma^2 \hat{w}(t)^T \hat{w}(t)] \; dt + V(t)\Big|_{t\to\infty} - V(t)\Big|_{t=0}$$
$$= \int_0^\infty [\hat{z}_e(t)^T \hat{z}_e(t) - \gamma^2 \hat{w}(t)^T \hat{w}(t) + \dot{V}(t)] \; dt$$

(or H_∞<0) means that the condition $\hat{z}_e(t)^T \hat{z}_e(t) - \gamma^2 \hat{w}(t)^T \hat{w}(t) + \dot{V}(t) < 0$ satisfies the H_∞ performance measure, and by applying Schur complement, one gets

$$\tilde{\Xi}_e < 0 \tag{19}$$

Then, we choose $P_3=\delta P_2$, $\delta \in R$, where δ is a tuning scalar parameter (which may be restrictive). Note that the matrix P_2 is non-singular due to the fact that the only matrix which can be negative definite in the second block on the diagonal of (19) is $\begin{bmatrix} \Xi_e & -D^s & D^a & O^a & -O^s \\ * & -S_d - R_{ds} & 0 & 0 & 0 \\ * & * & -3\bar{U}_1 & 0 & 0 \\ * & * & * & -3\bar{U}_2 & 0 \\ * & * & * & * & -H_d - T_{ds} \end{bmatrix} < 0$. Therefore, considering $-\delta\, sym(P_2) + U_{.}$, $i=1, 2, 3$ results in the LMI (6d).

Moreover, the condition $J_\infty<0$ for $w(t)\equiv 0$ implies $L_i = P_2^T K_i$ Then, we have $V(t)<V(0)$. From (7)-(8), one gets

$$\dot{V}(t) < 0. \tag{20}$$

Moreover, it is clear that

$$\begin{aligned}
V(0) &= e(0)^T P_1 e(0) + 2e(0)^T \int_{-\tau_1}^{0} Q(\xi)\, e(\xi)\ d\xi + \int_{-\tau_1}^{0} \hat{\psi}_2(e(s))^T Z_1 \hat{\psi}_2(e(s))\ ds + \int_{-\tau_2}^{0} \dot{e}(s)^T U \dot{e}(s)\ ds \\
&+ \int_{-\tau_1}^{0} e(\xi)^T S(\xi) e(\xi)\ d\xi + \int_{-\tau_1}^{0}\int_{-\tau_1}^{0} e(s)^T R(s,\xi)\, e(\xi)\ ds\, d\xi + \int_{-\tau_2}^{0} e(\xi)^T H(\xi) e(\xi)\ d\xi \\
&+ \int_{-\tau_2}^{0}\int_{-\tau_2}^{0} e(s)^T T(s,\xi)\, e(\xi)\ ds\, d\xi + \int_{-\tau_3}^{0} [\int_{s}^{0} \hat{\psi}_3(e(\theta))^T\ d\theta] Z_2 [\int_{s}^{0} \hat{\psi}_3(e(\theta))^T\ d\theta]\ ds \\
&+ \int_{0}^{\tau_3}\int_{-s}^{0} (\theta + s)\, \hat{\psi}_3(e(\theta))^T Z_2\, \hat{\psi}_3(e(\theta))\ d\theta\, ds \\
&\le \lambda_{\max}(P_1)\left\|\varsigma\right\|^2 + \tau_1 \lambda_{\max}(Q_p)_{p=0}^N \left\|\varsigma\right\|^2 + \tau_1 \lambda_{\max}(G^{+T} Z_1 G^{+})\left\|\varsigma\right\|^2 + \tau_2 \lambda_{\max}(U)\left\|\dot{\varsigma}\right\|^2 + \tau_1 \lambda_{\max}(S_p)_{p=0}^N \left\|\varsigma\right\|^2 \\
&+ \tau_1^2 \lambda_{\max}(R_{p,p})_{p=0}^N \left\|\varsigma\right\|^2 + \tau_2 \lambda_{\max}(H_p)_{p=0}^N \left\|\varsigma\right\|^2 + \tau_2^2 \lambda_{\max}(T_{p,p})_{p=0}^N \left\|\varsigma\right\|^2 + \frac{2}{3}\tau_3^3 \lambda_{\max}(H^{+T} Z_2 H^{+})\left\|\varsigma\right\|^2 \\
&= \Delta_1 \left\|\varsigma\right\|^2 + \Delta_2 \left\|\dot{\varsigma}\right\|^2
\end{aligned}$$

Therefore, we have

$$V(t) \ge e^{2\alpha t} e(t)^T P_1 e(t) \ge e^{2\alpha t} \lambda_{\min}(P_1)\left|e(t)\right|^2.$$

That is,

$$|e(t)|^2 \leq \frac{\max\{\Delta_1, \Delta_2\}}{\lambda_{\min}(P_1)} e^{-2\alpha t}[\|\varsigma\|^2 + \|\dot{\varsigma}\|^2].$$

which shows that the synchronization error network (4) with (5) is globally exponentially stable and has the exponential decay rate α. This completes the proof.

UNCERTAINTY CHARACTERIZATION

In this section, we will discuss the uncertainty characterization for the MSNNs (1)-(2) with different neutral, discrete and distributed delays.

Polytopic Uncertainty

The first class of uncertainty frequently encountered in practice is the polytopic uncertainty [Gu et al., 2003]. In this case, the matrices of the MSNNs (1)-(2) are not exactly known, except that they are within a compact set Ω denoting

$$|e(t)| \leq \sqrt{\frac{\max\{\Delta_1, \Delta_2\}}{\lambda_{\min}(P_1)}} e^{-\alpha t}[\|\varsigma\| + \|\dot{\varsigma}\|],$$

we assume that

$$\Omega = \begin{bmatrix} A & W_1 & W_2 & W_3 & W_4 & E \end{bmatrix} \tag{21}$$

for some scalars s_j satisfying

$$\Omega = \sum_{j=1}^{q} s_j \Omega_j \tag{22}$$

where the q vertices of the polytope are described by

$$0 \leq s_j \leq 1, \quad \sum_{j=1}^{q} s_j = 1 \tag{23}$$

In order to take into account the polytopic uncertainty in the exponential H_∞ synchronization problem of the MSNNs (1)-(2), we derive the following result from applying the same transformation that was used in deriving Theorem 1

Theorem 2. Let $h_i = \tau_i / N$, i=1,2, be given for any positive integer N. Under Assumptions 1 and 2, if the uncertainty set Ω is polytopic with vertices Ω_j, j = 1, 2,…, q, then the MSNNs described by (1)-(2)

and (21)-(23) are globally exponentially stable with the H_∞ performance level $\gamma > 0$ and an exponential decay rate $\alpha > 0$, if there exist some scalars $\Omega_j = \begin{bmatrix} A^{(j)} & W_1^{(j)} & W_2^{(j)} & W_3^{(j)} & W_4^{(j)} & E^{(j)} \end{bmatrix}$. ($i = 1,2,\ldots,$ N) matrices δ, $\sigma_i, \rho_i, \lambda_i$ P_2, L_1, L_2, L_3, Q_i, S_i, H_i, $R_{i,j} = R_{i,j}^T$, $T_{i,j} = T_{i,j}^T$ and positive-definite matrices $(i, j = 0, 1, \cdots, N)$ such that LMIs (6) are satisfied for all

$$P_1, Z_1, Z_2, \bar{U}_1, \bar{U}_2 \tag{24}$$

Then, the controller gains in (5) are given by

$\begin{bmatrix} A & W_1 & W_2 & W_3 & W_4 & E \end{bmatrix} = \begin{bmatrix} A^{(j)} & W_1^{(j)} & W_2^{(j)} & W_3^{(j)} & W_4^{(j)} & E^{(j)} \end{bmatrix}, \quad j = 1, 2, \cdots, q.$

$K_i = (P_2^T)^{-1} L_i$

Proof. It follows directly from the proof of Theorem 1 and using properties of (21)-(23).

Norm Bounded Uncertainty

There are also other uncertainties that cannot be reasonably modelled by a polytopic uncertainty set with a number of vertices. In such a case, it is assumed that the deviation of the system parameters of an uncertain system from their nominal values is norm bounded [Gu et al., 2003]. This kind of uncertainties often appear in modeled NNs mainly due to modeling error, external disturbance, and parameter fluctuation during the implementation; and such deviations and perturbations are usually bounded. To reflect such a reality, consider the MSNNs (1)-(2) with

$$(i = 1, 2, 3). \tag{25}$$

where the time-varying structured uncertainties $\Delta A(t)$, $\Delta W_i(t)$ and $\Delta E(t)$ are said to be admissible if the following form holds

$$A + \Delta A(t), \; W_i + \Delta W_i(t), \; E + \Delta E(t), \tag{26}$$

Where $[\Delta A(t) \quad \Delta W_i(t) \quad \Delta E(t)] = M_1 \Delta(t) [L_a \quad L_{w_i} \quad L_e]$ are constant matrices with appropriate dimensions; and $\Delta(t)$ is an unknown, real, and possibly time-varying matrix with Lebesgue measurable elements, and its Euclidean norm satisfies

$$L_a, L_{w_i}, L_e \tag{27}$$

In this section, we modify Assumption 1 in order to enable the application of Lyapunov's method for the stability of the uncertain MSNNs (1)-(2) with (25)-(27):

Assumption 3. Let the difference operator $\|\Delta(t)\| \leq 1, \quad \forall t.$ be delay-independently stable with respect to all delays and a sufficient condition is that all the eigenvalues of the matrix $W_3+\Delta W_3(t)$ lie inside the unit circle.

Theorem 3. Let $h_i=\tau_i/N$, i=1,2, be given for any positive integer N. Under Assumptions 2 and 3, the MSNNs described by (1)-(2) and admissible uncertainties (25)-(27) are globally exponentially stable with the H_∞ performance level γ >0 and an exponential decay rate α >0, if there exist some scalars $\nabla x_t = x(t) - (W_3 + \Delta W_3(t))\, x(t-\tau_2)$ matrices $\mu > 0, \delta, \ \sigma_i, \rho_i, \lambda_i (i = 1, 2, \cdots, N)$, $P_2, L_1, L_2, L_3, Q_i, S_i, H_i, R_{i,j} = R_{i,j}^T, T_{i,j} = T_{i,j}^T$ and positive-definite matrices $(i, j = 0, 1, \cdots, N)$ such that LMIs (6a)-(6c) and the following LMI are feasible

$$P_1, Z_1, Z_2, \bar{U}_1, \bar{U}_2 \tag{28}$$

where

$$\begin{bmatrix} \Pi & \Gamma_d & \mu \Gamma_e \\ * & -\mu I & 0 \\ * & * & -\mu I \end{bmatrix} < 0$$

$$\Gamma_d = [M_1^T P_2 \quad \delta\, M_1^T P_2 \quad \underbrace{0 \quad \cdots \quad 0}_{(4N+9)\, elements}]^T,$$

Then, the controller gains in (5) are given by

$$\Gamma_e = [-L_a \quad 0 \quad 0 \quad -\alpha\, e^{\alpha \tau_2} L_{w_3} \quad e^{\alpha \tau_2} L_{w_3} \quad L_{w_1} \quad 0 \quad 0 \quad L_{w_4} \quad L_{w_2} \quad L_e \quad \underbrace{0 \quad \cdots \quad 0}_{(4N)\ elements}].$$

Proof. If the state-space matrices A, $K_i = (P_2^T)^{-1} L_i (i = 1, 2, 3)$. and E in (6d) are replaced with $W_1, \cdots, W_4$, $A + M_1\, \Delta(t)\, L_a$ and $E+M_1\Delta(t)L_e$, respectively, then the inequality (6d) is equivalent to the following condition:

$$W_1 + M_1\, \Delta(t)\, L_{w_1}, \cdots, W_4 + M_1\, \Delta(t)\, L_{w_4} \tag{29}$$

By Lemma 1 (in Appendix), a necessary and sufficient condition for (29) is that there exists a scalar μ >0 such that

$$\Pi + sym(\Gamma_d^T \Delta(t)\ \Gamma_e) < 0. \tag{30}$$

then, applying Schur complements, we find that (30) is equivalent to (28).

Remark 4. The reduced conservatism of Theorems 1-3 benefit from the construction of the new DLKF in (8), introducing some free weighting matrices to express the relationship among the system matrices, utilizing a general form of the activation functions and neither the model transformation approach nor any bounding technique are needed to estimate the inner product of the involved crossing terms (see for instance [Cheng et al., 2006]). It can be easily seen that results of this paper is quite different from most existing results in the literature in the following perspective: theoretically exponential synchronization problem of MSNNs with mixed neutral, discrete and distributed delays is much more complicated, especially, for the case where the delays are different. In this paper, the derived sufficient conditions are convex and neutral-delay-dependent, discrete-delay-dependent and distributed-delay-dependent, which make the treatment in the present paper more general with less conservative in compare to most existing results in the literature which are independent of the neutral or distributed delays, see for instance the references [Cheng et al., 2006], [Wang & Cao, 2009], [Cao et al, 2008].

NUMERICAL RESULTS

Let us consider the MSNNs (1)-(2) with the following matrices:

$$\Pi + \mu^{-1}\Gamma_d^T\Gamma_d + \mu\Gamma_e^T\Gamma_e < 0,$$

where

$$A = I_2, W_1 = \begin{bmatrix} 1+\pi/4 & 20 \\ 0.1 & 1+\pi/4 \end{bmatrix}, W_2 = \begin{bmatrix} -1.3\sqrt{2}\,\pi/4 & 0.1 \\ 0.1 & -1.3\sqrt{2}\,\pi/4 \end{bmatrix}, W_3 = \begin{bmatrix} c & 0 \\ 0 & c \end{bmatrix},$$

$$W_4 = \begin{bmatrix} 2+\pi/2 & 40 \\ 0.2 & 2+\pi/2 \end{bmatrix}, C_1 = C_2 = C_3 = 1, E = \begin{bmatrix} 1 & 1 \end{bmatrix}^T, \tau_1 = 1, \tau_2 = 0.5, \tau_3 = 0.2,$$

Figure 1. x_1–x_2 plot

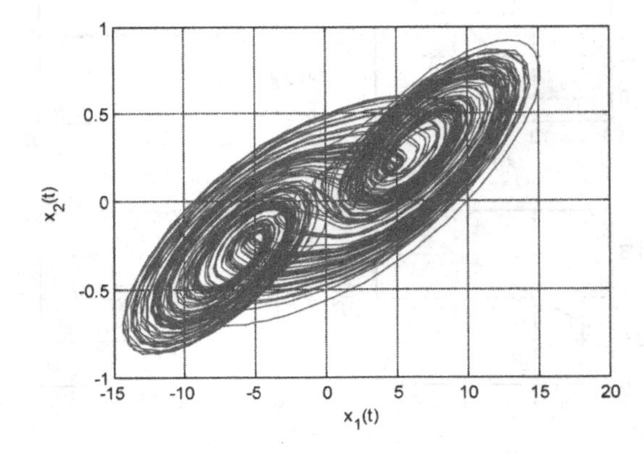

Table 1. Controller gains (with α =0.5) w.r.t. N

	N=1	*N*=2	*N*=3
K_1	$\forall s \in [-1, 0]$	$\begin{bmatrix} 294.4144 & 7.7231 \\ 31.9461 & 275.1222 \end{bmatrix}$	$\begin{bmatrix} 265.7394 & -27.0371 \\ -2.8888 & 252.2224 \end{bmatrix}$
K_2	$\begin{bmatrix} 264.8855 & -30.0154 \\ -8.4669 & 254.5174 \end{bmatrix}$	$\begin{bmatrix} -0.2117 & 0.2374 \\ 0.2341 & -0.2781 \end{bmatrix}$	$\begin{bmatrix} 0.3203 & -0.3582 \\ -0.3325 & 0.3653 \end{bmatrix}$
K_3	$\begin{bmatrix} -0.1742 & 0.1930 \\ 0.1850 & -0.2056 \end{bmatrix}$	$\begin{bmatrix} -1.0419 & 1.2039 \\ 1.1214 & -1.3026 \end{bmatrix}$	$\begin{bmatrix} -0.4198 & 0.4640 \\ 0.4237 & -0.4711 \end{bmatrix}$

Figure 2. Disturbance signal

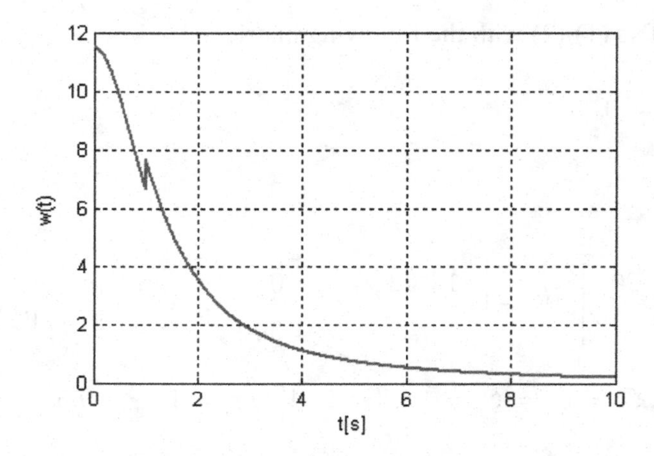

Figure 3. The phase trajectories of the MSNNs and the synchronization errors

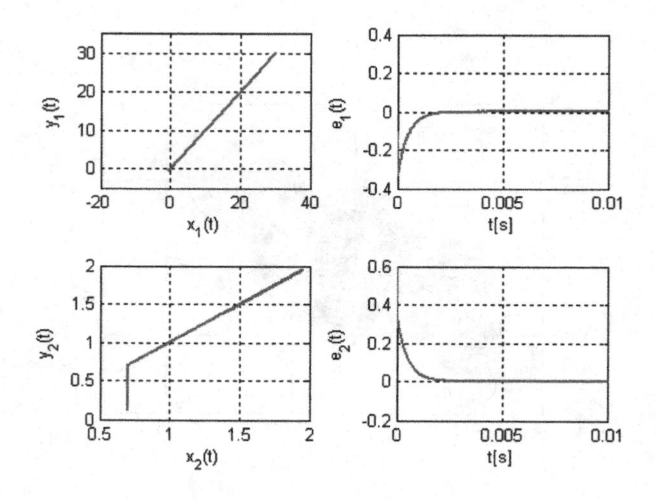

and

$0 \leq c \leq 1$

are monotonically increasing activation functions which is monotone increasing and globally Lipschitz continuous.

In the case of $f(x_i(t)) = g(x_i(t)) = h(x_i(t)) = 0.5\left(\left|x_i(t)+1\right| - \left|x_i(t)-1\right|\right)$ the chaotic behaviour of NN (1) with the parameters above has been investigated in [Cheng et al., 2006], [Karimi & Maass, 2009], [Cao et al., 2008]. The chaotic trajectory of the model, i.e. x_1–x_2 plane, for delay τ_1=0.95 with the initial condition $W_3 = W_4 = 0$, for $x(s) = [-0.1, 0.3]^T$ is plotted in Figure 1.

Assume that c =0.1. It is required to design a driving signal $\begin{bmatrix} -0.5580 & 0.6191 \\ 0.5751 & -0.6379 \end{bmatrix}$ of the form (5) such that the MSNNs (1)-(2) with the parameters above are exponentially synchronized with the H_∞ performance measure. To this end, in light of Theorem 1, LMIs (6) are solved using Matlab LMI Control Toolbox [Gahinet et al., 1995] for different values of the parameter N, i.e. $N \in \{1,2,3\}$ and the control gains $K_i(i=1,2,3)$ for α =0.5 are calculated and illustrated in Table 1.

For simulation purpose, an exogenous disturbance input, shown in Figure 2, is imposed on the system. Now, by considering N=3, initial conditions $\hat{u}(t)$ and applying the delayed feedback control (5) with the parameters available in Table 1, the phase trajectories of the network and the synchronization errors between the MSNNs are shown in Figure 3. It shows that the synchronization errors converge exponentially to zero. The simulation results imply that the MSNNs under consideration are globally exponentially synchronized.

CONCLUSION

This paper presented the exponential H_∞ synchronization problem for uncertain master and slave neural networks (MSNNs) with mixed time-delays, where the mixed delays comprise different neutral, discrete and distributed time-delays. Both the polytopic and the norm-bounded uncertainty cases were studied separately. An appropriate discretized Lyapunov-Krasovskii functional and some free weighting matrices were utilized to establish some delay-dependent sufficient conditions for designing a delayed state-feedback control as a synchronization law by convex optimization over LMIs under less restrictive conditions. It was shown that the synchronization law guaranteed the exponential H_∞ synchronization of the two coupled MSNNs regardless of their initial states. Detailed comparisons with existing results were made and numerical simulations were carried out to demonstrate the effectiveness of the established synchronization laws.

REFERENCES

Cai, C., & Chen, G. (2006). Synchronization of complex dynamical networks by the incremental ISS approach. *Physica A*, *371*, 754–766. doi:10.1016/j.physa.2006.03.052

Cao, J., Chen, G., & Li, P. (2008). Global synchronization in an array of delayed neural networks with hybrid coupling. *IEEE Trans. Systems, Man, and Cybernetics-Part B . Cybernetics*, *38*(2), 488–498.

Cao, J., & Liang, J. L. (2004). Boundedness and stability for Cohen-Grossberg neural network with time-varying delays. *Journal of Mathematical Analysis and Applications*, *296*(2), 51–65. doi:10.1016/j.jmaa.2004.04.039

Cao, J., & Wang, J. (2005). Global asymptotic and robust stability of recurrent neural networks with time delays. *IEEE Transactions on Circuits and Systems. I, Fundamental Theory and Applications*, *52*(2), 417–426. doi:10.1109/TCSI.2004.841574

Cao, J., Wang, Z., & Sun, Y. (2007). Synchronization in an array of linearly stochastically coupled networks with time delays. *Physica A*, *385*(2), 718–728. doi:10.1016/j.physa.2007.06.043

Cao, J., Yuan, K., & Li, H.-X. (2006). Global Asymptotical stability of generalized recurrent neural networks with multiple discrete delays and distributed delays. *IEEE Transactions on Neural Networks*, *17*(6), 1646–1651. doi:10.1109/TNN.2006.881488

Chen, G., & Dong, X. (1993). On feedback control of chaotic continuous-time systems. *IEEE Transactions on Circuits and Systems*, *40*, 591–601. doi:10.1109/81.244908

Cheng, C. J., Liao, T. L., Yan, J. J., & Hwang, C. C. (2006). Exponential synchronization of a class of neural networks with time-varying delays. *IEEE Trans. Sys., Man, and Cyber. Part B: Cybernetics*, *36*(1), 209–215. doi:10.1109/TSMCB.2005.856144

Feki, M. (2003). An adaptive chaos synchronization scheme applied to secure communication. *Chaos, Solitons, and Fractals*, *18*, 141–148. doi:10.1016/S0960-0779(02)00585-4

Fradkov, A. L., & Markov, A. Y. (1997). Adaptive synchronization of chaotic systems based on speed gradient method and passification. *IEEE Transactions on Circuits and Systems. I, Fundamental Theory and Applications*, *44*(10), 905–912. doi:10.1109/81.633879

Fradkov, A. L., Nijmeijer, H., & Markov, A. (2000). Adaptive observer-based synchronization for communications. *International Journal of Bifurcation and Chaos in Applied Sciences and Engineering*, *10*(12), 2807–2813. doi:10.1142/S0218127400001869

Fradkov, A. L., & Pogromsky, A. Y. (1996). Speed gradient control of chaotic continuous-time systems. *IEEE Transactions on Circuits and Systems. I, Fundamental Theory and Applications*, *43*(11), 907–913. doi:10.1109/81.542281

Fridman, E. (2006). Descriptor discretized Lyapunov functional method: Analysis and design. *IEEE Transactions on Automatic Control*, *51*(5), 890–897. doi:10.1109/TAC.2006.872828

Gahinet, P., Nemirovsky, A., Laub, A. J., & Chilali, M. (1995). *LMI control Toolbox: For use with Matlab*. Natick, MA: The MATH Works, Inc.

Gao, H., Chen, T., & Lam, J. (2008). A new delay system approach to network-based control. *Automatica, 44*(1), 39–52. doi:10.1016/j.automatica.2007.04.020

Gao, H., Lam, J., & Chen, G. (2006). New criteria for synchronization stability of general complex dynamical networks with coupling delays. *Physics Letters. [Part A], 360*(2), 263–273. doi:10.1016/j.physleta.2006.08.033

García-Valdovinos, L.-G., Parra-Vega, V., & Arteaga, M. A. (2007). Observer-based sliding mode impedance control of bilateral teleoperation under constant unknown time delay. *Robotics and Autonomous Systems, 55*(8), 609–617. doi:10.1016/j.robot.2007.05.011

Gu, K., Kharitonov, V. L., & Chen, J. (2003). *Stability of time-delay systems*. Boston: Birkhauser.

Hale, J., & Verduyn Lunel, S. M. (1993). *Introduction to functional differential equations*. New York: Springer Verlag.

He, Y., Wang, Q., Wu, M., & Lin, C. (2006). Delay-dependent state estimation for delayed neural networks. *IEEE Transactions on Neural Networks, 17*, 1077–1081. doi:10.1109/TNN.2006.875969

He, Y., Wang, Q., & Zheng, W. (2005). Global robust stability for delayed neural networks with polytopic type uncertainties. *Chaos, Solitons, and Fractals, 26*, 1349–1354. doi:10.1016/j.chaos.2005.04.005

Ho, D. W. C., Liang, J. L., & Lam, J. (2006). Global exponential stability of impulsive high-order BAM neural networks with time-varying delays. *Neural Networks, 19*, 1581–1590. doi:10.1016/j.neunet.2006.02.006

Hou, Y. Y., Liao, T. L., & Yan, J. J. (2007). H_∞ synchronization of chaotic systems using output feedback control design. *Physica A, 379*, 81–89. doi:10.1016/j.physa.2006.12.033

Huang, H., Feng, G., & Cao, J. (2008). Exponential synchronization of chaotic Luré systems with delayed feedback control. *Nonlinear Dynamics*. doi:.doi:10.1007/s11071-008-9454-z

Karimi, H. R., & Maass, P. (2009). Delay-range-dependent exponential H_∞ synchronization of a class of delayed neural networks. *Chaos, Solitons, and Fractals, 41*(3), 1125–1135. doi:10.1016/j.chaos.2008.04.051

Khargonekar, P. P., Petersen, I. R., & Zhou, K. (1990). Robust stabilization of uncertain linear systems: Quadratic stabilizability and H_∞ control Theory. *IEEE Transactions on Automatic Control, 35*, 356–361. doi:10.1109/9.50357

Liang, J., & Cao, J. (2004). Global asymptotic stability of bi-directional associative memory networks with distributed delays. *Applied Mathematics and Computation, 152*, 415–424. doi:10.1016/S0096-3003(03)00567-8

Liang, J., & Cao, J. (2006). A based-on LMI stability criterion for delayed recurrent neural networks. *Chaos, Solitons, and Fractals, 28*(1), 154–160. doi:10.1016/j.chaos.2005.04.120

Liang, J., & Cao, J. (2007). Global output convergence of recurrent neural networks with distributed delays. *Nonlinear Analysis Real World Applications, 8*(1), 187–197. doi:10.1016/j.nonrwa.2005.06.009

Liang, J., Wang, Z., & Liu, X. (2008). Exponential synchronization of stochastic delayed discrete-time complex networks. *Nonlinear Dynamics, 53*, 153–165. doi:10.1007/s11071-007-9303-5

Liao, T. L., & Tsai, S. H. (2008). Adaptive synchronization of chaotic systems and its application to secure communication. *Chaos, Solitons, and Fractals*, *11*(9), 1387–1396. doi:10.1016/S0960-0779(99)00051-X

Liu, Y., Wang, Z., & Liu, X. (2007). Design of exponential state estimators for neural networks with mixed time delays. *Physics Letters. [Part A]*, *364*(5), 401–412. doi:10.1016/j.physleta.2006.12.018

Lou, X. Y., & Cui, B. (2006). New LMI conditions for delay-dependent asymptotic stability of delayed Hopfield neural networks. *Neurocomputing*, *69*(16–18), 2374–2378. doi:10.1016/j.neucom.2006.02.019

Lu, H., & van Leeuwen, C. (2006). Synchronization of chaotic neural networks via output or state coupling. *Chaos, Solitons, and Fractals*, *30*, 166–176. doi:10.1016/j.chaos.2005.08.175

Mou, S., Gao, H., Lam, J., & Qiang, W. (2008a). A new criterion of delay-dependent asymptotic stability for Hopfield neural networks with time delay. *IEEE Transactions on Neural Networks*, *19*(3), 532–535. doi:10.1109/TNN.2007.912593

Mou, S., Gao, H., Qiang, W., & Chen, K. (2008b). New delay-dependent exponential stability for neural networks with time delay. *IEEE Trans. Systems, Man and Cybernetics-Part B . Cybernetics*, *38*(2), 571–576.

Park, J. H. (2006). Synchronization of Genesio chaotic system via backstepping approach. *Chaos, Solitons, and Fractals*, *27*, 1369–1375. doi:10.1016/j.chaos.2005.05.001

Park, P. (1999). A delay-dependent stability criterion for systems with uncertain time-invariant delays. *IEEE Transactions on Automatic Control*, *44*, 876–877. doi:10.1109/9.754838

Pecora, L. M., & Carroll, T. L. (1990). Synchronization in chaotic systems. *Physical Review Letters*, *64*, 821–824. doi:10.1103/PhysRevLett.64.821

Song, Q., & Wang, Z. (2008). Stability analysis of impulsive stochastic Cohen-Grossberg neural networks with mixed time delays. *Physica A: Statistical Mechanics and its Applications, 387*(13), 3314-3326.

Sun, Y., Cao, J., & Wang, Z. (2007). Exponential synchronization of stochastic perturbed chaotic delayed neural networks. *Neurocomputing*, *70*(13), 2477–2485. doi:10.1016/j.neucom.2006.09.006

Wang, L., & Cao, J. (2009). Global robust point dissipativity of interval neural networks with mixed time-varying delays. *Nonlinear Dynamics*, *55*(1-2), 169–178. doi:10.1007/s11071-008-9352-4

Wang, Y., Wang, Z., & Liang, J. (2008). A delay fractioning approach to global synchronization of delayed complex networks with stochastic disturbances. *Physics Letters. [Part A]*, *372*(39), 6066–6073. doi:10.1016/j.physleta.2008.08.008

Wang, Y. W., Wen, C., Soh, Y. C., & Xiao, J. W. (2006a). Adaptive control and synchronization for a class of nonlinear chaotic systems using partial system states. *Physics Letters. [Part A]*, *351*(1-2), 79–84. doi:10.1016/j.physleta.2005.10.055

Wang, Z., Ho, D. W. C., & Liu, X. (2005b). State estimation for delayed neural networks. *IEEE Transactions on Neural Networks*, *16*(1), 279–284. doi:10.1109/TNN.2004.841813

Wang, Z., Lauria, S., Fang, J., & Liu, X. (2007). Exponential stability of uncertain stochastic neural networks with mixed time-delays. *Chaos, Solitons, and Fractals*, *32*, 62–72. doi:10.1016/j.chaos.2005.10.061

Wang, Z., Liu, Y., Fraser, K., & Liu, X. (2006c). Stochastic stability of uncertain Hopfield neural networks with discrete and distributed delays. *Physics Letters. [Part A]*, *354*(4), 288–297. doi:10.1016/j.physleta.2006.01.061

Wang, Z., Liu, Y., & Liu, X. (2005a). On global asymptotic stability of neural networks with discrete and distributed delays. *Physics Letters. [Part A]*, *345*, 299–308. doi:10.1016/j.physleta.2005.07.025

Wang, Z., Liu, Y., Yu, L., & Liu, X. (2006d). Exponential stability of delayed recurrent neural networks with Markovian jumping parameters. *Physics Letters. [Part A]*, *356*(4), 346–352. doi:10.1016/j.physleta.2006.03.078

Wang, Z., Shu, H., Liu, Y., Ho, D. W. C., & Liu, X. (2006b). Robust stability analysis of generalized neural networks with discrete and distributed time delays. *Chaos, Solitons, and Fractals*, *30*(4), 886–896. doi:10.1016/j.chaos.2005.08.166

Wen, G., Wang, Q. G., Lin, C., Han, X., & Li, G. (2006). Synthesis for robust synchronization of chaotic systems under output feedback control with multiple random delays. *Chaos, Solitons, and Fractals*, *29*(5), 1142–1146. doi:10.1016/j.chaos.2005.08.078

Xu, S., Lam, J., & Ho, D. W. C. (2006). A new LMI condition for delay-dependent asymptotic stability of delayed Hopfield neural networks. *IEEE Transactions on Circuits and Wystems. II, Express Briefs*, *53*(3), 230–234. doi:10.1109/TCSII.2005.857764

Xu, S., Lam, J., Ho, D. W. C., & Zou, Y. (2005). Novel global asymptotic stability criteria for delayed cellular neural networks. *IEEE Transactions on Circuits and Wystems. II, Express Briefs*, *52*(6), 349–353. doi:10.1109/TCSII.2005.849000

Yan, J. J., Hung, M. L., Chiang, T. Y., & Yang, Y. S. (2006). Robust synchronization of chaotic systems via adaptive sliding mode control. *Physics Letters. [Part A]*, *356*(3), 220–225. doi:10.1016/j.physleta.2006.03.047

Yu, W., Cao, J., & Lu, J. (2008). Global synchronization of linearly hybrid coupled networks with time-varying delay. *SIAM Journal on Applied Dynamical Systems*, *7*(1), 108–133. doi:10.1137/070679090

Zhang, Q., Wen, X., & Xu, J. (2005). Delay-dependent exponential stability of cellular neural networks with time-varying delays. *Chaos, Solitons, and Fractals*, *23*, 1363–1369. doi:10.1016/j.chaos.2004.06.036

Zhao, H. (2004a). Global asymptotic stability of Hopfield neural network involving distributed delays. *Neural Networks*, *17*, 47–53. doi:10.1016/S0893-6080(03)00077-7

Zhao, H. (2004b). Existence and global attractivity of almost periodic solution for cellular neural network with distributed delays. *Applied Mathematics and Computation*, *154*, 683–695. doi:10.1016/S0096-3003(03)00743-4

APPENDIX

Lemma 1. [57] Given matrices $Y=Y^T$, D, E and F of appropriate dimensions with $\varphi(t) = [0.5 \quad -0.7]^T$, $\phi(t) = [-0.4 \quad 0.1]^T$ then the following matrix inequality

$$F^T F \leq I,$$

holds for all F if and only if there exists a scalar $Y + sym\,(D\,F\,E) < 0$ such that

$$\varepsilon > 0$$

Chapter 13
Adaptive Synchronization in Unknown Stochastic Chaotic Neural Networks with Mixed Time-Varying Delays

Jian-an Fang
Donghua University, China

Yang Tang
Donghua University, China

ABSTRACT

Neural networks (NNs) have been useful in many fields, such as pattern recognition, image processing etc. Recently, synchronization of chaotic neural networks (CNNs) has drawn increasing attention due to the high security of neural networks. In this chapter, the problem of synchronization and parameter identification for a class of chaotic neural networks with stochastic perturbation via state and output coupling, which involve both the discrete and distributed time-varying delays has been investigated. Using adaptive feedback techniques, several sufficient conditions have been derived to ensure the synchronization of stochastic chaotic neural networks. Moreover, all the connection weight matrices can be estimated while the lag synchronization and complete synchronization is achieved in mean square at the same time. The corresponding simulation results are given to show the effectiveness of the proposed method.

INTRODUCTION

It is widely believed that chaos synchronization has played a more and more significant role in nonlinear science (Chen & Dong, 1998; Pecora & Carroll, 1990; Tang & Fang, 2008; Tang et al, 2008; Tang et al, 2009a; Tang et al, 2009b; Tang et al, 2009c; Ojalvo & Roy, 2001; Yang & Chua, 1997; Boccaletti et al, 2002; Shahverdiev et al, 2002). Since Pecora and Carroll (Pecora & Carroll, 1990) synchronized two identical systems with different initial conditions, chaos synchronization has drawn much attention

DOI: 10.4018/978-1-61520-737-4.ch013

due to its potential applications in many fields, such as secure communication, pattern recognition, biological systems, and so on. So far, a wide variety of synchronization phenomena have been discovered such as complete synchronization, lag synchronization (Shahverdiev et al, 2002). Lag synchronization (LS) occurs as a coincidence of shifted-in-time states of two systems $y(t) \to x(t - \delta), t \to \infty$, where δ is a propagation delay. It is worth mentioning that, in many practical situations, a propagation delay will appear in the electronic implementation of dynamical systems. Therefore, it is very important and necessary to investigate the lag synchronization from the view of applications.

The past two decades have witnessed significant progress on the study of dynamical characteristics of neural networks because of their wide applications in many areas. There exist many works which are devoted to achieve the synchronization problems of neural networks. With respect to some recent representative works on this topic, we refer the reader to (Tang et al, 2009a; Tang et al, 2008; Tang et al, 2009b; Lu & Cao, 2007; Yu & Cao, 2007; Cao et al, 2006; He et al, 2008; Lou & Cui, 2008; Lu & Chen, 2004; Li & Chen, 2006; Li et al, 2007; Chen et al, 2004; Cheng et al, 2005) and references therein.

On the other hand, in the past few years, there has been an increasing interest in the research of neural networks with stochastic perturbations in the neural network community (Wang et al, 2007; Wang et al, 2006; Wan & Sun, 2005; Sun & Cao, 2007). It has been shown that, in real nervous systems, the transmission is a noisy process brought on by random fluctuations from the release of neurotransmitters and other probabilistic causes. Therefore, the synchronization and stability analysis problem for stochastic neural networks has been an important research issue. Recently, there have been some initial studies on the synchronization of neural networks (Yu & Cao, 2007; Sun & Cao, 2007). Very recently, in Ref. (Sun & Cao, 2007), the adaptive synchronization scheme has been developed to synchronize the stochastic neural networks with constant delay.

Recently, the distributed delay in neural networks has been an active subject nowadays. The reason for this is that the neural signal propagation is often distributed due to the presence of a multitude of parallel pathways with a variety of axon sizes and lengths. To be noted that distributed delay should be taken into consideration for modeling a realistic neural network (Ruan & Filfil, 2004). A number of works have been devoted to stability analysis issue for neural networks with distributed delay (Wang et al, 2007; Wang et al, 2006; Wang et al, 2008; Liu, 2006) based on linear matrix inequality (LMI) approach. However, there are few works about the synchronization of neural networks with time-varying delay and distributed delay. In Ref. (Wang et al, 2008), the sufficient conditions on the complete synchronization for two neural networks without noise perturbation which the parameters are known beforehand. But in real-life applications, this assumption is not realistic due to the stochastic perturbation widely existing in practical situations. Moreover, some systems' parameters cannot be exactly known in prior. The effect of these uncertainties will destroy the synchronization and even break it.

With the above motivations, our intention in this chapter is to study the synchronization for a class of stochastic chaotic neural networks with mixed time-delays and unknown parameters and consider more general coupling conditions, including states and outputs which result in different theoretical synchronization criteria. The mixed time-delays comprise the time-varying delay and distributed delay and the neural networks are subjected to stochastic disturbances described in terms of a Brownian motion. Via adaptive feedback approach (Huang, 2004; Huang, 2005; Huang, 2006), several simple criteria are developed to synchronize unknown neural networks with mixed time-delays and stochastic perturbations. The corresponding numerical simulations illustrate the feasibility of the presented method.

The organizations of this chapter are listed as follows. In Section 2, the synchronization for stochastic CNNs with mixed time-varying delays via state coupling is investigated. In Section 3, the synchronization for stochastic CNNs with discrete and distributed time-varying delays via output coupling is addressed. The simulations examples are provided in Section 4. The conclusions are given in Section 5.

Notations

Throughout this chapter, R^n and $R^{n\times m}$ denote, respectively, the n-dimensional Euclidean space and the set of all real matrices. The superscript 'T' denotes matrix transposition and the notation $X\geq Y$ (respectively, $X>Y$) where X and Y are symmetric matrices, means that $X-Y$ is positive semi-definite (respectively, positive definite). $\lambda_{max}(\cdot)$or $\lambda_{min}(\cdot)$denotes the largest or smallest eigenvalue of a matrix, respectively. Let $h>0$ and $C([-h,0; R^n])$ denote the family of continuous functions ϕ from $[-h,0]$ to R^n with the norm $\left\|\phi\right\|_2 = \sup_{-h\leq\theta\leq 0}\left\|\phi(\theta)\right\|$, where $\|\cdot\|$ is the Euclidean norm in R^n; $E\{\cdot\}$ stands for the mathematical expectation operator. Let $(\Omega,F,\{F_t\}_{t\geq 0},P)$ be a complete probability space with a filtration $\{F_t\}_{t\geq 0}$ satisfying the usual conditions (i.e., the filtration contains all P-null sets and is right continuous). Denote by $L^P_{F_0}([-h,0];\mathbb{R}^n)$ the family of all F_0-measurable $C([-h,0]; R^n)$ - valued random variables $\xi = \{\xi(\theta): -h \leq \theta \leq 0\}$ such that $\sup_{-h\leq\theta\leq 0} E\left\|\xi(\theta)\right\|^p < \infty$. I is an identity matrix.

LAG SYNCHRONIZATION OF STOCHASTIC UNKNOWN CHAOTIC NEURAL NETWORKS WITH MIXED TIME-VARYING DELAYS BY STATE COUPLING

Considering the following unknown neural networks with discrete and distributed time-varying delays:

$$dx_i(t) = [-c_i x_i(t) + \sum_{j=1}^{n} a_{ij}\tilde{f}_j(x_j(t)) + \sum_{j=1}^{n} b_{ij}\tilde{g}_j(x_j(t-\tau_1(t))) + \sum_{j=1}^{n} d_{ij}\int_{t-\tau_2(t)}^{t}\tilde{h}_j(x_j(s))ds + J_i]dt, \quad i = 1,2,\ldots,n \tag{1}$$

or equivalently,

$$dx(t) = [-Cx(t) + A\tilde{f}(x(t)) + B\tilde{g}(x(t-\tau_1(t))) + D\int_{t-\tau_2(t)}^{t}\tilde{h}(x(s))ds + J]dt, \tag{2}$$

where $x(t) = (x_1(t), x_2(t),\ldots, x_n(t))^T \in \mathbb{R}^n$ is the state vector associated with the neurons; $C = diag(c_1, c_2,\ldots, c_n)$ is an unknown positive matrix; $A = (a_{ij})_{n\times n}$, $B = (b_{ij})_{n\times n}$ and $D = (d_{ij})_{n\times n}$ denote the unknown connection weight matrix, the time-varying delayed connection weight matrix and the distributively delayed connection weight matrix, respectively; $J = [J_1, J_2,\ldots, J_n]^T$ is an external input vector; $\tau_1(t)$ is the time-varying delay; the scalar $\tau_2(t)$ is the distributed time-delay; $\tilde{f},\tilde{g},\tilde{h}$ are the activation functions of neurons, where

$$\tilde{f}(x(t)) = (\tilde{f}_1(x_1(t)), \tilde{f}_2(x_2(t)), ..., \tilde{f}_n(x_n(t)))^T \in \mathbb{R}^n,$$
$$\tilde{g}(x(t-\tau_1(t))) = (\tilde{g}_1(x_1(t-\tau_1(t))), \tilde{g}_2(x_2(t-\tau_1(t))), ..., \tilde{g}_n(x_n(t-\tau_1(t))))^T \in \mathbb{R}^n$$

and

$$\tilde{h}(x(t)) = (\tilde{h}_1(x_1(t)), \tilde{h}_2(x_2(t)), ..., \tilde{h}_n(x_n(t)))^T \in \mathbb{R}^n.$$

To realize lag synchronization, the noise-perturbed response system is given as

$$\begin{aligned} dy_i(t) = [&-\hat{c}_i y_i(t) + \sum_{j=1}^{n} \hat{a}_{ij} \tilde{f}_j(y_j(t)) + \sum_{j=1}^{n} \hat{b}_{ij} \tilde{g}_j(y_j(t-\tau_1(t))) + \sum_{j=1}^{n} \hat{d}_{ij} \int_{t-\tau_2(t)}^{t} \tilde{h}_j(y_j(s))ds \\ &+ J_i + \varepsilon_i(t) e_i(t)]dt + \sigma_i(t, y_i(t) - x_i(t-\delta), y_i(t-\tau_1(t)) - x_i(t-\delta-\tau_1(t)))d\omega_i(t), \\ &i = 1, 2, ..., n \end{aligned} \tag{3}$$

$$\begin{aligned} dy(t) = [&-\hat{C}y(t) + \hat{A}\tilde{f}(y(t)) + \hat{B}\tilde{g}(y(t-\tau_1(t))) + \hat{D} \int_{t-\tau_2(t)}^{t} \tilde{h}(y(s))ds + J + \varepsilon(t) \otimes e(t)]dt \\ &+ \sigma(t, y(t) - x(t-\delta), y(t-\tau_1(t)) - x(t-\delta-\tau_1(t)))d\omega(t), \end{aligned} \tag{4}$$

where $\hat{C}, \hat{A}, \hat{B}$ and $\hat{D}$ are the estimations of the unknown matrices C, A, B, D, respectively; $e(t) = y(t) - x(t-\delta) = (y_1(t) - x_1(t-\delta), ..., y_n(t) - x_n(t-\delta))^T \in \mathbb{R}^n$ denote the synchronization errors; δ is a propagation delay; $\varepsilon = (\varepsilon_1, \varepsilon_2, ..., \varepsilon_n)^T \in \mathbb{R}^n$ are the updated feedback gain; the mark $\otimes$ is defined as $\varepsilon \otimes e(t) = (\varepsilon_1 e_1(t), \ldots, \varepsilon_n e_n(t))$; $\omega(t)$ is a one dimensional Brownian motion satisfying $E\{d\omega(t)\}=0$ and $E\{[d\omega(t)]^2\}=dt$.

Let $\tilde{C} = C - \hat{C}, \tilde{A} = A - \hat{A}, \tilde{B} = B - \hat{B}$ and $\tilde{D} = D - \hat{D}$ be the estimation errors of the parameters C,A,B and D. Subtracting (2) from (4), yields the error dynamical system as follows

$$\begin{aligned} de(t) = [&-Ce(t) + Af(e(t)) + Bg(e(t-\tau_1(t))) + D \int_{t-\tau_2(t)}^{t} h(e(s))ds + \tilde{C}y(t) - \tilde{A}\tilde{f}(y(t)) \\ &- \tilde{B}\tilde{g}(y(t-\tau_1(t))) - \tilde{D} \int_{t-\tau_2(t)}^{t} \tilde{h}(y(s))ds + \varepsilon \otimes e(t)]dt + \sigma(t, e(t), e(t-\tau_1(t)))d\omega(t), \end{aligned} \tag{5}$$

where

$$f(e(t)) = \tilde{f}(e(t) + x(t-\delta)) - \tilde{f}(x(t-\delta));$$
$$g(e(t-\tau_1(t))) = \tilde{g}(e(t-\tau_1(t)) + x(t-\delta-\tau_1(t))) - \tilde{g}(x(t-\delta-\tau_1(t)));$$

and

$$h(e(t)) = \tilde{h}(e(t) + x(t-\delta)) - \tilde{h}(t-\delta).$$

Throughout this chapter, the following assumptions are needed:

(A_1) The activation functions $\tilde{f}_i(x)$ and $\tilde{g}_i(x)$ satisfy the Lipschitz condition, i.e., for all i=1,2,…,n, there exist constants $\lambda_i > 0$ and $\phi_i > 0$ such that

$$|\tilde{f}_i(x) - \tilde{f}_i(y)| \leq \lambda_i |x - y|, \qquad |\tilde{g}_i(x) - \tilde{g}_i(y)| \leq \phi_i |x - y|, \quad \forall x, y \in \mathbb{R}^n.$$

(A_2) The activation function $\tilde{h}(x)$ satisfies the Lipschitz condition of vector form:

$$\| h(x) - h(y) \| \leq \| H(x - y) \|, \quad \forall x, y \in \mathbb{R}^n,$$

where $H \in R^{n\times n}$ is a constant matrix.

$$(A_3). \quad \tilde{f}_i(0) = \tilde{g}_i(0) = \tilde{h}_i(0) = 0, \sigma(t, 0, 0) = 0.$$

(A_4) $\sigma(t,u,v)$ satisfies the Lipschitz condition. Furthermore, there exist known constant matrices of appropriate dimensions Σ_1 and Σ_2 and such that

$$trace[\sigma^T(t, u, v)\sigma(t, u, v)] \leq \| \Sigma_1 u \|^2 + \| \Sigma_2 v \|^2, \quad \forall (t, u, v) \in \mathbb{R}^+ \times \mathbb{R}^n \times \mathbb{R}^n.$$

(A_5). $\tau_1(t)$ and $\tau_2(t)$ are bounded and continuously differentiable functions, $0 < \tau_1(t) \leq \tau_1, 0 < \tau_2(t) \leq \tau_2$ and $\dot{\tau}_1(t) \leq \mu < 1, \dot{\tau}_2(t) \leq \rho < 1$,.

The initial conditions associated with system (1) are given as follows:

$$x_i(s) = \xi_i(s), -\tau^* \leq s \leq 0, i = 1, 2 \cdots, N,$$

where for any $\xi_i \in L^2_{F_0}([-\tau^*, 0], \mathbb{R}^n), \tau^* = \sup_{t\in\mathbb{R}} \in \{\tau_1(t), \tau_2(t)\}$, where $L^2_{F_0}([-\tau^*, 0], \mathbb{R}^n)$ is the family of all F_0-measurable $C([-\tau^*,0];R^n)$ -valued random variables satisfying that $\sup_{-\tau^* \leq \varphi \leq 0} E|\xi_i(\varphi)|^2 < \infty$, and $C([-\tau^*,0];R^n)$ denotes the family of all continuous R^n -valued functions $\xi_i(\varphi)$ on $[-\tau^*,0]$ with the norm $\|\xi_i\| = \sup_{-\tau^* \leq \varphi \leq 0} |\xi_i(\varphi)|$.

Definition 1. The master system (2) and slave system (4) are synchronized in the sense of lag synchronization, if the error system (5) is asymptotically stable in mean square

$$\lim_{t\to\infty} E \| e(t) \|^2 = 0. \tag{6}$$

The following lemmas will be essential in achieving the main results.

Lemma 1 (See Ref. (Xu et al,)). Let $\Omega_1 \Omega_2 \Omega_3$ be real matrices of appropriate dimensions with $\Omega_3 > 0$ .Given any vector x,y of appropriate dimensions, then the following inequality holds,

$2x^T\Omega_1^T\Omega_2 y \le x^T\Omega_1^T\Omega_3\Omega_1 x + y^T\Omega_2^T\Omega_3^{-1}\Omega_2 y.$

Lemma 2 (See Ref. (Wang et al, 2008)). For any positive definite matrix $N>0$, scalar $\nu > \nu(t) > 0$, vector function $w:[0,\nu] \to \mathbb{R}^n$ such that the integrations concerned are well defined, the following inequality holds:

$$\left(\int_0^{v(t)} w(s)ds\right)^T N\left(\int_0^{v(t)} w(s)ds\right) \le v(t)\left(\int_0^{v(t)} w^T(s)Nw(s)\right)ds.$$

Theorem 1. Under assumptions (A_1–A_5), the feedback strength $\varepsilon(t) = (\varepsilon_1(t), \varepsilon_2(t), \ldots, \varepsilon_n(t))^T$ and the estimated parameters $\hat{C}, \hat{A}, \hat{B}$ and $\hat{D}$ are adapted according to the following updated law, respectively,

$$\begin{cases} \dot{\varepsilon}_i = -\alpha_i e_i^2(t), & i = 1,2,\ldots,n, \\ \dot{\hat{c}}_i = \gamma_i e_i(t) y_i(t), & i = 1,2,\ldots,n, \\ \dot{\hat{a}}_{ij} = -\psi_{ij} e_i(t)\tilde{f}_j(y_j(t)), & i,j = 1,2,\ldots,n, \\ \dot{\hat{b}}_{ij} = -\beta_{ij} e_i(t)\tilde{g}_j(y_j(t-\tau_1(t))), & i,j = 1,2,\ldots,n, \\ \dot{\hat{d}}_{ij} = -\rho_{ij} e_i(t) \int_{t-\tau_2(t)}^{t} \tilde{h}_j(y_j(s))ds, & i,j = 1,2,\ldots,n, \end{cases} \tag{7}$$

in which $\alpha_i > 0, \gamma_i > 0, \psi_{ij} > 0, \beta_{ij} > 0$ and $\rho_{ij} > 0$ are arbitrary constants, respectively. Then the controlled response system (4) and drive system (2) can be synchronized lag synchronization in mean square.

Proof. For each $V \in C^{1,2}(\mathbb{R}^+ \times \mathbb{R}^n; \mathbb{R}^+)$, define an operator L associated with the error system acting on V by Friedman (1976)

$$\begin{aligned} LV(t,e(t)) = {} & V_t(t,e(t)) + V_e(t,e(t))[-Ce(t) + Af(e(t)) + Bg(e(t-\tau_1(t))) + D\int_{t-\tau_2(t)}^{t} h(e(s))ds \\ & + \tilde{C}y(t) - \tilde{A}\tilde{f}(y(t)) - \tilde{B}\tilde{g}(y(t-\tau_1(t))) - \tilde{D}\int_{t-\tau_2(t)}^{t} \tilde{h}(y(s))ds + \varepsilon \otimes e(t)] \\ & + \frac{1}{2} trace[\sigma^T(t,e(t),e_{\tau_1}(t))V_{ee}(t,e(t))\sigma(t,e(t),e_{\tau_1}(t))], \end{aligned} \tag{8}$$

where

$$V_t(t,e(t)) = \frac{\partial V(t,e(t))}{\partial t}, \quad V_e(t,e(t)) = \left(\frac{\partial V(t,e(t))}{\partial e_1}, \frac{\partial V(t,e(t))}{\partial e_2}, \ldots, \frac{\partial V(t,e(t))}{\partial e_n}\right),$$

$$V_{ee}(t,e(t)) = \left(\frac{\partial^2 V(t,e(t))}{\partial e_i \partial e_j}\right)_{n\times n}, \quad e(t) = y(t) - x(t-\delta), \quad e_{\tau_1} = y(t-\tau_1(t)) - x(t-\delta-\tau_1(t)).$$

Construct the following Lyapunov functional candidate:

$$V(t,e(t)) = \frac{1}{2}e^T(t)e(t) + \frac{1}{2}\int_{t-\tau_1(t)}^{t} e^T(s)Pe(s)ds + \frac{1}{2}\int_{t-\tau_2(t)}^{t}\int_{s}^{t} e^T(\theta)Qe(\theta)d\theta ds$$
$$+\frac{1}{2}\sum_{i=1}^{n}\left[\frac{1}{\alpha_i}(\varepsilon_i + l)^2 + \frac{1}{\gamma_i}\tilde{c}_i^{\,2} + \sum_{j=1}^{n}\frac{1}{\psi_{ij}}\tilde{a}_{ij}^{\,2} + \sum_{j=1}^{n}\frac{1}{\beta_{ij}}\tilde{b}_{ij}^{\,2} + \sum_{j=1}^{n}\frac{1}{\rho_{ij}}\tilde{d}_{ij}^{\,2}\right], \quad (9)$$

where P and Q are positive definite matrices and l is a constant to be determined in the following, respectively.

By the Ito-differential formula (Friedman, 1976), the stochastic derivative of V along the trajectory of the error system (5) can be obtained as follows,

$$dV(t,e(t)) = LV(t,e(t))dt + V_e(t,e(t))\sigma(t,e(t),e_{\tau_1}(t))d\omega(t),$$

where operator L is given as follows

$$\begin{aligned}
LV(t,e(t)) &= V_t(t,e(t)) + V_e(t,e(t))[-Ce(t) + Af(e(t)) + Bg(e(t-\tau_1(t))) + D\int_{t-\tau_2(t)}^{t} h(e(s))ds \\
&\quad + \tilde{C}y(t) - \tilde{A}\tilde{f}(y(t)) - \tilde{B}\tilde{g}(y(t-\tau_1(t))) - \tilde{D}\int_{t-\tau_2(t)}^{t}\tilde{h}(y(s))ds + \varepsilon\otimes e(t)] \\
&\quad + \frac{1}{2}trace[\sigma^T(t,e(t),e_{\tau_1}(t))V_{ee}(t,e(t))\sigma(t,e(t),e_{\tau_1}(t))] \\
&= e^T(t)[-Ce(t) + Af(e(t)) + Bg(e(t-\tau_1(t))) + D\int_{t-\tau_2(t)}^{t} h(e(s))ds \\
&\quad + \tilde{C}y(t) - \tilde{A}\tilde{f}(y(t)) - \tilde{B}\tilde{g}(y(t-\tau_1(t))) - \tilde{D}\int_{t-\tau_2(t)}^{t}\tilde{h}(y(s))ds + \varepsilon\otimes e(t)] \\
&\quad + \frac{1}{2}e^T(t)Pe(t) - \frac{1-\dot{\tau}_1(t)}{2}e^T(t-\tau_1(t))Pe(t-\tau_1(t)) + \frac{1}{2}\tau_2(t)e^T(t)Qe(t) \\
&\quad - \frac{1}{2}(1-\dot{\tau}_2(t))\int_{t-\tau_2(t)}^{t} e^T(s)Qe(s)ds - \sum_{i=1}^{n}(\varepsilon_i + l)e_i^2(t) - \sum_{i=1}^{n}\tilde{c}_i e_i(t)y_i(t) \\
&\quad + \sum_{i=1}^{n}\sum_{j=1}^{n}\tilde{a}_{ij}e_i(t)\tilde{f}_j(y_j(t)) + \sum_{i=1}^{n}\sum_{j=1}^{n}\tilde{b}_{ij}e_i(t)\tilde{g}_j(y_j(t-\tau_1(t))) \\
&\quad + \sum_{i=1}^{n}\sum_{j=1}^{n}\tilde{d}_{ij}e_i(t)\int_{t-\tau_2(t)}^{t}\tilde{h}_j(y_j(s))ds + \frac{1}{2}trace[\sigma^T(t,e(t),e_{\tau_1}(t))V_{ee}(t,e(t))\sigma(t,e(t),e_{\tau_1}(t))].
\end{aligned} \quad (10)$$

It can be obtained that

$$\begin{cases} e^T(t)(\varepsilon \otimes e(t)) = \sum_{i=1}^{n} \varepsilon_i e_i^{\,2}(t), \\ e^T(t)\tilde{C}y(t) = \sum_{i=1}^{n} \tilde{c}_i e_i(t) y_i(t), \\ e^T(t)\tilde{A}\tilde{f}(y(t)) = \sum_{i=1}^{n}\sum_{j=1}^{n} \tilde{a}_{ij} e_i(t)\tilde{f}_j(y_j(t)), \\ e^T(t)\tilde{B}\tilde{g}(y(t-\tau_1(t))) = \sum_{i=1}^{n}\sum_{j=1}^{n} \tilde{b}_{ij} e_i(t)\tilde{g}_j(y_j(t-\tau_1(t))), \\ e^T(t)\tilde{D}\int_{t-\tau_2(t)}^{t} \tilde{h}(y(s))ds = \sum_{i=1}^{n}\sum_{j=1}^{n} \tilde{d}_{ij} e_i(t)\int_{t-\tau_2(t)}^{t} \tilde{h}_j(y_j(s))ds. \end{cases} \tag{11}$$

By using Lemma 1, we can get

$$e^T(t)Af(e(t)) \le \frac{1}{2}e^T(t)AA^Te\;(t) + \frac{1}{2}f^T(e(t))f(e(t)) \tag{12}$$

and

$$e^T(t)Bg(e(t-\tau_1(t))) \le \frac{1}{2}e^T(t)BB^Te\;(t) + \frac{1}{2}g^T(e(t-\tau_1(t)))g(e(t-\tau_1(t))). \tag{13}$$

According to (11) -(13), we have

$$\begin{aligned} LV(t,e(t)) \le\; & -e^T(t)Ce(t) + \frac{1}{2}e^T(t)A^TAe(t) + \frac{1}{2}f^T(e(t))f(e(t)) + \frac{1}{2}e^T(t)BB^Te\;(t) \\ & + \frac{1}{2}g^T(e(t-\tau_1(t)))g(e(t-\tau_1(t))) + e^T(t)D\int_{t-\tau_2(t)}^{t} h(e(s))ds + \frac{1}{2}e^T(t)Pe(t) \\ & - \frac{1-\dot{\tau}_1(t)}{2}e^T(t-\tau_1(t))Pe(t-\tau_1(t)) + \frac{1}{2}\tau_2(t)e^T(t)Qe(t) \\ & - \frac{1}{2}(1-\dot{\tau}_2(t))\int_{t-\tau_2(t)}^{t} e^T(s)Qe(s)ds - le^T(t)e(t) \\ & + \frac{1}{2}trace[\sigma^T(t,e(t),e_{\tau_1}(t))\sigma(t,e(t),e_{\tau_1}(t))]. \end{aligned} \tag{14}$$

From (A_1), we can get

$$f^T(e(t))f(e(t)) = \sum_{i=1}^{n} f_i^2\left(e_i\left(t\right)\right) \le \sum_{i=1}^{n} \lambda_i^2 e_i^2\left(t\right) \le \Lambda e^T\left(t\right)e\;\left(t\right) \tag{15}$$

and

$$g^T(e(t))g(e(t)) = \sum_{i=1}^{n} g_i^2\left(e_i\left(t\right)\right) \leq \sum_{i=1}^{n} \phi_i^2 e_i^2\left(t\right) \leq \Phi e^T\left(t\right)e\left(t\right), \tag{16}$$

where $\Lambda = \max\{\lambda_i^2 : i = 1,2,\cdots,n\}$ and $\Phi = \max\{\phi_i^2 : i = 1,2,\cdots,n\}$.

According to the (A_5), we have

$$-\left(1-\dot{\tau}_1\left(t\right)\right) \leq -\left(1-\mu\right), \quad -\left(1-\dot{\tau}_2\left(t\right)\right) \leq -\left(1-\rho\right), \tag{17}$$

According to the (15) -(17), we have

$$\begin{aligned} LV(t,e(t)) \leq & -e^T(t)Ce(t) + \frac{1}{2}e^T(t)A^TAe(t) + \frac{1}{2}(\Lambda - lI)e^T(t)e(t) + \frac{1}{2}e^T(t)BB^Te\ (t) \\ & + \frac{1}{2}\Phi e^T(t-\tau_1(t))e(t-\tau_1(t)) + e^T(t)D\int_{t-\tau_2(t)}^{t} h(e(s))ds + \frac{1}{2}e^T(t)Pe(t) \\ & + \frac{1}{2}\tau_2 e^T(t)Qe(t) - \frac{1-\mu}{2}e^T(t-\tau_1(t))Pe(t-\tau_1(t)) \\ & - \frac{1}{2}(1-\rho)\int_{t-\tau_2(t)}^{t} e^T(s)Qe(s)ds \\ & + \frac{1}{2}trace[\sigma^T(t,e(t),e_{\tau_1}(t))\sigma(t,e(t),e_{\tau_1}(t))]. \end{aligned} \tag{18}$$

Let Ω_1, Ω_2, Ω_3 in Lemma 1 be the compatible dimensions identity matrix. Hence, we have

$$e^T(t)D\int_{t-\tau_2(t)}^{t} h\left(e\ \left(s\right)\right)ds \leq \frac{1}{2}e^T(t)DD^Te\ (t) + \frac{1}{2}\int_{t-\tau_2(t)}^{t} h^T\left(e\ \left(s\right)\right)ds\int_{t-\tau_2(t)}^{t} h\left(e\ \left(s\right)\right)ds. \tag{19}$$

From the assumption (A_2), the following inequality holds

$$\left\|h(e(t))\right\| \leq \left\|He(t)\right\|, \tag{20}$$

then $h^T(e(t))h(e(t)) \leq e^T(t)H^THe(t)$.

According to Lemma 2, we have

$$\begin{aligned} & \frac{1}{2}\int_{t-\tau_2(t)}^{t} h^T\left(e\ \left(s\right)\right)ds\int_{t-\tau_2(t)}^{t} h\left(e\ \left(s\right)\right)ds \\ \leq & \frac{\tau_2\left(t\right)}{2}\int_{t-\tau_2(t)}^{t} h^T\left(e\ \left(s\right)\right)h\left(e\ \left(s\right)\right)ds \\ \leq & \frac{\tau_2}{2}\int_{t-\tau_2(t)}^{t} e^T\left(s\right)H^THe\ \left(s\right)ds. \end{aligned} \tag{21}$$

It can be seen from assumption (A_4) that

$$\begin{aligned}&\frac{1}{2}trace[\sigma^T(t,e(t),e_{\tau_1}(t))\sigma(t,e(t),e_{\tau_1}(t))]\\&\leq\frac{1}{2}e^T(t)\Sigma_1{}^T\Sigma_1e(t)+\frac{1}{2}e^T(t-\tau_1(t))\Sigma_2{}^T\Sigma_2e(t-\tau_1(t)).\end{aligned}\tag{22}$$

Let $Q=\frac{\tau_2}{1-\rho}H^TH$ is a positive definite matrix. Using inequalities (19), (21) and (22), we obtain from (18) that

$$\begin{aligned}LV(t,e(t))&\leq e^T(t)\left[-C+\frac{1}{2}\tau_2Q+\frac{1}{2}AA^T+\frac{1}{2}BB^T+(\frac{\Lambda}{2}-l)I+\frac{1}{2}DD^T+\frac{1}{2}\Sigma_1{}^T\Sigma_1+\frac{1}{2}P\right]e(t)\\&\quad+e^T(t-\tau_1(t))[\frac{1}{2}\Sigma_2{}^T\Sigma_2+\frac{\Phi}{2}I-\frac{1-\mu}{2}P]e(t-\tau_1(t))\\&\leq e^T(t)[\lambda_{\max}(-C)+\lambda_{\max}(\frac{1}{2}AA^T)+\lambda_{\max}(\frac{1}{2}BB^T)+\lambda_{\max}(\frac{1}{2}DD^T)+\lambda_{\max}(\frac{1}{2}\tau_2Q)\\&\quad+\lambda_{\max}(\frac{1}{2}\Sigma_1{}^T\Sigma_1)+\frac{1}{2}\lambda_{\max}(P)+\frac{\Lambda}{2}-lI]e(t)\\&\quad+e^T(t-\tau_1(t))[\lambda_{\max}(\frac{1}{2}\Sigma_2{}^T\Sigma_2)+\frac{\Phi}{2}-\frac{1-\mu}{2}P]e(t-\tau_1(t)).\end{aligned}\tag{23}$$

Then, we take

$$\begin{aligned}l&=\lambda_{\max}(-C)+\lambda_{\max}(\frac{1}{2}AA^T)+\lambda_{\max}(\frac{1}{2}BB^T)+\lambda_{\max}(\frac{1}{2}DD^T)+\lambda_{\max}(\frac{1}{2}\tau_2Q)\\&\quad+\lambda_{\max}(\frac{1}{2}\Sigma_1{}^T\Sigma_1)+\frac{\Lambda}{2}+\frac{1}{2(1-\mu)}(\lambda_{\max}(\Sigma_2{}^T\Sigma_2)+\Phi)+1.\end{aligned}\tag{24}$$

$$P=\frac{1}{(1-\mu)}(\lambda_{\max}(\Sigma_2{}^T\Sigma_2)+\Phi)I.\tag{25}$$

Then, it can be obtained that

$$LV(t,e(t))\leq-e^T(t)e(t).\tag{26}$$

According to the invariant principle of stochastic differential equation proposed in (Mao, 2002), it can be derived that $E\,||\,e(t;\xi)\,||^2\rightarrow 0,\ \hat{C}\rightarrow C,\hat{A}\rightarrow A,\hat{B}\rightarrow B,\hat{D}\rightarrow D$ as $t\rightarrow\infty$. The unknown parameters of the drive system can be identified at the same time when the synchronization in mean square is achieved. This completes the proof of Theorem 1.

When the stochastic perturbation is removed from the response system, the response system is

$$dy(t) = [-\hat{C}y(t) + \hat{A}\tilde{f}(y(t)) + \hat{B}\tilde{g}(y(t-\tau_1(t))) + \hat{D}\int_{t-\tau_2(t)}^{t} \tilde{h}(y(s))ds + J + \varepsilon(t) \otimes e(t)]dt, \tag{27}$$

which can lead the following results.

Corollary 1. Under assumptions (A_1–A_3) and (A_5), the feedback strength $\varepsilon(t) = (\varepsilon_1(t), \varepsilon_2(t), ..., \varepsilon_n(t))^T$ and the estimated parameters $\hat{C}, \hat{A}, \hat{B}$ and $\hat{D}$ are adapted according to the following updated law, respectively,

$$\begin{cases} \dot{\varepsilon}_i = -\alpha_i e_i^2(t), & i = 1,2,...,n, \\ \dot{\hat{c}}_i = \gamma_i e_i(t) y_i(t), & i = 1,2,...,n, \\ \dot{\hat{a}}_{ij} = -\psi_{ij} e_i(t) \tilde{f}_j(y_j(t)), & i,j = 1,2,...,n, \\ \dot{\hat{b}}_{ij} = -\beta_{ij} e_i(t) \tilde{g}_j(y_j(t-\tau_1(t))), & i,j = 1,2,...,n, \\ \dot{\hat{d}}_{ij} = -\rho_{ij} e_i(t) \int_{t-\tau_2(t)}^{t} \tilde{h}_j(y_j(s))ds, & i,j = 1,2,...,n, \end{cases} \tag{28}$$

in which $\alpha_i > 0, \gamma_i > 0, \psi_{ij} > 0, \beta_{ij} > 0$ and $\rho_{ij} > 0$ are arbitrary constants, respectively. Then the controlled noiseless response system (4) and drive system (2) can be synchronized lag synchronization.

When the distributed delay is removed from the neural network, that is, D=0 in drive system (2) and response system (4) can be rewritten as follows

$$dx(t) = [-Cx(t) + A\tilde{f}(x(t)) + B\tilde{g}(x(t-\tau_1(t))) + J]dt, \tag{29}$$

and

$$\begin{aligned} dy(t) = {} & [-\hat{C}y(t) + \hat{A}\tilde{f}(y(t)) + \hat{B}\tilde{g}(y(t-\tau_1(t))) + J + \varepsilon(t) \otimes e(t)]dt \\ & + \sigma(t, y(t) - x(t-\delta), y(t-\tau_1(t)) - x(t-\delta-\tau_1(t)))d\omega(t). \end{aligned} \tag{30}$$

Corollary 2. Under assumptions (A_1),(A_3–A_5), the feedback strength $\varepsilon(t) = (\varepsilon_1(t), \varepsilon_2(t), ..., \varepsilon_n(t))^T$ and the estimated parameters $\hat{C}, \hat{A}$ and $\hat{B}$ are adapted according to the following updated law, respectively,

$$\begin{cases} \dot{\varepsilon}_i = -\alpha_i e_i^2(t), & i = 1,2,...,n, \\ \dot{\hat{c}}_i = \gamma_i e_i(t) y_i(t), & i = 1,2,...,n, \\ \dot{\hat{a}}_{ij} = -\psi_{ij} e_i(t) \tilde{f}_j(y_j(t)), & i,j = 1,2,...,n, \\ \dot{\hat{b}}_{ij} = -\beta_{ij} e_i(t) \tilde{g}_j(y_j(t-\tau_1(t))), & i,j = 1,2,...,n, \end{cases} \tag{31}$$

in which $\alpha_i > 0, \gamma_i > 0, \psi_{ij} > 0$ and $\beta_{ij} > 0$ are arbitrary constants, respectively. Then the response system and drive system can be synchronized lag synchronization in mean square.

When the matrices C,A,B and D are known in prior, the following corollary can be achieved by adaptive feedback approach.

Corollary 3. Under assumptions (A_1–A_5), the feedback strength $\varepsilon(t) = (\varepsilon_1(t), \varepsilon_2(t), ..., \varepsilon_n(t))^T$ is adapted according to the following updated law,

$$\dot{\varepsilon}_i = -\alpha_i e_i^2(t), \qquad i = 1, 2, ..., n, \tag{32}$$

in which $\alpha_i > 0$ are arbitrary constants. Then the response system and the known drive system can be synchronized lag synchronization in mean square.

3. SYNCHRONIZATION OF STOCHASTIC UNKNOWN CHAOTIC NEURAL NETWORKS WITH MIXED TIME-VARYING DELAYS BY OUTPUT COUPLING

Consider the following neural networks with mixed time-varying delays:

$$dx_i(t) = [-c_i x_i(t) + \sum_{j=1}^{n} a_{ij}\tilde{f}_j(x_j(t)) + \sum_{j=1}^{n} b_{ij}\tilde{f}_j(x_j(t-\tau_1(t))) + \sum_{j=1}^{n} d_{ij} \int_{t-\tau_2(t)}^{t} \tilde{f}_j(x_j(s))ds + J_i]dt,$$
$$i = 1, 2, ..., n, \tag{33}$$

or equivalently,

$$dx(t) = [-Cx(t) + A\tilde{f}(x(t)) + B\tilde{f}(x(t-\tau_1(t))) + D\int_{t-\tau_2(t)}^{t} \tilde{f}(x(s))ds + J]dt, \tag{34}$$

where $x(t) = (x_1(t), x_2(t), ..., x_n(t))^T \in \mathbb{R}^n$ is the state vector associated with the neurons; $C = diag(c_1, c_2, ..., c_n)$ is an unknown positive matrix; $A = (a_{ij})_{n\times n}$, $B = (b_{ij})_{n\times n}$ and $D = (d_{ij})_{n\times n}$ denote the unknown connection weight matrix, the time-varying delayed connection weight matrix and the distributively delayed connection weight matrix, respectively; $J = [J_1, J_2, ..., J_n]^T$ is an external input vector; $\tau_1(t)$ is the time-varying delay; the scalar $\tau_2(t)$ is the distributed time-delay; $\tilde{f}$ is the activation functions of neurons, where

$$\tilde{f}(x(t)) = (\tilde{f}_1(x_1(t)), \tilde{f}_2(x_2(t)), ..., \tilde{f}_n(x_n(t)))^T \in \mathbb{R}^n.$$

In order to achieve complete synchronization, the noise-perturbed response neural networks can be given as

$$\begin{aligned} dy_i(t) = [&-\hat{c}_i y_i(t) + \sum_{j=1}^{n} \hat{a}_{ij}\tilde{f}_j(y_j(t)) + \sum_{j=1}^{n} \hat{b}_{ij}\tilde{f}_j(y_j(t-\tau_1(t))) + \sum_{j=1}^{n} \hat{d}_{ij} \int_{t-\tau_2(t)}^{t} \tilde{f}_j(y_j(s))ds \\ &+ J_i + u_i(t)]dt + \sigma_i(t, y_i(t) - x_i(t), y_i(t-\tau_1(t)) - x_i(t-\tau_1(t)))d\omega_i(t), \\ & i = 1,2,\ldots,n \end{aligned} \tag{35}$$

$$\begin{aligned} dy(t) = [&-\hat{C}y(t) + \hat{A}\tilde{f}(y(t)) + \hat{B}\tilde{f}(y(t-\tau_1(t))) + \hat{D} \int_{t-\tau_2(t)}^{t} \tilde{f}(y(s))ds + J + U(t)]dt \\ &+ \sigma(t, y(t) - x(t), y(t-\tau_1(t)) - x(t-\tau_1(t)))d\omega(t), \end{aligned} \tag{36}$$

where $\hat{C}, \hat{A}, \hat{B}$ and $\hat{D}$ are the estimations of the unknown matrices C,A,B,D, respectively; $e(t) = y(t) - x(t) = (y_1(t) - x_1(t), \ldots, y_n(t) - x_n(t))^T \in \mathbb{R}^n$ denote the synchronization errors; $\omega(t)$ is a one dimensional Brownian motion satisfying $E\{d\omega(t)\} = 0$ and $E\{[d\omega(t)]^2\} = dt$.

The error state is $e(t) = y(t) - x(t)$. Subtracting Equation (34) from (36) yields the error dynamical system as follows:

$$\begin{aligned} de(t) = [&-Ce(t) + Af(e(t)) + Bf(e(t-\tau_1(t))) + D \int_{t-\tau_2(t)}^{t} f(e(s))ds + \tilde{C}y(t) - \tilde{A}\tilde{f}(y(t)) \\ &- \tilde{B}\tilde{f}(y(t-\tau_1(t))) - \tilde{D} \int_{t-\tau_2(t)}^{t} \tilde{f}(y(s))ds + U(t)]dt + \sigma(t, e(t), e(t-\tau_1(t)))d\omega(t), \end{aligned} \tag{37}$$

where $f(e(t)) = \tilde{f}(e(t) + x(t)) - \tilde{f}(x(t))$.

Definition 2. The master system (34) and response system (36) are synchronized in the sense of synchronization, if the error system (37) is asymptotically stable in mean square

$$\lim_{t\to\infty} E \| e(t) \|^2 = 0. \tag{38}$$

Theorem 2. Under assumptions (A_1),(A_3–A_5), let the controller be $U(t) = \Delta \otimes (\tilde{f}(y(t)) - \tilde{f}(x(t)))$. The mark $\otimes$ is defined as $\Delta \otimes (\tilde{f}(y(t)) - \tilde{f}(x(t))) = (\delta_1 f_1(e_1(t)), \delta_2 f_2(e_2(t)), \ldots, \delta_n f_n(e_n(t)))^T$, where feedback strength $\Delta(t) = (\delta_1(t), \delta_2(t), \ldots, \delta_n(t))^T$ and the estimated parameters $\hat{C}, \hat{A}, \hat{B}$ and $\hat{D}$ are adapted according to the following updated law, respectively,

$$\begin{cases} \dot{\delta}_i = -\alpha_i f_i^2(e_i(t)), & i = 1,2,\dots,n, \\ \dot{\hat{c}}_i = \gamma_i f(e_i(t)) y_i(t), & i = 1,2,\dots,n, \\ \dot{\hat{a}}_{ij} = -\psi_{ij} f(e_i(t)) \tilde{f}_j(y_j(t)), & i,j = 1,2,\dots,n, \\ \dot{\hat{b}}_{ij} = -\beta_{ij} f(e_i(t)) \tilde{f}_j(y_j(t-\tau_1(t))), & i,j = 1,2,\dots,n, \\ \dot{\hat{d}}_{ij} = -\rho_{ij} f(e_i(t)) \int\limits_{t-\tau_2(t)}^{t} \tilde{f}_j(y_j(s))ds, & i,j = 1,2,\dots,n, \end{cases} \tag{39}$$

in which $\alpha_i > 0, \gamma_i > 0, \psi_{ij} > 0, \beta_{ij} > 0$ and $\rho_{ij} > 0$ are arbitrary constants, respectively. Then the controlled response system (36) and drive system (34) can be synchronized synchronization in mean square.

Proof. Construct the following Lyapunov functional candidate:

$$\begin{aligned} V(t,e(t)) = {} & \sum_{i=1}^{n} \int_0^{e_i(t)} f_i(s)ds + \frac{1}{2} \int\limits_{t-\tau_1(t)}^{t} f^T(e(s))Pf(e(s))ds + \frac{1}{2} \int\limits_{t-\tau_2(t)}^{t} \int\limits_{s}^{t} f^T(e(\theta))Pf(e(\theta))d\theta ds \\ & + \frac{1}{2}\sum_{i=1}^{n}\left[\frac{1}{\alpha_i}(\delta_i + l)^2 + \frac{1}{\gamma_i}\tilde{c}_i^{\,2} + \sum_{j=1}^{n}\frac{1}{\psi_{ij}}\tilde{a}_{ij}^{\,2} + \sum_{j=1}^{n}\frac{1}{\beta_{ij}}\tilde{b}_{ij}^{\,2} + \sum_{j=1}^{n}\frac{1}{\rho_{ij}}\tilde{d}_{ij}^{\,2}\right], \end{aligned} \tag{40}$$

where P and Q are positive definite matrices and l is a constant to be determined. By the Ito-differential formula, the stochastic derivative of V along the trajectory of the error system (37) can be obtained as follows:

$$\begin{aligned} LV(t,e(t)) = {} & V_t(t,e(t)) + V_e(t,e(t))[-Ce(t) + Af(e(t)) + Bf(e(t-\tau_1(t))) \\ & + D\int\limits_{t-\tau_2(t)}^{t} f(e(s))ds + \tilde{C}y(t) - \tilde{A}\tilde{f}(y(t)) \\ & - \tilde{B}\tilde{f}(y(t-\tau_1(t))) - \tilde{D}\int\limits_{t-\tau_2(t)}^{t} \tilde{f}(y(s))ds + \Delta \otimes f(e(t))] \\ & + \frac{1}{2} trace[\sigma^T(t,e(t),e_{\tau_1}(t))V_{ee}(t,e(t))\sigma(t,e(t),e_{\tau_1}(t))]. \end{aligned} \tag{41}$$

Thus, we have

$$
\begin{aligned}
LV(t,e(t)) = {} & \sum_{i=1}^{n} f_i(e_i(t))[-c_i e_i(t) + \sum_{j=1}^{n} a_{ij} f_j(e_j(t)) + \sum_{j=1}^{n} b_{ij} f_j(e_j(t-\tau_1(t))) \\
& + \sum_{j=1}^{n} d_{ij} \int_{t-\tau_2(t)}^{t} f_j(e_j(s))ds + \tilde{c}_i y_i(t) - \sum_{j=1}^{n} \tilde{a}_{ij} \tilde{f}_j(y_j(t)) - \sum_{j=1}^{n} \tilde{b}_{ij} \tilde{f}_j(y_j(t-\tau_1(t))) \\
& - \sum_{j=1}^{n} \tilde{d}_{ij} \int_{t-\tau_2(t)}^{t} \tilde{f}_j(y_j(s))ds + \Delta \otimes f(e(t))] + \frac{1}{2} f^T(e(t))Pf(e(t)) \\
& - \frac{1}{2}(1-\dot{\tau}_1(t)) f^T(e(t-\tau_1(t)))Pf(e(t-\tau_1(t))) + \frac{1}{2}\tau_2(t) f^T(e(t))Qf(e(t)) \\
& - \frac{1}{2}(1-\dot{\tau}_2(t)) \int_{t-\tau_2(t)}^{t} f^T(e(s))Qf(e(s))ds - \sum_{i=1}^{n} (\delta_i + l) f_i^2(e_i(t)) - \sum_{i=1}^{n} \tilde{c}_i f_i(e_i(t)) y_i(t) \\
& + \sum_{i=1}^{n}\sum_{j=1}^{n} \tilde{a}_{ij} f_i(e_i(t)) \tilde{f}_j(y_j(t)) + \sum_{i=1}^{n}\sum_{j=1}^{n} \tilde{b}_{ij} f_i(e_i(t)) \tilde{f}_j(y_j(t-\tau_1(t))) \\
& + \sum_{i=1}^{n}\sum_{j=1}^{n} \tilde{d}_{ij} f_i(e_i(t)) \int_{t-\tau_2(t)}^{t} \tilde{f}_j(y_j(s))ds \\
& + \frac{1}{2} trace[\sigma^T(t,e(t),e_{\tau_1}(t))(diag(\dot{f}_1(e_1(t),\ldots,\dot{f}_n(e_n(t))))\sigma(t,e(t),e_{\tau_1}(t))].
\end{aligned} \tag{42}
$$

We easily have

$$
f^T(e(t))(\Delta \otimes f(e(t))) = \sum_{i=1}^{n} \delta_i f_i^2(e_i(t)). \tag{43}
$$

Using assumptions (A_1), we have

$$
| f_i(e_i(t)) | \le \lambda_i | e_i(t) |, \quad 0 \le e_i(t) f_i(e_i(t)) \le \lambda_i e_i^{\,2}(t), \quad i = 1,2,\ldots,n. \tag{44}
$$

It follows from Lemma 1 that

$$
\begin{aligned}
\sum_{i=1}^{n}\sum_{j=1}^{n} b_{ij} f_i(e_i(t)) f_j(e_j(t-\tau_1(t))) &= f^T(e(t))Bf(e(t-\tau_1(t))) \\
&\le \frac{1}{2} f^T(e(t))f(e(t)) + \frac{B^T B}{2} f^T(e(t-\tau_1(t)))f(e(t-\tau_1(t))).
\end{aligned} \tag{45}
$$

According to Lemma 1and Lemma 2, we have

$$\sum_{i=1}^{n}\sum_{j=1}^{n} d_{ij} f_i(e_i(t)) \int_{t-\tau_2(t)}^{t} f_j(e_j(s))ds$$
$$= f^T(e(t))D \int_{t-\tau_2(t)}^{t} f(e(t))ds \quad (46)$$
$$\leq \frac{1}{2} f^T(e(t))f(e(t)) + \frac{\tau_2}{2} D^T D \int_{t-\tau_2(t)}^{t} f^T(e(s))f(e(s))ds.$$

From assumption (A_4), it can be seen that

$$\frac{1}{2} trace[\sigma^T(t,e(t),e_{\tau_1}(t))(diag(\dot{f}_1(e_1(t),...,\dot{f}_n(e_n(t))))\sigma(t,e(t),e_{\tau_1}(t))]$$
$$\leq \frac{\Lambda}{2} f^T(e(t))\Sigma_1\Sigma_1^T f(e(t)) + \frac{\Lambda}{2} f^T(e(t-\tau_1(t)))\Sigma_2\Sigma_2^T f(e(t-\tau_1(t))). \quad (47)$$

Let the positive matrix Q be $\frac{\tau_2 DD^T}{1-\rho}$. Using (43) -(47), we obtain from (42) that

$$LV(t,e(t)) \leq f^T(e(t))[-\min_{1\leq i\leq n} \frac{c_i}{v_i} I + A + \frac{1}{2}\tau_2 Q + \frac{P}{2} + I + \frac{1}{2}\Lambda\Sigma_1\Sigma_1^T - lI]f(e(t))$$
$$+ f^T(e(t-\tau_1(t)))[\frac{1}{2}\Lambda\Sigma_2\Sigma_2^T + \frac{1}{2}BB^T - \frac{1-\mu}{2}P]f(e(t-\tau_1(t))). \quad (48)$$

$$l = -\min_{1\leq i\leq n} \frac{c_i}{v_i} + \lambda_{\max}(A) + \frac{1}{2}\tau_2\lambda_{\max}(Q) + \frac{1}{2}\lambda_{\max}(P) + 2 + \frac{1}{2}\Lambda\lambda_{\max}(\Sigma_1\Sigma_1^T), \quad (49)$$

and

$$P = \frac{1}{(1-\mu)}(\Lambda\Sigma_2\Sigma_2^T + BB^T). \quad (50)$$

Then, we get

$$LV(t,e(t)) \leq -f^T(e(t))f(e(t)). \quad (51)$$

According to the invariant principle of stochastic differential equation proposed in (Mao, 2002), it can be derived that $E\,||\,e(t;\xi)\,||^2 \rightarrow 0$, $\hat{C} \rightarrow C, \hat{A} \rightarrow A, \hat{B} \rightarrow B, \hat{D} \rightarrow D$ as $t\rightarrow\infty$. The unknown parameters of the drive system can be identified at the same time when the synchronization in mean square is achieved. This completes the proof of Theorem 2.

When the matrices *C,A,B* and *D* in drive system are known in priori, the following corollary can be achieved output coupling.

Corollary 4. Under assumptions $(A_1),(A_3 - A_5)$, if there exist arbitrary positive constants $\alpha_i (i = 1,2,...,n)$ such that feedback strength $\Delta(t) = (\delta_1(t),\delta_2(t),...,\delta_n(t))^T$ is adapted to the following update law:

$$\dot{\delta}_i(t) = -\alpha_i f^2(e_i(t)), \qquad i = 1,2,...,n, \tag{52}$$

then the known drive chaotic neural networks and the stochastic perturbed response system can be achieved in the mean square.

When the distributed delays are removed from the neural networks, i.e., $D = 0$ in the drive system (34) and response system (36) can be rewritten as follows:

$$dx(t) = [-Cx(t) + A\tilde{f}(x(t)) + B\tilde{f}(x(t - \tau_1(t))) + J]dt, \tag{53}$$

$$\begin{aligned} dy(t) = {} & [-\hat{C}y(t) + \hat{A}\tilde{f}(y(t)) + \hat{B}\tilde{f}(y(t - \tau_1(t))) + J + U(t)]dt \\ & + \sigma(t, y(t) - x(t), y(t - \tau_1(t)) - x(t - \tau_1(t)))d\omega(t), \end{aligned} \tag{54}$$

respectively. We can obtain the following results.

Corollary 5. Under assumptions $(A_1),(A_3 - A_5)$, let the controller be $U(t) = \Delta \otimes (\tilde{f}(y(t)) - \tilde{f}(x(t)))$. The mark $\otimes$ is defined as $\Delta \otimes (\tilde{f}(y(t)) - \tilde{f}(x(t))) = (\delta_1 f_1(e_1(t)), \delta_2 f_2(e_2(t)),...,\delta_n f_n(e_n(t)))^T$, where feedback strength $\Delta(t) = (\delta_1(t),\delta_2(t),...,\delta_n(t))^T$ and the estimated parameters $\hat{C},\hat{A},\hat{B}$ and $\hat{D}$ are adapted according to the following updated law, respectively,

$$\begin{cases} \dot{\delta}_i = -\alpha_i f_i^2(e_i\ (t)), & i = 1,2,...,n, \\ \dot{\hat{c}}_i = \gamma_i f(e_i(t))y_i(t), & i = 1,2,...,n, \\ \dot{\hat{a}}_{ij} = -\psi_{ij} f(e_i(t))\tilde{f}_j(y_j(t)), & i,j = 1,2,...,n, \\ \dot{\hat{b}}_{ij} = -\beta_{ij} f(e_i(t))\tilde{f}_j(y_j(t - \tau_1(t))), & i,j = 1,2,...,n, \end{cases} \tag{55}$$

in which $\alpha_i > 0, \gamma_i > 0, \psi_{ij} > 0$ and $\beta_{ij} > 0$ are arbitrary constants, respectively. Then the controlled response system (36) and drive system (34) can be synchronized synchronization in mean square.

NUMERICAL SIMULATIONS

In this section, we employ two examples to show the effectiveness of Theorems 1 and 2 obtained in the previous sections.

Examples

Example 1

We consider the following two-order stochastic chaotic neural network with mixed time-varying delays:

$$dx(t) = [-Cx(t) + A\tilde{f}(x(t)) + B\tilde{g}(x(t-\tau_1(t))) + D\int_{t-\tau_2(t)}^{t}\tilde{h}(x(s))ds + J]dt, \tag{56}$$

with

$$x(t) = (x_1(t), x_2(t))^T, \tilde{f}(x(t)) = \tilde{g}(x(t)) = \tilde{h}(x(t)) = \tanh(x(t)) = (\tanh(x_1), \tanh(x_2))^T,$$
$$\tau_1(t) = \frac{e^t}{e^t+1}, \tau_2(t) = 1, J = (0,0)^T,$$

and

$$C = \begin{pmatrix} 1 & 0 \\ 0 & 1 \end{pmatrix}, A = \begin{pmatrix} 1.8 & -0.15 \\ -5.2 & 3.5 \end{pmatrix}, B = \begin{pmatrix} -1.7 & -0.12 \\ -0.26 & -2.5 \end{pmatrix}, D = \begin{pmatrix} 0.6 & 0.15 \\ -2 & -0.12 \end{pmatrix},$$

respectively. Then, the neural network model has a chaotic attractor with initial values $x_1(t) = 0.2, x_2(t) = 0.5, \forall t \in [-1,0]$.

The noise-perturbed response system is designed as

$$dy(t) = [-\hat{C}y(t) + \hat{A}\tilde{f}(y(t)) + \hat{B}\tilde{g}(y(t-\tau_1(t))) + \hat{D}\int_{t-\tau_2(t)}^{t}\tilde{h}(y(s))ds + J + \varepsilon(t)\otimes e(t)]dt$$
$$+\sigma(t, y(t) - x(t-\delta), y(t-\tau_1(t)) - x(t-\delta-\tau_1(t)))d\omega(t), \tag{57}$$

with

$$\hat{C} = \begin{pmatrix} 1 & 0 \\ 0 & 1 \end{pmatrix}, \hat{A} = \begin{pmatrix} 1.8 & -0.15 \\ -5.2 & \hat{a}_{22} \end{pmatrix}, \hat{B} = \begin{pmatrix} -1.7 & -0.12 \\ -0.26 & \hat{b}_{22} \end{pmatrix}, \hat{D} = \begin{pmatrix} 0.6 & 0.15 \\ -2 & \hat{d}_{22} \end{pmatrix}.$$

The noise intensity can be chosen as

$$\sigma(t, e(t), e_{\tau_1}(t)) = \begin{pmatrix} a_1 e(t) + b_1 e_{\tau_1}(t) & 0 \\ 0 & a_2 e(t) + b_2 e_{\tau_1}(t) \end{pmatrix},$$

Where $w(t)$ is a one dimensional Brownian motion. From (A_2), we have $\Sigma_1 = diag(|a_1|, |a_2|)$ and $\Sigma_2 = diag(|b_1|, |b_2|)$.

In the simulations, the Euler-Maruyama numerical method is employed to simulate the drive system (56) and response system (57). The initial values of the response system is taken as

Figure 1. The chaotic dynamics of the response neural networks

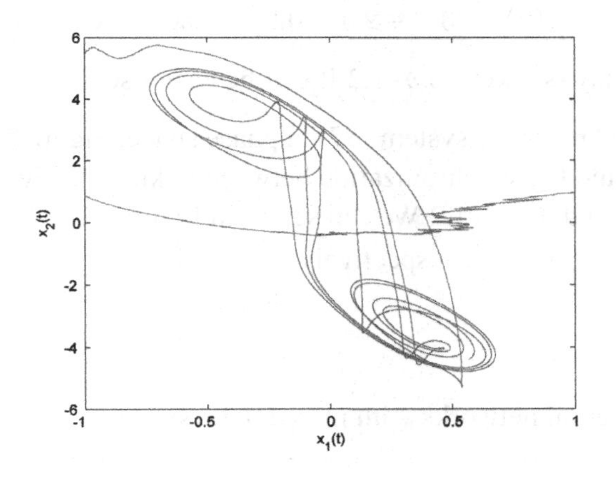

Figure 2. $t - e_1(t) - e_2(t)$.

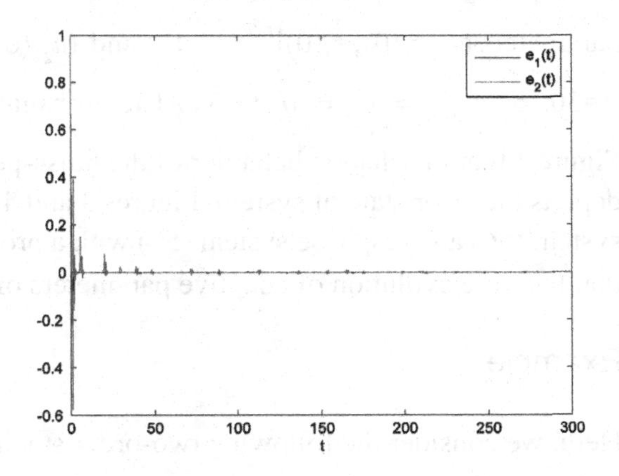

Figure 3. $t - x_1(t) - y_1(t)$.

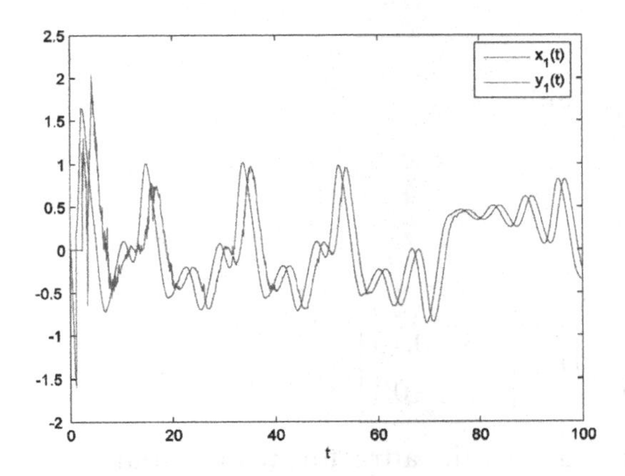

Figure 4. $t - x_2(t) - y_2(t)$.

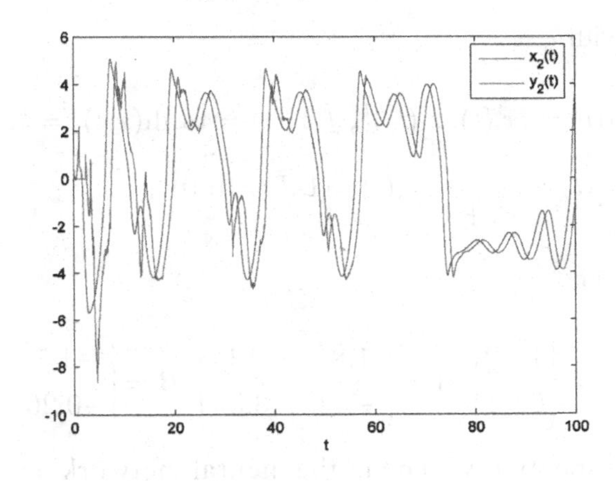

Figure 5. $t - \varepsilon_1(t) - \varepsilon_2(t)$.

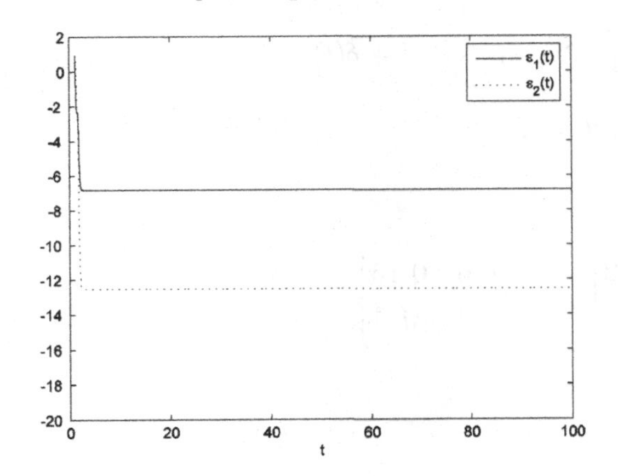

Figure 6. The time evolution of estimated parameters $\hat{a}_{22}, \hat{b}_{22}, \hat{d}_{22}$.

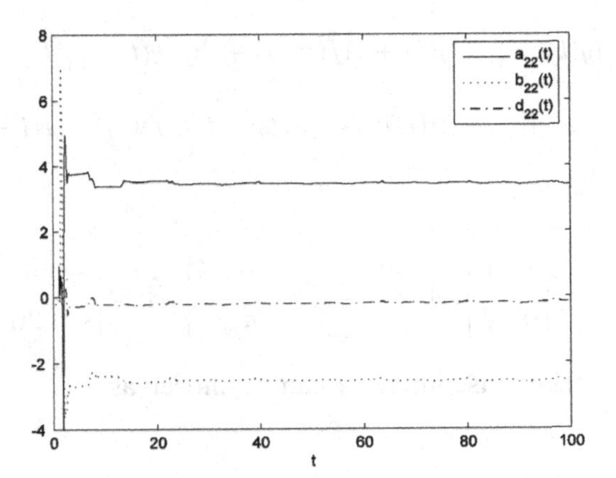

$y_1(t) = 1, y_2(t) = 1, \forall t[-1, 0]$. We take the initial conditions of the feedback strength and unknown parameters as $[\varepsilon_1(0), \varepsilon_2(0)]^T = [1, 1]^T$ and $(\hat{a}_{22}(0), \hat{b}_{22}(0), \hat{d}_{22}(0)) = [3.4, -2.4, -0.2]^T$, respectively, and a_i=30, $\gamma_i = \beta_{ij} = \psi_{ij} = \rho_{ij} = 50$. The propagation delay is taken as δ=1.2 It can be easily seen from Figure 1 that the chaotic behavior of the noise-perturbed response system (57) in phase space. Figure 2 depicts the error state of system. Figures 3 and 4 show the lag synchronization between unknown drive system (56) and response system (57) with a propagation delay δ=1.2 We can see from Figures 5 and 6 that the time evolution of adaptive parameters of $\varepsilon_1, \varepsilon_2, \hat{a}_{22}, \hat{b}_{22}, \hat{d}_{22}$, respectively.

Example 2

Here, we consider the following two-order stochastic neural networks with mixed delays:

$$dx(t) = [-Cx(t) + A\tilde{f}(x(t)) + B\tilde{f}(x(t - \tau_1(t))) + D \int_{t-\tau_2(t)}^{t} \tilde{f}(x(s))ds + J]dt, \tag{58}$$

with

$$x(t) = (x_1(t), x_2(t))^T, \ \tilde{f}(x(t)) = \tanh(x(t)) = (\tanh(x_1), \tanh(x_2))^T,$$
$$\tau_1(t) = \frac{e^t}{e^t + 1}, \tau_2(t) = 1, J = (0, 0)^T,$$

and

$$C = \begin{pmatrix} 1 & 0 \\ 0 & 1 \end{pmatrix}, A = \begin{pmatrix} 1.8 & -0.15 \\ -5.2 & 3.5 \end{pmatrix}, B = \begin{pmatrix} -1.7 & -0.12 \\ -0.26 & -2.5 \end{pmatrix}, D = \begin{pmatrix} 0.6 & 0.15 \\ -2 & -0.1 \end{pmatrix},$$

respectively. Then, the neural network model has a chaotic attractor with initial values $x_1(t) = 0.2, x_2(t) = 0.5, \forall t \in [-1, 0]$. We can see that $\lambda_1 = \lambda_2 = 1$.

The noise-perturbed response system is given as

$$\begin{aligned} dy(t) = {} & [-\hat{C}y(t) + \hat{A}\tilde{f}(y(t)) + \hat{B}\tilde{f}(y(t - \tau_1(t))) + \hat{D} \int_{t-\tau_2(t)}^{t} \tilde{f}(y(s))ds + J + \delta(t) \otimes f(e(t))]dt \\ & + \sigma(t, y(t) - x(t), y(t - \tau_1(t)) - x(t - \tau_1(t)))d\omega(t), \end{aligned} \tag{59}$$

with

$$\hat{C} = \begin{pmatrix} 1 & 0 \\ 0 & 1 \end{pmatrix}, \hat{A} = \begin{pmatrix} 1.8 & -0.15 \\ -5.2 & \hat{a}_{22} \end{pmatrix}, \hat{B} = \begin{pmatrix} -1.7 & -0.12 \\ -0.26 & \hat{b}_{22} \end{pmatrix}, \hat{D} = \begin{pmatrix} 0.6 & 0.15 \\ -2 & \hat{d}_{22} \end{pmatrix}.$$

The noise intensity can be chosen as

$$\sigma(t, e(t), e_{\tau_1}(t)) = \begin{pmatrix} a_1 e(t) + b_1 e_{\tau_1}(t) & 0 \\ 0 & a_2 e(t) + b_2 e_{\tau_1}(t) \end{pmatrix},$$

Where $w(t)$ is a one dimensional Brownian motion. From (A_2), we have $\Sigma_1 = diag(|a_1|, |a_2|) = diag(0.05, 0.05)$ and $\Sigma_2 = diag(|b_1|, |b_2|) = diag(0.06, 0.06)$.

In the simulations, the Euler-Maruyama numerical method is employed to simulate the drive system (58) and response system (59). The initial values of the response system is taken as $y_1(t) = 1, y_2(t) = 1, \forall t[-1, 0]$. We set the initial conditions of the feedback strength and unknown parameters as $[\delta_1(0), \delta_2(0)]^T = [1, 1]^T$ and $(\hat{a}_{22}(0), \hat{b}_{22}(0), \hat{d}_{22}(0)) = [3.4, -2.4, -0.2]^T$, respectively.

Note that the control method is the output coupling. The feedback strength and the parameters update laws can be listed as follows:

Figure 7. Chaotic dynamics of the drive chaotic neural networks

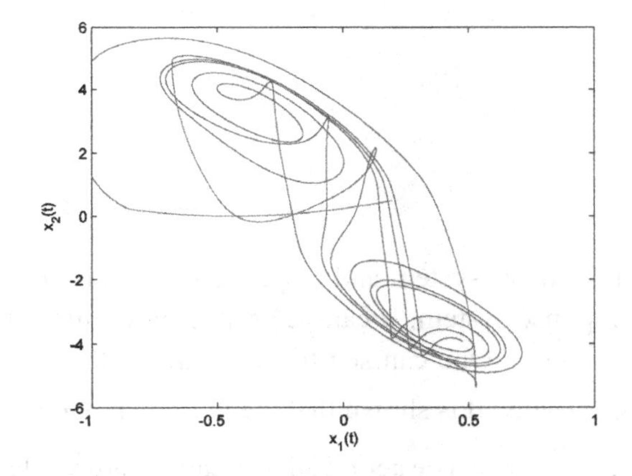

Figure 8. Chaotic dynamics of the response chaotic neural networks with stochastic perturbation

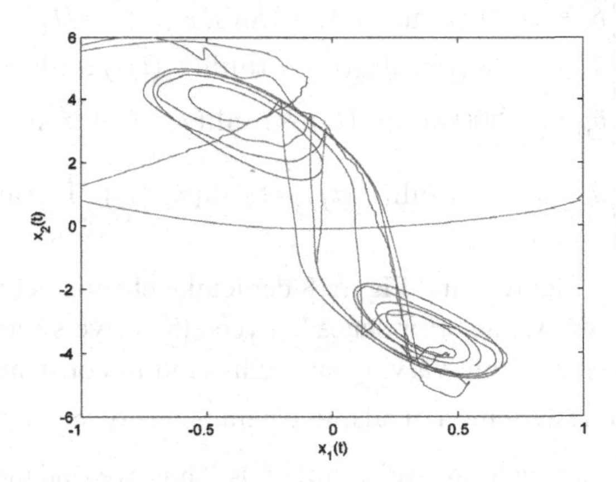

Figure 9. $t - e_1 - e_2$

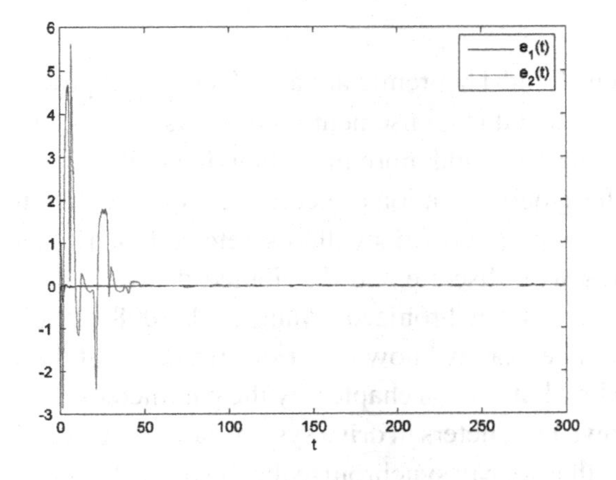

Figure 10. The time evolution of time-varying strength $\delta_1(t), \delta_2(t)$.

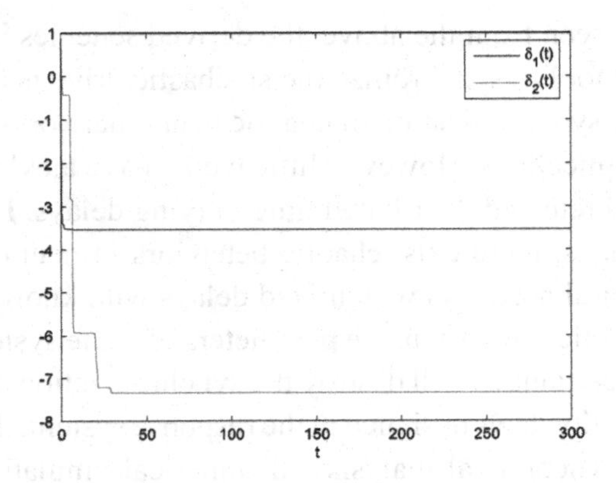

Figure 11. The time evolution of estimated parameters $\hat{a}_{22}, \hat{b}_{22}, \hat{d}_{22}$.

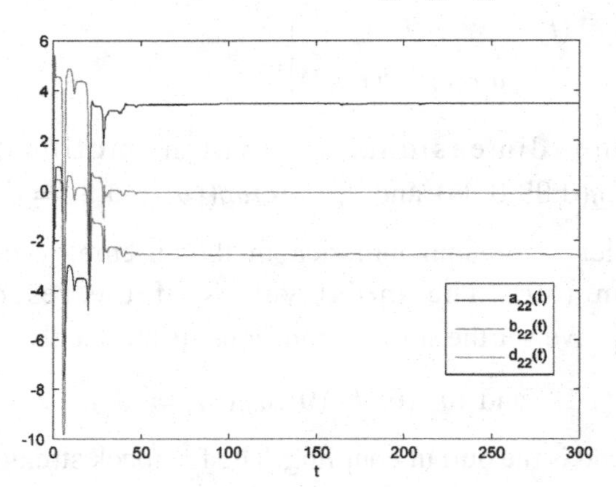

$$\begin{cases} \dot{\delta}_1 = -10(\tanh(y_1(t)) - \tanh(x_1(t)))(y_1(t) - x_1(t)), \\ \dot{\delta}_2 = -10(\tanh(y_2(t)) - \tanh(x_2(t)))(y_2(t) - x_2(t)), \\ \dot{\hat{a}}_{22} = -30(\tanh(y_2(t)) - \tanh(x_2(t)))\tanh(y_2(t)), \\ \dot{\hat{b}}_{22} = -30(\tanh(y_2(t)) - \tanh(x_2(t)))\tanh(y_2(t - \tau_1(t))), \\ \dot{\hat{d}}_{22} = -30(\tanh(y_2(t)) - \tanh(x_2(t))) \int_{t-\tau_2(t)}^{t} \tanh(y_2(s))ds. \end{cases} \tag{60}$$

Figure 7 and Figure 8 depict the chaotic behaviors of the drive system and response system. Figure 9 shows the error states between the drive system and response system. It can be seen from Figure 10 that the time-varying strengths tend to constants when $t \to \infty$. One can see from Figure 11 that the time evolution of adaptive parameters of $\hat{a}_{22}, \hat{b}_{22}, \hat{d}_{22}$, respectively. It is shown that the unknown parameters can be identified after 50s. Thus, we find the synchronization can be achieved in mean square while the unknown parameters in the drive system can be identified.

The Explanations of Applications

As seen from the above, the derived schemes in Theorem 1 and Theorem 2 are an effective and powerful tool to synchronize the stochastic drive neural networks and response neural networks. Nowadays, the synchronization of chaotic neural networks has become more and more important due to it's widely implications. However, little works have tackled with the synchronization of neural networks with both discrete and distributed time-varying delays. For chaotic neural networks with discrete and distributed delays, there exist chaotic behaviors in neural networks with discrete and distributed delays and the neural networks with mixed delays with known parameters is synchronized (Wang et al, 2008). But in practical situation, the parameters of some systems cannot be exactly known in priori; the effect of these uncertainties will destroy the synchronization and even break it. In this chapter, by the parameter identification laws designed in the response system, the unknown parameters in drive system can be estimated.

Theoretical analysis and numerical simulations show that we can synchronize the dynamical systems on stochastic neural networks. Most of the existing results are concerned with the synchronization of neural networks by state coupling. In this chapter, we consider more general coupling conditions, including outputs and state coupling which result in different theoretical synchronization criteria and are more useful in real-world applications.

Similar methods can be also applied to chaos synchronization, synchronization of complex networks and real-world communication networks. In many communication networks, synchronous transfers of digital or analog signals under noise perturbation are very important. Sometimes, one should synchronize the chaotic systems or complex networks under noise perturbation. Recently, many researchers have already obtained some elementary results that address this issue. It is now possible to synchronize the complex networks under stochastic perturbation by using adaptive feedback method. How to synchronize the complex networks under stochastic perturbation using adaptive feedback methods remains an important but challenging problem.

CONCLUSION

In this chapter, the problem of synchronization for a class of stochastic unknown chaotic networks with mixed time-delays by state coupling and output coupling is investigated in detail, where the mixed time-delays include time-varying delay and distributed delay. Via the adaptive state and output coupling, the theorems for lag synchronization and complete synchronization are derived. All the connection weight matrices can be identified while the complete synchronization is achieved at the same time. Two numerical examples are given to show the effectiveness of the developed method.

ACKNOWLEDGMENT

This research was partially supported by the National Natural Science Foundation of PR China (60874113), the Key Creative Project of Shanghai Education Community (09ZZ66), and the Research Fund for the Doctoral Program of Higher Education (200802550007) and Shanghai Natural Science Foundation (08ZR1400400).

REFERENCES

Boccaletti, S., Kurths, J., Osipov, G., Valladares, D. L., & Zhou, C. S. (2002). The synchronization of chaotic systems. *Physics Reports*, *366*, 1–101. doi:10.1016/S0370-1573(02)00137-0

Cao, J., Li, P., & Wang, W. (2006). Global synchronization in arrays of delayed neural networks with constant and delayed coupling. *Physics Letters. [Part A]*, *353*(4), 318–325. doi:10.1016/j.physleta.2005.12.092

Chen, G. R., & Dong, X. (1998). *From Chaos to Order: Methodologies, Perspectives, and Applications*. Singapore: World Science Publishing Company.

Chen, G. R., Zhou, J., & Liu, Z. R. (2004). Global synchronization of coupled delayed neural networks and applications to chaotic CNN models. *International Journal of Bifurcation and Chaos in Applied Sciences and Engineering*, *14*(7), 2229–2240. doi:10.1142/S0218127404010655

Cheng, C. J., Liao, T. L., & Hwang, C. C. (2005). Exponential synchronization of a class of chaotic neural networks. *Chaos, Solitons, and Fractals, 24*, 197–206.

Friedman, A. (1976). *Stochastic Differential Equations and Applications*. New York: Academic Press.

He, W., & Cao, J. (2008). Adaptive synchronization of a class of chaotic neural networks with known or unknown parameters. *Physics Letters. [Part A], 372*, 408–416. doi:10.1016/j.physleta.2007.07.050

Huang, D. B. (2004). Synchronization-based estimation of all parameters of chaotic systems from time series. *Physical Review E: Statistical, Nonlinear, and Soft Matter Physics, 69*, 067201. doi:10.1103/PhysRevE.69.067201

Huang, D. B. (2005). Simple adaptive-feedback controller for identical chaos synchronization. *Physical Review E: Statistical, Nonlinear, and Soft Matter Physics, 71*, 037203. doi:10.1103/PhysRevE.71.037203

Huang, D. B. (2006). Adaptive-feedback control algorithm. *Physical Review E: Statistical, Nonlinear, and Soft Matter Physics, 73*, 066204. doi:10.1103/PhysRevE.73.066204

Li, P., Cao, J. D., & Wang, Z. D. (2007). Robust impulsive synchronization of coupled delayed neural networks with uncertainties. *Physica A, 373*, 261–272. doi:10.1016/j.physa.2006.05.029

Li, Z., & Chen, G. (2006). Global synchronization and asymptotic stability of complex dynamical networks. *IEEE Transactions on Circuits and Systems, 53*, 28–33. doi:10.1109/TCSII.2005.854315

Liu, Y., Wang, Z., & Liu, X. (2006). Global asymptotic stability of generalized bi-directional associative memory networks with discrete and distributed delays. *Chaos, Solitons, and Fractals, 28*, 793–803. doi:10.1016/j.chaos.2005.08.004

Lou, X., & Cui, B. (2008). Synchronization of neural networks based on parameter identification and via output or state coupling. *Journal of Computational and Applied Mathematics, 222*(2), 440–457. doi:10.1016/j.cam.2007.11.015

Lu, J., & Cao, J. (2007). Synchronization-based approach for parameters identification in delayed chaotic neural networks. *Physica A, 384*, 432–443.

Lu, W., & Chen, T. (2004). Synchronization analysis of linearly coupled networks of discrete time systems. *Physica D. Nonlinear Phenomena, 198*, 148–168. doi:10.1016/j.physd.2004.08.024

Mao, X. (2002). A note on the LaSalle-type theorems for stochastic differential delay equations. *Journal of Mathematical Analysis and Applications, 268*, 125–142. doi:10.1006/jmaa.2001.7803

Ojalvo, J. G., & Roy, R. (2001). Spatiotemporal communication with synchronized optical chaos. *Physical Review Letters, 86*(22), 5204–5207. doi:10.1103/PhysRevLett.86.5204

Pecora, L. M., & Carroll, T. L. (1990). Synchronization in chaotic systems. *Physical Review Letters, 64*, 821–824. doi:10.1103/PhysRevLett.64.821

Ruan, S., & Filfil, R. (2004). Dynamics of a two-neuron system with discrete and distributed delays. *Physica D. Nonlinear Phenomena, 191*, 323–342. doi:10.1016/j.physd.2003.12.004

Shahverdiev, E. M., Sivaprakasam, S., & Shore, K. A. (2002). Lag-synchronization in time-delayed systems. *Physics Letters. [Part A], 292*, 320–324. doi:10.1016/S0375-9601(01)00824-6

Sun, Y., & Cao, J. (2007). Adaptive lag synchronization of unknown chaotic delayed neural networks with noise perturbation. *Physics Letters. [Part A]*, *364*, 277–285. doi:10.1016/j.physleta.2006.12.019

Tang, Y., Fang, J., & Miao, Q. (2009a). Synchronization of stochastic delayed neural networks with Markovian switching and its application. *International Journal of Neural Systems*, *19*, 43–56. doi:10.1142/S0129065709001823

Tang, Y., Fang, J., & Miao, Q. (2009b). On the exponential synchronization of stochastic jumping chaotic neural networks with mixed delays and sector-bounded nonlinearities. *Neurocomputing*, *72*, 1694–1701. doi:10.1016/j.neucom.2008.08.007

Tang, Y., & Fang, J. A. (2008). General methods for modified projective synchronization of hyperchaotic systems with known or unknown parameters. *Physics Letters. [Part A]*, *372*, 1816–1826. doi:10.1016/j.physleta.2007.10.043

Tang, Y., Qiu, R. H., Fang, J. A., Miao, Q., & Xia, M. (2008). Adaptive lag synchronization for unknown stochastic chaotic neural networks with discrete and distributed time-varying delays. *Physics Letters. [Part A]*, *372*, 4425–4433. doi:10.1016/j.physleta.2008.04.032

Tang, Y., Wang, Z., & Fang, J. (2009c). Pinning control of fractional-order weighted complex networks. *Chaos (Woodbury, N.Y.)*, *19*, 013112. doi:10.1063/1.3068350

Wan, L., & Sun, J. (2005). Mean square exponential stability of stochastic delayed Hopfield neural networks. *Physics Letters. [Part A]*, *343*(4), 306–318. doi:10.1016/j.physleta.2005.06.024

Wang, K., Teng, Z., & Jiang, H. (2008). Adaptive synchronization of neural networks with time-varying delay and distributed delay. *Physica A*, *387*, 631–647. doi:10.1016/j.physa.2007.09.016

Wang, Z., Fang, J. A., & Liu, X. (2008). Global stability of stochastic high-order neural networks with discrete and distributed delays. *Chaos, Solitons, and Fractals*, *36*, 388–396. doi:10.1016/j.chaos.2006.06.063

Wang, Z., Lauria, S., Fang, J. A., & Liu, X. (2007). Exponential stability of uncertain stochastic neural networks with mixed time-delays. *Chaos, Solitons, and Fractals*, *32*, 62–72. doi:10.1016/j.chaos.2005.10.061

Wang, Z., Liu, Y., Fraser, K., & Liu, X. (2006). Stochastic stability of uncertain Hopfield neural networks with discrete and distributed delays. *Physics Letters. [Part A]*, *354*, 288–297. doi:10.1016/j.physleta.2006.01.061

Xu, S., Chen, T., & Lam, J. (2003). Robust H Filtering for Uncertain Markovian Jump Systems With Mode-Dependent Time Delays. *IEEE Transactions on Automatic Control*, *48*, 900–907. doi:10.1109/TAC.2003.811277

Yang, T., & Chua, L. O. (1997). Impulsive stabilization for control and synchronization of chaotic system: theory and application to secure communication. *IEEE Trans. Circuits Syst. I*, *44*, 976–988. doi:10.1109/81.633887

Yu, W., & Cao, J. (2007). Synchronization control of stochastic delayed neural networks. *Physica A*, *373*, 252–260. doi:10.1016/j.physa.2006.04.105

Zhao, H. (2004). Global asymptotic stability of Hopfield neural network involving distributed delays. *Neural Networks*, *17*, 47–53. doi:10.1016/S0893-6080(03)00077-7

Chapter 14
Type-2 Fuzzy Sliding Mode Synchronization

Tsung-Chih Lin
Feng-Chia University, Taiwan

Ming-Che Chen
Feng-Chia University, Taiwan

Mehdi Roopaei
Islamic Azad University, Iran

ABSTRACT

This chapter presents an adaptive interval type-2 fuzzy neural network (FNN) controller to synchronize chaotic systems with training data corrupted by noise or rule uncertainties involving external disturbances. Adaptive interval type-2 FNN control scheme and sliding mode approach are incorporated to deal with the synchronization of non-identical chaotic systems. In the meantime, based on the adaptive fuzzy sliding mode control, the Laypunov stability theorem has been used to testify the asymptotic stability of the chaotic systems. The chattering phenomena in the control efforts can be reduced and the stability analysis of the proposed control scheme will be guaranteed in the sense that all the states and signals are uniformly bounded and the external disturbance on the synchronization error can be attenuated. The simulation example is included to confirm validity and performance of the advocated design methodology.

INTRODUCTION

In general, the synchronization phenomenon is happened when two, or more, chaotic oscillators are coupled, or when a chaotic oscillator drives another chaotic oscillator. In the other word in the synchronization problem the output of the drive system is used to control the response system so that the output of the response system follows the output of the drive system asymptotically. Although chaotic systems have deterministic behavior, they are extremely sensitive to initial conditions and difficult to predict. Motivated by potential applications in chaos synchronization, such as communication theory, biological engineering, pattern recognition and information processing, control chaotic dynamics has received and

DOI: 10.4018/978-1-61520-737-4.ch014

increasing interest. The OGY method, a model-free chaos control method, was proposed to stabilize one of the unstable periodic orbits by perturbing an accessible system parameter over (Ott, Grebogi, & Yorke, 1990). Besides, many chaos control strategies have been presented based on feedback control technologies as well as sliding mode control (SMC) (Chang, 2001; Kim, Yang, & Hong, 2003; Leu, Lee, & Wang, 1999; Li & Tong, 2003; S. C. Lin & Chen, 1994; Palm, 1992; Sastry & Bodson, 1989; Chi Hsu Wang, Lin, Lee, & Liu, 2002; Chi Hsu Wang, Liu, & Lin, 2002; L. X. Wang, 1993, 1994; L. X. Wang & Mendel, 1992; W. Y. Wang, Chan, Hsu, & Lee, 2002; Yoo & Ham, 1998; Zheng, Liu, Tong, & Li, 2009). Recently, the study of chaos synchronization has become a hot spot in the nonlinear dynamics field and researchers in this field have explored a variety of problems on chaos synchronization, such as the stability conditions for chaos synchronization, the realization for a successful synchronization and the applications of chaos synchronization.

In recent years, some chaos synchronizations based on fuzzy systems have been proposed (Noroozi, Roopaei, Balas, & Lin, 2009; Noroozi, Roopaei, Karimaghaee, & Safavi, 2010; Noroozi, Roopaei, & Zolghadri Jahromi, 2009; Mehdi Roopaei & Jahromi, 2008; Mehdi Roopaei, Zolghadri Jahromi, & Jafari, 2009; Mehdi Roopaei, Zolghadri, John, & Lin, in press; Mehdi Roopaei, Zolghadri, & Meshksar, 2009; Zadeh, 1965). The fuzzy set theory was initiated by Zadeh (Chen, Lee, & Chang, 1996). Recently, intelligent control approach has been done on applications of FNNs, which combine the capability of fuzzy reasoning to handle uncertain information and the capability of artificial neural networks to learn from processes. The FNNs do not require mathematical models and have the ability to approximate nonlinear and uncertainties systems. Therefore, there were many researches using FNNs to represent complex plants and construct advanced controllers (Chen, Tseng, & Uang, 2000; Golea, Golea, & Benmahammed, 2003; Hojati & Gazor, 2002; Kim et al., 2003; Kosko, 1994; Kovacic, Balenovic, & Bogdan, 1998; C. C. Lee, 1990; H. Lee & Tomizuka, 2001; Leu et al., 1999; Li & Tong, 2003; J. M. Mendel, 2004; Nguang & Shi, 2003; Sastry & Bodson, 1989; Tseng & Chen, 2001; C. H. Wang, T. C. Lin et al., 2002; C. H. Wang, H. L. Liu et al., 2002; J. S. Wang & Lee, 2002; L. X. Wang, 1993, 1994, 1997; L. X. Wang & Mendel, 1992; Yang & Zhou, 2005; Zheng et al., 2009) based on the back propagation algorithm. Currently, there were only few works to analyze and simulate the type-2 FNN (Hsiao, Li, Lee, Chao, & Tsai, 2008; Karnik, Mendel, & Liang, 1999; Kheireddine, Lamir, Mouna, & Hier, 2007; Tsung Chih Lin, 2009; Tsung Chih Lin, Kuo, & Hsu, in press; Tsung Chih Lin, Liu, & Kuo, 2009; J.M Mendel, 2007; Jerry M. Mendel, John, & Liu, 2006; J. M. Mendel & John, 2000; C. H. Wang, Cheng, & Lee, 2004)

As the membership functions for the type-1 fuzzy sets contain no uncertainty information, the control problem cannot be directly handled if the nonlinear system has the rule uncertainty will be existed in following three possible ways, (i1 the words that are used in antecedents and consequents of rules can mean different things to different people; (2) consequents obtained by polling a group of experts will often be different for the same rule because the experts will not necessary be in agreement; and (3) noisy training data. Type-2 FLSs, are very useful where it is difficult to determine an exact membership function, and there are measurement uncertainties. It is known that type-2 fuzzy sets enable modeling and minimizing the effects of uncertainties in rule-based fuzzy logic systems. Type-2 fuzzy sets are able to model such uncertainties because their membership functions are themselves fuzzy. A type-1 fuzzy set is a special case of a type-2 fuzzy set. A type-2 FLS is again characterized by IF-THEN rules, but its antecedent or consequent sets are now of type-2 (Hsiao et al., 2008; Karnik et al., 1999; Kheireddine et al., 2007; Tsung Chih Lin, 2009; J.M Mendel, 2007; Jerry M. Mendel et al., 2006; J. M. Mendel & John, 2000; C. H. Wang et al., 2004). The type-2 FLS has been successfully applied to fuzzy neural network, VLSI testing (Tsung Chih Lin, 2009) and fuzzy controller designs.

This chapter systematically investigates a synchronization scheme consisting of two non identical chaotic systems by an interval type-2 adaptive fuzzy sliding mode controller. The performance of the proposed scheme has been compared with the same methodology with type-1 and it is shown that type-2 can handle better synchronization while it guarantees the synchronization performance and asymptotical stability regards to Lyapunov theory.

This chapter is organized as follows. Problem formulation and the system description are given in section 2. A brief of the interval type-2 fuzzy logic system is presented in section 3. Adaptive interval type-2 fuzzy sliding mode control for chaos synchronization between two different chaotic systems is presented in section 4. Simulation example to demonstrate the performances of the proposed method is provided in section 5. Section 6 gives the conclusions of the advocated design methodology.

PROBLEM FORMULATION AND SYSTEM DESCRIPTION

Consider the two nth-order chaotic systems of the form.

The master system:

$$\begin{cases} x_i = x_{i+1}, 1 \le i \le n-1, \\ x_n = g(x,t) \end{cases} x = \left[x_1, x_2, \cdots, x_n\right] \in \Re^n \tag{1}$$

The slave system:

$$\begin{cases} y_i = y_{i+1}, 1 \le i \le n-1, \\ y_n = f(x,t) + d(t) + u(t) \end{cases} y = \left[y_1, y_2, \cdots, y_n\right] \in \Re^n \tag{2}$$

where $g(x,t)$ and $f(y,t)$ are unknown but bounded functions; $x_i \in R$ and $y_i \in R$, $i=1,2\cdots,n$ are state outputs of the master system and the slave system, respectively; $u(t) \in R$ is the control input and $d(t)$ *is the* external bounded disturbance, $|d(t) \le D$.

The second system is driven by the first system but the behavior of the first system is not affected by the second system. The control objective is to design an appropriate control input $u(t)$ in slave system to synchronize both systems, i.e., master and slave systems, under the constraint that all signals involved must be bounded.

To begin with, the synchronization errors between the master system and the slave system will be defined as

$$e_i = y_i - x_i,\ i=1,2,\cdots,n \tag{3}$$

Therefore, the error dynamics of the driven system and the response system can be obtained as

$$\begin{cases} e_1 = e_2, \\ e_2 = e_3, \\ \vdots \\ e_{n-1} = e_n, \\ e_n = f(y,t) - g(x,t) + d(t) + u(t) \end{cases} \tag{4}$$

$e = [e_1, \ddot{e}_1, \cdots e_1^{(n-1)}]$

In general, in the space of the error state a sliding surface is defined by

$$s(e,t) = -ke = -\left(k_1 e + k_2 \dot{e} + L + k_{n-1} e^{(n-2)} + e^{(n-1)}\right) \quad (5)$$

where in which the k_i's are all real and are chosen such that $h(r) = \sum_{i=1}^{i=1} k_i r^{(i-1)}, k_n = 1$ is a Hurwitz polynomial where r is a Laplace operator. The synchronizing problem will be considered as the state error vector e remaining on the sliding surface $s(e,t)=0$ for all $t \geq 0$. The sliding mode control process can be classified into two phases, the approaching phase with $s(e,t) \neq 0$ and the sliding phase with $s(e,t)=0$ for initial error $e(0)=0$. In order to guarantee that the trajectory of the state error vector e will translate from the approaching phase to the sliding phase, the sufficient condition

$$s(e,t) \cdot s(e,t) \leq -\eta |s|, \ \eta > 0 \quad (6)$$

must be satisfied. Two type of control law must be derived separately for those two phases described above. In the sliding phase, it implies $s(e,t)=0$ and $\dot{s}(e,t) = 0$. In order to force the system dynamics to stay on the sliding surface, the equivalent control can be derived as follows:

If $f(y,t)$ and $g(x,t)$ are known and free of external disturbance, i.e., $d(t) = 0$, taking the derivative of the sliding surface with respective to time, we get

$$\dot{s} = -\left(k_1 \dot{e}_1 + k_2 \ddot{e}_1 + \cdots + k_{n-1} e_1^{(n-1)} + e_1^{(n)}\right) = -\left(\sum_{i=1}^{n-1} k_i e_1^{(i)} + e_1^{(n)}\right)$$

$$= -\left(\sum_{i=1}^{n-1} k_i e_1^{(i)} + f(y,t) - g(x,t) + u\right)$$

$$= -\sum_{i=1}^{n-1} k_i e_1^{(i)} - f(y,t) + g(x,t) - u = 0 \quad (7)$$

Therefore, the equivalent control can be obtained as

$$u_{eq} = -\sum_{i=1}^{n-1} k_i e_1^{(i)} - f(y,t) + g(x,t) \quad (8)$$

On the contrary, in the approaching phase,, a approaching-type control must be added in order satisfy the sufficient condition (6) and the complete sliding mode control will be expressed as

$$u = u_{eq} + u_r, \ u_r = \eta_\Delta \operatorname{sgn}(s) \quad (9)$$

where $\eta_\Delta \geq \eta > 0$.

To obtain the sliding mode control (9), the system functions $f(y,t)$, $g(x,t)$ and switching parameter η_Δ must be known in advance. However, $f(y,t)$, $g(x,t)$ and η_Δ are unknown and $d(t) \neq 0$ in our problem, it is impossible to obtain the control (8). The purpose of this paper is to approximate $f(y,t)$, $g(x,t)$ and u_r by interval type-2 fuzzy logic system (FLS). Furthermore, the adaptive laws to adjust parameters will be derived.

BRIEF DESCRIPTION OF INTERVAL TYPE-2 FUZZY LOGIC SYSTEM

Due to the complexity of the type reduction, the general type-2 FLS becomes computationally intensive. In order to make things simpler and easier to compute meet and join operations, the secondary MFs of an interval type-2 FLS are all unity which leads finally to simplify type reduction. The 2-D interval type-2 Gaussian membership function (MF) with uncertain mean $m \in [m_1, m_2]$ and a fixed deviation is shown in Figure 1.

$$\mu_{\tilde{A}}(x) = \exp\left[-\frac{1}{2}\left(\frac{x-m}{\sigma}\right)^2\right], m \in [m_1, m_2] \tag{10}$$

It is obvious that the type-2 fuzzy set is in a region bounded by an upper MF and a lower MF denoted as $\overline{\mu}_{\tilde{A}}(x)$ and $\underline{\mu}_{\tilde{A}}(x)$, respectively, and is called a foot of uncertainty (FOU). In the meantime, the firing strength for the *i*th rule can be an interval type-1 set expressed as

$$F^i = \begin{bmatrix} \underline{f}^i \\ \overline{f}^i \end{bmatrix} \tag{11}$$

where

$$\underline{f}^i = \underline{\mu}_{\tilde{F}_1^i}(x_1) * \cdots * \underline{\mu}_{\tilde{F}_1^i}(x_n) = \prod_{j=1}^{n} \underline{\mu}_{\tilde{F}_j^i}(x_j) \tag{12}$$

Figure 1. Interval type-2 fuzzy set with uncertain mean

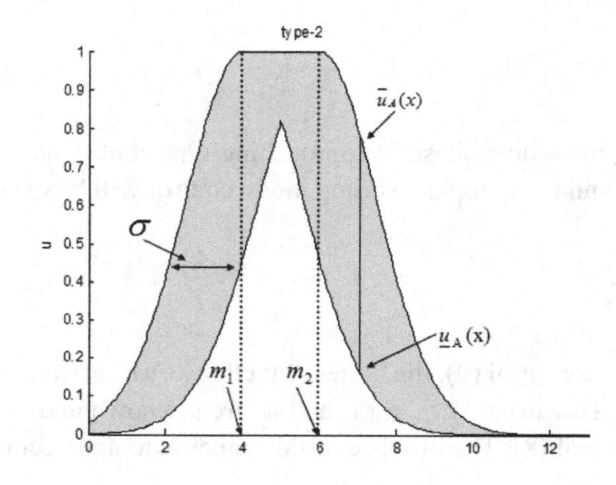

$$\overline{f}^{i} = \overline{\mu}_{\tilde{F}_1^i}(x_1) * \cdots * \overline{\mu}_{\tilde{F}_n^i}(x_n) = \prod_{j=1}^{n} \overline{\mu}_{\tilde{F}_j^i}(x_j) \tag{13}$$

Based on the center of sets type reduction, the defuzzified crisp output from an interval type-2 FLS is the average of y_l and y_r, i.e.,

$$y(\underline{x}) = \frac{y_l + y_r}{2} = \frac{1}{2}\left[\underline{\xi_r^T}\ \underline{\xi_l^T}\right]\begin{bmatrix}\underline{\Theta}_r \\ \underline{\Theta}_l\end{bmatrix} = \underline{\xi}^T \underline{\Theta} \tag{14}$$

where and are the left most and right most points of the interval type-1 set which can be obtained as

$$y_l = \frac{\sum_{i}^{M} f_l^i w_l^i}{\sum_{i}^{M} f_l^i} = \frac{\sum_{i=1}^{L} \overline{f}^i w_l^i + \sum_{i=L+1}^{M} \underline{f}^i w_l^i}{\sum_{i=1}^{L} \overline{f}^i + \sum_{i=L+1}^{M} \underline{f}^i}$$

$$= \sum_{i=1}^{L} \overline{q}_l^i y_l^i + \sum_{i=L+1}^{M} \underline{q}_l^i y_l^i = \left[\underline{Q}_l \overline{Q}^l\right]\begin{bmatrix}\underline{y}_l \\ \overline{y}^l\end{bmatrix} = \underline{\xi}_l^T \underline{\Theta}_l \tag{15}$$

where $\underline{q}_r^i = \underline{f}^i / D_r, \overline{q}_r = \overline{f}^i / D_r$ and $D_r = \left(\sum_{i=1}^{R} \underline{f}^i + \sum_{i=R+1}^{M} \overline{f}^i\right)$. In the meantime, we have $\underline{Q}_r = [\underline{q}_r^1, \underline{q}_r^2, \cdots, \underline{q}_r^R], \overline{Q}^r = [\overline{q}_r^1, \overline{q}_r^2, \cdots, \overline{q}_r^R], \underline{\xi}_r^T = [\underline{Q}_r \overline{Q}^r]$ and $\underline{\Theta}_r^T = [\underline{y}_r \overline{y}^r]$.

Moreover

$$y_r = \frac{\sum_{i}^{M} f_r^i w_r^i}{\sum_{i}^{M} f_r^i} = \frac{\sum_{i=1}^{R} \underline{f}^i w_r^i + \sum_{i=R+1}^{M} \overline{f}^i w_r^i}{\sum_{i=1}^{R} \underline{f}^i + \sum_{i=R+1}^{M} \overline{f}^i}$$

$$= \sum_{i=1}^{R} \underline{q}_r^i y_r^i + \sum_{i=R+1}^{M} \overline{q}_r^i y_r^i = \left[\underline{Q}_r \overline{Q}^r\right]\begin{bmatrix}\underline{y}_r \\ \overline{y}^r\end{bmatrix} = \underline{\xi}_r^T \underline{\Theta}_r \tag{16}$$

where $\underline{q}_r^i = \underline{f}^i / D_l, \overline{q}_l = \overline{f}^i / D_l$ and $D_l = \left(\sum_{i=1}^{L} \overline{f}^i + \sum_{i=L+1}^{M} \underline{f}^i\right)$. In the meantime, we have $\underline{Q}_l = [\underline{q}_l^1, \underline{q}_l^2, \cdots, \underline{q}_l^R], \overline{Q}^l = [\overline{q}_l^1, \overline{q}_l^2, \cdots, \overline{q}_l^R], \underline{\xi}_l^T = [\overline{Q}_r \underline{Q}^r], \underline{\Theta}_l^T = [\overline{y}_l \underline{y}^l]$, and M is the total number of rules in the rule base of the interval type-2 fuzzy neural network. The weighting factors w_l^i and w_r^i of the consequent part represents the centroid interval set of the consequent type-2 fuzzy set of the $i$the rule.

In the meantime, R and L can be determined by using the iterative Karnik-Mendel procedure (Karnik et al., 1999).

A type-2 fuzzy logic system (FLS) is very similar to a type-1 FLS as shown in Figure 2, the major structure difference being that the defuzzifier block of a type-1 FLS is replaced by the output processing block in a typ-2 FLS which consists of type-reduction followed by defuzzification.

There are five main parts in a type-2 FLS: fuzzifier, rule base, inference engine, type-reducer and defuzzifier. A type-2 FLS is a mapping f:$R^p \rightarrow R^1$. After defuzzification, fuzzy inference, type-reduction and defuzzification, a crisp output can be obtained.

Consider a type-2 FLS having p inputs $x_1 \in X_1, \cdots, x_p \in X_p$ and one output $y \in Y$. The type-2 fuzzy rule base consists of a collection of IF-THEN rules. As in the type-1 case, we assume there are M rules and the rule of a type-2 relation between the input space $X_1 \times X_2 \times \cdots \times X_p$ and the output space Y can be expressed as

$$R^l : \text{IF } x_1 \text{ is } \tilde{F}_1^l \text{ and } \cdots \text{ and } x_p \text{ is } \tilde{F}_p^l, \text{ THEN } y \text{ is } \tilde{G}^l \cdots l = 1, 2, \cdots, M \tag{17}$$

where $\tilde{F}_j^l$ s are antecedent type-2 sets (j=1,2,⋯,p) and $\tilde{G}^l$ s are consequent type-2 sets.

The inference engine combines rules and gives a mapping from input type-2 fuzzy sets to output type-2 fuzzy sets. To achieve this process, we have to compute unions and intersection of type-2 sets, as well as compositions of type-2 relations. The output of inference engine block is a type-2 set. By using the extension principle of type-1 defuzzification method, type-reduction takes us from type-2 output sets of the FLS to a type-1 set called the "type-reduced set". This set may then be defuzzified to obtain a single crisp value.

SYNCHRONIZATION USING INTERVAL TYPE-2 FUZZY SLIDING MODE CONTROL

To begin with, our objective is to use interval type-2 FNN as shown in Figure 3 to approximate the nonlinear functions $f(y,t)$, $g(x,t)$ and u_r. Based on the Lyapunov approach, the adaptive laws can be developed to adjust the parameters of FNN to attenuate the synchronization error and external disturbance.

Figure 2. The structure of Type-2 fuzzy logic system

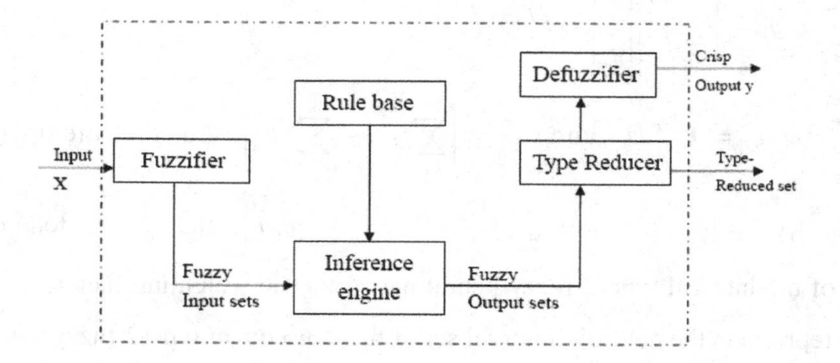

We replace $f(y,t)$, $g(x,t)$ and u_r in (1), (2) and (9) by the interval type-2 FNN $f(y \mid \underline{\theta}_f)$ $g(x \mid \underline{\theta}_g)$ and $h(s \mid \underline{\theta}_h)$, respectively, as in (14)

$$f(y \mid \underline{\theta}_f) = \frac{1}{2}\begin{bmatrix} \underline{\xi}_{fr}^T & \underline{\xi}_{fl}^T \end{bmatrix}\begin{bmatrix} \underline{\theta}_{fr} \\ \underline{\theta}_{fl} \end{bmatrix} = \underline{\xi}_f^T \underline{\theta}_f \tag{18}$$

$$g(x \mid \underline{\theta}_g) = \frac{1}{2}\begin{bmatrix} \underline{\xi}_{gr}^T & \underline{\xi}_{gl}^T \end{bmatrix}\begin{bmatrix} \underline{\theta}_{gr} \\ \underline{\theta}_{gl} \end{bmatrix} = \underline{\xi}_g^T \underline{\theta}_g \tag{19}$$

$$h(s \mid \underline{\theta}_h) = \frac{1}{2}\begin{bmatrix} \underline{\xi}_{hr}^T & \underline{\xi}_{hl}^T \end{bmatrix}\begin{bmatrix} \underline{\theta}_{hr} \\ \underline{\theta}_{hl} \end{bmatrix} = \underline{\xi}_h^T \underline{\theta}_h \tag{20}$$

Therefore, if the control input is chosen

$$u = -\sum_{i=1}^{n-1} k_i e_1^{(i)} - f(\underline{y} \mid \underline{\theta}_f) + g(\underline{x} \mid \underline{\theta}_g) + h(s \mid \underline{\theta}_h) \tag{21}$$

with the assumption that $|h(s|\underline{\theta}_h)|=D+\eta_\Delta+\omega_{\max}$.

In order to adjust the parameters in the fuzzy logic systems, we have to derive adaptive laws. Hence, the optimal parameter estimations $\underline{\theta}_f^*$, $\underline{\theta}_g^*$ and $\underline{\theta}_h^*$ are defined as

$$\underline{\theta}_f^* = \arg\min_{\underline{\theta}_f \in \Omega_f} \left[\sup_{y \in \Omega_y} | f(y \mid \underline{\theta}_f) - f(y,t) | \right] \tag{22}$$

Figure 3. The structure of interval type-2 fuzzy neural network

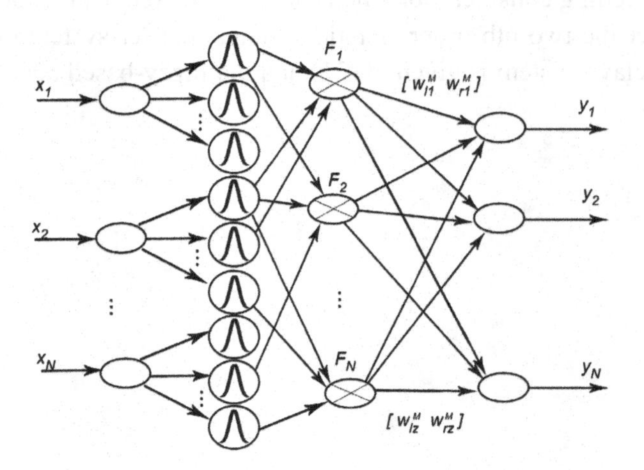

$$\underline{\theta}_g^* = \arg\min_{\underline{\theta}_g \in \Omega_g} \left[\sup_{x \in \Omega_x} | g(x \mid \underline{\theta}_g) - g(x,t) | \right] \tag{23}$$

$$\underline{\theta}_h^* = \arg\min_{\underline{\theta}_h \in \Omega_h} \left[\sup_{s \in \Omega_s} | h(s \mid \underline{\theta}_h) - u_r | \right] \tag{24}$$

where Ω_f, Ω_g, Ω_h, Ω_x, Ω_y, and Ω_s are constraint sets of suitable bounds on $\underline{\theta}_f$, $\underline{\theta}_g$, $\underline{\theta}_h$, y x and s, respectively and they are defined as $\Omega_f = \left\{ \underline{\theta}_f \mid \left| \underline{\theta}_f \right| \le M_f \right\}$, $\Omega_g = \left\{ \underline{\theta}_g \mid \left| \underline{\theta}_g \right| \le M_g \right\}$, $\Omega_y = \left\{ y \mid \left| y \right| \le M_y \right\}$ $\Omega_x = \left\{ x \mid \left| x \right| \le M_x \right\}$ and $\Omega_s = \left\{ s \mid \left| s \right| \le M_s \right\}$, where M_f, M_g, M_h,, and M_s are positive constants.

Define the minimum approximation errors as

$$\omega = -\left(f(y,t) - f(y \mid \underline{\theta}_f^*) \right) + \left(g(x,t) - g(x \mid \underline{\theta}_g^*) \right) \tag{25}$$

The time derivative of the sliding surface is

$$\dot{s} = -\sum_{i=1}^{n-1} k_i e_1^{(i)} - f(y,t) + g(x,t) - u$$

$$\begin{aligned}
&= f(y \mid \underline{\theta}_f) - f(y,t) - g(x \mid \underline{\theta}_g) + g(x,t) - h(s \mid \underline{\theta}_h) - d(t) + f(y \mid \underline{\theta}_f^*) \\
&+ g(x \mid \underline{\theta}_g^*) - g(x \mid \underline{\theta}_g^*) + h(s \mid \underline{\theta}_h^*) - h(s \mid \underline{\theta}_h^*) \\
&= \omega + \left(f(y \mid \underline{\theta}_f) - f(y \mid \underline{\theta}_f^*) \right) - \left(g(x \mid \underline{\theta}_g) - g(x \mid \underline{\theta}_g^*) \right) - \left(h(s \mid \underline{\theta}_h) - h(s \mid \underline{\theta}_h^*) \right) \\
&- d(t) - h(s \mid \underline{\theta}_h^*) \\
&= \omega + \tilde{\theta}_f \xi(y) - \tilde{\theta}_g \xi(x) - \tilde{\theta}_h \phi(s) - h(s \mid \underline{\theta}_h^*) - d(t)
\end{aligned} \tag{26}$$

where $\tilde{\underline{\theta}}_f = \underline{\theta}_f - \underline{\theta}_f^*$, $\tilde{\underline{\theta}}_g = \underline{\theta}_g - \underline{\theta}_g^*$, and $\tilde{\underline{\theta}}_h = \underline{\theta}_h - \underline{\theta}_h^*$.

Therefore, from proceeding consideration, the following theorem can be obtained.

***Theorem* 1:** Consider the two nth-order chaotic systems, master system (1) and slave system (2), the control input of the clave system is given in (21) and the fuzzy-based adaptive laws are chosen as

$$\dot{\underline{\theta}}_f = -r_f s \xi(y) \tag{27}$$

$$\dot{\underline{\theta}}_g = -r_g s \xi(x) \tag{28}$$

$$\dot{\underline{\theta}}_h = -r_h s \phi(s) \tag{29}$$

Then, the overall adaptive scheme guarantees the global stability of the resulting closed-loop system in the sense that all signals involved are uniformly bounded and the synchronization error will converge to zero asymptotically. The overall control scheme is shown in Figure 4.

Proof: In order to analyze the closed-loop stability, the Lyapunov function candidate is chosen as

$$V = \frac{1}{2}s^2 + \frac{1}{2r_f}\tilde{\underline{\theta}}_f^T\tilde{\underline{\theta}}_f + \frac{1}{2r_g}\tilde{\underline{\theta}}_g^T\tilde{\underline{\theta}}_g + \frac{1}{2r_h}\tilde{\underline{\theta}}_h^T\tilde{\underline{\theta}}_h \tag{30}$$

Taking the derivative of the Equation (30) with respect to time, we get

$$\dot{V} = s\dot{s} + \frac{1}{r_f}\tilde{\underline{\theta}}_f^T\dot{\tilde{\underline{\theta}}}_f + \frac{1}{r_g}\tilde{\underline{\theta}}_g^T\dot{\tilde{\underline{\theta}}}_g + \frac{1}{r_h}\tilde{\underline{\theta}}_h^T\dot{\tilde{\underline{\theta}}}_h \tag{31}$$

By substituting (26) into (31) and using (25) yields

$$\dot{V} = s\left\{\omega + \tilde{\theta}_f\xi(y) - \tilde{\theta}_g\xi(x) - \tilde{\theta}_h\phi(s) - h(s \mid \theta_h^*) - d(t)\right\} + \frac{1}{r_f}\tilde{\underline{\theta}}_f^T\dot{\tilde{\underline{\theta}}}_f + \frac{1}{r_g}\tilde{\underline{\theta}}_g^T\dot{\tilde{\underline{\theta}}}_g + \frac{1}{r_h}\tilde{\underline{\theta}}_h^T\dot{\tilde{\underline{\theta}}}_h$$

$$= s\omega + s\tilde{\theta}_f\xi(y) - s\tilde{\theta}_g\xi(x) - s\tilde{\theta}_h\phi(s) - sh(s \mid \theta_h^*) - sd(t) + \frac{1}{r_f}\tilde{\underline{\theta}}_f^T\dot{\tilde{\underline{\theta}}}_f + \frac{1}{r_g}\tilde{\underline{\theta}}_g^T\dot{\tilde{\underline{\theta}}}_g + \frac{1}{r_h}\tilde{\underline{\theta}}_h^T\dot{\tilde{\underline{\theta}}}_h$$

$$= s\omega + \frac{1}{r_f}\tilde{\underline{\theta}}_f^T\left(\dot{\tilde{\underline{\theta}}}_f + r_f s\xi(y)\right) + \frac{1}{r_g}\tilde{\underline{\theta}}_g^T\left(\dot{\tilde{\underline{\theta}}}_g + r_g s\xi(x)\right) + \frac{1}{r_h}\tilde{\underline{\theta}}_h^T\left(\dot{\tilde{\underline{\theta}}}_h + r_h s\phi(s)\right)$$
$$-sh(s \mid \underline{\theta}_h^*) - sd(t)$$

$$\leq \frac{1}{r_f}\tilde{\underline{\theta}}_f^T\left(\dot{\tilde{\underline{\theta}}}_f + r_f s\xi(y)\right) + \frac{1}{r_g}\tilde{\underline{\theta}}_g^T\left(\dot{\tilde{\underline{\theta}}}_g - r_g s\xi(x)\right) + \frac{1}{r_h}\tilde{\underline{\theta}}_h^T\left(\dot{\tilde{\underline{\theta}}}_h - r_h s\phi(s)\right) - (D + \eta_\Delta)\,\mathrm{sgn}(s)$$

Figure 4. Overall scheme of the adaptive interval type-2 fuzzy sliding mode control system

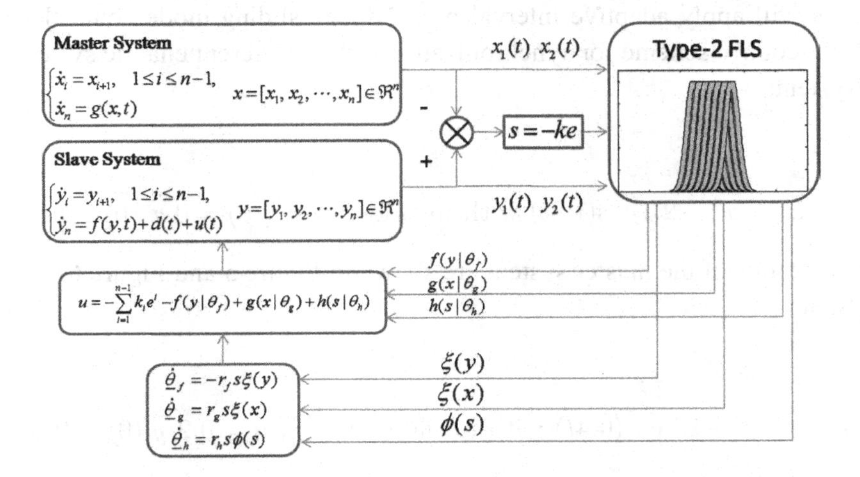

$$< \frac{1}{r_f}\tilde{\underline{\theta}}_f^T\left(\dot{\tilde{\underline{\theta}}}_f + r_f s\underline{\xi}(y)\right) + \frac{1}{r_g}\tilde{\underline{\theta}}_g^T\left(\dot{\tilde{\underline{\theta}}}_g - r_g s\underline{\xi}(x)\right) + \frac{1}{r_h}\tilde{\underline{\theta}}_h^T\left(\dot{\tilde{\underline{\theta}}}_h - r_h s\underline{\phi}(s)\right) - |s|\,\eta_\Delta$$

where $\dot{\tilde{\underline{\theta}}}_f = \dot{\underline{\theta}}_f$, $\dot{\tilde{\underline{\theta}}}_g = \dot{\underline{\theta}}_g$, and $\dot{\tilde{\underline{\theta}}}_h = \dot{\underline{\theta}}_h$. Substitute (27)-(29), we have

$$\begin{aligned} \dot{V} &< -s\eta_\Delta \operatorname{sgn}(s) \\ &< -|s|\,\eta_\Delta \end{aligned} \tag{33}$$

Since $k=[k_1,k_2,\cdots,k_{n-1},1]$ and $e=[e_1,\dot{e}_1,\cdots,e_1^{(n-1)}]$ in which the k_i's are all real and are chosen such that $h(r) = \sum_{i=1}^{i=1} k_i r^{(i-1)}, k_n = 1$ is a Hurwitz polynomial where r is a Laplace operator, we have $\lim_{t\to\infty} |e(t)| = 0$ and $\lim_{t\to\infty} |s(e,t)| = 0$ (Sastry & Bodson, 1989).The proof is completed.

The overall scheme of the adaptive interval type-2 fuzzy sliding mode control is shown in Figure 4.

To summarize above analysis, the design algorithm for adaptive interval type-2 fuzzy sliding mode control is proposed as follows:

- Step 1: Specify the desired coefficients k_i, such that $h(r)$ is a Hurwitz polynomial and obtain the sliding surface.
- Step 2: Define the membership functions $\mu_{F_i^l}(x)$ $\mu_{F_i^l}(y)$ and $\mu_{F_i^l}(s)$ for and compute the fuzzy basis functions $\underline{\xi}(x)$ $\underline{\xi}(y)$ and $\underline{\phi}(s)$.
- Step 3: Select suitable the adaptive parameters r_f, r_g, r_h.
- Step 4: Obtain the control and apply to the plant, then compute the adaptive law (27)-(29) to adjust the parameter vector $\underline{\theta}_f$ $\underline{\theta}_g$ and $\underline{\theta}_h$. Then control input of the $u(t)$ slave system can be constructed as (21).

SIMULATION EXAMPLE

In this section, we will apply adaptive interval type-2 fuzzy sliding mode controller to evaluate the performance of our control scheme for synchronization of two different chaotic systems as follows:

The master system:

$$\begin{cases} \dot{x}_1 = x_2 \\ \dot{x}_2 = -0.4x_2 - 1.1x_1 - x_1^3 - 2.1\cos(1.8t) \text{ with initial states:} x_1(0) = 0, x_2(0) = 0 \end{cases}$$

The simulation results of the master system are shown in Figure 5 and Figure 6.

The slave system:

$$\begin{cases} \dot{y}_1 = x_2 \\ \dot{y}_2 = 1.8y_1 - 0.1y_2 - y_1^3 - 1.1\cos(0.4t) \text{ with initial states:} y_1(0) = 0.2, y_2(0) = 0.2 \end{cases}$$

The simulation results of the slave system are shown in Figure 7 and Figure 8.

In order to force the slave system to track the master system, the adaptive interval type-2 fuzzy sliding mode control $u(t)$ is added into the slave system as follows:

The slave system:

$$\begin{cases} \dot{y}_1 = y_2 \\ \dot{y}_2 = 1.8y_1 - 0.1y_2 - y_1^3 - 1.1\cos(0.4t) + d(t) + u(t) \end{cases}$$

where $d(t) = 0.2\sin(2t)$ and $|d(t)| \leq 0.2 = D$. According to the design procedure, the design is given in the following steps:

- Step 1: The feedback gain matrix is chosen as $k=[k_2 k_1]=[15,1]$,and the sliding surface is obtained as $s = -(k_1\dot{e}_1 + k_2 e_1)$.
- Step 2: Specify the design parameters $r_f=30, r_g=30, r_h=30$, in adaptive laws (27)-(29), simulation time $t_f=20$ second and the step size $h=0.01$.

Figure 5. Time responses of states $x_1(t)$ and $x_2(t)$

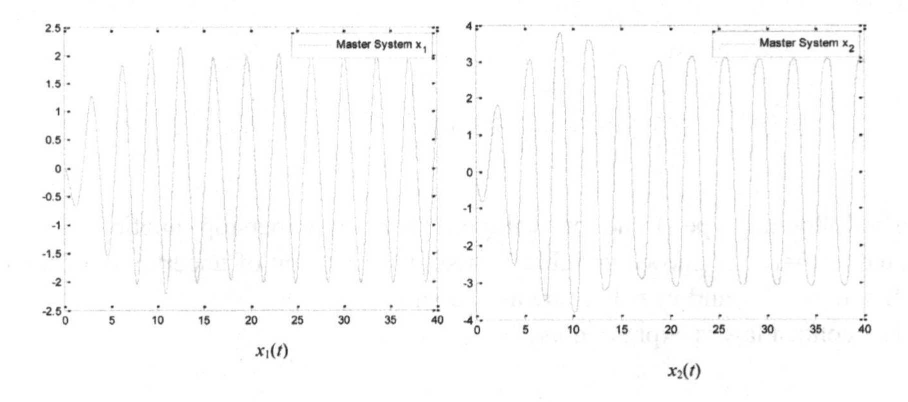

Figure 6. Phase-plane trajectory of the master system

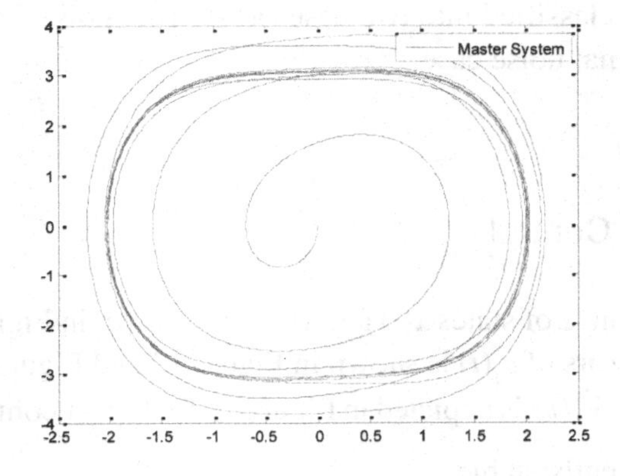

Figure 7. Time responses of states $y_1(t)$ and $y_2(t)$

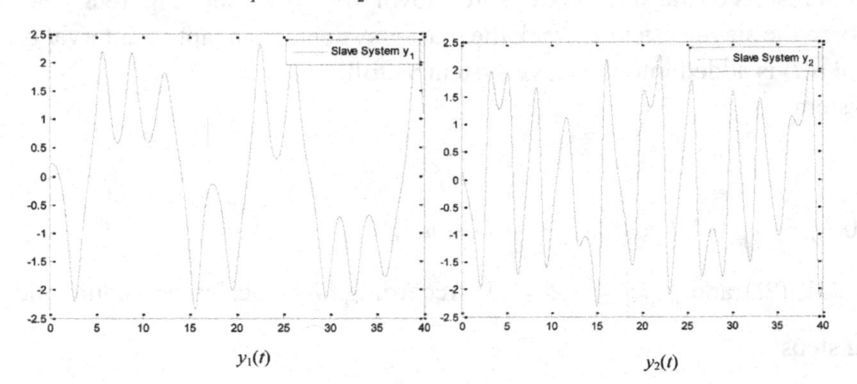

Figure 8. Phase-plane trajectory of the slave system

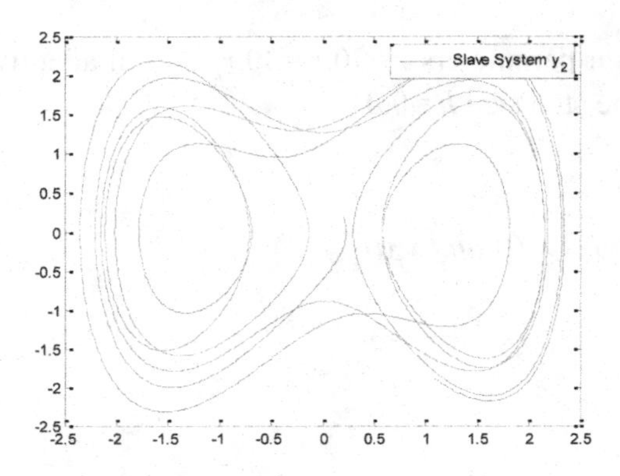

- Step 3: The following type -1 and interval type-2 fuzzy membership functions for x_i and y_i $i = 1, 2$ are selected as, j=1,...,7 shown in Table 1. Also, the footprint of uncertainty of the type-2 membership function for x_i and y_i $i = 1, 2$, are as shown in Figure 9.
- Step 4: The control law is expressed as

$$u = -\sum_{i=1}^{n-1} k_i e^i - f(y \mid \theta_f) + g(x \mid \theta_g) + h(s \mid \theta_h)$$

The simulation results are classified into two cases as flollws: noise free case and training data corrupted with SNR=20 dB internal noise case.

Part I: Noise Free Case

A. Adaptive Type-1 FNN Control

The synchronization performance of states $x(t)$ and $y(t)$ is as shown in Figure 10 and Figure 11, respectively. The synchronization errors of $e_1(t)$ is shown in Figure 12 and Figure 13 shows control input $u(t)$.

Furthermore, the trajectory $\dot{V}(t)$ is depicted in Figure 14 for type-1 controller, respectively, is always negative defined and consequently stable.

Figure 9. The interval type-2 Gaussian membership functions for x_i and y_i $i = 1, 2$

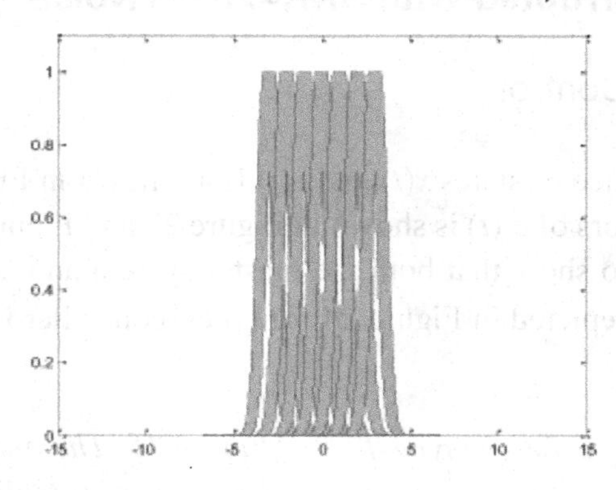

B. Adaptive Interval Type-2 FNN Control

The synchronization performance of states $x(t)$ and $y(t)$ is as shown in Figure 15 ad Figure 16, respectively. The synchronization errors of $e_1(t)$ is shown in Figure 17 and Figure 18 shows control input $u(t)$.

Moreover, the trajectory $\dot{V}(t)$ Equation (35) depicted in Figure 19 for type-2 controller, respectively, is always negative defined and consequently stable.

Figure 10. The output trajectories of y_1 (dashed line) and $x_1(t)$ (solid line)

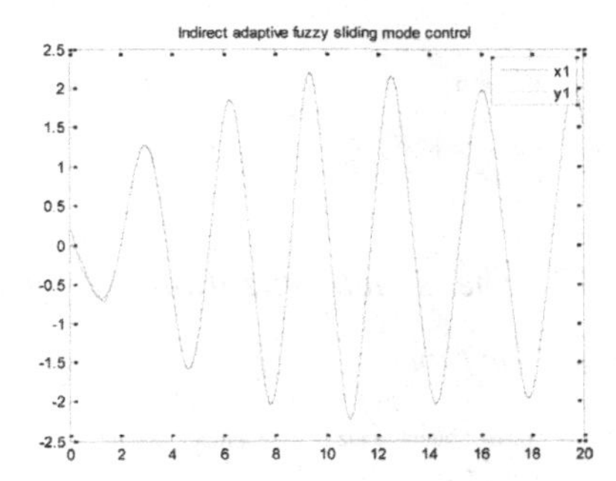

Figure 11. The output trajectories of y_2 (dashed line) and $x_2(t)$ (solid line)

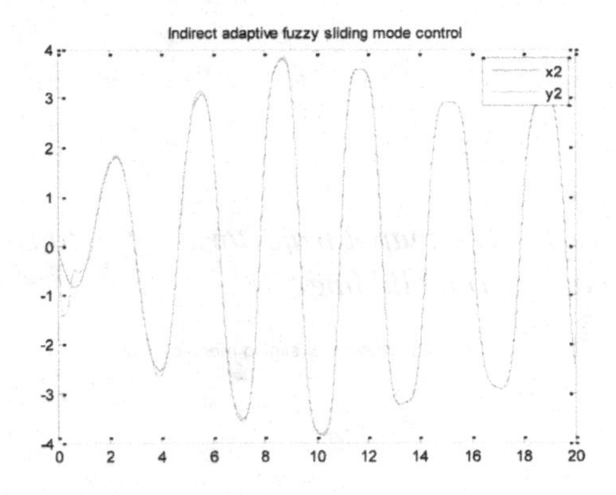

Figure 12. The synchronization performance $\int_0^{20} e_1^2(t)dt$ for type-1

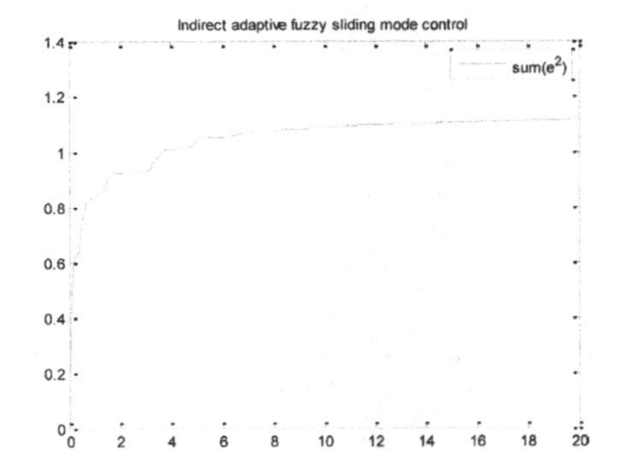

Figure 13. Control input u(t)

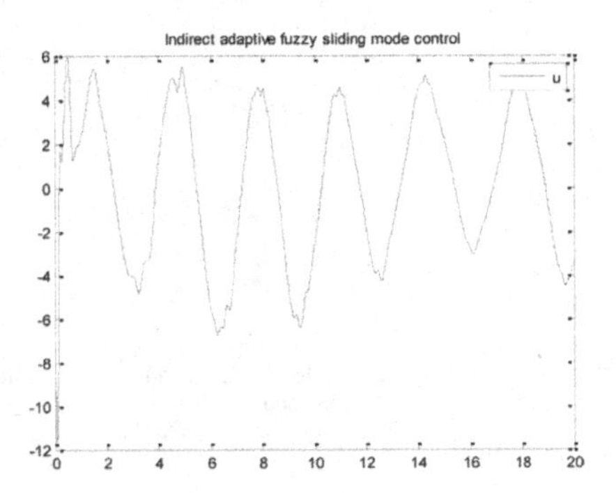

Part II: Training Data Corrupted With SNR=20 dB Noise

A. Adaptive Type-1 FNN Control

The synchronization performance of states $x(t)$ and $y(t)$ is as shown in Figure 20 ad Figure 21, respectively. The synchronization errors of $e_1(t)$ is shown in Figure 22 and Figure 23 shows control input $u(t)$.

In the meantime, in order to show that both the master system and slave system are consequently stable, the trajectory $\dot{V}(t)$ is depicted in Figure 24 for type-1controller is always negative defined.

Figure 14. The trajectory of $\dot{V}(t)$ for the type-1 fuzzy controller

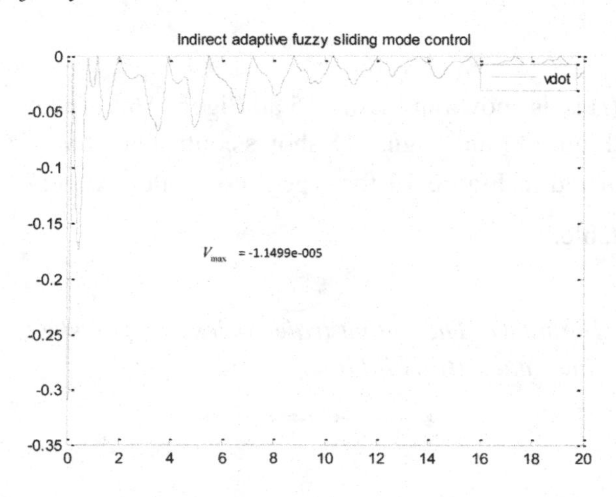

Figure 15. The output trajectories of y_1 (dashed line) and $x_1(t)$ (solid line)

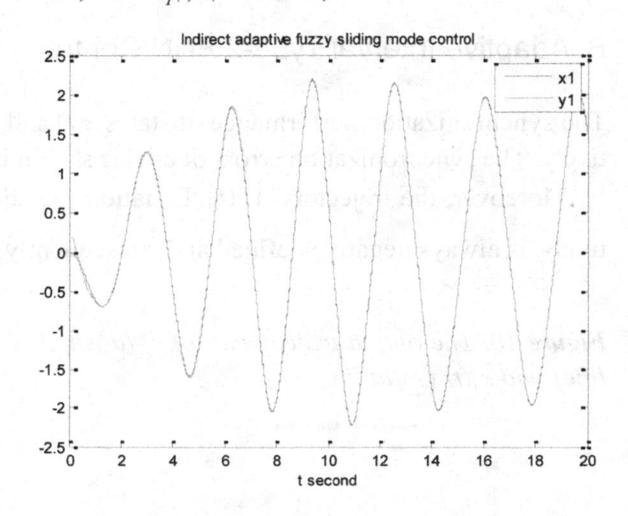

Figure 16. The output trajectories of y_2 (dashed line) and $x_2(t)$ (solid line)

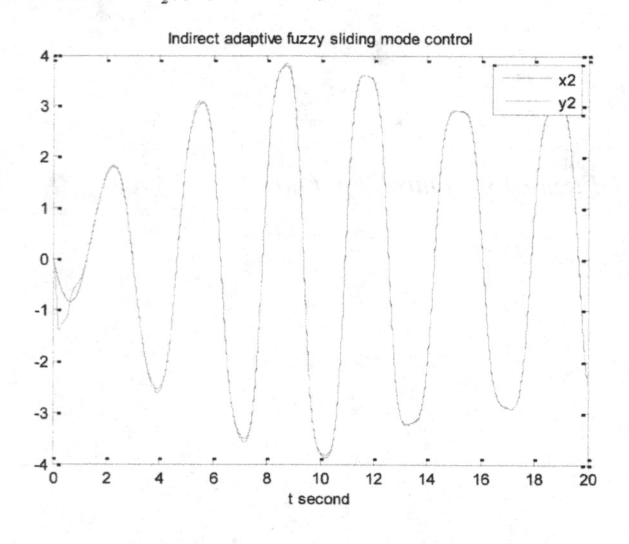

Figure 17. The synchronization performance $\int_0^{20} e_1^2(t)dt$ for type-2

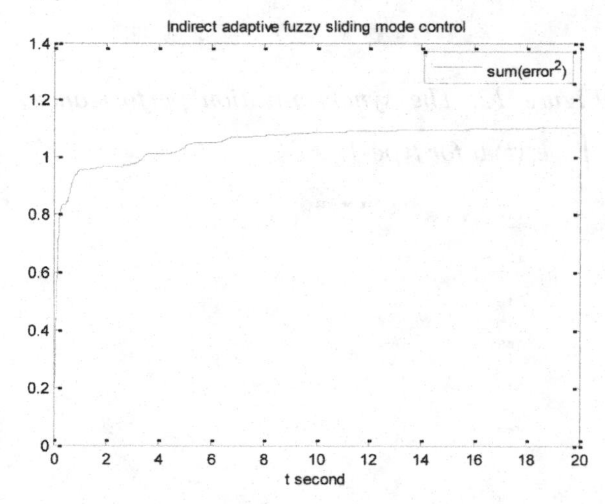

Figure 18. Control input u(t)

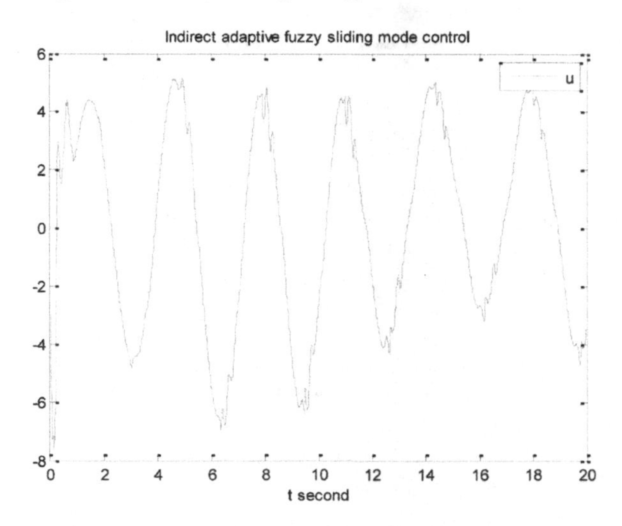

Figure 19. The trajectory of $\dot{V}(t)$ for the type-2 fuzzy controller

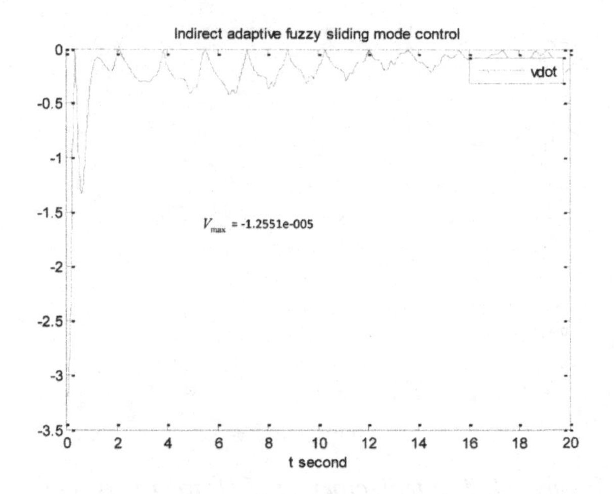

B. Adaptive Interval Type-2 FNN Control

The synchronization performance of states $x(t)$ and $y(t)$ is as shown in Figure 25 and Figure 26, respectively. The synchronization errors of $e_1(t)$ is shown in Figure 27 and Figure 28 shows control input $u(t)$.

In the meantime, in order to show that both the master system and slave system are consequently stable, the trajectory Equation (35) depicted in Figure 29 for type-1controller is always negative defined.

Table 2 manifests that the proposed interval type-2 fuzzy controller gives the better performance compared with type-1 fuzzy controller as system internal uncertainties (noisy training data) and external disturbance appeared. Moreover, the adaptive type-1 fuzzy controller must expend more control effort to deal with noisy training data.

Figure 20. The output trajectories of y_1 (dashed line) and $x_1(t)$ (solid line)

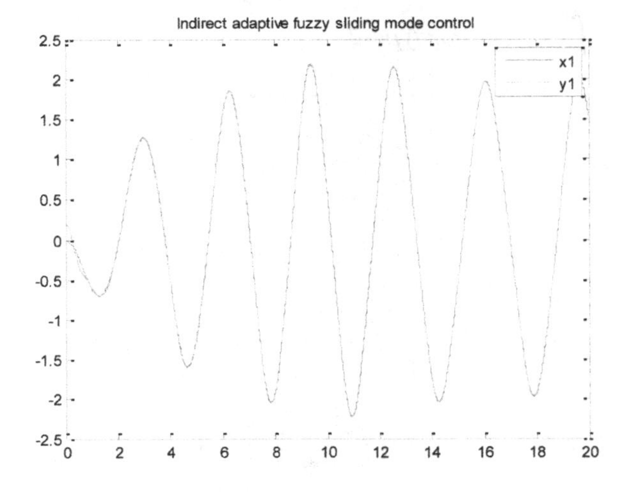

Figure 21. The output trajectories of y_2 (dashed line) and $x_2(t)$ (solid line)

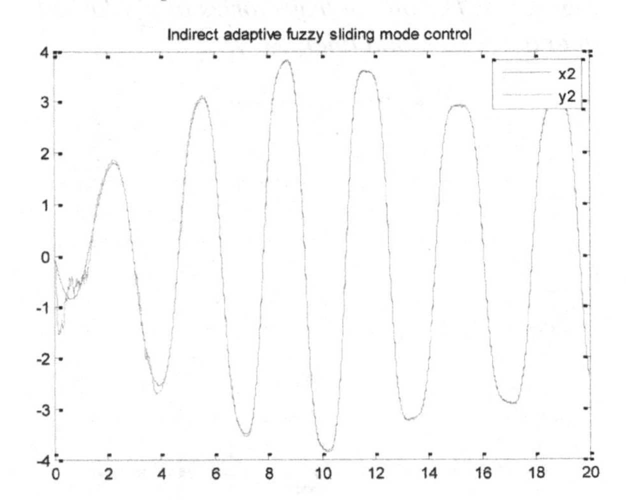

Figure 22. The synchronization performance $\int_0^{20} e_1^2(t)dt$ *for type-1*

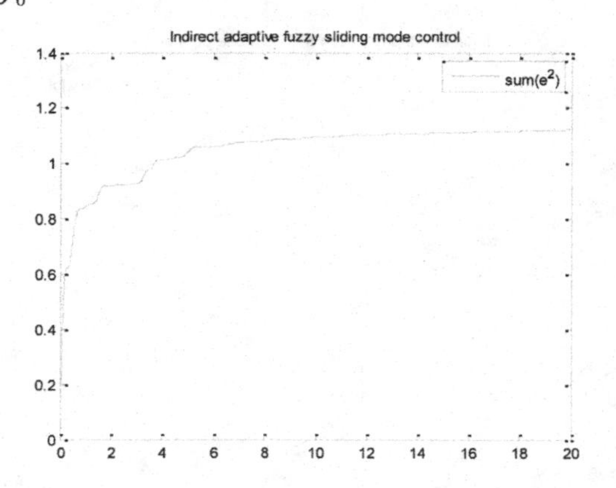

Figure 23. Control input u(t)

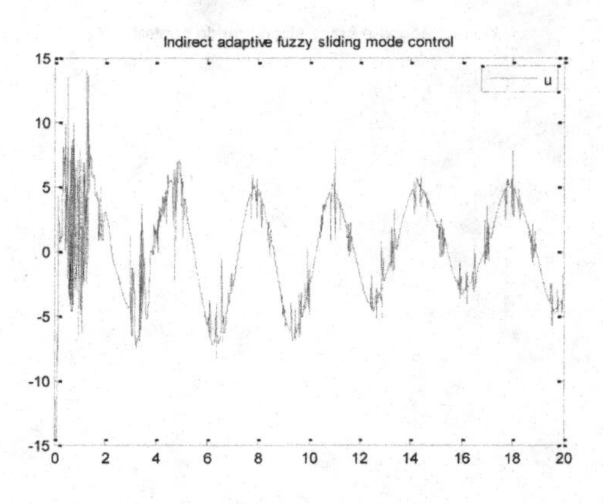

Figure 24. The trajectory of $\dot{V}(t)$ *for the type-1 fuzzy controller*

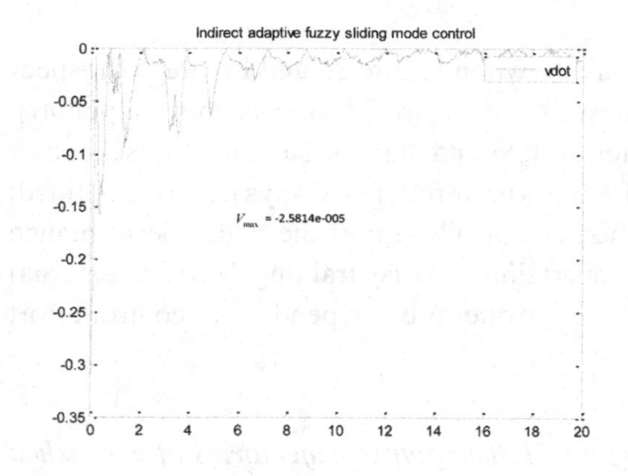

Figure 25. The output trajectories of y_1 *(dashed line) and* $x_1(t)$ *(solid line)*

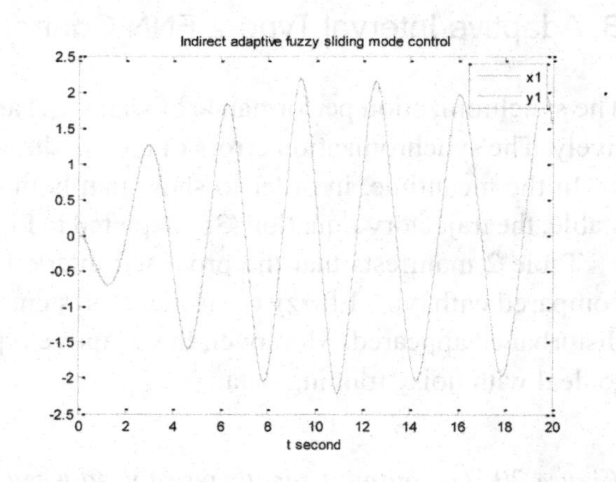

Figure 26. The output trajectories of y_2 *(dashed line) and* $x_2(t)$ *(solid line)*

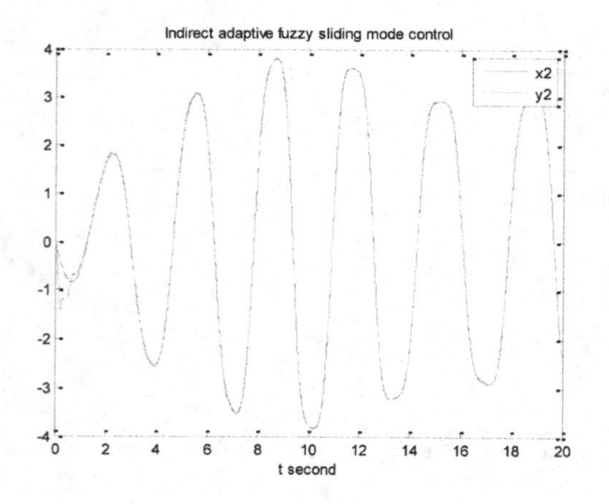

Figure 27. The synchronization performance $\int_0^{20} e_1^2(t)dt$ *for type-2*

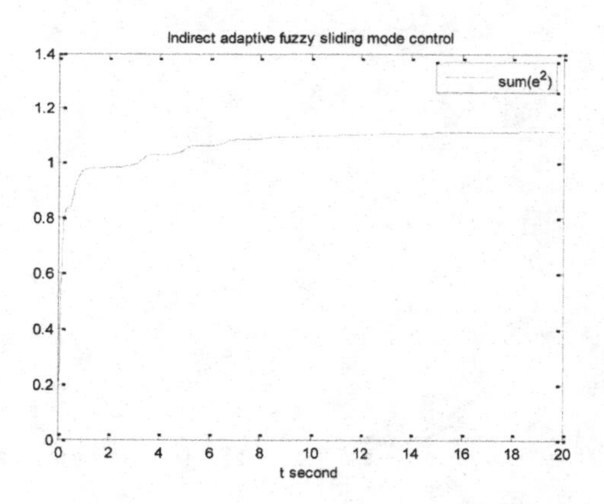

Figure 28. Control input u(t)

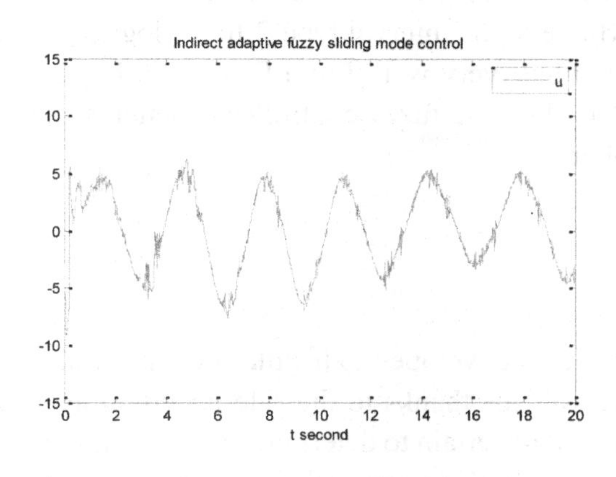

Figure 29. The trajectory of $\dot{V}(t)$ for the type-2 fuzzy controller

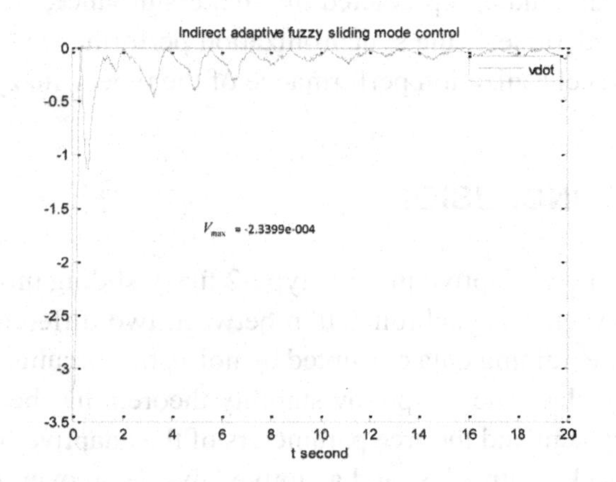

Table 1. Interval type-2 and type -1 fuzzy membership functions for x_i and i = 1, 2

	Variance(σ)	Mean(m)				Variance(σ)	Mean(m)		
		m_1	m_2	m(type-1)			m_1	m_2	m(type-1)
$\mu_{F_i^1}(x_i)$	0.5	-3.5	-2.5	-3	$\mu_{F_i^5}(x_i)$	0.5	0.5	1.5	1
$\mu_{F_i^2}(x_i)$	0.5	-2.5	-1.5	-2	$\mu_{F_i^6}(x_i)$	0.5	1.5	2.5	2
$\mu_{F_i^3}(x_i)$	0.5	-1.5	-0.5	-1	$\mu_{F_i^7}(x_i)$	0.5	2.5	3.5	3
$\mu_{F_i^4}(x_i)$	0.5	-0.5	0.5	0					

Table 2. Comparison of control efforts and synchronization errors of type-1 and interval type-2 controllers applied to slave system

$T = tf / h = 2000$				
	$\sum_{i=1}^{T} \lvert u_i \rvert$		$\int_0^{20} (x_1 - y_1)^2 dt$	
	Type-1	*Type-2*	*Type-1*	*Type-2*
Noice free	6196	6137	1.1193	1.0997
Noice 20dB	6497	6150	1.1436	1.1017

From above simulation results, we can see that in order to deal with noisy training data the type-1 fuzzy controller must expend more control effort. Nevertheless, the interval type-2 fuzzy logic system can handle unpredicted internal disturbance, data uncertainties, very well. From Figure 27, Figure 22 and Table 1, the synchronization performance of the interval type-2 fuzzy controller is better than the synchronization performance of the type-1 fuzzy controller.

CONCLUSION

A new adaptive interval type-2 fuzzy sliding mode controller is developed to handle such uncertainties for chaos synchronization between two different chaotic systems thanks to the rule uncertainties and the training data corrupted by noise, the circumstances are too uncertain to determine exact membership grades. The Laypunov stability theorem has been used to testify the asymptotic stability of the whole system and the free parameters of the adaptive fuzzy controller can be tuned on-line by an output feedback control law and adaptive laws. Moreover, the chattering phenomena in the control efforts can be reduced. From the simulation results, it is obvious that the synchronization performance obtained from interval type-2 fuzzy controller is better than the synchronization performance obtained from type-1 fuzzy controller. Besides, in order to minimize the influence from external disturbance and noisy training data, type-1 fuzzy controller must expend more control effort. The overall adaptive interval type-2 control scheme guarantees the stability of the resulting whole system in the sense that all the states and signals are uniformly bounded and synchronization performance can be achieved.

REFERENCES

Chang, Y. C. (2001). Adaptive Fuzzy-Based Tracking Control for Nonlinear SISO Systems via VSS and H∞ Approaches. *IEEE Transactions on Fuzzy Systems*, *9*, 278–292. doi:10.1109/91.919249

Chen, B. S., Lee, C. H., & Chang, Y. C. (1996). H1 tracking design of uncertain nonlinear SISO systems: adaptive fuzzy approach. *IEEE Transactions on Fuzzy Systems*, *4*(1), 32–43. doi:10.1109/91.481843

Chen, B. S., Tseng, C. S., & Uang, H. J. (2000). Mixed H2/H1 fuzzy output feedback control design for nonlinear dynamical systems: an LMI approach. *IEEE Transactions on Fuzzy Systems*, *8*(3), 249–265. doi:10.1109/91.855915

Golea, N., Golea, A., & Benmahammed, K. (2003). Fuzzy Model Reference Adaptive Control. *Fuzzy Sets and Systems*, *137*(3), 353–366. doi:10.1016/S0165-0114(02)00279-8

Hojati, M., & Gazor, S. (2002). Hybrid Adaptive Fuzzy Identification and Control of Nonlinear Systems. *IEEE Transactions on Fuzzy Systems*, *10*(2), 198–210. doi:10.1109/91.995121

Hsiao, M. Y., Li, T. H. S., Lee, J. Z., Chao, C. H., & Tsai, S. H. (2008). Design of interval type-2 fuzzy sliding-mode controller. *Information Science*, *178*, 1696–1716. doi:10.1016/j.ins.2007.10.019

Karnik, N. N., Mendel, J. M., & Liang, Q. (1999). Type-2 fuzzy logic systems. *IEEE Transactions on Fuzzy Systems*, *7*, 643–658. doi:10.1109/91.811231

Kheireddine, C., Lamir, S., Mouna, G., & Hier, B. (2007). Indirect adaptive interval type-2 fuzzy control for nonlinear systems. *International Journal of Modeling, Identification and Control, 2*(2), 106-119.

Kim, D., Yang, H., & Hong, S. (2003). An Indirect Adaptive Fuzzy Sliding-Mode Control for Decoupled Nonlinear Systems. *IEEE Transactions on Fuzzy Systems*, *6*(2), 315–321.

Kosko, B. (1994). Fuzzy systems are universal approximators. *IEEE Transactions on Computers*, *43*(11), 1329–1333. doi:10.1109/12.324566

Kovacic, Z., Balenovic, M., & Bogdan, S. (1998). Sensitivity based self learning fuzzy logic control for a servo system. *IEEE Control Systems Magazine*, *18*(3), 41–51. doi:10.1109/37.687619

Lee, C. C. (1990). Fuzzy logic in control system: Fuzzy logic controller- Parts I, II. *IEEE Trans. Syst., Man, Cybern 20, 20*, 404-435.

Lee, H., & Tomizuka, M. (2001). Robust Adaptive Control using a Universal Approximator for SISO Nonlinear Systems. *IEEE Transactions on Fuzzy Systems*, *8*, 95–106.

Leu, Y. G., Lee, T. T., & Wang, W. Y. (1999). Observer-based adaptive fuzzy-neural control for unknown nonlinear dynamical systems. *IEEE Transactions on Systems, Man, and Cybernetics*, *29*, 583–591. doi:10.1109/3477.790441

Li, H. X., & Tong, S. (2003). A Hybrid Adaptive Fuzzy Control for A Class of Nonlinear MIMO Systems. *IEEE Transactions on Fuzzy Systems*, *11*(1), 24–34. doi:10.1109/TFUZZ.2002.806314

Lin, S. C., & Chen, Y. Y. (1994). *Design of adaptive fuzzy sliding mode for nonlinear system control.* Paper presented at the Int. Conf. Fuzzy Syst., Orlando, FL.

Lin, T. C. (2009). Analog Circuit Fault Diagnosis under Parameter Variations Based on Type-2 Fuzzy Logic Systems. *International Journal of Innovative Computing, Information and Control., accepted to be published.*

Lin, T. C., Kuo, M. J., & Hsu, C. H. (in press). Robust Adaptive Tracking Control of Multivariable Nonlinear Systems Based on Interval Type-2 Fuzzy approach. *International Journal of Innovative Computing . Information and Control.*

Lin, T. C., Liu, H. L., & Kuo, M. J. (2009). Direct adaptive interval type-2 fuzzy control of multivariable nonlinear systems. *Engineering Applications of Artificial. Intelligence*, *22*, 420–430.

Mendel, J. M. (2004). Computing Derivatives in Interval Type-2 Fuzzy logic Systems. *IEEE Transactions on Fuzzy Systems*, *12*(1), 84–98. doi:10.1109/TFUZZ.2003.822681

Mendel, J. M. (2007). Type-2 fuzzy sets and systems: an overview. *Computational. Intelligence Magazine, IEEE*, *2*(1), 20–29. doi:10.1109/MCI.2007.380672

Mendel, J. M., John, R. I., & Liu, F. (2006). Interval Type-2 Fuzzy Logic Systems Made Simple. *IEEE Transactions on Fuzzy Systems*, *14*(6), 808–821. doi:10.1109/TFUZZ.2006.879986

Mendel, J. M., & John, R. I. B. (2000). Type-2 fuzzy sets made simple. *IEEE Transactions on Fuzzy Systems*, *10*, 117–127. doi:10.1109/91.995115

Nguang, S. K., & Shi, P. (2003). H1 fuzzy output feedback control design for nonlinear systems: an LMI approach. *IEEE Transactions on Fuzzy Systems*, *11*(3), 331–340. doi:10.1109/TFUZZ.2003.812691

Noroozi, N., Roopaei, M., Balas, V. E., & Lin, T. C. (2009). *Observer-based adaptive variable structure control and synchronization of unknown chaotic systems.* Paper presented at the SACI 2009 5th International Symposium on Applied Computational Intelligence and Informatics.

Noroozi, N., Roopaei, M., Karimaghaee, P., & Safavi, A. A. (2010). Simple adaptive variable structure control for unknown chaotic systems *Communications in Nonlinear Science and Numerical Simulation, 15*(3), 707-727.

Noroozi, N., Roopaei, M., & Zolghadri Jahromi, M. (2009). Adaptive fuzzy sliding mode control scheme for uncertain systems. *Communications in Nonlinear Science and Numerical Simulation, 14*(11), 3978-3992.

Ott, E., Grebogi, C., & Yorke, J. A. (1990). Controlling chaos. *Physical Review Letters*, *64*(11), 1196–1199. doi:10.1103/PhysRevLett.64.1196

Palm, R. (1992). *Sliding mode fuzzy control.* Paper presented at the Int. Conf. Fuzzy Syst., San Diego, CA.

Roopaei, M., & Jahromi, M. Z. (2008). Synchronization of two different chaotic systems using novel adaptive fuzzy sliding mode control. *Chaos (Woodbury, N.Y.)*, *18*, 033133. doi:10.1063/1.2980046

Roopaei, M., Zolghadri, M., John, R., & Lin, T.-C. (in press). Unknown Nonlinear Chaotic Gyros Synchronization Using Adaptive Fuzzy Sliding Mode Control with Unknown Dead-Zone Input *Communications in Nonlinear Science and Numerical Simulation.*

Roopaei, M., Zolghadri, M., & Meshksar, S. (2009). Enhanced adaptive fuzzy sliding mode control for uncertain nonlinear systems. *Communications in Nonlinear Science and Numerical Simulation, 14*(9), 3670-3681.

Roopaei, M., Zolghadri Jahromi, M., & Jafari, S. (2009). Adaptive gain fuzzy sliding mode control for the synchronization of nonlinear chaotic gyros. *Chaos 19:013125*

Sastry, S., & Bodson, M. (1989). *Adaptive Control Stability, Convergence, and Robustness*. Englewood Cliffs, NJ: Prentice-Hall.

Tseng, C. S., & Chen, B. S. (2001). H1 decentralized fuzzy model reference tracking control design for nonlinear interconnected systems. *IEEE Transactions on Fuzzy Systems*, *9*(6), 795–809. doi:10.1109/91.971729

Wang, C. H., Cheng, C. S. C. S., & Lee, T. T. (2004). Dynamical Optimal Training for Interval Type-2 Fuzzy Neural Network (T2FNN). *IEEE Trans. Systems, Man and Cybernetics, . Part B*, *34*(3), 1462–1477.

Wang, C. H., Lin, T. C., Lee, T. T., & Liu, H. L. (2002). Adaptive Hybrid Intelligent Control for Uncertain Nonlinear Dynamical Systems. *IEEE Trans. Syst., Man . Cybern.*, *32*(5), 583–597.

Wang, C. H., Liu, H. L., & Lin, T. C. (2002). Direct Adaptive Fuzzy-Neural Control with State Observer and Supervisory Control for Unknown Nonlinear Dynamical Systems. *IEEE Transactions on Fuzzy Systems*, *10*(1), 39–49. doi:10.1109/91.983277

Wang, J. S., & Lee, C. S. G. (2002). Self-Adaptive Neuro-Fuzzy Inference Systems For Classification Application. *IEEE Transactions on Fuzzy Systems, 10*(6), 790–802. doi:10.1109/TFUZZ.2002.805880

Wang, L. X. (1993). Stable adaptive fuzzy control of nonlinear systems. *IEEE Transactions on Fuzzy Systems, 1*(1), 146–155. doi:10.1109/91.227383

Wang, L. X. (1994). *Adaptive Fuzzy Systems and Control: Design and Stability Analysis*. Englewood Cliffs, NJ: Prentice-Hall.

Wang, L. X. (1997). *A Course in Fuzzy Systems and Control*. Englewood Cliffs, NJ: Prentice Hall.

Wang, L. X., & Mendel, M. (1992). Fuzzy basis functions universal approximation, and orthogonal least squares learning. *IEEE Transactions on Neural Networks, 1*(3), 804–814.

Wang, W. Y., Chan, M. L., Hsu, C. C. J., & Lee, T. T. (2002). H∞ Tracking-Based Sliding Mode Control for Uncertain Nonlinear Systems via Adaptive Fuzzy-Neural Approach. *IEEE Transactions on Systems, Man, and Cybernetics, 32*, 483–492. doi:10.1109/TSMCB.2002.1018767

Yang, Y., & Zhou, C. (2005). Adaptive fuzzy H1 stabilization for strict-feedback canonical nonlinear systems via backstepping and small-gain approach. *IEEE Transactions on Fuzzy Systems, 13*(1), 104–114. doi:10.1109/TFUZZ.2004.839663

Yoo, B., & Ham, W. (1998). Adaptive Fuzzy Sliding Mode Control of Nonlinear System. *IEEE Transactions on Fuzzy Systems, 6*(2), 315–321. doi:10.1109/91.669032

Zadeh, L. A. (1965). Fuzzy Sets. *Information and Control, 8*(3), 338–353. doi:10.1016/S0019-9958(65)90241-X

Zheng, Y. Q., Liu, Y. J., Tong, S. C., & Li, T. S. (2009). *Combined Adaptive Fuzzy Control for Uncertain MIMO Nonlinear Systems.* Paper presented at the American Control Conference.

Section 3
Cryptographic Applications

Chapter 15
Secure Transmission of Analog Information Using Chaos

A.S. Dmitriev
Inst. of Radio Eng. & Electr. of Russian Academy of Sciences, Russia

E.V. Efremova
Inst. of Radio Eng. & Electr. of Russian Academy of Sciences, Russia

L.V. Kuzmin
Inst. of Radio Eng. & Electr. of Russian Academy of Sciences, Russia

A.N. Miliou
Aristotle University of Thessaloniki, Greece

A.I. Panas
Inst. of Radio Eng. & Electr. of Russian Academy of Sciences, Russia

S.O. Starkov
Inst. of Radio Eng. & Electr. of Russian Academy of Sciences, Russia

ABSTRACT

In this work the authors present a thorough experimental study of a practical realization of a complex analog signal transmission system using dynamic chaos. It is demonstrated that the chaotic synchronous response could be used as a basis for the design of secure communication channels. The results presented in this work confirm the possibility of secure wireless communications in RF band, while they allow the authors to analyze in detail the restrictions and problems connected with the quality of synchronization of the transmitter and the receiver of the wireless communication systems. The effect of the perturbing factors on the transmission quality is investigated theoretically. It is shown that the main reason of the transmission's quality degradation is the chaotic response desynchronization associated with the phenomenon of "on-off" intermittency. It is found that under the effect of the perturbing factors, the level of information signal fed to the transmitter must be increased in order to obtain qualitative information transmission. However, in order to provide secure communication, one must decrease the information signal level. A compromise on these contradictory requirements provides an improvement of the quality of the synchronous chaotic response in the receiver.

DOI: 10.4018/978-1-61520-737-4.ch015

1. INTRODUCTION

A number of approaches to the design of communication systems with chaos was proposed at the beginning of 1990s (Kocarev et al., 1992; Belsky & Dmitriev, 1993; Cuomo et al., 1993; Dedieu et al., 1993; Halle et al., 1993; Volkovskii & Rul'kov, 1993, Bohme et al., 1994). Some of them were designed for transmission of analog signals. There was also experimental data that dealt mainly with transmission of simple analog signals, such as sinusoidal signals (Kocarev et al., 1992, Halle et al., 1993, Volkovskii & Rul'kov, 1993).

The main element of the above reported systems is a chaotic module devoted to generate chaotic signal and to introduce information into the transmitting signal, and at the same time, to retrieve the information in the receiver. One of the most important operation requirements to the majority of the known systems is the identity of the chaotic module parameters in the transmitter and receiver (Belsky & Dmitriev, 1995). This is associated with the necessity to obtain in the receiver an exact copy of the signal formed in the transmitter (synchronous response) (Pecora & Carroll, 1990). A question arises then: can the synchronous response, hence, information retrieval, be achieved in real conditions, i.e., using chaotic modules in RF communication systems? The problem is that in such systems the signal undergoes a number of additional manipulations (modulation, frequency conversion, detection, amplification, etc.) which can lead to distortions, regarding the complex structure of the formed signal and essentially to nonlinear characteristics of the system's functional elements.

This chapter describes an experimental study of a practical realization of a complex analog signal transmission system using dynamic chaos (Dmitriev et al., 1997, 2000).

The structure of the chapter is as follows; First, the system and its operating principles are described, then the fundamental information on the system's hardware implementation is provided, a summary of experimental results on transmission of music and speech signals in low frequency range is presented. Thereafter, the structure of a communication system in RF band employing dynamic chaos is described. The mathematical model of the communications system is presented in the next section and computer modeling results are analyzed. The last section is devoted to experiments on communications in RF band.

2. BASE BAND EXPERIMENTS ON SPEECH AND MUSIC SIGNAL TRANSMISSION USING CHAOS

The block diagram of the communication system is shown in Figure 1a. It includes a transmitter, a receiver and a communication channel. This system is based on the idea of nonlinear signal mixing and subsequent restoration of the synchronous chaotic response as suggested in (Volkovskii & Rul'kov, 1993).

Both the transmitter and the receiver are based on the decomposition of Chua's circuit (Madan, 1993) into two subsystems (RLC and RCN_R) (Figures 1b and 1c) which are connected in a closed loop (Figure 1d). Subsystem RLC is a band pass filter and subsystem RCN_R is a low-frequency first-order filter loaded at a nonlinear resistance N_R with three-segment of piecewise-linear voltage-current characteristics (Madan, 1993):

$$I_N(V_{C1}) = G_b V_{C1} + 1/2(G_b - G_a)[|V_{C1} + E| - |V_{C1} - E|], \quad (1)$$

Figure 1. (a) Block-diagram of the communication system. It consist of a transmitter (1), a receiver (2) and a communication channel (3). The transmitter contains an adder (+), and subsystems RLC and RCN_R. The receiver contains a subtractor (-), and subsystems RLC and RCN_R. The triangular symbols are voltage buffers (op amp). S - input information signal, S' - recovered information signal, N - transmitted signal. (b) RLC subsystem. (c) RCN_R subsystem. N_R - nonlinear element with a three-segment piecewise - linear V-I characteristic. (d) Block-diagram of the chaotic generator. It includes the RLC and RCN_R subsystems.

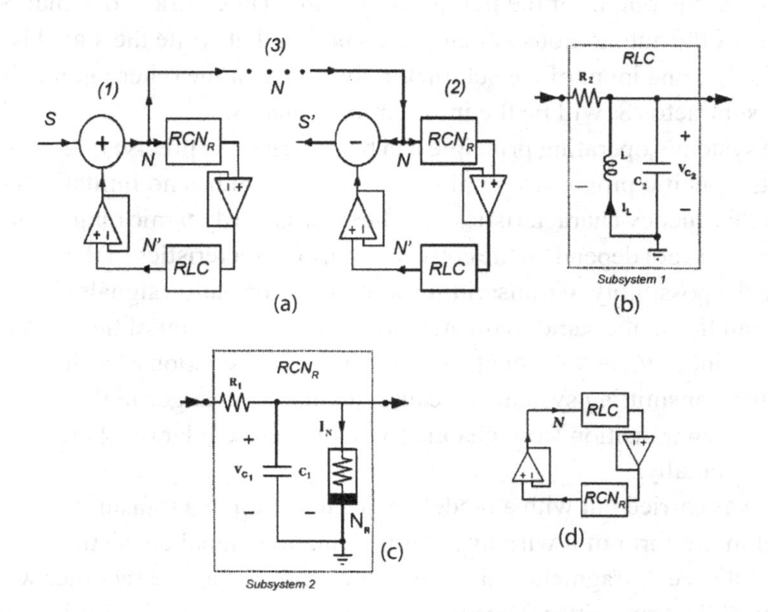

where I_N and V_{C1} are the current and the voltage across the resistance, and G_a, G_b and E are constants. Buffer amplifiers (unit gain, high input and low output resistances) between the blocks play the role of insulators and provide unidirectional feedback loop in the generator.

Similar decomposition was used before in the design of the receiver in a masking communication system and in the transmission of binary signals (Kocarev et al., 1992, Parlitz et al., 1992). Our system uses such a decomposition in both the receiver and the transmitter in order to organize the nonlinear mixing of information to the chaotic signal, and to send the resulting signal into the channel.

Let us consider the operating principle of the communication system presented in Figure 1a.

A voltage signal from the output of the nonlinear subsystem 2 (RCN_R), denoted as V_{c1} in Figure 1c, is applied to one of the inputs of the transmitter's adder where it is combined with an information signal S applied to the other input terminal. The resulting signal is transmitted over the communication channel to the receiver and is applied to the input of the first subsystem (RLC), and to one of the two inputs of the subtractor, as shown in Figure 1a. The adder is designed so as not to break the feedback loop. In other words, the system generates chaotic oscillations and sends them over the channel irrespectively of the presence or the absence of the information signal applied to the input of the adder.

Let us consider the operation of the system in the absence of an information signal. The signal applied to the input of the first subsystem (RLC) will be denoted as N, while the signal taken from the output of the second subsystem (RLC_R) as N'. When S = 0 the signals are equal N ≡ N' (the gain of the buffer stages is equal to unity). If the parameters in the transmitter and the receiver are identical, the signal coming from the channel, after being processed in subsystems 1 and 2 of the receiver, induces a synchronous response N' ≡ N. Thus the same signal is applied to both inputs of the receiver's subtractor which produces a zero output voltage.

The presence of an information signal S breaks the static process in the transmitter. Now, the signal N' is observed at the input of the nonlinear subsystem 2 while the signal at the input of subsystem 1 is the combination of signals S + N'=N. Thus the application of signal S + N' to the input of subsystem 1 induces the signal N' at the output of the nonlinear system 2 (in contrast to signal N', signal N' + S is not the "eigenmode" of the autonomous system). This means that while the signal N' + S, coming from the channel, is applied to one input of the subtractor, the signal at the other input is N'. Thus, the signal at the output of the subtractor S' will be the information signal S.

The fact that the system's operating principle can be described in precise relationships, without mentioning the elements' specific properties, implies that our system has no limitations on the information signal amplitude and frequency characteristics. In other words the dynamic range of transmitted message and the frequency band do not depend on the corresponding characteristics of the original chaotic oscillations. This opens up the possibility of transmitting complex information signals (e.g., speech and music), which power spectrum lies in the same frequency range as the spectrum of the autonomous oscillations.

From a practical point of view it is not clear whether the introduction of a signal, having sufficiently high amplitude, to the transmitting system can cause qualitative changes in the operation of the generator; for example a sudden transition from chaotic to modulated regular oscillations. The answer can be obtained only experimentally.

The experiment was carried out with a model device made up of a transmitter, a receiver and a communication channel in the form of a wire line. An information signal consisting of a sinusoidal signal, or recorded music and speech fragments taken from the output of a tape recorder was applied to one of the two adder inputs of the transmitter. The output signal from the receiver's subtractor was reproduced by an acoustic system.

Our experimental unit allowed the visualization of the signal waveforms and signal power spectra at any point of the system, while listening to the corresponding audio fragments. The system's transmitter provided an autonomous generation of chaotic oscillations in the range of 0 - 5 kHz. The mean ratio of the amplitude of the information signal to the transmitter's output signal varied from 0.1 to 0.3. The parameter values in the transmitter's elements are: $L = 18$ mH, $C_2 = 47$ nF, $C_1 = 5100$ pF, $R_1 = R_2 = 1.65$ kΩ, with a tolerance less than 1%. The nonlinear element with three-segment characteristics (Eq. 1) was based on Kennedy's two-operational-amplifier (op amp) circuit (Kennedy, 1992). IC KR1401UD2a was used with 4 identical op amps on the chip. The parameter values for the circuit are: $G_b = -0.459$ mS, $G_a = -0.757$ mS, and $E = 1.56$ V.

The main results of our experiments are summarized in Figures 2 - 5.

Our first experiment was performed with a 1 kHz tone (harmonic) signal. The typical results are presented in Figure 2 showing the oscilloscope traces of the information signals at the transmitter's input (top trace) and the receiver's output (bottom trace). The results indicate that the reproduction of the message transmitted through the communication channel is accurate enough. This gives hope that our system can be successfully applied for transmitting more complex analog (continuous time) signals.

Typical results of the experiments on transmitting complex analog signals are presented in Figures 3 - 5. Comparing Figure 3a and Figure 3b (phase portraits), Figure 4 (power spectra) and Figure 5 (signal waveforms), consisting of a music fragment, one could conclude:

Figure 2. Input signal versus recovered signal. The top trace is the sinusoidal signal S. The bottom trace is the recovered signal S'.

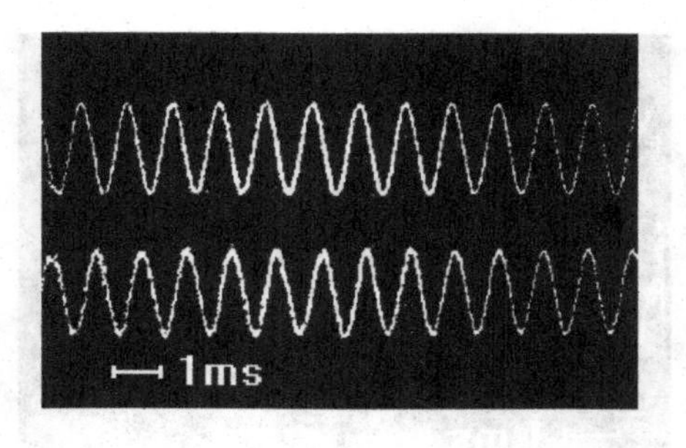

Figure 3. Phase portraits of the transmitter's signal in the (V_{C1}, V_{C2}) plane. V_{C1} - horizontal axis; V_{C2} - vertical axis. (a) With zero input signal. (b) In the presence of a musical signal input.

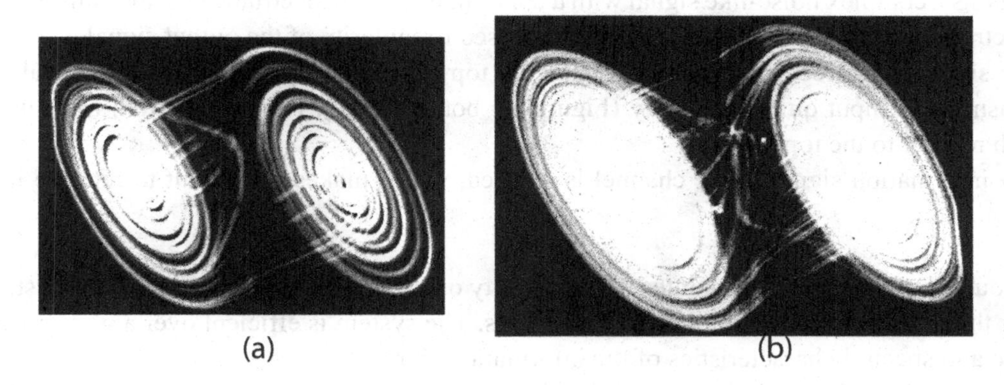

Figure 4. (a) The power spectrum of the transmitted signal with zero input. (b) The spectrum of the musical signal input. (c) The spectrum of the transmitted oscillations with input signal.

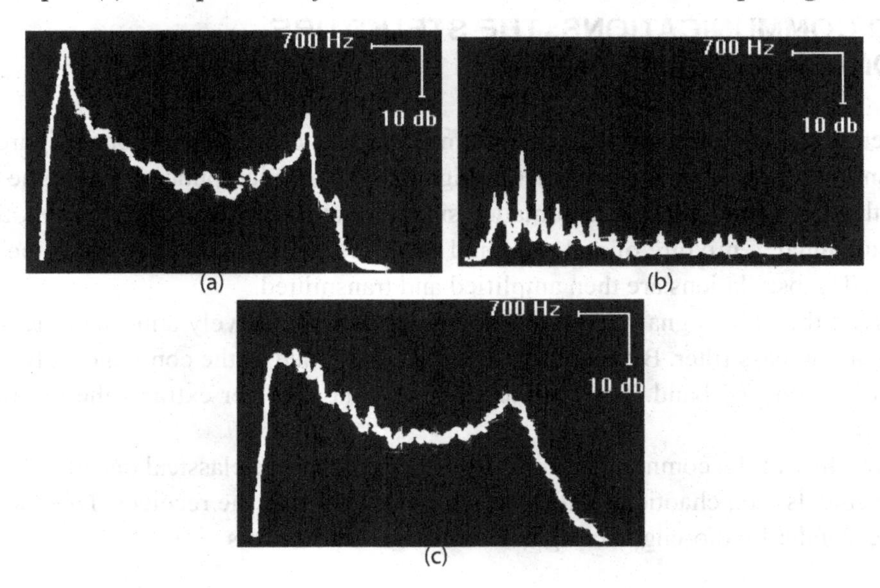

Figure 5. Input signal versus recovered signal. The top trace is the musical input signal S. The bottom trace is the recovered signal S'.

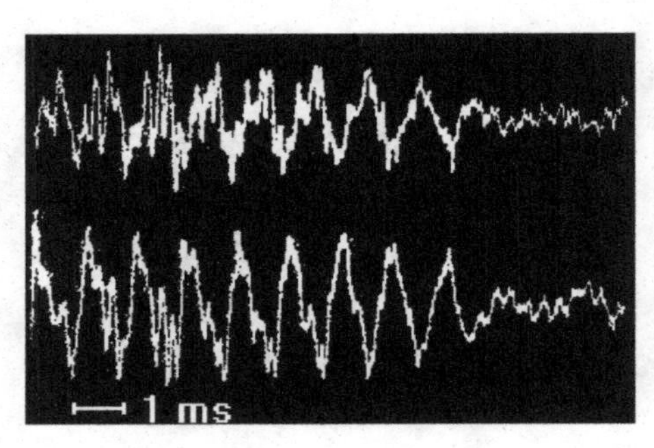

1. After the information signal was applied to the transmitter, the signal at the transmitter's output remains a complex noise-like signal with a continuous power spectrum. The non-uniformity of the spectrum diminishes which indicates an increased irregularity of the output signal.
2. The signal at the receiver's output (Figure 5 - top trace) displays the recovered signal from the transmitter's input quite accurately (Figure 5 - bottom trace). Here the bottom trace is inverted with respect to the top trace.
3. The information signal in the channel is masked, which makes it difficult to retrieve it without permission.

Thus our present study demonstrates the possibility of a practical implementation of a system using chaos for the transmission of complex analog signals. The system is efficient over a wide range of the amplitude and spectral characteristics of the information signal.

Since the study was completed in baseband, it allows us to proceed to investigate the conditions for secure communications in RF band.

3. RF BAND COMMUNICATIONS: THE STRUCTURE OF THE COMMUNICATION SYSTEM

The structure chosen for the communication system was close to the classical one (Figure 6).

In the transmitter, low-frequency information signal from the microphone is fed to the chaotic module's input. In the chaotic module, the information signal is nonlinearly mixed to the chaotic signal. The module's output signal, a mixture of a chaotic and an information signal, modulates the amplitude of RF oscillations. The oscillations are then amplified and transmitted.

In the receiver, the radio signal received by the antenna is selectively amplified, demodulated and passed through a low-pass filter. By filtering, the signal is cleaned of the components lying outside the chaotic module's frequency band. The chaotic module at the receiver extracts the information signal from the mixture.

Thus, the structure of the communication system differs from the classical one only by the presence of additional elements, i.e., chaotic modules, in the transmitter and the receiver. This fact allows us to extensively use standard radio-engineering devices in the experiments.

Figure 6. The structure of the communication system. It includes transmitter (1), receiver (2), microphone (3), low-frequency amplifier (4), chaotic module (5), generator of RF carrier oscillations (6), modulator (7), power amplifier (8), antenna (9), selective amplifier (10), demodulator (11), low-frequency filter (12), information signal S.

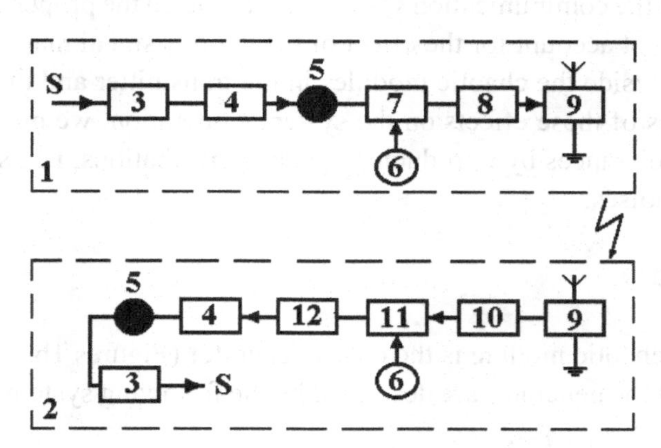

Formally, in a chaotic AM communication system, if one will transmit a mixture of an information signal with low-frequency chaotic signals, instead of the information signal using, e. g., a chaotic transmission module (Dmitriev et al., 1994, 1995), then in the receiver one can extract the information component by means of the synchronous response of the chaotic receiving module. In practice, however, there are difficulties. As it is known, in the communication systems based on synchronous response in the receiver (an example of such a system is the system with nonlinear mixing of information that is discussed in this chapter), tough conditions should be fulfilled on the equality of the chaotic module's parameters in the receiver and the transmitter.

In low-frequency experiments, this problem is solved by careful selection and tuning of the circuits' elements. However, when signals are carried up to RF band and back, they undergo a set of additional manipulations (amplification, modulation, filtering, demodulation, etc.). Each of these manipulations leads to additional distortions, and inhibits the formation, in the receiver, of an exact copy of the signal formed in the transmitter. Hence, the transmission of speech signals imposes tougher restrictions on the accuracy of forwards and backwards manipulations, than in a low-frequency system. Practically, the total signal distortion along the entire manipulation path should not exceed 1-2%.

Another distorting factor is the presence of additive noises in the channel.

In the next section a model of communication system is discussed and the effect of various perturbing factors on its characteristics is estimated.

4. MATHEMATICAL MODEL AND COMPUTER SIMULATION OF THE SYSTEM

There are two frequency scales in the presented communication system: low frequencies (to several kilohertz), characteristic of the information signals and chaotic module oscillations, and high frequencies (tens of MHz), characteristic of RF oscillations. Low-frequency signals modulate high-frequency oscillations. By the analysis of the dynamic properties of standard radio systems, "slow variables" describing the high-frequency signal envelope are introduced, and the corresponding differential equations are investigated.

In our case, let us consider a low-frequency model as a mathematical model of the communication system, i.e., a model describing the dynamics of the transmitting and the receiving chaotic modules, both in the absence and presence of the information signal. Such a model allows us to analyze the main dynamic characteristics of the communication system depending on the properties of the chaotic module. At the same time, it does not account for the effect of the various signal manipulations on the system's dynamics that are made outside the chaotic modules in the transmitter and the receiver. In order to estimate the potential results of these effects on the system's operation, we model the manipulation and communication channel tolerances by introducing special perturbations, i.e., signal filtering, nonlinear distortions, and additive noises.

4.1 Chaotic Modules

The basic element of the chaotic module is the chaos generator (Figures 1b, 1c and 1d).

The dynamic modes of the generator are described by the following system of differential equations:

$$C_1(dV_{C1}/dt) = (V_{C2} - V_{C1})/R_1 - I_N(V_{C1})$$

$$C_2(dV_{C2}/dt) = (V_{C1} - V_{C2})/R_2 + I_L \qquad (2)$$

$$L(dI_L/dt) = - V_{C2}.$$

Resistances R_1 and R_2 are the control parameters of the generator's oscillation modes. With $R_1 = R_2 = R$, Eq. 2 coincide with the equations for the canonical Chua's circuit (Madan, 1993).

The "double scroll" attractor's mode (Figure 7) was taken as a basic mode for the following analysis. The mode takes place when the parameter values are set at $G_b = -0.714$ mS, $G_a = -1.143$ mS, $E = 1$ V, $L = 0.0625$ H, $C_2 = 1$ F, $C_1 = 0.10204$ F, and $1/R_1 = 1/R_2 = 1$ S.

Figure 7. Basic attractor's mode of the chaotic generator. (a) Chaotic signal (V_{C2}). (b) Phase portrait of the generator's signal in the (V_{C1}, V_{C2}, I_L) plane. (c) The power spectrum of the chaotic signal V_{C2}.

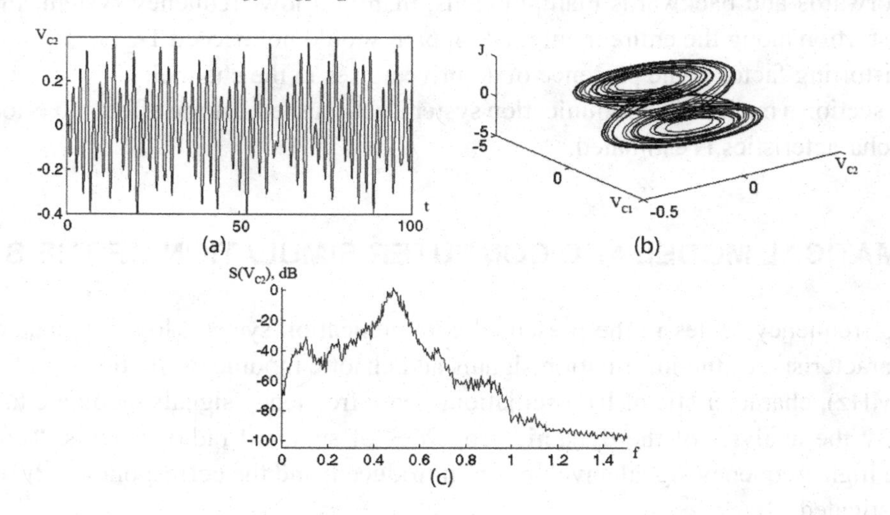

The structure of the communication system is presented in Figure 1a. It comprises of a transmitter (transmitting chaotic module) 1, a receiver (receiving chaotic module) 2, and a communications channel 3. The transmitting chaotic module is built from the initial chaos generator (Figure 2d) by means of including an adder (+) into the feedback loop. The receiving chaotic module is based on a copy of the same generator, with the feedback loop disconnected and a subtractor added (-). In the case of identical elements in the transmitter and the receiver, the system's operation is described by the following equations:

$$C_1(dV_{C1}/dt) = (V_{C2} + S - V_{C1})/R_1 - I_N(V_{C1})$$

$$C_2(dV_{C2}/dt) = (V_{C1} - V_{C2})/R_2 + I_L$$

$$L(dI_L/dt) = - V_{C2}.$$

$$C_1(dV'_{C1}/dt) = (V_{C2} + S - V'_{C1})/R_1 - I_N(V'_{C1}) \quad (3)$$

$$C_2(dV'_{C2}/dt) = (V'_{C1} - V'_{C2})/R_2 + I'_L$$

$$L(dI'_L/dt) = - V'_{C2}.$$

where S is an information signal fed to the transmitter, and $S' = V_{C2} + S - V'_{C2}$ is a signal retrieved in the receiver, (V_{C1}, V_{C2}, I_L) and (V'_{C1}, V'_{C2}, I'_L) are voltages across the capacitances C_1 and C_2 and the current through inductance L in the transmitter and the receiver respectively, while $I_N(V_{C1})$ is the piecewise-linear current-voltage characteristic (Eq. 1) of the nonlinear element N_R.

The first three equations are referring to the transmitter, while the other three to the receiver.

If the conditions for synchronous chaotic response are satisfied in the system then $V_{C2} = V'_{C2}$ (Figure 8b) and $S' = V_{C2} + S - V'_{C2} = S$. If the response is desynchronized (Figure 8a), the output signal S' is a complex mixture of the information and the chaotic signal.

Figure8. Chaotic response in the (V_{C2}, V'_{C2}) plane. (a) Desynchronized response. (b) Synchronous response.

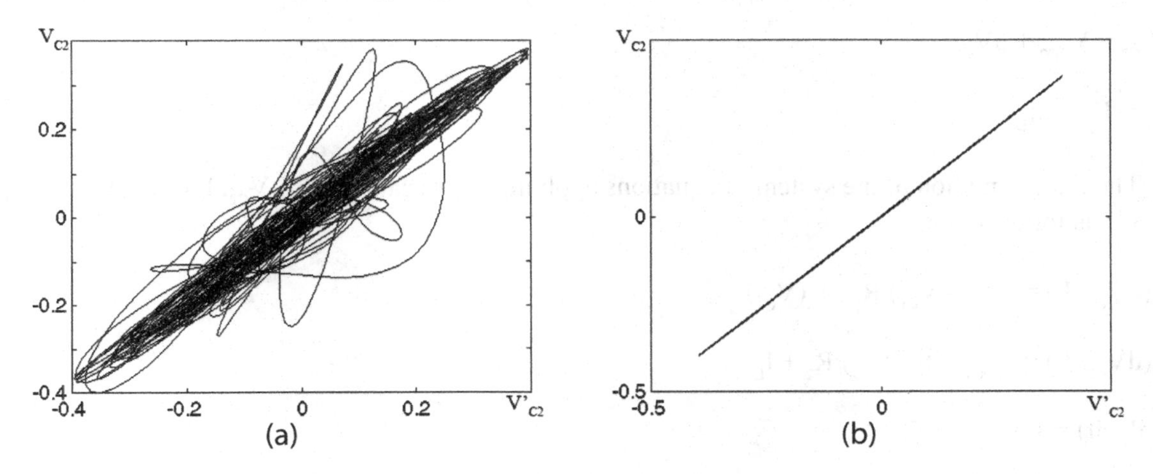

4.2 Synchronous Response and On-Off Intermittency

The necessary and sufficient conditions for synchronous chaotic response in the receiver are:

1. The trajectory $(V_{C1}, V_{C2}, I_L) = (V'_{C1}, V'_{C2}, I'_L)$ should exist;
2. The system's motion (Eq. 3) along this trajectory should be stable with respect to any small perturbations, transversal to the attractor.

A trajectory satisfying the first condition is chaotic and belongs to the chaotic attractor. The motion at the attractor is characterized with exponential, in average, divergence of the trajectories. At the same time, the attractor has a basin of attraction, and all trajectories starting from this basin attract to it. Consequently, sufficiently small perturbations do not destroy the attractor, i.e., it is stable with respect to small perturbations.

A necessary condition for the stability of the attractor is the negative value of the second Lyapunov exponent of the system (Eq. 3): $\lambda_2 < 0$. This condition guarantees that the trajectory, averaged over the attractor, will attract to it. Direct calculations show that the system (Eq. 3) in "double scroll" mode has $\lambda_2 = -0.498 < 0$, i.e., the necessary condition for the existence of the synchronization mode attractor is fulfilled. However, there can be some special trajectories in the attractor where the condition $\lambda_2 < 0$ is not satisfied. Therefore, the attractor can prove to be unstable by small transversal perturbations. This potential danger is inevitable, e.g., if the attractor's attraction basin has a positive Lesbesque measure but it is not an open set. In this case, deviations of the receiver's parameters with respect to the transmitter's parameters or a low-amplitude noise added to the signal at the receiver's input will lead to the effect of on-off intermittency (Platt et al., 1993; Heagy et al., 1994). Calculations show that in the case of equal parameters in the transmitter and the receiver and in the absence of additive noise, ideal synchronization is established in the system. However, in the case of perturbed parameters or additive noise in the channel, on-off intermittency actually occurs (Figure 8a; Figure 11a).

In order to rationalize this phenomenon, we consider the system's dynamics (Eq. 3) near the synchronization attractor, whose trajectory satisfies the above conditions. Assume the trajectories (V_{C1}, V_{C2}, I_L) and (V'_{C1}, V'_{C2}, I'_L) to be close, and we introduce new variables:

$$V''_{C1} = V'_{C1} + \delta V_{C1}$$

$$V''_{C2} = V'_{C2} + \delta V_{C2}$$

$$I''_L = I'_L + \delta I_L$$

The transformation of the system of equations applying the variables (V_{C1}, V_{C2}, I_L) and $(V''_{C1}, V''_{C2}, I''_L)$ is as follows:

$$C_1(dV_{C1}/dt) = (V_{C2} - V_{C1})/R_1 - I_N(V_{C1})$$

$$C_2(dV_{C2}/dt) = (V_{C1} + S - V_{C2})/R_2 + I_L$$

$$L(dI_L/dt) = - V_{C2}.$$

$$C_1(d\delta V_{C1}/dt) = (V_{C2} + S - \delta V_{C1})/R_1 - I_N(V_{C1}) \tag{4}$$

$$C_2(d\delta V_{C2}/dt) = (\delta V_{C1} - \delta V_{C2})/R_2 + \delta I_L$$

$$L(d\delta I_L/dt) = -\delta V_{C2}.$$

The first three equations of (Eq. 4) coincide with the first three of the system represented by Eq. 3. The other equations, linear with variable coefficients, describe the small deviations of the trajectories of the "transmitting-receiving" system from the synchronization attractor. If a trajectory belongs to the attractor, $\delta V_{C1} = \delta V_{C2} = \delta I_L = 0$.

Consider the stability conditions for the solution $\delta V_{C1} = \delta V_{C2} = \delta I_L = 0$. The problem can be reduced to the calculation of the coefficients' matrix eigenvalues for the last three linear equations of the system represented by Eq. 4. Formally,

$$M = \begin{pmatrix} -\frac{1}{C1}\left(\frac{1}{R1} + \frac{\partial I_N(V'_{C1})}{\partial V'_{C1}}\right) & 0 & 0 \\ \frac{1}{R2 \cdot C2} & -\frac{1}{R2 \cdot C2} & \frac{1}{C2} \\ 0 & -\frac{1}{L} & -\frac{R0}{L} \end{pmatrix} \tag{5}$$

is a matrix with variable coefficients, which are determined by the nonlinear element characteristic. Since the characteristic is piecewise-linear, the coefficients could only have values from two fixed sets:

$$M_a = \begin{pmatrix} -\frac{(1/R1 + G_a)}{C1} & 0 & 0 \\ \frac{1}{R2 \cdot C2} & -\frac{1}{R2 \cdot C2} & \frac{1}{C2} \\ 0 & -\frac{1}{L} & -\frac{R0}{L} \end{pmatrix} \tag{6}$$

and

$$M_b = \begin{pmatrix} -\frac{(1/R1 + G_b)}{C1} & 0 & 0 \\ \frac{1}{R2 \cdot C2} & -\frac{1}{R2 \cdot C2} & \frac{1}{C2} \\ 0 & -\frac{1}{L} & -\frac{R0}{L} \end{pmatrix} \tag{7}$$

Matrix eigenvalues for the first set (corresponding to the phase space part containing the origin) are $\alpha_1 = 1.4014$, $\alpha_2 = -0.5000 + 3.9686i$, and $\alpha_3 = -0.5000 - 3.9686i$, i.e., have both positive and negative real parts. For the second set (phase space part related to the outer segments of the nonlinear element characteristic) $\alpha_1 = -2.8028$, $\alpha_2 = -0.5000 + 3.9686i$, and $\alpha_3 = -0.5000 - 3.9686i$. Thus, the system trajectory is stable with respect to small transversal perturbations when it passes through the phase space part corresponding to matrix M_b, and is not stable when it is in the phase space region corresponding to matrix M_a. Hence, the synchronization attractor is not absolutely stable with respect to transversal perturbations, and this is the cause of the effect of on-off intermittency occurring by deviation in the values of the transmitter's parameters compared to those of the receiver's.

4.3 Transmission of Analog Information

Calculations of the system (Eq. 3) with identical parameters in the transmitter and receiver were performed with one-frequency (tone) signals,

$$S(t) = A\cos(2\pi f t), \tag{8}$$

and multi-frequency (frequency-modulated, FM) signals,

$$S(t) = A\sin(2\pi f_0 t - \psi_m \cos(2\pi F t)), \tag{9}$$

where f_0 is the signal spectrum mean frequency, ψ_m is the FM index, and $F = 0.1 f_0$.

a. Tone Signal

A communication system was simulated with sinusoidal oscillations which are used as information signals. The ratio of the information signal power P_S to the chaotic signal power P_N

$$\mu = P_S / P_N \tag{10}$$

was varied in the range of 10^{-6} to 10^{-2}, and the frequency f was varied from 0.1 to 0.6. The frequency $f = 0.6$ corresponds to the normalized upper bound frequency of the chaotic signal (Figure 7c).

When the sinusoidal signal was mixed in at $\mu > 10^{-2}$, the system's trajectory, as a rule, went rapidly to infinity. So, the value of $\mu \approx 10^{-2}$ sets an upper admissible limit of the introduced sinusoidal signal power.

The calculations have shown that within the entire range of the parameter's variation, the signal is retrieved in the receiver without distortions. In particular, "perturbation" of the transmitter's dynamics by mixing the information signal does not lead to on-off intermittency.

In Figures 9a-c the power spectra of the mixture of chaotic and information signals, fed to the receiver's input, are presented. As it follows from these figures, the form of the spectrum is very sensitive to the introduced signal. For example, if the frequency of the introduced signal is far from the characteristic peaks of the chaotic signal, then its presence becomes visible at $\mu \approx 10^{-4}$ (Figure 9b). If the frequency of the introduced signal is close to a characteristic spectral peak of the chaotic oscillations, then its presence becomes noticeable at $\mu \approx 10^{-5}$ (Figure 9c) due to the effect of resonant amplification of the information signal.

Figure 9. The power spectra of the transmitted signal with input information signal. (a), (b), (c) Sinusoidal signal $S(t) = A\cos(2\pi f t)$ is the input signal. ((a): $A=2x10^{-2}$, $f=0.4$; (b): $A= 2.5x10^{-2}$, $f=0.4$; (c): $A=2x10^{-2}$, $f=0.5$). (d) Frequency modulated signal $S(t) = A\sin(2\pi f_0 t - \psi_m \cos(2\pi F t))$ (where $A=2.5x10^{-2}$, $f_0=0.4$ is the signal's spectrum mean frequency, $\psi_m=0.3$ is the FM index, and $F = 0.1f_0$) is input signal.

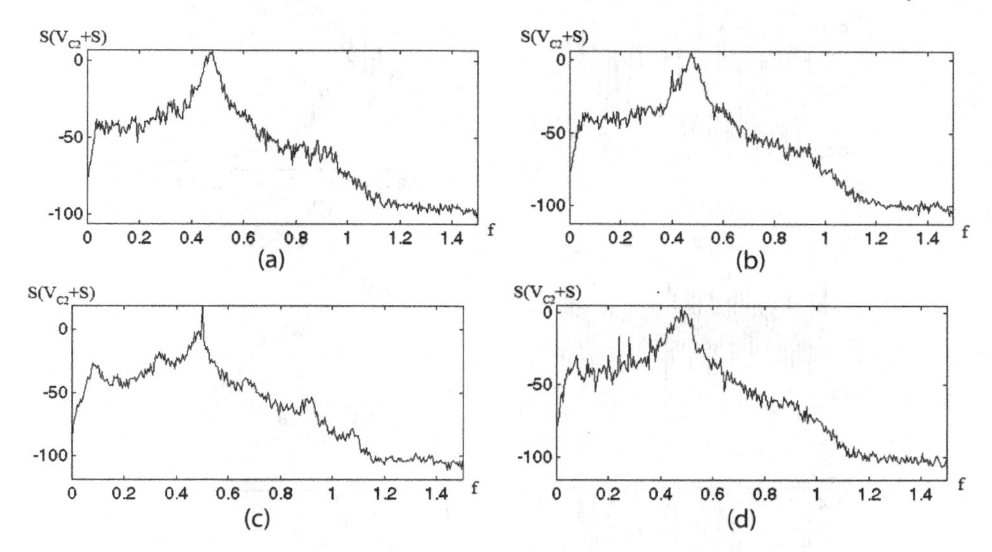

b. FM Signal

FM signal has a discrete power spectrum, so it imitates better the structure of the speech signal than the sinusoidal signal.

The calculations have shown that the maximum power of the introduced information signal, at which the system's trajectory still remains in a finite zone of the phase space, is $\mu \approx 4 \cdot 10^{-2}$, i.e., approximately four times higher than in the case of the sinusoidal signal.

As in the previous case, the information signal is retrieved without distortions in the receiver (Figure 10).

The power spectra of the mixture of chaotic and information signal, at the receiver's input, are less sensitive to the introduced signal than in the case of the tone signal, i.e., the components of the spectrum of information signal become noticeable only at $\mu \geq 2 \cdot 10^{-2}$ (Figure 9d).

4.4 Estimation of the Quality of Communication

Non-ideality (desynchronization) of the chaotic response leads to a noise at the receiver's output. This noise is the main cause of the received signal degradation. Therefore, in order to estimate the quality of communication, we use the ratio of the information signal power at the receiver's output, P_S, to the desynchronization noise power, P_D, (signal/noise ratio, SNR), defined by the expression:

$$SNR = P_S/P_D. \tag{11}$$

Figure 10. FM signal transmission. (a) FM input signal ($A=2x10^{-2}$, $f_0=0.4$, $\psi_m=0.3$). (b) The power spectrum of the FM signal. (c) The transmitted signal ($V_{C2}+S$). (d) The power spectrum of the transmitted signal. (e) The recovered FM signal (S'). (f) The power spectrum of the recovered FM signal.

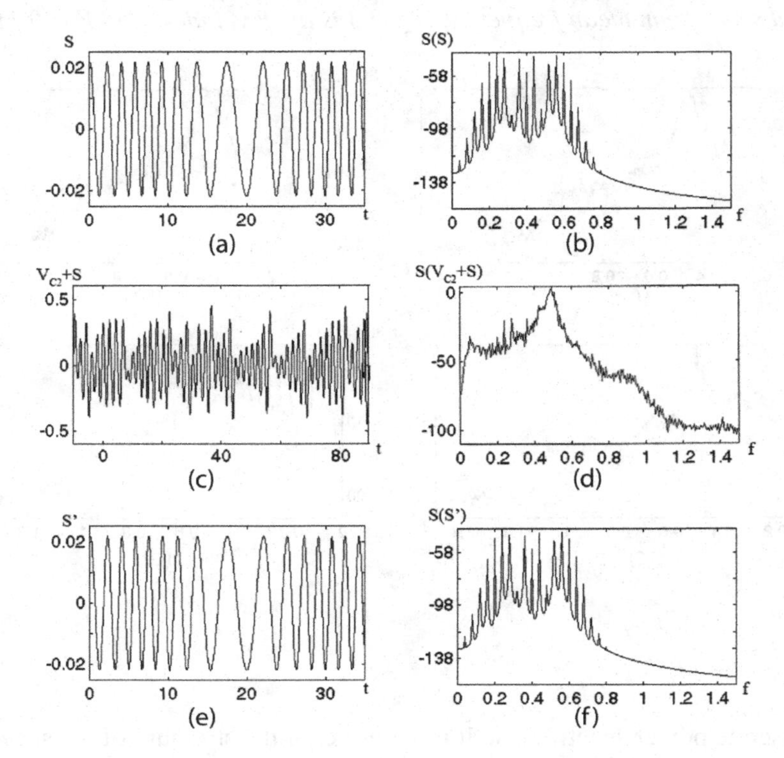

Figure 11. The receiver's output with zero input information signal and non-identity of elements. (a) The fragments of the desynchronization signal at the receiver's output (V_{C2}-V'_{C2}). (b) The power spectra of the receiver's output, (1) 0.1% deviation of the resistance's value, R, in the receiver and the transmitter. (2) 1% deviation of the resistance' value, R, in the receiver and the transmitter.

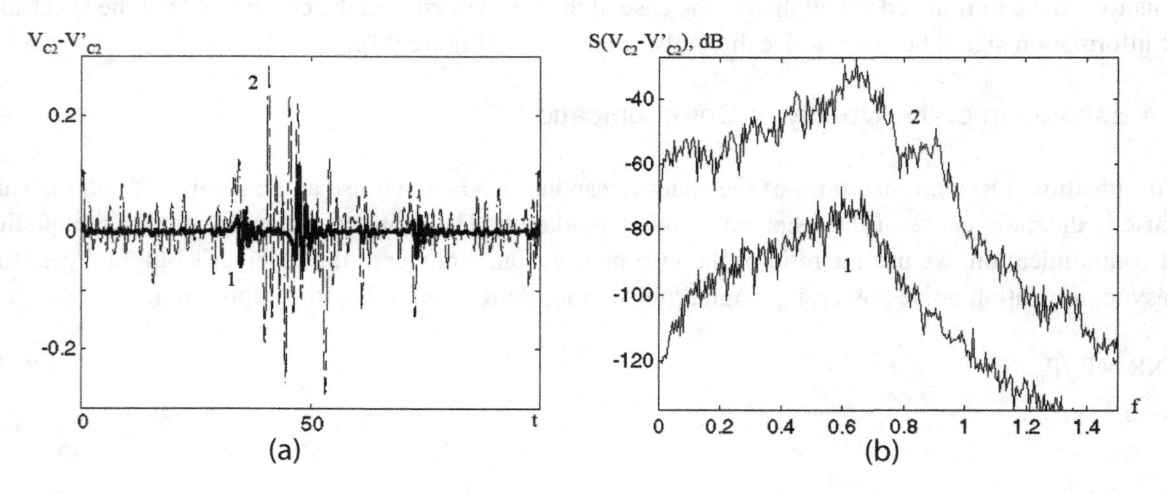

The signal power in the channel depends weakly on the presence of the information signal in the transmitter, because the information signal power does not exceed by several percents the chaotic signal power, hence it is convenient for the calibration. It is used below as a normalization coefficient for the noise power. Namely, instead of the absolute power of the desynchronization noise, its relative value is considered:

$$\eta = P_D/P_N. \quad (12)$$

Obviously, SNR can be expressed by μ and η as:

$$SNR = P_S/P_D = (P_S/P_N)\cdot(P_N/P_D) = \mu/\eta. \quad (13)$$

In the ideal case of exact equality in the transmitter's and the receiver's parameters and the absence of distortions in the channel, $\eta = 0$ and SNR $=\infty$. Practically η is always different from zero, since it characterizes the degree of distortion of the transmitted signal. For example, if the relative information signal power is equal to $\mu = 10^{-2}$ and the relative desynchronization noise power is $\eta = 10^{-4}$, then SNR $= 10^2$.

4.5 Effect of Perturbations

Typical perturbing factors that decrease the quality of communication are: (a) the non-identical elements in the transmitter and the receiver, (b) the nonlinear distortions, (c) the non-uniformity of amplitude-frequency response of the functional system's elements, and (d) the external noise in the communication channel. We consider the effect of these factors on the quality of the synchronous chaotic response at the receiver's output.

a. Non-Identity of Elements

In practice, deviation of the parameters of the circuit's element in the transmitter and the receiver is inevitable. Deviation of at least one element transmitter's circuit with respect to the corresponding receiver's element leads to desynchronization signal at the receiver's output. Characteristic fragments of the desynchronization signals and their power spectra are presented in Figure 11 for deviations of the value of the resistance, R, in the receiver and the transmitter by 0.1% (curve 1) and 1% (curve 2).

As it was noted in Section 4.2, non-identical parameters in the transmitter's and receiver's circuits lead to "on-off" intermittency. So, in the waveforms in Figure 11, besides the low-amplitude desynchronization signal, proportional to the value of deviation, one can observe high-amplitude irregular bursts comparable to the amplitude of the chaotic oscillations at the receiver's input. The average frequency of these bursts increase with increasing parameter deviation, and leads to a sharp increase of the relative mean power of the desynchronization signal η at the receiver's output.

b. Non-Uniformity of the Amplitude-Frequency Response

The non-uniformity of the amplitude-frequency response of the functional elements in the transmitter and the receiver leads to signal filtering and transformations of its spectral characteristics. As a rule, this effect is caused by the RC-circuits that play the role of low-pass filters. Therefore, the problem of the effect of the amplitude-frequency response non-uniformity is discussed here on the example of the low-pass filter characteristics.

Assume that the parameters in the transmitter and the receiver are identical, the transmitter's output signal is passed through a first-order low pass filter, and the only reason for the desynchronization of the response is the signal distortion by the filter. The signal transformation in the filter is described by the following equation:

$$Tx + x = y, \tag{14}$$

where y is the signal at the transmitter's output (V_{C2}), x is the signal at the receiver's input and time constant T defines the low-pass filter cutoff frequency, $f = 1/T$.

The analysis shows that in this case the quality of the synchronous response is determined by the relation of the filter's cutoff frequency, f, and upper bound frequency, F, of the signal power spectrum at the transmitter's output. At $f >> F$, the presence of the filter does not destroy the system's operation. Since $F \approx 0.6$ (see Figure 7c), at $f > 4$ the communication system doesn't experience the presence of the filter. From the other hand, comparable values of F and f lead to degradation of the synchronous response quality, and this tendency pertains as f decreases.

Desynchronization noise relative power η and SNR (at $\mu=1$) are presented in Figure 12 as a function of the cutoff frequency, f. As it is obvious from the figure, the mean square error rapidly decreases and SNR quickly increases with increasing f.

c. Nonlinear Distortions in the Communication Channel

The effect of nonlinear distortions is investigated using as an example the cubic nonlinearity.

$$x = y \cdot (1 - a \cdot y^2), \tag{15}$$

Figure 12. Non-uniformity of the amplitude-frequency characteristic of the communication channel. (a) Desynchronization noise relative power η and (b) SNR (at $\mu=1$) as a function of the cutoff frequency f (low-frequency filter).

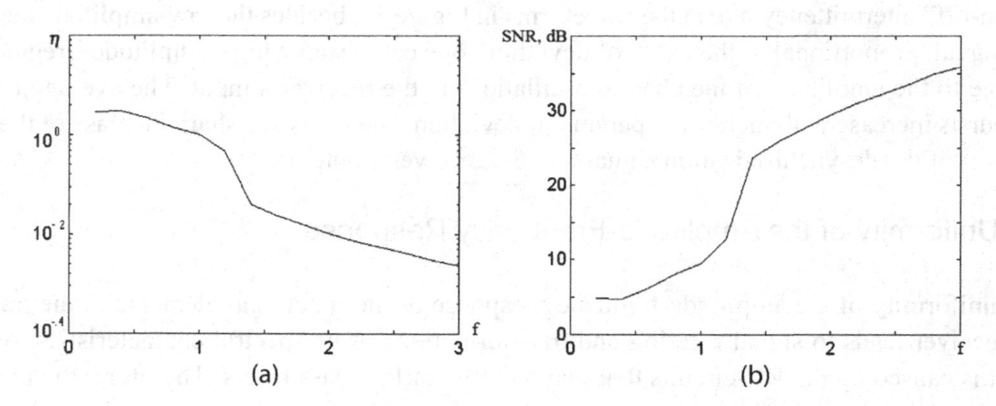

where y is the signal at the transmitter's output, and x is the signal at the receiver's input. The parameter a is varied in the range of 0.001 to 0.03 that corresponds to a variation of the nonlinear distortion level of the signal amplitude at the transmitter's output from 0.1% to 3%. SNR (at μ=1) at the receiver's output is presented in Figure 13a as a function of the nonlinear distortions. It has the same tendency as in the case of the amplitude-frequency distortions. In particular, in both cases the maximum achievable SNR is equal to 35-40 dB.

d. Effect of Additive Normally-Distributed Noise

The effect of the external noise on the system's operation in wireless communications is considered here. The noise frequency bandwidth is restricted with a first-order low-pass filter, and the noise power is fixed at the filter's output.

The signal to noise ratio at μ=1 is presented in Figure 13b as a function of the cutoff frequency for noise levels of -40 dB, -50 dB, and -60 dB. The analysis of these dependencies indicates that the presence of external noise within the carrier signal frequency band decreases the SNR at the receiver's output (here, by 13-15 dB) with respect to the case in the channel (or receiver's input). This is observed at different noise power levels, and is explained by quality losses in the retrieval of the information signal from the mixture with the chaotic.

5. EXPERIMENTS

The main purpose of the experiments is to realize the conditions for transmitting analog information (e.g., speech) in RF band via dynamic chaos.

The idea is to use of the low-frequency chaotic modules discussed in section 2 together with the functional elements of one of the standard RF communication systems.

Figure 13. SNR (at μ=1). (a) Nonlinear distortions in the communication channel (a is the parameter of the nonlinear distortion). (b) Additive noise in the communication channel (f is the cutoff frequency of the noise level). (1) noise level -60 dB. (2) noise level -50 dB (uncertainty zone associated with the finite length of the time series and the irregular occurrence of desynchronization bursts). (3) noise level -40 dB.

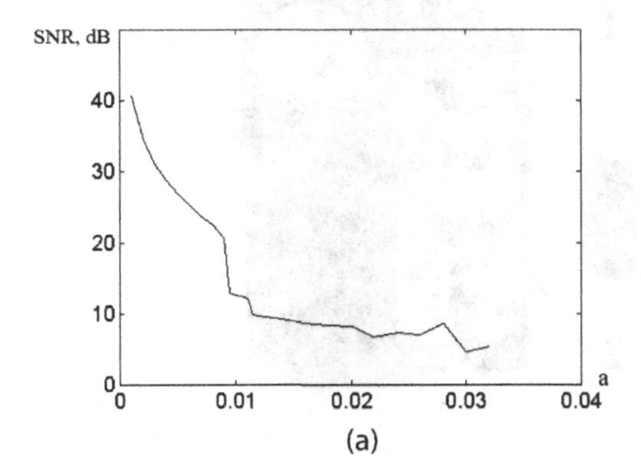

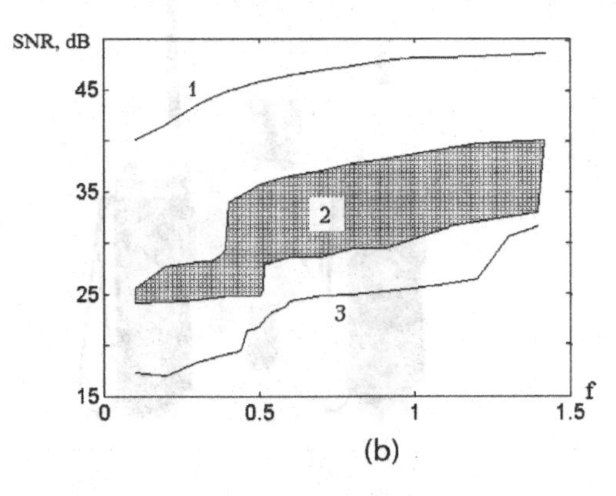

5.1 Experimental Devices and Their Characteristics

a. The Basic Communications System

Portable pocket-size AM transceiver stations ("walkie-talkie"), with carrier frequency 27 MHz, capable of transmitting speech signals with the bandwidth 2 kHz, were used as a basic system (Figure 14). The transceivers perform the following operations in the signal.

In the transmission mode, the speech signal from the built-in microphone is amplified and fed to the modulator. The modulator is a transistor amplifier; a crystal generator signal (27 MHz) is fed to its input, and low-frequency speech (modulating) signal is added to the amplifier bias voltage, thus modulating its gain. Then, the radio signal is amplified and fed to the system's antenna.

In the receiver mode, the radio signal from the antenna is amplified and fed to a demodulator. The demodulator is a mixer (multiplier), made of two pairs of differential transistors. The crystal generator signal is fed to one of its inputs, while the radio signal is fed to the other input. The demodulator's output signal is filtered, and only the low-frequency component is left. Then the signal is passed through an amplifier with automatic gain control, and is sent to the speaker.

b. Chaotic Modules in the Transmitter and the Receiver

The chaotic modules of the experimental devices had the same configuration (Figure 1a) as the ones used in the experiments on speech and music signal transmission in the frequency range of 0 to 5 kHz (section 2). Since the used transceivers were capable to transmit signals with bandwidth up to 2 kHz, the chaotic module band was correspondingly decreased. For this reason, the values of the elements L, C_1 and C_2 were increased (L = 40 mH, C_1 = 0.015 μF, C_2=0.1 μF, R_1=R_2 = 1.9 kΩ). Thus, the chaotic module in the transmitter in the autonomous mode was generating chaotic oscillations in the range of 0 to 2 kHz (Figure 15a).

Figure 14. The basic communication system. (a) The appearance of the transceivers and the chaotic module. (b) The chaotic module built-in transceiver. (1) transceiver. (2) chaotic module.

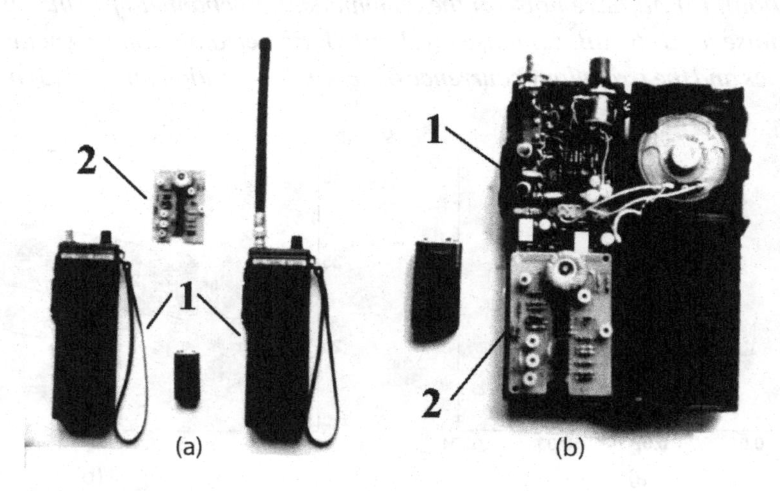

Figure 15. (a) The power spectrum of the transmitter's chaotic module in the autonomous mode. (b) The spectrum of the speech signal.

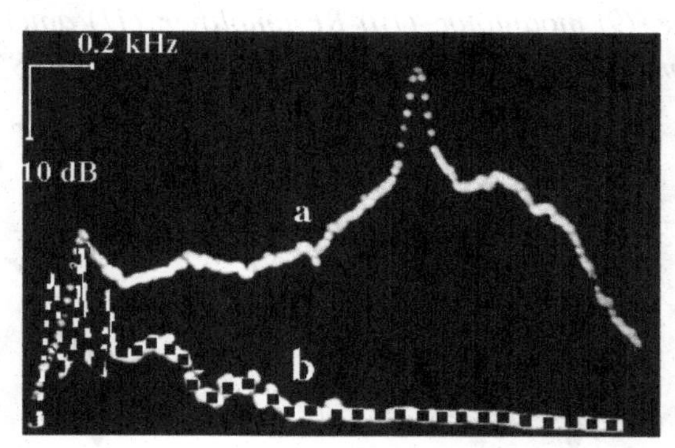

The function of the receiver's chaotic module (Figure 1a) is to extract the speech signal from the mixture with the chaotic one. The module is a passive device and it consists of the same elements (with the same values) as the ones at the transmitter's module. The mixture of chaotic and information signals is fed to an amplifier with tunable gain, and then to the module. The purpose of the amplifier is to restore the input signal level (by means of gain control) and to make it equal to the level of the transmitter's chaotic module output signal, which is a necessary condition for restoring the speech signal at the receiver's module output.

c. Communication Devices

Three transceiver stations are used in the experiments (Figure 16). The first one, with an added chaotic module, plays the role of the communication's system transmitter. The speech signal from its built-in microphone is fed to the chaotic module input as an information signal. At the same time, the module's output signal, a mixture of speech and chaotic signals, plays the role of an information signal with respect to the station: i.e., after amplification, it modulates the carrier amplitude (27 MHz) and then through the antenna is going through the communication channel (Figure 17).

The role of the second station is to receive the radio signal from the communication channel, to amplify and demodulate it, to filter and then to amplify the low-frequency mixture of the speech and chaotic signals, and then to retrieve the information component from the mixture. The last operation is performed by a chaotic module also added to the station. The signal after an amplifier with automatic gain control is fed to the module's input, and its output signal is sent to the station speaker. Thus, the station plays the role of the receiver at the communication system.

Finally, the third station has no chaotic modules and plays the role of a reference receiver.

5.2 Transmission of Speech Information in RF Band by Cable

The first group of experiments is devoted to the transmission of speech information in radio frequency band by cable lines. The communication channel is represented by a long piece of a radio cable with an attenuator with total attenuation up to 70 dB. The experiments have to resolve the following problems:

Figure 16. Block-diagram of the communication system: (1) transmitter, (2) receiver, (3) attenuator, (4) comparison unit, (5) built-in microphone, (6) low frequency amplifier, (7) transmitter's chaotic module, (8) receiver's chaotic module, (9) modulator, (10) RF amplifier, (11) antenna, (12) demodulator, (13) filter, (14) automatic gain control amplifier, S - input information signal, S' - recovered signal.

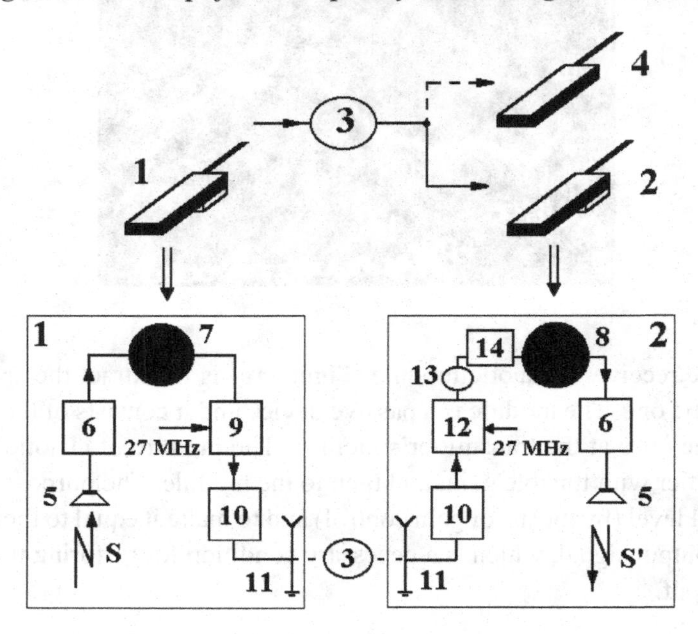

Figure 17. The transmitter's RF output signal in the presence of a speech signal input.

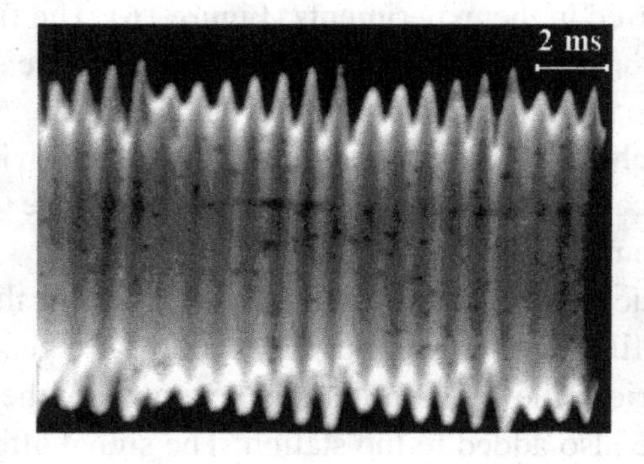

- Investigation of the effects associated with the transfer of low-frequency signals to RF band and back.
- Acquire the synchronous chaotic response and speech information transmission.
- Tuning of the chaotic modules and minimization of the transmitted signal distortions in the radio channel, preparing for the wireless communication experiments.

Just before the chaotic modules were connected to the stations, they were carefully tuned in order to provide equality of their parameters with accuracy 0.5 - 0.8%. This is necessary in order to facilitate the signal's propagation in the radio channel, but is not sufficient to obtain synchronous response in the receiver.

As a result of the tuning of the chaotic module and further low-frequency experiments, the synchronous response with $\eta \approx 9 \cdot 10^{-4}$ is obtained. However, when the modules are built into the stations, in RF band experiments, η increased to 10^{-2}. This increase in η is explained by the signal's distortions in the radio channel. In order to minimize the distortions, the amplitude-frequency response of the functional elements of the basic communication system is corrected. For this purpose, the input and output signals of the functional elements are analyzed, and if there is any distortion in the structure of the low-frequency signal component it is additionally adjusted. This procedure allows us to obtain the amplitude-frequency response of the entire radio channel practically independent of frequency within the range of the chaotic carrier's signal spectrum (Figure 15a).This has led to a decrease in η down to $1.6 \cdot 10^{-3}$.

At the next stage, experiments on speech signal transmission are performed. In the transmitter, the ratio μ of the speech signal level (power) to the chaotic module signal level is varied. At $\mu < 2.5 \cdot 10^{-3}$ the speech signal in the receiver is hardly observable, and is absolutely imperceptible in the reference station. An increase in μ leads to an improvement in the quality of the retrieved speech signal in the receiver. Waveform fragments of the initial speech signal and the signal at the receiver's output are shown in Figure 18 for $\mu = 2.25 \cdot 10^{-2}$ (SNR $\approx$ 11.5 dB in the receiver). Finally, at $\mu > 4 \cdot 10^{-2}$ the signal becomes perceptible, though hardly, in the reference station. The receiver's chaotic module allows us to increase μ up to $1.6 \cdot 10^{-1}$ without a change in the chaotic mode. SNR at the receiver's module output is then equal to 20 dB.

5.3 Wireless Transmission of Speech Information in RF Band

In the experiments on wireless transmission of speech signals in RF band, the cable connecting the transmitter's and the receiver's stations is removed, and the radio signal is radiated and received by the antennas of the transmitter and receiver, respectively. The distance between the stations is varied within the range of the laboratory room (10-15 m). In the absence of the information signal, desynchronization signal level at the receiver's subtractor output is measured. An increase in this level up to the value of $\eta = 1.2 \cdot 10^{-2}$ is found. We were not able to obtain a better response in the receiver by means of tuning the

Figure 18. Information retrieval in the receiver and the comparison unit: (a) speech input signal S (top trace) versus recovered signal S' (bottom trace), (b) speech input signal S (top trace) versus low frequency output signal of the comparison unit (bottom trace).

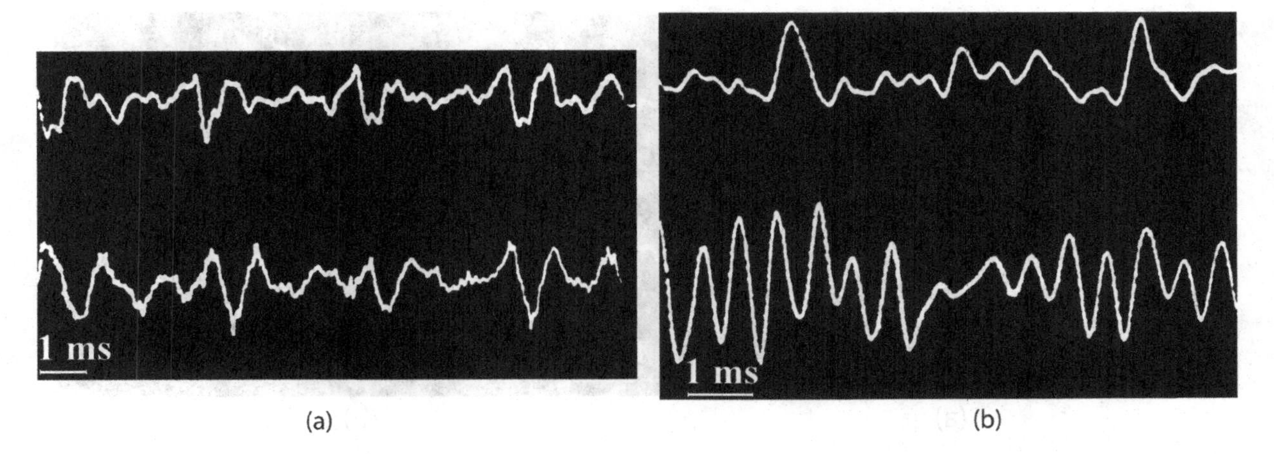

module or restricting the bandwidth of the information channel. The main reasons for the noise level increase are the strong external noise levels in the frequency range of the transceivers (27 MHz), and reflections of the radiated signal within the laboratory room, that appear due to the operation in the near zone of the employed transceivers.

In order to investigate the effect of external factors on the transmission's quality, experiments with sinusoidal signal are performed. In Figure 19a, the waveforms of the initial and retrieved 600 kHz tone signals are presented, $\mu = 4 \cdot 10^{-2}$. As it is obvious, the test signal is visible, but some of its fragments are essentially distorted. Thus, with the increased η taken into account, in order to retrieve a legible signal, the speech signal level at the transmitter's chaotic module has to be increased to the value of $\mu = (1.2 \ldots 1.6) \cdot 10^{-1}$ (Figure 19b). This level of information signal grants a SNR $\approx$ 11 dB, but also leads to partial speech legibility in the reference station, where SNR = -10 dB.

6. CONCLUSION

The above study demonstrates that the chaotic synchronous response could be used as a basis for the design of secure communication channels. From one hand, the results presented in this chapter confirm the possibility of secure wireless communications in RF band, while they allow us to analyze in detail the restrictions and problems connected, first of all, with the quality of synchronization of the transmitter and the receiver of the wireless communication systems.

The effect of the perturbing factors on the transmission quality is investigated theoretically. It is shown that the main reason of the transmission's quality degradation is the chaotic response desynchronization associated with the phenomenon of "on-off" intermittency.

It is found that under the effect of the perturbing factors, the level of information signal fed to the transmitter must be increased in order to obtain qualitative information transmission. However, in order to provide secure communication, one must decrease the information signal level. A compromise on

Figure 19. Information retrieval in the receiver in the wireless communication. (a) The top trace is a test signal S. The bottom trace is the recovered signal S'. (b) Speech input signal S (top trace) versus recovered signal S' (bottom trace).

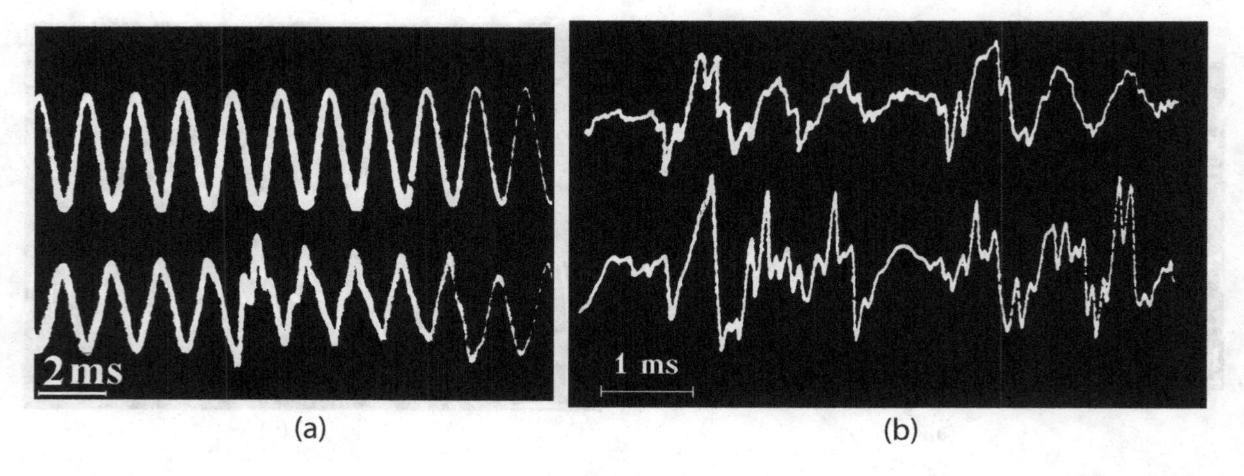

these contradictory requirements provides an improvement of the quality of the synchronous chaotic response in the receiver (lower η).

7. REFERENCES

Belsky, Y. L., & Dmitriev, A. S. (1993). Communication by means of dynamic chaos. *Radiotekhnika I Elektronika, 38*(7), 1310–1315.

Belsky, Y. L., & Dmitriev, A. S. (1995). The effect of Perturbing Factors on the operation of an Information Transmission System with a Chaotic Carrier. *Radiotekhnika I Elektronika, 40*(2), 265–281.

Bohme, F., Feldman, U., Schwartz, W., & Bauer, A. (1994 July). Information transmission by chaotizing. In *Proc. of Workshop NDES'94,* Krakov, Poland, (pp. 163-168).

Cuomo, M. K., Oppenheim, A. V., & Strogatz, S. H. (1993). Synchronization of Lorenz-based chaotic circuits with applications to communications. *IEEE Transactions on Circuits and Systems, 40*(10), 626–632. doi:10.1109/82.246163

Dedieu, H., Kennedy, M. P., & Hasler, M. (1993). Chaos Shift Keying: Modulation and Demodulation of a Chaotic Character Using Self-Synchronizing Chua's Circuits. *IEEE Transactions on Circuits and Systems, 40*(10), 634–642. doi:10.1109/82.246164

Dmitriev, A., Panas, A., Starkov, S., & Kuzmin, L. (1997). Experiments on RF band communications using chaos. *International Journal of Bifurcation and Chaos in Applied Sciences and Engineering, 7*(11), 2511–2527. doi:10.1142/S0218127497001680

Dmitriev, A. S., & Panas, A. I. (2000). *Dynamic chaos: novel type of information carrier for communication systems*. Moscow: Fiz.-Mat. Lit.

Dmitriev, A. S., Panas, A. I., & Starkov, S. O. (1994). Experiments on transmission of speech and musical signals using dynamical chaos. In *Preprint IRE RAS* N12(600), Moscow, Russia (pp. 1-42).

Dmitriev, A. S., Panas, A. I., & Starkov, S. O. (1995). Experiments on speech and music signals transmission using chaos. *International Journal of Bifurcation and Chaos in Applied Sciences and Engineering, 5*(4), 1249–1254. doi:10.1142/S0218127495000910

Halle, K. S., Wu, C. W., Itoh, M., & Chua, L. O. (1993). Spread spectrum communication through modulation of chaos. *International Journal of Bifurcation and Chaos in Applied Sciences and Engineering, 3*(2), 469–477. doi:10.1142/S0218127493000374

Heagy, J. F., Platt, N., & Hammel, S. M. (1994). Characterization of on-off intermittency. *Physical Review E: Statistical Physics, Plasmas, Fluids, and Related Interdisciplinary Topics, 49*(2), 1140–1150. doi:10.1103/PhysRevE.49.1140

Kennedy, M. P. (1992). Robust op amp realization of Chua's circuit. *Frequuenz, 46*(3-4), 66–80.

Kocarev, L., Halle, K. S., Eckert, K., Chua, L. O., & Parlitz, U. (1992). Experimental demonstration of secure communication via chaotic synchronization. *International Journal of Bifurcation and Chaos in Applied Sciences and Engineering, 2*(3), 709–713. doi:10.1142/S0218127492000823

Madan, R. (1993). *Chua's Circuit: A Paradigm for Chaos*. Singapore: World Scientific.

Parlitz, U., Chua, L., Kocarev, L., Halle, K., & Shang, A. (1992). Transmission of digital signals by chaotic synchronization. *International Journal of Bifurcation and Chaos in Applied Sciences and Engineering, 2*(4), 973–977. doi:10.1142/S0218127492000562

Pecora, L. M., & Carroll, T. L. (1990). Synchronization in chaotic systems. *Physical Review Letters, 64*, 821–824. doi:10.1103/PhysRevLett.64.821

Platt, N., Spiegel, E. A., & Tresser, C. (1993). On-off intermittency a mechanism for bursting. *Physical Review Letters, 70*(3), 279–282. doi:10.1103/PhysRevLett.70.279

Volkovskii, A. R., & Rul'kov, N. V. (1993). Synchronous chaotic response of a nonlinear oscillating system as the detection principle of chaos informational component. *Pis'ma v GTF, 19*(3), 71-75.

Chapter 16
Control-Theoretical Concepts in the Design of Symmetric Cryptosystems

Gilles Millérioux
University Henri Poincaré of Nancy, France & Research Center on Automatic Control of Nancy (CRAN), France

José María Amigó
University Miguel Hernandez of Elche (UMH), Spain & Centro de Investigación Operativa, Spain

ABSTRACT

In this chapter, it is shown how control-theoretical concepts can be useful in the design of symmetric cryptosystems. The authors first provide some background on cryptography with special emphasis on symmetric ciphering and, more specifically, on stream ciphers. It is explained how some permutation or substitution primitives can be derived from chaotic dynamical systems for cryptographic purposes. After a review of the most popular synchronization-based cryptosystems, a comparative study between these chaotic cryptosystems and the conventional symmetric ciphers, specifically stream ciphers, is carried out. In particular, it is shown that message-embedded chaotic ciphers and conventional self-synchronizing stream ciphers are structurally equivalent under the so-called flatness condition, a condition borrowed from control theory.

1 INTRODUCTION

The considerable progress in communication technology during the last decades has led to an increasing need for security in information exchanges. In this context, cryptography plays a major role as information is mostly conveyed through public networks. The main objective of cryptography is, precisely, to conceal the content of messages transmitted through insecure channels, to unauthorized users or, in other words, to guarantee privacy and confidentiality in the communications. Since the early 1960s, cryptography has no longer been restricted to military or governmental concerns, what has spurred an unprecedented development of it. At the same time, this development benefited very much from the ad-

DOI: 10.4018/978-1-61520-737-4.ch016

vances in digital communication technology in form of new and efficient ways of designing encryption schemes. Despite the diversity of cryptographic techniques, two major classes are typically distinguished: *public-key* ciphers and *symmetric-key* ciphers (also called private-key ciphers).

Let us shortly recall that modern cryptography originates in the works of Claude Shannon after World War II (Shannon, 1949). Shannon's ideas substanciated in form of substitution-permutation networks, that are at the heart of the Lucifer encryption algorithm, designed by IBM in the late 1960s and early 1970s (Pieprzyk et al, 2003). One of the key dates in the recent history of cryptography is 1977, when the symmetric cipher Data Encryption Standard (DES) was adopted by the U.S. National Bureau of Standards (now the National Institute of Standards and Technology ---NIST), for encrypting unclassified information. DES is now in the process of being replaced by the Advanced Encryption Standard (AES), a new standard adopted by NIST in 2001. Another milestone is 1978, marked by the publication of RSA, the first full-fledged public-key algorithm. This discovery not only solved the key-exchange problem of symmetric cryptography but, most importantly, did it open new whole areas (like authentication and electronic signature) in modern cryptology. Among symmetric-key ciphers, stream ciphers are of special interest for high speed encryption, like in satellite communications, private TV channels broadcasting, and networked embedded systems. They are mainly based on generators of complex sequences, which must be synchronized at the transmitter and receiver sides. Stream ciphers have received much attention recently. Two European projects have influenced this evolution: The project NESSIE within the Information Society Technologies (IST) Programme of the European Commission, which started in 2000 and ended in 2004, followed by ECRYPT[1], launched on February 1st, 2004. Sponsored by ECRYPT, ESTREAM is a multi-year effort to identify promising software- and hardware-oriented algorithms with the aid of proposals from industry and academia.

Chaotic behavior is one of the most complex dynamics a nonlinear system can exhibit. One of the formal definitions of chaos is due to R.L. Devaney (Devaney, 1989). A dynamical system is said to be chaotic in the sense of Devaney if it fulfills two properties: Transitivity and density of periodic points. It can be shown that sensitivity to the initial condition, which is the property mostly associated with chaotic behavior, is actually a consequence of those two other properties. Roughly speaking, a system is said to be sensitive to initial conditions if a small change in the initial condition drastically changes the behavior of a system in the long run, thus making long-term predictions unfeasible in practice. Complex dynamics had its beginnings in the work of the French mathematician Henri Poincaré (1854-1912), who also recognized the practical unpredictability of such systems. Sensitive dependent phenomena were highlighted by Edward Lorenz in 1963 while simulating a simplified model of convection. But it was the paper "Period Three Implies Chaos" by Li and Yorke in 1975 (Li, & Yorke, 1975), where the word "chaos" was coined in the framework of dynamical systems, which triggered a tremendous interest in this kind of phenomena. Since then a great number of applications have been proposed in such disparate areas as mechanics, physics, biology, economy, engineering, avionics or weather forecasting.

Signals generated by chaotic systems are broadband, noiselike and present random-like statistical properties, in spite of being deterministic. This makes them a very convenient tool to implement the principles of confusion and diffusion required by Shannon for cryptosystems (Shannon, 1949; Massey, 1992). The first proposals in this direction were made around 1990 (Matthews, 1989; Habutsu et al, 1991). In 1990 entered the scene chaos synchronization (Carroll, & Pecora, 1991), a new technique for analog communication (Parlitz et al, 1993) whose potential for cryptography was soon exploited (Cuomo et al, 1993). Since then, many schemes have been proposed to mask information using the properties of chaos, either based on discrete chaotic maps or on the synchronization of signals. This new approach to encryption is commonly called chaos-based (or "chaotic') cryptography; see (Amigó et al, 2007a) for a general view of digital chaotic cryptography. To illustrate the high activity in this field, let us refer to the many special issues that have been already published in international journals like IEEE Transactions on Circuits and Systems, or International Journal on Bifurcations and Chaos (see (Alvarez & Li, 2006; Hasler, 1998; Millérioux et al, 2008; Ogorzalek, 1993; Yang, 2004) for some surveys). Furthermore, numerous invited sessions on chaos-based cryptography have been organized at international conferences, e.g., the International Symposium on Circuits And Systems (ISCAS), the International Symposium on NOnLinear Theory and Applications (NOLTA), and many others. Let us also mention the existence of an IEEE Action Group, called "Chaos, Synchronization and Control", under the aegis of the IEEE Control Systems Society Technical Committee on Computer Aided Control System Design (CACSD).

The contribution of this chapter is twofold. First, we provide a state-of-the-art of the structures involved in chaotic cryptographic schemes and compare them with conventional ciphers with a special treatment of the symmetric ones. Then, we discuss some relevant issues which deserve, to our opinion, a deeper study. We show how control-theoretical concepts are useful for both carrying out the comparative study and suggesting new approaches in the design of so-called self-synchronizing stream ciphers. We are going to restrict to digital encryption, and so to discrete-time dynamical systems (maps).

The layout of this chapter is the following. Section 2 is devoted to a background on cryptography with special emphasis on symmetric ciphering and, more specifically, on stream ciphers. In section 3 it is explained how some permutation or substitution primitives can be derived from chaotic dynamical systems for cryptographic purposes. Section 4 reviews the most popular synchronization-based cryptosystems. Section 5 is devoted to a comparative study between these chaotic cryptosystems and the conventional symmetric ciphers, specifically, stream ciphers. Further issues that deserve deeper investigation are detailed. In particular, it is shown that message-embedded chaotic ciphers and conventional self-synchronizing stream ciphers are equivalent under the so-called flatness condition, a condition borrowed from control theory. As a result, this kind of chaotic ciphers may be an interesting alternative for the design of Self-Synchronizing Stream Ciphers.

2 BACKGROUND ON CRYPTOGRAPHY

2.1 Generalities

A general encryption mechanism, also called cryptosystem or cipher, is illustrated in Fig. 1. We are given an alphabet A, that is, a finite set of basic elements named symbols. On the *transmitter* part, a plaintext m (also called information or message) consisting of a string of symbols $m_k \in A$ is encrypted according to an encryption function e which depends on the key $k^e \in K$ (the key space). The resulting ciphertext c, a string of symbols c_k from an alphabet usually identical to A, is conveyed through a public channel to the *receiver*. At the receiver side, the ciphertext c is decrypted according to a decryption function d which depends on the key $k^d \in K$. The encryption scheme corresponding to the pair (e,d) must be designed in an appropriate way so as it is a hard task for an eavesdropper to retrieve the plaintext m. Therefore, there must exist a unique pair (k^e,k^d) such that $d(k^d,c)$ where $c=e(k^e,m)$.

Among a wide variety of cryptographic techniques, two major classes can be typically distinguished: *public-key* ciphers and *symmetric-key* ciphers. Public-key ciphers are largely based upon computationally demanding mathematical problems, for instance, prime factorization. One of the best well-known public-key ciphers is the RSA algorithm (Menezes et al, 1996).

Figure 1. General encryption mechanism

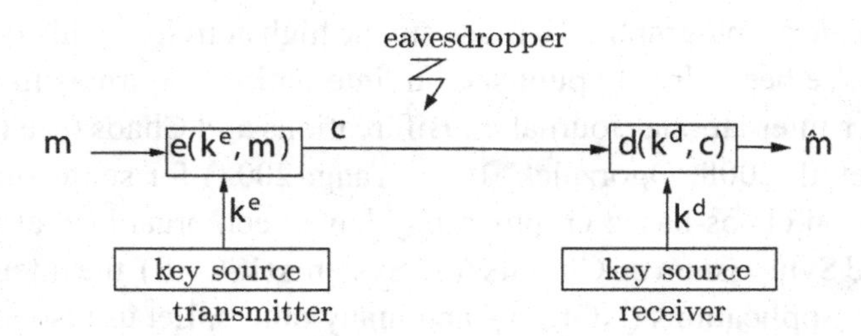

In contrast to public-key ciphers, symmetric-key ciphers are characterized by an encryption scheme (e,d) for which the determination of the key k^d can be easily done from the knowledge of k^e. Hence, not only k^d must be kept secret but the key k^e as well. It is customary that both keys are identical, that is k^d=k^e. There are two classes of symmetric-key encryption schemes which are commonly distinguished: block ciphers and stream ciphers.

A block cipher is an encryption scheme that breaks up the plaintext messages into strings (called blocks) of a fixed length over an alphabet and encrypts one block at a time. Block ciphers usually involve *substitution* or *transposition* operations or compositions of them.

Next we describe stream ciphers in more detail.

2.2 Stream Ciphers

In the case of stream ciphers, the *plaintext* is broken up into blocks of the same length, called symbols and denoted by m_k. A major distinction with respect to the block ciphers lies in that the encryption function e can change for each symbol because it depends on a time-varying key K_k also called *running key*. The sequence $\{K_k\}$ is called the *keystream*. This being the case, stream ciphers are generally well appropriate and their use can even be compulsory when buffering is limited or when only one symbol can be processed at a time: the field of telecommunications is subject to such constraints. They benefit from smaller footprint (gates, power consumption,...) in low-end hardware implementation, high encryption speed, small input/output delay and simple protocols for handling variable sized inputs. They are efficient and compact in constrained devices (e.g. RFID).

Stream ciphers require a keystream generator. When the plaintext m_k and the ciphertext c_k are binary words, the most widely adopted function e is the bitwise XOR operation. If the keystream $\{K_k\}$ is truly random and never used again, the encryption scheme is called *one-time pad*—the only cipher provably secure so far. However, in order to decrypt the ciphertext, the recipient party would have to know the random keystream and, thus, would require again a secure transmission of the key; this is the so-called key-exchange problem which can be solved in different ways, notably via public-key cryptography. However, for the *one-time pad* cipher, the key should be as long as the plaintext which drastically increases the difficulty of the key distribution. As an alternative to such an ideal encryption scheme, one can resort to pseudo-random generators. Indeed, for such generators, the keystream is produced by a deterministic function while its statistical properties look random. Generally, keystreams are generated iteratively by feedback shift registers (or compositions of them) since they produce pseudo-random sequences in a very efficient way (Knuth, 1998). There are two classes of stream ciphers, the difference lying in the way the keystream is generated: the *synchronous* stream ciphers (SSC) and the *self-synchronizing* stream ciphers (SSSC).

At the transmitter side, Synchronous Stream Ciphers are described by the equations

$$\begin{cases} K_k = \sigma_\theta^s(K_{k-1}), \\ c_k = e(K_k, m_k). \end{cases} \tag{1}$$

The running key K_k is generated by a function σ_θ^s parameterized by θ, this parameter acting as the secret (or master) key. c_k is generated by the encryption function e, which depends on the time-varying key K_k.

At the transmitter side, the Self-Synchronizing Stream Ciphers are described by the equations

$$\begin{cases} K_k = \sigma_\theta^{ss}(c_{k-l}, \ldots, c_{k-l-M}), \\ c_k = e(K_k, m_k). \end{cases} \tag{2}$$

where σ_θ^{ss} is also a function parameterized by the secret key θ, the same that generates the keystream $\{K_k\}$. σ_θ^{ss} depends on a fixed number of past values of c_k. The nonnegative integer l stands for a possible delay, and M is called the *delay of memorization*. Similarly to Synchronous Stream Ciphers, c_k is generated by the encryption function e, and this depends on the time-varying key K_k.

Synchronization Issues and Definition of the Secret Key

For stream ciphers, the generators at both sides of the communication channel have the same generator function (σ_θ^s for SSC and σ_θ^{ss} for SSSC), and synchronization of keystreams $\{K_k\}$ and $\{\hat{K}_k\}$, produced respectively at the transmitter and receiver sides, is a condition for correct decryption.

For SSC, the generators are not coupled to each other. Consequently, the only way to guarantee synchronization of the keystreams is to share the seed (the initial running key $K_0 = \hat{K}_0$). This being the case, the secret key θ is nothing but the seed K_0. For SSSC, since the generator function σ_θ^{ss} at the transmitter and receiver sides, shares the same quantities, namely the past ciphertexts, it is clear that the generators synchronize automatically after a finite transient time of length M. The secret key θ is the parametrization of the function σ_θ^{ss}.

It has been pointed out in the introduction that, because chaotic signals are broadband, noiselike and exhibit random-like statistical properties, they are specially well suited to implement the principles of confusion and diffusion put forward by Shannon to guarantee secrecy. We present in Section 3 and Section 4 the main chaos-based ciphers proposed since the 90's.

3 CHAOS-BASED PERMUTATIONS AND SUBSTITUTIONS

A first method for scrambling information consists in computing a great number of iterations of a chaotic map, using a digital message as initial data. The resulting transformation on the original data must be a permutation, a substitution or a composition of both.

One of the techniques used to design permutations is the discretization of chaotic maps. The reader is referred to (Fridrich, 1998; Schmitz, 2001) and references therein for details. This is basically also the strategy provided in (Amigó et al, 2005;Szczepanski et al, 2005). The discretization is obviously required since chaotic maps take on values in a continuum, whereas digital messages are intrinsically discrete. The intuition that the resulting permutations may have different diffusion and mixing properties have been embodied in an approach called discrete chaos. Any discrete approximation of a chaotic system (*X*, *f*) in form of a permutation $F_M : \{0,1,...,M-1\} \to \{0,1,...,M-1\}$ is called a chaotic cryptographic primitive; we say that a cryptographic algorithm is chaotic if some of its building blocks is a chaotic cryptographic primitive. Usually, the state space *X* is a compact metric space (like an n-dimensional interval or an n-torus). Chaos-based primitives takes advantage of the ergodic properties of dynamical systems. The theoretical framework of discrete chaos was presented in (Kocarev et al, 2006). The main tool of discrete chaos is the discrete Lyapunov exponent.

Let $\mathsf{X} = \{\xi_0,...,\xi_{L-1}\}$ be a linearly ordered finite set, $\xi_i < \xi_{i+1}$, endowed with a metric $d(\cdot,\cdot)$, and $F : \mathsf{X} \to \mathsf{X}$ be a bijection or, equivalently, a permutation of X. We define the *discrete Lyapunov exponent* (DLE) of *F* as

$$\lambda_F = \frac{1}{L}\sum_{i=0}^{L-1} \log \frac{d\left(F(\xi_{i+1}), F(\xi_i)\right)}{d\left(\xi_{i+1}, \xi_i\right)}, \tag{3}$$

where the definition of ξ_L, the right neighbor of ξ_{L-1}, depends on the 'topology' of X (see below for a choice). We will use natural logarithms to calculate λ_F. Observe that λ_F depends both on the order (that determines whose neighbor is who) and on the metric *d*, but it is invariant under rescaling and, furthermore, has the same invariances that *d* might have.

In the examples and applications we will consider below, *F* will be a permutation of the subset X $= \{0,...,L-1\}$ of R endowed with the Euclidean distance $d(\xi_i,\xi_j) = |\xi_i - \xi_j|$ and the standard order. In this case, we will refer to X as a linear set and choose $\xi_L \equiv \xi_{L-2}$ to be the 'right' neighbor of the last (or greatest) state ξ_{L-1} in the definition (3); we have $\lambda_F \geq 0$.

The justification for calling λ_F the discrete Lyapunov exponent of *F* is as follows. Let $x_{j+1} = f(x_j)$, $j = 0,1,...,L-1$, be a typical trajectory of length *L* of a chaotic self-map *f* of a one-dimensional interval *I*, such that $x_{j+1} \neq x_j$ for all *j* and $|x_{L-1} - x_0| < \varepsilon$. We define $f(x_{L-1}) = x_0$ and order $x_0, x_1, ..., x_{L-1}$ in *I* to obtain $x_{n_0} < x_{n_1} < ... < x_{n_{L-1}}$, so that x_{n_i} and $x_{n_{i+1}}$ are neighbors. Furthermore, set $\xi_i = \lfloor x_{n_i} N \rfloor$, where *N* is chosen such that $\xi_i \neq \xi_j$ for all $i \neq j$. The map *f* induces then the obvious permutation

$$F(\xi_i) = \xi_j \quad if \quad f(x_{n_i}) = x_{n_j}$$

on $\{\xi_0, ..., \xi_{L-1}\}$. Then,

Proposition 1. *Let I be a one-dimensional interval and* $f : I \to I$ a chaotic map with piecewise continuous derivative. Then $\lim_{L\to\infty} \lambda_{F_L} = \lambda_f$, where λ_f is the Lyapunov exponent of f.

In (Kocarev et al, 2006) the reader can find a proof of Proposition 1 and also a generalization to higher dimensions.

Examples of chaotic primitives include the finite-state tent map, the finite-state Chebyshev map and the finite-state *n*-dimensional torus automorphisms. Affine transformations on the *n*-torus in chaos synchronization-based cryptography have been studied in (Rosier et al, 2006). These maps have the nice property that the precision of the initial point does not degrade along its orbit.

We consider next some examples of permutations on linear sets.

Example 1. *For the right shift modulo L, defined on* X $= \{0, ..., L-1\}$ as $\theta_L(\xi) = \xi + 1$ for $\xi = 0, 1, ..., L-2$ and $\theta_L(L-1)=0$, *we find*

$$\lambda_{\theta_L} = \frac{2}{L} \ln(L-1).$$

Example 2. *Define the permutation*

$$F_{2l}^{\max}(\xi) = \begin{cases} l+k & if\ \xi = 2k \quad\ \ 0 \le k \le l-1, \\ k & if\ \xi = 2k+1 \quad 0 \le k \le l-1, \end{cases}$$

on S $=\{0, ..., 2l-1\}$. The DLE of $F_{2l}^{\max}$ is easily seen to be

$$\lambda_{F_{2l}^{\max}} = \frac{l+1}{2l} \ln l + \frac{l-1}{2l} \ln(l+1) \equiv \lambda_{2l}^{\max}. \tag{4}$$

For large *l* we have:

$$\lambda_{\theta_{2l}} \approx \frac{1}{l} \log 2l \quad and \quad \lambda_{2l}^{\max} \approx \log l.$$

It was proved in (Amigó et al, 2007b) that $\lambda_{2l}^{\max}$ is the greatest DLE a bijection on the linear set $\{0, ..., 2l-1\}$ may have. This makes possible to gauge the 'diffusivity' of a permutation of an even number of elements.

Example 3. *The Advanced Encryption Standard (AES) or Rijndael is a symmetric cipher designed for 128, 192 and 256 bit block lengths (*Pieprzyk et al, 2003*); for simplicity, we consider here the first implementation only. In order to calculate the DLE for the AES, we assign to each 128 bit block an integer in* $\{0, 1, ..., 2^{128}-1\}$ *via its binary representation. The computation of the DLE has been performed* *on 7000 iterations of the AES map obtaining DLE* = 20.93 *after the first round and DLE* = 87.22 *after the second and subsequent rounds (to be compared to* $\lambda_{2^{128}}^{\max} = 88.03$) (Amigó et al, 2008*).*

Example 4. *Serpent handles 128-bit messages using a key that can be either 128-, 192-, or 256-bits long (*Pieprzyk et al, 2003*). The encryption proceeds basically in 32 rounds, using 8 S-boxes* S_i . In the simplest version, the input to the *ith round is first XORed with the round key* K_i, *next each 4-bit subblock is input in parallel into 32 copies of the same S-box* $S_{i \bmod 8}$, and finally (except in the last round) the output of the S-boxes is submitted to linear transformations. In order to measure the diffusion property of the whole algorithm, we followed the orbit of a sample of 128-bit random messages. The result is DLE =84.16 *after the first round and DLE* = 87.22 *(the same as for AES) after the second and subsequent rounds (*Amigó et al, 2008*).*

4 SYMMETRIC CHAOTIC CRYPTOSYSTEMS

The second approach for scrambling information views a chaotic system with dynamic *f* as a complex sequence generator. *f* is often specified by a state representation with corresponding state vector $x_k \in X$, the dimension of the system being *n*. *f* is parametrized by a vector θ of dimension *L* which is expected to act as the secret key. Only a part of the state vector x_k obtained via a function *h*, possibly parameterized by θ as well, called the "output" and denoted by $y_k \in Y$, is conveyed through the public channel. y_k is usually of low dimension and should be unidimensional in the ideal case. In what follows, we will assume that y_k is a scalar (dimension 1), the transmitter being thus restricted to a so-called Single Input Single Output (SISO) system. The receiver is a dynamical system with dynamics $\tilde{f}$ and with output function $\tilde{h}$, both parameterized by $\hat{\theta}$. The state vector is denoted $\hat{x}_k$.

Both functions $\tilde{f}$ and $\tilde{h}$ must be properly chosen to recover the message m_k at the receiver side. A first condition is that $\hat{\theta} = \theta$. For most of the chaos-based cryptosystems proposed so far, the recovering of the message m_k is performed in two steps, namely *chaos synchronization* and *static inversion*.

i) Chaos Synchronization

Let *T* be a constant matrix of appropriate dimension, and *U* a non empty set of initial conditions. There are two main concepts of synchronization:

Definition 1. *Asymptotical synchronization obeys the following condition:*

$$\lim_{k \to \infty} \left\| Tx_k - \hat{x}_k \right\| = 0 \quad \forall \hat{x}_0 \in U. \tag{5}$$

Definition 2. *Finite time synchronization obeys the following condition:*

$$\exists k_f < \infty, \ \left\| Tx_k - \hat{x}_k \right\| = 0 \quad \forall \hat{x}_0 \in U \ \text{ and } \ \forall k \geq k_f. \tag{6}$$

In practice, however, measurements have finite accuracy, so the error of an asymptotical synchronization can be considered to be zero after a finite transient time. Synchronization can be viewed as a state reconstruction. In 1997 several papers (Grassi & Mascolo, 1997; Itoh et al, 1997; Millérioux, 1997; Nijmeijer & Mareels, 1997) brought out this connection. As a result, the receiver often consists in an observer. If only a part of the components are reconstructed, the observer is a reduced observer and $rank(T) < n$. If all the components of the state vector are reconstructed, the observer is a full observer and T is the identity matrix.

ii) Static Inversion

Static inversion involves a "static" function d that depends on the internal state $\hat{x}_k$ and the output y_k. The function d delivers a quantity $d(\hat{x}_k, y_k) = \hat{m}_k$, and must verify

$$d(\hat{x}_k, y_k) := \hat{m}_k = m_k \quad if \quad \hat{x}_k = x_k.$$

Various cryptosystems, corresponding to distinct ways of scrambling a message, have drawn the attention of the researchers over the years. They are reviewed in the following subsections. Let us point out that we are going to restrict to discrete-time systems (maps) having in mind the comparison with digital conventional cryptography, but most of these chaotic cryptosystems can also be found in the literature for the continuous time.

4.1 Additive Masking

This scheme was first suggested in (Cuomo et al, 1993) and (Wu & Chua, 1993). The information m_k to be concealed is merely added to the output y_k of the *transmitter* (Figure 2):

$$\begin{cases} x_{k+1} &= f_\theta(x_k), \\ y_k &= h_\theta(x_k) + m_k. \end{cases} \tag{7}$$

The generic equations of the *receiver* read:

$$\begin{cases} \hat{x}_{k+1} &= \tilde{f}_\theta(\hat{x}_k, y_k), \\ \hat{y}_k &= \tilde{h}_\theta(\hat{x}_k). \end{cases} \tag{8}$$

The quantity y_k which appears in (8) reveals the unidirectional coupling between both the transmitter and the receiver systems. Provided that synchronization (5) or (6) can be achieved, the recovering of the information is performed by the static inversion

$$\hat{m}_k = y_k - \tilde{h}_\theta(\hat{x}_k).$$

Figure 2. Additive masking

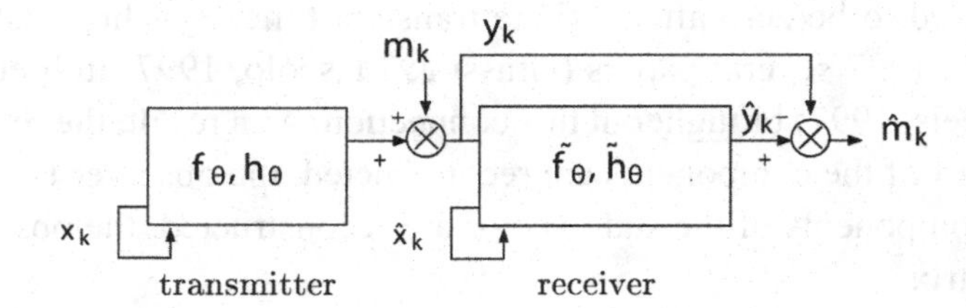

Unfortunately, more often than not the information cannot be exactly retrieved. Indeed, m_k acts as a disturbance on the channel and precludes the *receiver* from being exactly synchronized; neither (5) nor (6) can be exactly fulfilled. As a result, $\hat{x}_k \neq x_k$, $\hat{y}_k \neq y_k$ and, finally, $\hat{m}_k \neq m_k$ for any *k*.

4.2 Modulation

4.2.1 Chaotic Switching

Chaotic switching is also referred to as chaotic modulation or chaos shift keying. Such a technique has been mostly proposed in the digital communications context. A description with deep insights can be found in (Kolumban et al, 1998), even though the method was proposed a couple of years before, say, in 1993 (Dedieu et al, 1993). Basically, at the *transmitter* side, to each symbol $m_k = m^i$ belonging to a finite set $\{m^1,...,m^N\}$, it is assigned a chaotic signal emanating from the dynamic f_θ^i with output function h_θ^i (i=1,...,N). Therefore, in the transmitter description, the index *i* depends on m_k.

$$\begin{cases} x_{k+1} &= f_\theta^{i(m_k)}(x_k), \\ y_k &= h_\theta^{i(m_k)}(x_k). \end{cases} \tag{9}$$

The simplest case involves binary-valued information and only two different chaotic dynamics f^1, f^2 are needed. Then, according to the current value of the symbol m_k at times $k = jK$ ($j \in \mathbb{N}$), a switch is periodically triggered on every *K* samples. During the interval of time $[jK,(j+1)K-1]$, m_k is assumed to be constant and the chaotic signal y_k of the system which has been switched on is conveyed through the channel (Figure 3).

The objective at the *receiver* end is to decide which chaotic system f_θ^i is most likely to have produced the sequence $\{y_k\}_{jK,...,(j+1)K-1}$. To this end, the receiver part is composed of as many systems, say *N*, as at the transmitter part:

$$\begin{cases} \hat{x}_{k+1} &= \tilde{f}_\theta^i(\hat{x}_k,[y_k]), \\ \hat{y}_k &= \tilde{h}_\theta^i(\hat{x}_k). \end{cases} \tag{10}$$

Figure 3. Chaotic switching

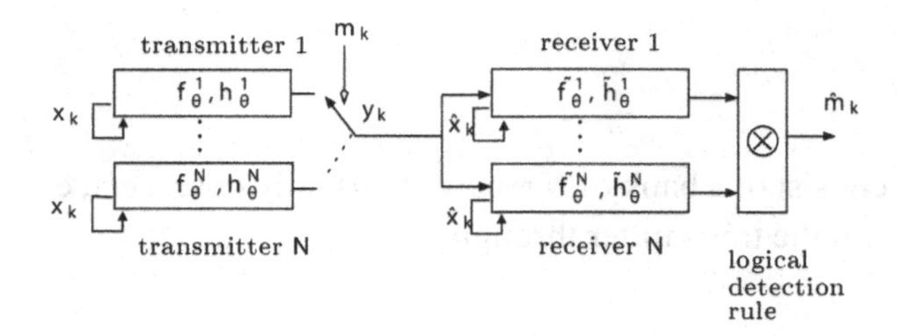

The symbol $[\cdot]$ distinguishes two methods: The coherent and the non coherent detections. Non coherent detection involves statistical approaches mainly based upon correlation operations between the transmitted signal y^k and the estimated signal $\hat{y}_k$. In this case, the receivers are autonomous systems with dynamics $\tilde{f}^i_\theta$, and y^k must be omitted in (10). Coherent methods require the synchronization of both the transmitter and the receiver. The synchronization (5) or (6) (where x_0 must be replaced by x_{jK}) is obtained by unidirectional coupling through the variable y^k which is thereby involved in $\tilde{f}^i_\theta$ of Eq. (10). Only one of the N receivers (observers, for instance) can be synchronized according to the value of m_k which is constant within the interval of time $[jK,(j+1)K-1]$. A simple logical decoder permits to retrieve the original information when analyzing the residuals r^i_k, where

$$r^i_k = y_k - \tilde{h}^i_\theta(\hat{x}_k).$$

When multi-valued information is considered (Palaniyandi & Lakshmanan, 2001), the number of receivers increases and a sophisticated logical mechanism, located after the bank of receivers, is required.

Regarding a noisy context, the modulation technique is appealing because it benefits from some immunity properties. In a noise-free context though, it is much less attractive because it suffers from the fact that each switch of m_k causes a transient in the synchronization process. That motivates the requirement that m_k must be constant within an interval of time. Unfortunately, that prevents from high throughput transmissions. To the lack of efficiency adds the mounting number of receivers when N becomes larger.

4.2.2 Parameter Modulation

Basically, there are two kinds of modulations: The discrete and the continuous one.

The setup corresponding to a discrete parameter modulation (Dedieu et al, 1993;Parlitz et al, 1993) is depicted in Fig. 4a. In such a case, a parameter λ (different from the key parameter θ) of a single chaotic system, takes values $\lambda(m_k) = \lambda^i$ according to a prescribed rule over a finite set $\{\lambda^1,\ldots,\lambda^N\}$. For binary messages, the parameter of the *transmitter* only takes two distinct values λ^1, λ^2. During the interval of time $[jK,(j+1)K-1]$, m_k is assumed to be constant and the chaotic signal y_k is conveyed through the channel:

$$\begin{cases} x_{k+1} &= f_\theta^{\lambda(m_k)}(x_k), \\ y_k &= h_\theta^{\lambda(m_k)}(x_k). \end{cases} \tag{11}$$

The receiver part can consist of a bank of N receivers, usually observers, each of them being coupled in a unidirectional way with the transmitter through y_k:

$$\begin{cases} \hat{x}_{k+1} &= \tilde{f}_\theta^{\lambda^i}(\hat{x}_k, y_k), \\ \hat{y}_k &= \tilde{h}_\theta^{\lambda^i}(\hat{x}_k). \end{cases} \tag{12}$$

Only one observer, set with the same value λ^i of the transmitter which has actually delivered the sequence $\{y_k\}_{jK,\ldots,(j+1)K-1}$, can be synchronized in the form (5) or (6) (where x_0 must be replaced by x_{jK}) within the time interval $[jK,(j+1)K-1]$. Thus, again, a simple logical decoder permits to retrieve the original information when analyzing the residuals

$$r_k^i = y_k - \tilde{h}_\theta^{\lambda^i}(\hat{x}_k).$$

For the continuous modulation (Figure 4b), the information m_k takes values over an uncountable set M. Consequently, an infinite number of units at the receiver side would be required. As a matter of fact, for the recovering of λ_k and of m_k, we usually resort to adaptive techniques and identification procedures (Anstett et al, 2004; Dedieu & Ogorzalek, 1997; Fradkov & Markov, 1997; Huijberts et al, 2004). The estimation must fulfill $\hat{\lambda}_k = \lambda_k$ after a transient as short as possible.

For both discrete and continuous modulation, the function delivering $\lambda(m_k)$ must be bijective so that m_k can be recovered in a unique way.

Figure 4. Parameter modulation

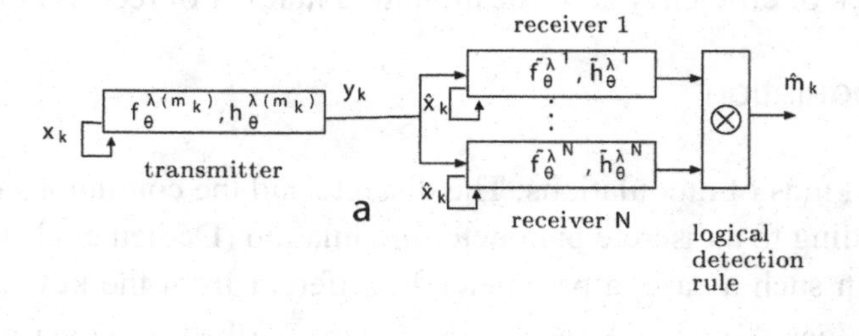

Nevertheless, for the parameter modulation and similarly to the chaotic switching, the information must be constant during a prescribed interval of time (or at least slowly time-varying in a bounded range) to cope with the transients induced by the adaptive or the identification process. As happened with chaotic switching, this technique severely limits high throughput purposes and, therefore, it does not seem very appealing for encryption.

4.3 Two-Channel Transmission

For a two-channel transmission (Figure 5), a first channel is used to convey the output y_k of an autonomous chaotic system with dynamic f_θ and output function h_θ. Besides, a function ν_e, depending on a time-varying quantity, say, the state vector x_k of the chaotic system, encrypts the information m_k and delivers $u_k = \nu_e(x_k, m_k)$. Then, the signal u_k is transmitted via a second channel. The set of equations governing the transmitter is

$$\begin{cases} x_{k+1} = f_\theta(x_k), \\ y_k = h_\theta(x_k), \\ u_k = \nu_e(x_k, m_k). \end{cases} \tag{13}$$

At the receiver end, since the chaotic signal y_k is information-free (and so not disturbed), a perfect synchronization fulfilling (5) or (6) can be achieved by resorting to an observer. As a consequence, the information m_k can be correctly recovered by:

$$\begin{cases} \hat{x}_{k+1} = \tilde{f}_\theta(\hat{x}_k, y_k), \\ \hat{y}_k = \tilde{h}_\theta(\hat{x}_k), \\ \hat{m}_k = \nu_d(\hat{x}_k, u_k). \end{cases} \tag{14}$$

The decrypting function ν_d is defined by

$$\hat{m}_k = \nu_d(\hat{x}_k, u_k) = m_k \text{ when } \hat{x}_k = x_k. \tag{15}$$

Figure 5. Two-channel transmission

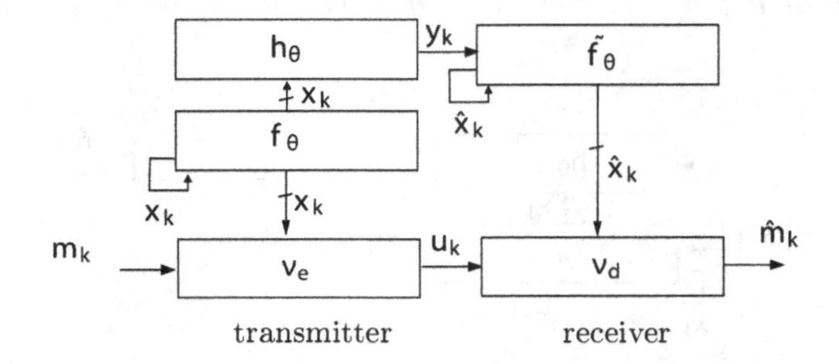

This technique has been proposed, for example, in (Jiang, 2002; Millérioux & Mira, 1998). The advantage lies in that, unlike modulation-based approaches, m_k is allowed to switch every discrete times k without inducing synchronization transients for each symbol. The recovering is wrong only for a finite number of first symbols of the message. On the other hand, a transmission involving two channels may be unsatisfactory for throughput purposes.

4.4 Message-Embedding

First of all, the reader is cautioned that different but equivalent terminologies can be encountered in the literature referring to the same technique: embedding (Lian &. Liu, 2000; Millérioux & Daafouz, 2004), non autonomous modulation (Yang, 2004) or direct chaotic modulation (Hasler, 1998). The reasons for this diversity are the following. At the transmitter part, the information $m_k \in A$ (the alphabet) is directly injected (or, as it is also usually put, embedded) in a chaotic dynamic f_θ. The resulting system turns into a non autonomous one since the information acts as an exogenous input. So, the system involves a state vector $x_k \in X$ and m_k. Injecting m_k into the dynamic could be considered as a "modulation" of the phase space. Only the output y_k of the system is transmitted. It is assumed that y_k belongs to the same set A as m_k.

A message-embedded cryptosystem (depicted on Figure 6) obeys:

$$\begin{cases} x_{k+1} = f_\theta(x_k, m_k), \\ y_k = h_\theta(x_k, m_k). \end{cases} \tag{16}$$

In order to understand the way how the plaintext m_k can be retrieved at the receiver part, we recall at this point two important definitions and one structural property, borrowed from control theory.

Iterated Functions

The internal state $x_{k+i} \in X$ of (16) at time $k+i$ depends on the state $x_k \in X$ and on the sequence of i input symbols $m_k \cdots m_{k+i-1} \in A^i$, by means of the so called i-order iterated next-state function, $f^{(i)}: X \times A^i \to X$, defined for $i \geq 1$, and recursively obeying for $k \geq 0$,

Figure 6. Message-embedding. When $r = 0$, m_k is embedded into f_θ and h_θ. When $r > 0$, m_k is only embedded into f_θ

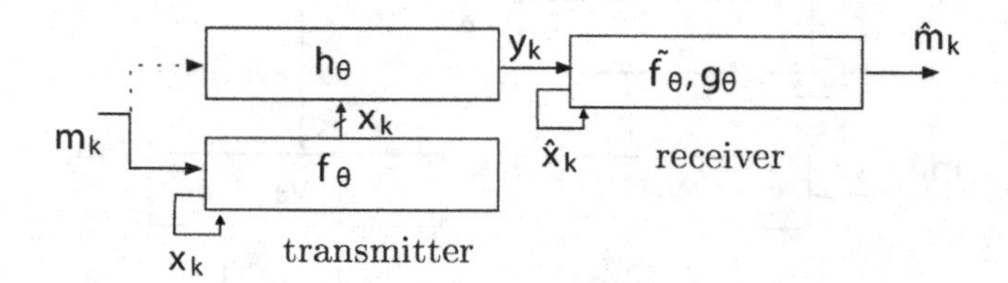

$$\begin{cases} f^{(1)}(x_k, m_k) = f(x_k, m_k), \\ f^{(i+1)}(x_k, m_k \cdots m_{k+i}) = f(f^{(i)}(x_k, m_k \cdots m_{k+i-1}), m_{k+i}) \quad for\ i \geq 1. \end{cases}$$

Similarly, the output y_{k+i} of (16) at time $k+i$ depends on the state $x_k \in X$ and on the sequence of $i+1$ input symbols $m_k \cdots m_{k+i} \in A^{i+1}$, by means of the so-called *i*-order iterated output function $h^{(i)}: X \times A^{i+1} \to A$, defined for $i \geq 0$, and recursively obeying for $k \geq 0$,

$$\begin{cases} h^{(0)}(x_k, m_k) = h(x_k, m_k). \\ h^{(i)}(x_k, m_k \ldots m_{k+i}) = h(f^{(i)}(x_k, m_k \cdots m_{k+i-1}), m_{k+i}) \quad for\ i \geq 1. \end{cases}$$

Relative Degree

Definition 3. *The relative degree of the dynamical system (16) is the quantity denoted r with*

- $r = 0$ if $\exists x_k \in X, \exists m_k, m_k^{'} \in A$ with $h(x_k, m_k) \neq h(x_k, m_k^{'})$
- $r > 0$ if for any sequence $m_{k+i} \cdots m_{k+r}$ $(i > 0)$ of input symbols

$$\exists x_k \in X, \exists m_k, m_k^{'} \in A\ with\ h^{(r)}(x_k, m_k m_{k+1} \cdots m_{k+r}) \neq h^{(r)}(x_k, m_k^{'} m_{k+1}^{'} \cdots m_{k+r}^{'}),$$

In other words, the relative degree of the dynamical system (16) is the minimum number of iterations such that the output at time $k + r$ is influenced by the input at time k and

- if the relative degree r of (16) equals zero, there exists a state $x_k \in X$ and two distinct input symbols $m_k, m_k^{'} \in A$ that lead to different values of the output,
- if the relative degree is $r > 0$, then for $i < r$, the iterated output function $h^{(i)}$ only depends on x_k while for $i \geq r$, it depends both on x_k and on the sequence of $i - r + 1$ input symbols $m_k \cdots m_{k+i-r}$. In particular, for $i = r$, the iterated output function depends both on m_k and on x_k, that is, there exists a state $x_k \in X$ and two distinct input symbols $m_k \in A$ and $m_k^{'} \in A$ that lead to different values of the output, for any sequence $m_{k+1} \cdots m_{k+r}$ of input symbols.

Consequently, for $r > 0$, the r-order output function $h^{(r)}$ may be considered as a function on $X \times A$, and thereby one has for $r \geq 0$:

$$y_{k+r} = h^{(r)}(x_k, m_k). \tag{17}$$

Remark 1. *We do not consider the case when r may depend on m_k. Hereafter, the relative degree will be constant for (16). The case when the relative degree is time-varying is addressed in (*Millérioux & Daafouz, 2009*).*

Remark 2. *For Single Input Single Output (SISO) linear systems, the relative degree r corresponds to the difference between the degree of the denominator and the degree of the numerator in their transfer function*

Left Invertibility

Definition 4. *The dynamical system (16) is left invertible if there exists a nonnegative integer* $R < \infty$, called inherent delay, such that for any two inputs $m_k \in A$ and $m_k^{'} \in A$ the following implication holds:

$$\begin{aligned} &\forall x_k \in X \\ &h^{(0)}(x_k, m_k) \cdots h^{(R)}(x_k, m_k \cdots m_{k+R}) = h^{(0)}(x_k, m_k^{'}) \cdots h^{(R)}(x_k, m_k^{'} \cdots m_{k+R}^{'}) \\ &\Rightarrow m_k = m_k^{'}. \end{aligned} \tag{18}$$

The left invertibility property means that the input m_k is uniquely determined by the knowledge of the state x_k and of the output sequence $y_k, \ldots, y_{k+R}$.

Remark 3. Hereafter, it will be assumed that the inherent delay and the relative degree coincide, that is $r = R$.

Another interpretation of left invertibility is that for any internal state $x_k \in X$, the map

$$h_{x_k} : \begin{array}{ccc} A & \rightarrow & A \\ m_k & \mapsto & h^{(r)}(x_k, m_k) \end{array}$$

is a permutation, where $r \geq 0$ is the relative degree of (16). The output function $h(r)$ may be considered as a family of permutations, indexed by the set S of the internal states, or at least by a subset.

Having in mind those definitions, let us turn back to the design of the receiver.

If the dynamical system (16) is used at the transmitter part, then the receiver must be able to recover the input only from the output. From the dynamical system theory, it actually implies the so-called left invertibility. x_k can be reconstructed by a synchronization mechanism called input independent synchronization followed by a static inversion. To this end, we resort to a dynamical system with dynamics $\hat{f}$ and with output function g, both parametrized by $\hat{\theta} = \theta$. The state vector is denoted $\hat{x}_k$. Input independent synchronization corresponds to the following definitions.

Let T be a constant matrix of appropriate dimension and U a non empty set of initial conditions. We distinguish two concepts.

Definition 5. *Asymptotical input independent synchronization-*

$$\lim_{k \to \infty} \left\| T x_k - \hat{x}_k \right\| = 0 \ \ \forall \hat{x}_0 \in U, \ \ \forall m_k \in A. \tag{19}$$

Definition 6. *Finite time input independent synchronization*

$$\exists k_f < \infty, \ \left\| T x_k - \hat{x}_k \right\| = 0 \ \ \forall \hat{x}_0 \in U, \ \ \forall m_k \in A \ \ and \ \ \forall k \geq k_f. \tag{20}$$

Two mechanisms have been proposed in the literature: the inverse system approach (Feldmann et al, 1996) and the unknown input observer (UIO) approach (Boutat-Baddas et al, 2004; Boutayeb et al, 2002; Inoue & Ushio, 2001; Millérioux & Daafouz, 2003; Millérioux & Daafouz, 2004; Millérioux & Daafouz, 2006). As a matter of fact, UIO is nothing else but an inverse system slightly modified by adding some extra terms, for the sake of robustness in noisy environments. The generic equations governing an inverse system or an UIO for (16) are:

$$\begin{cases} \hat{x}_{k+1} = \tilde{f}_\theta(\hat{x}_k, y_k, \ldots, y_{k+r}), \\ \hat{m}_k = g_\theta(\hat{x}_k, y_{k+r}), \end{cases} \tag{21}$$

With g such that

$$\hat{m}_k = g_\theta(\hat{x}_k, y_{k+r}) = m_k \quad when \quad \hat{x}_k = x_k. \tag{22}$$

This technique follows the same spirit than the observer-based techniques described in Subsect. 4.1, 4.2 and 4.3. Nevertheless, a major difference lies in that the receiver, unlike a mere observer, must reconstruct the state x_k to guarantee the synchronization without the knowledge of m_k.

5 COMPARATIVE STUDY AND CONNECTION

The aim of this section is to highlight the connections between chaotic ciphers and symmetric conventional ones. Relevant issues which call for a closer look are also discussed. It has been already pointed out that a first condition for proper recovering of the information is that the transmitter and the receiver shares the same master key θ. As a result, all the chaotic cryptosystems belong to the class of symmetric ciphers.

5.1 Continuous vs. Discrete Dynamics

A major and obvious difference between chaotic dynamics and pseudo-random sequences lies in that a chaotic generator is assumed to produce an aperiodic sequence $\{x_k\}$ ranging in a dense set, while the pseudo-random generators produce discrete sequences. Yet, observe that when chaotic generators are implemented in machines with finite precision (say, a computer) for cryptographic purposes, the sequences $\{x_k\}$ are not really chaotic. Indeed, since the set in which the x_k's take values has finite cardinality, such sequences will obviously get trapped in a loop, called *cycle*, of finite period. We can expect this period to be not too short and the degree of 'randomness' of the sequence to be high (as measured e.g. by standard statistical tests). In this regard, the following points deserve further investigation.

- The recourse to discretized chaotic systems for keystream generators in Synchronous Stream Ciphers is questionable and seems a poor alternative because the discretization is usually machine dependent, and/or its effect is in general not completely predictable. Put in mild terms, guaranteeing sequences with good cryptographic properties requires some caution. Important contributions to this issue can be found in (Kocarev et al, 2006)

- Periodic approximations of chaotic automorphisms can be theoretically used to define substitutions (so-called S-boxes) resistant to linear and differential cryptanalysis. S-boxes are the most important entities involved in block ciphers. Whether this proposal, which is non-constructive, can be applied in practice to the design of block ciphers, is an open issue.

5.2 Additive Masking and Two-Channels Transmission vs. Synchronous Stream Ciphers

As far as only structural considerations are concerned, regardless the complexity of the dynamics, a natural connection between additive masking and SSC can be made. In fact, the transmitter of the respective schemes has exactly the same structure. The sequences $\{x_k\}$ for chaotic cryptosystems (*resp.* $\{K_k\}$ for SSC) are independent from the plaintext m_k and the ciphertext y_k (*resp.* c_k). For a SSC, the same initialization is required at both ends to guarantee the synchronization. And yet, sharing the same seed is sufficient to produce synchronized sequences because the generators operate on integers. For additive masking and assuming that the generator is really chaotic, synchronization would be inevitably lost within a very short time window due to sensitivity to initial conditions. To handle such a problem, a synchronization at the receiver part usually requires observers for state reconstruction, as mentioned in Section 4.1. Nevertheless, as already pointed out, the information to be masked acts as a disturbance, preventing an exact synchronization. This renders the additive masking not very appealing compared to a conventional SSC.

To circumvent the bad synchronization properties of additive masking, we can resort to the two-channels transmission setup. Such a scheme is also highly connected to the SSC. Indeed, it involves two entities: the keystream generator on one hand, and the encrypting function on the other hand. But similarly to the additive masking, synchronization would be inevitably lost within a very short time window without any coupling between the generators at both ends. Thus, an additional channel must be used to convey the state vector for ensuring exact synchronization. Again, the resulting scheme is a poor solution and certainly not an alternative to conventional SSC.

5.3 Message-Embedding vs. Self-Synchronizing Stream Ciphers

In this subsection it is shown that there is an interesting connection between the message-embedded setup and the Self-Synchronizing Stream Cipher.

Let us recall the equations (see (16)) governing a message-embedded cryptosystems:

$$\begin{cases} x_{k+1} = f_\theta(x_k, m_k), \\ y_k = h_\theta(x_k, m_k). \end{cases}$$

The results stated in this subsection are based on the notion of *flatness* (see (Fliess et al, 1995) for an introductory theory)

Definition 7. *An output for (16) is said to be flat if all system variables of (16) can be expressed as a function of y_k and a finite number of its forward/backward iterates. In particular, there exists two functions F and G and integers* $t_1 < t_2$ and $t_1' < t_2'$ such that

$$
\begin{aligned}
x_k &= F(y_{k+t_1}, \cdots, y_{k+t_2}) \\
m_k &= G(y_{k+t_1'}, \cdots, y_{k+t_2'})
\end{aligned} \tag{23}
$$

Definition 8. *The dynamical system (16) is said to be flat if it admits a flat output.*
We define the *flatness characteristic number* as the quantity $t_2 - t_1 + 1$.

We are now in a position of bringing out a structural connection with self-synchronizing stream ciphers. Assume that the dynamical system (16) is left invertible with inherent delay R and, according to the Remark 3, has finite relative degree $r = R$. Besides, assume that (16) is flat with flat output y_k and a flatness characteristic number $t_2 - t_1 + 1$. Thus both equations (17) and (23) hold respectively.

Identification of (17) and (23) with the respective equations in (2), leads to the following Proposition:
Proposition 2. *The system (16) is strictly equivalent to the transmitter part of a self-synchronizing stream cipher of the form (2) with the correspondences (explicited for short by the symbol $\leftrightarrow$)*

- a keystream generator $\sigma_\theta^{ss} \leftrightarrow F$
- a running key $K_k \leftrightarrow x_k$
- a ciphertext c_{k+r} corresponding to the plaintext $m_k \leftrightarrow y_{k+r}$
- a ciphering function $e \leftrightarrow h^{(r)}$

Let us stress that, when the relative degree r is strictly greater than zero, there is a delay r between the plaintext m_k and the corresponding ciphertext y_{k+r}. Actually, this is what typically happens in conventional Self-Synchronizing Stream Ciphers when the output function is pipelined (see, for example, the algorithm called MOUSTIQUE (Daemen & Paris, 2004), or the comparative study (Vo Tan et al, 2007)).

The equivalent representation of the message-embedded cryptosystem (16) is depicted on Figure 7 as a canonical representation of a SSSC.

Remark 4. *It is worthwhile noticing that the function G in (23) gives explicitly the way how the plaintext m_k can be obtained at the receiver side. Even more is true: the knowledge of the keystream is no longer useful. Nevertheless, it is usually more convenient to implement a recursive left inversion which performs in two steps (see Subsection 4.4): Computation of the keystream $\hat{x}_k$ (fulfilling $\hat{x}_k = x_k$* with a finite transient time) and recovering of the plaintext through a static function d.

Figure 7. Self-synchronizing message-embedded stream cipher

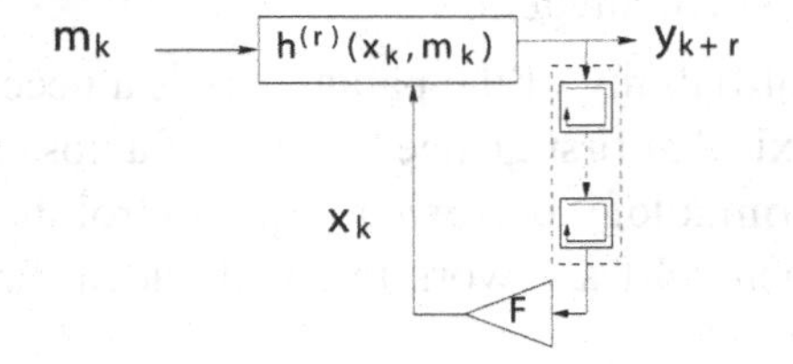

The ability of SSSC to self-synchronizing provides many advantages. First, if a ciphertext is deleted, inserted or flipped, the SSSC will automatically resume proper decryption after a short, finite and predictable transient time. Hence, SSSC does not require any additional synchronization flags or interactive protocols for recovering lost synchronization. Secondly, the self-synchronizing mechanism also enables the receiver to switch at any time into an ongoing enciphered transmission. Third, any modification of ciphertext symbols by an active eavesdropper causes incorrect decryption for a fixed number of next symbols. As a result, an SSSC prevents active eavesdroppers from undetectable tampering with the plaintext: Message authenticity is guaranteed. Finally, since each plaintext symbol influences a fixed number of following ciphertexts, the statistical properties of the plaintext are thereby diffused through the ciphertext. Hence, SSSC are very efficient against attacks based on plaintext redundancy.

When looking into the open literature, it turns out that less attention has been paid on this special class of symmetric ciphers, namely the self-synchronizing stream ciphers. Actually, these ciphers have been just touched on but not really explored. As a result, they deserve deep investigation and the message-embedded setup could be revealed as a promising alternative for the design of conventional SSSC.

5.4 Security, Algebraic Attacks vs. Identifiability and Identification

An essential issue for the validation of cryptosystems is cryptanalysis, that is, the study of attacks against cryptographic schemes in order to reveal their possible weakness. A fundamental assumption in cryptanalysis, first stated by A. Kerkhoff in (Delfs & Knebl, 2002), is that the eavesdropper knows all the details of the cryptosystem, including the algorithm and its implementation, except the secret key, on which the security of the cryptosystem must be entirely based. As a result, the security is directly related to the complexity of the parameters recovering task.

First of all, the eavesdropper may be assumed (depending on the type of the attack) to have different privileges, like access to pairs of inputs and outputs of the cipher. Yet, observe that a cryptosystem must resist the most basic attack, i.e. the brute force attack. This attack consists in trying exhaustively every possible parameter value in the parameter space of the secret key (which is, in practice, a finite space). The quicker the brute force attack succeeds on average, the weaker the cryptosystem. Consequently, the worst situation for the eavesdropper and the best for the security arises when, for known plaintexts (*resp.* ciphertexts) and corresponding ciphertext sequences (*resp.* plaintext sequences), only one solution in the parameters of the cipher exists. As it turns out, the unicity is directly related to the notion of parameter identifiability.

Indeed, let us recall that a parameter θ of a discrete-time dynamical system is *identifiable* if θ can be rewritten as a unique function φ of the input, the output and their iterates

$$\theta = \phi(y_k, \ldots, y_{k+M}, m_k, \ldots, m_{k+M'}) \tag{24}$$

with $M < \infty$ and $M' < \infty$ some positive integers.

As a result, we conclude that identifiability of the parameters is a necessary condition for security. Such a result might appear as paradoxical at first glance because of a possible misunderstanding on the meaning of "identifiable", a usual terminology borrowed from control theory. Actually, identifiability means unicity in the parameters. A general framework to test the identifiability has been proposed in (Anstett et al, 2006).

Besides, the algebraic attack is one of the newest and most efficient form of cryptanalytic attacks. Algebraic attacks have had an important impact on cryptanalysis. First paper which introduced algebraic attack is the paper by Kipsis and Shamir (Kipnis & Shamir, 1999). Its principle relies on the algebraic model of the cipher. The objective of an algebraic attack is to find out a set of algebraic equations which can be solved efficiently. Actually, the idea of breaking cipher as solving a system of equations is very old with a paper written by Shannon (Shannon, 1949). An efficient algebraic attack is one for which the complexity is below the complexity of an exhaustive search. One of the main tool for that purpose is the elimination technique, in particular, by means of Grobner bases. The security with respect to algebraic attacks is directly related to the complexity of the parameters (secret key) recovering task.

In automatic control theory, the parameters recovering task is nothing else but identification. Thus, the security is related to the complexity of the identification procedure required for retrieving the parameter(s) used as the secret key(s). An interesting issue is the choice of a relevant class of dynamical systems for which identification is a hard task. We expect hybrid systems to fall into this class as they involve heterogeneity. Further work is needed to answer such a conjecture.

6 CONCLUSION

Throughout this chapter, we have reviewed the main architectures of chaotic cryptosystems proposed since the 1990's in the open literature. Next we have established a formal parallelism with conventional ciphers and pointed out some relevant issues to be addressed in the near future. The main conclusion of this study is that the so-called message embedding seems the most relevant chaotic cryptosystem proposed so far; moreover, it could be used as an alternative for the design of conventional self-synchronizing stream ciphers. The design highly rests on control theoretical concepts. On the other hand, the chaos-based algorithms proposed so far belong more to the field of steganography than to pure cryptography, and unfortunately they always appear as poor cryptographic solutions.

REFERENCES

Alvarez, G., & Li, S. (2006). Some basic cryptographic requirements for chaos-based cryptosystems. *International Journal of Bifurcation and Chaos in Applied Sciences and Engineering*, *16*(8), 2129–2151. doi:10.1142/S0218127406015970

Amigó, J. M., Kocarev, L., & Szczepanski, J. (2007). Theory and practice of chaotic cryptography. *Physics Letters. [Part A]*, *366*, 211–216. doi:10.1016/j.physleta.2007.02.021

Amigó, J. M., Kocarev, L., & Szczepanski, J. (2007). Discrete lyapunov exponents and resistance to differential cryptanalysis. *IEEE Transactions on Circuits and Systems, II*, 54, 882–886.

Amigó, J. M., Szczepanski, J., & Kocarev, L. (2005). A chaos-based approach to the design of crytographically secure substitutions. *Physics Letters. [Part A]*, *343*, 55–60. doi:10.1016/j.physleta.2005.05.057

Amigó, J. M., Szczepanski, J., & Kocarev, L. (2008). On some properties of the discrete Lyapunov exponent. *Physics Letters. [Part A]*, *372*, 6265–6268. doi:10.1016/j.physleta.2008.07.076

Anstett, F., Millérioux, G., & Bloch, G. (2004 May). Global adaptive synchronization based upon polytopic observers. In *Proc. of IEEE International symposium on circuit and systems, ISCAS'04*, Vancouver, Canada (pp. 728 - 731).

Anstett, F., Millérioux, G., & Bloch, G. (2006 December). Chaotic cryptosystems: Cryptanalysis and identifiability. *IEEE Trans. on Circuits and Systems: Regular papers, 53*(12), 2673-2680.

Boutat-Baddas, L., Barbot, J. P., Boutat, D., & Tauleigne, R. (2004, May). Sliding mode observers and observability singularity in chaotic synchronization. *Mathematical Problems in Engineering, 1*, 11–31. doi:10.1155/S1024123X04309038

Boutayeb, M., Darouach, M., & Rafaralahy, H. (2002, March). Generalized state-space observers for chaotic synchronization and secure communications. *IEEE Transactions on Circuits and Systems. I, Fundamental Theory and Applications, 49*(3), 345–349. doi:10.1109/81.989169

Carroll, T. L., & Pecora, L. M. (1991, April). Synchronizing chaotic circuits. *IEEE Transactions on Circuits and Systems, 38*(4), 453–456. doi:10.1109/31.75404

Cuomo, K. M., Oppenheim, A. V., & Strogatz, S. H. (1993). Synchronization of lorenz-based chaotic circuits with applications to communications. *IEEE Trans. Circuits. Syst. II: Anal. Digit. Sign. Process, 40*(10), 626–633. doi:10.1109/82.246163

Daemen, J., & Paris, K. (2004). *The self-synchronzing stream cipher moustique*. Technical report, e-Stream Project. Retrieved from http://www.ecrypt.eu.org/stream/p3ciphers/mosquito/mosquito-p3.pdf

Dedieu, H., Kennedy, M. P., & Hasler, M. (1993). Chaos shift keying: modulation and demodulation of a chaotic carrier using self-synchronizing Chua's circuits. *IEEE Trans. Circuits. Syst. II: Anal. Digit. Sign. Process, 40*, 634–642. doi:10.1109/82.246164

Dedieu, H., & Ogorzalek, M. (1997). Identification of chaotic systems based on adaptive synchronization. In *Proc. ECCTD'97*, Budapest, Sept. 1997 (pp. 290-295).

Delfs, H., & Knebl, H. (2002). *Introduction to cryptography*. Berlin: Springer-Verlag.

Devaney, R. L. (1989). *An introduction to Chaotic Dynamical Systems*. Redwood City, CA: Addison-Wesley.

Feldmann, U., Hasler, M., & Schwarz, W. (1996). Communication by chaotic signals:the inverse system approach. *Int. J. of Circuit Theory Appl., 24*, 551–579. doi:10.1002/(SICI)1097-007X(199609/10)24:5<551::AID-CTA936>3.0.CO;2-H

Fliess, M., Levine, J., Martin, P., & Rouchon, P. (1995). Flatness and defect of non-linear systems: introductory theory and examples. *International Journal of Control, 61*(6), 1327–1361. doi:10.1080/00207179508921959

Fradkov, A. L., & Markov, A. Y. (1997, October). Adaptive synchronization of chaotic systems based on speed-gradient method and passification. *IEEE Transactions on Circuits and Systems. I, Fundamental Theory and Applications, 44*(10), 905–912. doi:10.1109/81.633879

Fridrich, J. (1998, June). Symmetric ciphers based on two-dimensional chaotic maps. *International Journal of Bifurcation and Chaos in Applied Sciences and Engineering*, *8*(6), 1259–1284. doi:10.1142/S021812749800098X

Grassi, G., & Mascolo, S. (1997, October). Nonlinear observer design to synchronize hyperchaotic systems via a scalar signal. *IEEE Transactions on Circuits and Systems. I, Fundamental Theory and Applications*, *44*(10), 1011–1014. doi:10.1109/81.633891

Habutsu, T., Nishio, Y., Sasase, I., & Mori, S. (1991). A secret key cryptosystem by iterating a chaotic map. In *Proc. of EuroCrypt'91* (LNCS 0547, pp. 127-140). Berlin: Springer Verlag.

Hasler, M. (1998, April). Synchronization of chaotic systems and transmission of information. *International Journal of Bifurcation and Chaos in Applied Sciences and Engineering*, *8*(4). doi:10.1142/S0218127498000450

Huijberts, H. J. C., Nijmeijer, H., & Willems, R. (2000). System identification in communication with chaotic systems. *IEEE Transactions on Circuits and Systems. I, Fundamental Theory and Applications*, *47*(6), 800–808. doi:10.1109/81.852932

Inoue, E., & Ushio, T. (2001). Chaos communication using unknown input observers. *Electronics and Communications in Japan (Part III Fundamental Electronic Science)*, *84*(12), 21–27. doi:10.1002/ecjc.1053

Itoh, M., Wu, C. W., & Chua, L. O. (1997). Communications systems via chaotic signals from a reconstruction viewpoint. *International Journal of Bifurcation and Chaos in Applied Sciences and Engineering*, *7*(2), 275–286. doi:10.1142/S0218127497000194

Jiang, Z. P. (2002, January). A note on chaotic secure communication systems. *IEEE Transactions on Circuits and Systems. I, Fundamental Theory and Applications*, *49*(1), 92–96. doi:10.1109/81.974882

Kipnis, A., & Shamir, A. (1999). Cryptanalysis of the hfe public key cryptosystem by relinearization. *Lecture Notes in Computer Science*, *1666*, 19–30. doi:10.1007/3-540-48405-1_2

Knuth, D. E. (1998). *The Art of Computer Programming* (*Vol. 2*). Reading, MA: Addison-Wesley.

Kocarev, L., Szczepanski, J., Amigó, J. M., & Tomosvski, I. (2006). Discrete chaos: part I. *IEEE Transactions on Circuits and Systems*, *I*, 53.

Kolumban, G., Kennedy, M. P., & Chua, L. O. (1998, November). The role of synchronization in digital communications using chaos - part ii: Chaotic modulation and chaotic synchronization. *IEEE Transactions on Circuits and Systems. I, Fundamental Theory and Applications*, *45*, 1129–1140. doi:10.1109/81.735435

Li, T. Y., & Yorke, J. A. (1975). Period three implies chaos. *The American Mathematical Monthly*, *82*, 985–992. doi:10.2307/2318254

Lian, K. Y., & Liu, P. (2000). Synchronization with message embedded for generalized lorenz chaotic circuits and its error analysis. *IEEE Transactions on Circuits and Systems. I, Fundamental Theory and Applications*, *47*(9), 1418–1424. doi:10.1109/81.883341

Massey, J. L. (1992). *Contemporary cryptology: an introduction*. New York: IEEE Press Ed.

Matthews, R. (1989). On the derivation of a chaotic encryption algorithm. *Cryptologia, 13*, 29–41. doi:10.1080/0161-118991863745

Menezes, A. J., Oorschot, P. C., & Vanstone, S. A. (1996 October). *Handbook of Applied Cryptography*. Boca Raton, FL: CRC Press.

Millérioux, G. (1997). Chaotic synchronization conditions based on control theory for systems described by discrete piecewise linear maps. *International Journal of Bifurcation and Chaos in Applied Sciences and Engineering*, *7*(7), 1635–1649. doi:10.1142/S0218127497001266

Millérioux, G., Amigó, J. M., & Daafouz, J. (2008 July). A connection between chaotic and conventional cryptography. *IEEE Trans. on Circuits and Systems I: Regular Papers*, 55(6).

Millérioux, G., & Daafouz, J. (2003, October). An observer-based approach for input independent global chaos synchronization of discrete-time switched systems. *IEEE Transactions on Circuits and Systems. I, Fundamental Theory and Applications*, *50*(10), 1270–1279. doi:10.1109/TCSI.2003.816301

Millérioux, G., & Daafouz, J. (2004, April). Unknown input observers for message-embedded chaos synchronization of discrete-time systems. *International Journal of Bifurcation and Chaos in Applied Sciences and Engineering, 14*(4), 1357–1368. doi:10.1142/S0218127404009831

Millérioux, G., & Daafouz, J. (2006). *Chaos in AutomatiControl-theoretical concepts in the design of symmetric cryptosystems c Control*. Boca Raton, FL: CRC Press.

Millérioux, G., & Daafouz, J. (2009). *Flatness of switched linear discrete-time systems*. IEEE Trans. on Automatic Control.

Millérioux, G., & Mira, C. (1998). Coding scheme based on chaos synchronization from noninvertible maps. *International Journal of Bifurcation and Chaos in Applied Sciences and Engineering, 8*(10), 2019–2029. doi:10.1142/S0218127498001674

Nijmeijer, H., & Mareels, I. M. Y. (1997, October). An observer looks at synchronization. *IEEE Transactions on Circuits and Systems. I, Fundamental Theory and Applications*, *44*, 882–890. doi:10.1109/81.633877

Ogorzalek, M. J. (1993). Taming chaos - part I: Synchronization. *IEEE Transactions on Circuits and Systems. I, Fundamental Theory and Applications*, *40*(10), 693–699. doi:10.1109/81.246145

Palaniyandi, P., & Lakshmanan, M. (2001). Secure digital signal transmission by multistep parameter modumation and alternative driving of transmitter variables. *International Journal of Bifurcation and Chaos in Applied Sciences and Engineering, 11*(7), 2031–2036. doi:10.1142/S021812740100319X

Parlitz, U., Chua, L. O., Kocarev, L., Halle, K. S., & Shang, A. (1993). Transmission of digital signals by chaotic synchronization. *International Journal of Bifurcation and Chaos in Applied Sciences and Engineering, 3*(2), 973–977.

Pieprzyk, J., Hardjono, T., & Seberry, J. (2003). *Fundamentals of Computer Security*. Berlin: Springer Verlag.

Rosier, L., Millérioux, G., & Bloch, G. (2006). Chaos synchronization for a class of discrete dynamical systems on the n-dimensional torus. *Systems & Control Letters*, *55*, 223–231. doi:10.1016/j.sysconle.2005.07.003

Schmitz, R. (2001). Use of chaotic dynamical systems in cryptography. *Journal of the Franklin Institute*, *338*, 429–441. doi:10.1016/S0016-0032(00)00087-9

Shannon, C. E. (1949). Communication theory of secrecy systems. *The Bell System Technical Journal*, *28*, 657–715.

Szczepanski, J., Amigó, J. M., Michalek, T., & Kocarev, L. (2005, February). Crytographically secure substitutions based on the approximation of mixing maps. *IEEE Transactions on Circuits and Systems. I, Regular Papers*, *52*(2), 443–453. doi:10.1109/TCSI.2004.841602

Vo Tan, P., Millérioux, G., & Daafouz, J. (2007 August). A comparison between the message-embedded cryptosystem and the self-synchronous stream cipher mosquito. In *Proc. of IEEE 18th European Conference on Circuits Theory and Design, ECCTD'07*, Sevilla, Spain.

Wu, C. W., & Chua, L. O. (1993). A simple way to synchronize chaotic systems with applications to secure communications systems. *International Journal of Bifurcation and Chaos in Applied Sciences and Engineering*, *3*(6), 1619–1627. doi:10.1142/S0218127493001288

Yang, T. (2004). A survey of chaotic secure communication systems. *Int. J. of Computational Cognition*. Retrieved from http://www.YangSky.com/yangijcc.htm

ENDNOTE

[1] available at http://www.ecrypt.eu.org/stream/

Chapter 17
Unmasking Optical Chaotic Cryptosystems Based on Delayed Optoelectronic Feedback

Silvia Ortín
Instituto de Física de Cantabria (CSIC-Universidad de Cantabria), Spain

Luis Pesquera
Instituto de Física de Cantabria (CSIC-Universidad de Cantabria), Spain

ABSTRACT

The authors analyze the security of optical chaotic communication systems. The chaotic carrier is generated by a laser diode subject to delayed optoelectronic feedback. Transmitters with one and two fixed delay times are considered. A new type of neural networks, modular neural networks, is used to reconstruct the nonlinear dynamics of the transmitter from experimental time series in the single-delay case, and from numerical simulations in single and two-delay cases. The authors show that the complexity of the model does not increase when the delay time is increased, in spite of the very high dimension of the chaotic attractor. However, it is found that nonlinear dynamics reconstruction is more difficult when the feedback strength is increased. The extracted model is used as an unauthorized receiver to recover the message. Therefore, the authors conclude that optical chaotic cryptosystems based on optoelectronic feedback systems with several fixed time delays are vulnerable.

INTRODUCTION

Chaotic signals typically have broadband spectrum. This property is desirable for applications that require robustness against interference, jamming and low detection probability. Those issues have been addressed by traditional communication systems by using spread spectrum and frequency hopping modulations. In chaos-based communications the broadband chaotic signal is generated at the physical layer instead of algorithmically. Additionally, chaotic carriers offer a certain degree of intrinsic privacy in the data

DOI: 10.4018/978-1-61520-737-4.ch017

transmission. In chaotic communication systems (Cuomo et al., 1993b; Colet & Roy, 1994) the masking of the message is performed at the physical layer by embedding the signal within a chaotic carrier in the emitter. The recovery of the message is based on the synchronization phenomenon (Ashwin, 2003) by which a receiver, quite similar to the transmitter, is able to reproduce the chaotic part of the transmitted signal. After synchronization occurs, the decoding of the message is straightforward by comparing the input and output of the receiver. Privacy in chaotic communication systems results from the fact that the eavesdropper must have the proper hardware and parameter settings in order to recover the message. The suitability of chaos-based optical communication systems for encrypting gigabit signals has been recently demonstrated in an installed optical network infrastructure of approximately 120 km that covers the metropolitan area of Athens (Argyris et al., 2005). However, the security of these systems remains the key issue to be addressed.

In conventional encryption techniques a key is used to alter the information symbols. The transmitter and the receiver share the key so that the information can be recovered. In a chaotic communication system the transmitter generates a time-evolving chaotic waveform that is used to mask the message. The cryptographic key relies on structural characteristics of the hardware as well as on the set of operating parameters chosen for the system. The message can be recovered with a receiver such that its configuration and parameter settings are matched to those of the transmitter. Encryption is achieved by encoding at the physical layer, providing full compatibility to conventional software encryption techniques. Dynamical encoding with a chaotic waveform can then be considered as an additional layer of encryption.

Chaos cryptography is a recent encryption technique (the idea was proposed in the early 90s), and it will take some time for its security analysis to mature. Some rules have been suggested to achieve a reasonable degree of security (Alvarez & Li, 2006). Methods to quantify the cryptanalysis of chaotic encryption schemes have been also proposed (Tenny & Tsimring, 2004). However, more research needs to be done to develop a systematic cryptographic approach for the analysis of the security of different chaotic communication systems. Many chaos-based encryption schemes have been proposed, and many of those schemes have been broken later on.

Some chaotic encryption systems were broken even without reconstructing the transmitter's chaotic dynamics, that is, without searching for the secret key that was used to encrypt the message. This kind of attack is usually applicable if the statistical properties of the ciphertext change as a result of changing the transmitted plaintext. Return maps (Perez & Cerdeira, 1995) and spectral analysis (Yang et al., 1998a) of the transmitted ciphertext have been used to decode the message eliminating the need to reconstruct the secret dynamics.

Another type of attacks relies on partial knowledge of the chaotic dynamics. If the unauthorized receiver knows the type of attractor used for the transmission and reception, but ignores the precise value of the parameters, generalized synchronization (Rulkov et al., 1995) can be used to extract the message (Yang et al., 1998b). In this case a generalized synchronization between transmitter and unauthorized receiver with a different set of parameters occurs, and the message is decoded using variations in the synchronization error. In another case the unauthorized receiver knows that the transmitter is an erbium-doped fiber-ring laser with two delay loops (Geddes et al., 1999). The dynamics of this chaotic transmitter is high dimensional (dimension greater than 10). However, using a simplified lower dimensional model with four parameters, the message can be decoded by estimating the model parameters. In this case the laser dynamics was governed almost entirely by the modulation signal, which echoed in the two loops, and nonlinear effects could be neglected to a good approximation.

In another type of attacks the unauthorized receiver attempts to reconstruct the secret chaotic dynamics of the transmitter without having any a-priori knowledge about the type of dynamics used. Nonlinear dynamics forecasting techniques have been used to extract the chaotic carrier signal (Short, 1994; 1996; Short & Parker, 1998). The message signal can be obtained by removing the carrier signal from the transmitted ciphertext signal.

Two important factors to privacy considerations in chaotic communication systems are the effort required to obtain the necessary parameters for a matched receiver and the dimensionality of the chaos. Important aspects of receiver design are the number of parameters that have to be matched for information recovery and the precision required for parameter matching. Higher dimensional signals, especially those involving hyperchaotic dynamics, are likely to provide improved security. Chaotic communication systems were originally implemented with electronic circuits with low dimensional chaos (Kocarev et al., 1992; Cuomo & Oppenheim, 1993a). Unfortunately, messages encoded with low dimensional chaotic carriers can be often extracted using standard nonlinear techniques (Perez & Cerdeira, 1995; Short & Parker, 1998). Nonlinear systems with delayed feedback can have many positive Lyapunov exponents and chaotic attractors whose dimension increases with the delay time, reaching very high values (Farmer, 1982). It is then computationally difficult to reconstruct the nonlinear dynamics of these systems with time-series analysis techniques based on the standard embedding approach. However, it is possible to recover the equation that describes a single-variable time-delay system without using embedding techniques by exploiting the particular structure of these systems. In that approach it is assumed that the structure of that equation is known, and only the functions and parameters are unkown. In this way chaotic cryptosystems based on time-delay systems can be unmasked in spite of the very high dimension of the chaotic carrier (Zhou & Lai, 1999; Ponomarenko & Prokhorov, 2002; Udaltsov et al., 2003; Robilliard et al., 2006).

When the structure of the equations that govern a time-delay system is unknown, an embedding-like approach has been used to recover the dynamics with a local linear model (Hegger et al., 1998; Bünner et al., 2000a; 2000b). This method works with a special embedding space that includes both short time and feedback-time delayed values of the system variable. In this way the dynamics is reconstructed in a space with a dimension much smaller than the attractor´s dimension. The dimension of this space is independent of the delay time. In this chapter we use this special embedding space to reconstruct the nonlinear dynamics of semiconductor lasers subject to delayed optoelectronic feedback. A global nonlinear model based on neural networks will be used for this reconstruction.

Optical systems provide simple ways of generating very high dimensional chaotic carriers that offer the possibility of high transmission rates (Uchida et al., 2005; Donati & Mirasso, 2002). Generation of chaotic signals with high dimension and high information entropy can be achieved in diode lasers by means of delayed feedback. Two schemes based on all-optical and electro-optical feedback have been considered (Argyris et al., 2005). In both systems the dimension of the chaotic attractor increases linearly with the delay time, reaching very high values (Vicente et al., 2005). However, the entropy of systems with optoelectronic feedback increases with the feedback strength, whereas it saturates for all-optical feedback systems (Vicente et al., 2005). Therefore the behaviour can be more unpredictable for chaotic carriers based on optoelectronic feedback. In this work we consider optoelectronic chaos generators with fixed values of the feedback delay time (Goedgebuer et al., 1998a; Larger et al., 1998a). We have recently shown that the nonlinear dynamics of those optical chaotic carriers can be reconstructed by using time series obtained from numerical simulations (Ortín & Pesquera, 2006; Ortín et al., 2007c) and from experimental data (Ortín et al., 2007a; 2007b). The security of chaotic cryptosystems based on optoelectronic feedback with fixed delay time is then compromised.

We study in this chapter the security of optical chaotic communication systems based on a chaotic carrier generated by a laser diode subject to delayed optoelectronic feedback. Two different cases with one and two fixed delay times are considered. The chapter is structured as follows. The first Section is devoted to the study of the one fixed delay time case through numerical simulations. We introduce the model for a chaotic generator based on a semiconductor laser with delayed electro-optical feedback (Larger et al., 1998a; Goedgebuer et al., 1998a). The value of the delay time is extracted from the numerical time series. This step is crucial to work in the special embedding space that includes both short time and feedback-time delayed values of the system variable. The nonlinear dynamics of the chaotic carrier is reconstructed from numerical simulations by using a new type of modular neural networks. Using this model as an unauthorized receiver, it is shown that the message can be extracted. In the second Section we apply the previous techniques to recover the message from experimental time series. The experiments were performed in the group of Prof. Laurent Larger (Université de Franche-Comté, Besançon, France). The third Section is devoted to the study of the two fixed delay times case by using numerical simulations. Two different configurations, serial and parallel, are considered. It is shown that in all the cases this cryptosystem can be broken. Finally, we present the conclusions.

ONE-DELAY CHAOTIC SIGNAL GENERATOR: NUMERICAL SIMULATIONS

The generator of the chaotic wavelength beam considered in this work consists of an electrically tunable DBR multielectrode laser diode with a feedback loop formed by a delay line and an optically birefringent plate whose peculiarity is to exhibit a nonlinearity in wavelength (Goedgebuer et al., 1998a ; 1998b; Larger et al., 1998a; 1998b). The wavelength of the chaotic carrier can be described by the following time-delay differential equation:

$$\lambda(t) + T\frac{d\lambda(t)}{dt} = \beta_{\lambda}\sin^2\left(\frac{\pi D}{\Lambda_0^2}\lambda(t-\tau) - \Phi_0\right), \quad (1)$$

where λ is the wavelength deviation from the center wavelength Λ_0, D is the optical path difference of the birefringent plate which constitutes the nonlinearity, Φ_0 is the feedback phase, τ is the delay time, T is the response time of the feedback loop, implemented by a first-order low-pass filter and β_{λ} is the feedback strength. Eq. (1) is the Ikeda equation (Ikeda et al., 1982) and once normalized it takes the form

$$x(t) + T\frac{dx(t)}{dt} = \beta\sin^2(x(t-\tau) - \varphi_0), \quad (2)$$

where $x = \pi D \lambda / \Lambda_0^2$ and $\beta = \pi D \beta_{\lambda} / \Lambda_0^2$. The feedback strength can be adjusted through the gain of an amplifier in the loop. The regime of oscillations in wavelength depends on the value of the parameter β, which determines the strength of the feedback as well as the strength of the nonlinearity. The system can display chaotic behavior for $\beta > 2.1$, but this threshold value depends on the feedback phase. As the feedback strength β increases the influence of the phase decreases and for $\beta > 5$ the dynamics is independent of ϕ_0 (Vicente et al., 2005). The number of extreme values of the sin^2 nonlinear function increases also with β.

The dependence of some chaotic indicators with operating parameters has been investigated in detail for this electro-optical feedback laser system (Vicente et al., 2005). The number of positive Lyapunov exponents grows linearly with the delay time. The Kaplan-Yorke dimension also increases linearly with the delay time. Therefore, chaotic attractors with very large dimensions can be achieved. However, the Lyapunov exponents that become positive as the delay time is increased have a very small magnitude. This, together with the fact that the largest positive Lyapunov exponent decreases as the delay time increases, yields saturation in the Kolmogorov-Sinai entropy. Therefore, although the system has a larger dimensionality when increasing the delay, its behavior does not become more unpredictable. These results suggest that increasing the delay time beyond the value at which the entropy saturates will neither yield a better masking nor improve the security. Unlike all-optical feedback systems, in these electro-optical systems the feedback is nonlinear while the laser operates in the linear regime. The number of positive Lyapunov exponents as well as their value increases with the feedback strength in a linear way. Therefore, the Kaplan-Yorke dimension and the Kolmogorov-Sinai entropy grow also linearly with the feedback strength. Then a way to achieve a better masking and more secure encoding is to increase the nonlinear feedback strength.

A chaotic communication system has been demonstrated experimentally (Goedgebuer et al., 1998a; Bavard et al., 2007) using the chaotic generator described by Eq. (1). A schematic diagram of the chaotic communication system is presented in Figure 1. In this type of communication protocol the message *m(t)* is injected into the feedback loop and thus participates actively to the dynamics. Two different injection points have been considered in experiments (Goedgebuer et al., 1998a; Bavard et al., 2007). Here we consider that the message is injected before the low-pass filter (see Figure 1). The open-loop receiver is a replica of the transmitter (in the sense that it is formed by the same elements). The message is recovered by the receiver upon synchronization with the emitter. The receiver can be viewed as performing a nonlinear filtering process, intended to generate locally a message-free chaotic signal, which is subtracted from the transmitted signal to extract the message.

To unmask this optical chaotic cryptosystem we construct a model that reproduces the nonlinear dynamics of the chaotic carrier. The model is used as an unauthorized receiver to recover the message. Several methods have been used to extract the nonlinear dynamics of time-delay systems. Most of them assume that the structure of the equation ruling the chaotic carrier dynamics is known. Here we consider that no a-priori knowledge is available about that equation. Nonlinear dynamics reconstruction is performed in a special embedding space that includes both short time and feedback-time delayed values of the system variable $x(t)$. In this way the dynamics is reconstructed in a space with a dimension much smaller than the attractor's dimension (Bünner et al., 2000a; 2000b). The dimension of the embed-

Figure 1. Schematic diagram of the chaotic communication system. LD: laser diode, NL: nonlinear element, delay line: τ, amplifier: β and low-pass filter.

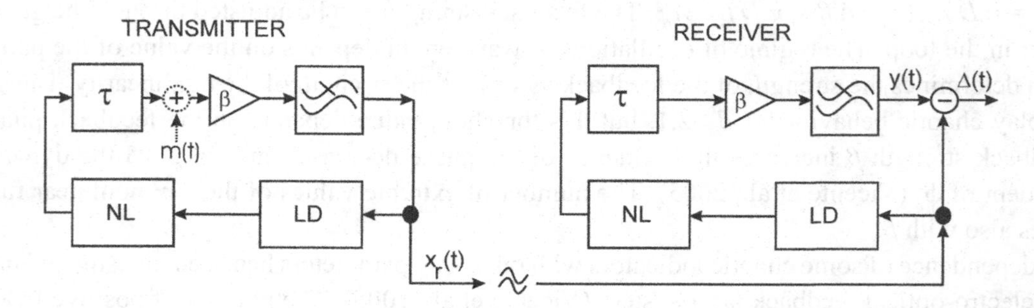

ding space turns out to be independent of the delay time and thus of the dimensionality of the attractor dynamics. In this Section we use time series of $x(t)$ obtained from numerical simulations of Eq. (2) to reconstruct the nonlinear dynamics of one-delay chaotic carriers.

We have carried out numerical simulations of Eq. (2) using the Adams-Bashforth-Moulton predictor-corrector scheme (Press et al., 1992) with an integration step of 0.01. To study the effect of the nonlinearity strength we have considered several values of β in the range from $\beta = 5$ (low nonlinearity strength) to $\beta = 50$ (high nonlinearity strength). The maximum value of β that has been achieved in experiments is around 22 (Goedgebuer et al., 1998a). Several values of T and τ have been also considered. The feedback phase is $\phi_0 = 0.26\,\pi$.

Time-Delay Extraction

The first step to reconstruct the nonlinear dynamics is the estimation of the delay time. Several methods have been proposed to recover the delay time from time series. Here we consider the delayed mutual information and the filling factor method. The delayed mutual information (DMI) (Abarbanel, 1996) is a statistical function that measures some nonlinear link between the chaotic signal $x(t)$ and a delayed version of it, $x(t-v)$. Since these variables are connected to each other when $v = \tau$ (see Eq. (2)), the DMI function versus v exhibits a peak located at $v = \tau$. Figure 2 shows the DMI function obtained from numerical time series with a delay time $\tau = 100$. Two values of the feedback strength that correspond to low, $\beta = 5$, and high nonlinearity, $\beta = 50$, are considered. Short, $T = 1$, and long, $T = 40$, response times are also considered. The sampling time is $T/100$. For short response times a sharp peak is obtained close to $v = 100$ (see the insets in Figures $2a_1$ and $2b_1$). However the DMI always overestimates the delay time due to the finite reaction of the system, which is given by the response time T. Therefore, the mutual information is not a valid method to identify the delay time for long response times, in particular for low nonlinearity strengths (see the inset in Figure $2a_2$) (Ortín et al., 2007b). Smaller peaks are also obtained at multiples of the delay time. The amplitude of the peaks decreases when increasing the nonlinearity strength (Locquet et al., 2006). This corresponds to a decrease in the statistical link given by the DMI when the nonlinearity is increased.

The filling factor method exploits also the functional relationship between $x(t)$ and $x(t-\tau)$ (Bünner et al., 1997). The trajectory will collapse to a line in the space ($x(t)$, $x(t-v)$, dx/dt) when $v = \tau$. By taking only the extreme values such that $dx(t_{ext})/dt = 0$, we cover the space ($x(t_{ext})$, $x(t_{ext}-v)$) with squares. The filling factor is the number of squares that are visited by the trajectory normalized to the total number of squares. The filling factor is computed under variation of v. A minimum in the filling factor appears when $v = \tau$ because the trajectory collapses into the surface with the minimum area. The filling factor is plotted in Figure 3. A minimum is obtained at the correct value of the delay time, $v = 100$. Minima with smaller amplitude are also obtained at multiples of the delay time. The amplitude of the valleys decrease when increasing the nonlinearity strength, just as happens for the DMI. The filling factor method is not limited by the response time if we use only the data that correspond to extreme values. However, this method is more sensitive to noise than the delayed mutual information.

Nonlinear Dynamics Reconstruction with Modular Neural Networks

Artificial Neural Networks (NNs) have been successfully applied for extracting the dynamics of nonlinear systems with low dimensionality. NNs are trained with an input vector given by $(x(t-t_e), \ldots, x(t-mt_e))$,

Figure 2. Delayed mutual information obtained from numerical simulation of Eq. (2) with a delay time $\tau = 100$ for low ($\beta = 5$: a_1, a_2) and high ($\beta = 50$: b_1, b_2) nonlinearity strength, and short ($T = 1$: a_1, b_1) and long response time ($T = 40$: a_2, b_2).

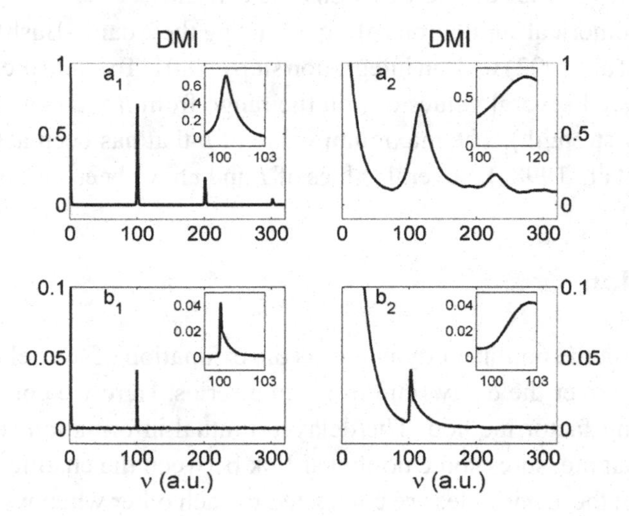

Figure 3. Filling factor obtained from numerical simulation of Eq. (2) with a delay time $\tau = 100$ for low ($\beta = 5$: a_1, a_2) and high ($\beta = 50$: b_1, b_2) nonlinearity strength, and short ($T = 1$: a_1, b_1) and long response time ($T = 40$: a_2, b_2).

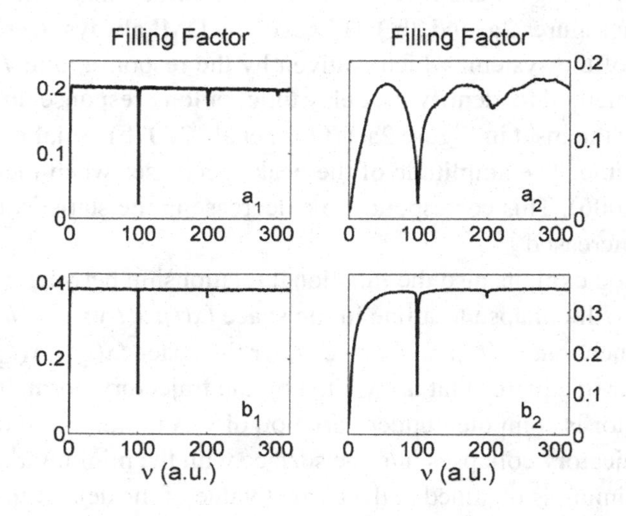

where t_e is the embedding delay and m the dimension of the embedding space. However, the dimension of the embedding space is very high for large delay times and standard embedding techniques can not be applied. It has been shown that nonlinear dynamics of time-delay systems can be reconstructed in a special embedding space with a dimension much smaller than the attractor´s dimension (Hegger et al., 1998; Bünner et al., 2000a; 2000b). This space includes both short time and feedback-time delayed values of the system variable $x(t)$. We consider non-uniform input vectors given by data delayed by t_e, $x_{nf} = (x(t-t_e),..,x(t-m_1t_e))$, and by the feedback time, $x_f = (x(t-\tau +m_2t_e),.., x(t -\tau),.., x(t-\tau- m_2t_e))$. The numbers of non-feedback and feedback inputs are m_1 and $2m_2+1$, respectively. We have shown that the nonlinear dynamics of time-delay systems can be reconstructed in this special embedding space by using standard feed-forward NNs (Ortín et al., 2005; 2007a). However, standard NNs have a rigid structure of fully connected layers with many degrees of freedom that may overfit the data, train slowly, or converge to local minima (see Figure 4 left). Modular neural networks (MNNs) have been recently introduced to obtain more flexible models that require a smaller number of parameters than the standard NNs (Ortín et al., 2005; 2007a). According to the structure of the time-delay system, the MNN has two modules, one for the non-feedback part with input data $\boldsymbol{x}_{nf}$, and a second one for the feedback part with input data $\boldsymbol{x}_f$. A feed-forward NN is used for each of the modules. The value given by the MNN is $x_{nn}(t+t_e)=f_{nn}(\boldsymbol{x}_{nf}) + g_{nn}(\boldsymbol{x}_f)$, where f_{nn} and g_{nn} correspond to the non-feedback and feedback modules, respectively (see Figure 4 right).

We have used time series obtained from numerical simulations of Eq. (2) to train the MNN model with one input for the non-feedback module (m_1= 1), and three inputs (m_2= 1) for the feedback module with an embedding time t_e =0.01. Different values of the feedback strength in the range from $\beta = 5$ to $\beta = 50$ and two values of the delay time, $\tau = 10$ and 100, have been considered. The response time is $T = 1$. A single hidden layer with one neuron has been used for the non-feedback module. Two layers with a and b neurons, denoted by $a{:}b$, are used for the feedback module. We take 3000 training points chosen to cover uniformly the chaotic attractor. The resulting MNN model was tested with 100000 data. The root mean square error (RMSE) is calculated as the mean of the best five test errors obtained out of ten models trained starting at different initial weights. This error is normalized to the standard deviation of the chaotic carrier. We show in Figure 5 the RMSE as a function of the feedback strength for two values of the delay time. For a given value of β the error decreases as the number of parameters of the model increases. However, no significant error reduction is obtained when the MNN has a large number of parameters. It is found that the model error increases with the feedback strength, but not with the delay time. This result is in agreement with the fact that the entropy increases with β, but not with τ (Vicente et al., 2005). A similar error is obtained by using a MNN model with the same degree of complexity for different values of τ, although the attractor dimension increases with the delay time. An estimate of the dimension of the chaotic attractor is given by $0.4\beta\tau/T$ (Vicente et al., 2005) that leads to values as large as 2000 for $\beta = 50$ and $\tau = 100$.

Figure 4. Topology of a standard neural network (left) and modular neural network (right).

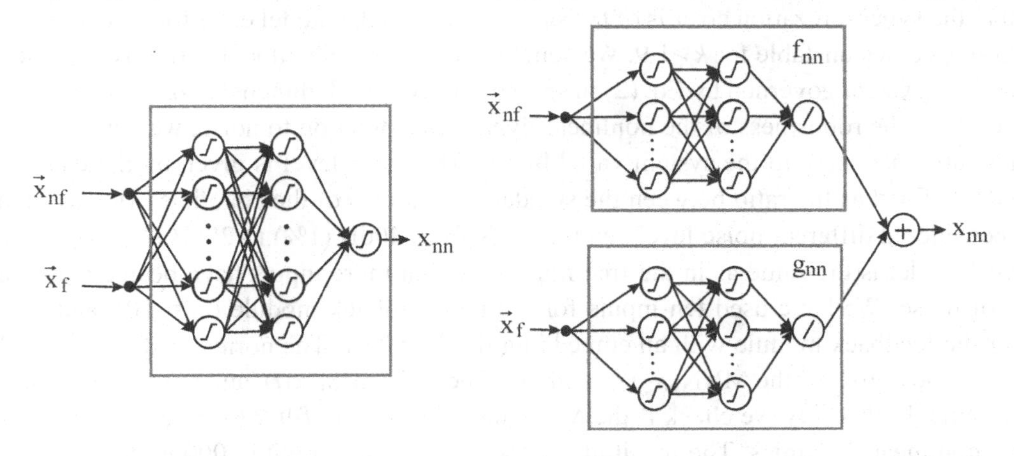

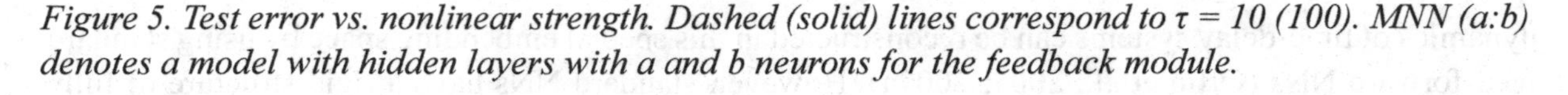

Figure 5. Test error vs. nonlinear strength. Dashed (solid) lines correspond to τ = 10 (100). MNN (a:b) denotes a model with hidden layers with a and b neurons for the feedback module.

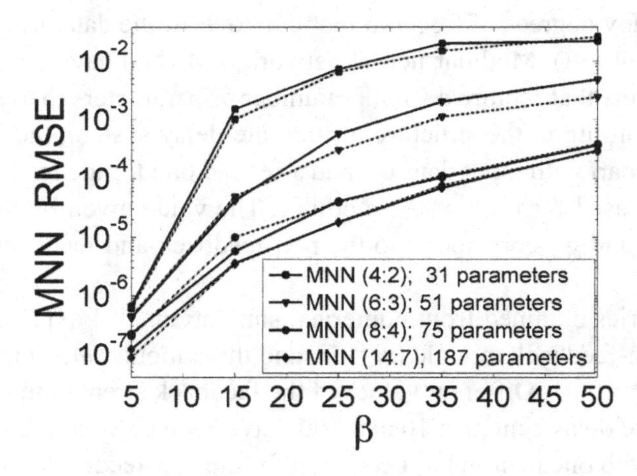

Thanks to the modularity of the model, it is possible to represent the non-feedback, f_{nn}, and feedback, g_{nn}, functions given by the MNN model. We show in Figure 6 these functions for three different values of the feedback strength, β =5, 25 and 50, and a delay time, τ = 100. The MNN model has 31, 51 and 75 parameters for β= 5, 25 and 50, respectively. The functions plotted are similar to the linear and nonlinear terms of a discrete-time version of Eq. (2). However, note that the function g_{nn} depends on the vector $\boldsymbol{x}_f$ and not only on $x(t\text{-}\tau)$. The results show that the MNN model is able to extract the nonlinear feedback function.

Although the results shown in Figure 5 indicate a good accuracy in one-step ahead prediction, it does not mean that the MNN model can reproduce the dynamics described by Eq. (2) when iterated in time (Principe et al., 1992). Synchronization is a good test to validate models of chaotic systems (Brown et al., 1994; Aguirre et al., 2006). In this test it is assumed that valid models can synchronize with the data when a diffusive coupling between the system and the MNN model is used. A term $k(x\text{-}x_{nn})$ is added to the MMN variable x_{nn} in the input of the non-feedback module. We plot in Figure 7 the synchronization error, η, divided by the RMSE as a function of the coupling constant, *k*, for three different values of the feedback strength, β =5, 25 and 50, and a delay time, τ = 10. Similar results are obtained for τ = 100. It is found that the synchronization error is of the same order than the model error for k >0.4 until the synchronization becomes unstable for k >1.9. We conclude that the MNN model correctly reconstructs the dynamics of the system governed by Eq. (2) in spite of the very high dimension of the chaotic attractor.

To investigate the robustness of the nonlinear dynamics extraction to noise, we have added a zero-mean Gaussian noise, $n(t)$, to the system variable $x(t)$. The noise level is given by the chaos to noise ratio (CNR), defined as the ratio between the standard deviations of the chaotic carrier $x(t)$ and noise. We have considered different noise levels with a CNR from 40dB (1%) to 20dB (10%). The structure of the MNN model is the same as in the free-noise case, but more inputs have been taken to average the effect of noise. We have used ten inputs for the non-feedback module (m_1= 10), and five inputs (m_2= 2) for the feedback module with an embedding time t_e =0.01. The normalized RMSE is obtained by comparing the output of the MNN, $x_{nn}(t)$, with the free-noise data, $x(t)$, and not with the input of the MNN, $x(t)+n(t)$. In this way we check if the MNN model is able to filter the noise and to recover the underlying nonlinear dynamics. The resulting MNN model is tested with 100000 data. The test error

Figure 6. Left (middle): non-feedback (feedback) module function of the MNN model. Right: nonlinear feedback function of Eq. (2). Different values of β are considered: 5 (top), 25 (middle row) and 50 (bottom). The delay time is τ =100.

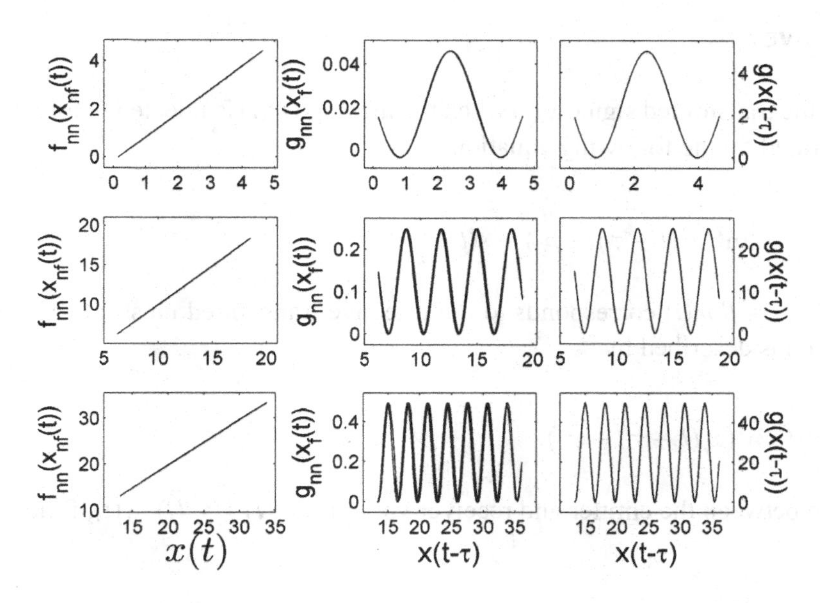

Figure 7. Synchronization error divided by MNN model error vs. k for τ = 10 and three values of β: 5 (dash-dotted line with circles), 25 (solid line with squares) and 50 (dashed line with triangles). The MNN has 31, 51 and 75 parameters for β= 5, 25 and 50, respectively.

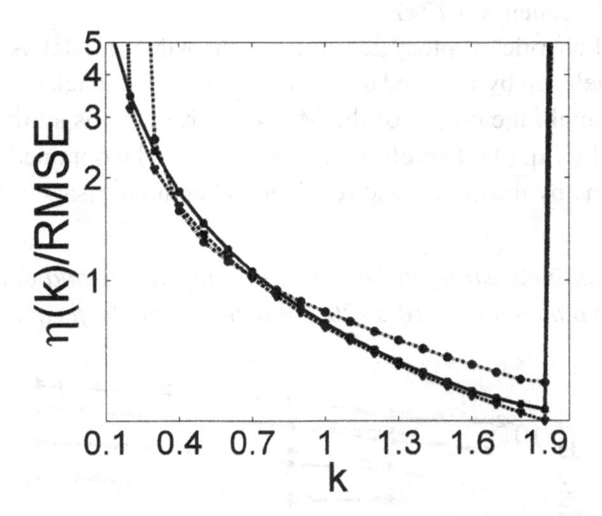

is plotted in Figure 8 as a function of the feedback strength for different noise levels. The delay time is τ = 100. Two different feedback modules have been considered with *10:5* (131 parameters) and *20:10* (351 parameters) neurons, which have been trained with 8000 and 14000 points, respectively. Figure 8 shows that the MNN model is not improved by increasing the number of parameters of the model. As expected, the error increases with the noise level. For a low noise level a clear increase with β is also observed. However, when the CNR decreases similar errors are obtained for low and high feedback strengths. The error is always smaller than the noise level. Therefore, we conclude that the MNN model filters to some extent the noise.

Message Recovery

The dynamics of the transmitted signal $x_r(t)$ when the message $m(t)$ is injected before the low-pass filter (see Figure 1) is ruled by the following equation

$$x_r(t) + T\frac{dx_r(t)}{dt} = \beta \sin^2(x_r(t-\tau) - \varphi_0) + s(t), \tag{3}$$

where the signal $s(t) = \beta\, m(t)$ corresponds to the effective transmitted message. The dynamics of the receiver system $y(t)$ is described by

$$y(t) + T\frac{dy(t)}{dt} = \beta \sin^2(x_r(t-\tau) - \varphi_0). \tag{4}$$

The difference between the emitter and receiver variables, $\Delta(t) = x_r(t) - y(t)$, follows the equation

$$T\frac{d\Delta(t)}{dt} + \Delta(t) = s(t). \tag{5}$$

Therefore the difference between the emitter and receiver variables gives the message filtered by a first-order filter of cut-off frequency $1/(2\pi T)$.

To unmask this optical chaotic cryptosystem we use the MNN model as an unauthorized receiver. The transmitted signal x_r delayed by τ is used as the input for the feedback module of the MNN receiver. When the model error is small, the output of the MNN receiver, y_{nn}, is similar to the output of the authorized receiver, y, ruled by Eq. (4). Therefore, the message can be obtained from $\Delta_{nn}(t) = x_r(t) - y_{nn}(t)$ in the same way that is done by the authorized receiver. When no message is transmitted Δ_{nn} is zero for

Figure 8. Test error vs. nonlinear strength for different noise levels and delay time $\tau = 100$. Dashed (solid) lines correspond to a model with 10:5 (20:10) neurons for the feedback module.

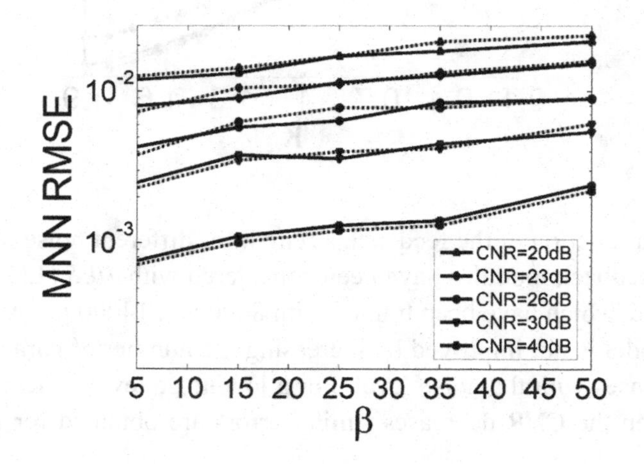

a perfectly matched receiver. We define the decoding error ε_{nn} as the root mean square amplitude of Δ_{nn} when no message is transmitted. This error is normalized to the standard deviation of the chaotic carrier. The decoding error is always greater than the model error.

We consider a chaotic carrier with a response time $T = 1$ and a delay time $\tau = 100$. The feedback strength is $\beta = 50$ that is the most difficult case for an eavesdropper, since the model error increases with β. When the chaotic carrier without message $x(t)$ is used to reconstruct the nonlinear dynamics, the normalized RMSE of the MNN model is around 5×10^{-4} and the decoding error is $\varepsilon_{nn} = 10^{-3}$. We first consider the case of a periodic square signal $s(t)$. The signal frequency is chosen as 0.04 to avoid the detection in the power spectrum of $x_r(t)$ (see Figure 9). Regarding message amplitude, we define the message to chaos ratio (MCR) as the ratio between the standard deviations of the signal $s(t)$ and that of the chaotic carrier $x(t)$. In order to achieve a high signal to noise ratio (SNR) at the receiver, the amplitude of $s(t)$ is chosen as the highest one that avoids the detection of the signal in the spectrum of $x_r(t)$. This signal amplitude corresponds to a MCR of –20dB. We plot in Figure 9 the recovered message given by $\Delta_{nn}(t)$. The slight difference between the recovered Δ_{nn} and the original signal $s(t)$ is mainly due to the fact that $\Delta_{nn}(t)$ corresponds to the filtered message (see Eq. (5)). A high SNR =52 dB is obtained by using the MNN model as an unauthorized receiver.

We now consider that the nonlinear dynamics is extracted from the transmitted signal $x_r(t)$, which includes the message. The message can be viewed as a perturbation. The MNN is able to filter the message to some extent in the same way than for noisy data. However, the model error will increase with respect to the previous case (training without message), in particular when the message amplitude is not negligible. A pseudorandom NRZ (non return to zero) message of frequency 0.04 is injected into the feedback loop. The amplitude of $s(t)$ is chosen as the highest one that avoids the detection of the signal in the spectrum of $x_r(t)$. In this case the maximum allowed amplitude corresponds to a MCR of –9dB. We use the same MNN structure, *8:4* neurons for the feedback module and 75 parameters, as the one used for the time series without message. The resulting MNN model has an error around 0.02 and the decoding error is $\varepsilon_{nn} = 0.1$. It is shown in Figure 10 (top left) that the message is recovered.

We study the robustness of message recovery to noise by adding a zero-mean Gaussian noise, $n(t)$, to the transmitted signal $x_r(t)$. Different noise levels with a CNR from 40dB (1%) to 20dB (10%) are considered. The structure of the MNN model is the same as the one used for noisy data with a feedback module with *10:5* neurons and 131 parameters. The results plotted in Figure 10 show the message can be recovered in spite of the noise.

Figure 9. Top: transmitted signal and its spectrum. Bottom: recovered message (solid line) and its spectrum. Dashed line: original message

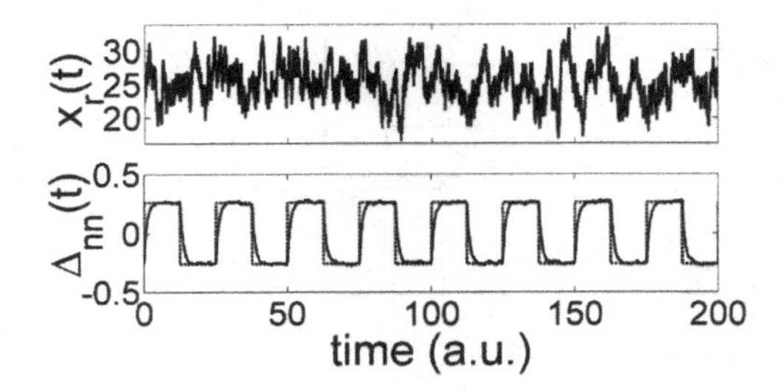

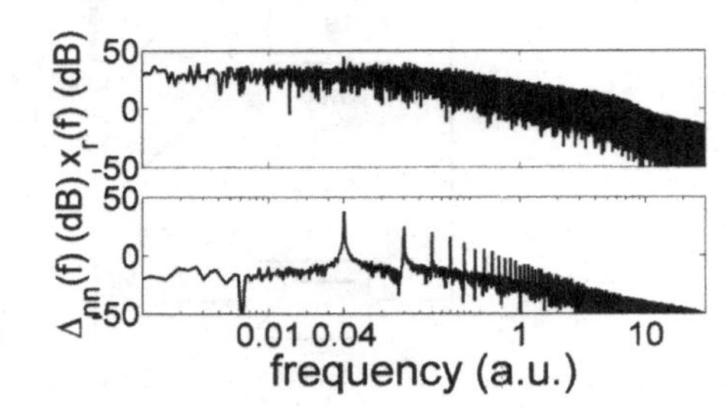

Figure 10. MNN model trained with message. Dashed lines: original message. Solid lines: recovered message without noise (top left), and with noise: CNR =40 dB (bottom left), CNR =30 dB (top right) and CNR =20 dB (filtered recovered message: bottom right)

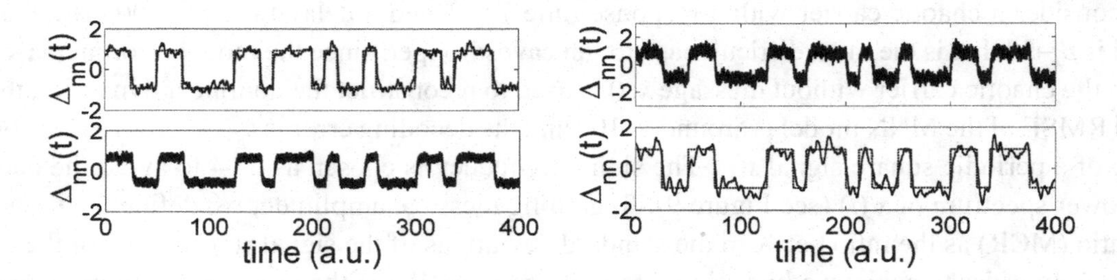

ONE-DELAY CHAOTIC SIGNAL GENERATOR: EXPERIMENTS

In this Section we apply the same methods used with numerical simulations to recover the message from experimental time series. The generator of the chaotic wavelength beam that is used to encrypt the information is depicted in Figure 11 (Goedgebuer et al., 1998a ; 1998b; Larger et al., 1998a; 1998b). It is formed by a Distributed Bragg Reflector (DBR) two-section tunable laser diode at 1.55 μm with a nonlinear feedback loop containing a birefringent plate, a photodetector, a delay line, an amplifier and a first-order low-pass filter. An electronic corrector is also included in the loop to achieve constant wavelength-independent optical power at the laser output. The nonlinearity in wavelength is induced by the birefringent plate whose fast and slow axes are oriented at 45° to two crossed polarizers. The intensity detected by the photodetector is a sin^2 nonlinear function of the wavelength emitted by the laser diode. The cut-off frequency of the low-pass filter determines the response time of the feedback loop.

The delay module consists of a First-In First-Out (FIFO) memory that stores 2048 data in queue order so the first input element goes out the first. A sampling clock fixes the delay value. The clock frequency is limited to a maximum value of 5 MHz by the analog-to-digital and digital-to-analog converters. A minimum clock frequency of 1 MHz is considered to avoid aliasing (Bavard et al., 2007). Therefore the delay time can take values between 0.4 and 2 ms.

The working regime of the device is set by the value of the feedback strength β_λ, which can be adjusted through the gain of the amplifier. The number of extreme values of the sin^2 nonlinear function increases with β_λ. The system parameters are set to operate in the chaotic regime. The maximum number

Figure 11. Experimental set-up of the chaos generator used as transmitter

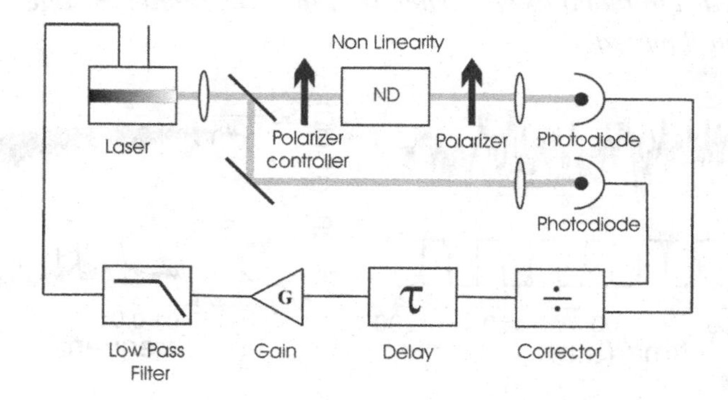

of extrema of the nonlinear function reached with the present set-up when the loop is closed is around 6. In this case the chaotic carrier has a Gaussian probability density. The standard deviation normalized to the mean value decreases when β_λ is increased. For the maximum feedback strength the normalized standard deviation is 0.23. The value of the normalized parameter β can be estimated by comparing the results obtained from numerical simulations of Eq. (2) with the experimental results. If we compare the normalized standard deviation and the number of extrema of the nonlinear function, the maximum feedback strength corresponds to a value of the normalized parameter β around 15.

Time-Delay Extraction

We extract the delay time from the experimental time series with the delayed mutual information and the filling factor method. Two different cases are considered: low nonlinearity with 2-3 extreme values and moderate nonlinearity with 6 extreme values of the nonlinear function. We have also studied the effect of the response time in the time delay identification (Ortín et al., 2007b). Two values of the cut-off frequency of the low-pass filter have been considered: 20 kHz and 200 Hz that lead to response times of $T = 8$ and 800 μs, respectively. A sampling time of 0.1 and 10 μs have been used when $T = 8$ and 800 μs, respectively. Time series with one million points have been recorded. Data acquisition is performed for all the cases in this work with a 8 *bits* resolution oscilloscope. A delay time greater than the largest response time has been considered. The value of the delay time has been estimated as 1.65 ms from the operating parameters of the delay module.

Figure 12 shows the DMI function obtained from experimental time series. For short response times a sharp peak is obtained close to $\nu = 1.65$ ms (see the insets in Figures 12a_1 and 12b_1). However the DMI always overestimates the delay time due to the non-zero response time of the system T. Therefore, the mutual information is not a valid method to identify the delay time for long response times, in particular for low nonlinearity strength (see the inset in Figure 12a_2) (Ortín et al., 2007b). Smaller peaks are also

Figure 12. Delayed mutual information obtained from experimental time series with a delay time τ = 1.65 ms for low (a_1, a_2) and moderate (b_1, b_2) nonlinearity strength, and short (T = 8 μs: a_1, b_1) and long response time (T = 800 μs: a_2, b_2).

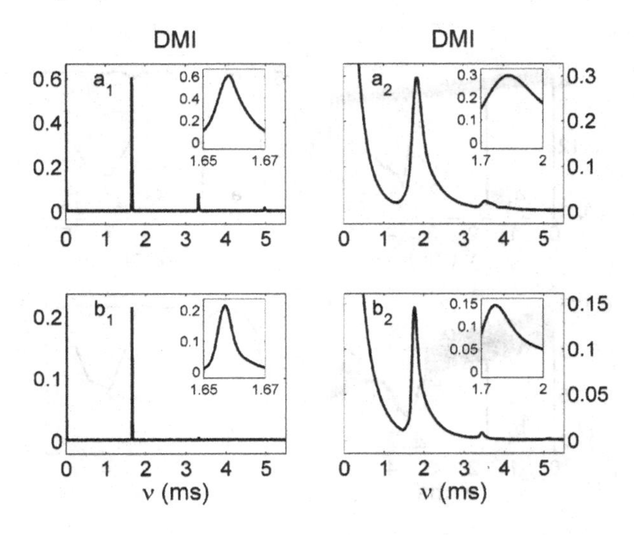

obtained at multiples of the delay time for low nonlinearity strength. These results are similar to those obtained by using numerical simulations.

For noise-free data the filling factor method is not affected by the response time if we use only the data that correspond to extreme values. However, in the presence of noise spurious extrema appear that leads to an overestimation of the delay time for long response times. A rough estimation of the noise is obtained by comparing the recorded time series with a smoothed version of that series. A moving average over 30 μs (3 μs) in the long (short) response time case leads to a 0.028 (0.024) noise level. The noise level is normalized to the standard deviation of the chaotic carrier. In the smoothed time series the number of spurious extrema is reduced. Therefore, the filling factor obtained by using the smoothed time series leads to the correct delay time for both, short and long response times. The resulting filling factor is plotted in Figure 13. A minimum is obtained at the correct value of the delay time. The amplitude of the valleys decreases when increasing the nonlinearity strength, just as happens for the DMI. These results are again similar to those obtained by using numerical simulations.

We have shown (Ortín et al., 2007b) that the correct value of delay time can be also extracted from the recorded experimental time series by using a method based on the forecasting error. In this method a neural network model is trained from the experimental time series by using a non-uniform input vector given by $(x(t-t_e), x(t-v), x(t-v-t_e), x(t-v-2t_e))$, where t_e is the sampling time. The forecasting error of the NN is calculated under variation of v. A minimum in the error appears at the correct value of the delay time for both, short and long response times. The main drawback of this method is its high computational cost. However, this method seems to be the most appropriate when the noise level is not small.

Nonlinear Dynamics Reconstruction with Modular Neural Networks

We have shown with numerical simulations that the most difficult case for an eavesdropper corresponds to the maximum nonlinear strength. Here we consider the case of a nonlinear function with 5 extrema that

Figure 13. Filling factor obtained from the smoothed experimental time series with a delay time τ = 1.65 ms for low (a_1, a_2) and moderate (b_1, b_2) nonlinearity strength, and short (T = 8 μs: a_1, b_1) and long response time (T = 800 μs: a_2, b_2).

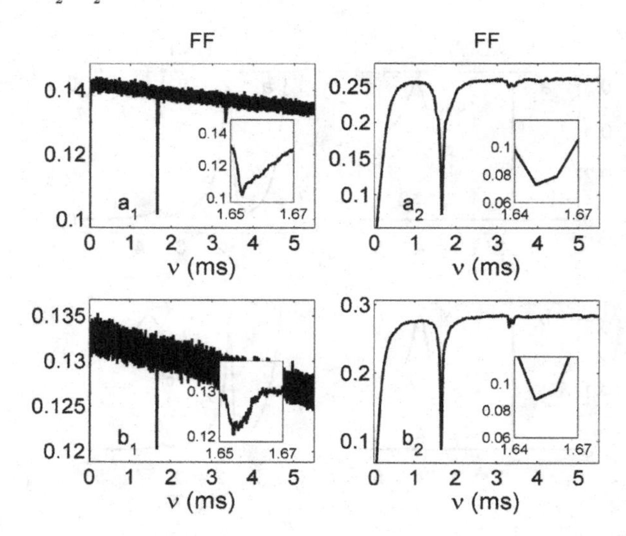

is close to the maximum nonlinear strength reached with the present experimental set-up. The estimated feedback strength β is greater than 12. The cut-off frequency of the low-pass filter is 20 kHz that leads to a response time of 8 μs. The value of the delay time, 0.476 ms, is extracted from the experimental time series.

Modular neural networks are used to reconstruct the nonlinear dynamics from experimental time series, $x_{exp}(t)$. We use four inputs, $\mathbf{x}_{nf} = (x_{exp}(t-t_e),..,x_{exp}(t-m_1 t_e))$ with $m_1 = 4$, for the non-feedback module, and five inputs, $\mathbf{x}_f = (x_{exp}(t-\tau+m_2 t_e),.., x_{exp}(t-\tau),.., x_{exp}(t-\tau-m_2 t_e))$ with $m_2 = 2$, for the feedback module with an embedding time $t_e = 1$ μs. We take 25000 training points chosen to cover uniformly the chaotic attractor. The MNN has one linear neuron for the non-feedback module, and two layers with 6 and 3 neurons for the feedback module. The resulting MNN model is tested with 100000 data. The normalized test error (RMSE) is around 0.06 and the estimated normalized noise level is 0.024. The normalized RMSE is obtained by comparing the output of the MNN, $x_{nn}(t)$, with the experimental data, $x_{exp}(t)$. Therefore, the test error is always greater than the normalized noise level. We plot in Figure 14 the functions f_{nn} and g_{nn} extracted by the MNN model that correspond to the non-feedback and feedback modules, respectively. It can be seen that the MNN model recovers the shape of the sin^2 nonlinear function.

We have used the synchronization test (Aguirre et al., 2006) to validate the MNN model extracted from the experimental time series. A coupling term $k(x_{exp} - x_{nn})$ is added to the model variable x_{nn} in the input of the non-feedback module. When the dynamics of the model is similar to that of the experimental system, chaotic synchronization between these two systems is achieved. We plot in Figure 15 the synchronization error, η, divided by the RMSE as a function of the coupling constant, *k*. It is found that the synchronization error is of the same order that the model error for $k>0.4$ until the synchronization becomes unstable for $k>1.4$. These results are similar to those obtained with numerical simulations (see Figure 7). We conclude that the MNN model correctly reproduces the dynamics of the experimental system. We have checked that these results also hold for longer delay times such as 1.23 and 2.05 ms. An estimate of the dimension of the chaotic attractor is given by $0.4\beta\tau/T$ (Vicente et al., 2005) that leads to dimensions from 280 for $\tau = 0.476$ ms to 1200 for $\tau = 2.05$ ms. We conclude that the MNN model correctly reconstructs the dynamics of the experimental system in spite of the very high dimension of the chaotic attractor, that can be estimated to be greater than 200 in all the cases.

Figure 14. Left (right): non-feedback (feedback) module function extracted by the MNN model from the experimental time series for a response time of 8 μs. The delay time is 0.476 ms.

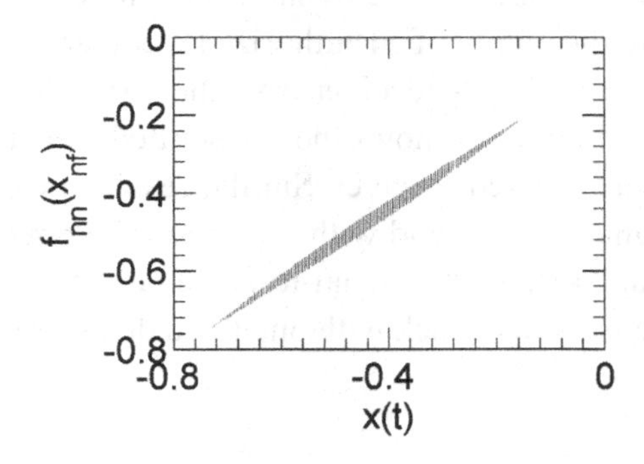

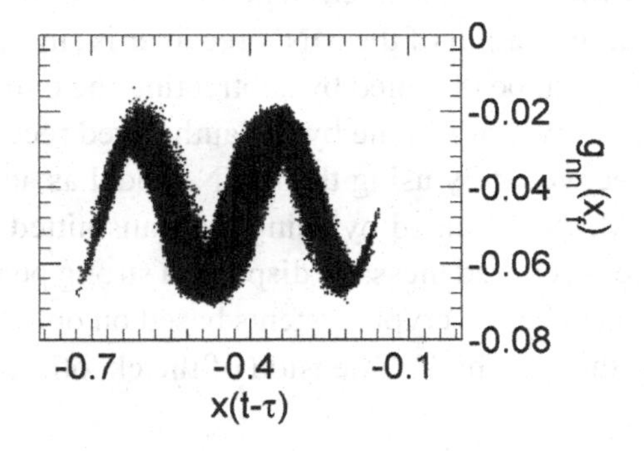

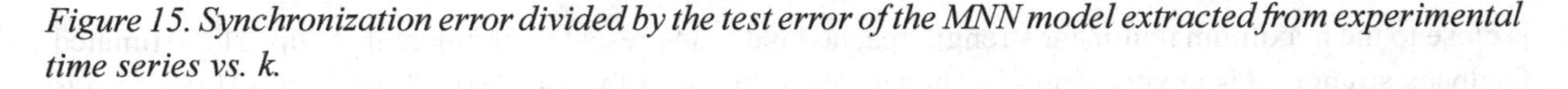

Figure 15. Synchronization error divided by the test error of the MNN model extracted from experimental time series vs. k.

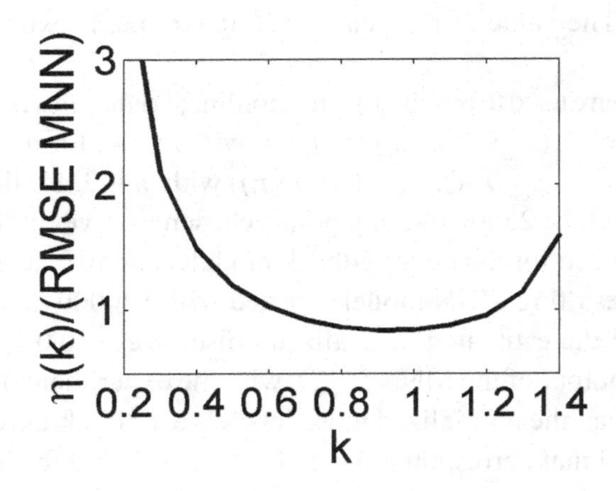

Message Recovery

An optical chaotic cryptosystem has been demonstrated experimentally (Goedgebuer et al., 1998a; Bavard et al., 2007) using the chaotic generator shown in Figure 11. In this system the message is injected into the feedback loop and thus participates actively to the dynamics. The receiver is formed by the same elements than the transmitter, but the loop is open (see Figure 1). Message recovery is based on the synchronization phenomenon by which the receiver is able to reproduce the chaotic part of the transmitted signal. The message is recovered by subtracting the receiver signal from the transmitted signal. Here we consider that the message is injected before the low-pass filter (Bavard et al., 2007) (see Figure 1). To unmask this optical chaotic cryptosystem we use the MNN model extracted from experimental time series as an unauthorized receiver.

A periodic square function message with a message-to-chaos ratio of –8dB is injected into the feedback loop. The signal frequency, 2 kHz, is chosen to avoid the detection in the power spectrum of the transmitted signal (see Figure 16). The message amplitude is similar to the one used in the experimental chaotic communication system to achieve a good signal to noise ratio (SNR) at the receiver (Goedgebuer et al., 1998a; Bavard et al., 2007). This signal amplitude is the highest one that avoids the detection of the signal in the spectrum of the transmitted signal (see Figure 16).

The recorded time series from experimental transmitted signal is used as the input for the feedback module of the MNN receiver. When the model reproduces the chaotic dynamics of the experimental transmitter, the output of the MNN receiver is similar to the output of an authorized receiver. Therefore, the message can be obtained by subtracting the output of the MNN receiver from the transmitted signal in the same way that is done by the authorized receiver. Figure 16 shows the transmitted signal and the message recovered by using the MNN model as an unauthorized receiver. Similar results are obtained when the MNN is trained by using the transmitted signal without and with message. The power spectrum of the recovered message displays a strong peak at 2 kHz with a signal-to-noise ratio of 27 dB. We conclude that chaotic cryptosystems based on optoelectronic feedback with one fixed delay are not safe in spite of the very high dimension of the chaotic carrier.

Figure 16. Left: Experimental transmitted signal with message (top) and its power spectrum (bottom). Middle (right): recovered filtered message (top) and its power spectrum (bottom) when the MNN model is extracted from the experimental time series without (with) message.

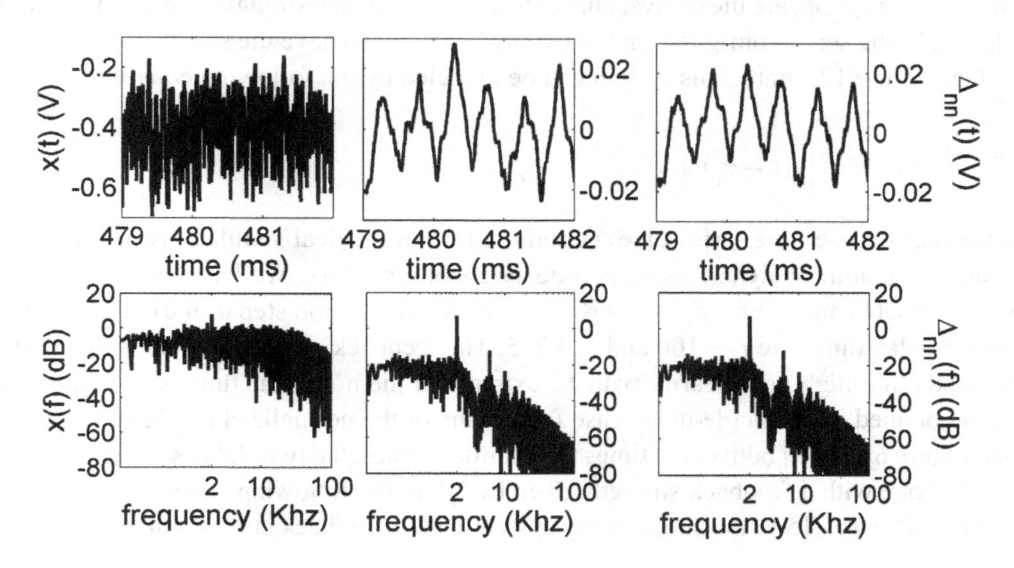

TWO-DELAY CHAOTIC SIGNAL GENERATOR: NUMERICAL SIMULATIONS

It has been proposed to use chaotic generators with multiple delayed feedback loops to improve the security of optical chaotic communication systems (Lee et al., 2004; 2005). In this Section we study the security of an optoelectronic chaos generator with two fixed delay times through numerical simulations. Two different configurations, parallel and serial, are considered. In the parallel configuration two feedback loops with different delay modules and nonlinear elements are put in parallel. The block-diagram of this configuration is represented in Figure 17 (left). The elements of this diagram are the same ones used in the emitter of wavelength chaos with one feedback loop. The delay-differential equation describing the dynamics of the system in the parallel configuration is

Figure 17. Left (right): Block diagram of the parallel (serial) configuration of chaotic emitter with two feedback loops

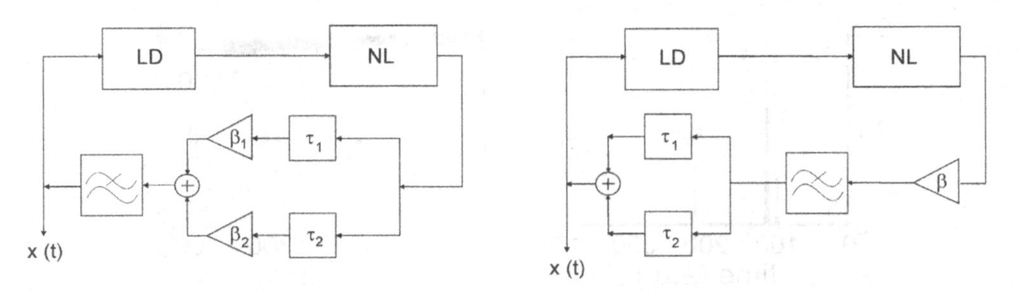

$$x(t) + T\frac{dx(t)}{dt} = \beta_1 \cdot \sin^2\left[x(t-\tau_1) - \varphi_1\right] + \beta_2 \cdot \sin^2\left[x(t-\tau_2) - \varphi_2\right], \qquad (6)$$

where τ_1, β_1, ϕ_1 and τ_2, β_2, ϕ_2 are the delays, phase shifts, and feedback strengths, respectively, of the two feedback loops. In the serial configuration the two feedback loops have the same nonlinearity module, $\beta_1 = \beta_2 = \beta$ (see Figure 17 right). This system can be modeled by the following equation

$$x(t) + T\frac{dx(t)}{dt} = \beta \cdot \sin^2\left[x(t-\tau_1) + x(t-\tau_2) - \varphi\right]. \qquad (7)$$

In this Section we use time series of $x(t)$ obtained from numerical simulations of Eqs. (6) and (7) to reconstruct the nonlinear dynamics of two-delay chaotic carriers. We have carried out numerical simulations of Eqs. (6) and (7) for $\beta_1 = \beta_2 = \beta = 15$ with an integration step of 0.01. The response time is T=1 and the delay times are $\tau_1 = 100$ and $\tau_2 = 215$. The feedback phase is $\phi_1 = \phi_2 = \phi = 0.26\,\pi$. This case corresponds to a high nonlinearity with 12 extrema in the nonlinear function. A similar number of extrema is obtained in the single-delay case for a value of the normalized feedback strength greater than 25 (see Figure 6). When both delay times take similar values, the two-delay system is equivalent to the single-delay one with a feedback strength given by 2β. In the following we compare the two-delay system with $\beta_1 = \beta_2 = \beta = 15$ to the single-delay system with a feedback strength of 30.

Time-Delay Extraction

The first step to extract the nonlinear dynamics is the estimation of the delay times. In the parallel configuration it is possible to identify the delay values using the same methods that work for the single-delay case. Figure 18 represents the delayed mutual information and the filling factor for the system described by Eq. (6). We observe that it is possible to identify clearly the two delays. Two peaks or valleys of similar amplitude appear at the correct delay times. An additional peak or valley is obtained at $\tau_2 - \tau_1 = 115$, due to the existence of a linear relationship between $x(t)$ and $x(t-(\tau_2 - \tau_1))$ when $\phi_2 - \phi_1 = 0$ or $\pi/4$. We have checked that these results hold regardless the values of the delays and of the feedback strengths (Locquet et al., 2006). Stronger feedbacks may require the use of smaller sampling times, but it is always possible to identify the delays with the same techniques that work for the single-delay system.

Figure 18. Delayed mutual information (left) and filling factor (right) obtained from numerical simulation of Eq. (6), parallel configuration, with delay times $\tau_1 = 100$ and $\tau_2 = 215$.

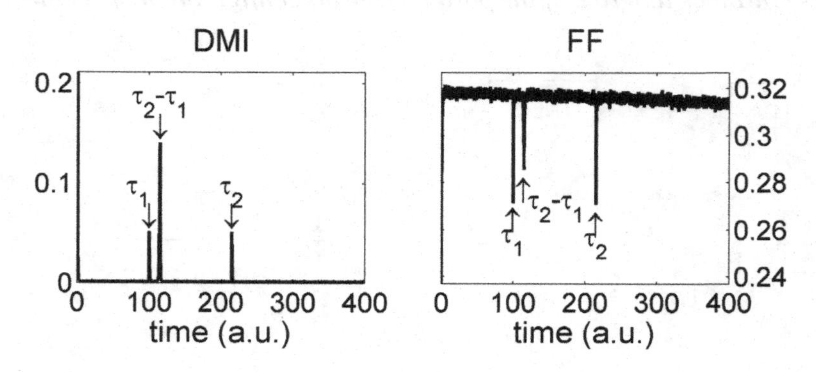

In the serial configuration described by Eq. (7) the methods applied to the single-delay case fail unless the feedback strength is small (Locquet et al., 2006). We should remember that the techniques used in Figure 18 to identify the delays try to measure the presence of a structure in the plane $(x(t),x(t-v))$, in the case of the mutual information, and in the space $(x(t_{ext}), x(t_{ext}-v))$, in the case of the filling factor. Eqs. (6) and (7) for the two-delay systems define, however, a constraint in the four-dimensional space $(dx(t)/dt, x(t), x(t-\tau_1), x(t-\tau_2))$, or in the three-dimensional space $(x(t_{ext}), x(t_{ext}-\tau_1), x(t_{ext}-\tau_2))$ when only the extrema points are considered. It is not guaranteed, however, that a projection on a lower-dimensional space leads to an identifiable object. This depends on the system considered. This is the case for the parallel configuration, but not for the serial one. We propose to generalize the filling factor method to the case of systems with two delays in the serial configuration, by working in the space $(x(t_{ext}), x(t_{ext}-\tau_1), x(t_{ext}-\tau_2))$. The results obtained with this modified method are shown in Figure 19, where the two delays are clearly identified.

We have shown (Locquet et al., 2006) that the correct value of the delay times can be also extracted in the serial configuration by using the method based on the forecasting error. Two models, based on local linear fits and neural networks, have been applied to obtain the forecasting error under variation of the two delay times. The minimum in the error leads to the correct value of the two delay times. The main drawback of all these methods that consider different values for both delay times is the high computational cost. Therefore, we conclude that time-delay identification is more difficult in the serial configuration.

Nonlinear Dynamics Reconstruction with Modular Neural Networks

In this Section we apply modular neural networks to reconstruct the nonlinear dynamics from numerical simulations of feedback systems with two delays (Ortín & Pesquera, 2006; Ortín et al., 2007a). We first use the same MNN structure as in the single-delay case. Hence, the MNN has two modules, one for the non-feedback part with input data delayed by $t_e=0.01$, $\boldsymbol{x}_{nf}=(x(t-t_e),..,x(t-m_1t_e))$, and a second one for the feedback part. The feedback module has an input vector with data delayed by both feedback times, $\boldsymbol{x}_f=(x(t-\tau_1+m_2t_e),..,x(t-\tau_1),..,x(t-\tau_1-m_2t_e),x(t-\tau_2+m_2t_e),..,x(t-\tau_2),..,x(t-\tau_2-m_2t_e))$. The numbers of non-feedback and feedback inputs are $m_1=1$ and $2(2m_2+1)=6$, respectively. A single hidden layer with one neuron is used for the non-feedback module. Two layers with a and b neurons, a:b, are used for the feedback

Figure 19. Filling factor obtained from numerical simulation of Eq. (7), serial configuration, with delay times $\tau_1 = 100$ and $\tau_2 = 215$.

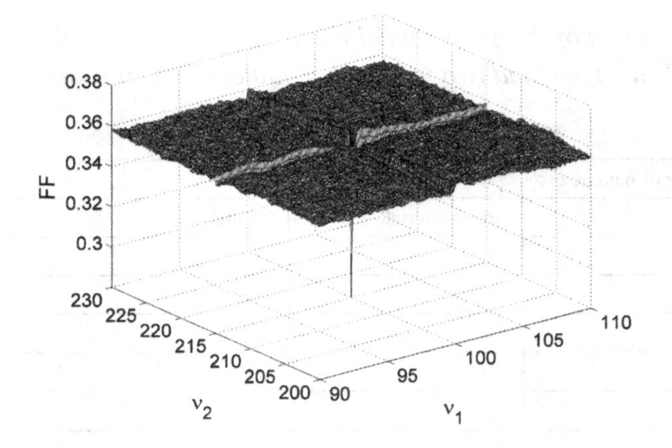

module. We take 3000 training points chosen to cover uniformly the chaotic attractor. The resulting MNN model was tested with 100000 data. In Table 1 we show the resulting normalized RMSE for both serial and parallel configurations with $\beta_1 = \beta_2 = \beta = 15$ and delay times $\tau_1 = 100$ and $\tau_2 = 215$. These results are compared with the results obtained for the single-delay case with a feedback strength of 30 and $\tau = 215$.

It is found that similar test errors are obtained for the two-delays serial configuration and the single-delay case. Therefore, the difficulty to extract the nonlinear dynamics is very similar in both cases. The only difference is that the number of parameters of the model is greater in the two-delay case due to the inclusion of the data delayed by the second feedback time in the input vector of the feedback module (see Table 2). However, the error obtained with one feedback module for the parallel configuration is two orders of magnitude higher than for the serial configuration and the single-delay cases. We have checked that no better results are obtained for the parallel configuration when the number of training points or the complexity of the neural network is increased (see Table 1). Adapting the MNN model to the characteristics of the parallel configuration by including a second feedback module can reduce this error. The input data for each feedback module, $x_{fi} = (x(t-\tau_i + m_2 t_e),.., x(t-\tau_i),.., x(t-\tau_i - m_2 t_e))$, are only delayed by the corresponding delay time τ_i. In this model the output of the MNN is given by $x_{nn}(t+t_e) = f_{nn}(x_{nf}) + g_{nn,1}(x_{f1}) + g_{nn,2}(x_{f2})$, where f_{nn}, $g_{nn,1}$ and $g_{nn,2}$ correspond to the non-feedback, first and second feedback modules, respectively. The results presented in Table 1 show that the error is clearly reduced for the parallel configuration when this MNN model with two feedback modules is used. However, the number of parameters of this model increases due to the inclusion of the second feedback module (see Table 2).

Table 1. Test error of the MNN model for the single-delay system with a feedback strength of 30 and $\tau = 215$, two-delay system in the serial and in the parallel (with one and two feedback modules) configurations with $\beta_1 = \beta_2 = \beta = 15$, $\tau_1 = 100$ and $\tau_2 = 215$. In the first column (a:b) denotes a model with hidden layers with a and b neurons for the feedback module.

	MNN RMSE			
NN feedback	1 delay	2 delays serial	Parallel 1 module	Parallel 2 modules
6: 3	2.33 10^{-3}	2 10^{-3}	0.015	1.7 10^{-3}
8: 4	6.62 10^{-5}	3.63 10^{-5}	9.7 10^{-3}	4.11 10^{-5}
10: 5	6.12 10^{-5}	3.57 10^{-5}	4.7 10^{-3}	2.53 10^{-5}
12: 6	4.03 10^{-5}	3.21 10^{-5}	2.4 10^{-3}	2.39 10^{-5}
14: 7	4.03 10^{-5}	1.74 10^{-5}	1 10^{-3}	1.62 10^{-5}

Table 2. Number of parameters of the MNN model for the single-delay system, two-delay system in the serial and in the parallel (with one and two feedback modules) configurations. The parameters are the same as in Table 1.

	Number of parameters			
NN feedback	1 delay	2 delays serial	Parallel 1 module	Parallel 2 modules
6: 3	51	69	69	100
8: 4	75	99	99	148
10: 5	103	133	133	204
12: 6	135	171	171	268
14: 7	187	229	229	372

The nonlinear functions extracted by the MNN model are plotted in Figures 20 and 21 for the serial and parallel configurations, respectively. MNN model with two feedback modules has been used for the parallel configuration. Each feedback module has *14:7* neurons. Good agreement is achieved between the original and recovered functions for both configurations. Note that the original nonlinear functions depend on both delayed variables, whereas in the parallel configuration each feedback module is a function of only one delayed variable. Therefore, the functions extracted by the MNN model are sharper than the original ones in Figure 21.

We have validated the MNN models by using the synchronization test. Chaotic synchronization between the data obtained from numerical simulations of Eqs. (6) and (7) and the MNN models is achieved. The synchronization error is of the same order that the model error.

Finally, we investigate the robustness of the nonlinear dynamics extraction to noise. The same approach as in the single-delay case is followed. We add a zero-mean Gaussian noise to the system variable $x(t)$. The structure of the MNN model is the same as in the free-noise case, but with ten inputs for the non-feedback module (m_1= 10), and ten inputs (m_2= 2) for the feedback modules. The model with two

Figure 20. Serial configuration. Left: nonlinear feedback function of Eq. (7). Right: feedback module function of the MNN model. The feedback module has 14:7 neurons.

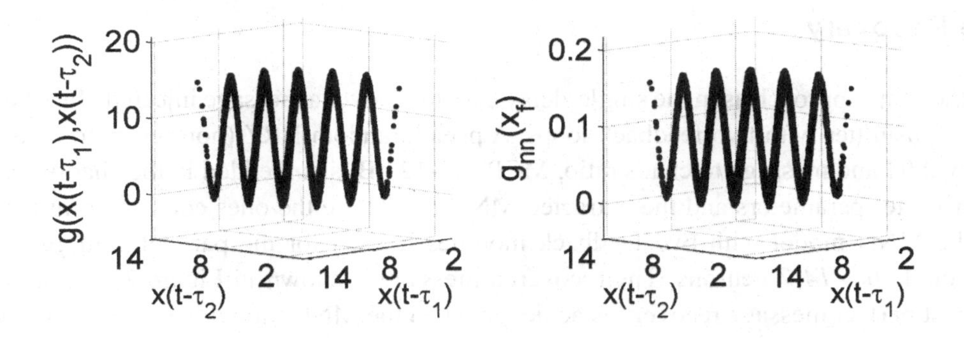

Figure 21. Parallel configuration. Left: nonlinear feedback function of Eq. (6). Right: feedback module functions of the MNN model. Each feedback module has 14:7 neurons.

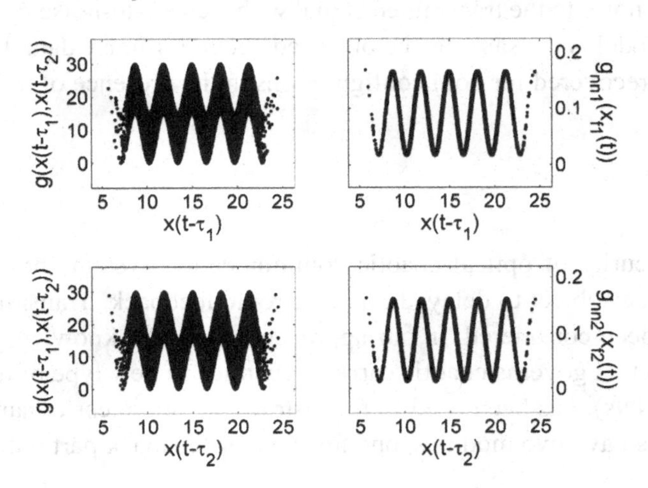

Table 3. Test error of the MNN model for the two-delay system in the parallel (with two feedback modules) and in the serial configurations with $\beta_1 = \beta_2 = \beta = 15$, $\tau_1 = 100$ and $\tau_2 = 215$ for different values of the chaos-to-noise ratio (CNR). Each feedback module has 14:7 neurons.

	MNN RMSE	
CNR (dB)	2 delays parallel	2 delays serial
40	$1.6\ 10^{-3}$	$1.9\ 10^{-3}$
30	$4.83\ 10^{-3}$	$9.1\ 10^{-3}$
25	0.011	0.015
20	0.023	0.031

feedback modules has been used for the parallel configuration. Each feedback module has *14:7* neurons. The normalized RMSE is obtained by comparing the output of the MNN, $x_{nn}(t)$, with the free-noise data, $x(t)$, as in the one delay case. The normalized test errors are presented in Table 3 for different noise levels. The error increases with the noise level for the two configurations, but noise affects more to the serial configuration. It is found that the error is always smaller than the noise level. Therefore, we conclude that the MNN model filters to some extent the noise.

Message Recovery

We follow the same approach as in the single-delay case to recover a message injected after the low-pass filter in the transmitter with two feedback loops. A pseudorandom NRZ (non return to zero) message of frequency 0.04 and message to chaos ratio, MCR, of -12 dB is embedded in the chaotic carrier. The chaotic transmitter parameters and the extracted MNN model are the ones considered in the previous analysis. The MNN model with two feedback modules is used for the parallel configuration. Each feedback module has *14:7* neurons. The recovered message is shown in Figure 22. In the noise-free case an almost perfect message recovery is achieved when the MNN model is trained from the chaotic carrier without message. In this case the model error is around 10^{-5}. When the transmitted signal, which includes the message, is used to train the MNN the error increases to around $9\text{x}10^{-4}$. Figure 22 shows that the message can be also recovered when the MNN model is extracted from the transmitted signal. Finally we have added a noise to the transmitted signal with a chaos-to-noise ratio, CNR, of 30 dB. The structure of the MNN model is the same as the one used above for noisy data. It is shown in Figure 22 that the message is also recovered for both configurations in the presence of noise.

CONCLUSION

We have studied the security of optical chaotic communication systems based on a chaotic carrier generated by a laser diode subject to delayed optoelectronic feedback. Transmitters with one and two fixed delay times have been considered. In our approach no a-priori knowledge is assumed about the structure of the equation that governs chaotic carrier dynamics. A new type of neural networks, modular neural networks (MNNs), has been used to reconstruct the nonlinear dynamics of the transmitter. Modular neural networks have two modules, one for the non-feedback part with input data delayed by

Figure 22. Top (bottom): recovered message for the serial (parallel) configuration. Right (middle): recovered message without noise when the MNN model is extracted from the time series without (with) message. Left: recovered message with noise, CNR =30 dB, when the MNN model is extracted from the time series without message.

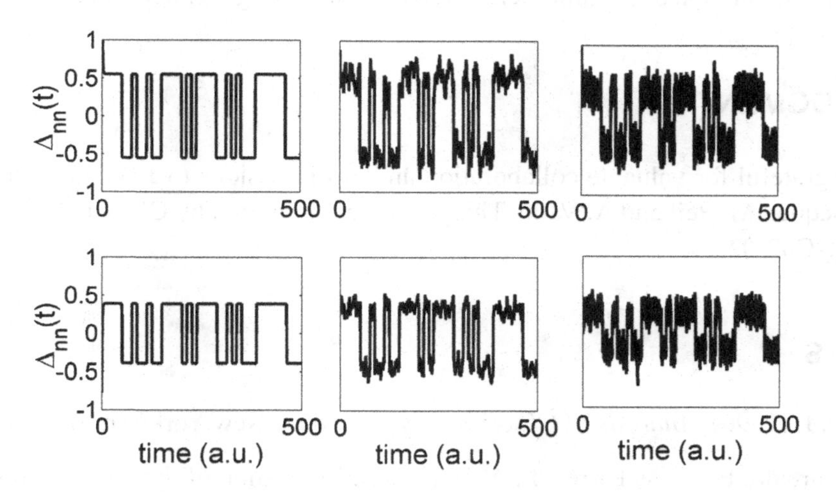

the embedding time, and a second one for the feedback part with input data delayed by the delay time. Nonlinear dynamics have been reconstructed from experimental time series in the single-delay case, and from numerical simulations in the one and two-delay cases. Two different configurations, serial and parallel, have been considered in the two-delay case. In the parallel configuration adapted MNNs with two feedback modules, each for one of the delay times, are required to reduce the model error. Therefore a more complex model with more parameters is needed to extract the nonlinear dynamics in the parallel configuration. We have shown that the extracted MNN model can be used in all considered cases as an unauthorized receiver to recover the message.

It is found that the complexity of the neural network required to achieve a low model error does not increase when the delay time is increased, in spite of the very high dimension of the chaotic attractor. Nonlinear dynamics of chaotic attractors with dimensions greater than 200 have been reconstructed from both experimental and numerical time series. However, it is found that nonlinear dynamics extraction is more difficult when the feedback strength is increased. These results are in agreement with the fact that the entropy increases with the feedback strength, but not with the delay time.

It is crucial to identify the delay in order to reconstruct the nonlinear dynamics of time-delay systems. Several methods have been used to extract the delay in the single-delay case. In the two-delay case with parallel configuration the delay times can be extracted with the methods used in the single-delay case. However, adapted methods with high computational cost are necessary to identify the delay times in the serial configuration.

The main conclusion of this chapter is that optical chaotic cryptosystems based on optoelectronic feedback systems with several fixed time delays are vulnerable. It has been proposed to use a variable time delay to avoid its identification by an eavesdropper, and then to enhance the security (Kye et al., 2004a). However, we have recently shown that periodic time delay functions can be extracted from experimental time series (Ortín et al., 2008). Therefore, it can be expected that the retrieval of the time delay function will allow nonlinear dynamics reconstruction by working in a low-dimensional projection of the phase-space. These optoelectronic feedback systems with periodic time delay are then vulnerable. A possible way to enhance the security of these systems is to use a chaotic time delay (Kye et al., 2004b), obtained as a function of the time-delay system variable or generated by another chaotic system.

ACKNOWLEDGMENT

The authors are grateful for valuable collaboration and/or discussions to J. M. Gutiérrez, M. Jacquot, L. Larger, A. Locquet, M. Peil and A. Valle. This work was supported by CICYT (Spain) under Project TEC2009-14581-C02-02.

REFERENCES

Abarbanel, H. D. I. (1996). *Analysis of Observed Chaotic Data*. New York: Springer-Verlag.

Aguirre, L. A., Furtado, E. C., & Tôrres, L. A. B. (2006). Evaluation of dynamical models: dissipative synchronization and other techniques. *Physical Review E: Statistical, Nonlinear, and Soft Matter Physics*, *74*, 019612-1–16. doi:10.1103/PhysRevE.74.066203

Alvarez, G., & Li, S. (2006). Some Basic Cryptographic Requirements for Chaos-Based Cryptosystems. *International Journal of Bifurcation and Chaos in Applied Sciences and Engineering*, *16*, 2129–2151. doi:10.1142/S0218127406015970

Argyris, A., Syvridis, D., Larger, L., Annovazzi-Lodi, V., Colet, P., & Fischer, I. (2005). Chaos-based communications at high bit rates using commercial fibre-optic links. *Nature*, *438*, 343–346. doi:10.1038/nature04275

Ashwin, P. (2003). Synchronization from chaos. *Nature*, *422*, 384–385. doi:10.1038/422384a

Bavard, X., Locquet, A., Larger, L., & Goedgebuer, J. P. (2007). Influence of digitisation on master–slave synchronisation in chaos-encrypted data transmission. *IET Optoelectron.*, *1*, 3–8. doi:10.1049/iet-opt:20060022

Brown, R., Rulkov, N. F., & Tracy, E. R. (1994). Modeling and synchronizing chaotic systems from time-series data. *Physical Review E: Statistical Physics, Plasmas, Fluids, and Related Interdisciplinary Topics*, *49*, 3784–3800. doi:10.1103/PhysRevE.49.3784

Bünner, M. J., Ciofini, M., Giaquinta, A., Hegger, R., Kantz, H., Meucci, R., & Politi, A. (2000a). Reconstruction of systems with delayed feedback: Theory. *The European Physical Journal D*, *10*, 165–176. doi:10.1007/s100530050538

Bünner, M. J., Ciofini, M., Giaquinta, A., Hegger, R., Kantz, H., Meucci, R., & Politi, A. (2000b). Reconstruction of systems with delayed feedback: Application. *The European Physical Journal D*, *10*, 177–187. doi:10.1007/s100530050539

Bünner, M. J., Meyer, T., Kittel, A., & Parisi, J. (1997). Recovery of the time-evolution of time-delay systems from time series. *Physical Review E: Statistical Physics, Plasmas, Fluids, and Related Interdisciplinary Topics*, *56*, 5083–5089. doi:10.1103/PhysRevE.56.5083

Colet, P., & Roy, R. (1994). Digital communication with synchronized chaotic lasers. *Optics Letters*, *19*, 2056–2058. doi:10.1364/OL.19.002056

Cuomo, K. M., & Oppenheim, A. V. (1993a). Circuit implementation of synchronized chaos with applications to communications. *IEEE Journal of Quantum Electronics*, *38*, 65–68.

Cuomo, K. M., Oppenheim, A. V., & Strogatz, S. H. (1993b). Synchronization of Lorenz-based chaotic circuits with applications to communications. *IEEE Trans. Circuits Systems II*, 40, 626-633.

Donati, S., & Mirasso, C. R. (Eds.). (2002). Feature Section on Optical Chaos and Applications to Cryptography. *IEEE J. of Quantum Electron.*, 38.

Farmer, J. (1982). Chaotic attractors of an infinte dimenisonal system. *Physica D. Nonlinear Phenomena*, *4*, 366–393. doi:10.1016/0167-2789(82)90042-2

Geddes, J. B., Short, K. M., & Black, K. (1999). Extraction of signals from chaotic laser data . *Physical Review Letters*, *83*, 5389–5392. doi:10.1103/PhysRevLett.83.5389

Goedgebuer, J. P., Larger, L., & Porter, H. (1998a). Optical cryptosystem based on synchronization of hyperchaos generated by a delayed feedback tunable laser diode. *Physical Review Letters*, *80*, 2249–2252. doi:10.1103/PhysRevLett.80.2249

Goedgebuer, J. P., Larger, L., Porter, H., & Delorme, F. (1998b). Chaos in wavelength with a feedback tunable laser diode. *Physical Review E: Statistical Physics, Plasmas, Fluids, and Related Interdisciplinary Topics*, *57*, 2795–2798. doi:10.1103/PhysRevE.57.2795

Hegger, R., Bünner, M. J., Kantz, H., & Giaquinta, A. (1998). Identifying and modeling delay feedback systems. *Physical Review Letters*, *81*, 558–561. doi:10.1103/PhysRevLett.81.558

Ikeda, K., Kondo, K., & Akimoto, O. (1982). Successive higher-harmonic bifurcations in systems with delayed feedback. *Physical Review Letters*, *49*, 1467–1470. doi:10.1103/PhysRevLett.49.1467

Kocarev, L., Halle, K. S., Eckert, K., Chua, L. O., & Parlitz, U. (1992). Experimental demonstration of secure communications via chaotic synchronization. *International Journal of Bifurcation and Chaos in Applied Sciences and Engineering*, *2*, 709–713. doi:10.1142/S0218127492000823

Kye, W. H., Choi, M., Kim, M. W., Lee, S. Y., Rim, S., Kim, C. M., & Park, Y. J. (2004a). Synchronization of delayed systems in the presence of delay time modulation. *Physics Letters. [Part A]*, *322*, 338–343. doi:10.1016/j.physleta.2004.01.046

Kye, W. H., Choi, M., Rim, S., Kurdoglyan, M. S., Kim, C. M., & Park, Y. J. (2004b). Characteristics of a delayed system with time-dependent delay time. *Phys. Rev. E*, 69, 055202(R)1-4.

Larger, L., Goedgebuer, J. P., & Delorme, F. (1998a). Optical encryption system using hyperchaos generated by an optoelectronic wavelength oscillator. *Physical Review E: Statistical Physics, Plasmas, Fluids, and Related Interdisciplinary Topics*, *57*, 6618–6624. doi:10.1103/PhysRevE.57.6618

Larger, L., Goedgebuer, J. P., & Merolla, J. M. (1998b). Chaotic Oscillator in Wavelength: A New Setup for Investigating Differential Difference Equations Describing Nonlinear Dynamics. *IEEE Journal of Quantum Electronics*, *34*, 594–601. doi:10.1109/3.663432

Lee, M. W., Larger, L., Udaltsov, V., Génin, E., & Goedgebuer, J. P. (2004). Demonstration of chaos generator with two delays. *Optics Letters*, *29*, 325–327. doi:10.1364/OL.29.000325

Lee, M. W., Rees, P., Shore, K. A., Ortín, S., Pesquera, L., & Valle, A. (2005). Dynamical characterization of laser diode subject to double optical feedback for chaotic optical communications. *IEE Proceedings. Optoelectronics*, *152*, 97–102. doi:10.1049/ip-opt:20045025

Locquet, A., Ortín, S., Udaltsov, V., Larger, L., Citrin, D. S., Pesquera, L., & Valle, A. (2006). Delay-time identification in chaotic optical systems with two delays. In D. Lenstra, M. Pessa, I. H. White (Eds.) *SPIE Proceedings: Vol. 6184. Semiconductor Lasers and Laser Dynamics II.* (pp. 173-184). Bellingham, WA: SPIE.

Ortín, S., Gutierrez, J. M., Pesquera, L., & Vasquez, H. (2005). Nonlinear dynamics extraction using modular neural network: synchronization and prediction. *Physica A*, *351*, 133–141. doi:10.1016/j.physa.2004.12.015

Ortín, S., Jacquot, M., Pesquera, L., Peil, M., & Larger, L. (2007a). Nonlinear dynamics reconstruction of chaotic cryptosystems based on delayed optoelectronic feedback. In CLEO-E/IQEC (Ed.), *Proceedings CLEO-E/IQEC*(JSI1-3-THU). Munich, Germany: CLEO-E/IQEC.

Ortín, S., Jacquot, M., Pesquera, L., Peil, M., & Larger, L. (2007b). Security analysis of chaotic communication systems based on delayed optoelectronic feedback . In Zubia, J., Aldabaldetreku, G., Durana, G., Arrue, J., & Jiménez, F. (Eds.), *Proceedings 5ª Reunión Española de Optoelectrónica* (pp. 429–434). Bilbao, Spain: Escuela Técnica Superior de Ingeniería.

Ortín, S., Jacquot, M., Pesquera, L., Peil, M., & Larger, L. (2008). Time delay extraction in chaotic cryptosystems based on optoelectronic feedback with variable delay. In K. P. Panajotov, M. Sciamanna, A. A. Valle, R. Michalzik. (Eds.) *SPIE Proceedings: Vol. 6997. Semiconductor Lasers and Laser Dynamics III.* (pp. 69970E-1-12). Bellingham, WA: SPIE.

Ortín, S., & Pesquera, L. (2006). Cracking Chaos-based Encryption Systems Based on Laser Diodes with Optoelectronic Feedback with Two Delays. In D. Lenstra, J. M. Dudley, J. Rarity, C. R. Mirasso (Eds.), *Proceedings ECOC: Vol. 6. CLEO Focus meeting* (Tu3_1_2, pp. 15-16). Cannes, France: European Conference on Optical Communications.

Ortín, S., Pesquera, L., Gutiérrez, J. M., Valle, A., & Cofiño, A. (2007c). Nonlinear dynamics reconstruction with neural networks of chaotic communication time-delay systems. In J. Marro, P. L. Garrido, J. J. Torres (Eds.), *AIP Proceedings: Vol. 887. Cooperative Behavior in Neural Systems.* (pp. 249-258). New York: American Institute of Physics.

Perez, G., & Cerdeira, H. (1995). Extracting messages masked by chaos. *Physical Review Letters*, *74*, 1970–1073. doi:10.1103/PhysRevLett.74.1970

Ponomarenko, V. I., & Prokhorov, M. D. (2002). Extracting information masked by the chaotic signal of a time-delay system. *Physical Review E: Statistical, Nonlinear, and Soft Matter Physics*, *66*, 026215-1–7. doi:10.1103/PhysRevE.66.026215

Press, W. H., Flannery, B. P., Teukolsky, S. A., & Vetterling, W. T. (1992). *Numerical Recipes*. Cambridge, UK: Cambridge University Press.

Principe, J. C., Rathie, A., & Kuo, J. M. (1992). Prediction of chaotic time series with neural networks and the issue of dynamic modelling. *International Journal of Bifurcation and Chaos in Applied Sciences and Engineering*, *2*, 989–996. doi:10.1142/S0218127492000598

Robilliard, C., Huntington, E. H., & Webb, J. G. (2006). Enhancing the Security of Delayed Differential Chaotic Systems With Programmable Feedback. *IEEE Transactions on Circuits and Wystems. II, Express Briefs*, *53*, 722–726. doi:10.1109/TCSII.2006.876405

Rulkov, N. F., Sushckik, M. M., Tsimring, L. S., & Abarbanel, H. D. I. (1995). Generalized syncronization of chaos in directionally coupled chaotic systems. *Physical Review E: Statistical Physics, Plasmas, Fluids, and Related Interdisciplinary Topics*, *51*, 980–994. doi:10.1103/PhysRevE.51.980

Short, K. M. (1994). Steps toward unmasking secure communications. *International Journal of Bifurcation and Chaos in Applied Sciences and Engineering*, *4*, 959–977. doi:10.1142/S021812749400068X

Short, K. M. (1996). Unmasking a modulated chaotic communications scheme. *International Journal of Bifurcation and Chaos in Applied Sciences and Engineering*, *6*, 367–375. doi:10.1142/S0218127496000114

Short, K. M., & Parker, A. T. (1998). Unmasking a hyperchaotic communication scheme. *Physical Review E: Statistical Physics, Plasmas, Fluids, and Related Interdisciplinary Topics*, *58*, 1159–1162. doi:10.1103/PhysRevE.58.1159

Tenny, R., & Tsimring, L. S. (2004). Steps towards cryptanalysis of chaotic active/passive decomposition encryption schemes using average dynamics estimation. *International Journal of Bifurcation and Chaos in Applied Sciences and Engineering*, *14*, 3949–3968. doi:10.1142/S0218127404011727

Uchida, A., Rogister, F., García-Ojalvo, J., & Roy, R. (2005). Synchronization and communication with chaotic laser systems . In Wolf, E. (Ed.), *Progress in Optics* (*Vol. 48*, pp. 203–341). Amsterdam, The Netherlands: Elsevier B. V.

Udaltsov, V. S., Goedgebuer, J. P., Larger, L., Cuenot, J. B., Levy, P., & Rhodes, W. T. (2003). Cracking chaos-based encryption systems ruled by nonlinear delay differential equations. *Physics Letters. [Part A]*, *308*, 54–60. doi:10.1016/S0375-9601(02)01776-0

Vicente, R., Daudén, J., Colet, P., & Toral, R. (2005). Analysis and Characterization of the Hyperchaos Generated by a Semiconductor Laser Subject to a Delayed Feedback Loop. *IEEE Journal of Quantum Electronics*, *41*, 541–548. doi:10.1109/JQE.2005.843606

Yang, T., Yang, L. B., & Yang, C. M. (1998a). Breaking chaotic secure communication using a spectrogram. *Physics Letters. [Part A]*, *247*, 105–111. doi:10.1016/S0375-9601(98)00560-X

Yang, T., Yang, L. B., & Yang, C. M. (1998b). Breaking chaotic switching using generalized synchronization: Examples. *IEEE Trans. Circuits and Systems I*, 45, 1062-1067.

Zhou, C., & Lai, C. (1999). Extracting messages masked by chaotic signals of time-delay systems. *Physical Review E: Statistical Physics, Plasmas, Fluids, and Related Interdisciplinary Topics*, *60*, 320–323. doi:10.1103/PhysRevE.60.320

KEY TERMS AND DEFINITIONS

Chaotic Communication Systems: Communication systems that use a chaotic carrier.

Synchronization: Phenomenon by which two coupled systems starting from different initial conditions follow the same trajectory after a transient.

Time-Delay Systems: Dynamical systems whose evolution depends of a delayed value of the system variable.

Feedback: Describes the situation when output from an event or phenomenon in the past will influence the same event/phenomenon in the future.

Artificial Neural Network: A model based on biological neural networks. It consists of an interconnected group of artificial neurons and processes information using a connectionist approach to computation. The network behaviour is stored in the connections between neurons in values called weights, which represent the strength of each link. In most cases an artificial neural network is an adaptive system that changes its structure based on external or internal information that flows through the network during the learning phase.

Chapter 18
Encryption of Analog and Digital Signals Through Synchronized Chaotic System

Kehui Sun
Central South University, China

ABSTRACT

Chaos is characterized by aperiodic, wideband, random-like, and ergodicity. Chaotic secure communication has become one of the hot topics in nonlinear dynamics since the early 1990s exploiting the technique of chaos synchronization. As distinguished by the type of information being carried, chaos-based communication systems can be categorized into analogy and digital, including four popular techniques such as Chaos Masking, Chaos Shift Keying, Chaos Modulation, and Chaos Spreading Spectrum. In this chapter, the principles of these schemes and their modifications are analyzed by theoretical analysis as well as dynamic simulation. In addition, chaos-based cryptography is a new approach to encrypt information. After analyzing the performances of chaotic sequence and designing an effective chaotic sequence generator, the authors briefly presented the principle of two classes of chaotic encryption schemes, chaotic sequence encryption and chaotic data stream encryption.

1 INTRODUCTION

Chaos is an aperiodic long-term behavior in a deterministic system that exhibits sensitively dependent on initial conditions. There are three keywords in this definition, *i.e.* "aperiodic long-term behavior", "deterministic" and "sensitive dependence on initial conditions". "Aperiodic long-term behavior" means that the system's trajectory in phase space does not settle down to any fixed points, periodic orbits, or quasi-periodic solutions as time tends to infinity. "Deterministic" means the irregular behavior of chaotic systems arises from intrinsic nonlinearities rather than noise. "Sensitive dependence on initial conditions" means that trajectories originating from very nearly identical initial conditions will diverge exponentially quickly.

DOI: 10.4018/978-1-61520-737-4.ch018

Actually, two important features, sensitivity and randomness, give the way to application of secure communication. For a chaotic system, on one hand, the state space trajectories originating from two closely spaced initial conditions diverge exponentially. So its behaviors are unpredictable. It is useful for effective bits confusion and diffusion. Confusion and Diffusion are two principles in cryptology. "Confusion" is standing for substitution operations, and "diffusion" standing for transposition or permutation operations. These two principles are still actively used in modern ciphers. On the other hand, chaos is a kind of stochastic signals produced by deterministic model, so it has internal random, and has no relationship with external factors. It is useful for producing output signals with satisfactory statistics. This means the probability of symbols "1" and "0" for a chaotic sequence is equal.

As a kind of signal, chaos signals have some especial characters. For example, chaos signals are aperiodic, noise-like in time domain, and chaos signals are continuous and wideband in frequency domain. Chaos signals can be easily generated by deterministic nonlinear dynamic systems (ODE or iterative map). For chaos signals, its autocorrelation is impulse-like, delta function shown in Figure 1(a), and its cross-correlation is zero (low) shown in Figure 1(b).

Communication systems are designed primarily for conveying useful information from one place to another. Typical communication systems including amplitude modulated (AM), frequency modulated (FM), phase modulated (PM), cellular mobile communication and wireless local area networks. The simplest communication consists of only four components, namely an information source, a modulator, a channel and a demodulator, as represented by the function block diagram as shown in Figure 2 (Stavroulakis, 2005).

Since the late 1980s, chaos-based cryptography has attracted more and more attention from researchers in many different areas. It has been found that chaotic systems and cryptosystems share many similar properties. For instance, chaotic systems are sensitive to the initial conditions, which corresponds to the diffusion property of good cryptosystems (for a comparison of chaos and cryptography, see Table 1 (Alvarez & Li, 2006).

Figure 1. Normalized correlation functions of chaotic signals against normalized time delay

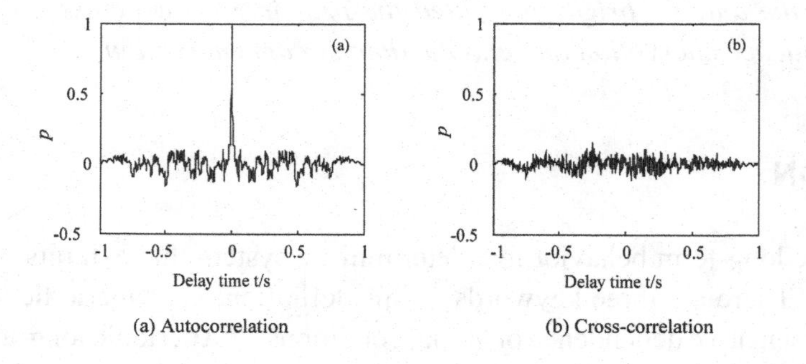

Figure 2. Block diagram of a simple communication system

Table 1. Comparison between chaos properties and cryptography properties

Chaotic properties	Cryptographic properties	Description
Ergodicity	Confusion	The output has the same distribution for any input.
Sensitivity to initial conditions / control parameter	Diffusion with a small change in the plaintext /secret key	A small deviation in the input can cause a large change at the output.
Mixing property	Diffusion with a small change in one plain-block of the whole plaintext	A small deviation in the local area can cause a large change in the whole space.
Deterministic dynamics	Deterministic pseudo-randomness	A deterministic process can cause a random-like (pseudo-random) behavior.
Structure complexity	Algorithm (attack) complexity	A simple process has a very high complexity.

As with some traditional technologies, chaotic communication is also reliant on synchronization. Many synchronization approaches have been proposed since Pecora and Carroll reported the first synchronization scheme in 1990 (Pecora & Carroll, 1990), such as master-slaver synchronization (Kocarev & Parlitz, 1995) co-coupling synchronization (Roy & Scott, 1994; Sugawara, 1994), continuous variable feedback synchronization (Pyragas & Tamasevicius, 1993), adaptive synchronization (John, 1994), impulse synchronization, and dead-bead synchronization (Angeli et al, 1995).

To illustrate the concept of synchronization and illustrate the Pecora-Carroll synchronization technique, the Lorenz system is applied in Figure 3. The drive system is given by Lorenz equations,

$$\begin{cases} \dot{x}_1 = \sigma(x_2 - x_1) \\ \dot{x}_2 = rx_1 - x_2 - x_1x_3 \\ \dot{x}_3 = x_1x_2 - bx_3 \end{cases} \quad (1)$$

where, $\sigma = 10$, $b = 8/3$, and $r = 28$. This system is decomposed so that the x_2 variable is coupled to the response system. The response system, therefore, is given by

$$\begin{cases} \dot{y}_1 = \sigma(x_2 - y_1) \\ \dot{y}_2 = ry_1 - x_2 - y_1y_3 , \\ \dot{y}_3 = y_1x_2 - by_3 \end{cases} \quad (2)$$

Figure 3. Drive-response synchronization (y_2 in response system is replaced by x_2 in drive system)

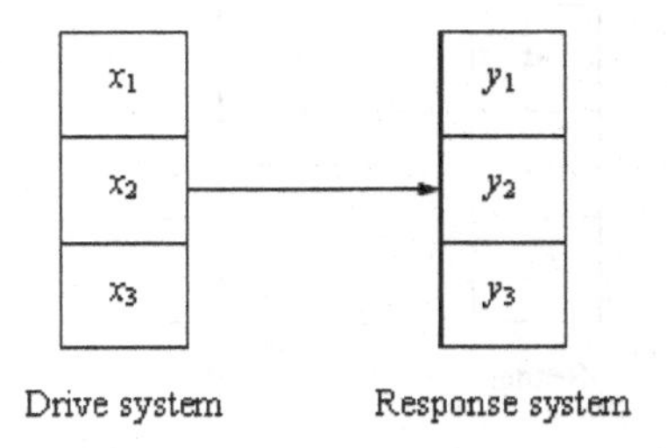

Figure 4. Synchronization phase figure of $x_1 - y_1$ with initial conditions of (15, 8, -10) and (5, -5, 20)

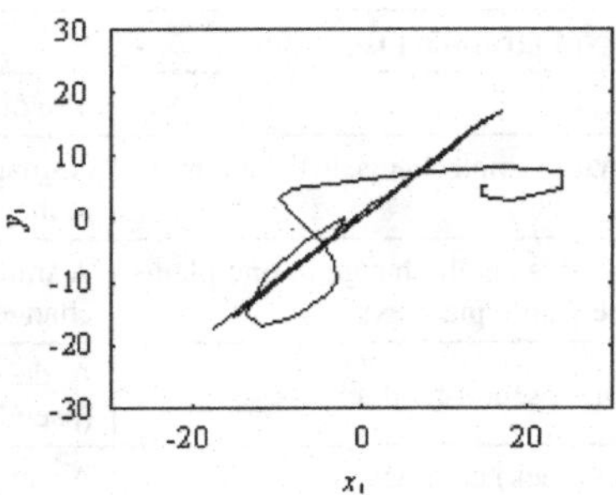

where the variable y_2 has been completely replaced by x_2. By simulation on Simulink, the synchronization phase figure on $x_1 - y_1$ plane is presented in Figure 4, which shows that two Lorenz systems with initial conditions of (15, 8, -10) and (5, -5, 20) synchronize quickly.

2 CHAOS-BASED ANALOG AND DIGITAL COMMUNICATION TECHNIQUES

2.1 Chaos Masking

In 1992, Chaotic masking was proposed by Kocarev *et al* (Kocarev et al, 1992), in which the analog message signal *m*(*t*) is added to the output of the chaos generator, *x*(*t*), within the transmitter. The principle of chaotic masking, using the Lorenz system and drive – response synchronization as its basis, is shown in Figure 5. As illustrated in the figure, a small-amplitude signal *m*(*t*) scaled by *k* is added to the much larger chaotic fluctuation of the x_1 variable and transmitted to the receiver. Because the amplitude of the signal is smaller than that of the chaotic signal, it is effectively hidden, or masked, by the chaotic signal during transmission. The other communication channel is used to transmit the x_2 variable to the receiver where drives the response system to synchronize the drive system. This synchronization enables the original signal to be recovered from the masking signal *s*(*t*) by simply subtracting y_1 from the masked transmission, $x_1 + km(t)$.

Figure 5. Two channels chaotic masking communication scheme based on drive-response synchronization

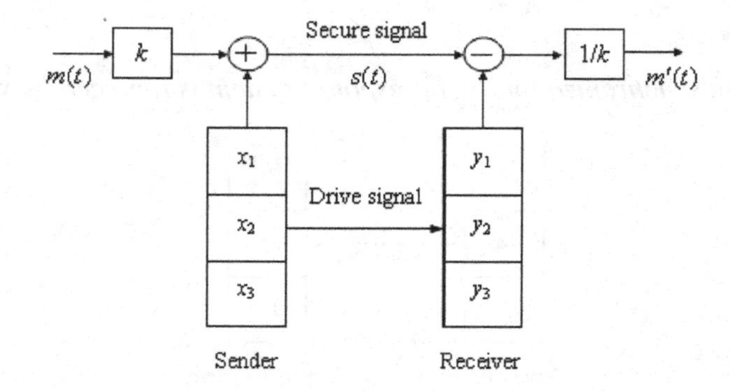

At the sender of the secure communication system, the modulator can be write as

$$\begin{cases} \dot{x}_1 = \sigma(x_2 - x_1) \\ \dot{x}_2 = -x_1 x_3 + r x_1 - x_2 \\ \dot{x}_3 = x_1 x_2 - b x_3 \\ s(t) = km(t) + x_1 \end{cases} \tag{3}$$

At the receiver, the demodulator are given by

$$\begin{cases} \dot{y}_1 = \sigma(x_2 - y_1) \\ \dot{y}_2 = -y_1 y_3 + r y_1 - y_2 \\ \dot{y}_3 = y_1 x_2 - b y_3 \\ m'(t) = (s(t) - y_1) \,/\, k \end{cases} \tag{4}$$

To save the channel resource, a modification scheme was presented as shown in Figure 6. In this scheme, the drive signal is the sum of information signal and chaotic signal, so only one channel is needed to realize secure communication. In this scheme, the equations at sender and receiver are as follows.

$$\begin{cases} \dot{x}_1 = \sigma(x_2 - x_1) \\ \dot{x}_2 = -x_1 x_3 + r x_1 - x_2 \\ \dot{x}_3 = x_1 x_2 - b x_3 \\ s(t) = km(t) + x_1 \end{cases} \tag{5}$$

$$\begin{cases} \dot{y}_1 = \sigma(s(t) - y_1) \\ \dot{y}_2 = -y_1 y_3 + r y_1 - y_2 \\ \dot{y}_3 = y_1 s(t) - b y_3 \\ m'(t) = (s(t) - y_1) \,/\, k \end{cases} \tag{6}$$

One feature of this chaos masking scheme is that the message signal disturbs the driving signal, therefore chaos synchronization cannot be achieved exactly and the message signal cannot be recovered exactly. Another obvious feature of this chaotic masking scheme is that the message signal does not influence the dynamics of the master system at all. The security of chaotic masking is questionable against various attacks, mainly due to the fact that an attacker can always obtain some information from the driving signal to construct (at least part of) the dynamics of the master system. To improve the security performance of this system, another modification scheme (Mianovic & Zaghloul, 1996) is presented as shown in Figure 7. In this scheme, the sum of information signal and chaotic signal is feedback to the drive system. The sender and receiver have the same dynamic behaviors, so it can be synchronized completely, and the received signal can be demodulated immediately. The information is melted into the

Figure 6. One channel chaotic masking communication scheme based on drive-response synchronization

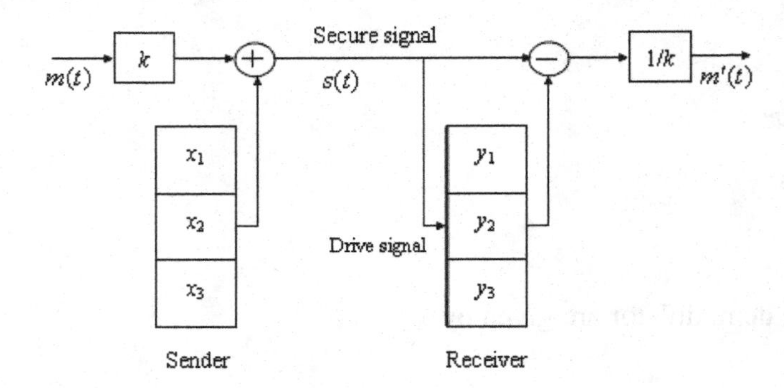

chaotic system, so it is more difficult to extract the message from the channel an attacker. The sender and receiver can be described respectively as

$$\begin{cases} \dot{x}_1 = \sigma(s(t) - x_1) \\ \dot{x}_2 = -x_1 x_3 + r x_1 - x_2 \\ \dot{x}_3 = x_1 s(t) - b x_3 \\ s(t) = k m(t) + x_2 \end{cases} \tag{7}$$

$$\begin{cases} \dot{y}_1 = \sigma(s(t) - y_1) \\ \dot{y}_2 = -y_1 y_3 + r y_1 - s(t) \\ \dot{y}_3 = y_1 s(t) - b y_3 \\ m'(t) = (s(t) - y_2) \,/\, k \end{cases} \tag{8}$$

Figure 7. One channel modified chaotic masking communication scheme based on drive-response synchronization

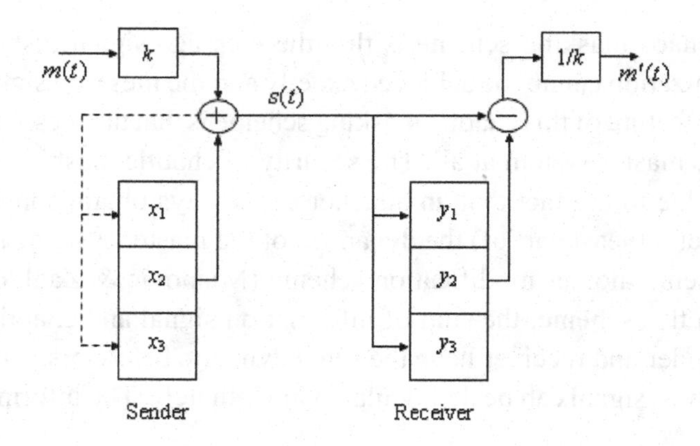

Figure 8. Series chaotic masking communication scheme based on drive-response synchronization

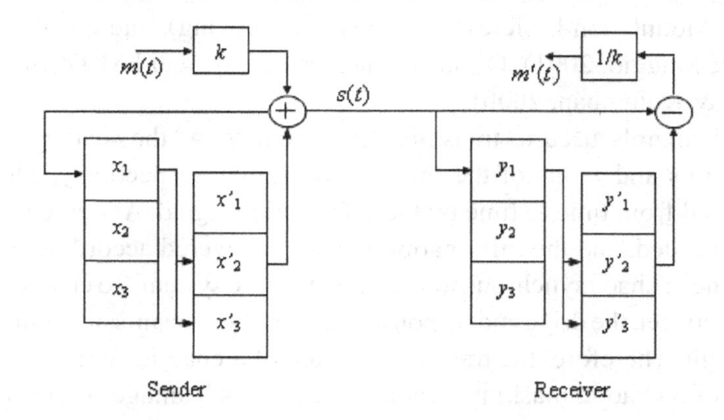

To improve secure performance further, Carroll *et al* (Carroll, 1996) extended their synchronization scheme to high order series synchronization to transmit information, and its principle block diagram is shown in Figure 8. In this scheme, there are two chaotic systems in both of the sender and receiver.

The sender and receiver can be described respectively as

$$\begin{cases} \dot{x}_1 = \sigma(s(t) - x_1) \\ \dot{x}_2 = -x_1x_3 + rx_1 - x_2 \\ \dot{x}_3 = x_1x_2 - bx_3 \\ \dot{x}'_1 = \sigma(x'_2 - x'_1) \\ \dot{x}'_2 = -x_1x'_3 + rx_1 - x'_2 \\ \dot{x}'_3 = x_1x'_2 - bx'_3 \\ s(t) = km(t) + x'_2 \end{cases} \tag{9}$$

$$\begin{cases} \dot{y}_1 = \sigma(s(t) - y_1) \\ \dot{y}_2 = -y_1y_3 + ry_1 - y_2 \\ \dot{y}_3 = y_1y_2 - by_3 \\ \dot{y}'_1 = \sigma(y'_2 - y'_1) \\ \dot{y}'_2 = -y_1y'_3 + ry_1 - y'_2 \\ \dot{y}'_3 = y_1y'_2 - by'_3 \\ m'(t) = (s(t) - y'_2) / k \end{cases} \tag{10}$$

2.2 Chaos Shift Keying

Chaos shift keying (CSK) was proposed by Parlitz *et al* (Parlitz, 1992) and Dedieu *et al* (Dedieu, 1993). Then, it was improved by Kolumban G (Kolumbán et al, 1996; Kolumbán et al, 1997). There are several schemes, including COOK (Chaotic On-Off Keying), DCSK (Differential Chaos Shift Keying), FM-DCSK (Frequency Modulation Differential Chaos Shift Keying), and QCSK (Quadrature Chaos-Shift Keying) (Galias & Maggio, 2001). Of these schemes, DCSK and FM-DCSK have the best noise performance (Kennedy & Kolumbán, 2000).

Chaos shift keying is mainly used to transmit digital signals. At the sender, two different chaotic systems are used for 0-bits and 1-bits of the information signal, respectively. That is, the employed chaotic system is switched from time to time by the information signal. At the receiver, only one of the two chaotic systems is needed, and the information bits are recovered according to whether or not the response system can achieve chaos synchronization with the drive system. To ensure the establishment of chaos synchronization between the drive and response systems, the transmission time of each information bit should be long enough. Therefore, the transmission rate of a chaotic switching system is generally much slower than that of a chaotic masking system. The main advantage of chaotic switching is that the information signal can exactly be recovered as long as the level of the signal-to-noise is not too low.

1) COOK

COOK is the simplest chaos shift keying scheme. Figure 9 is the basic structure of COOK. In this scheme, $\{b_i\}$ is the binary message signal, and it is a series of "1" and "-1". When b_i =1, the switch is on; When b_i =-1, the switch is off. The message is recovered by correlative algorithm. If symbols of the input signal are equal probability, and the average bit energy of the digital sequence is E_b, then the symbol energy is $2E_b$ and 0 for "1" and 1 for "-1" respectively. Obviously, the bigger the bit energy difference is, the better the performance against noise is. The main advantage of this scheme is that it needs only one chaos oscillator, but security of the simplest chaotic switching system is not high.

2) CSK

A scheme of chaos shift keying based on two chaos generators is presented here. At the sender, the modulator structure is shown in Figure 10.

Here, $\{b_i\}$ is the information signal, and it is a sequence of "1" and "-1". So the transmitted signal in channel is chaotic, like noise.

$$s_i(t) = \begin{cases} g_1(t) & b_i = 1 \\ g_2(t) & b_i = -1 \end{cases} \quad (11)$$

Figure 9. Block diagram of noncoherent COOK modulation / demodulation scheme

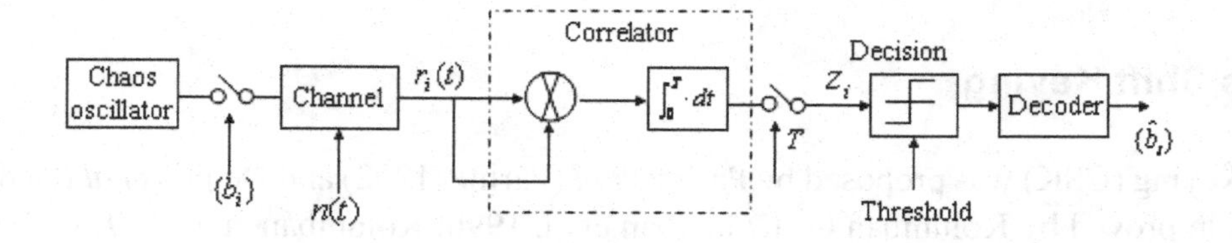

Figure 10. Block diagram of a CSK modulator

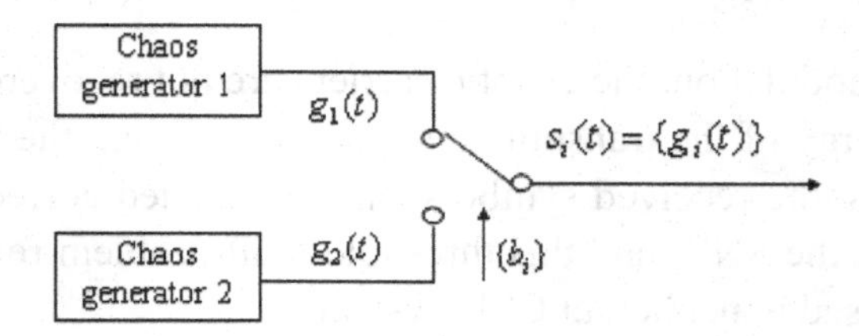

At the receiver, two coherent demodulators are employed to recover the information signal, and the coherent demodulation structure is shown in Figure 11, and the noncoherent demodulator structure is shown in Figure 12.

For CSK coherent demodulator, the output of the correlator is

$$z_{i1} = \int_{T_s}^{T} r_i(t)\hat{g}_1(t)dt = \int_{T_s}^{T} [g_i(t)+n(t)]\hat{g}_1(t)dt = \int_{T_s}^{T} g_i(t)\hat{g}_1(t)dt + \int_{T_s}^{T} n(t)\hat{g}_1(t)dt \tag{12}$$

$$z_{i2} = \int_{T_s}^{T} r_i(t)\hat{g}_2(t)dt = \int_{T_s}^{T} [g_i(t)+n(t)]\hat{g}_2(t)dt = \int_{T_s}^{T} g_i(t)\hat{g}_2(t)dt + \int_{T_s}^{T} n(t)\hat{g}_2(t)dt \tag{13}$$

If $b_i=1$, $s_i(t)=g_1(t)$, $r_i(t)=g_1(t)+n(t)$, then $z_{i1}>z_{i2}$, and we get $\hat{b}_i=1$. If $b_i=-1$, $s_i(t)=g_2(t)$, $r_i(t)=g_2(t)+n(t)$, then $z_{i1}<z_{i2}$, and we get $\hat{b}_i=-1$

For CSK noncoherent demodulator, the output of correlator is

Figure 11. Block diagram of a coherent CSK receiver

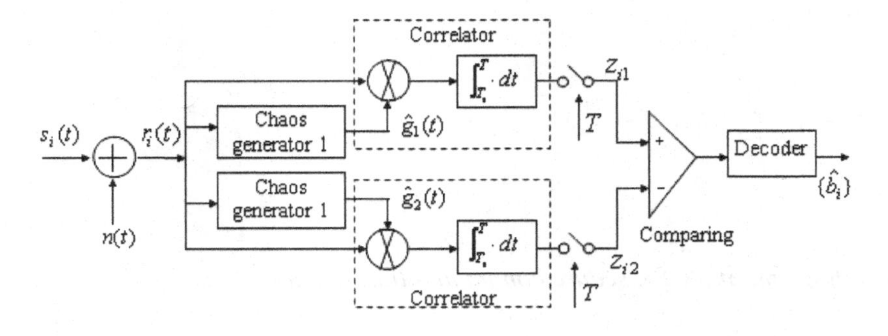

Figure 12. Block diagram of a non-coherent CSK receiver

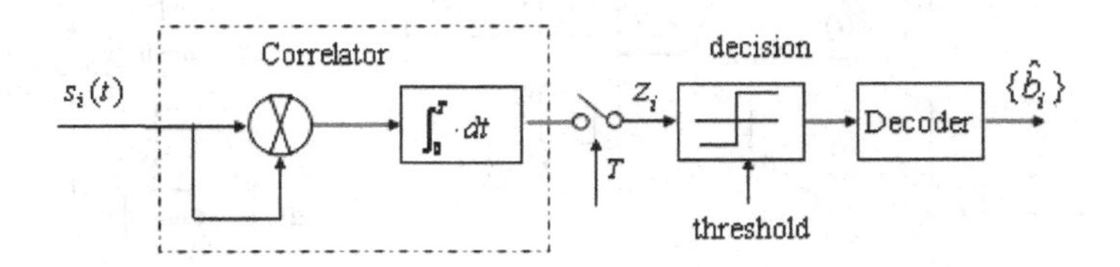

$$z_i = \int_0^T s_i^2(t)dt = \int_0^T [g_i(t) + n(t)]^2 dt = \int_0^T g_i^2(t)dt + 2\int_0^T g_i(t)n(t)dt + \int_0^T n^2(t)dt \tag{14}$$

In noncoherent CSK demodulation, the chaotic carriers are not recovered at the receiver. Detection is done based on the bit energy of the transmitted signals. By setting the threshold mid-way between the two average bit energies, the received symbols can be decoded correctly. Obviously, the optimal threshold is dependent upon the SNR, and this threshold-shift problem remains one of the drawbacks of this type of bit-energy-based noncoherent CSK system.

A CSK secure communication scheme based on chaotic synchronization and coherent demodulation is presented in Figure 13. Two unified chaotic systems (UCS1 and UCS2) with different parameters are employed. At the sender, different chaotic signal will be sent by different symbols, and the signal is recovered by coherent demodulation at the receiver.

At the sender, the transmitted signal $s(t)$ is

$$s(t) = x_1(t) \cdot b(t) + (1 - b(t)) \cdot x_2(t) = \begin{cases} x_1(t) & b(t) = 1 \\ x_2(t) & b(t) = -1 \end{cases} \tag{15}$$

At the receiver, the outputs of correlator demodulator are

$$z_1 = \int_T r(t)x_{21}(t)dt = \begin{cases} \int_T (x_1(t) + n(t))x_{21}(t)dt = E_b + \int_T n(t)x_{21}(t)dt & b(t) = 1 \\ \int_T (x_2(t) + n(t))x_{21}(t)dt = \rho E_b + \int_T n(t)x_{21}(t)dt & b(t) = -1 \end{cases} \tag{16}$$

$$z_2 = \int_T r(t)x_{22}(t)dt = \begin{cases} \int_T (x_1(t) + n(t))x_{22}(t)dt = \rho E_b + \int_T n(t)x_{22}(t)dt & b(t) = 1 \\ \int_T (x_2(t) + n(t))x_{22}(t)dt = E_b + \int_T n(t)x_{22}(t)dt & b(t) = -1 \end{cases} \tag{17}$$

Where,

$$\rho = (\int_T x_{11}(t)x_{12}(t)dt) \, / \, E_b \tag{18}$$

$$E_b = \int_T x_{11}^2(t)dt = \int_T x_{12}^2(t)dt \tag{19}$$

Figure 13. Block diagram of a CSK secure communication scheme

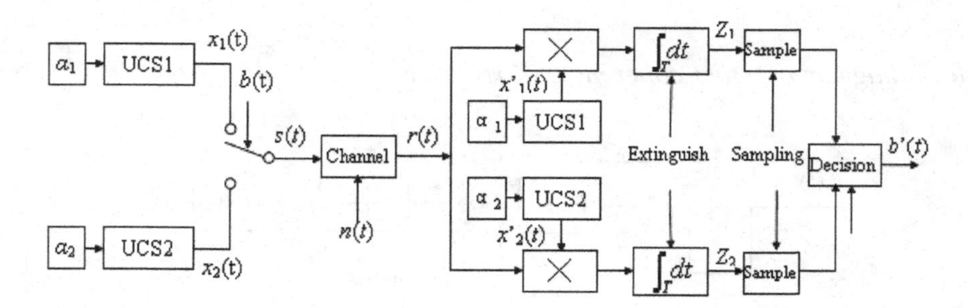

$$b'(t) = \begin{cases} \text{"1"} & z_1 > z_2 \\ \text{"}-1\text{"} & z_1 < z_2 \end{cases} \tag{20}$$

Just as analyzing previously, detection is done based on the bit energy of the transmitted signals. Obviously, the optimal threshold is dependent upon the SNR.

The simulation model of CSK digital secure communication on Matlab / Simulink 7.0 is shown in Figure 14. The signal is generated by Bernoulli random binary generator. The parameters of two unified chaotic systems are 1 and 0.3 respectively. The channel is typical AWGN. The results can be observed by scope, which are shown in Figure 15. The transmitted signals $s(t)$ in the channel are chaotic, so it is secure. The signals can be recovered by coherent demodulator and chaotic synchronization completely.

Figure 14. Dynamic simulation model of CSK

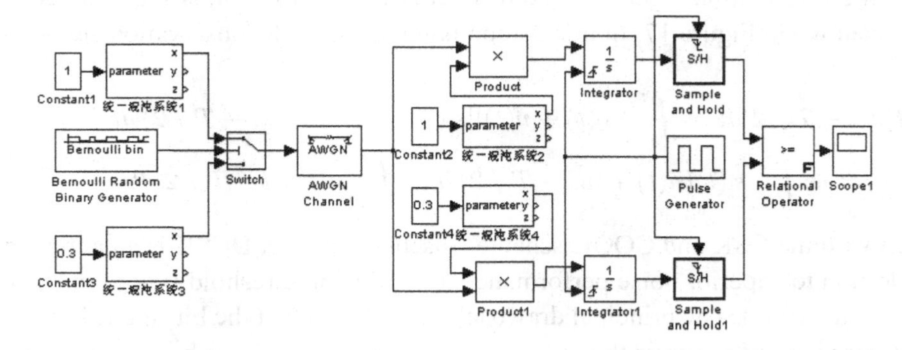

Figure 15. Simulation waves in different point of CSK (a) Clock (b) Information signal (c) The signal transmitted in channel (d) The recovered signals

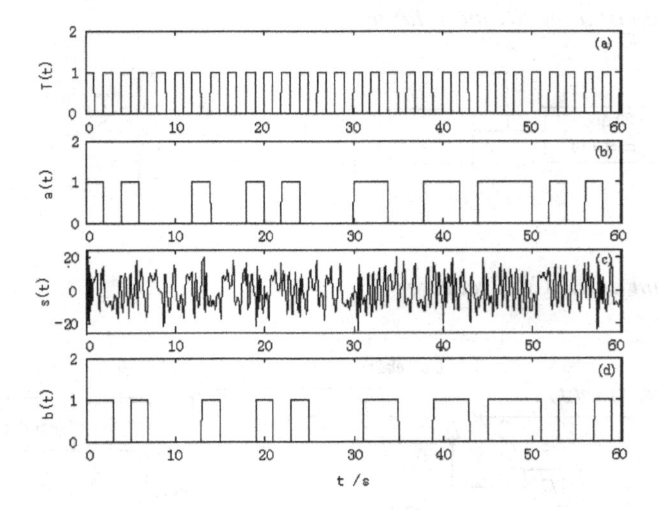

3) DCSK

In differential chaos shift keying (DCSK) (Kolumbán et al, 1996), every bit to be transmitted is represented by two chaotic sample functions. A block diagram of a DCSK modulator is shown in Figure 16. The first sample function serves as a reference, while the second one carries the information. Bit "1" is sent by transmitting a reference signal provided by a chaos generator twice in succession, while for bit "0", the reference chaotic signal is transmitted, followed by an inverted copy of the same signal. Thus, the signal transmitted in the sender is

$$s_i(t) = \begin{cases} x(t) & t_k \le t \le t_k + T/2 \\ +x(t - T/2) & t_k + T/2 < t \le t_k + T \end{cases} \quad for \quad b_i = 1 \tag{21}$$

$$s_i(t) = \begin{cases} x(t) & t_k \le t \le t_k + T/2 \\ -x(t - T/2) & t_k + T/2 < t \le t_k + T \end{cases} \quad for \quad b_i = 0 \tag{22}$$

Since each bit is mapped to the correlation between successive segments of the transmitted signal of length T/2, the information signal can be recovered by a correlator. A block diagram of a DCSK demodulator is shown in Figure 17. In this demodulator scheme, the observation signal has the form

$$\begin{aligned} z_i &= \int_{T/2}^{T} r_i(t) r_i(t - T/2) dt = \int_{T/2}^{T} [s_i(t) + n(t)][s_i(t - T/2) + n(t - T/2)] dt \\ &= \int_{T/2}^{T} s_i^2(t)\, dt + \int_{T/2}^{T} s_i(t)(n(t) + n(t - T/2)) dt + \int_{T/2}^{T} n(t) n(t - T/2) dt \end{aligned} \tag{23}$$

By contrast with the CSK and COOK schemes discussed above, DCSK is an antipodal modulation scheme. In addition to superior noise performance, the decision threshold is zero independently of E_b/N_0 (Kolumbán *et al*, 1996). The principal drawback of DCSK is that the bit-rate is low, and the estimation has a non-zero variance even in the noise-free case; this puts a lower bound on the bit duration and thereby limits the data rate. To solve the estimation problem, FM-DCSK is proposed by Kolnmbán G *et al* (Kolumbán *et al*, 1997).

Figure 16. Block diagram of a DCSK modulator

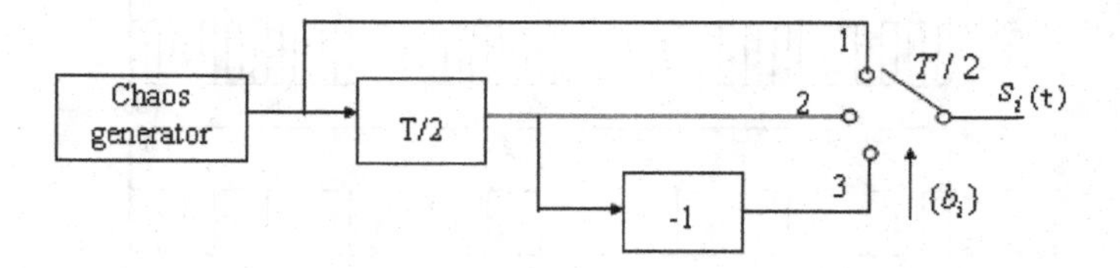

Figure 17. Block diagram of a DCSK receiver

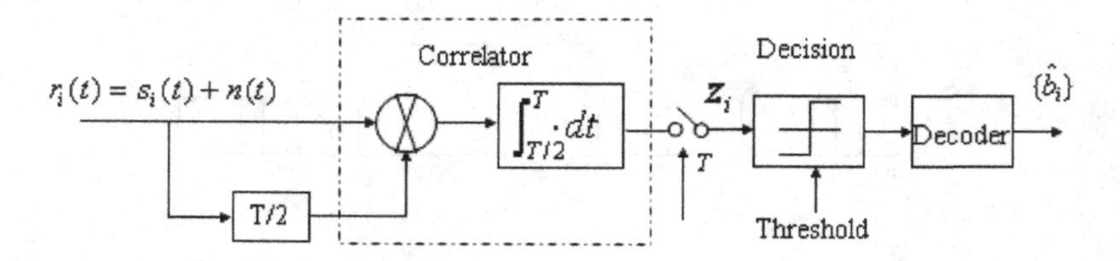

4) FM-DCSK

The operation of the modulator is the same as in DCSK, but the FM modulated signal is the input to the DCSK modulator, which is shown in Figure 18. The input of the FM modulator is a chaotic signal, which can be generated by a chaotic analog phase-locked loop (APLL).

The demodulator of an FM-DCSK system is the same as in a DCSK receiver. The only difference is that, instead of low-frequency chaotic signals, the FM signals are correlated directly, as shown in Figure 19. The main advantage of FM-DCSK modulation over CSK, COOK, and DCSK is that the data rate is not limited by the properties of the chaotic signal.

2.3 Chaos Modulation

Chaos modulation was proposed in 1993 (Halle et al, 1993; Cuomo, 1993). Different from chaos masking and chaos switching schemes, in a chaos modulation scheme the message signal m(t) is injected into the sender system so that its dynamics is changed by the message signal continuously. In this case, generally, an adaptive controller (which can also be considered as an extra dynamical system bidirectionally coupled with the sender system) is added at the response system according to some rule such that its output m'(t) asymptomatically converges to m(t). To follow the drive system's dynamics, generally the controller's output (*i.e.*, m(t)) should be injected into the response system in the same way as in the sender.

There are two different types of chaos modulation: (1) parameter modulation, in which the message signal m(t) modulates the values of one or more control parameter. (2) direct modulation, in which m(t) is injected into one or more variables of the driving system without changing the value of any control parameter. In some chaos modulation schemes, the message signal is also embedded into the driving signal, which can be regarded as a modified version of chaos masking as shown in Figure 7 and Figure 8.

Ref. (Yang & Chua, 1996) designed chaotic parameter modulation based on Chua's circuit. As shown in Figure 20, Chua's circuit consists of a linear inductor L, a linear resistor R, two linear capacitors C_1 and C_2, and a nonlinear resistor – the Chua's diode N_R. Four different modulation rules were presented

Figure 18. Block diagram of an FM-DCSK modulator

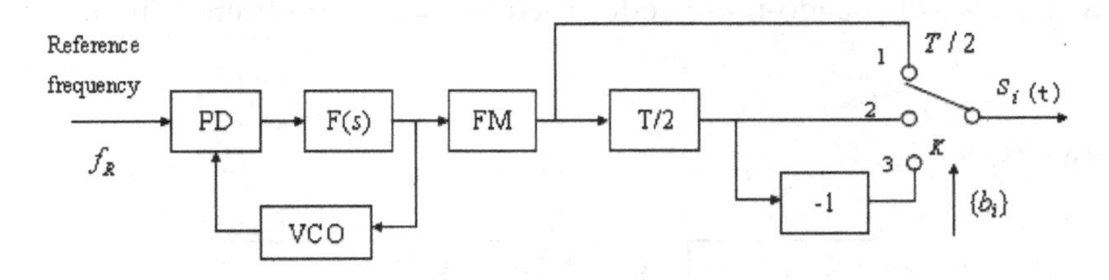

Figure 19. Block diagram of an FM-DCSK demodulator

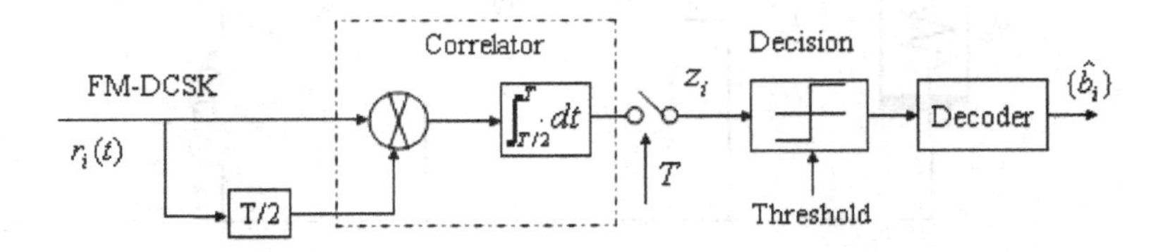

according to different parameters modulation in this paper, *i.e.* G-modulation, C_1-modulation, C_2-modulation, and L-modulation, and four different adaptive controllers were used at the receiver to maintain synchronization by continuously tracking the changes in the different modulated parameter.

Compared with chaos masking schemes, chaos modulation schemes can exactly recover the message signal (in an asymptotical manner) if some conditions are satisfied. Considering that chaos switching systems can only transmit digital signals, chaos modulation also has a better performance than chaos switching. In fact, by being designed carefully, the chaotic modulation technique can even be used to transmit more than on message signal. One possible way for this is to modulate n control parameters of the master system with n message signals, respectively.

The main disadvantage of chaos modulation is that the controller depends on the drive and response systems' structure, which means that different controllers needs to be designed for different drive systems. Controllers may not even exist in certain cases for essential defects of the drive/response chaotic systems (Li *et al*, 2007).

2.4 Chaos Direct-Sequence Spreading Spectrum

At present, spread spectrum is the basis for many popular and emerging wireless communications standards, including portable telephones, global positioning systems, Bluetooth, and some cellular telephone network schemes. The two most widely used versions of spread spectrum are frequency hopping spread spectrum (FHSS) and direct sequence spread spectrum (DSSS). Here, the DSSS was taken as an example to explain how to use chaos to implement spread spectrum communication.

Chaos direct sequence spread spectrum is proposed by Li and Haykin in 1995 (Li & Haykin, 1995), and Itoh extended this application on to frequency hopping spread spectrum system (Itoh, 1998). In spread spectrum communications, the transmitted signal is "spread" over a much wider bandwidth compared with that of a narrow band signal. When the signal spreads, the average power spectral density becomes much lower and the signal can be concealed in the background noise.

The motivation to use chaos for spread spectrum communication is mainly due to its wideband nature and its noise-like appearance. Compared with common spread-spectrum techniques, chaotic signals provide robustness against frequency selective fading in multipath channels and narrowband interference. Owing to the intricate dynamic of chaos, the security of communications can be significantly increased in comparison with standard pseudo-noise codes used in spread spectrum. Without knowledge of the

Figure 20. Chua's circuit

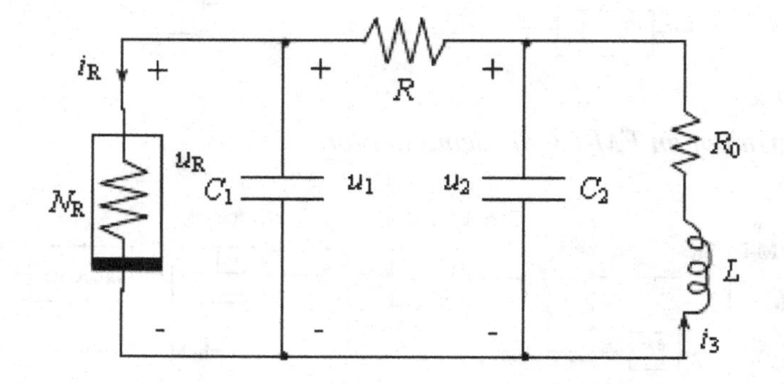

type of nonlinearity on which the transmission is based, it will be extremely difficult for the unauthorized user aware of the transmission to access the information.

Figure 21 shows a functional block diagram of the direct sequence spread spectrum with binary shift keying communication (DS-SS BPSK) analyzed (Itoh, 1998). The binary data input is transformed in NRZ (nonreturn to zero) signaling pulse format. The spectrum of this signal is spread, in baseband, by NRZ-chaos spread sequence. A modulator BPSK, which shifts the spectrum to the assigned frequency range, follows the resulting spread signal. The modulated signal is then amplified and sent through the radio frequency (RF) channel, *i.e.*, a wireless communication channel, which has a limited bandwidth. It is assumed that the channel introduces only additive white Gaussian noise (AWGN). At the receiving end, the signal is amplified and filtered at the desired RF bandwidth and is fed to the demodulator, which shifts the spectrum to baseband. The resulting signal is despread by NRZ-chaos spread sequence. The NRZ chaos spread sequence generated at the receiver end must be in synchronization with the chaos sequence in the sender. The synchronizer does this. After this, the binary output information is detected correctly.

3 CHAOS-BASED CRYPTOGRAPHY

Modern telecommunication networks, especially the Internet and mobile-phone networks, have tremendously extended the limits and possibilities of communications and information transmissions. Associated with this rapid development, there is a growing demand of cryptographic techniques, which has spurred a great deal of intensive research activities in the study of chaotic cryptography.

3.1 Chaotic Sequence Generation

Chaos exhibits uncertainty behavior, the intrinsic randomness, sensitivity to initial conditions, and long-term unpredictability. therefore, the pseudo-random sequence constructed by the chaotic systems has good randomness, correlativity, complexity, and the number of the sequence is very large. The sequence is renewable, difficult to reconstruction and prediction.

Generally, chaotic sequence is divided into three different sequences, *i.e.* real numerical sequence, binary sequence, bit sequence (Kocarev *et al*, 2003). Bitary sequence and bit sequence can be got from real numerical as shown in Figure 22.

Figure 21. Chaotic DS-SS communication scheme

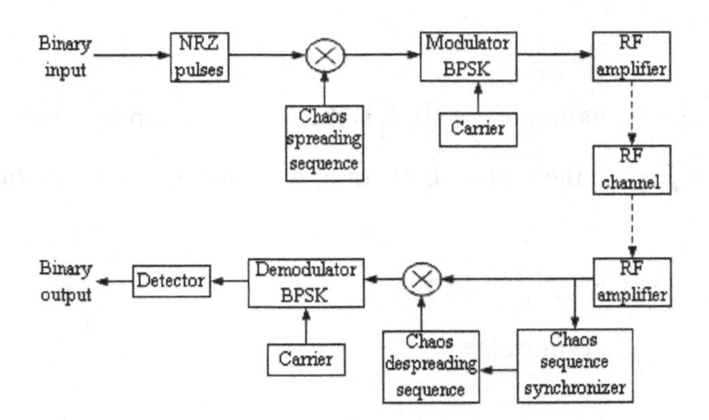

Figure 22. Sequence flow generating

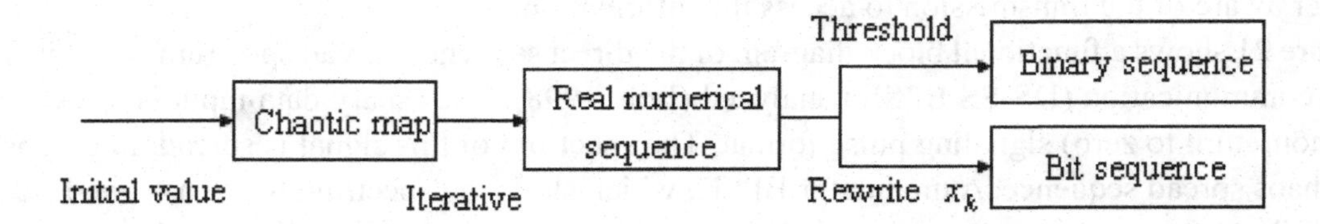

(1) Real numerical sequence, $\{x_k, k = 0,1,2\ldots\}$, is a sequence formed by the track points of the chaotic map. Generally, the distribution of the chaotic systems isn't uniform, so a real numerical sequence should not be applied to encryption directly.
(2) Binary sequence, it can be got by defining a threshold function $T(x_k)$ for the above real numerical sequence

$$T(x_k) = \begin{cases} 1 & x_k \geq c \\ 0 & x_k < c \end{cases}, \tag{24}$$

where c is the threshold value. If the threshold selection is unreasonable, then the performance of chaotic sequence is not good. For a balanced binary sequence, c should be chosen so that the likelihood of $x_k < c$ is equal to that of $x_k \geq c$. This sequence can be expressed as $\{T(x_k), k = 0,1,2\cdots\}$.

(3) Bit sequence, compared with the binary sequence, is also got from a real numerical sequence. What the difference is that it rewrites the x_k to the form of L-bit floating-point numbers, as $|x_k| = 0.b_1b_2\cdots b_i\cdots, b_i = 0or1, i = 1,2,\cdots l$ $b_i = 0$ or 1, $i = 1,2,\cdots l$. For each real value x_k, one can extract some or all of the binary bits from its L-bit to constitute a pseudo-random sequence.

Generally, the existing chaotic sequence generator is mainly based on traditional types of chaotic system, such as Logistic map, improved Logistic map, Tent map, Chebyshev map.

To analyze the performance of chaotic sequence, Logistic map is taken as an example. Logistic map is one of the simplest, and is most studied nonlinear discrete chaotic system. Logistic map is given by

$$x_{k+1} = \mu x_k(1 - x_k), \tag{25}$$

where $x_k \in (0,1)$ is the system variable, $\mu \in [0,4]$ is the system parameter, and k is the number of iterations. When $3.5699\cdots < \mu \leq 4$, the system is chaotic, its probability density function is given by

$$\rho(x) = \begin{cases} \dfrac{1}{\pi\sqrt{x(1-x)}} & 0 < x < 1, \mu = 4 \\ 0 & otherwise \end{cases} \tag{26}$$

Using the knowledge of the probability theory, we present the statistical characteristics of the logistic map system as follows.

(1) Average value

$$\overline{x} = \lim_{N\to\infty} \frac{1}{N} \sum_{i=1}^{N} x_i = \frac{1}{2} \tag{27}$$

(2) Auto-correlation function

$$R_{xx}(\tau) = \lim_{N\to\infty} \sum_{n=0}^{N-1} x_n x_{n+\tau} = \begin{cases} 0.125 & \tau = 0 \\ 0 & \tau \neq 0 \end{cases} \tag{28}$$

(3) Cross-correlation function

$$R_{xy}(\tau) = \lim_{N\to\infty} \sum_{n=0}^{N-1} x_n y_{n+\tau} = 0 \tag{29}$$

According to Eq. (27), (28), (29), the average value of the chaotic sequence is a constant, and auto-correlation function is delta function, and cross-correlation function equal to 0. It is similar to white noise statistical properties. These statistical properties describe quantitatively the sensitivity and randomness of the chaotic signal. Thus, it is applied to chaotic cryptography widely.

With in-depth study of chaos theory, people found that, at present, the encryption systems, which use traditional types of chaotic systems, are not safe enough. The pseudo-random sequence based on the structure of these chaotic systems still exit security risks. Many researchers put forward improvement measures for having more or less shortcomings of each kind of chaotic sequence generator. There are mainly two methods to resolve this problem.

(1) **Improve the traditional chaotic sequence generator:** There are two methods to improve the realization algorithm of the chaotic sequence. One of them is improve the individual traditional chaotic sequence generator by employing improved Logistic map, which is proposed in Ref. (Zheng et al, 2003). The other is to design chaotic sequence generator by using multi-chaotic systems, a combination of Logistic and Chebyshev, which was studied in Ref. (Zhao & Gan, 2005).
(2) **Use a new chaotic system in chaotic sequence generator:** Recently, a simple pseudo-random number generator (PRNG) is proposed with parallel TD-ERCS (Sheng *et al*, 2005). In this generator, users' keys are no longer fixed, and can be chosen in the interval (264, 2672) arbitrarily. By testing the basic statistic characteristics such as equilibrium, runs, and correlation of the binary pseudo-random sequences (viz. TD-ERCS sequences) being generated from the PRNG, and comparing with m-sequences, the logistic sequences, Chebyshev sequences and SCQC sequences, the experimental results show that TD-ERCS sequences have better statistic characteristics.

3.2 Chaotic Sequence Encryption

In last section, we have talked about the generation of chaotic sequence. Now, we will apply the chaotic sequence to encrypt information. As we know, Chaos is characterized by ergodicity, sensitive dependence on initial conditions, and random-like behaviors, which are used for the diffusion and confusion process in cryptography. The main aim of cryptography is to ensure security by changing messages unavailable to anyone else except for the legal receiver. Encryption is also defined as the science of using mathematics to encrypt and decrypt the data, the text mode message, or the image. It allows us to store sensitive information and transmit it across insecure channel, so that it cannot be achieved by anyone else (Yin & Wang, 2008). According to the definition of the encryption form in traditional cryptology, the chaos encryption may be divided into sequence encryption and data stream encryption. In this section, we will discuss about the chaotic sequence encryption.

As shown in Figure 23, the sender applies an encryption algorithm to encrypt a plain text message. Similarly, the receiver uses the decryption algorithm to decrypt the received encrypted message. Both sender and receiver must agree on a common algorithm. The most important thing about the encryption is to find a reliable approach to accomplish the encryption in sender, and it must be safe enough to resist any attack. A safe key is used during the encryption and decryption process. To achieve the secret key, the chaos and chaotic sequence, such as Logistic map and Henon map, are introduced in the field of the encryption (Zhu *et al*, 2003; Yen & Guo, 2000). Chaotic systems have been testified very suitable for data message encryption.

The encoder in the chaotic secure communication system can be described by encoding equation $Y = E(Z, X)$, and the decoder can be described by decoding equation $X = E^{-1}(Z,Y)$, where $X, Z \in R^n$ are the plaintext and the cryptograph respectively, $Y \in R$ is the cipher, and E, E^{-1} are the encryption algorithm and the decryption algorithm (Zhu *et al*, 2003).

The process of chaotic sequence encryption can be described by Figure 24. Firstly, the sender should translate the message (such as text, data, images) into binary sequence using quantization coding technology. Therefore, we assume that the sequence in plaintext sequence space, cipher text space and key space are composed of binary digit sequence set. The six elements (M, C, K, E_k, D_k, Z) are used to describe the chaotic sequence encryption system, where M is the plaintext sequence space; C is the cipher text sequence space; K is the key space; Z is the algorithm, which is used to generate the cipher test sequence. Here we use chaotic sequence as the cipher text sequence, so Z is generated by a chaotic sequence generator. $E_k(m)$ and $D_k(c)$ are the encryption rule and the decryption rule, and the XOR operator is run to generate the cipher text sequence, *i.e.* $C_i = M_i \oplus K_i$.

Figure 23. Block diagram of secure communication system

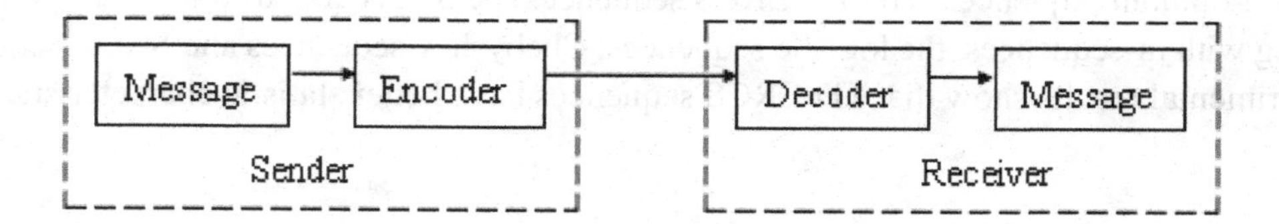

The chaotic sequence encryption process can be summarized as follows:

(1) Translate the message we want to send into binary digital sequence

$$M = m_{b0} m_{b1} m_{b3} \cdots m_{bn-1} m_{bn} \tag{30}$$

(2) Use the chaotic sequence generator to generate a group of cipher sequence with the encryption key source $K = k_0 k_1 k_2 \cdots$

$$Z = z_{b0} z_{b1} z_{b2} \cdots z_{bn-1} z_{bn} \tag{31}$$

(3) Encrypt the plaintext sequence messages by the encryption rule we have chose. To get the cipher sequence by the operation of $c_i = m_i \oplus z_i \ for \quad i = 0, 1, 2 \cdots n-1, n$.

$$C = E_k(m) = c_0 c_1 c_2 \cdots c_{n-1} c_n \tag{32}$$

(4) Transmit the cipher sequence $C = c_0 c_1 c_2 \cdots c_{n-1} c_n$ to the receiver through public signal channel, and transmit the encryption key source $K = k_0 k_1 k_2 \cdots$ to the receiver through private channel.

(5) The receiver use the encryption key source $K = k_0 k_1 k_2 \cdots$ and the chaotic sequence generator to generate a group of chaotic sequence

$$Z' = z'_{b0} z'_{b1} z'_{b2} \cdots z'_{bn-1} z'_{bn} \tag{33}$$

Z' and Z must be same, otherwise, the receiver can not get the plaintext sequence the sender have send. The chaotic sequence generator in two sides is required to be synchronous.

Figure 24. Chaotic sequence encryption system

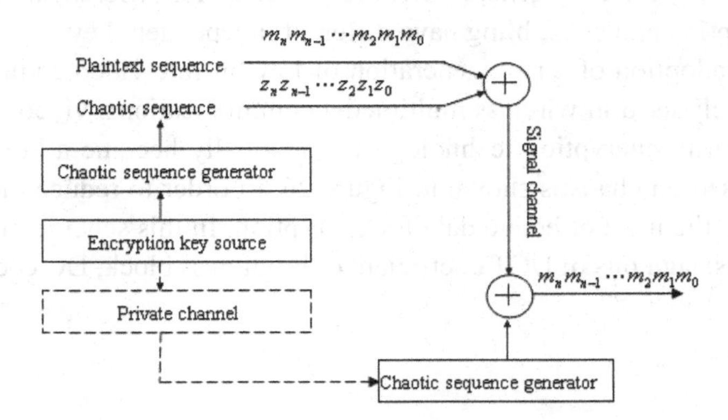

(6) Receiver decrypt the cipher sequence $C = c_0 c_1 c_2 \cdots c_{n-1} c_n$ with the chaotic sequence generated by $Z' = z'_{b0} z'_{b1} z'_{b2} \cdots z'_{bn-1} z'_{bn}$,. To get the plaintext sequence by the operation of $m_i = z'_{bi} \oplus c_i$, $for \quad i = 0,1,2 \cdots n-1, n$

$$M = m_{b0} m_{b1} m_{b3} \cdots m_{bn-1} m_{bn} \tag{34}$$

3.3 Chaotic Data Stream Encryption

In last section, we have introduced the chaotic sequence encryption. Now, we'll talk about chaotic data stream encryption. As we know, with the development of information and social progress, people's demands for information are growing. Recently, image and video are more valuable for people's concern for its directness and vividness, and they shared the characteristics of large amount of data. As a result, the security of the relative information is becoming more and more important, especially in transmission. In the following content, we'll mainly talk about video encryption.

Video data are highly relevant. That is to say. There is a lot of redundancy information. Data compression technology is to remove redundant information. What commonly used video compression algorithms are the MPEG series based on DCT (discrete cosine transform) and the H26X series. Here, a Chaos-based MPEG series video encryption algorithm is presented (Shang *et al*, 2008).

Practical MPEG video encryption algorithm should meet some basic requirements, such as invariant compression formats, invariant compression ratio, low calculated amount and widely adjustable security level. On the whole, the entire MPEG video encryption scheme consists of the following two parts:

1. Do encryption to fixed-length code data including intra DC coefficients, non-intra DC coefficients and AC coefficients, ESCAPE DCT coefficients, motion vector sign bit and residue.
2. Scramble the macroblock bit stream, and distribute a different scrambling table for each frame.

Since all the operations are completed on the video streams, the entire encryption algorithm has no effect on compression ratio. Stream cipher and scrambling table are generated by two chaotic maps, discrete PLCM and cat map. Figure 25 shows the complete chaotic MPEG video encryption process. In order to reduce the chaotic iteration, we used a discrete cat map to generate a stream cipher and the replacement table. Of course, in order to strengthen the security, the chaotic maps with different initial values and parameters can be used to generate chaotic sequence, for different encryption process. In this case, the stream encryption and scrambling have different independent key.

H.264 is the latest adoption of a new generation of low bit-rate video coding standard (Wiegand, 2003). It has been widely used in wireless multimedia communications, video conferencing, and thus the corresponding security encryption technology has gradually become a hot topic. A H.264 video encryption scheme based on chaos is shown in Figure 26. In order to reduce the data needed encryption, we should choose the most effective data for encryption. In this scheme, the selected content for encryption is as follows: sign bits of DCT coefficient of brightness block, DC coefficient of color block and motion vector.

Figure 25. Block diagram of MPEG encryption algorithm based on chaos

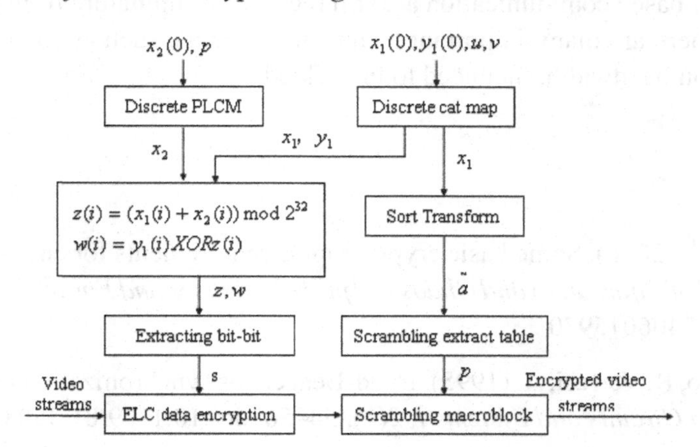

4 CONCLUSION

This chapter presents some proposed applications of chaotic communications through synchronized chaotic system. It first introduces a number of properties of chaotic systems that are appealing for chaotic communications systems. Then, it illustrates the principles of chaos-based analog and digital communication schemes. Finally, a brief introduction of chaos-based cryptography is presented. It should be

Figure 26. H.264 encryption process diagram

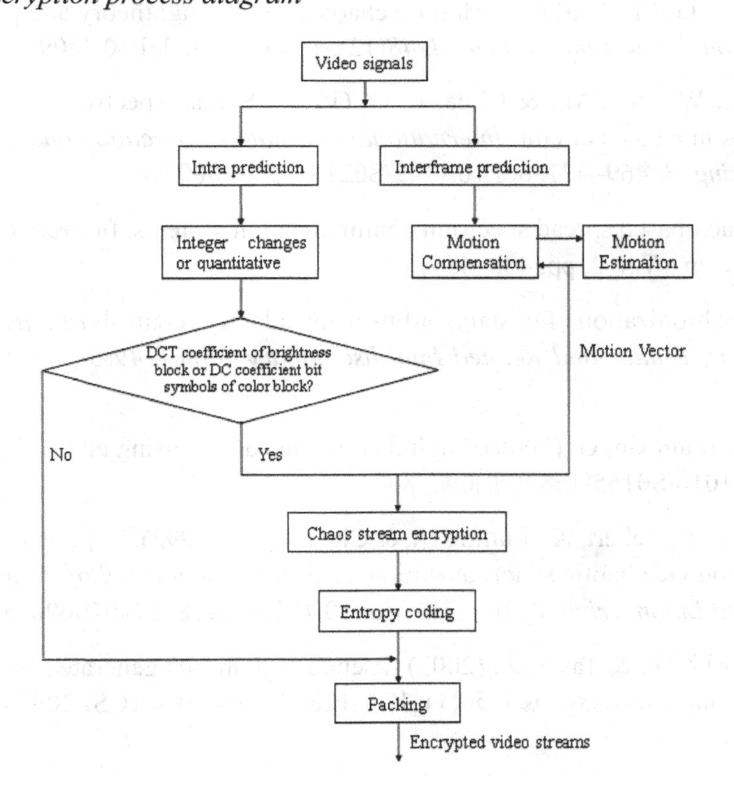

pointed out that chaos-based communication are still regarded as immature from the practical engineering standpoint, and there are many important technical problems, such as robust synchronization and control of transmission bandwidth, that need to be tackled.

REFERENCES

Alvarez, G., & Li, S. J. (2006). Some basic cryptographic requirements for chaos-based cryptosystems. *International Journal of Bifurcation and Chaos in Applied Sciences and Engineering, 16*(8), 2129–2151. doi:10.1142/S0218127406015970

Angeli, A. D., Genesio, R., & Tesi, A. (1995). Dead-Beat chaos synchronization in discrete-time system. *IEEE Transactions on Circuits and Systems I, 2*(1), 54–56. doi:10.1109/81.350802

Carroll, T. C., Heagy, J. F., & Pecora, L. M. (1996). Transforming signals with chaotic synchronization. *Physical Review E: Statistical Physics, Plasmas, Fluids, and Related Interdisciplinary Topics, 54*(5), 4676–4680. doi:10.1103/PhysRevE.54.4676

Cuomo, K. M., Openheim, A. V., & Strogatz, S. H. (1993). Synchronization of Lorenz-based chaotic circuits with applications to communications. *IEEE Transactions on Circuits and Systems-I, 40*, 426–633.

Dedieu, H., Kennedy, M. P., & Hasler, M. (1993). Chaos shift keying: Modulation and Demodulation of a chaotic carrier using self-synchronizing. *IEEE Transactions on Circuits and Systems-I, 40*, 634–641. doi:10.1109/82.246164

Galias, Z., & Maggio, G. M. (2001). Quadrature chaos–shift keying: theory and performance analysis. *IEEE Transactions on Circuits and Systems-I, 48*(12), 1510–1519. doi:10.1109/TCSI.2001.972858

Halle, K. S., Wu, C. W., Itoh, M., & Chua, L. O. (1993). Spread spectrum communication through modulation of chaos in Chua's circuit. *International Journal of Bifurcation and Chaos in Applied Sciences and Engineering, 3*, 469–477. doi:10.1142/S0218127493000374

Itoh, M. (1998). Chaos-based spread spectrum communication systems. In *Proc. Industrial Electronics ISIE 98 IEEE Int. Symp.* (Vol. 2, pp. 430–435).

John, K. (1994). synchronization of unstable orbits using adaptive control. *Physical Review E: Statistical Physics, Plasmas, Fluids, and Related Interdisciplinary Topics, 49*(6), 4843–4848. doi:10.1103/PhysRevE.49.4843

Kennedy, M. P., & Kolumbán, G. (2000). Digital communications using chaos. *Signal Processing, 80*, 1307–1320. doi:10.1016/S0165-1684(00)00038-4

Kocarev, L., Halle, K. S., Eckert, K., Piulitz, U., & Chua, L., O. (1992). Experimental demonstration of secure communication via chaotic synchronization. *International Journal of Bifurcation and Chaos in Applied Sciences and Engineering, 2*, 709–713. doi:10.1142/S0218127492000823

Kocarev, L., Jakimoski, G., & Tasev, Z. (2003). Pseudorandom bits generated by chaotic maps. *IEEE Transactions on Circuits and Systems-I, 50*(1), 123–126. doi:10.1109/TCSI.2002.804550

Kocarev, L., & Parlitz, U. (1995). General approach for chaotic synchronization with application to communication. *Physical Review Letters*, *74*(25), 5028–5031. doi:10.1103/PhysRevLett.74.5028

Kolumbán, G., Kis, G., Kennedy, M., et al. (1997). FM-DCSK: A new and robust solution to chaos communications. In *Proc. Int. Symposium on Nonlinear Theory and Its Applications*, Hawaii (pp. 117-120).

Kolumbán, G., Vizvári, B., Schwarz, W., & Abel, A. (1996). Differential chaos shift keying: a robust coding for chaos communication. In *Proc. Int. Workshop on Nonlinear Dynamic of Electronic Systems*, Sevilla, Spain (pp. 87-92).

Kolumbán, G., Vizvári, B., Schwarz, W., & Abel, A. (1996). Differential chaos shift keying: A robust coding for chaotic communication. In *Proceedings of the Fourth International Workshop on Nonlinear Dynamics of Electronic Systems*, Sevilla (pp. 87-92).

Li, B. X., & Haykin, S. (1995). A new pseudo-noise generator for spread spectrum communications. In *ICASSP-95, Int. Conf. Acoustics, Speech, and Signal Process.* (Vol. 5, pp. 3603–3606).

Li, S. J., Alvarez, G., Li, Z., & Halang, W. A. (2007). Analog chaos-based secure communications and cryptanalysis: a brief survey. In *Proceedings of 3rd International IEEE Scientific Conference on Physics and Control*, Potsdam, Germany (pp. 92-98).

Mianovic, V., & Zaghloul, M. E. (1996). Improved masking algorithm chaotic communication systems. *Electronics Letters*, *1*, 11–12. doi:10.1049/el:19960004

Parlitz, U., Chua, L. O., Kocarev, L., Halle, K. S., & Shang, A. (1992). Transmission of digital signals by chaotic synchronization. *International Journal of Bifurcation and Chaos in Applied Sciences and Engineering*, *2*, 973–977. doi:10.1142/S0218127492000562

Pecora, L. M., & Carroll, T. C. (1990). Synchronization in chaotic system. *Physical Review Letters*, *64*(8), 821–824. doi:10.1103/PhysRevLett.64.821

Pyragas, K., & Tamasevicius, A. (1993). Experimental control of chaos by delayed self-controlling feedback. *Physics Letters. [Part A]*, *180*, 99–102. doi:10.1016/0375-9601(93)90501-P

Roy, R., & Scott, K. (1994). Experimental synchronization of chaos. *Physical Review Letters*, *72*(13), 2009–2012. doi:10.1103/PhysRevLett.72.2009

Shang, F., Sun, K. H., & Cai, Y. Q. (2008). A new efficient MPEG video encryption system based on chaotic maps. In *Proceedings-1st International Congress on Image and Signal Processing* (Vol. 3, pp. 12-16).

Sheng, L. Y., Cao, L. L., & Sun, K. H. (2005). Pseudo-random number generator based on TD-ERCS chaos and it's statistic characteristic s analysis. *ACTA PHYSICA SINICA*, *54*(9), 4031–4037.

Stavroulakis, P. (2005). *Chaos applications in telecommunications*. Philadelphia, PA: Taylor & Francis Group. doi:10.1201/9780203025314

Sugawara, T., Tachikawa, M., Tsukamoto, T., & Shimizu, T. Y. (1994). Observation of synchronization in Laser chaos. *Physical Review Letters*, *72*(22), 3502–3505. doi:10.1103/PhysRevLett.72.3502

Wang, X. Y., & Yu, Q. (2009). A block encryption algorithm based on dynamic sequences of multiple chaotic systems. *Communications in Nonlinear Science and Numerical Simulation*, *14*, 574–581. doi:10.1016/j.cnsns.2007.10.011

Wiegand, T., Sullivan, G. J., Bjontegaard, G., & Luthra, A. (2003). Overview of the H.264/AVC video coding standard. *IEEE Trans. Circuits and Systems for Video Technology*, *7*, 560–576. doi:10.1109/TCSVT.2003.815165

Yang, T., & Chua, L. O. (1996). Secure communication via chaotic parameter modulation. *IEEE Transactions on Circuits and Systems-I*, *43*, 817–819. doi:10.1109/81.536758

Yen, J. C., & Guo, J. I. (2000). A new chaotic key-based design for image encryption and decryption. In *IEEE International Symposium on ISCAS*, Geneva, IV-49-IV-52

Yin, M., & Wang, L. W. (2008). A new study in encryption based on fractional order chaotic system. *Journal of Electronic Science and Technology of China*, *6*(3), 238–241.

Zhao, H. X., & Gan, J. (2005). An improved algorithm of generate chaotic sequence. *Computer Applications*, *25*(B12), 78–79.

Zheng, S. H., Li, C. D., & Liao, X. F. (2003). Pseudo-random bit generator based on coupled Logistic maps and its applications in chaotic stream-cipher algorithms. *Computer Science*, *30*(12), 95–98.

Zhu, G. B., Cao, C. X., & Hu, Z. Y. (2003). An image scrambling and encryption algorithm based on affine transformation. *Journal of Computer-Aided Design & Computer Graphics*, *15*(6), 711–715.

Chapter 19
Digital Information Transmission Using Discrete Chaotic Signal

A. N. Anagnostopoulos
Aristotle University of Thessaloniki, Greece

A. N. Miliou
Aristotle University of Thessaloniki, Greece

S. G. Stavrinides
Aristotle University of Thessaloniki, Greece

A. S. Dmitriev
Inst. of Radio Eng. & Electr. of Russian Academy of Sciences, Russia

E. V. Efremova
Inst. of Radio Eng. & Electr. of Russian Academy of Sciences, Russia

ABSTRACT

In this work the authors thoroughly investigated a digital information transmission system using discrete chaotic signal over cable. As an example in their work the authors consider the non-autonomous 2nd order non-linear oscillator system presented in Tamaševičious, Čenys, Mycolaitis, and Namajunas (1998) which is particularly suitable for digital communications and present the experimental results regarding synchronization. The effect of noise (internal or external) on the synchronization of the drive-response system (unidirectional coupling between two identical systems) is analyzed and since in every practical implementation of a communication system, the transmitter and receiver circuits (although identical) operate under slightly different conditions the case of the mismatch between the parameters of the transmitter and the receiver is considered. Moreover, there is a study of the robustness of the system with reference to the desired security, proposing a more sophisticated approach, which combines the simplicity in the implementation of a chaotic system with an enhanced encoding scheme that will overall increase security.

DOI: 10.4018/978-1-61520-737-4.ch019

1. INTRODUCTION

The significance of private and secure communications is very clear in a world, which increasingly relies on rapid transmission of large amounts of information. The current solutions for secure communications are the public key cryptosystems using software techniques to achieve computational complexity while quantum cryptography has the potential to render such techniques obsolete. However, hardware complexity is another method of increasing security in communications by hiding or masking the message on a chaotic carrier (Pecora, et. al. 1990; Ott, et. al. 1990; Carroll, et. al. 1993; Chua et. al. 1996; Wu et. al. 1996; Kolumban, et. al. 1998; Tamaševičious, et. al. 1998; Mycolaitis, et. al. 1999; Pikovsky, et. al. 2003; Yang, 2004).

The introduction of non-linear chaos theories has offered several new applications and performance enhancements to existing communication systems. A chaotic generator can produce non-linear and non-repeating sequences and it is very hard to predict chaotic patterns and sequences even when the chaotic function is known to the interceptors. This is because different estimation of the initial condition will lead to a very different chaotic sequence.

Chaotic communication systems are simpler, by means of circuit engineering implementation, as compared to traditional spread spectrum systems. Two key features of chaos are a noise-like time series and sensitive dependence on initial conditions and control parameters, which cause chaotic transmissions to have low probability of detection as an information-bearing signal and low probability of interception, respectively (Chambers, & Frey, 1993; Ogorzalek, 1993; Frey, 1993; Dedieu, et. al. 1993; Short, 1994; Wu, & Chua, 1994; Murali, & Lakshmanan, 1994; Chua, et. al. 1996; Yang, & Chua, 1996; Yang, Wu & Chua, 1997; Yang, & Chua, 1997; Yang, & Chua, 1997; Yang, et. al. 1997; Yang, & Yang, 1997; Yang, Sharuz, 1997; He, & Vaidya, 1998; Chien, & Liao, 2005).

The key fact in chaos-based communication systems is that the procedure used by the transmitter to generate the chaotic waveform is deterministic; the knowledge of this procedure by an authorized receiver allows him to replicate, or synchronize, the chaotic waveform, and then to recover the message by subtracting the chaotic carrier (Frey, 1993; Cuomo, & Oppenheim, 1993; Cuomo, Oppenheim, & Strogatz, 1993; Wu, & Chua, 1994; Chua, et. al. 1996; Yang, & Yang, 1997; Yang, Sharuz, 1997; Tamaševičious, et. al. 1998; Mycolaitis, et. al. 1999; Chien, & Liao, 2005). The confidentiality of the encryption technique is based on the difficulty to reproduce the chaotic carrier signal if an intruder does not know the particular dynamical system used.

In this chapter, we present a digital information transmission system using discrete chaotic signal over cable. The chapter is structured as follows: The first section is a sort general theoretical presentation of synchronization while the next ones are devoted to the description of the circuit and experimental results regarding synchronization. The last sections are presenting the synchronization of the system (transmitter - receiver) under noisy channel conditions, as well as, the case where different noise levels are added in the transmitter and the receiver (internal noise) due to electronics. Moreover, there is a study of the robustness of the system with reference to the desired security, proposing a more sophisticated approach, which combines the simplicity in the implementation of a chaotic system with an enhanced encoding scheme that will overall increase security.

2. SYNCHRONIZATION

The term "Synchronization" comes from the Greek complex word "*συν-χρονισμός*" which means "*sharing a common timing*". This very meaning has remained unchanged until nowadays, as an association of different processes in the time domain (Hornby, 1984).

Synchronization of oscillating systems was first observed in systems with periodic oscillations. The first time that the phenomenon was reported, was in 1665 by Huygens. It was referred to motion synchronization between two clocks, which were weakly coupled by hanging them from the same wooden rod, and it was described as "*a phenomenon of sympathy between two clocks*" (Huygens, 1986; V)

Later on, it was revealed that synchronization also existed in the case of systems with irregular or chaotic oscillations. Nonlinear oscillator synchronization is a process that is frequently encountered in nature, explaining relevant phenomena. The ability that nonlinear dynamical systems possess, to be synchronized, is a significant property. As a consequence, chaotic system synchronization is encountered in a variety of scientific fields from astronomy and electronic engineering to social sciences.

In general, tuning up the oscillations of two or more dynamical systems, that are weakly interacting, is called synchronization. More specifically, chaotic synchronization is a process in which two or more chaotic dynamical systems adjust their oscillations or a given property of their oscillations, to a common behavior, due to a coupling or to a forcing. The synchronized systems could be either equivalent or nonequivalent (Boccaletti, et. al. 2002; Pikovsky, Rosenblum, & Kurths, 2003).

Chaotic synchronization between nonlinear dynamical systems is an incompatible idea, taking into account their main characteristic, that of sensitive dependence on the initial conditions. This feature implies exponential grow of initial state differences, even in the case of identical dynamical systems, getting them uncorrelated in the course of time (Cuomo, & Oppenheim, 1993; Cuomo, Oppenheim, & Strogatz, 1993; Dedieu, et. al. 1993; Frey, 1993; Ogorzalek, 1993; Chambers, & Frey, 1993; Wu, & Chua, 1994; Murali, & Lakshmanan, 1994; Yang, & Chua, 1996; Chua, et. al. 1996; Yang, & Chua, 1997; Yang, et. al. 1997; Yang, & Yang, 1997; Yang, Sharuz, 1997; He, & Vaidya, 1998; Boccaletti, et. al. 2002; Chien, & Liao, 2005). However, it has been experimentally shown that synchronization is possible even for chaotic dynamical systems (Stavrinides, 2007).

There are two coupling configurations, namely *unidirectional coupling* and *bidirectional coupling*. Each one leads both the coupled systems to synchronized states, by using quite different mechanisms. These mechanisms are different so that, till now, no method has ever been proposed to link formally the two cases or to reduce one process to another (Rulkov, 1996; Pikovsky, Rosenblum, & Kurths, 2003).

The first case is a typical master-slave configuration, where two subsystems form a global system. One subsystem acts as the drive and the other as the response; meaning that the driving system evolves freely, driving the evolution of the response system. Consequently, the response system is obliged to follow the dynamics of the driving system. In this kind of coupling, each subsystem dynamics are not influenced by the coupling. It should be noticed that the master subsystem acts as an external chaotic forcing for the response system and as a result, one could say that in this case *external synchronization* exists.

In the second case, that of bidirectional coupling, both subsystems are coupled with each other in such a way that they mutually influence their behavior. The coupling factor induces an adjustment of the rhythms onto a common synchronized manifold, thus inducing a mutual synchronization behavior in both subsystems.

Both couplings are used in electronic chaotic circuits with typical applications in communications with chaos or cryptography. However, the bidirectional coupling is encountered very often in nonlinear optics e.g. coupled laser systems and biology, e.g. between interacting neurons.

Synchronization ranges from complete agreement of trajectories, where coordinates of different subsystems coincide, to locking of phases. During the last twenty years, many different synchronization states have been reported and studied such as *Complete* and *Generalized Synchronization* or *Phase* and *Lag Synchronization* (Pikovsky, Rosenblum, & Kurths, 2003; Schuster, & Just, 2005).

Complete Synchronization was the first discovered and is the simplest form of synchronization in chaotic systems. This type of behavior can only be expected when identical subsystems are involved. In this case, although, each subsystem operates in a chaotic way, there is complete agreement of trajectories and the coordinates of different subsystems coincide. This is achieved by means of a coupling signal, in such a way that they remain in step with each other in the course of the time. This mechanism was first shown to occur when two identical chaotic systems are coupled unidirectionally, provided that the conditional Lyapunov exponents of the subsystem to be synchronized are all negative (Pecora, & Carroll, 1990).

Generalized Synchronization, as a notion, goes further in using different physical systems. The idea of Generalized Synchronization has been introduced in 1995, aiming in treating synchronization with non-identical subsystems. In this case the output of one system is associated to a given function of the output of the other system and this functional relation holds for all time (Rulkov, et. al. 1995). Again, the stability of the synchronized state is expressed in terms of the appropriate Lyapunov exponents.

Phase Synchronization appears when coupled non-identical oscillatory systems can reach an intermediate mode of operation, where a locking of the phases is established, while correlation in the amplitudes remains weak. Transition to Phase Synchronization was first reported in two coupled Rössler dynamical subsystems (Pikovsky, Rosenblum, & Kurths, 2003).

Lag Synchronization is an intermediate stage between Phase and Complete Synchronization. The two outputs of each subsystem lock their phases and amplitudes, but with the presence of a time lag. It implies the asymptotic boundedness of the difference between the output of one system at a specific instant and the output of the other, shifted in time of a lag time space (Schuster, & Just, 2005).

Another aspect in the synchronization of chaotic systems is the way it is established. When the coupling factor falls short of a critical value, the synchronized state becomes unstable and characteristic intermittent dynamics are observed. The difference between two signals from each subsystem blows occasionally from its almost zero value, demonstrating instant desynchronization. This kind of transition is termed *On-Off Intermittency* and its dynamics are quite different from those of the three classic types (scenarios) for chaotic systems (Toniolo, Provenzale, & Spiegel, 2002). The mechanism behind the intermittent synchronization of chaotic system is the antagonism between the instability of trajectories, due to chaotic elements and the synchronization inclination, due to coupling. As a result the strength of coupling plays a crucial role in the evolution of the phenomenon of on-off intermittency (Chen, Hong, Chen, & Hill, 2006).

In electronic circuits, simple resistive coupling is the most common way of synchronizing circuits, thus this kind of coupling is bidirectional. Unidirectional coupling is established by using buffers or amplifiers in the coupling branch. Direct coupling is another case that is used in unidirectional coupling and is applied in synchronization of chaotic electronic circuits, through cable or RF. As a result numerous applications in communications with the use of coupled chaotic circuits have appeared over the last 15 years.

3. CIRCUIT FOR CHAOTIC SYGNAL TRANSMISSION

3.1 Circuit Description and Synchronization Properties

A circuit capable of synchronized chaotic communication, suitable for transmission of digital signal is presented. It should be noted that the transmitted chaotic signal is not analog but discrete one. The transmitter and the receiver are identical circuits similar to those in (Mycolaitis, et. al. 1999). The circuits include an integrator based second order *RC* resonance loop, a comparator *H* (the circuit's non-linear element), an exclusive OR gate, with an input *M* for the external source and a buffer to avoid overloading of the XOR gate. *M(t)* can be a sequence of square pulses of period $T = 2\pi/\omega$ or a more complex signal if one wants to encode an arbitrary message, for example.

The transmitter-receiver system is shown in Figure 1. The principle of operation is demonstrated below. Here the chaotic pulses $U^*(t) \propto F(y_1, t)$ drive both the resonance loop of the transmitter and the resonance loop of the receiver.

The transmitter-receiver system is governed by the following set of equations:

Figure 1. Schematic diagram of the transmitter-receiver system

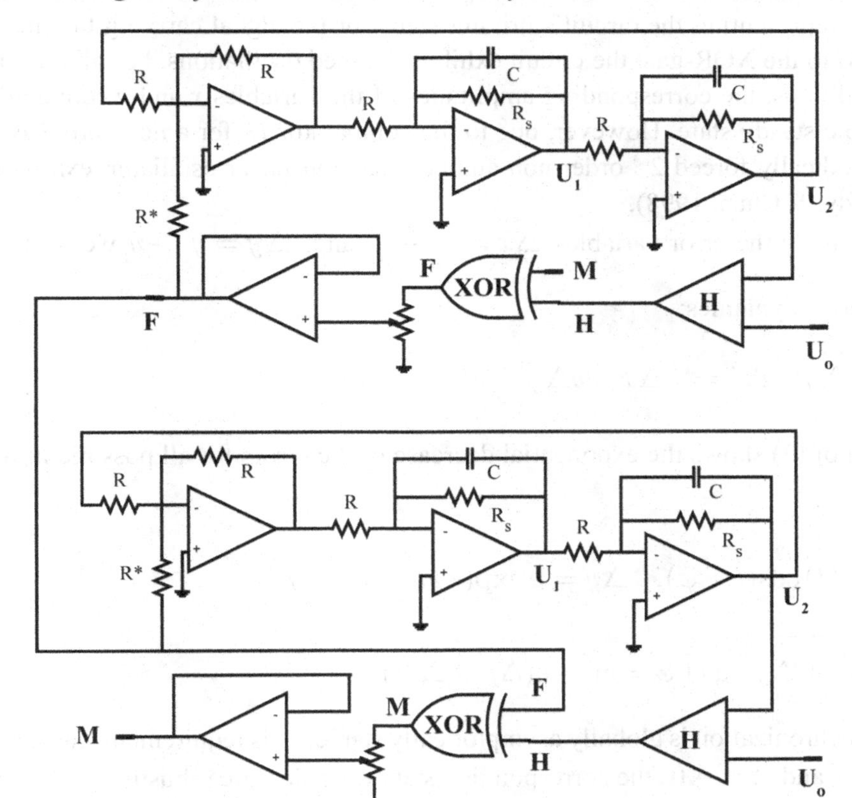

$$\begin{aligned} \dot{x}_1 &= aF(y_1,t) - bx_1 + y_1 \\ \dot{y}_1 &= -x_1 - by_1 \\ F(y_1,t) &= H(y_1) \oplus M(t) \\ \dot{x}_2 &= aF(y_1,t) - bx_2 + y_2 \\ \dot{y}_2 &= -x_2 - by_2 \end{aligned} \tag{1}$$

Subscripts '1' and '2' at the state variables specify the transmitter and the receiver, respectively. Note, the same driving term $F(y_1,t)$ in the equations for the transmitter and the receiver which represents the system's coupling factor. The following substitutions have been used in the previous system of equations since the parameters are usually written in a dimensionless form:

$$\begin{aligned} x &= \frac{U_1}{U_o}, \quad y = \frac{U_2}{U_o}, \quad t = \frac{t}{RC} \\ \pm &= \frac{U^* \cdot R}{U_o \cdot R^*}, \quad b = \frac{R}{R_s} \\ \omega &= \omega_M RC \end{aligned} \tag{2}$$

The shifted Heaviside function $H(y)$=$H(-y-1)$ has the following values $H(-y>1)=1$ and $H(-y\leq 1)=0$, while the symbol Å denotes the exclusive-OR operation and $M(t)$ denotes the normalized square pulses of period $2\pi/\omega$, representing the circuit's driving signal or the signal carrying the message. For zero external drive M to the XOR-gate the circuit exhibits damped oscillations. For all reasonable ($x_o^2 + y_o^2 < 1$) initial conditions, the corresponding amplitudes of the variables x and y converge exponentially ($\mu\ e^{-bt}$) to a stable steady state. However, due to the comparator H, for a non-zero drive M the circuit becomes a periodically forced 2nd order non-autonomous non-linear oscillator, exhibiting chaos (Kolumban, Kennedy, & Chua, 1998).

Introducing in (1) the error variables $\Delta x = x_2 - x_1$ and, $\Delta y = y_2 - y_1$ we obtain the equations governing the error dynamics:

$$\Delta \dot{x} = b\Delta x + \Delta y, \quad \Delta \dot{y} = -\Delta x - b\Delta y \tag{3}$$

The solution of (3) shows the exponential decrease of the errors for all possible initial errors Δx_0 and Δy_0 :

$$\Delta x = a\exp(-bt)\cos(t+\varphi), \quad \Delta y = a\exp(-bt)\sin(t+\varphi) \tag{4}$$

where $a = \sqrt{\Delta x_0^2 + \Delta y_0^2}$ and $\varphi = \arctan(\Delta y_0 / \Delta x_0)$

Thus, the synchronization is globally asymptotically stable. This requirement leads to the conclusion that for $\Delta x \to 0$ and $\Delta y \to 0$, the corresponding state variables, are robustly synchronized ($x_2 \to x_1$

and $y_2 \rightarrow y_1$). Consequently, the non-linear functions behave in a synchronous way $H(y_2) \rightarrow H(y_1)$ as well. This result suggests an extremely simple technique of recovering the signal $M(t)$ at the receiver end. The received signal $F(y_1, t) \propto U^*(t)$ is applied to the XOR unit of the receiver. Due to the sum mod2 property, the signal $M(t)$ can be recovered from the chaotic one $F(y_1, t)$ without any errors, according to:

$$\begin{aligned} &F(y_1, t) \oplus H(y_2) = H(y_1) \oplus M(t) \oplus H(y_2) \rightarrow \\ &\rightarrow H(y_1) \oplus M(t) \oplus H(y_1) = M(t) \end{aligned} \tag{5}$$

The above circuit was extensively studied and characterized, both theoretically as well as experimentally, and the rich structures of chaotic signals produced are reported in Refs. (Stavrinides, Kyritsi, Deliolanis, Anagnostopoulos, Tamaševičious, Čenys, 2004; Stavrinides, Deliolanis, Miliou, Laopoulos, & Anagnostopoulos, 2008; Stavrinides, Deliolanis, Miliou, Laopoulos, & Anagnostopoulos, 2008).

3.2 Synchronization: Experimental Results

Synchronization between transmitter and receiver was experimentally verified. Both sub-circuits remained synchronized under different conditions regarding the circuit parameters as well as the driving frequency f_M that was provided by a digital signal generator (HM8130). It should be noted that the driving frequency represents the external digital information that is fed to the communication system.

In this section a typical modulation-demodulation procedure through chaotic synchronization is presented. All signals were monitored by a digital storage oscilloscope (HP54603B), further connected to a PC for recording and analysis purposes, so that the proper characterization of the circuit behavior could be achieved. Appropriate software built in NI's LabView environment was used in order to control all digital instruments used and process the signals acquired (Stavrinides, Laopoulos, & Anagnostopoulos, 2005). In Figures 2-8 the following diagrams are presented:

- Digital input signal $M(t)$ and its power spectrum
- transmitter's chaotic output signal F and its power spectrum
- receiver's recovered output signal $M^*(t)$ and its power spectrum
- phase portrait $U_2 - U_3$ of transmitter's operation

These diagrams regard a typical case of chaotic synchronized transmission of a simple pulse series $M(t)$, in the case of $f_M = 6{,}222$ *KHz* ($U_o = 350$ *mV*, $U^* = 4$ $V_{p\text{-}p}$).

Transmitter's phase portraits in Figure 8 as well as the power spectrum (Figure 4) of the output signal $F(t)$ confirm the circuit chaotic mode of operation. Synchronization between the transmitter and the receiver is ascertained by the almost perfect recover of the input signal $M(t)$ at the output of the receiver $M^*(t)$, as proved by comparing either the signals $M(t)$ and $M^*(t)$ [Figures 2 and 6] or their power spectrums [Figures 3 and 7]. Glidges are an issue that arises in all systems using digital signals and are confronted by various proposed methods in the relevant literature (Weste, & Eshraghian, 1993).

Figure 2. Digital Input Signal M(t)

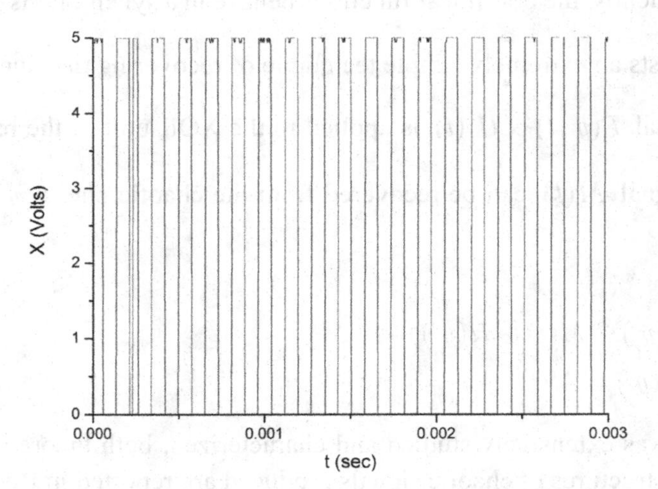

Figure 3. Power spectrum (FFT) for input signal M(t)

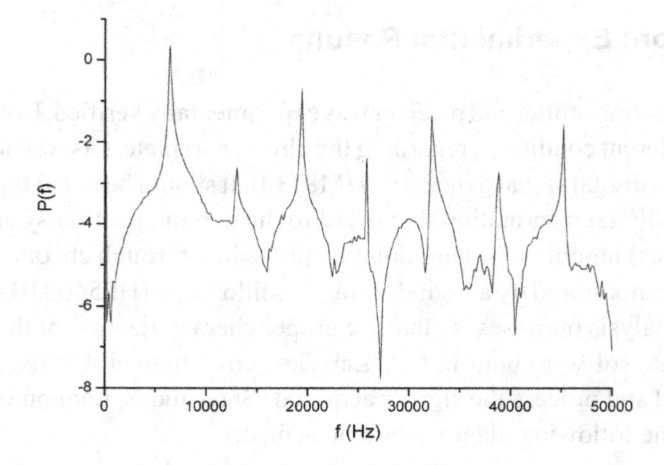

Figure 4. Chaotic Signal F(t) at transmitter's output

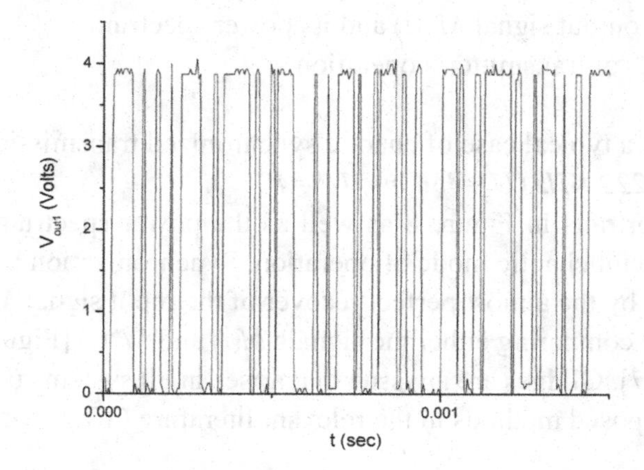

Figure 5. Power spectrum (FFT) for output signal F(t)

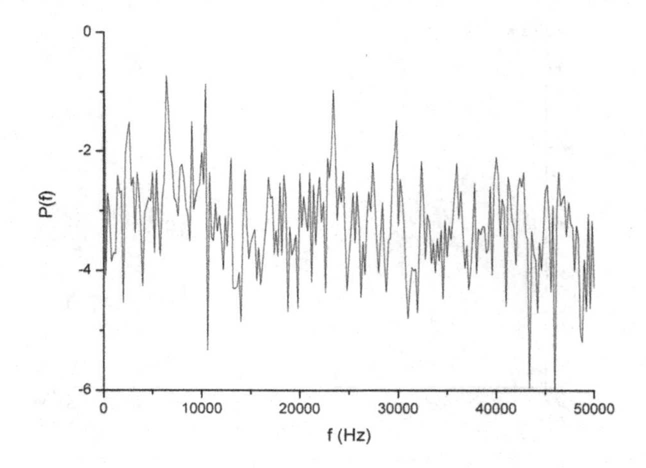

Figure 6. The recovered signal M(t) at the receiver's output*

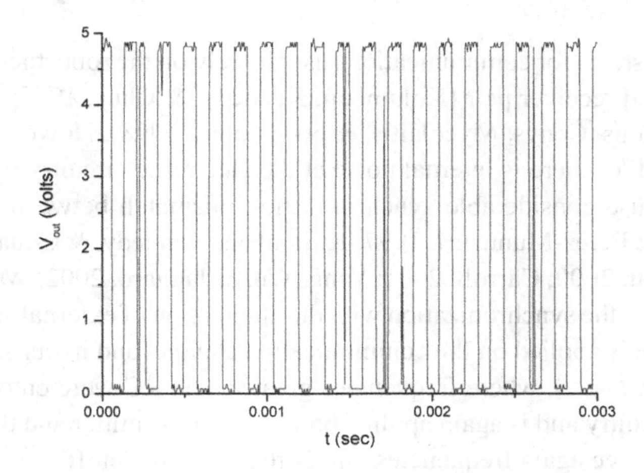

Figure 7. Power spectrum (FFT) for the recovered signal M(t)*

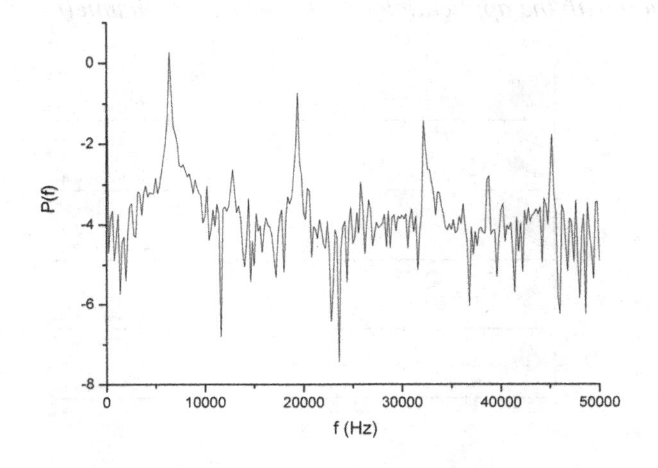

Figure 8. The phase portrait $U_2 - U_3$ characterizing transmitter's chaotic operation

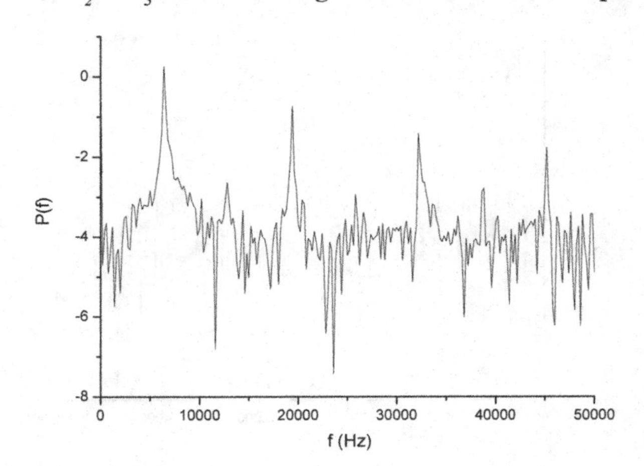

3.3 The Effect of Noise and Parameter Mismatch in the System

Synchronized chaotic systems for communications usually rely on the robustness of the synchronization within the transmitter and receiver pair (Kolumban, Kennedy, & Chua, 1997; Kolumban, Kennedy, & Chua, 1998; Tamaševičious, Čenys, Mycolaitis, & Namajunas, 1998). However, if the communication channel is imperfect and /or there is internal noise at the electronic circuitry the distorted signal at the receiver input might cause considerable synchronization mismatch between the transmitter-receiver pair (Sanchez, Matias, & Perez-Munuzuri, 1997; Kolumban, Kennedy, & Chua, 2000; Lorenzo, Perez-Munuzuri, & Perez-Villar, 2000; Carroll, 2001; Yang, Wu, & Jaggard, 2002; Wang, Hou, & Xin, 2005).

Figures 9 and 10 depict the synchronization with the application of external and internal noise respectively. The external noise is applied on the communication channel and in our simulation is represented by white noise added on $F(y_1,t)$, where frequencies greater than RC were cutoff. The internal noise is due to the electronic circuitry and is again applied both on the transmitter and the receiver (added on x_1, y_1 and x_2, y_2 variables). Once again frequencies higher than RC are cutoff.

Figure 9. Synchronization with the application of external noise (channel)

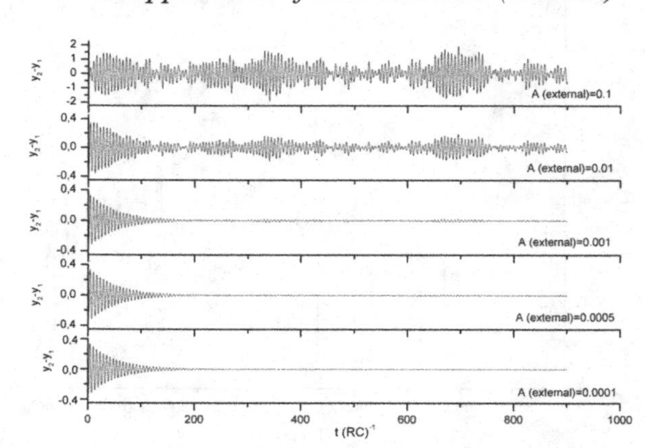

Figure 10. Synchronization with the application of internal noise (electronics)

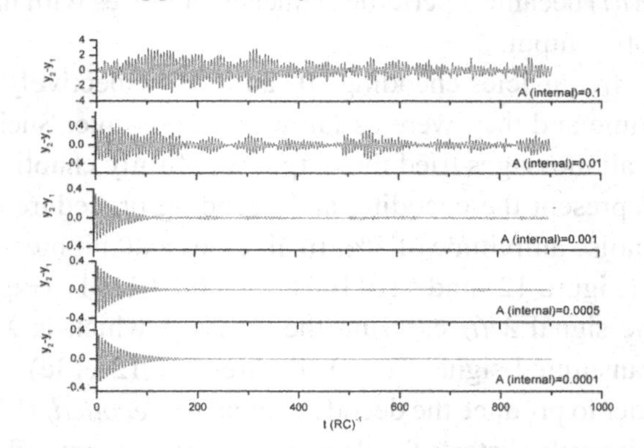

Different noise amplitudes A, have been utilized ranging from 0.01% to 50% of the mean signal amplitude. As the noise amplitude A is increased, the synchronization of the system continuously deteriorates and is practically destroyed in both cases (external and internal noise) above a certain noise level.

The artificial noise added at the simulation was produced as follows: A pseudorandom number generator produces an array of random numbers in the interval [0, 1]. The random numbers are equal to the total number of simulation steps. Then the Fourier Transform of this series is obtained by standard procedures and amplitudes for frequencies larger than a particular cutoff value are zeroed. The inverse Fourier Transform is employed in order to produce the noise series to be used in the subsequent simulation. In our simulation we cutoff frequencies larger than the characteristic frequency of the system RC. By this "noise filtering" procedure we avoid the dependence of the generated noise on the simulation step. Noise was added at every simulation step, to the signal $F(y_1,t)$ coming out of the transmitter (external noise) or to each of the dynamic variables x_1, y_1 and x_2, y_2 of the transmitter and receiver in respect (internal noise). In the latter case, four different noise series were used one for each dynamic variable.

There are several options on how to construct the signal $M(t)$ in order to load a specific binary message onto the transmitted signal ($F(y_1,t)$). However, since for security reasons the system must operate in the chaotic regime which strongly depends on the driving signal $M(t)$, the encoding scheme must be such that this requirement is met for all possible messages. Therefore, we attempted two different encoding schemes: in scheme A, "1" and "0" were represented each by a square pulse of a single period duration, of the same frequency but with opposite phases, while in scheme B, "1" and "0" were represented by square pulses of single period duration, of the same phase but of two distinct frequencies.

In both scheme we tried three types of messages: (i) a sequence of several "1", (ii) an alternating sequence of "1" and "0" and (iii) a random binary string. The messages were loaded on the signal $M(t)$ starting at $t>t_{synch}$, where t_{synch} was the characteristic time for the synchronization to be achieved. For times t outside the time window where the message was loaded, $M(t)$ was effectively "padded" with zeros (essentially a periodic train of square pulses).

In realizing scheme A it was very difficult to produce chaotic output of the system *for all*types of messages that we tried. The reason was that encoding a "0" and "1" by a pulse of opposite phase leads, in certain combinations of "0" and "1", to a an effective change in the frequency of the signal $M(t)$ so that the system was carried outside of the chaotic regime. For example, if one was encoding a message of alternating "0" and "1" (101010101...) using a frequency value ω=1.1 for each pulse (which lies in the chaotic regime), the signal $M(t)$ became a periodic sequence of pulses with half the frequency (ω=0.55), which did not produce chaotic output.

In scheme B, for the two frequencies encoding "0" and "1" respectively, we picked two values that each lied in the chaotic regime and they were as far apart as possible. Such values were ω_1=0.75 and ω_2=1.15. We found that for all messages tried the output was firmly chaotic.

Figures 11, 12 and 13 represent the encoding and decoding procedure for the three types of messages tried and for internal noise amplitude of 5%. In all cases a 20 bit message was used: "1111...111" (Figure 11), "1010...1010" (Figure 12) and "10110011101000101101" (Figure 13). In each figure the first one from the top is the signal $M(t)$ carrying the message which is XORed with $H(y_1)$ (Figures 11b,12b,13b) to give the transmitted signal $F(y_1,t)$ (Figures 11c,12c,13c) which is XORed with $H(y_2)$ (Figures 11d,12d,13d) in order to produce the decoded signal *M(decoded)* (Figures 11e,12e,13e). We see that internal noise although it partly distorts the decoded signal does not affect the bit error rate because the correct bits can be recognized and restored by a suitable decision algorithm.

Figure 14 demonstrates the square mean mismatch $\sqrt{<\Delta y^2>}$ over the initial $\Delta y = y_2 - y_1$ (=0.3) and the mean difference $\left[MD = \frac{1}{t}\int_0^t (M(t) - M_d(t))dt \right]$ between signals $M(t)$ and $M_d(t)$ in the time interval of the simulation versus the noise amplitude, for external and internal noise. The signal $M(t)$ in this case was simply a periodic sequence of square pulses of frequency ω=1.1. As expected, both system synchronization and *MD* (message mismatch) become worst as noise amplitude increases and internal noise, is affecting the system's synchronization more than external noise of the same amplitude level.

Figure 15 illustrates the quality of synchronization using a different measure, namely the fractional deviation of the mean square difference of $\sqrt{<\Delta y^2>}$ from the initial Δy (=0.3). This quantity is equal to unity for perfect synchronization. The internal noise, is affecting the system's synchronization more than external noise of the same amplitude level.

Figure 11. The effect of internal noise on the received (decoded) signal $M_d(t)$ for a message "1111...1111"

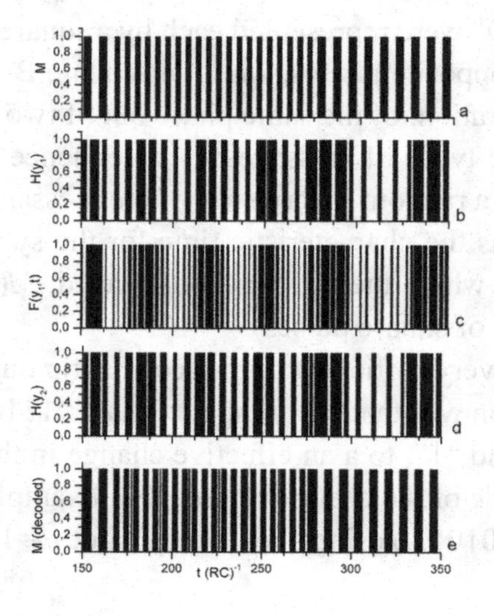

Figure 12. The effect of internal noise on the received (decoded) signal $M_d(t)$ for a message "1010...1010"

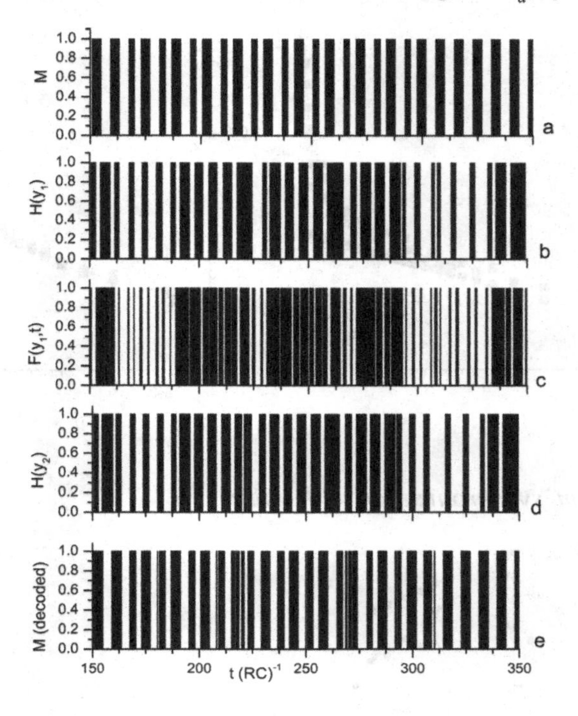

Figure 13. The effect of internal noise on the received (decoded) signal $M_d(t)$ for a message "10110011101000101101"

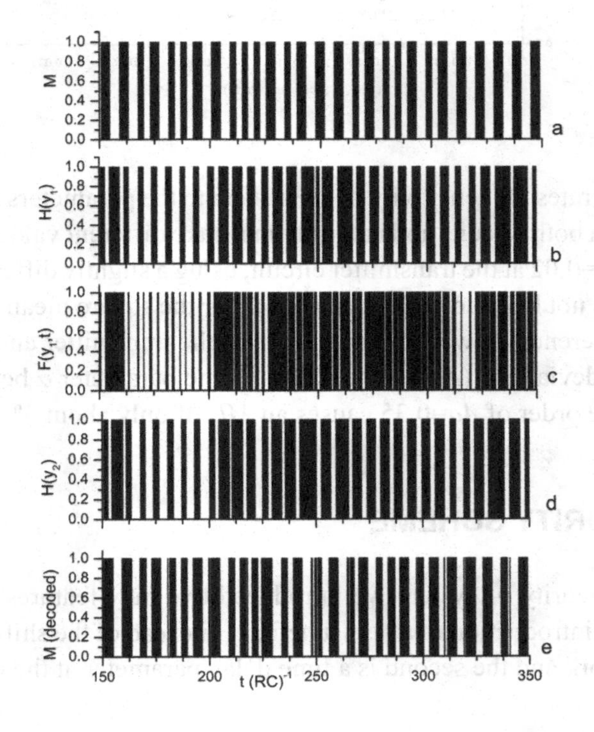

Figure 14. The effect of noise on the square mean mismatch of the y parameters and MD

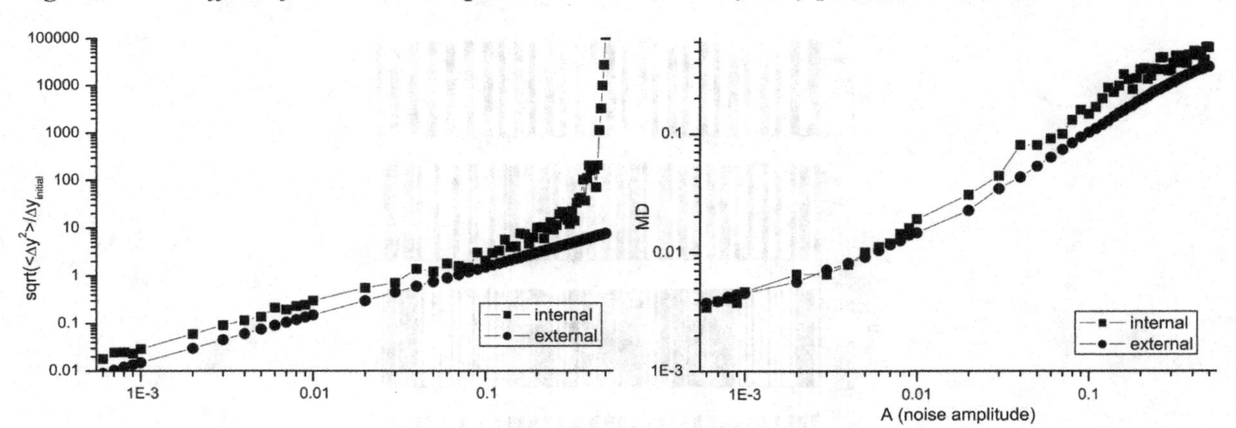

Figure 15. Quality of system' synchronization with noise

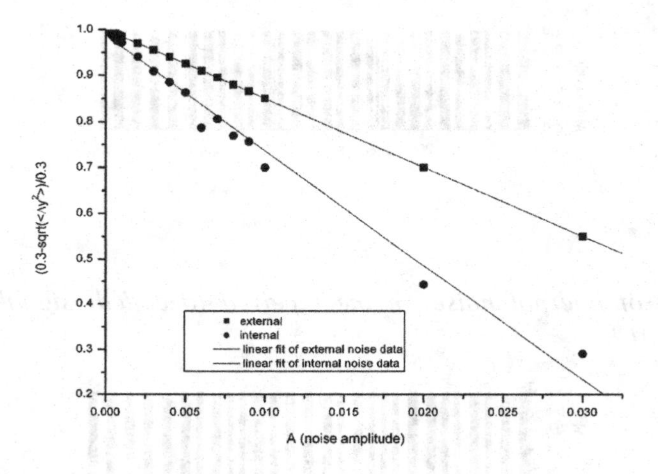

Finally, Figure 16 illustrates the sensitivity of the system to the parameters α and b. These parameters depend on resistors used in both circuits and naturally may take different values between transmitter and receiver. For α=2.65 and b=0.02 at the transmitter circuit, using a slightly different value of b (b=0.0199) at the receiver circuit does not have any effect on the *MD* or the square mean mismatch of the y parameters regardless of the difference between α parameters of the transmitter and the receiver. The system is also robust in terms of deviations ($\Delta\alpha$) of the values of the parameter α between the transmitter and receiver; a deviation in the order of $\Delta\alpha$=0.35 causes an *MD* of only about 3%.

4. ENHANCED SECURITY SCHEME

In order to enhance the security of system we introduced two new features in the original electronic circuit. The first one is the introduction of a new function H instead of the shifted Heaviside proposed in the third section of this work and the second is a time delay parameter at the non-linear driving term in

Figure 16. Sensitivity of the system to parameters α and b

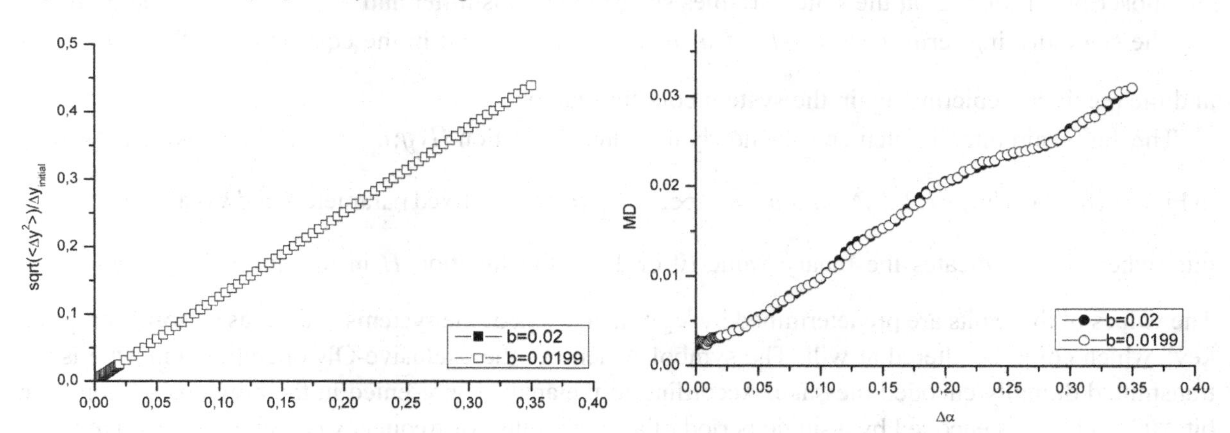

order to improve the chaotic nature of the dynamical system. Figure 17 presents the proposed security-enhanced non-linear electronic circuit.

The new transmitter-receiver system is governed by the following set of equations:

$$\begin{aligned}
\dot{x}_1 &= aF(y_1(t-t_d),t;a_1,a_2,...,a_{N_a},k) - bx_1 + y_1 \\
\dot{y}_1 &= -x_1 - by_1 \\
F(y_1(t-t_d),t;a_1,a_2,...,a_{N_a},k) &= H(y_1(t-t_d);a_1,a_2,...,a_{N_a},k) \oplus M(t) \\
\dot{x}_2 &= aF(y_1(t-t_d),t;a_1,a_2,...,a_{N_a},k) - bx_2 + y_2 \\
\dot{y}_2 &= -x_2 - by_2
\end{aligned} \tag{6}$$

Figure 17. Schematic diagram of the transmitter-receiver system

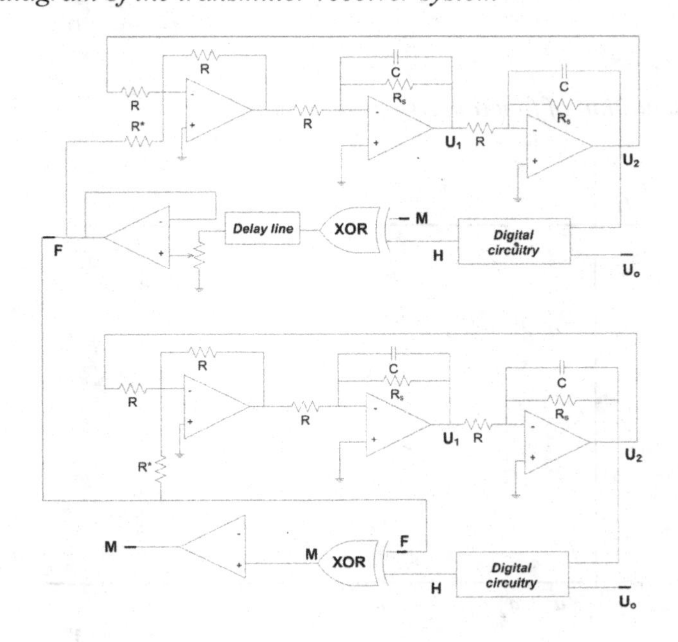

Subscripts '1' and '2' at the state variables specify the transmitter and the receiver, respectively. The same driving term $F(y_1(t-t_d),t;a_1,a_2,...,a_{N_a},k)$ is used in the equations for the transmitter and the receiver, depicting again the system coupling factor.

The digital circuitry, implements the newly introduced function $H(y;a_1,a_2,...,a_{N_a},k)$, which is shown in Figure 18. The a_j's ($j=1,2,...,N_a$ and $a_o=-\infty, a_{N_a+1}=\infty$) are fixed parameters and k is a set of N_a+1 bits, where bit j indicates the binary value (0 or 1) of the function H in the interval $a_j < y < a_{j+1}$. The values of these bits are predetermined by legitimate users of the systems and act as a "soft encryption key" which could be altered at will. The symbol Å denotes the exclusive-OR operation and $M(t)$ is the transmitted digitally encoded message. According to the analysis presented in the previous section each bit, "0", and "1", is encoded by a single period of a square pulse of frequency ω_1 and ω_2 respectively.

The message, $M(t)$, is XORed with the value of the function $H(y(t-t_d))$ yielding the transmitted *digital* signal F which drives the dynamical system of the receiver. If the system of the transmitter and the receiver are identical, the dynamic variables are synchronized enabling the decoding of the message.

As shown in [3] the synchronization is globally asymptotically stable. Thus the corresponding state variables are robustly synchronized ($x_2 \to x_1$ and $y_2 \to y_1$). Consequently, the non-linear functions behave in a synchronous way $H(y_2;a_1,a_2,...,a_{N_a},k) \to H(y_1;a_1,a_2,...,a_{N_a},k))$ as well. This result suggests an extremely simple technique for recovering the signal $M(t)$ at the receiver end. The received signal $F(y_1(t-t_d),t;a_1,a_2,...,a_{N_a},k) \propto U^*(t-t_d)$ is applied to the XOR unit of the receiver. Due to the sum mod2 property, the signal $M(t)$ can be recovered from the chaotic one $F(y_1(t-t_d),t;a_1,a_2,...,a_{N_a},k)$ without any errors, according to:

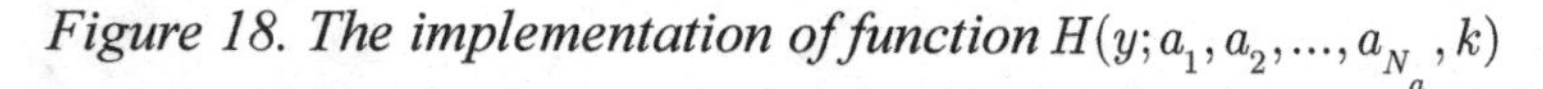

Figure 18. The implementation of function $H(y;a_1,a_2,...,a_{N_a},k)$

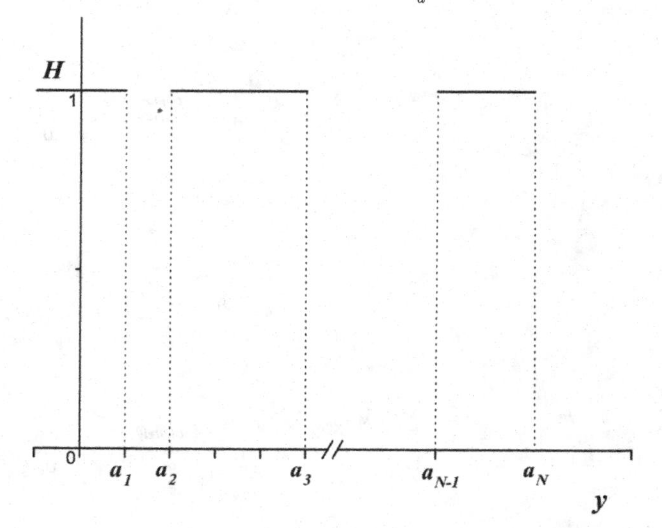

$$F(y_1(t-t_d),t;a_1,a_2,...,a_{N_a},k)\oplus H(y_2(t-t_d);a_1,a_2,...,a_{N_a},k)=$$
$$H(y_1(t-t_d);a_1,a_2,...,a_{N_a},k)\oplus M(t)\oplus H(y_2(t-t_d);a_1,a_2,...,a_{N_a},k)\to \tag{7}$$
$$\to H(y_1(t-t_d);a_1,a_2,...,a_{N_a},k)\oplus M(t)\oplus H(y_1(t-t_d);a_1,a_2,...,a_{N_a},k)=M(t)$$

The division of the y range in many regions, instead of only two presented in [3] and the fact that there is the encryption key, *k*, determining the value of *H* in the corresponding region (either 0 or 1) introduces an uncertainty as to how well the intruder could predict the values of *H*(*y*) even if he manages to perfectly synchronize his system with the transmitter. The intruder must know the values of *all* the limits of the regions (a_j) as well as all bits of the key *k* in order to accurately retrieve the message. The larger the number of a_js and the longer the key, the hardest it is to break the system by brute force.

Moreover, the introduction of a time delay parameter, t_d, in the system improves the chaotic nature of the dynamical system. For example, Ref (Dorizzi, et. al. 1987) has shown that the introduction of a time delay in a dissipative system with a non-linear driving term increases the attractor dimension and yields a probability density function which tends asymptotically to a Gaussian. The high degree of complexity thus obtained makes such systems good candidates for signal encryption; an intruder that is able to intercept the transmitted signal without possessing an identical dynamical system will have a much greater difficulty in separating the message from the chaotic carrier. Even if one possesses an identical dynamical system and manages to synchronize with the transmitter, one will not be able to decode the message if the exact value of the time delay t_d is not known.

Equations (6) have been numerically integrated. The chaotic behavior of the system depends also on the value of the time delay parameter, the position of the a_j and the key *k*. Strong chaotic signals were obtained for time delay values ranging from t_d=20 to 50.

In order to test the security of the newly proposed encoding scheme, we first assumed that an intruder knows the exact value of the key *k* and the values of the a_j with some uncertainty. By varying the uncertainty in the values of a_j we investigated its effect on the Bit Error Rate (BER) suffered by the intruder at the decoding stage. We define two measures of the BER; the Raw BER (RBER) which is the average mismatch between the actual message *M*(*t*) and the decoded message $M_d(t)$

$$RBER=\frac{1}{t}\int^{t}\left|M(t)-M_d(t)\right|dt, \tag{8}$$

and the Relative BER (RelBER) which is RBER divided by the fractional difference of the two encoding frequencies ω_1 and ω_2

$$\text{RelBER}=\frac{RBER}{Z},\quad \text{where}\quad Z=\frac{\left|\omega_1-\omega_2\right|}{\frac{1}{2}\left(\omega_1+\omega_2\right)} \tag{9}$$

The rational for the second definition is that the smaller the difference between the encoding frequencies, the larger the uncertainty in distinguishing between values "0" and "1" when there is a mismatch between the decoded signal M_d and the actual *M*. In all our simulations we used the values ω_1=0.75 and ω_2=1.15 (in RC units).

In Figure 19, we plot the RelBER versus Δa, the fractional mismatch between the values of each respective a_j of the transmitter and the receiver (intruder).

Figure 19. Relative BER vs. fractional mismatch Δa

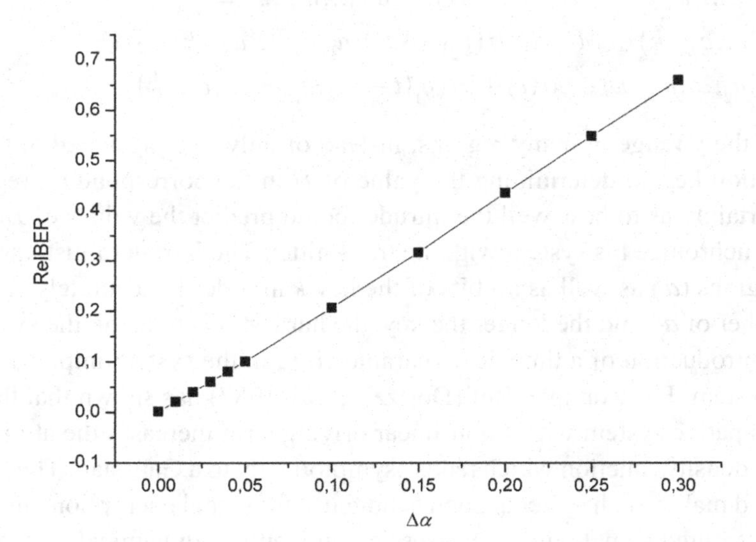

We notice that RelBER rises linearly with Δa and for a relative low mismatch of Δa ~22% the intruder has a RelBER of about 0.5 which means almost 100% uncertainty in the knowledge of *M*. This is because a Bit Error Rate of 50% means that for each bit there is a probability of ~50% to get the correct value, which is essentially as good as a guess. In the present simulation we used only 5*a*'s (N_a=5).

Next we assumed that the intruder knows the exact values of all the *a*'s and all the bits of the key *k* except of one. This means that the decoded signal will be incorrect for all values of y that lie between the *a*'s corresponding to the unknown bit. Thus, we expect that the BER will be proportional to the size, y_{mis}, of the interval in *y* where the key mismatch occurs.

In Figure 20, we plot the dependence of the *Bit Uncertainty* versus the total mismatch y_{mis} for different positions of the incorrect bit in the key *k*, where Bit Uncertainty is defined as

$$\text{Bit Uncertainty} = 2 \cdot RBER \qquad (10)$$

As expected by the ergodicity of the system in *y* the BER depends linearly on the size of the mismatch.

The above result is further demonstrated in Figure 21 where we plot the Bit Uncertainty versus y_{mis} for three different values of N_a (5, 9 and 11) and assuming that the intruder knows all bits of the key *k* in each case. The fractional mismatch Δa in each a_j is the same for all three cases and equal to 0.05. By increasing the N_a, the total y_{mis} increases.

In all the above results we presume that the intruder is able to synchronize perfectly with the transmitter, i.e. he knows exactly the system dynamic parameters. Also, that he knows all the a_j to a good precision and all or almost all the bits of the key *k*. In actuality, the intruder is much more disadvantaged as the N_a increases and the quality of the synchronization is not perfect (he doesn't possess an identical system to the transmitter).

Finally, we investigated the dependence of Bit Uncertainty on the mismatch in the time delay between the receiver and the transmitter. We assume that the intruder perfectly matches all system parameters (including the a_j and the key *k*), except for the time delay. The time delay t_d of the transmitter is 50 $(RC)^{-1}$.

Figure 20. Bit Uncertainty vs. y_{mis} for different positions of the incorrect bit in the key k. The line corresponds to a linear least squares fit

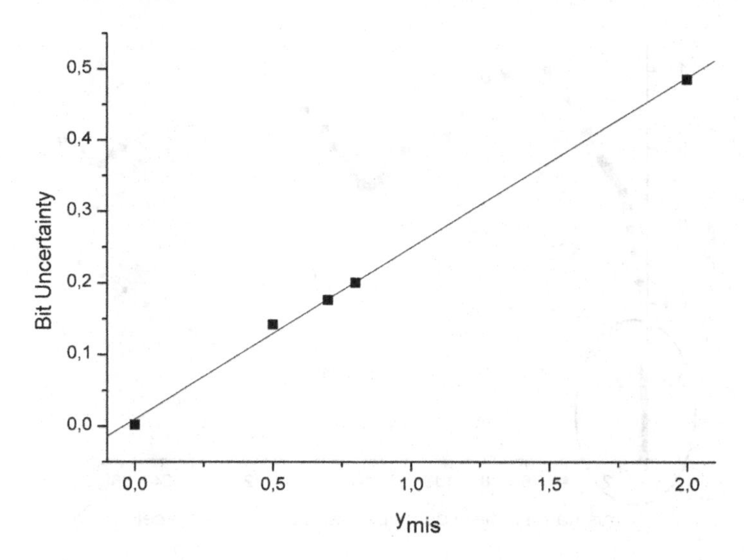

Figure 21. Bit Uncertainty vs. y_{mis} for three different values of N_a. The line corresponds to a linear least squares fit

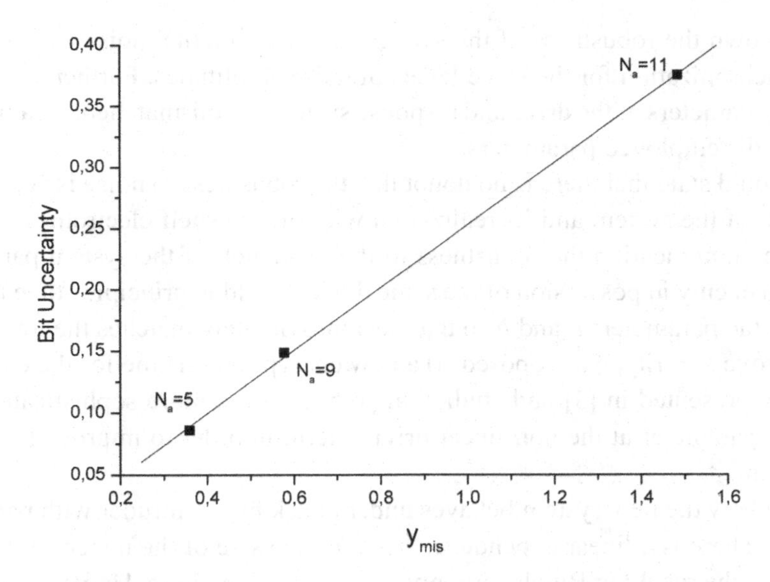

In Figure 22 we plot the Bit Uncertainty versus the mismatch in time delay between the receiver (intruder) and the transmitter. We notice that Bit Uncertainty is very sensitive to the time delay mismatch; for a mismatch of only 0.1 $(RC)^{-1}$ (=0.2%) the Bit Uncertainty is as high as ~27%. The Bit Uncertainty grows linearly for small values of mismatch while it reaches its maximum value of ~1 for a mismatch of ~2.5 $(RC)^{-1}$ (=5%). For larger values of mismatch the Bit Uncertainty oscillates with a period of ~6.85 $(RC)^{-1}$. The rise and fall of the Bit Uncertainty is probably attributed to the relation of the delay mismatch with the encoding frequencies ω_1 and ω_2 .

Figure 22. Bit Uncertainty vs. mismatch in time delay between the receiver and the transmitter. The inset figure is a blow-up of circled region

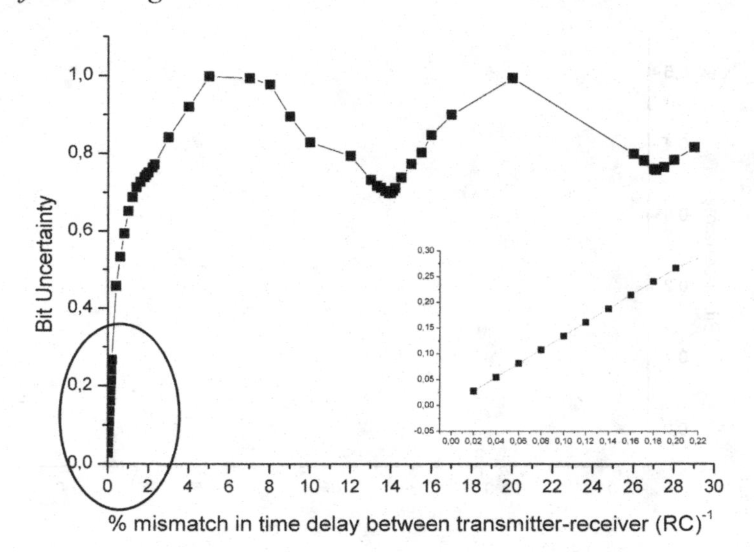

5. CONCLUDING REMARKS

The results have shown the robustness of the system to noise. Internal noise has more influence than external on the synchronization for the same levels of noise amplitudes. Furthermore, synchronization occurs, even if the parameters of the drive and response system are mismatched, meaning that the system is robust regarding the employed parameters.

Moreover, we could state that there is no doubt that the robustness to noise is very advantageous for the synchronization of the system and its realization with off the shelf electronics. However, in terms of the security of communication the robustness to the mismatch of the system parameters would be destructive since an enemy in possession of the same device could in principle "tune in" the transmitted signal by adjusting the parameters a and b so that he approximately matches the transmitter's values.

In order to improve security, we proposed (i) a new encryption scheme for the chaotic communication system already presented in [3] and studied in [], based on a more sophisticated process; (ii) we added a time delay parameter at the non-linear driving term in order to improve the chaotic nature of the dynamical system.

We investigated how the new system behaves under attack by an intruder with partial knowledge of system parameters. There is a linear dependence between the size of the uncertainty in the encryption parameters (a_j,k) and the resulting Bit Uncertainty at the decoding phase. However, small uncertainties in the encryption parameters lead to high Bit Uncertainties.

Furthermore, the Bit Uncertainty is very sensitive to time delay mismatch between the transmitter and the receiver a fact that produces an extra disadvantage for the potential intruder.

Weak points of the proposed system are: (i) the fact that by changing the values and the number of the encryption parameters one often has to retune the other system parameters (time delay, encoding frequencies, etc) in order to achieve a good chaotic signal, (ii) the encryption scheme is vulnerable under a known-message attack in the case of that an intruder knows all parameters of the dynamical system but not the encryption parameters and he possesses both the encrypted and decrypted versions on unknown message. This is because he could easily XOR the encrypted signal with the known message and restores the encrypting function H which could be used to decode all subsequent communications. Therefore, valid users must alter the encryption key k at each communication.

The great advantage of the system still remains the fact that it transmits a digital chaotic signal which gives all the advantages of digital technology (for example filtering of channel noise, etc).

REFERENCES

Boccaletti, S., Kurths, J., Osipov, G., Valladares, D. L., & Zhou, C. S. (2002). The synchronization of chaotic systems. *Physics Reports, 366*, 1–101. doi:10.1016/S0370-1573(02)00137-0

Carroll, T. L. (2001). Noise-resistant chaotic synchronization. *Physical Review E: Statistical, Nonlinear, and Soft Matter Physics*, *64*, 1–4. doi:10.1103/PhysRevE.64.015201

Carroll, T. L., & Pecora, L. M. (1993). Synchronizing nonautonomous chaotic circuits. *IEEE Trans. Circuits and Systems II*, 40, 646-650.

Chambers, W. G., & Frey, D. (1993). Comments on Chaotic digital encoding: an approach to secure communication and reply. *IEEE Trans. On Circuits and Systems II*, 46, 1445–1448.

Chen, Q., Hong, Y., Chen, G., & Hill, D. J. (2006). Intermittent phenomena in switched systems with high coupling strengths. *IEEE Trans Circuits and Systems I*, 53, 2692-2704.

Chien, T., & Liao, T.-L. (2005). Design of secure digital communication systems using chaotic modulation, cryptography and chaotic synchronization. *J. of Chaos Solitons and Fractals*, *24*, 241–255.

Chua, L. O., Yang, T., Zhong G. Q., & Wu, C. W. (1996). Synchronization of Chua's circuits with time-varying channels and parameters. *IEEE Trans. Circuits and Systems I*, 43, 862-868.

Chua, L. O., Yang, T., Zhong, G. Q., & Wu, C. W. (1996). Adaptive synchronization of Chua's Oscillators. *International Journal of Bifurcation and Chaos in Applied Sciences and Engineering*, *6*, 189–201. doi:10.1142/S0218127496001946

Cuomo, K. M., & Oppenheim, A. V. (1993). Circuit implementation of synchronized chaos with applications to communications. *Physical Review Letters*, *71*, 65–68. doi:10.1103/PhysRevLett.71.65

Cuomo, K. M., Oppenheim A. V., & Strogatz, S. H. (1993). Synchronization of Lorenz based chaotic circuits with applications to communications. *IEEE Trans. Circuits and Systems I*, 40, 626-633.

Dedieu, H., Kennedy M. P., & Hasler, M. (1993). Chaos shift keying: modulation and demodulation of a chaotic carrier using self-synchronizing Chua's circuits. *IEEE Trans. Circuits and Systems I*, 40, 634-642.

Dorizzi, B., Grammaticos, B., Le Berre, M., Pomeau, Y., Ressayre, E., & Taller, A. (1987). Statistics and dimension of chaos in differential delay systems. *Physical Review A.*, *35*, 328–339. doi:10.1103/PhysRevA.35.328

Frey, D. (1993). Chaotic digital encoding: an approach to secure communication. *IEEE Trans. Circuits and Systems II*, 40, 660–666.

He, R., & Vaidya, P. G. (1998). Implementation of chaotic cryptograph with chaotic synchronization. *Physical Review E: Statistical Physics, Plasmas, Fluids, and Related Interdisciplinary Topics*, *57*, 1532–1535. doi:10.1103/PhysRevE.57.1532

Hornby, A. S. (1984). *Oxford Advanced Learner's Dictionary*. Oxford, UK: Oxford University Press.

Huygens, C. (1986). *Horologium Oscillatorium* [The Pendulum clock]. Iowa City, IA: Iowa State University Press. (Original manuscript published in 1673).

Kolumban, G., Kennedy, M. P., & Chua, L. O. (1997). The role of synchronization in digital communications using chaos – part I: fundamentals of digital communications, *IEEE Trans. Circuits and Systems I*, 44, 927-936.

Kolumban, G., Kennedy, M. P., & Chua, L. O. (1998). The role of synchronization in digital communications using chaos – part II: chaotic modulation and chaotic synchronization. *IEEE Trans. Circuits and Systems I*, 45, 1129-1140.

Kolumban, G., Kennedy, M. P., & Chua, L. O. (2000). The role of synchronization in digital communications using chaos– part II: Chaotic Modulation and Chaotic Synchronization. *IEEE Trans. Circuits and Systems I*, 47, 1129-1140.

Lorenzo, M. N., Perez-Munuzuri, V., & Perez-Villar, V. (2000). Noise Performance of a synchronization scheme through compound chaotic signal. *International Journal of Bifurcation and Chaos in Applied Sciences and Engineering*, *10*, 2863–2870. doi:10.1142/S0218127400001912

Murali, M., & Lakshmanan, M. (1994). Drive-response scenario of chaos synchronization in identical nonlinear systems. *Physical Review E: Statistical Physics, Plasmas, Fluids, and Related Interdisciplinary Topics*, *49*, 4882–4885. doi:10.1103/PhysRevE.49.4882

Mycolaitis, G., Tamaševičious, A., Čenys, A., Namajunas, A., Navionis, K., & Anagnostopoulos, A. N. (1999). Globally synchronizable non-autonomous chaotic oscillator. In *Proc. of 7th International Workshop on Nonlinear Dynamics of Electronic Systems,* Denmark (pp. 277-280).

Ogorzalek, M. J. (1993). Taming chaos—Part I: Synchronization. *IEEE Trans. Circuits and Systems I*, 40, 693–699.

Ott, E., Grebogi, C., & Yorke, J. A. (1990). Controlling chaos. *Physical Review Letters*, *64*, 1196–1199. doi:10.1103/PhysRevLett.64.1196

Pecora, L. M., & Carroll, T. L. (1990). Synchronization in chaotic systems. *Physical Review Letters*, *64*, 821–823. doi:10.1103/PhysRevLett.64.821

Pikovsky, A., Rosenblum, M., & Kurths, J. (2003). *Synchronization: A universal concept in nonlinear sciences*. Cambridge, UK: Cambridge University Press.

Rulkov, N. F. (1996). Images of synchronized chaos: experiments with circuits. *Chaos (Woodbury, N.Y.)*, *6*, 262–279. doi:10.1063/1.166174

Rulkov, N. F., Sushchik, M. M., Tsimring, L. S., & Abarbanel, H. D. I. (1995). Generalized synchronization of chaos in directionally coupled chaotic systems. *Physical Review E: Statistical Physics, Plasmas, Fluids, and Related Interdisciplinary Topics*, *51*, 980–994. doi:10.1103/PhysRevE.51.980

Sanchez, E., Matias, M. A., & Perez-Munuzuri, V. (1997). Analysis of synchronization of chaotic systems by noise: an experimental study. *Physical Review E: Statistical Physics, Plasmas, Fluids, and Related Interdisciplinary Topics*, *56*, 4068–4071. doi:10.1103/PhysRevE.56.4068

Schuster, H. G., & Just, W. (2005). *Deterministic Chaos: An Introduction*. Berlin: Wiley-VCH Verlag. doi:10.1002/3527604804

Short, K. M. (1994). Steps toward unmasking secure communications. *International Journal of Bifurcation and Chaos in Applied Sciences and Engineering*, *4*, 959–977. doi:10.1142/S021812749400068X

Stavrinides, S. G. (2007). *Characterization of the behavior of a nonlinear electronic oscillator producing chaotic signals.* Unpublished doctoral dissertation, Aristotle University of Thessaloniki.

Stavrinides, S. G., Deliolanis, N. C., Miliou, A. N., Laopoulos, T., & Anagnostopoulos, A. N. (2008). Internal Crisis in a Second-Order Non-Linear Non-Autonomous Electronic Oscillator. *J. of Chaos Solitons and Fractals*, *36*, 1055–1061. doi:10.1016/j.chaos.2006.07.025

Stavrinides, S. G., Kyritsi, K. G., Deliolanis, N.C., Anagnostopoulos, A. N., Tamaševičious, A., & Čenys, A. (2004). The period doubling route to chaos of a second order non-linear non-autonomous chaotic oscillator – Part I. *J. of Chaos Solitons and Fractals*, 20, 849-854.

Stavrinides, S. G., Kyritsi, K. G., Deliolanis, N. C., Anagnostopoulos, A. N., Tamaševičious, A., & Čenys, A. (2004). The period doubling route to chaos of a second order non-linear non-autonomous chaotic oscillator – Part II. *J. of Chaos Solitons and Fractals*, *20*, 843–847. doi:10.1016/j.chaos.2003.09.008

Stavrinides, S. G., Laopoulos, T., & Anagnostopoulos, A. N. (2005). An Automated Acquisition setup for the analysis of chaotic systems. In *Proc. IEEE IDAACS 2005*, Sofia, Bulgaria (pp. 628-632).

Stavrinides, S. G., Miliou, A. N., Laopoulos, Th., & Anagnostopoulos, A. N. (2008). The Intermittercy Route to Chaos of an Electronic Digital Oscillator. *International Journal of Bifurcation and Chaos in Applied Sciences and Engineering*, *18*, 1561–1566. doi:10.1142/S0218127408021178

Tamaševičious, A., Čenys, A., Mycolaitis, G., & Namajunas, A. (1998). Synchronizing hyperchaos in infinite-dimensional dynamical systems. *J. of Chaos Solitons and Fractals*, *9*, 1403–1408. doi:10.1016/S0960-0779(98)00072-1

Toniolo, C., Provenzale, A., & Spiegel, E. A. (2002). Signature of on-off intermittency in measured signals. *Physical Review E: Statistical, Nonlinear, and Soft Matter Physics*, *66*(066209), 1–066209.

Wang, M., Hou, Z., & Xin, H. (2005). Internal noise-enhanced phase synchronization of coupled chemical chaotic oscillators. *Journal of Physics. A. Mathematical Nuclear and General*, *38*, 145–152. doi:10.1088/0305-4470/38/1/010

Weste, N., & Eshraghian, K. (1993). *Principles of CMOS VLSI Design: A systems Prospective*. AT&T.

Wu, C. W., & Chua, L. O. (1994). A simple way to synchronize chaotic systems with applications to secure communication systems. *International Journal of Bifurcation and Chaos in Applied Sciences and Engineering, 3*, 1619–1627. doi:10.1142/S0218127493001288

Wu, C. W., Yang, T., & Chua, L. O. (1996). On adaptive synchronization and control of nonlinear dynamical systems. *International Journal of Bifurcation and Chaos in Applied Sciences and Engineering, 6*, 455–471. doi:10.1142/S0218127496000187

Yang, L. B., & Yang, T. (1997). Synchronization of Chua's circuits with parameter mismatching using adaptive model-following control. [English Version]. *Chinese Journal of Electronics, 6*, 90–96.

Yang, T. (2004). A survey of chaotic secure communication systems. *International Journal of Bifurcation and Chaos in Applied Sciences and Engineering, 2*, 81–139.

Yang, T., & Chua, L. O. (1996). Secure communication via chaotic parameter modulation. *IEEE Trans. Circuits and Systems I*, 43, 817-19.

Yang, T., & Chua, L. O. (1997). Impulsive control and synchronization of nonlinear dynamical systems and application to secure communication. *International Journal of Bifurcation and Chaos in Applied Sciences and Engineering, 7*, 645–664. doi:10.1142/S0218127497000443

Yang, T., & Chua, L. O. (1997). Impulsive stabilization for control and synchronization of chaotic systems: Theory and application to secure communication. *IEEE Trans. Circuits and Systems I*, 44, 976-988.

Yang, T., & Sharuz, S. (1997). Channel-independent chaotic secure communication system using general chaotic synchronization. *Telecommunications Review, 7*, 240–254.

Yang, T., Wu C. W., & Chua, L. O. (1997). Cryptography based on chaotic systems. *IEEE Trans. Circuits and Systems I*, 44, 469-472.

Yang, T., Yang, L. B., & Yang, C. M. (1997). Impulsive synchronization of Lorenz systems. *Physics Letters. [Part A], 226*, 349–354. doi:10.1016/S0375-9601(97)00004-2

Yang, T., Yang, L. B., & Yang, C. M. (1998). Breaking Chaotic Switching Using Generalized Synchronization: Examples. *IEEE Trans. Circuits and Systems I*, 45, 1062-1067.

Yang, X., Wu, T. X., & Jaggard, D. L. (2002). Synchronization recovery of chaotic wave through an imperfect channel. *IEEE Antennas and Wireless Propagation Letters, 1*, 154–156. doi:10.1109/LAWP.2002.807569

Chapter 20
Mathematical Treatment for Constructing a Countermeasure Against the One-Time Pad Attack on the Baptista Type Cryptosystem

M.R.K. Ariffin
Universiti Putra Malaysia, Malaysia

M.S.M.Noorani
Universiti Kebangsaan Malaysia, Malaysia

ABSTRACT

In 1998, M.S. Baptista proposed a chaotic cryptosystem using the ergodicity property of the simple lowdimensional and chaotic logistic equation. Since then, many cryptosystems based on Baptista's work have been proposed. However, over the years research has shown that this cryptosystem is predictable and vulnerable to attacks and is widely discussed. Among the weaknesses are the non-uniform distribution of ciphertexts and succumbing to the one-time pad attack (a type of chosen plaintext attack). In this chapter the authors give a mathematical treatment to the phenomenon such that the cryptosystem would no longer succumb to the one-time pad attack and give an example that satisfies it.

1.0 INTRODUCTION

The relationship between chaos and cryptography makes it natural to employ chaotic systems to design new cryptosystems. It is based on the facts that chaotic signals are usually noise-like and chaotic systems are very sensitive to initial conditions. Their sensitivity to initial conditions and their spreading out of trajectories over the whole interval seems to be a model that satisfies the classic Shannon requirements of confusion and diffusion (Shannon, 1949). From 1989 onwards, many different chaotic encryption systems have been proposed. The most celebrated cryptosystems based on the ergodicity property of

DOI: 10.4018/978-1-61520-737-4.ch020

chaotic maps is presented by Baptista (1998) and has received more and more attentions in the past literature (Grassi & Mascolo, 1998; Alvarez, et.al., 1999; Chu & Chang 1999; Alvarez, et.al., 2000; Jakimoski, & Kocarev, 2001; Li, et.al., 2001; Wong, et.al., 2001; Garcia & Jimenez, 2002; Wong, 2002; Palacios & Juarez, 2002; Alvarez, et.al., 2003; Pareek, et.al., 2003; Li, et.al., 2003; Wong, 2003; Wong, et.al., 2003; Alvarez, et.al., 2004; Li, et.al., 2004; Alvarez & Li 2006). Researchers in this field have also constructed chaotic cryptosystems without using chaotic synchronization (most are designed for implementation on digital circuits or computers (Jakimoski, & Kocarev, 2001; Alvarez, et.al., 2003)) and secure communications based on chaotic synchronization of analog circuits (Baptista, 1998; Alvarez, et.al., 1999; Alvarez, et.al., 2000).

In 1998, M.S. Baptista proposed a chaotic cryptosystem using the ergodicity property of the simple low-dimensional and chaotic logistic equation $X_{n+1} = bX_n\left(1 - X_n\right)$ where X_0 and b are the secret keys. This cryptosystem has the ability to produce various ciphers responding to the same message input. In other words, this type of cryptosystem is a dynamic cryptosystem due to mathematical considerations and not due to computer programming methods. Since the ciphertexts are small integers, they are suitable to be transmitted through today's public digital networks. In Baptista's original work, in order to avoid statistical and differential cryptanalysis, a random number is generated each time the chaotic trajectory has reached the desired region. If it is greater than a threshold η, the current number of iterations will be transmitted as the ciphertext. Otherwise, the iteration will continue.

Motivated by the interest in chaotic cryptosystem, and by Baptista's ergodic cipher, numerous algorithms based on variations of Baptista's have been proposed. However, over the years research has shown that this cryptosystem is predictable and vulnerable to attacks and is widely discussed. Among the weaknesses are the non-uniform distribution of ciphertexts and succumbing to the one-time pad attack (a type of chosen plaintext attack).

Wong, et.al., (2001), examined the system and came out with two major drawbacks with Baptista's approach. First, the resultant ciphertext is usually concentrated at the smaller number of iterations (i.e. the distribution of the ciphertext is non-uniform). Second, a sequence of random numbers may have to be generated for a single block of message text. After examining the problems, Wong proposed a remedy that gave a flatter distribution of ciphertext, with single random number generation for each block of message text. Wong states that, the tradeoff between the spread of the distribution of ciphertext and the encryption time can be controlled by a single parameter. Wong also used the logistic map in illustrating the remedy.

Wong (2002) proposed a fast chaotic cryptographic scheme based on iterating the logistic map whereby no random numbers are needed to be generated. Wong proposed the use of a dynamical look-up table instead of a static one. This means that the table for looking up the ciphertext and plaintext is no longer fixed during the whole encryption and decryption processes. Instead, it depends on the plaintext and will continuously be updated in encryption and decryption. The dynamical table updating process is performed until the end of the input source. By doing so, Wong claims that the relationship between consecutive ciphertext becomes dynamic and it is much more difficult for cryptanalysis. Wong performed decryption using values of X_0 and b which differ from the correct value by 10^{-9} and found that even the first decrypted block is incorrect.

Alvarez, et.al., (2003), examined Baptista's system. He presented three types of cryptanalytic attacks: one-time pad attacks, entropy attacks and key recovery attacks. The one-time pad attack is based on the chosen plain text attack scenario. However, it is noted here in Alvarez's attack, it is assumed that the S

association between the $S\varepsilon$ -intervals and the units of some alphabet is part of the algorithm. Hence, the association is known to the cryptanalyst. In Baptista's work, this element is part of the secret key. If kept secret would result in the cryptanalyst opting for the brute force attack. Through the entropy attack a table consisting of the frequency of each symbol source $S_n = \{s_i\}_{i=1}^{n}$ is established. Due to weaknesses of the original Baptista cryptosystem, the information in the table allows one to perform a ciphertext only attack. The key recovery attack (more of an academic attack, since the assumption is $N_0 = 0, \eta = 0$) is to obtain X_0 when b is known and vice-versa.

In 2003 Alvarez presented 4 different cryptanalytic attacks after a complete study of Baptista' algorithms: one-time pad attacks, entropy attacks and key recovery attacks, comprised of parameter and initial point estimation. Among the 4 types of attacks, the one-time pad attack is the most efficient. It is a type of known plaintext attack. Without the knowledge of the secret parameters and the initial condition, the one-time pad attack can decrypt any message encrypted via this method. The one-time pad attack capitalizes on the fact that the ergodic cipher put forward by Baptista behaves as a one-time pad which reuses its key, and as a result, easy to break. Alvarez's attack is based on symbolic dynamics of 1-D quadratic maps, such as the logistic map. In 2004 Alvarez, et.al., cryptanalyzed a variation of Baptista by Wong (2002). Alvarez stated that the look-up table updating method is most unfortunate, since it allows the attacker to easily predict the new positions of the symbols even without the exact knowledge of X_0 and b. He states that, it is not necessary to know the exact value of X_0 and b to predict the next update, it suffices to know the subinterval where X_0 lands.

The focus of our research is to overcome the one-time pad attack. As pointed out by Alvarez, obtaining the one-time pad is as good as knowing the key (i.e. X_0 and b), making the system 100% vulnerable. We give a formal treatment for the one-time pad attack. We derive definitions and give mathematical explanations for this phenomenon. Finally, we give a theorem, if satisfied by a *"counter measure"* method, would result in this cryptosystem being invulnerable against the one-time pad attack. Finally, after deriving the mathematical counter measure treatment, we will demonstrate an example of a counter measure that satisfies the mathematical treatment.

2.0 BACKGROUND: THE BAPTISTA TYPE CRYPTOSYSTEM

Using the simple one-dimensional logistic map

$$X_{n+1} = bX_n\left(1 - X_n\right) \tag{1}$$

where X∈[0, I], choose a control parameter b set to make (1) have a chaotic behavior, Baptista suggested a quick and safe encryption process for further transmission. Baptista proposed that the encryption of some character is the number of iterations applied in Eq. (1) to make its trajectory, departing from an initial condition X_0, reach an ε -interval associated with that character.

Each interval, or site, is the range $\left[X_{min} + \left(S-1\right)\varepsilon, X_{min} + S\varepsilon\right]$, where, in this work, S = 256, $\varepsilon = \left(X_{max} - X_{min}\right) / S$, and $\left[X_{max}, X_{min}\right]$ is a portion of the attractor (it can be the whole attractor). The number of iterations (the ciphertext) is used together with the secret keys: the S associations between the $S\varepsilon$ -intervals and the S units of some alphabet, the first initial condition X_0, and the control parameter b (thus, we work with $S+2$ secret keys), allowing the receiver to decrypt the ciphertext (recovering the original character) by iterating Eq. (1) as much times as indicated by the ciphertext. The position of the final point, with respect to the $S\varepsilon$ -intervals, points out the original character to the receiver.

3.0 BACKGROUND: THE ONE-TIME PAD ATTACK

The Baptista ergodic cipher behaves as a modified version of the one-time pad. A one-time pad is perfectly secure under certain conditions. It is crucial to the security of the one-time pad that the key be as long as the message and never be reused, thus preventing 2 different messages encrypted with the same portion of the key being intercepted or generated by an attacker. Alvarez proved that the ergodic cipher put forward by Baptista behaves as a one-time pad which reuses its key, and as a result, is easy to break. The method of attack is based on the symbolic dynamics of one dimensional quadratic map.

Alvarez proved that the one-time pad attack capitalizes on the fact that the ergodic cipher put forward by Baptista behaves as a one-time pad which reuses its key, and as a result, easy to break. Alvarez's attack is based on symbolic dynamics of 1-D quadratic maps, such as the logistic map given by (1).

Since it is a known plaintext attack, a few pairs of cipher and plain texts are requested. Every pair of cipher and plain text allows recovering exactly 50% of the one-time pad. After analyzing one pair, the probability of finding any correct symbol of the one-time pad is exactly $1 - \frac{1}{2} = 0.5$. If a second pair is analyzed, and assuming that plain texts are perfectly random and mutually independent, the probability of finding any symbol of the one-time pad is $1 - \frac{1}{2^2} = 0.75$. With p different plaintext messages, the probability of finding a correct symbol approaches $1 - \frac{1}{2^p}$. This is the maximum rate of convergence to the one-time pad for small values of p, independent of the key length.

Alvarez used the following illustration for conducting the known plaintext attack as follows. Assume that unknown to the cryptanalyst, the keys are $X_0 = 0.232323$ and $b = 3.78$, using the interval [0.2,0.8]. Let us use a 4-symbol source $S_4 = \{s_1, s_2, s_3, s_4\}$. Under these assumptions, in a ***known plaintext attack*** scenario, we request the ciphertext of messages consisting of all their symbols set to s_1, s_2, s_3 and s_4, respectively (see Figure 1).

We begin by requesting the ciphertext for a plaintext consisting of only s_1: $P = (s_1, s_1, s_1, s_1, s_1, s_1, \ldots)$. The corresponding ciphertext is $E_1 = (5, 9, 5, 3, 4, 5, \ldots)$. We are now going to construct the one-time pad which indicates the position of X_n. Examining the ciphertext, we know for sure that the 6th symbol in the one-time pad is s_1 (or equivalently, we can say that X_6 belongs to the cell that encodes s_1), that the 14th symbol is another s_1, and the 20th, and the 23rd and so on. After considering the whole message, we get the following partial sequence for the one-time pad O=xxxxx$s_1$xxxxxxxx$s_1$xxxxx$s_1$xx$s_1$xxxx$s_1$xxxxxs_1.... Since we

Figure 1.

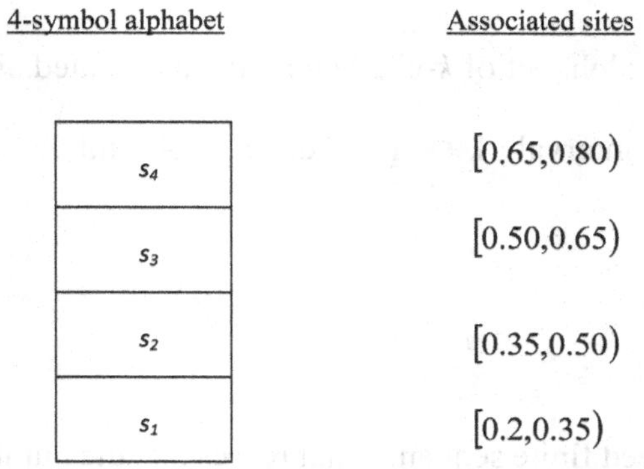

4-symbol alphabet	Associated sites
s_4	$[0.65,0.80)$
s_3	$[0.50,0.65)$
s_2	$[0.35,0.50)$
s_1	$[0.2,0.35)$

chose the interval [0.2,0.8] instead of the whole attractor, the letters marked x could correspond to either an iteration below the lower bound 0.2 or beyond the upper bound 0.8 or it could also correspond to the other symbols from the 4-symbol source.

For s_2, the corresponding ciphertext is $E_2 = (8,2,2,5,7,8,\ldots)$. We can complete more gaps in the sequence $O{=}xxxxxs_1xxs_2xs_2xs_2xs_1xxs_2xs_1xxs_1xs_2xs_1xxs_2xs_1x\ldots$. For s_3, $E_3 = (3,3,14,7,\ldots)$ and $O{=}xxxs_3xs_1s_3x\,s_2xs_2xs_2xs_1xxs_2xs_1s_3xs_1xs_2xs_1s_3xs_2xs_1x\ldots$. Lastly for s_4, the ciphertext is given by $E_4 = (1,14,17,1,\ldots)$, and we are able to construct $O{=}xs_4xs_3xs_1s_3xs_2xs_2xs_2xs_1s_4xs_2xs_1s_3xs_1xs_2xs_1s_3xs_2xs_1s_4\ldots$

Hence, the one-time pad is constructed. The symbol x correspond to points outside the boundaries and cannot be used for encryption. It is important to note that knowing the one-time pad generated by a certain key (X_0 and b) is entirely equivalent to knowing the key.

As an example, let us first construct the ε-intervals with its corresponding source. Since $n = 4$, we have $\varepsilon = 0.15$, and s_1 is associated with the interval $[0.2, 0.35)$, s_2 with the interval $[0.35, 0.5)$, s_3 with the interval $[0.5, 0.65)$ and finally s_4 with the interval $[0.65, 0.8)$. Given the following ciphertext $C^* = (1,4,3,2,2,3,5)$, it is easily derived from the one-time pad that the corresponding plaintext is $P = s_4, s_1, s_2, s_2, s_2, s_4, s_3$.

4.0 CONDITIONS FOR COUNTER MEASURES AGAINST THE ONE-TIME PAD ATTACK

In this section we give a formal treatment for the one-time pad attack. We derive definitions and give mathematical explanations for this phenomenon. Finally, we give a theorem, if satisfied by a *"counter measure" method*, will relieve this cryptosystem which has the ability to produce various ciphers responding to the same message input from the one-time pad attack.

Definition 3.1 (Associated Sites)

Let $A = \left(a,b\right) \subset \left[0,1\right]$. For an alphabet of *k*-characters, the associated site for each member of the alphabet $s_i, i = 1,2,3,\ldots,k$ is an interval $A_i \subset A$, where $s_i \in A_i$ and $A_i = \left(a + \left(i-1\right)\varepsilon, a + i\varepsilon\right)$ where $\varepsilon = \dfrac{b-a}{k}$.

Definition 3.2

Let $E_i = \left\{s_{i,m}\right\}_{m=1}^{n}$ be an ordered finite sequence that represents the number of iterations a chaotic map on a pre-defined initial condition x_0, needs to arrive at an associated site of a designated alphabet s_i. From definition 3.2, let

$$\alpha_{i,1} = \sum_{m=1}^{1} s_{i,m}$$

$$\alpha_{i,2} = \sum_{m=1}^{2} s_{i,m}$$

$$\vdots$$

$$\alpha_{i,n} = \sum_{m=1}^{n} s_{i,m}$$

Definition 3.3

Let E_i^* be the cumulative set where $E_i^* = \left\{\alpha_{i,1}, \alpha_{i,2}, \ldots, \alpha_{i,n}\right\}$.

Motivated by observation and evidence provided by Alvarez, we derive the following definition.

Definition 3.4

A chaotic map on a predefined initial condition x_0, behaves as a OTP that reuses its key when $\bigcap_{i=1}^{j} E_i^* = \varnothing$.

From computational evidence, for the logistic map, the following statement holds.

Statement

If a finite sequence of identical s_i's is substituted arbitrarily by a sequence of s_r's where $r \in \left\{1,2,\ldots,k\right\}$ then $\bigcap_{i=1}^{j} E_i^* \neq \varnothing$. That is, the chaotic map that generates each E_i^* no longer behaves as an OTP that reuses its key.

Theorem 3.1

The one-time pad attack on Baptista type chaotic cryptosystem fails if and only if $\bigcap_{i=1}^{j} E_i^* \neq \varnothing$.

Proof:

$(\Rightarrow)$

First, let us conduct the OTP attack by requesting the ciphertexts of each alphabet. We will be given the following sets:

$$E_1 = \{s_{1,1}, s_{1,2}, \ldots, s_{1,n}\}$$
$$\vdots$$
$$E_i = \{s_{i,1}, s_{i,2}, \ldots, s_{i,n}\}$$

where $i = 1, 2, \ldots, k$.

We will then obtain the cumulative sets;

$$E_i^* = \{\alpha_{1,1}, \alpha_{1,2}, \ldots, \alpha_{1,n}\}$$
$$\vdots$$
$$E_i^* = \{\alpha_{i,1}, \alpha_{i,2}, \ldots, \alpha_{i,n}\}$$

where $i = 1, 2, \ldots, k$.

The OTP attack fails when there exists $\alpha_{j,l} \in E_j^*$ such that for some $p = 1, 2, \ldots, k$ and $p \neq j$, $\alpha_{j,l} \in E_p^*$. That is, $\bigcap_{i=1}^{j} E_i^* \neq \varnothing$.

$(\Leftarrow)$

If $\bigcap_{i=1}^{j} E_i^* \neq \varnothing$, there exists $\alpha_{j,l} \in E_j^*$ such that for some $p = 1, 2, \ldots, k$ and $p \neq j$, $\alpha_{j,l} \in E_p^*$. For a OTP attack on a Baptista type chaotic cryptosystem to be successful, elements from distinct sets E_i^* where $i = 1, 2, \ldots, k$, must not overlap (hence the term OTP). Thus, the OTP attack fails.

5.0 EXAMPLE OF THE COUNTERMEASURE: THE MODIFIED BAPTISTA TYPE CHAOTIC CRYPTOSYSTEM VIA MATRIX SECRET KEY

We will present an example that satisfies the counter measure method as explained in section 4. It will be presented in stages.

5.1 Encryption

5.1.1 Preparing the associated sites choosing a chaotic map and secret keys.

i. Assume that we construct a look-up table consisting of $n\varepsilon$-intervals. The sites are numbered 0,1,2,…,*n*-1.
ii. Represent each site with its decimal representation.
iii. The minimum value of the first interval is 0, and the upper bound of the final interval is 1.
iv. Choose a one-dimensional chaotic map with the onto property (Alvarez & Li, 2006).

Example 1
The logistic map, $x\left(n+1\right)=bx\left(n\right)\left(1-x\left(n\right)\right)$ for $b=4$.

Example 2
The skewed Piece Wise Linear Tent Map

$$F\left(x\right)=\begin{cases}x/p & x\in\left[o,p\right]\\ \dfrac{1-x}{1-p} & x\in\left(p,1\right]\end{cases}$$

where $p\in\left(0,1\right)$ is the control parameter.

v. The secret keys are:-
   a. A 128-bit initial condition, X_0 (used to iterate the chaotic map)
   b. The secret matrix key (used to prepare the distorted plaintext -further discussed in 5.1.2)

5.1.2 Preparing the Matrix Secret Key

Say we have a plaintext consisting of *m*-characters ($m\leq pn$) where *pn* is any *p*-multiple of *n* (*p* integer).

i. If *m*<*pn*, pad it with zeroes (after the last character in the plaintext), so that the length of the plain-text is *pn*.

Choose an integer *k*, such that $k\mid pn$ (*k* divides *pn*).

Generate a *k* X *k* invertible matrix ($\left[A\right]_{k\times k}$) and its inverse $\left[A\right]^{-1}_{k\times k}$. This is our secret key. As for the strength of a secret key, it is widely discussed in cryptographic literatures. For our example we used a 2×2 invertible matrix. If observed this is just a 32-bit key, which is a very week secret key. The condition for a secret key to be strong is that it has to be at least (with current technology) 128-bit key (for non-government data). To create a 128-bit key with a matrix, one has to generate a 4×4 invertible matrix which consists of 16 digits. Each digit is an 8-bit number. Hence, a 128-bit key. To further enhance, one can design a 5×5, 6×6 invertible matrices and so on.

5.1.3 Preparing the Plaintext

i. Determine each plaintext character decimal representation from the look-up table.

From left-to-right, group the decimal numbers in groups of *k*'s and represent the group of decimal numbers in a matrix $k \times 1$ ($\left[B\right]_{k\times 1}$).

5.1.4 Preparing Distorted Plaintext

For each matrix constructed in 3.1.3 (ii) do the following matrix multiplication: $\left\{\left\{\left[C\right]_{k\times 1}\right\}_l\right\}_{l=1}^{r} = \left\{\left[A\right]_{k\times k} \cdot \left\{\left[B\right]_{k\times 1}\right\}_l\right\}_{l=1}^{r}$ where $r = \dfrac{pn}{k}$.

Use each integer in the resultant matrix $\left\{\left\{\left[C\right]_{k\times l}\right\}_l\right\}_{l=1}^{r}$, to iterate the chosen chaotic map accordingly (i.e. the number of iteration is the integer in the matrix).

i. From the iteration value, determine the corresponding character.

5.1.5 Encrypt via Baptista Cryptographic Method

i. For each character acquired from 5.1.4 (iii) encrypt via Baptista cryptographic method.
ii. The iteration numbers are the ciphertexts.

5.2 Decryption

5.2.1 Use the ciphertext to iterate chaotic map.
5.2.2 Would result in corresponding character (refer to look-up table).
5.2.3 Start iterating the chaotic map until it falls in the corresponding phase space of the first character. Obtain the number of iterations required.
5.2.4 Continue iterating the chaotic map until it falls in the corresponding phase space of the second character. Continue until the final character. Each time, obtain the number of iterations required.
5.2.5 From left-to-right group the decimal numbers (acquired from steps 5.2.3 and 5.2.4) in groups of *k*'s and represent the group of decimal numbers in a matrix $k \times 1$ ($\left[B\right]_{k\times 1}^{\prime}$).

5.2.6 Multiply with $[A]^{-1}_{k\times k}$. We will get ($\left\{\left\{[C]_{k\times 1}\right\}_l\right\}_{l=1}^{r} = \left\{[A]^{-1}_{k\times k}\cdot\left\{[B]_{k\times 1}\right\}_l\right\}_{l=1}^{r}$).

5.2.7 From left-to-right, assign character to each decimal obtained from step 5.2.6 (refer to look-up table).

5.2.8 Get back the plaintext.

6.0 An Example

Let us choose the chaotic logistic map as given in 5.1.1 (iv). For illustrative purposes the key is $X_0 = 0.232323$. Assume we are using a 4-symbol source and the plaintext is given by $P = s_3, s_1, s_2, s_2, s_4, s_1, s_4, s_3$.

6.0.1 Preparing associated sites.

Let us use a 4-symbol source $S_4 = \{s_1, s_2, s_3, s_4\}$. It is illustrated in Figure 2.

6.0.2 Preparing matrix key

i. Observe that m=8. Choose k=2 (since 2|8).

ii. Let $A = \begin{bmatrix} 2 & 3 \\ 1 & 2 \end{bmatrix}$.

iii. Decimal representation of plaintext: 2,0,1,1,3,0,3,2.

6.0.3 Preparing distorted plaintext

$$\begin{bmatrix} 2 & 3 \\ 1 & 2 \end{bmatrix}\begin{bmatrix} 2 \\ 0 \end{bmatrix} = \begin{bmatrix} 4 \\ 2 \end{bmatrix},\ \begin{bmatrix} 2 & 3 \\ 1 & 2 \end{bmatrix}\begin{bmatrix} 1 \\ 1 \end{bmatrix} = \begin{bmatrix} 5 \\ 3 \end{bmatrix},\ \begin{bmatrix} 2 & 3 \\ 1 & 2 \end{bmatrix}\begin{bmatrix} 3 \\ 0 \end{bmatrix} = \begin{bmatrix} 6 \\ 3 \end{bmatrix},\ \begin{bmatrix} 2 & 3 \\ 1 & 2 \end{bmatrix}\begin{bmatrix} 3 \\ 2 \end{bmatrix} = \begin{bmatrix} 12 \\ 7 \end{bmatrix}$$

i. From the matrix multiplication, we have the following decimal numbers: 4,2,5,3,6,3,12,7.

ii. Iterate the chaotic map accordingly. First 4-times, second 2-times and so on.

iii. List down the corresponding characters: $s_2, s_2, s_4, s_2, s_4, s_2, s_1, s_2$

6.0.4 Encrypt each character in 6.0.3(iii) via Baptista cryptographic method (in this example we are using the logistic map and $b = 4, X_0 = 0.232323$)

Figure 2.

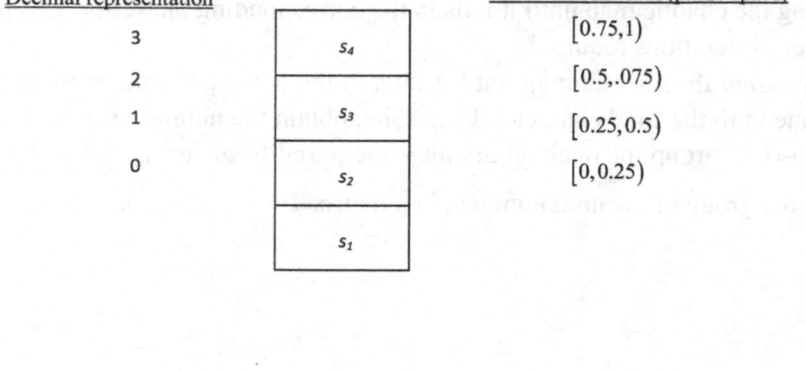

Decimal representation		Associated interval (phase space)
3	s_4	[0.75,1)
2		[0.5,.075)
1	s_3	[0.25,0.5)
0	s_2	[0,0.25)
	s_1	

We have the following cipher:

8,2,5,2,15,18,2,5

Deciphering the ciphertext is trivial process.

7.0 CRYPTANALYSIS USING ALVAREZ'S ONE-TIME PAD ATTACK (CHOSEN PLAINTEXT ATTACK)

We will now proceed to conduct an attack using Alvarez's method.

7.0.1 We start by requesting the ciphertext for s_1 we will get (5,3,1,1,1,3,11,1,1,1,1,...).
7.0.2 For $s_{2,}$ we have (5,3,5,3,1,1,7,7,5,1,...).
7.0.3 For s_3 we have (5,2,1,4,20,10,1,7,3,2,1,1,...).
7.0.4 For s_4 we have (6,1,8,17,2,4,3,2,7,3,1,2,...).
7.0.5 With only the above information we will construct a partial one-time pad which has 2 possibilities at 7 places and 3 possibilities at 3 places. The partial pad is as follows (see Figure 3)

With only the above information a cryptanalysts has to try $2^7 \times 3^3 = 10,368$ possibilities of combinations.

7.0.6 The larger the number of characters involved in the "alphabet", one cannot rule out the high possibility of constructing a one-time pad which has more than 2^{1024} possibilities of combinations.
7.0.7 Hence, making this type of attack unfeasible to our modified Baptista cryptographic method.

8.0 RESULT AND ITS SIGNIFICANCE

From the results, one can clearly see the advantages of this method. Assume a cryptanalyst employs the one-time pad attack against our modified Baptista type chaotic cryptosystem. The cryptanalyst will obtain a one-time pad with more than one possible combination. In our 4-character "alphabet" example, there are at least 10,368 combinations to try.

The number of possibilities to construct the correct one-time pad sequence increases exponentially as the number of characters in an "alphabet" increases. This fact ensures the security of our modified Baptista type chaotic cryptosystem against the one-time pad attack.

Figure 3.

O=XXXXX $s_1s_2s_3$ s_4 s_3s_4 $s_1s_2s_3$ s_1 s_1 s_1 s_3 s_2 s_1 s_4 s_2 s_2 s_2 XXXXXX s_1s_2 s_1 s_1 s_1 s_1 XX $s_2s_3s_4$ X s_4 XX s_2
s_2s_4 XX s_4s_3 s_3s_4 XXXXXX s_3s_4 XX s_3s_4 s_4 s_3 s_3s_4 s_3...

ACKNOWLEDGMENT

This research was supported by the Science Fund, Ministry of Science, Technology and Innovation, Malaysia under grant number 04-01-02-SF0177.

REFERENCES

Alvarez, E., Fernandez, A., Garcia, P., Jimenez, J., & Marcano, A. (1999). New approach to chaotic encryption. *Physics Letters. [Part A]*, *263*, 373–375. doi:10.1016/S0375-9601(99)00747-1

Alvarez, G., & Li, S. (2006). Some basic cryptographic requirements. for chaos-based cryptosystems. *International Journal of Bifurcation and Chaos in Applied Sciences and Engineering*, *16*(8), 2129–2151. doi:10.1142/S0218127406015970

Alvarez, G., Montoya, F., Romera, M., & Pastor, G. (2000). Cryptanalysis of a chaotic encryption system. *Physics Letters. [Part A]*, *276*, 191–196. doi:10.1016/S0375-9601(00)00642-3

Alvarez, G., Montoya, F., Romera, M., & Pastor, G. (2003). Cryptanalysis of a chaotic secure communication system. *Physics Letters. [Part A]*, *306*, 200–205. doi:10.1016/S0375-9601(02)01502-5

Alvarez, G., Montoya, F., Romera, M., & Pastor, G. (2003). Cryptanalysis of an ergodic chaotic cipher. *Physics Letters. [Part A]*, *311*, 172–179. doi:10.1016/S0375-9601(03)00469-9

Alvarez, G., Montoya, F., Romera, M., & Pastor, G. (2003). Cryptanalysis of a discrete chaotic cryptosystem using external key. *Physics Letters. [Part A]*, *319*, 334–339. doi:10.1016/j.physleta.2003.10.044

Alvarez, G., Montoya, F., Romera, M., & Pastor, G. (2004). Cryptanalysis of dynamic look-up table based chaotic cryptosystems. *Physics Letters. [Part A]*, *326*, 211–218. doi:10.1016/j.physleta.2004.04.018

Baptista, M. S. (1998). Cryptography with chaos. *Physics Letters. [Part A]*, *240*, 50–54. doi:10.1016/S0375-9601(98)00086-3

Chu, Y. H., & Chang, S. (1999). Dynamic data encryption system based on synchronized chaotic systems. *Electronics Letters*, *35*, 271–273. doi:10.1049/el:19990148

Garcia, P., & Jimenez, J. (2002). Communication through chaotic map systems. *Physics Letters. [Part A]*, *298*, 35–40. doi:10.1016/S0375-9601(02)00382-1

Grassi, G., & Mascolo, S. (1998). Observer Design for Cryptography based on Hyperchaotic Oscillators. *Electronics Letters*, *34*, 1844–1846. doi:10.1049/el:19981285

Jakimoski, G., & Kocarev, L. (2001). Analysis of some recently proposed chaos-based encryption algorithms. *Physics Letters. [Part A]*, *291*, 381–384. doi:10.1016/S0375-9601(01)00771-X

Li, S., Chen, G., Wong, K. W., Mou, X., & Cai, Y. (2004). Baptista-type chaotic cryptosystems: Problems and countermeasures. *Physics Letters. [Part A]*, *332*, 368–375. doi:10.1016/j.physleta.2004.09.028

Li, S., Mou, X., & Cai, Y. (2001). Improving security of a chaotic encryption approach. *Physics Letters. [Part A]*, *290*, 127–133. doi:10.1016/S0375-9601(01)00612-0

Li, S., Mou, X., Ji, Z., Zhang, J., & Cai, Y. (2003). Performance analysis of Jakimoski–Kocarev attack on a class of chaotic cryptosystems. *Physics Letters. [Part A]*, *307*, 22. doi:10.1016/S0375-9601(02)01659-6

Palacios, A., & Juarez, H. (2002). Crytography with cycling chaos. *Physics Letters. [Part A]*, *303*, 345–351. doi:10.1016/S0375-9601(02)01323-3

Pareek, V., Patidar, N. K., & Sud, K. K. (2003). Discrete chaotic cryptography using external key. *Physics Letters. [Part A]*, *309*, 75–82. doi:10.1016/S0375-9601(03)00122-1

Shannon, C. E. (1949). Communication theory of secrecy systems. *The Bell System Technical Journal*, *28*, 656–715.

Wong, K. W. (2002). A fast chaotic cryptographic scheme with dynamic look-up table. *Physics Letters. [Part A]*, *298*, 238–242. doi:10.1016/S0375-9601(02)00431-0

Wong, K. W. (2003). A combined chaotic cryptographic and hashing scheme. *Physics Letters. [Part A]*, *307*, 292–298. doi:10.1016/S0375-9601(02)01770-X

Wong, K. W., Ho, S. W., & Yung, C. K. (2003). A chaotic cryptography scheme for generating short ciphertext. *Physics Letters. [Part A]*, *310*, 67–73. doi:10.1016/S0375-9601(03)00259-7

Wong, W. K., Lee, L. P., & Wong, K. W. (2001). A modified chaotic cryptographic method. *Computer Physics Communications*, *138*, 234–236. doi:10.1016/S0010-4655(01)00220-X

Chapter 21
Chaos Synchronization with Genetic Engineering Algorithm for Secure Communications

Sumona Mukhopadhyay
Army Institute of Management, India

Mala Mitra
Camellia School of Engineering and Technology, India

Santo Banerjee
Politecnico di Torino, Italy & Techfab, Italy

ABSTRACT

In this chapter, the authors have attempted to explore an association between chaos synchronization with genetic engineering algorithm. The authors developed a cryptographic technique using keys generated by time delayed synchronized systems by modifying the selection mechanism of the basic genetic algorithm. This has resulted into a proposed genetic engineering algorithm for cryptography which successfully deciphers the message. The observed results support the stated aim of the proposed genetic engineering algorithm that it is a reliable, effective and computationally cheaper substitute to the cryptography also developed here with (μ /ρ, λ)- selection scheme of Evolutionary strategy. A detail study is also presented with the analysis.

1. CHAOS SYNCHRONIZATION AND CRYPTOGRAPHY

The turning point in traditional schemes of cryptography was brought about by the revolutionary work of Pecora and Carroll (Pecora & Carroll, 1990). This opened a gateway to varied areas of applications of chaos synchronization namely secure communication, chaos generators design, chemical reactions, biological systems, information science (Hramov & Koronovskii, 2005a ; Garcia – Ojalio, & Roy, 2001; Yang & Chu 1997; Bowong, 2004; Kittel et al, 1994) and many more significant applications.

DOI: 10.4018/978-1-61520-737-4.ch021

Pecora and Carroll experimented on two systems producing their phenomenal work. The first is known as a driver and the other a response system. They used a novel idea of generating a chaotic signal to drive a non linear dynamic system such that the state of the second system is governed by the state of the driving system. This is possible if the two systems are coupled by a proper controlling function. However, it is to be noted that the behavior and system parameters of the response system is dependent on the behavior of the driving system alone, similar to a master-slave relationship. When subjected to suitable conditions, the response system will exhibit a chaotic pattern which is in sync with the driver system. This phenomenon is known as chaos synchronization. Till now many different types of synchronizations have been analyzed in ordinary and time delayed dynamical systems such as complete synchronization (Fujisaka. & Yamada, 1983), generalized synchronization (GS) (Rulkov et al, 1995; Kocarev & Parlitz, 1996; Hramov & Koronovskii, 2005a), generalized projective synchronization of a unified chaotic system (Yan and Li 2005), anticipated synchronization (AS) (Masoller, 2001), lag synchronization (LS) (Rosenblum et al, 1997; Zhan et al, 2002), phase synchronization (PS) (Rosenblum et al, 1996; Koronovskii & Hramov, 2004), antiphase synchronization (APS), time scale synchronization (Hramov et al, 2005b), intermittent generalized synchronization in which the authors (Hramov et al, 2005c) detected that before the transition of unidirectionally coupled chaotic oscillators to generalized synchronization, in some time intervals a non-synchronous behavior occurs and functional synchronization (FS) (Banerjee & Chowdhury, 2009).

Recent research on coupled non linear systems proved that the properties of chaotic dynamics such as ergodicity(Caponetto et al, 2003) and sensitivity on initial conditions can be tapped for potential applications in secure communications around which this chapter primarily focuses on. The information signal is masked by a chaotic signal which forms a carrier and transmits the information signal in a secure way. The masking is removed at the receiver end. Even if the communication is intercepted, the information signal cannot be deciphered unless total information of the chaos signal generating system is at the disposal of the interpreter. A proper choice for the chaotic system must be made accurately for secure communication. Simple chaotic systems such as those having one positive Lyapunov exponent should be avoided for cryptographic purposes (Perez. & Cerdeira, 1995). L. Pecora (Pecora, 1996) suggested that a chaotic system with at least two positive Lyapunov exponents should be an ideal choice. Such a chaotic system is known as a hyperchaotic system (Rossler, 1979), characterized by higher randomness and unpredictability which are vital for secure transmission of signals. Chaos theory has found its roots in secure communication due to its fundamental characteristics of sensitivity to initial conditions, ergodicity and mixing and several works have been discussed on combining features of chaos using both circuits and software simulations (Alvarez et al, 1999a; Alvarez et al, 1999b; Dachselt & Schwarz, 2001; Kocarev et al, 1998; Kocarev, 2001). The most attractive feature of chaotic system that makes it apt for novel engineering applications is its unpredictable and random nature. These are the key properties which are explored in communication theory for secure transmission of data as pointed out by Shannon (Shannon, 1949). To quote Shannon, " *In a good mixing transformation functions are complicated, involving all variables in a sensitive way. A small variation of any one (variable) changes the outputs considerably*." This implied that small changes may result into large changes and if this be mapped to chaos theory it will mean that chaotic behavior will continue to manifest the effects of any minuscule disturbances. However, it is to be noted that systems used in chaos are defined on real numbers only (Gickenheimer & Holmes, 1983) whereas cryptographic systems operate on finite number of integers (Schneier, 1996). The two disciplines can form a symbiotic relationship in cryptography which is explained by the works of many researchers in the following paragraph.

T. Habutsu *et al* (Habutsu et al, 1991) proposed an algorithm that uses the iterations of the chaotic tent map which was then generalized by Kotulski (Kotulski & Szczepanski, 1997). Later in 1999, an enhanced version of their previous encryption algorithm was presented using multiple iterations of a dynamical chaotic system based on gas dynamic models (Kotulski et al,1999). Also a two dimensional chaotic map was used by J. Fridrich to develop a new symmetric block encryption scheme (Fridrich, 1998). Baptista (Baptista, 1998) showed that it is possible to have an increased number of encryption iterations ranging from 250 to 65536 in the transformation function used for encoding the message. This overcame the limitation of the number of iterations the traditional encryption methods were subjected to which is usually less than 30. Another encryption algorithm based on synchronized chaotic system was proposed in (Alvarez et al, 1999b). Here the transmitter sends a string of binary symbols constituting a binary file assuming that the usage of a d- dimensional chaotic dynamical rule, which generates sequences of real numbers, has already been agreed upon before communication between the two parties commences. The authors in (Chu and Chang, 1999) used a different chaotic attractor originating from synchronized chaotic systems to encrypt each byte of the message. In (Jakimoski & Kocarev, 2001) a block encryption cipher using exponential and logistic chaotic maps defined on the unit interval by $x \rightarrow a^x \bmod 1$ and $x \rightarrow 4x(1\text{-}x)$ respectively was proposed and cryptanalysed. The authors in [Parker & Short, 2001) highlighted the weakness of the chaotic cryptographic system developed by Yang et al (Yang et al, 1997) in which eventhough the key stream was not transmitted but by observing the transmitted message could give a clue about the cipher function . An analysis of the geometric information in the transmission could help to extract an estimate of the keystream highlighting a major security vulnerability. Then in (Li et al, 2007) a spatiotemporal chaotic system was used to generate pseudo random bits to form the transaction numbers. A spatiotemporal chaotic system is a spatially extended system which can exhibit chaotic motion in both dimensions of space and time. This system is generally modeled by either a partial differential equation (PDE), a coupled ordinary differential equation (CODE) or a coupled map lattice (CML). In their work they used a one-way CML consisting of logistic maps to devise the system which yielded low computational expenditure. In (Banerjee et al, 2008a), the authors used a time delayed multiplexed synchronized chaotic system in chaotic cryptography which successfully communicated encrypted messages with several receiver systems simultaneously with different plaintexts. Again in (Banerjee & Chowdhury, 2008), the authors proposed a new method of synchronization of non linear systems based upon Lyapunov function and parameter estimation and effectively applied in cryptography by using different systems as transmitter and receiver thereby injecting enhance security. Again in 2008 Xiyin Liang et al (Liang et al, 2008) suggested that a 1 –D discrete time control system with a Delta modulated feedback control mechanism can be utilized to improve security.

This chapter mainly deals with the fundamental application of chaos synchronization in Cryptography. The benefit of chaos synchronization can be coupled with those of evolutionary algorithms since both are dynamical and probabilistic in nature and utilize the advantages accompanied with computer simulation. The work explained in subsequent sections is based on a modified selection scheme for our application giving rise to a genetic engineering algorithm(GEA). Section 2 presents a background knowledge on Evolutionary algorithm. Other pioneering works of many researchers combining chaos synchronization with Genetic Algorithm has also been highlighted. The generic Genetic Algorithm is explained in subsection 2.2. SubSection 2.3 highlights the motivation for the proposed application along with the Chaotic Synchronized system. SubSection 2.4 explains the design of the Partial Fitness based Roulette Wheel Replacement strategy (PFRWR) for selection of the individual in our application of cryptography. Section 3 shows the simulation result and observations concluding with Section 4 in which we have attempted to list some vital points to be kept in mind for applying evolutionary algorithm especially genetic algorithm together with the future prospect of our scheme.

2. CHAOS AND EVOLUTIONARY ALGORITHM BASED SECURE COMMUNICATION

2.1 Brief Introduction of Evolutionary Algorithm

Evolutionary algorithm mimics the metaphor of natural biological model of evolution. Evolutionary algorithms(EA) have been stated to be optimized search methods (Goldberg, 1994; Srinivas & Patnaik, 1994). The principle of natural evolution revolves around the fact that individuals are always competing for survival. Adaptation to the environment determines the fitness of the individual to be selected for taking part in reproduction. They propagate their features to their offsprings usually by exchange of genetic material between two individuals by a process known as crossover. The diversity among species is maintained by another stochastic operator known as - *mutation*. Randomness inherent in nature can bring stableness or disorder however, evolution maintains its stability where individuals identify other members of the population by associating them with special "tags" (Banzhaf et al,1998; Koza, 1992). Evolutionary algorithm, also known as Evolutionary computation (Abraham, 2005), is an umbrella term encompassing

- Evolution strategies(ES) (Rechenberg, 1973; Rechenberg, 1994; Schwefel,1977; Schneier, 1996),
- Evolutionary Programming(EP) (Fogel et al, 1966),
- Genetic Algorithms(GA) (Holland, 1975)and
- Genetic Programming(GP) (Koza, 1992).

ES basically solves technical optimization problem of finding optimal parameter setting including continuous, discrete and combinatorial search spaces. Fogel et al(Fogel et al, 1966), had been working with mutation operator to change states of Finite state Automata when they stumbled upon the concept of EP. EP begins with an initial population consisting of solutions of Finite State Machines(FSM) that changes its state based on the input and the current state. In EP, mutation is the primary operator for creating offsprings by manipulating its states and symbols. The fitness assigned to each FSM is determined if the symbols produced as output agreed with its input symbols. But, EP differs strictly from ES in the selection scheme. The former conducts a stochastic tournament to determine solutions to be retained whereas, the latter is more inclined towards deterministic selection. Selection scheme in ES (Beyer, 1995; Beyer & Schwefel, 2002; Beyer, 2001; Hansen & Ostermeier, 2001; Hansen et al, 2003; Rechenberg, 1994; Schwefel, 1995) are of four types- (μ, λ) (better known as Comma based strategy), $(\mu + \lambda)$ (Plus strategy) and other two canonical versions for self adaptations $(\mu /\rho, \lambda)$and $(\mu /\rho+ \lambda)$.In the first type of (μ, λ) -ES Selection, the μ parents produce λ children using mutation and $\mu \leq \lambda$. After this fitness values are calculated for each of the λ children and the best children become next generation parents. In $(\mu + \lambda)$, the μ parents produce λ children using mutation operator. The fitness values are calculated for each of the λ children and the best μ individuals which are to become parents for next generation are then selected from both the old parents and their λ offspring. In the last two selection mechanisms, the μ parents produce λ children using mutation and recombination where ρ denotes the number of individuals involved in crossover. In (μ /ρ, λ) selection strategy, from among the μ individuals ρ number of parents are selected for reproduction by crossover and mutation operation to yield λ number of offsprings and their ρ parents are discarded. Out of the λ children just produced, the best is selected to become the μ parents for next generation. In the (μ /ρ+ λ) selection, from among the μ individuals ρ number of parents are selected for reproduction by crossover and mutation operation to yield λ number of offsprings. Each of the λ children are then subjected to fitness evaluation and the best μ individuals from both the parent pool and children become parents for the next generation. Apart from using an additional crossover operator these methods are exactly the same as (μ, λ) strategy and (μ+ λ) strategy.

Genetic Programming(GP) is a technique for machine learning where a population of computer programs evolve to improvise themselves based on the training set (Mitchell, 1996). These programs are expressed in GP as parse trees, rather than as lines of code (Banzhaf et al, 1998; Koza, 1992). GP breeds such a population of candidate solutions which are variable length computer programs usually represented in a tree structure (Banzhaf et al,1998; Koza, 1992) thereby achieving automatic programming using the principles of Darwinian natural selection and biologically inspired operations. The operations include most of the techniques discussed already. A GP works best for problems that do not have a single best solution and for problems with dynamically changing variables. However, recently there has been some concern over the robustness of GP solutions (Kushchu, 2002). A GP program is labeled to be robust if it arrives at a good solution when applied to an environment similar to the one it had been evolved for. Many researchers have successfully used GP on a wide variety of problems, including automated synthesis of controllers, circuits, antennas, genetic networks, and metabolic pathways. Koza (Koza,1994) and Koza et al. (Koza, et al. 1999; Koza et al, 2003a; Koza et al, 2003b.) provided a wide range of examples, computer code, and video of GP applications.The main difference between GP and GA is the representation of the solution. GA represents solutions in fixed length binary string format or a problem dependent representation and the genetic operators are chosen which facilitates unrestricted navigation of the search space.

Genetic algorithm(GA) - a subset of Evolutionary Algorithm, developed by John Holland (Holland, 1975), his colleagues and students, K. DeJong, D. Goldberg, at the University of Michigan(1970's), USA, exploits Darwin's principle of evolution theory. It is a well proven powerful stochastic optimization technique which evaluates a population of individuals iteratively transforming the current population $G(t)$ at iteration t into a new population $G(t+1)$ of offspring. A new generation of population is produced by applying the genetic operators of reproduction and gene mutation which serves as the transformation operators to transform the previous set of population to a fitter, goal-oriented solution. The basic prerequisites for developing GA are a proper representation of the possible solution, known as chromosomes, as a string of symbols and choosing of the three genetic operators which manipulate the population – reproduction, crossover and mutation (Goldberg, 1989). Darwinian survival of the fittest individual is simulated in GA by a fitness function which is used to evaluate individuals and reproductive success is proportional to fitness of such individuals. Each chromosome competes for reproduction but only the best are chosen based on its fitness value and offsprings are produced which are added to the new gene pool. Thereby crossover or mixing of pairs of chromosomes occurs replacing the worst chromosomes with the better ones, or without replacement of the best chromosome by a technique known as elitism. Figure 1 shows two of the most common crossover schemes. The first subpart a and the other subpart b shows a schematic diagram of one point and two point crossover respectively. In Figure 1a, crossover occurs with the elements of parents selected after the crossover point and in Figure 1b, crossover occurs with the elements lying in between Crossover point1 and Crossover point2.

Figure 1. One point and two point crossover in genetic algorithm

a: One Point Crossover in Genetic Algorithm

b: Two Point Crossover in Genetic Algorithm

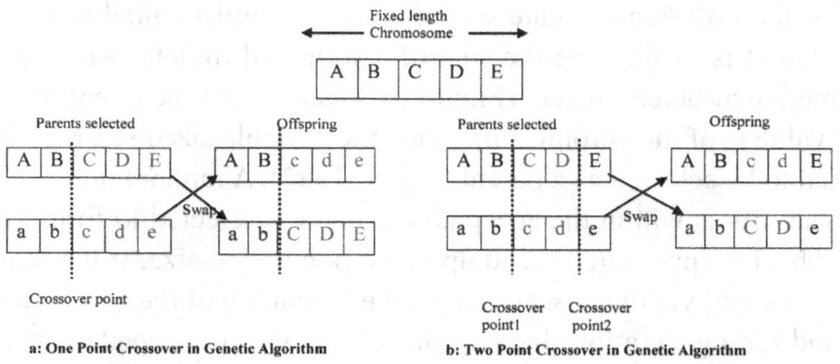

In order to increase the search space offsprings are mutated with very low probability. Such a single cycle of evolutionary algorithm constitutes a generation and is efficient enough to perform well on a wide variety of problems. However, better results may be obtained by multiple iterations of subpopulation. In GA, the progress of the population is measured by a commonly used term "selection intensity" (Bulmer, 1980; Blickle & Thiele, 1995b). This throws light on the average fitness of the population before and after selection and leads to a normalized and dimensionless measure. Selection intensity basically depends on the fitness distribution and for a particular selection method it is defined mathematically as the difference between the mean selection criterion of those individuals selected to be parents and the average selection criterion of all potential parents divided by the mean variance of the fitness of population. The features of GA which makes it apt for novel engineering applications are listed below

- Less computational effort
- Finds an optimal solution with continuous or discrete variables
- Scalability
- Modularity
- Adaptive, self guided by using a fitness measure
- Less chance of getting stuck with a non optimal solution(local optima) than back propagation neural network by applying the genetic operator: mutation

An interesting feature worth mentioning is the variety of selection schemes at disposal with GA. The ubiquitous question as to which chromosome will contest as a parent for mating can be obtained by choosing a suitable one out of the basic selection mechanisms as discussed below depending on the application.

1. *Fitness Proportionate selection:* This is the simplest selection scheme and is popularly referred to as Stochastic sampling with replacement. In this method the probability of selecting a parent p_i for mating is governed by their fitness and is defined by the formula

$$p_i = \frac{f_i}{\sum_{j=1}^{n} f_j}$$

Fitness proportionate selection results in choosing several solutions from the population by repeated random sampling. This kind of proportionate selection can be implemented in a number of ways. The simplest way to implement is to simulate the spin of a weighted roulette wheel (Goldberg, 1989; De Jong, 1775). If we imagine a roulette wheel (Figure 2), wherein each segment of it is occupied by the proportionate fitness value p_i of the chromosome, then the variable sized segment is equal to the probability of the individual to be selected as a parent (Baker, 1987). A random number is generated and an individual whose probability is within the acceptable criteria is selected to form the mating pool. The number of times the wheel is spun will depend upon the population size. If the search for the location of the chosen slot is performed via linear search from the beginning of the list, each selection requires a complexity of $O(N)$ and for a generation since n spins of the wheel are needed the algorithm will have a total complexity of $O(N^2)$.

Proportionate reproduction can also be performed by stochastic remainder selection (Brindle, 1981; Booker, 1982) with two options - replacement or without replacement. Here the expected number of copies e of an individual is calculated by $e = p_i$ * population size.

In stochastic remainder sampling with replacement, the integer portions of this count e, are assigned deterministically to select the individuals based on this computed expected number of trials. Each string is allocated samples according to the integer part followed by the population being sorted according to the fractional part of the calculated e value. The remainder of the strings needed to fill in the mating pool are then drawn from the top of the sorted list. In remainder stochastic sampling with replacement the fractional parts of the expected number values are used to calculate weights in a roulette wheel selection procedure that is then used to fill the remaining population slots. The remainders are used to size the slots of the roulette wheel. The time complexity taken by this scheme is $O(N^2)$ because in each spin only a single individual gets extracted and a checking for N individuals for their fractional part will be performed till the population size is achieved. Stochastic sampling without replacement minimizes the stochastic error present in the Roulette Wheel selection. This requires only a single spin of the roulette wheel to assign space to the N individuals based on the integer part of e and hence incurs a time complexity of $O(N)$. Then the rest of the space in the mating pool is filled using the idea of tossing a weighted coin (Golberg & Deb, 1991). This decides the success probability for an individual getting its copy in the mating pool and this has a time complexity of $O(1)$. So overall, the time complexity of Stochastic sampling without replacement is $O(N)$. The fractional part of e is hence used probabilistically, to fill the mating pool.

Figure 2. Schematic approach of Roulette wheel based on fitness of the individual

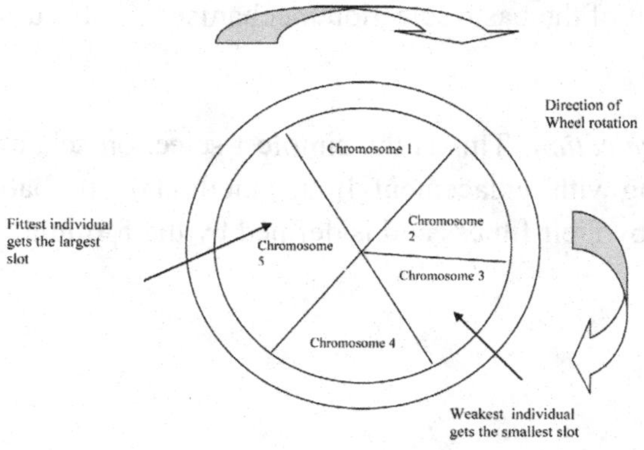

2. *Rank based fitness selection:* Baker (Baker, 1985) introduced the notion of ranking selection. In rank based selection, the individuals are sorted by their rank and assigned a fitness accordingly. Over here the population is sorted from best to worst and assigned the number of copies that each individual would receive according to a non-increasing assignment function. After this, proportionate selection is performed based on that assignment. The chance of an individual to be selected is based upon the rank assigned to it (Bäck & Hoffmeister, 1991). This scheme is more robust than the previous method discussed because it prevents the premature convergence encountered in proportional fitness assignment. Rank based selection mechanism requires access to all the individuals and their fitness values to sort them rank wise. Actually the Ranking based selection method comprises of two basic steps. At first, all the individuals must be sorted in accordance to their fitness values, and next these are used in some form of proportionate selection. So, the calculation of the time complexity of ranking requires the cumulative consideration of sorting, which takes $O(N \log N)$ steps to sort, plus an additional time to select the parents and proportionate selection has a time complexity which lies in between $O(N)$ and $O(N^2)$.
3. *Tournament Selection*: Brindle's doctoral dissertation (Brindle, 1981) studies explains that in this process out of a large population size, a sample of individuals is selected at random(known as the tournament size). From this sample, the best individual depending upon the objective function is selected as a parent and mated with the next fitter individual and the offspring is added in the new gene pool. This process is repeated till the sample size is achieved. Sometimes the competition is held between two individuals in which case it becomes a binary tournament but this selection can be extended to any tour size. The calculation of the time complexity of tournament selection is straightforward. Each competition in the tournament requires the random selection of a constant number of individuals from the population. The comparison among those individuals can be performed in constant time, and N such competitions are required to fill a generation. Thus, the time complexity for tournament selection is $O(N)$. Often there is no sorting of the population and implementation is also simpler but a time complexity of $O(N)$ is required. (Blickle & Thiele, 1995a; Blickle & Thiele, 1995b)
4. *Truncation selection:* In truncation selection individuals are sorted according to their fitness. Only the best individuals are selected for parents. The parameter for truncation selection is the truncation threshold *Trunc*. This threshold indicates that only a fraction of the best population will be selected as parents. Individuals below the truncation threshold will not participate in mating. Truncation selection leads to a much higher loss of diversity for the same selection intensity compared to ranking and tournament selection. This selection is more likely to replace less fit individuals with fitter offspring, because all individuals below a certain fitness threshold do not have a probability to be selected. Ranking and tournament selection seem to behave similarly (Blickle & Thiele, 1995b). This option of selection has a time complexity of $O(N \log N)$ incurred due to sorting of the N individuals.
5. *Stochastic Universal sampling(SUS):* Introduced by Baker (Baker, 1987; Grefenstette & Baker, 1989), it exhibits no biasing and a minimal spread. The concept of implementation is borrowed from Rowlette Wheel where in every individual is evenly placed on contiguous segments of a line, such that each individual's segment is equal in size to its fitness. Uniformly spaced pointers equal to the number of individuals (say N) to be selected are placed over an imaginary line. So, there will be N pointers. The distance between each pointer will be $1/N$ and the position of the first pointer is given by a randomly generated number in the range $[0, 1/N]$. SUS uses a single random

value to sample all of the solutions by choosing them at evenly spaced intervals and the number of copies for an individual is obtained by checking the individuals fitness which spans the position of the pointers. The algorithm has a time complexity of $O(N)$, because only a single pass is needed through the list after the sum of the function values is calculated.

2.2 Detail Steps of General Genetic Algorithm

This section gives the working steps of the basic Genetic Algorithm.

- **Step 1: Initial Population Generation:** Randomly generate an initial population of G_p chromosomes where p=1,2,…p_size where p_size is the size of the population. The chromosome is a string of symbols(in case of binary it is a series of bits) $G_p = \{G_{p1}, G_{p2}, \ldots, G_{pn}\}$ where n is the fixed length of each chromosomes. The symbols $G_{p1}, G_{p2}, \ldots, G_{pn}$ are the elements or genes of the chromosomes $G_{p.}$
- **Step 2: Fitness Calculation:** Compute and save the fitness measure fitness($G_p(t)$) for each individual G_p at generation t where t stands for the number of generations. This tests each chromosome to see how "good" it is at solving the problem at hand and assign a score accordingly.
- **Step 3: Parent Selection:**
  - Define selection probabilities Select($G_p(t)$) for each individual chromosome G_p in the current generation $G_p(t)$ so that Select($G_p(t)$) is directly proportional to fitness($G_p(t)$)
  - Select parents with a suitable selection scheme.
  - Then choose two members from the current population $G_p(t)$. Chance of being selected is proportional to the fitness of the chromosome, with those chromosomes having a higher fitness value being selected more often.
- **Step 4: Crossover:**
  - Define a crossover rate.
  - For each pair of parents identified in step 3, perform crossover operation. Select type of crossover.
  - If no crossover takes place then form two offsprings that are exact copies of their parents.
- **Step 5: Replacement and Mutation:**
  - Generate the offspring population $G_p(t+1)$ until some convergence criteria is reached.
  - For the bit-based representation, mutate the individual bits G_{pi}'s (i=1,2….,n) with probability *Pm*.
  - One has the option of choosing whether to make the elitist chromosomes candidates for mutation.
- **Step 6: Termination:**
  - Compute fitness of new generation G_p (t+1) of G_p number of chromosomes.
  - If stopping criteria is met, terminate the algorithm. Else return to step 3, 4, 5 until a new population of G_p members have been created.

2.3 Motivation and Proposed Scheme

The genesis of chaos theory stands on the fact that minute perturbations in a dynamical system can have an enormous effect on the behavior of the chosen system. The chaotic behavior is itself probabilistic in nature and the transformation operations used in GA are also of similar nature. The objective in our proposed cryptography is to extend these features and use the probabilistic operators of natural evolution viz selection of parents, reproduction via the operator known as crossover and the diversity creation mutation operator to choose a cipher text (CT) of better quality, a subjective term with relative to the rest of the candidate solutions in search space, by using Genetic Engineering Algorithm (GEA). Evolution theory is a dynamical process similar to the dynamical behavior of chaos. Chaos theory models the complex non linear system and after several iterations reveals a pattern of a strange attractor whose behavior can be monitored. GA is also subjected to several iterations of the individuals represented as chromosomes, to manipulate themselves into fitter solutions and reach a goal guided by a heuristic function known as fitness. Due to the genetic operator mutation, these individuals bring diversity and the solution does not converge prematurely. In GA the recombination operator crossover cuts and joins pieces of information from different individuals to get a better overall solution. So, a mixing of the information results in the search to be directed to reach a globally optimal solution. Due to this linkage and the third probability operator of GA- *mutation* the overall fitness gets influenced. In effect we are dealing with a dynamical system wherein the influence of the GA operators can either stabilize or destabilize the search space. This is similar to what a minuscule disturbance can do in a non linear dynamical system. Chaos theory shows that a small disturbance results in drastic changes manifested as butterfly effect and a small change in the initial condition results in wide changes in the attractor domain after some time t.

Another vital link is chaos, EA and GA can be simulated. Simulation of random and dynamic behavior inherent in organic evolution can be realized in EA by using a pseudo-random number generator combined with selection based on the decision metric *fitness*, simulated version of reproduction by *crossover* and the diversity creator operator- *mutation.* When the random changes of individuals and fitness based selection are combined then a computerized system will be able to solve the problem faster than a random search. *This section covers a brief comprehensive overview of the interdisciplinary work between chaos and evolutionary theory.*

Evolutionary algorithms (EA) are meta heuristic search technique and not a directed search technique because biological evolution itself is not purposeful or directed. GA is probabilistic because initial population is generated randomly, parents selected for participation in reproduction are chosen randomly, fitness determination can also be probabilistic, even the crossover points and mutation rate are determined randomly.

Many developments have taken place in data encryption techniques (Wenbo, 2004). Recent trend shows a deviation from these traditional schemes towards techniques which are self learning and have less overhead in implementation. The stochastic properties of GA are also finding its roots in cryptography (Husainy & Muhammed, 2006) and cryptanalysis (Gorodilov & Morozenko, 2008; Spillman, 1993; Toemeh & Arumugam, 2007). ICIGA (Improved Cryptography Inspired by Genetic Algorithm) developed by A.Tragha et al. (Tragha et al, 2005) is an improved version of GIC (Genetic Algorithms Inspired Cryptography) (Tragha et al, 2006). Another vital application of chaotic synchronization in cryptography is to design pseudo-random number generators (PRNGs). In 2003 an important contribution by Riccardo Caponetto(Caponetto, 2003) uncovered the significance of the chaotic generators in enhancing the convergence speed of EA. The optimizing property of GA has been used in optimization of the PID parameter by merging properties of GA with chaotic dynamics which accelerated the convergence of GA to a global optimal solution (Wu et al, 2008). Klein et al (Klein et al, 2005) demonstrated that a secure key exchange protocol can be established between the driver which acts as the transmitter and the response system which can be treated as the receiver. Therefore, the parameters and the keys transmitted are secure since the synchronized dynamical system does not necessitate the transmission of keys over the communication channel. The random sequence obtained from chaotic generator further transforms the process into a powerful stochastic method of searching the solution space in varied directions for an optimal solution escaping points of local optima.Thus the usage of synchronized chaotic systems for key generation in secure communication, which is also adopted for our chapter, has a two fold computational edge mainly due to the chaotic generator properties and the evolvable property of EA. Hence, attempt has been made in this chapter to use these compatible features in secure communication with Evolutionary algorithm. There has also been significant amount of successful attempt to reap the benefits of the dynamical features of evolution with chaos theory (Guoa et al, 2008; Yan et al, 2003). Another successful attempt was made in (Yiqiang et al, 2009) by the authors who exploited the mentioned features of chaotic dynamics with GA to improve the accuracy of remote sensing image classification. An emerging derivative of GA known as Genetic engineering(GE) has been applied effectively for solving complex goal oriented problems like synthesis problem (Jozwiak and Postula, 2002). In their paper the authors emphasized that for complex problems such as synthesis of a minimal input support, it was imperative to substitute the basic operators of genetic algorithm with deterministic operators goal-oriented in the form of a Genetic engineering algorithm (GEA). In 2001, the authors(Gero & Kazakov, 2001) came out with a solution to their problem by extending the underlying principle of GA in the form of GEA which is tailor made to fit into specific class of problems and which also provided a computational edge. In GA the selection mechanism and both transformation operators are probabilistic. Such property can be usefully tapped to perform a blind search of the solution space in varied directions to obtain an optimal solution, but randomicity sometimes can destabilize the system also. Moreover, there is no guarantee that it will yield an improved offspring. Furthermore, convergence speed can be rather slow for many complex optimization problem.

In our application of cryptography, we want to choose the best solution in comparison to the available candidate solutions as the cipher text (CT) which should be recovered. GA is rendered dynamical by the genetic operators viz selection scheme, crossover and mutation which bring drastic changes which can yield to a near optimal or no solution at all. Again, for most optimization problem, a good solution needs to be stable and optimal. To suit the purpose of our scheme, such randomness needed to be controlled by injecting certain deterministic features into the probabilistic operator - *selection* of the basic GA. Thus a genetic engineering algorithm(GEA) inspired from GA has been proposed for the purpose of cryptography. The basic selection mechanism in GA has been re-designed to suit our purpose. The way selection is done leads to two solution of cryptography application. The first method Selection1 renders the GEA deterministic and the second method Selection2 makes the GEA pseodo deterministic. The former is a self learning and crossover is guided with a "tag" concept along with the heuristic function *fitness* which measures the quality of how "good" the CT is. Before, delving into the proposed scheme some points of the most commonly used selection mechanism of GA needs to be re-iterated. In the roulette wheel selection method, it is observed that since the qualifying criteria for a chromosome to become a parent is its highest fitness value compared to the rest of the individuals, the individuals having similar fitness values tend to be selected more often. This eliminates a fair chance for the other individuals of the same population to participate in reproduction since the fittest individual dominates over the rest. The Rank based scheme chooses candidate solution as parents based on the rank index and not directly on the fitness. In the proposed scheme, instead of sorting the initial population based on their increasing order of fitness, a search is performed to find the individual with the highest fitness. This is similar to a spin of the Roulette wheel where all the individuals in the population are checked.

A "tag" concept is used which serves as a tool for learning to assist the rest of the candidate solutions to identify a solution which showed some promise. A chromosome with the highest fitness measure is named as $Chromosome_{best}$ which is chosen as one parent and crossed over with replacement option with the remaining chromosomes which are also chosen randomly in the next turn of the Roulette wheel. The offspring so produced is passed on to the next generation even if it is better or worse to the second parent in fitness. The selection operator has been modified for the purpose of cryptography to successfully converge to the exact goal oriented solution to the problem with minimum computation and simple design of the algorithm. The selection scheme is not purely based on the greedy method of choosing the fittest individual always as a parent which occurs in GA selection mechanism explained already especially in Rowlette Wheel. Hence a tradeoff is made and a partial fitness based selection scheme emerging from the concept of Roulette wheel with replacement strategy (known as PFRWR in short) has been proposed for our case giving rise to GEA. The evolutionary process of the proposed GEA makes the individuals of sub generations distributed ergodically in the defined space and leads to the convergence towards the goal - successful selection of a "good" quality CT relative to the rest and convergence of decryption with the output of recovered plain text (PT). To trigger crossover in both the solutions for cryptography, the fittest candidate solution is selected as a parent from all the neighbouring solutions. This also happens in Stochastic hill climbing, a variant of hill climbing approach. When the candidate solution with highest fitness is chosen during both encryption and decryption process, it implies that all the solutions are compared first and then a decision is made to move to the one which is closest to the goal or to move to another sample point akin to a spin of the Roulette wheel where all the individuals in the population are checked for the contesting individual to become the first parent. Our aim is to extend the benefit of synchronization of two different time delayed systems in cryptography, (Banerjee et al, 2008b) using GEA in secure communication.

The driving system is an autonomous continuous time difference-differential equation with one variable expressed as,

$$x = ax(t-\tau) - b[x(t-\tau)]^3. \quad (1)$$

Here a and b represent positive parameters and τ is the time delay. The system is proposed as a chaos generator and is studied in (Ucar, 2003). This model is a special class of of functional differential equation (Hale, 1977), and it is characterized by an infinite dimensional solution space (Farmer, 1982) with initial condition as any continuous function defined on the closed interval $[-\tau, 0]$.

For the numerical simulation, the system parameters and initial condition can be chosen as

$$a = 1.7, \quad b = 1.0, \quad \tau = 1.0, \quad x_0 = 0.1. \quad (2)$$

Again the dynamic equation of the Ikeda system is given by

$$y = -cy + m_1 \sin(y(t-\tau_1)) = m_2 \sin(y(t-\tau_2)), \quad (3)$$

where τ_1, τ_2 are the time delays and c, m_1, m_2 are the system parameters. We consider the above system (3) is the response system. The system (3) had been introduced to describe the dynamics of an optical bistable resonator, it plays an important role in electronics and physiological studies and is wellknown for delay-induced chaotic behavior (Voss, 2000; Masoller, 2001).

The system (3) exhibits chaotic behavior at values c=1.8, m_1 =5.0, m_2 =1.0, τ_1=2.0, τ_2=1.0, y_0 = 2.0.

Now, we describe the process of synchronization between two systems. For this a control function u in system (3) has been introduced. So re-writing eq (3) as

$$y = -cy + m_1 \sin(y(t - \tau_1)) + m_2 \sin(y - \tau_2)) + u(t). \tag{4}$$

The control functions for synchronization between systems (1) and (3) is obtained by subtracting equation (1) from (3) to result

$$\dot{e} = -cy + m_1 \sin(y(t - \tau_1)) + \mathrm{m}_2 \sin(y(t - \tau_2)) - ax(t - \tau) + b[x(t - \tau)]^3 + u(t), \tag{5}$$

where the relative error $e = y - x$.

The active control function u can be defined as

$$u(t) = cx + ay(t - \tau) - m_1 \sin(y(t - \tau 1)) - m_2 \sin(y(t - \tau_2)) - b[x(t - \tau)]^3. \tag{6}$$

Therefore the error system (5) becomes

$$\dot{e} = -ce + ae(t - \tau). \tag{7}$$

It is obvious that $e = 0$ is the trivial solution of eq. (7) for any value of time delay τ. To study the stability of synchronization manifold $x = y$, we can use the Krasovskii-Lyapunov functional approach. According to Pyragas (Pyragas,1998) the sufficient condition for the stability of the trivial solution e = 0 for the time-delayed system

$$\dot{e} = -r(t)e + s(t)e_\tau \tag{8}$$

is $r(t)>|s(t)|$. The Krasovskii-Lyapunov functional (similar to the Lyapunov function in the case of ODE) is

$$V(t) - \frac{e^2}{2} + \mu \int_{-\Gamma}^{0} e^2(t + \theta) d\theta$$

where μ>0 is an arbitrary positive parameter. The solution e= 0 is stable if the derivative of the functional $V(t)$ along eq. (8),

$$\dot{V}(t) = -r(t)e^2 + s(t)e\ e_\tau + \mu\ e^2 - \mu\ e^2_\tau,$$

is negative.

$\dot{V}(t)$ will be negative if $4(r - \mu) > s^2$ and $r > \mu > 0$. The asymptotic stability condition for $e = 0$ is given for

$$\mu = \frac{|\, s(t)\,|}{2} \text{ and } r(t) > |\, s(t)\,| \,. \tag{9}$$

If this inequality holds for all $t > 0$, the identity synchronization manifold of the equation is asymptotically stable. By using the Krasovskii-Lyapunov functional approach we obtain that the sufficient stability condition for the synchronization manifold y = x can be written as

$$c > a. \tag{10}$$

As eq. (7) is valid for small e, the stability condition (10) found above holds locally. Numerically, simulations fully support the analytical results. We select the parameters of the prototype model as $a = 1.7$, $b = 1.0$, $\tau = 1.0$, and the parameters of the Ikeda system as $c = 1.8$, $\tau_1 = 2.0$, $\tau_2 = 1.0$, $m_1 = 5.0$, $m_2 = 1.0$. The time variation of synchronization error is shown in Figure 3.

After the system in consideration becomes stable and coupled, the response system behaves as a slave to the driving system. This feature can be used to generate a sequence of real numbers which can be converted to integer numbers to be used as keys for the cryptosystem. For our scheme, the key sets are generated from the synchronized systems (1) and (3) with the proper choice of the controller (6). The sender use the x-variable of the system (1) and the receiver use the corresponding of the system (3). They both pre determine to choose the corresponding numerical values at $t = 200, 201, 202$ respectively. To convert the keys into integer, the sender and receiver both pre determine to [1000x] (Mod 38) and [1000 x_1] (Mod 38) respectively.Therefore the first key k_1=12 (see Table 1) is the value of x at $t = 200$ with the transformation [1000x] Mod(38) and so on.

2.4 Application: Proposed Cryptography

This section explains a GEA scheme for the purpose of cryptography inspired from GA. The key sets used for encoding and decoding are derived from the previous section. In our scheme GEA for encryption, initialization of the population by breeding the different encoded versions of the cipher text to generate the initial population is also an essential operator apart from the rest of the GA operators. In the present

Figure 3. The time variation of synchronization error

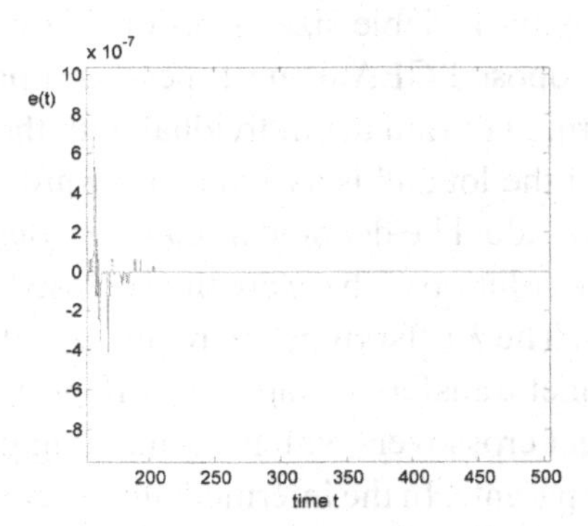

Table 1. Key set

Keys	k_1	k_2	k_3	k_4	k_5	k_6	k_7	k_8	k_9	k_{10}	k_{11}	k_{12}
Value	12	1	29	19	11	4	37	32	30	28	29	31
Keys	k_{13}	k_{14}	k_{15}	k_{16}	k_{17}	k_{18}	k_{19}	k_{20}	k_{21}	k_{22}	k_{23}	k_{24}
Value	34	1	8	17	27	1	15	30	9	28	10	32
Keys	k_{25}	k_{26}	k_{27}	k_{28}	k_{29}	k_{30}	k_{31}	k_{32}	k_{33}	k_{34}	k_{35}	k_{36}
Value	18	5	32	23	15	9	4	0	37	36	37	2
Keys	k_{37}	k_{38}	k_{39}	k_{40}	k_{41}	k_{42}	k_{43}	k_{44}	k_{45}			
Value	5	10	17	24	33	5	17	29	5			

manuscript, we have used the features of natural biological evolution and its transformation operators viz selection, crossover and mutation by adhering to its inherent meaning but the way they subsequently mix their information is modified to increase reliability and decrease the uncertainty of recovery of the message. The discrete solution space is represented by encoded and decoded chromosomes in the encryption and decryption process respectively. The objective is to choose a strongly encrypted string of better quality relative to the available candidate solutions. This choice is guided by the fitness measure and the selection schemes discussed in this section. Here, a combination of rank based selection scheme and Roulette wheel is adopted since the reproduction range is limited. Selection process is governed by a maximization objective function *fitness*. The selection process has also been closely monitored by implementing the proposed selection operator which is partially dependent on fitness for choosing the ρ parents. Subsequent sections explain the proposed GEA for cryptography. A predetermined number of individuals n are generated from the cipher text (CT) using a generator function G_n to produce valid solutions based on CT. The initial population $G_n(0)$ of encrypted chromosomes is populated using integer values obtained from the system described in previous section. They are then evaluated with a heuristic function -fitness. The design of the selection operator are discussed first followed by the algorithm for cryptography. We have considered that in a population of μ candidate solutions, always two parents participate for crossover(ρ) producing two λ offsprings.

Selection1: Deterministic GE Algorithm

To instantiate reproduction(mixing), the initial population of encrypted chromosomes $G_n(t)$ ($t = 0$ for initial population) are not sorted according to their fitness measure that is they are not assigned any rank. They are arranged randomly occupying variable sized slots of a Rowlette Wheel. This section explains the selection mechanism for the proposed GEA using fitness as a partial guiding metric along with a concept of "tag". A search is performed to find the individual with the highest fitness value.Out of the ρ parents, one of them is the fittest of the lot and is assigned a reward of being the best and marked with a tag of *best*. Let this privileged individual be denoted as $Chromosome_{best}$. The other parent is randomly selected from rest of the μ candidate solutions. These are then crossed over producing λ offspring having mixed traits of both their ρ parents. The λ offsprings are retained and their ρ parents are discarded. But $Chromosome_{best}$ before getting extinct transfers its tag to its offspring and the other offspring is passed onto the next generation. Subsequent crossovers with the remaining μ individuals is performed with the tagged chromosome as one of the ρ parents. In the intermediate generation, this tagged chromosome gets replaced due to crossover but passes its tag to its offspring which replaces it. Here, elitism is switched off. Therefore, the λ offspring replace their parents irrespective of their quality. However, this mechanism guarantees that every individual of the current population gets a chance of reproducing. The tag serves as a learning tool to guide the μ individuals to crossover with a chromosome as it shows some promise in yielding a better candidate solution because this was derived from a candidate solution which was the fittest in its generation. Since with the best chromosome the rest of the chromosomes in the population are mated, the best features are mixed increasing the chance of a better breed of chromosomes in the new population. This is in turn governed by the population size and the formulation of the fitness metric. The same strategy is followed for subsequent generations till a complete -spin of the Rowlette Wheel after which they are passed on to the next generation. Hence for the intermediate crossovers, decision to select a parent is not governed by resorting to the highest fitness scheme. Thus the name of partial fitness based Rowlette Wheel selection with replacement (PFRWR) has been coined as an abbreviation for the selection criteria and ths proposed selection mechanism for our application can be used in two ways yielding two varieties of GEA for cryptography- Deterministic and Pseudo-Deterministic GEA for cryptography with chaos synchronization playing a vital role as a key generator. Figure 4 shows a diagram explaining this process. In the figure, information is a collective term used to denote the PT and CT at transmitter and receiver section respectively.

Selection2: Pseudo-Deterministic GE Algorithm

This selection mechanism differs from the previous selection mechanism as described under. After the first crossover $\mathit{Chromosome}_{best}$ releases its tag. The ρ parents are replaced by their λ offsprings which are passed onto the intermediate population. They also can again participate in reproduction, similar to (μ / ρ, λ) selection scheme, mentioned in the previous sections, but with a difference that when λ are produced there is no fitness evaluation for every λ th offspring to choose a ρ parent for next crossover in the same generation. This reduces considerable computation in cryptography which when applied with (μ / ρ, λ)-ES selection scheme would have account to

Figure 4. Schematic diagram of selection 1: Selection method for cryptography with deterministic GEA

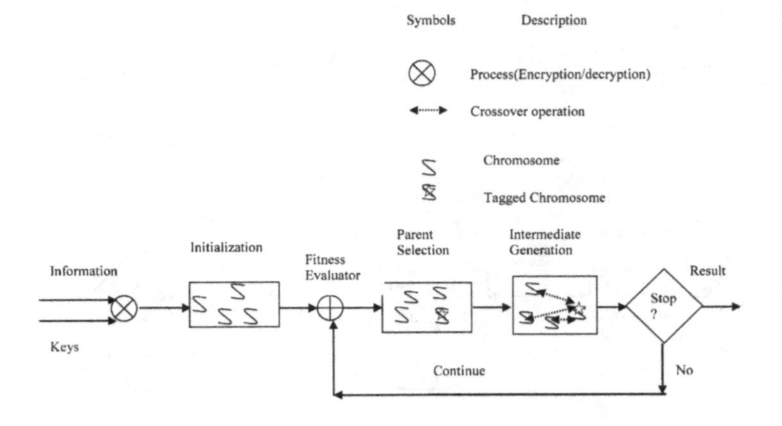

Number of times fitness is evaluated for parent selection at the transmitter section = sample size number of iterations at encryption process,*

and

Number of times fitness is evaluated for parent selection at the receiver's section = sample size number of iterations required to recover the message.*

Whereas, for our application with both the proposed selection methods the number of times fitness evaluation occurs in both the transmitter and receiver sections will be equal to the number of iterations the population of individuals is subjected to in both the ends. Say if the number of generations the transmitter side is subjected to is 5, then only 5 times fitness is called, one time for every generation. Thus, a lot of computation time is saved and the simulation results in the subsequent sections show that the proposed scheme achieves exploitation and exploration oriented optimal solution. Figure 5 illustrates this process.

So for our purpose for the subsequent crossovers within the same generation, the decision of parent selection is not guided by the fitness measure and chromosomes from the variable slots of the Rowlette Wheel are randomly picked up.Hence, the randomness property of GA is maintained. However, the same individual cannot self reproduce.This cycle is repeated for the entire chromosomes in the initial population preparing the same number of individuals as the former one to be passed into the next generation $G_n(t+1)$.After all the μ individuals have been subjected to crossover, they have their fitness updated. Here all chromosomes get a fair chance to compete even more than once and the problem of dominance of the fittest in subsequent crossovers within the intermediate population is controlled. This evolution scheme of switching from a deterministic to a random parent selection leads to an enhanced exploration of search space. The following section shows the application of the above mentioned modified selection scheme (PFRWR) in cryptography and a comparison of the proposed GEA with PFRWR selection has been made with the $(\mu / \rho, \lambda)$ -ES selection mechanism in cryptography. The algorithm for the test system has also been developed.

The key sets keys=$\{k_1, k_2,\ldots, k_n\}$ used are of the same length n as that of M to generate the CT $C=\{c_0, c_1,\ldots, c_{n-1}\}$ where $c_i= m_i \oplus k_i$, $0 \leq i \leq n-1$. Initially, each letter of the plaintext M is encrypted. M is XO-

Figure 5. Schematic diagram of selection 2 scheme for cryptography with pseudo-deteministic GEA

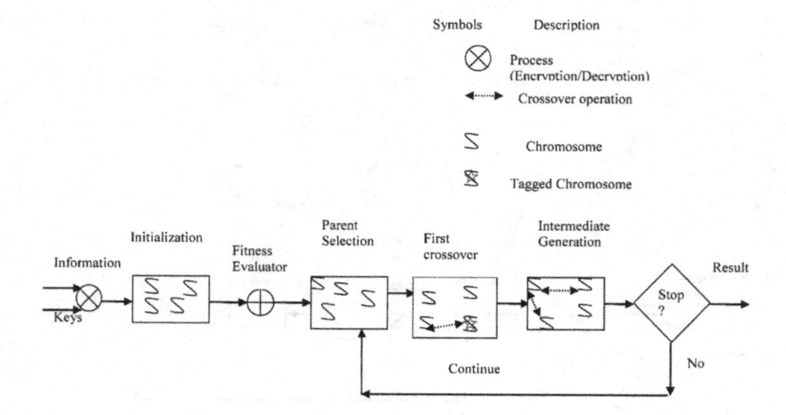

Red with n* population size(denoted as *p_size*) number of keys obtained from synchronization of chaotic systems described in the previous section.. The encryption technique adopted here is Vernam cipher. The message to be transmitted $M=\{m_0, m_1,..., m_{n-1}\}$ is XORed with randomly generated keys from synchronization of chaotic systems. The key sets $K=\{k_1, k_2,..., k_n\}$ used are of the same length n as that of M to generate the cipher text. If the stream of data is truly random and used only once then this is the one-time pad and hence theoretically unbreakable (Menezes et al, 1997). An example is illustrated with two point crossover. In our case, the heuristic function is defined as sum of how many steps or characters away each encrypted element is from the message M.

The encoding and decoding resulting in change of an element of a chromosome is synonymous to mutation in evolution. In evolution theory, mutation is defined as a change in the DNA sequence of a gene (Aminetzach et al, 2005; Bertram, 2000; Burrus & Waldor, 2004; Watson et al, 1987). This creates diversity which can be deleterious or beneficial if passed onto the progeny. There are six basic types of mutation explained briefly as under:

1. *Point mutation*: In this kind of mutation, one change in the base of the gene sequence is changed. This can be thought like changing of a letter in a word.
2. *Frame-shift*: This mutation results in a deletion or addition of one or more bases in the DNA resulting in shorter protein. For example, in a three letter word a letter is removed and that word is shifted to the corresponding word resulting in a changed sentence.
3. *Deletion*: this mutation happens when an entire section of the DNA is chopped off, resulting in a shorter one. This can be thought of randomly removing a word anywhere from a sentence.
4. *Inversion*: over here, an entire section of DNA is reversed. It is similar to cutting off a part of a sentence and pasting it in reverse order.
5. *Insertion*: In this kind, a DNA is added which also results in a frame-shift mutation. This is quite similar to adding a three letter word anywhere in a sentence.
6. *DNA expression mutation*: In biological mutation, the inversion mutation causes a whole section of a person's DNA to be reversed while in an expression mutation, it is not just the protein that may be changed but the location where it is made or the amount of the protein produced gets affected.

For the purpose of cryptography, we can consider the process of encryption as subjected to several point mutations wherein each element of M is changed during encoding. Similarly, decryption can be considered to inversion mutation scheme in which every element of the CT undergoes an inverse of the encryption scheme adopted. For bit based representation, point mutation will result in flipping the selected gene. Since it is a problem specific GEA, a certain amount of deterministic factor is injected and we have assumed that all successive crossovers till the termination of the scheme occurs with a fixed crossover point C_p and another pair of crossover points- C_{p1} and C_{p2} generated randomly for one point and two point crossover respectively.

The algorithm for Cryptography with Deterministic and Pseudo-Deterministic GEA performed with the modified selection mechanism is presented below

1. TRANSMITTER SECTION

1.1. Initial Population

Initialization: Decide a predefined number of generations t and the *p_size*

Create initial population $G_p(t)$ of encrypted chromosomes, where p=2,3,.., *p_size* and t=0 of fixed length n from message $M=\{m_0, m_1,\ldots,m_{n-1}\}$ using the series of keys $K=\{k_1, k_2,\ldots, p\_size\}$ by performing an XOR operation on M.

```
Also
Set counter for incrementing keys k=0;
For i= 1,2,…, p_size and
      For j= 1,2,.., n
     Chromosome i,j = M j ⊕ keys k++
```

1.2. Evaluate Fitness

Set fitness for all individuals fitness $_i$ =0

```
For each i=1,2,…, p_size and
For each j=1,2,…, n
  Calculate fitness i = fitness i + abs (M j - Chromosome i, j)
Find best fitted chromosome Chromosome best
```

1.3. Parent Selection and Crossover

Select parents based on either

(a) Selection1-The Deterministic GEA
(b) Selection2 -The Pseudo-Deterministic GEA

Select type of crossover scheme. The system has been tested effectively with one-point and two-point crossover scheme and results have been shown in subsequent section with two-point crossover.

For i=1,2,…, p_size perform crossover between $Chromosome_{best}$ and $Chromosome_i$. It is to be noted that no crossover occurs with itself.

1.4. Termination of Encoding: Terminate the Encryption Algorithm after a Fixed Number of Iterations Depending upon the Parameters/User Choice

The ciphertext $Chromosome_{best}$ and the crossover points are transmitted which may be encoded by the Vernam cipher technique. For simplicity, we have not encoded the crossover points.

2. RECOVERY OF PLAINTEXT

2.1. *Storing the Fitness of the CT:Chromosome*$_{best}$ gets evaluated into *fitness*$_{cipher}$
2.2. *Initialization*: Breed the same number of individuals by performing inverse mutation which is decoding the CT with the key set used in encoding by performing XOR operation.
2.3. *Evaluate fitness*: Evaluate fitness of decoded chromosomes as discussed in step 1.2
2.4. *Parent selection and Crossover*: This step is the same as 1.3.
2.5. *Termination of the process*: Tune the fitness of decoded chromosomes to achieve the fitness of *fitness*$_{cipher}$

For each i=1,2,…,p_size

Terminate when *fitness*$_{cipher}$ becomes equal to *fitness*$_{cipher}$ else continue with the steps starting from 2.3.

The proposed scheme has been tested with encryption with the (μ /ρ, λ)Selection scheme. The test system is also developed and explained below in a nutshell. All the steps mentioned in Transmitter and Receiver end are the same except for a change in Step 1.3 and 2.3. For cryptography with (μ /ρ, λ) Selection and crossover is as follows:

a. Choose the fittest chromosome and denote it as *Chromosome*$_{best}$ and another parent randomly from the μ individuals.
b. Perform crossover replacing the ρ parents with λ offsprings.
c. Evaluate fitness of the λ offsprings. Select the λ offspring with higher fitness as one of the ρ parent for remaining crossovers in the current generation.This selected offspring now becomes *Chromosome*$_{best}$.
d. Return the other offspring into the next generation pool.

During decryption, the fitness measure is used to direct the individuals to adjust their fitness value to match with that of the calculated fitness *fitness*$_{cipher}$. Hence, this heuristic operator gives an estimate of how far we are from the goal of successfully recovery of PT. This is an essential property of evolution that individuals are self learning and adapt themselves to solve a problem in hand. The algorithm is simulated with one point and two point recombination scheme and a comparison of successful recovery with the (μ /ρ, λ) apart from other observations have been discussed in the following subsections.

A flowchart for the proposed cryptography is illustrated in Figure 6.

3. SIMULATION RESULT AND ANALYSIS

Let us illustrate the above steps with a small example. Suppose the sender wants to send the message *The globe* to the receiver. The population size is considered to be small in the presented example. A small population size of 5 chromosomes and two point recombination scheme has been opted. Therefore 45 number of keys will be needed for encoding the message. The experimental observations are subdivided into two segments. The key sets displayed in Table 1 are chosen and Table 2 and 3 shows a working example of proposed cryptography and cryptography with the (μ / ρ, λ) respectively.

Figure 6. Flowchart for the proposed cryptography

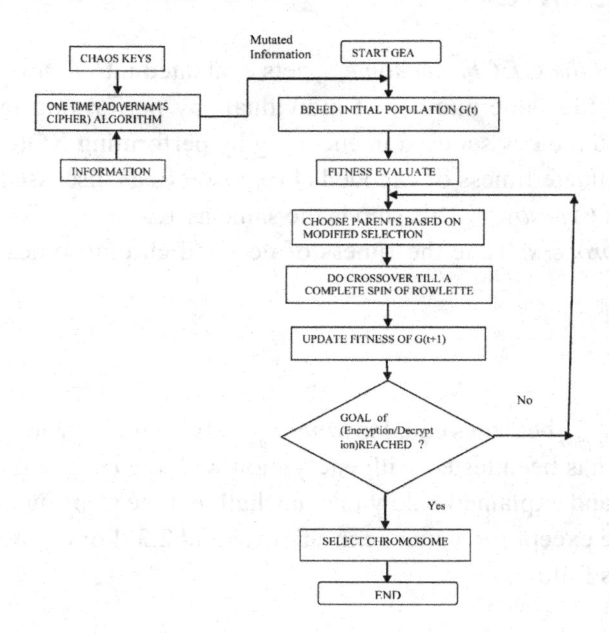

In Table 2 it is observed that the initial population (Generation =0) by breeding several different versions of the cipher text is the same for both the flavors of cryptography. The encoded strings are scrambled enough to keep things from prying eyes.

Table 2. Simulation result of the proposed cryptography

	Cryptography with Deterministic GEA	*Cryptography with Pseudo-Deterministic GEA*
Generation	Population Fitness	Population Fitness
0	Xix3lhJB{ 143 Huz1fd~yd 124 [vl<mL}gE 145 Cgl$gIKGg 129 Qbt8Fi~` 133	Xix3lhJB{ 143 Huz1fd~yd 124 [vl<mL}gE 145 Cgl$gIKGg 129 Qbt8Fi~` 133
1	Xil<mL}B{ 146 Hux3lhJyd 133 [vt8Fi~gE 148 Cgz1fd~Gg 122 Qbl$gIK ` 125	Xit8Fi~B{ 149 Huz1fd~yd 124 [vl$gIKgE 140 Cgl<mL}Gg 134 Qbx3lhJ ` 127
2	Xit8Fi~B{ 149 Hul<mL}yd 136 [vl$gIKgE 140 Cgx3lhJGg 131 Qbz1fd~` 118	Xil<mL}B{ 146 Hul$gIKyd 131 [vt8Fi~gE 148 Cgx3lhJGg 131 Qbz1fd~` 118
3	Xiz1fd~B{ 134 Hut8Fi~yd 139 **[vl<mL}gE 145** Cgl$gIKGg 129 Qbx3lhJ` 127	**Xit8Fi~B{ 149** Hul<mL}yd 136 [vx3lhJgE 142 Cgz1fd~Gg 122 Qbl$gIK ` 125

Table 3. Cryptography with (μ / ρ, λ)

Generation	Population	Fitness
1	Xiz1fd~B{	134
	Hux3lhJyd	133
	[vt8Fi~gE	148
	Cgl<mL}Gg	134
	Qbl$gIK `	125
2	Xiz1fd~B{	134
	Hut8Fi~yd	139
	[vl$gIKgE	140
	Cgx3lhJGg	131
	Qbl<mL} `	130
3	Xiz1fd~B{	134
	Hul$gIKyd	131
	[vl<mL}gE	145
	Cgt8Fi~Gg	137
	Qbx3lhJ `	127

First time crossover occurs with the best fitted parent with fitness 149 producing the next generation(Generation 1). From this step onwards the behavior and fitness of candidate solution differs in both the cases as seen from observation. At the end of the process, the cipher text transmitted from cryptography with Deterministic GEA is of fitness value 145. On the other hand the cipher text obtained from cryptography with Deterministic GEA is considerably of higher fitness measure 149. Although, the difference in their quality is significantly small due to small sample size, still it shows that randomicity enhances the searching and mixing in search space in a less restrictive manner and may sometimes lead to a better solution. But uncontrolled randomness can lead to a solution which may be incoherent with the original message. Hence, this concern has to be addressed leading to answers which are suited for goal oriented purpose. Coincidently, for this example both the cryptographic versions decipher the cipher text after three generations. But cryptography with (μ / ρ, λ) selection scheme takes 6 iterations/ generations to recover the original message along with the overhead of intermediate computation of fitness between the λ offsprings competing to participate in crossover.

Figure 7 and Figure 8 represents a comparison between cryptography with deterministic and pseudo-deterministic GEA with the probabilistic data points corresponding to the population size 20. From these figures we can see that the deterministic GEA has a maxima at transmitter generation 6 and after that the peak is below 10 throughout the generations. On the other hand the pseudo deterministic GEA has a similar curve upto transmitter generation 20 and then it attains its maxima at transmitter generation 25 for the rest of the interval. At generation 25 it has a high peak (y=40) which shows that the randomness takes more iterations to stabilize and come to a conclusion. Here we can see that the pseudo deterministic GEA takes more iterations to recover the plaintext, but it yields the best cipher text at the same generation 25 than the deterministic GEA.

Figure 9 and Figure 10 for population size 40 shows a slightly different observation. The pseudo deterministic GEA attains a maxima at transmitter generation 30 but it also yields the best cipher text

Figure 7. Comparison of convergence in cryptography with deterministic and pseudo-deterministic GEA for population size 20

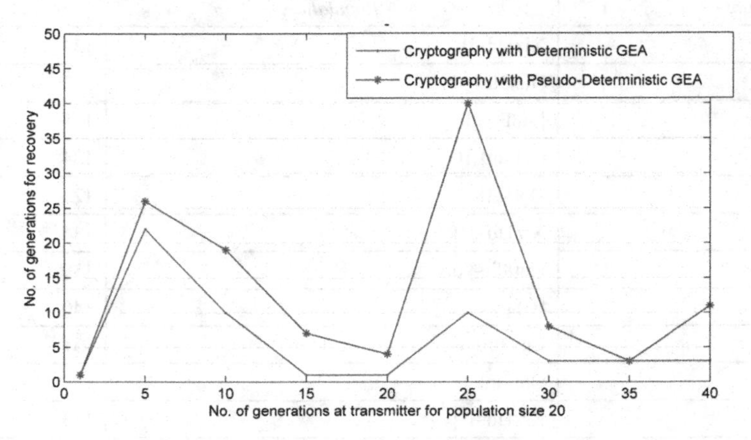

Figure 8. Comparison of fitness of cipher text for population size 20

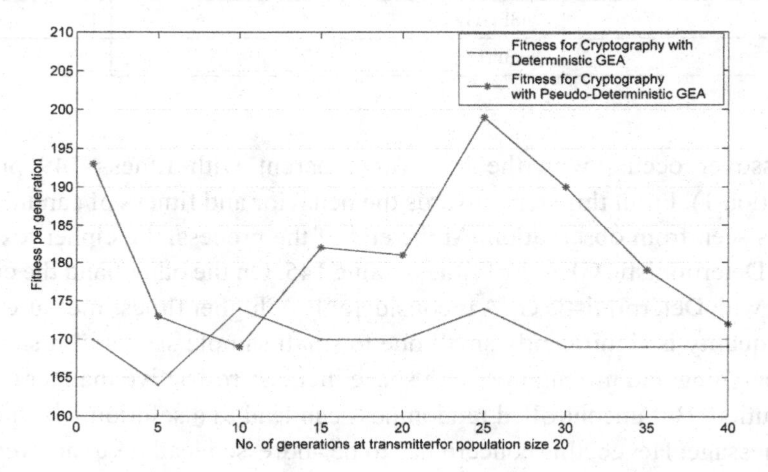

in the early generation at 15. Further, pseudo deterministic option takes more iterations to recover the plaintext for transmitter generation 40 and the cipher text obtained from it is of poor quality than the deterministic GEA. But this covers more of the solution space than the Deterministic GEA, evolving with better candidate solution. It is interesting to note how both the schemes escape form getting stuck with a cipher text in a generation and gradually self adapt themselves to yield a better solution. Overall, the behavior is unpredictable which is sometimes advantageous in applications such as this. Figure 11 through Figure 14 shows a comparison between the proposed cryptography and cryptography with (μ / ρ, λ) selection. We can see that the Deterministic solution has overall low peaks after transmitter generation 15 and 1 in Figure 11 and Figure 13 respectively. Whereas there are abrupt maxima and minima for cryptography with Pseudo deterministic and (μ / ρ, λ) selection respectively.

Figure 12 shows that cryptography with Pseudo deterministic and (μ / ρ, λ) selection exhibit more or less the same fitness contour. Pseudo deterministic GEA yields a solution of fitness 193 at transmitter generation 1 whereas cryptography with (μ / ρ, λ) selection yields a candidate solution of 170 fitness

Figure 9. Comparison of convergence in cryptography with deterministic and pseudo-deterministic GEA for population size 40

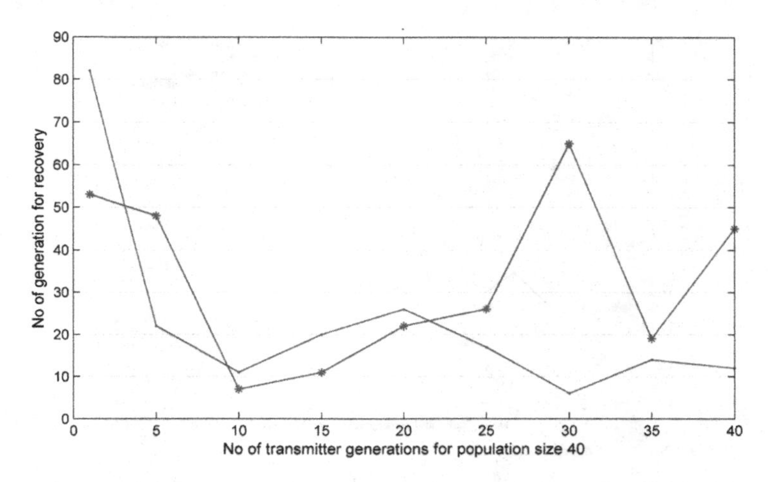

Figure 10. Comparison of fitness of cipher text for population size 40

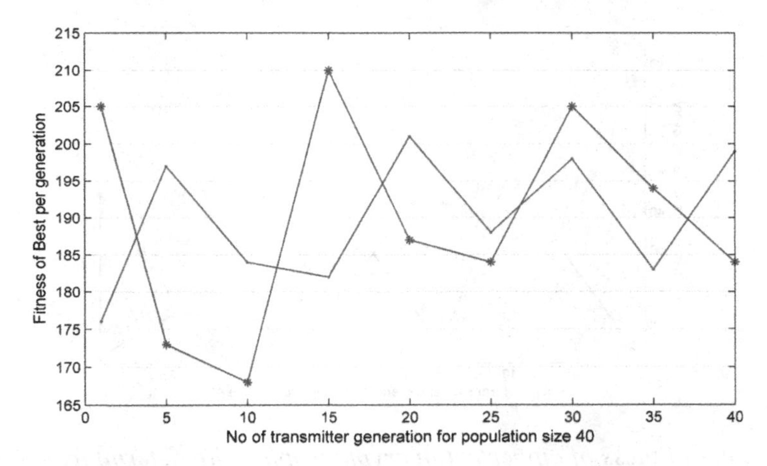

Figure 11. Comparison of convergence in cryptography with deterministic GEA, pseudo-deterministic GEA and (μ / ρ, λ) selection mechanism for population size 20

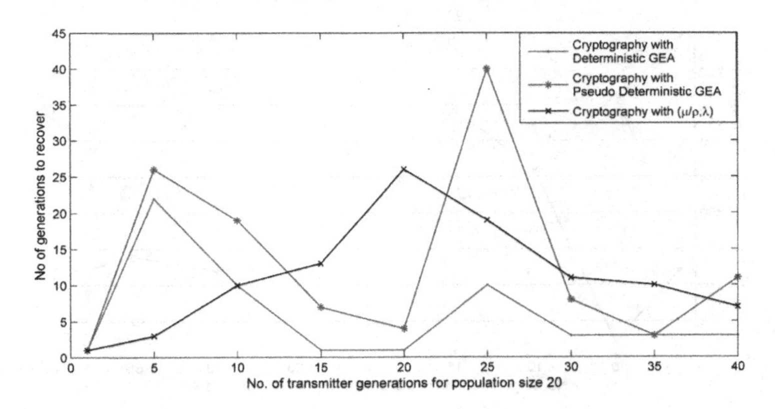

Figure 12. Comparison of fitness of ciphertext on cryptography with deterministic GEA, pseudo-deterministic GEA and (μ / ρ, λ) selection mechanism for population size 20

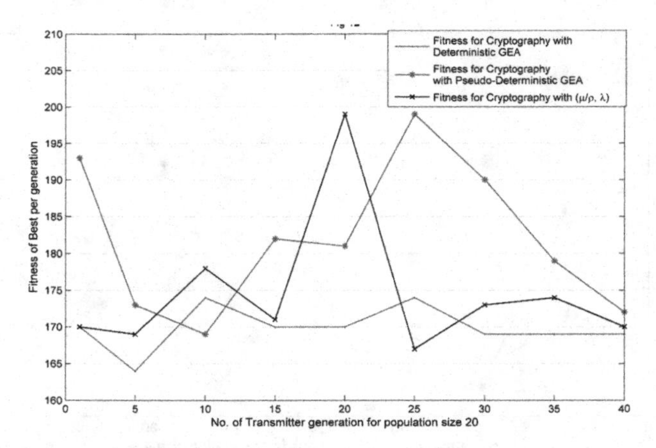

Figure 13. Comparison of convergence in cryptography with deterministic GEA, pseudo-deterministic GEA and (μ / ρ, λ) selection mechanism for population size 40

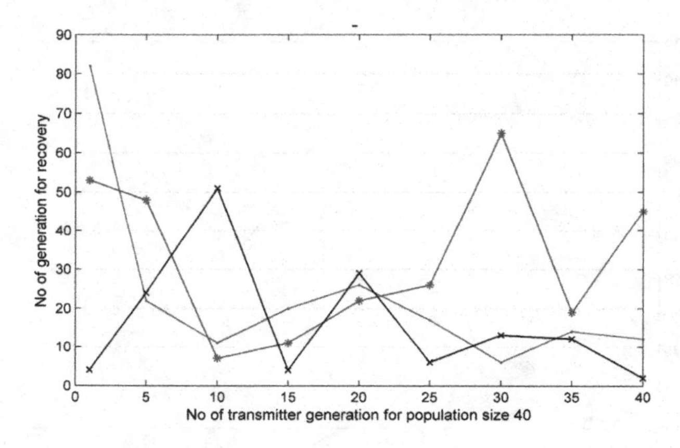

Figure 14. Comparison of fitness of ciphertext in cryptography with deterministic GEA, pseudo-deterministic GEA and (μ / ρ, λ) selection mechanism for population size 40

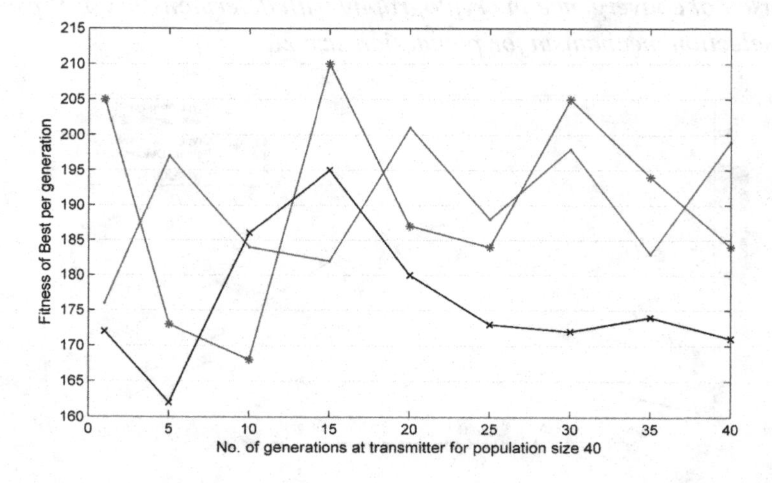

value. However, over here the Pseudo deterministic GEA explores more search area in contrast to (μ / ρ, λ). On the other hand, Deterministic GEA is not impressive in navigating the search space in comparison. Figure 14 again highlights that fitness of transmitted cipher text in Pseudo deterministic GEA is of overall good quality in comparison to the other two. Also the fitness landscape in cryptography with (μ / ρ, λ) shows that it does not evolve as effectively after generation 20 in comparison to the proposed cryptography. This demonstrates that the proposed scheme is a self adaptable and efficient option than the test system.

4. CONCLUSION AND FUTURE WORK

A few thumb rules for choosing and setting parameters are presented here although the list is not exhaustive for the parameters and their usage are application dependent.

1. *Population size*: This is an important parameter to be carefully chosen to begin automation of genetic algorithm for any application. If a large population size is chosen then subsequent generation of new populations of the same size would incur more time. Furthermore, such big populations have more genetic diversity, explore more points in the search domain thereby reducing the number of evaluations required for finding a solution. Many of the Genetic Programming researchers have argued over the maximum number of generations to be 50 because nothing significant development happens after these number of offsprings have been produced [Banzhaf et al 1998). In multipoint crossover due to more diversity in exchange of information, the evolution may stagnate after some more iterations in comparison to one point crossover. For the given problem domain, positive results have been achieved for almost all different sized population.
2. *Fitness function*: Fitness function determines the selection criteria which is problem specific. There are a number of different selection algorithms generally followed in basic GA listed in Table 4.

Fitness function is the evaluation metric which guides the individuals of the population towards an effective solution determining how well a program has learned to predict the outputs from the inputs. The fitness function's computation time t is problem dependent and often excessively long. So it should be designed to provide some ordering of the individuals so that there is a continuous feedback to the self-evolving algorithm regarding which individuals should have a higher probability of being allowed to multiply and reproduce and subsequently be removed from the population. Generally, probability of

Table 4. Different algorithms/methods of selection employed in various steps of genetic algorithm

Step	Algorithm
Creation	Uniform
Fitness evaluation	Rank-based, proportional, multi-objective, Monte Carlo
Selection	Uniform, Roulette wheel, Stochastic Uniform Selection, tournament selection, rank based, truncation
Crossover	One-point, two-point, Uniform, Arithmetic, Heuristic
Mutation	Uniform, Adaptive feasible, Gaussian
Termination criteria	Generation number, evolution time, fitness threshold, fitness convergence, gene convergence

individuals in a generation G of being chosen for participating in evolution is $p(G)\alpha f(t)$. In this chapter, the authors have chosen a continuous fitness function and the algorithm converges to a stopping criteria.

Selection is without elitism since it has been observed that if the best chromosome is also replaced in a population then there is a chance of another chromosome to become better than the former after crossover else the GEA stagnates with the same chromosome being always chosen as the best for encoding.

3. *Termination criteria*: This criteria has to be closely monitored otherwise there may be an explosive growth which can diminish the possibility of an effective convergence.

To summarize, this is a preliminary attempt to exploit the benefits and features of GA with a hybrid version of it in cryptography with keys generated from the system in consideration. In the Deterministic selection GEA, the "tag" remains with the successive offsprings of the fittest parent initially chosen to start crossover. In Pseudo-Deterministic GEA, the tag is released after the first crossover in both the transmitter and receiver's end and for the rest of the crossovers the process wanders randomly through the discrete search space which contains both good and bad solutions. The end result is a successful termination with the global optimal solution. In a way the tag in this case transforms the Pseudo Deterministic solution into a simulated annealing algorithm after the first crossover in a generation. On closer examination,the proposed PFRWR selection scheme is quite similar to Simulated Annealing but with a major difference. In Simulated Annealing, emphasis is laid on modifying the solution by one move without any population of individuals. This is quite similar to the function of mutation operation in GA. Whereas, in GA, a solution is achieved by taking combinations of existing solutions. In general, the major drawback of premature convergence by settling with a poor offspring in GA is solved by using PFRWR selection scheme as a selection operator for the proposed cryptography. Both the versions of the GEA for cryptography guarantees an initial population of unique candidate solutions, however there is no assurance that a generation will not get regenerated again after some iterations since, the mixing of information is restricted within a fixed problem oriented solution space. The Deterministic option guarantees that every individual will get a fair chance to participate in reproduction but in the case of its counterpart, it may happen that an individual may be passed without participating in crossover or even after repeatative crossovers the original individual appears. In general, a comparison of cryptography with these two varieties of proposed selection mechanism for some values of μ does not lead to a consensus about the decision regarding which is a better option for cryptography since experiments show with several test cases that the behavior varies with different population sizes, crossover point and randomness inherent in the underlying principle of GA. In this present work, we have substituted the random and probabilistic selection operator of GA with problem specific operator to design the cryptosystem to control the inherent random behavior otherwise it would lead to a solution which is uncorrelated with the original message and may also lead to loss of information. The way selection has been modified leads to two versions of the proposed genetic engineering algorithm for cryptography. Simulation results demonstrates that both the flavors of the proposed cryptography successfully recover the message. The example produced is with a text but there is an equal scope to extend this scheme for image cryptography. Further work can be done to test and employ the discussed scheme by using other types of crossover mechanisms. The interdisciplinary work between chaos and evolutionary theory is still emerging and there is ample scope to reap the benefits of both in various applications of science and engineering.

REFERENCES

Abraham, A. (2005). Evolutionary Computation . In Sydenham, P., & Thorn, R. (Eds.), *Handbook for Measurement, Systems Design* (pp. 920–931). London: John Wiley and Sons Ltd. doi:10.1002/0471497398.mm423

Alvarez, E., Fernandez, A., Garcia, P., Jimenez, J., & Marcano, A. (1999b). New approach to chaotic encryption. *Physics Letters. [Part A]*, 373–375. doi:10.1016/S0375-9601(99)00747-1

Alvarez, G., Montoya, G., Pastor, F., & Romera, M. (1999a). Chaotic cryptosystems. In *Proc. IEEE Int. Carnahan Conf. Security Technology* (pp. 332-338).

Aminetzach, Y. T., Macpherson, J. M., & Petrov, D. A. (2005). Pesticide resistance via transposition-mediated adaptive gene truncation in Drosophila. *Science, 309*(5735), 764–767. doi:10.1126/science.1112699

Bäck, T., & Hoffmeister, F. (1991). Extended Selection Mechanisms in Genetic Algorithms. In R. K. Belew & L. B. Booker (Eds.), *Proceedings of the Fourth International Conference on Genetic Algorithms* (pp. 92-99). San Mateo, CA: Morgan Kaufmann Publishers.

Baker, J. E. (1985). Adaptive selection methods for genetic algorithms. In *Proceedings of an International Conference on Genetic Algorithms and Their Applications* (pp. 100-111).

Baker, J. E. (1987). Reducing Bias and Inefficiency in the Selection Algorithm. In J. J. Grefenstette (Ed.), *Proceedings of the Second International Conference on Genetic Algorithms and their Application* (pp. 14-21). Hillsdale, NJ: Lawrence Erlbaum Associates.

Banerjee, S., & Chowdhury, A. R. (2008). *Lyapunov function, parameter estimation, synchronization and chaotic cryptography*. Commun. Nonlinear Sci Numer Simulat.

Banerjee, S., & Chowdhury, A. R. (2009). Functional synchronization and its application to secure communications. *International Journal of Modern Physics B, 23*(9), 2285–2295. doi:10.1142/S0217979209052157

Banerjee, S., Ghosh, D., & Chowdhury, A. R. (2008a). Multiplexing synchronization and its applications in cryptography. *Physica Scripta*, 78.

Banerjee, S., Ghosh, D., Ray, A., & Chowdhury, A. R. (2008b). Synchronization between two different time-delayed systems and image encryption. *EPL, 81*, 20006. doi:10.1209/0295-5075/81/20006

Banzhaf, W., Nordin, P., Keller, E. R., & Francone, D. F. (1998). *Genetic Programming:An Introduction On the Automatic Evolution of Computer Programs and Its Applications*. San Francisco, CA: Morgan Kaufmann Publishers.

Baptista, M. S. (1998). Cryptography with chaos. *Physics Letters. [Part A], 240*, 50–54. doi:10.1016/S0375-9601(98)00086-3

Bertram, J. (2000). The molecular biology of cancer. *Molecular Aspects of Medicine, 21*(6), 167–223. doi:10.1016/S0098-2997(00)00007-8

Beyer, H.-G. (1995). Toward a theory of evolution strategies: On the benefit of sex- the (μ/μ, λ)-strategy. *Evolutionary Computation, 3*(1), 81–111. doi:10.1162/evco.1995.3.1.81

Beyer, H.-G. (2001). *The Theory of Evolution Strategies*. Berlin: Springer.

Beyer, H.-G., & Schwefel, H. P. (2002). Evolution Strategies: A Comprehensive Introduction. *Natural Computing, 1*(1), 3–52. doi:10.1023/A:1015059928466

Blickle, T., & Thiele, L. (1995a). A Mathematical Analysis of tournament selection. In L. Eshelman (Ed.), *Proceedings of the Sixth International Conference on Genetic Algorithms (ICGA95)*. San Fransisco, CA: Morgan Kaufmann Publishers.

Blickle, T., & Thiele, L. (1995b). A Comparison of Selection Schemes used in Genetic Algorithms (2nd Ed.). (TIK Report No. 11). Zürich, Switzerland: Swiss Federal Institute of Technology (ETH), Computer Engineering and Communication Networks Lab (TIK).

Booker, L. B. (1982). Intelligent behavior as an adaptation to the task environment. (Doctoral dissertation). *Dissertation Abstracts International, 43*(2), 469B.

Bowong, S. (2004). Stability analysis for the synchronization of chaotic systems. *Phys. Rev. Lett. A, 326*, 102–113.

Brindle, A. (1981). *Genetic algorithms for function optimization* (Doctoral dissertation and Technical Report TR81-2). Edmonton, Canada: University of Alberta, Department of Computer Science.

Bulmer, M. G. (1980). *The Mathematical Theory of Quantitative Genetics*. Oxford, UK: Clarendon Press.

Burrus, V., & Waldor, M. (2004). Shaping bacterial genomes with integrative and conjugative elements. *Research in Microbiology, 155*(5), 376–378. doi:10.1016/j.resmic.2004.01.012

Caponetto, R., Fortuna, L., Fazzino, S., & Xibilia, M. G. (2003). Chaotic sequences to improve the performance of evolutionary algorithms. *IEEE Transactions on Evolutionary Computation, 7*(3), 289–304. doi:10.1109/TEVC.2003.810069

Chu, Y. H., & Chang, S. (1999). Dynamical cryptography based on synchronized chaotic systems. *Electronics Letters, 35*, 97–975.

Dachselt, F., & Schwarz, W. (2001). Chaos and Cryptography. *IEEE Transactions on Circuits and Systems, 48*(12), 1498–1509. doi:10.1109/TCSI.2001.972857

De Jong, K. A. (1975). An analysis of the behavior of a class of genetic adaptive systems. (Doctoral dissertation, University of Michigan). *Dissertation Abstracts International, 36*(10), 5140B.

Farmer, J. D. (1982). Chaotic Attractors of an Infinite-Dimensional Dynamical System. *Physica D. Nonlinear Phenomena, 4*, 366. doi:10.1016/0167-2789(82)90042-2

Fogel, L. J., Owens, A. J., & Walsh, M. J. (1966). *Artificial Intelligence Through Simulated Evolution*. New York: John Wiley and Sons.

Fridrich, J. (1998). Symmetric ciphers based on two-dimensional chaotic maps. *International Journal of Bifurcation and Chaos in Applied Sciences and Engineering, 8*, 1259–1284. doi:10.1142/S021812749800098X

Fujisaka, H., & Yamada, T. (1983). Stability theory of synchronized motion in coupled-oscillator systems. *Progress of Theoretical Physics*, *69*, 32–47. doi:10.1143/PTP.69.32

Garcia-Ojalio, J., & Roy, R. (2001). Spatiotemporal Communication with Synchronized Optical Chaos. *Physical Review Letters*, *86*, 5204. doi:10.1103/PhysRevLett.86.5204

Gero, J. S., & Kazakov, V. (2001). A Genetic Engineering Approach to Genetic Algorithms Evolutionary Computation. *Evolutionary Computation*, *9*(1), 71–92. doi:10.1162/10636560151075121

Ghosh, D., Banerjee, S., & Chowdhury, A. R. (2007). Synchronization between variable time-delayed systems and cryptography. *Europhysics Letters*, *80*, 30006. doi:10.1209/0295-5075/80/30006

Gickenheimer, J., & Holmes, P. (1983). *Nonlinear Oscillations, Dynamical Systems and Bifurcations of Vector Fields*. Berlin, Germany: Springer.

Goldberg, D. E. (1989). *Genetic Algorithms in Search, optimization and machine learning*. Reading, MA: Addison-Wesley.

Goldberg, D. E. (1994). Genetic and Evolutionary Algorithms Come of Age. *Communications of the ACM*, *37*(3), 13–119. doi:10.1145/175247.175259

Goldberg, D. E., & Kalyanmoy, D. (1991). A Comparative Analysis of Selection Schemes Used in Genetic Algorithms . In Rawlins, G. J. E. (Ed.), *Foundations of Genetic Algorithms*. San Mateo, CA: Morgan Kaufmann Publishers.

Gorodilon, A., & Morozenko, V. (2008). Genetic Algorithm for finding the Key's length and Crypt-analysis of the Permutation Cipher. *International Journal Information Theories and Applications, 15*.

Grefenstette, J. J., & Baker, J. E. (1989). How genetic algorithms work: A critical look at implicit parallelism. In *Proceedings of the Third International Conference on Genetic Algorithms* (pp. 20-27). San Mateo, CA: Morgan Kaufmann Publishers.

Guo, S. M., Liu, K. T., Tsai, J. S. H., & Shieh, L. S. (2008). An observer-based tracker for hybrid interval chaotic systems with saturating inputs: The chaos-evolutionary-programming approach. *Computers & Mathematics with Applications (Oxford, England)*, *55*, 1225–1249. doi:10.1016/j.camwa.2007.06.024

Habutsu, T., Nishio, Y., Sasane, I., & Mori, S. (1991). A secret key cryptosystem by iterating a chaotic map . In *Proc. Advances in Cryptology – EUROCRYPT '91* (pp. 127–140). Berlin, Germany: Springer-Verlag.

Hale, J. K. (1977). *Theory of Functional Differential Equations*. Berlin: Springer-Verlag.

Hansen, N., Müller, S. D., & Koumoutsakos, P. (2003). Reducing the Time Complexity of the Derandomized Evolution Strategy with Covariance Matrix Adaptation (CMA-ES). *Evolutionary Computation*, *11*(1). doi:10.1162/106365603321828970

Hansen, N., & Ostermeier, A. (2001). Completely Derandomized Self-Adaptation in Evolution Strategies. *Evolutionary Computation*, *9*(1), 159–195. doi:10.1162/106365601750190398

Holland, J. H. (1975). *Adaptation in natural and artificial systems*. Ann Arbor, MI: The University of Michigan Press.

Hramov, A. E., & Koronovskii, A. A. (2005a). Generalized synchronization: A modified system approach. *Physical Review E: Statistical, Nonlinear, and Soft Matter Physics*, *71*, 067201. doi:10.1103/PhysRevE.71.067201

Hramov, A. E., & Koronovskii, A. A. (2005c). Intermittent generalized synchronization in unidirectionally coupled chaotic oscillators. *Europhysics Letters*, *70*, 169–175. doi:10.1209/epl/i2004-10488-6

Hramov, A. E., Koronovskii, A. A., & Levin, Y. I. (2005b). Synchronization of Chaotic Oscillator Time Scales. *Journal of Experimental and Theoretical Physics*, *100*, 784. doi:10.1134/1.1926439

Husainy, A., & Muhammed, A. F. (2006). Image Encryption Using Genetic Algorithm. *Information Technology Journal*, *5*(3), 516–519. doi:10.3923/itj.2006.516.519

Jakimoski, G., & Kocarev, L. (2001). Chaos and Cryptography: Block Encryption Cipher Based on Chaotic Maps. *IEEE Transactions on Circuits and Systems. I, Fundamental Theory and Applications*, *48*(2). doi:10.1109/81.904880

Jozwiak, L., & Postula, A. (2002). Genetic engineering versus natural evolution Genetic algorithms with deterministic operators. *Journal of Systems Architecture*, *48*, 99–112. doi:10.1016/S1383-7621(02)00094-2

Kittel, A., Pyragas, K., & Richter, R. (1994). Pre-recorded history of a system as an experiment tool to control chaos. *Physical Review E: Statistical Physics, Plasmas, Fluids, and Related Interdisciplinary Topics*, *50*, 262–268. doi:10.1103/PhysRevE.50.262

Klein, E., Mislovaty, R., Kanter, I., & Kinzel, W. (2005). Public-channel cryptography using chaos synchronization. *Physical Review E: Statistical, Nonlinear, and Soft Matter Physics*, *72*, 016214. doi:10.1103/PhysRevE.72.016214

Kocarev, L. (2001). Chaos-based cryptography: A brief overview. *IEEE Circuits and Systems Magazine*, *1*(3), 6–21. doi:10.1109/7384.963463

Kocarev, L., Jakimoski, G., Stojanovski, T., & Parlitz, U. (1998). From chaotic maps to encryption schemes. In *Proc. IEEE Int. Symposium Circuits and Systems, 4*, 514-517.

Kocarev, L., & Parlitz, U. (1996). Generalized synchronization, predictability, and equivalence of unidirectionally coupled dynamical systems. *Physical Review Letters*, *76*, 1816–1819. doi:10.1103/PhysRevLett.76.1816

Koronovskii, A. A., & Hramov, A. E. (2004). Chaotic Phase Synchronization Studied by Means of Continuous Wavelet Transform. *Technical Physics Letters*, *30*(7), 587–590. doi:10.1134/1.1783411

Kotulski, Z., & Szczepanski, J. (1997). Discrete chaotic cryptography. *Annals of Physics*, *6*, 381–394.

Kotulski, Z., Szczepanski, J., Grski, K., Paszkiewicz, A., & Zugaj, A. (1999). Application of discrete chaotic dynamical systems cryptography – DCC method. *International Journal of Bifurcation and Chaos in Applied Sciences and Engineering*, *9*, 1121–1135. doi:10.1142/S0218127499000778

Koza, J. R. (1992). *Gentic Programming: On the Programming of computers by Means of natural selection*. Cambridge, MA: The MIT Press.

Koza, J. R. (1994). *Genetic Programming II: Automatic Discovery of Reusable Programs*. Cambridge, MA: The MIT Press.

Koza, J. R., Andre, D., Bennett, F. H., & Keane, M. A. (1999). *Genetic Programming III: Darwinian Invention and Problem Solving*. San Francisco: Morgan Kaufmann.

Koza, J. R., Keane, M. A., & Streeter, M. J. (2003). Evolving inventions. *Scientific American, 288*, 52–59. doi:10.1038/scientificamerican0203-52

Koza, J. R., Keane, M. A., Streeter, M. J., Mydlowec, W., Yu, J., & Lanza, G. (2003). *Genetic Programming IV: Routine Human-Competitive Machine Intelligence*. Hingham, MA: Kluwer Academic Press.

Kushchu, I. (2002). Genetic programming and evolutionary generalization. *IEEE Transactions on Evolutionary Computation, 6*, 431–442. doi:10.1109/TEVC.2002.805038

Li, P., Li, Z., Fetinger, S., Mao, Y., & Halang, W. A. (2007). Application of Chaos – based Pseudo – Random – Bit Generators in Internet-based Online Payments. [SCI]. *Studies in Computational Intelligence, 37*, 667–685. doi:10.1007/978-3-540-37017-8_31

Liang, X., Zhang, J., & Xia, X. (2008, July). Improving the Security of Chaotic Synchronization With a Δ-Modulated Cryptographic Technique. *IEEE Transactions on Circuits and Wystems. II, Express Briefs, 55*(7), 680–684. doi:10.1109/TCSII.2008.921585

Masoller, C. (2001). Anticipation in the Synchronization of Chaotic Semiconductor Lasers with Optical Feedback. *Physical Review Letters, 86*(13), 2782. doi:10.1103/PhysRevLett.86.2782

Menezes, A. J., Oorschot, P. V., & Vanstone, S. (1997). *Handbook of Applied Cryptography*. Boca Raton, FL: CRC Press.

Mitchell, T. (1996). *Machine Learning*. New York: McGraw-Hill.

Parker, A. T., & Short, K. M. (2001). Reconstructing the keystream from a chaotic encryption Scheme. *IEEE Trans on Circuits and Syst- I, 48*(5), 624-630.

Pecora, L. (1996). Hyperchaos harnessed. *Physics World, 9*(51), 17.

Pecora, L. M., & Carroll, T. L. (1990). Synchronization in Chaotic systems. *Physical Review Letters, 64*, 821–824. doi:10.1103/PhysRevLett.64.821

Perez, G., & Cerdeira, H. A. (1995). Extracting Messages Masked by Chaos. *Physical Review Letters, 74*, 1970. doi:10.1103/PhysRevLett.74.1970

Pyragas, K. (1998). Synchronization of coupled time-delay systems: Analytical estimations. *Physical Review E: Statistical Physics, Plasmas, Fluids, and Related Interdisciplinary Topics, 58*, 3067. doi:10.1103/PhysRevE.58.3067

Rechenberg, I. (1973). *Evolutionsstrategie: Optimierung technischer Systeme nach Prinzipien der biologischen Evolution*. Stuttgart, Deutschland: Fromman-Holzboog.

Rechenberg, I. (1994). *Evolutionsstrategie '94*. Stuttgart, Deutschland: Frommann-Holzboog.

Rosenblum, M. G., Pikovsky, A. S., & Kurths, J. (1996). Phase synchronization of chaotic oscillators. *Physical Review Letters, 76*, 1804–1807. doi:10.1103/PhysRevLett.76.1804

Rosenblum, M. G., Pikovsky, A. S., & Kurths, J. (1997). From Phase to Lag Sychronization in Coupled Chaotic Oscillators. *Physical Review Letters, 78*, 4193. doi:10.1103/PhysRevLett.78.4193

Rossler, O. E. (1979). An equation for hyperchaos. *Physics Letters. [Part A], 71*, 155–157. doi:10.1016/0375-9601(79)90150-6

Rulkov, N. F., Sushchik, M. M., Tsimring, L. S., & Abarbanel, H. D. I. (1995). Generalized synchronization of chaos in directionally coupled chaotic systems. *Physical Review E: Statistical Physics, Plasmas, Fluids, and Related Interdisciplinary Topics, 51*, 980. doi:10.1103/PhysRevE.51.980

Schneier, B. (1996). *Applied Cryptography: Protocols, Algorithm, and Source Code in C*. New York: Wiley.

Schwefel, H. P. (1977). *Numerische Optimierung von Computermodellen mittels der Evolutionsstrategie*. Basel: Birkhaeuser.

Schwefel, H. P. (1995). *Evolution and Optimum Seeking*. New York: Wiley.

Shannon, C. E. (1949). Communication theory of secrecy systems. *The Bell System Technical Journal, 28*, 656–715.

Spillman, R. (1993, April). Use of a genetic algorithm in the cryptanalysis of simple substitution ciphers. *Cryptologia, 17*(1), 187–201. doi:10.1080/0161-119391867746

Srinivas, M., & Patnaik, L. M. (1994). Genetic Algorithms: A Survey. *IEEE Computer,* 17-26.

Tigan, G. (2004). Analysis of a dynamical system derived from the Lorenz system. In *Mathematics in engineering and numerical physics* (pp. 265-272).

Toemeh, R., & Arumugam, S. (2007). Breaking Transposition Cipher with Genetic Algorithm. *Electronics and Electrica Engineering, 7*(79).

Tragha, A., Omary, F., & Kriouile, A. (2005). Genetic Algorithms Inspired Cryptography. *A.M.S.E Association for the Advancement of Modeling & Simulation Techniques in Enterprises, Series D:Computer Science and Satistics*.

Tragha, A., Omary, F., & Moloudi, A. (2006). ICIGA: Improved Cryptography Inspired by Genetic Algorithms. In *Proceedings of the International Conference on Hybrid Information Technology (ICHIT'06)* (pp. 335-341).

Ucar, A. (2003). On the chaotic behaviour of a prototype delayed dynamical system. *Chaos, Solitons, and Fractals, 16*(2), 187. doi:10.1016/S0960-0779(02)00160-1

Voss, H. U. (2000). Anticipating chaotic synchronization. *Physical Review E: Statistical Physics, Plasmas, Fluids, and Related Interdisciplinary Topics, 61*, 5115. doi:10.1103/PhysRevE.61.5115

Watson, J. D., Hopkins, N. H., Roberts, J. W., Steitz, J. A., & Weiner, A. M. (1987). *Molecular Biology of the Gene*. Menlo Park, CA: Benjamin Cummings.

Wenbo, M. (2004). *Modern Cryptography: Theory and Practice*. Upper Saddle River, NJ: Prentice Hall.

Wu, T., Cheng, Y., Tan, J., & Zhou, T. (2008). The Application of Chaos Genetic Algorithm in the PID Parameter Optimization. In *Proceedings of 3rd International Conference on Intelligent System and Knowledge Engineering.*

Yan, J., & Li, C. (2005, November). Generalized projective synchronization of a unified chaotic system. *Chaos, Solitons, and Fractals*, *26*(4), 1119–1124. doi:10.1016/j.chaos.2005.02.034

Yan, X. F., Chen, D. Z., & Hu, S. X. (2003). Chaos-genetic algorithms for optimizing the operating conditions based on RBF-PLS model. *Computers & Chemical Engineering*, *27*, 1393–1404. doi:10.1016/S0098-1354(03)00074-7

Yang, T., Wu, C. W., & Chua, L. O. (1997, May). Cryptography based on chaotic systems. *IEEE Trans. Circuits Syst. I.*, *44*, 69.

Yiqiang, G., Yanbin, W., Zhengshan, J., Jun, W., & Luyan, Z. (2009). *Remote sensing image classification by the Chaos Genetic Algorithm in monitoring land use changes*. Mathematical and Computer Modelling.

Zhan, M., Wei, G. W., & Lai, C. H. (2002). Transition from intermittency to periodicity in lag synchronization in coupled Rössler oscillators. *Physical Review E: Statistical, Nonlinear, and Soft Matter Physics*, *65*, 036202–036205. doi:10.1103/PhysRevE.65.036202

Compilation of References

Aasen, T., Kugiumtzis, D., & Nordahl, S. H. G. (1997). Procedure for estimating the correlation dimension of optokinetic nystagmus signals. *Computers and Biomedical Research, an International Journal, 30*, 95–116. doi:10.1006/cbmr.1997.1441

Abarbanel, H. D. I., Rulkov, N. F., & Sushchik, M. M. (1996). Generalized synchronization of chaos: The auxiliary system approach. *Physical Review E: Statistical Physics, Plasmas, Fluids, and Related Interdisciplinary Topics, 53*, 4528–4535. doi:10.1103/PhysRevE.53.4528

Abarbanel, H. D. I. (1996). *Analysis of Observed Chaotic Data*. New York: Springer-Verlag.

Abraham, A. (2005). Evolutionary Computation . In Sydenham, P., & Thorn, R. (Eds.), *Handbook for Measurement, Systems Design* (pp. 920–931). London: John Wiley and Sons Ltd.doi:10.1002/0471497398.mm423

Agiza, H. N., & Matouk, A. E. (2006). Adaptive synchronization of Chua's circuits with fully unknown parameters. *Chaos, Solitons, and Fractals, 28*, 219–227. doi:10.1016/j.chaos.2005.05.055

Agiza, H. N., & Yassen, M. T. (2001). Synchronization of Rossler and Chen chaotic dynamical systems using active control. *Physics Letters. [Part A], 278*(4), 191–197. doi:10.1016/S0375-9601(00)00777-5

Aguilar, R., Martínez-Guerra, R., & Maya-Yescas, R. (2003). State Estimation for Partially Unknown Nonlinear Systems: A Class of Integral High Gain Observers. *IEE Proceedings. Control Theory and Applications, 150*(3), 240–244. doi:10.1049/ip-cta:20030400

Aguilar, R., Poznyak, A., Martínez-Guerra, R., & Maya-Yescas, R. (2002). Temperature control in catalytic cracking reactors via robust PID controller. *Journal of Process Control, 12*(6), 695–705. doi:10.1016/S0959-1524(01)00034-8

Aguilar, R., Martínez, R., & Poznyak, A. (2001). PI observers for uncertainty estimation in continuous chemical reactors. In *IEEE Conference on Control Applications/ISIC* (pp. 1037-1041).

Aguilar-Ibañez, C., Suarez, C. M. S., Fortunato Flores, A., Martínez-Guerra, R., & Garrido, M. R. (2006). Reconstructing and Identifying the Rossler system by using a high gain observer. *Asian Journal of Control, 8*(4), 401–407.

Aguilar-López, R., & Alvarez-Ramírez, J. (2002). Sliding-mode control scheme for a class of continuous chemical reactors. *IEE Proceedings. Control Theory and Applications, 149*(4), 263–268. doi:10.1049/ip-cta:20020558

Aguilar-López, R., & Martínez-Guerra, R. (2008). Synchronization of a coupled Hodgkin-Huxley neurons via high order sliding-mode feedback. *Chaos, Solitons, and Fractals, 37*, 539–546. doi:10.1016/j.chaos.2006.09.029

DOI: 10.4018/978-1-61520-737-4.chcrf

Aguilar-López, R., Martínez-Guerra, R., Puebla H., & Hernandez Suarez, R. (2008). High order sliding-mode dynamic control for chaotic intracellular calcium oscillations. *Journal of Nonlinear Analysis-B: Real World Applications*.

Aguirre, L. A., Furtado, E. C., & Tôrres, L. A. B. (2006). Evaluation of dynamical models: dissipative synchronization and other techniques. *Physical Review E: Statistical, Nonlinear, and Soft Matter Physics*, *74*, 019612-1–16. doi:10.1103/PhysRevE.74.066203

Ahmad, W. M., & El-Khazali, R. (2007). Fractional-order dynamical models of love. *Chaos, Solitons, and Fractals*, *33*, 1367–1375. doi:10.1016/j.chaos.2006.01.098

Ahmed, E., & Elgazzar, A. S. (2007). On fractional order differential equations model for nonlocal epidemics. *Physica A*, *379*, 607–614. doi:10.1016/j.physa.2007.01.010

Aislam, T., & Edwards, J. A. (1996). Secure communications using chaotic digital encoding. *Electronics Letters*, *32*, 190–191. doi:10.1049/el:19960107

Allaria, E., Arecchi, F. T., Garbo, A., & Di, ., & Meucci, M. (2001). Synchronization of Homoclinic Chaos. *Physical Review Letters*, *86*(5), 791–794. doi:10.1103/PhysRevLett.86.791

Alvarez, G., Montoya, F., Romera, M., & Pastor, G. (2004). Breaking parameter modulated chaotic secure communication system. *Chaos, Solitons, and Fractals*, *21*, 783–787. doi:10.1016/j.chaos.2003.12.041

Alvarez, J., & Hernandez, H. (2003). Robust estimation of contiunuous nonlinear plants with discrete measurements. *Journal of Process Control*, *13*(1), 69–89. doi:10.1016/S0959-1524(02)00010-0

Alvarez, G., & Li, S. J. (2006). Some basic cryptographic requirements for chaos-based cryptosystems. *International Journal of Bifurcation and Chaos in Applied Sciences and Engineering*, *16*(8), 2129–2151. doi:10.1142/S0218127406015970

Alvarez, E., Fernandez, A., Garcia, P., Jimenez, J., & Marcano, A. (1999). New approach to chaotic encryption. *Physics Letters. [Part A]*, *263*, 373–375. doi:10.1016/S0375-9601(99)00747-1

Alvarez, G., Montoya, F., Romera, M., & Pastor, G. (2000). Cryptanalysis of a chaotic encryption system. *Physics Letters. [Part A]*, *276*, 191–196. doi:10.1016/S0375-9601(00)00642-3

Alvarez, G., Montoya, F., Romera, M., & Pastor, G. (2003). Cryptanalysis of a chaotic secure communication system. *Physics Letters. [Part A]*, *306*, 200–205. doi:10.1016/S0375-9601(02)01502-5

Alvarez, G., Montoya, F., Romera, M., & Pastor, G. (2003). Cryptanalysis of an ergodic chaotic cipher. *Physics Letters. [Part A]*, *311*, 172–179. doi:10.1016/S0375-9601(03)00469-9

Alvarez, G., Montoya, F., Romera, M., & Pastor, G. (2003). Cryptanalysis of a discrete chaotic cryptosystem using external key. *Physics Letters. [Part A]*, *319*, 334–339. doi:10.1016/j.physleta.2003.10.044

Alvarez, G., Montoya, F., Romera, M., & Pastor, G. (2004). Cryptanalysis of dynamic look-up table based chaotic cryptosystems. *Physics Letters. [Part A]*, *326*, 211–218. doi:10.1016/j.physleta.2004.04.018

Alvarez, E., Fernandez, A., Garcia, P., Jimenez, J., & Marcano, A. (1999b). New approach to chaotic encryption. *Physics Letters. [Part A]*, 373–375. doi:10.1016/S0375-9601(99)00747-1

Alvarez, G., Montoya, G., Pastor, F., & Romera, M. (1999a). Chaotic cryptosystems. In *Proc. IEEE Int. Carnahan Conf. Security Technology* (pp. 332-338).

Amigó, J. M., Kocarev, L., & Szczepanski, J. (2007). Theory and practice of chaotic cryptography. *Physics Letters. [Part A]*, *366*, 211–216. doi:10.1016/j.physleta.2007.02.021

Amigó, J. M., Kocarev, L., & Szczepanski, J. (2007). Discrete lyapunov exponents and resistance to differential cryptanalysis. *IEEE Transactions on Circuits and Systems, II*, 54, 882–886.

Amigó, J. M., Szczepanski, J., & Kocarev, L. (2005). A chaos-based approach to the design of crytographically secure substitutions. *Physics Letters. [Part A], 343*, 55–60. doi:10.1016/j.physleta.2005.05.057

Amigó, J. M., Szczepanski, J., & Kocarev, L. (2008). On some properties of the discrete Lyapunov exponent. *Physics Letters. [Part A], 372*, 6265–6268. doi:10.1016/j.physleta.2008.07.076

Aminetzach, Y. T., Macpherson, J. M., & Petrov, D. A. (2005). Pesticide resistance via transposition-mediated adaptive gene truncation in Drosophila. *Science, 309*(5735), 764–767. doi:10.1126/science.1112699

Andronova, I. A., & Malykin, G. B. (2002). Physical problems of fiber gyroscopy based on the Sagnac effect. *Phys. Usp., 45*, 793. Retrieved from doi: 10.1070/PU2002v045n08ABEH001073

Angeli, A. D., Genesio, R., & Tesi, A. (1995). Dead-Beat chaos synchronization in discrete-time system. *IEEE Transactions on Circuits and Systems I, 2*(1), 54–56. doi:10.1109/81.350802

Anstett, F., Millérioux, G., & Bloch, G. (2004 May). Global adaptive synchronization based upon polytopic observers. In *Proc. of IEEE International symposium on circuit and systems, ISCAS'04*, Vancouver, Canada (pp. 728 - 731).

Anstett, F., Millérioux, G., & Bloch, G. (2006 December). Chaotic cryptosystems: Cryptanalysis and identifiability. *IEEE Trans. on Circuits and Systems: Regular papers, 53*(12), 2673-2680.

Argyris, A., Hamacher, M., Chlouverakis, K. E., Bogris, A., & Syvridis, D. (2008). Photonic integrated device for chaos applications in communications. *Physical Review Letters, 100*, 194101. doi:10.1103/PhysRevLett.100.194101

Argyris, A., Syvridis, D., Larger, L., Annovazzi-Lodi, V., Colet, P., & Fischer, I. (2005). Chaos-based communications at high bit rates using commercial fibre-optic links. *Nature, 437*, 343–346. doi:10.1038/nature04275

Ashwin, P. (2003). Synchronization from chaos. *Nature, 422*, 384–385. doi:10.1038/422384a

Bäck, T., & Hoffmeister, F. (1991). Extended Selection Mechanisms in Genetic Algorithms. In R. K. Belew & L. B. Booker (Eds.), *Proceedings of the Fourth International Conference on Genetic Algorithms* (pp. 92-99). San Mateo, CA: Morgan Kaufmann Publishers.

Bagley, R. L., & Calico, R. A. (1991). Fractional order state equations for the control of viscoelastically damped structures. *Journal of Guidance, Control, and Dynamics, 14*, 304–311. doi:10.2514/3.20641

Bai, E. W., & Lonngren, K. E. (1997). Synchronization of two Lorenz systems using active control. *Chaos, Solitons, and Fractals, 8*, 51–58. doi:10.1016/S0960-0779(96)00060-4

Bai, E.-W., & Lonngren, K. E. (2000). Sequential synchronization of two Lorenz systems using active control. *Chaos, Solitons, and Fractals, 11*(7), 1041–1044. doi:10.1016/S0960-0779(98)00328-2

Bai, E.-W., Lonngren, K. E., & Sprott, J. C. (2002). On the synchronization of a class of electronic circuits that exhibit chaos. *Chaos, Solitons, and Fractals, 13*(7), 1515–1521. doi:10.1016/S0960-0779(01)00160-6

Baker, J. E. (1985). Adaptive selection methods for genetic algorithms. In *Proceedings of an International Conference on Genetic Algorithms and Their Applications* (pp. 100-111).

Baker, J. E. (1987). Reducing Bias and Inefficiency in the Selection Algorithm. In J. J. Grefenstette (Ed.), *Proceedings of the Second International Conference on Genetic Algorithms and their Application* (pp. 14-21). Hillsdale, NJ: Lawrence Erlbaum Associates.

Banerjee, S., & Chowdhury, A. R. (2009). Lyapunov function, parameter estimation, synchronization and chaotic cryptography. *Communications in Nonlinear Science and Numerical Simulation, 14*, 2248–2254. doi:10.1016/j.cnsns.2008.06.006

Banerjee, S., Saha, P., & Chowdhury, A. R. (2004). On the application of adaptive control and phase synchronization in non-linear fluid dynamics. *International Journal of Non-linear Mechanics*, *39*, 25–31. doi:10.1016/S0020-7462(02)00125-7

Banerjee, S., & Chowdhury, A. R. (2009). Functional synchronization and its application to secure communications. *International Journal of Modern Physics B*, *23*(9), 2285–2295. doi:10.1142/S0217979209052157

Banerjee, S., Ghosh, D., & Chowdhury, A. R. (2008a). Multiplexing synchronization and its applications in cryptography. *Physica Scripta*, 78.

Banerjee, S., Ghosh, D., Ray, A., & Chowdhury, A. R. (2008b). Synchronization between two different time-delayed systems and image encryption. *EPL*, *81*, 20006. doi:10.1209/0295-5075/81/20006

Banks, J., Brooks, J., Cairns, G., Davis, G., & Stacey, P. (1992). On Devaney's definition of chaos. *The American Mathematical Monthly*, *99*(4), 332–334. doi:10.2307/2324899

Banzhaf, W., Nordin, P., Keller, E. R., & Francone, D. F. (1998). *Genetic Programming:An Introduction On the Automatic Evolution of Computer Programs and Its Applications*. San Francisco, CA: Morgan Kaufmann Publishers.

Baptista, M. S. (1998). Cryptography with chaos. *Physics Letters. [Part A]*, *240*, 50–54. doi:10.1016/S0375-9601(98)00086-3

Bavard, X., Locquet, A., Larger, L., & Goedgebuer, J. P. (2007). Influence of digitisation on master–slave synchronisation in chaos-encrypted data transmission. *IET Optoelectron.*, *1*, 3–8. doi:10.1049/iet-opt:20060022

Behzad, M., Salarieh, H., & Alasty, A. (2008). Chaos synchronization in noisy environment using nonlinear filtering and sliding mode control. *Chaos, Solitons, and Fractals*, *36*(5), 1295–1304. doi:10.1016/j.chaos.2006.07.058

Belsky, Y. L., & Dmitriev, A. S. (1993). Communication by means of dynamic chaos. *Radiotekhnika I Elektronika*, *38*(7), 1310–1315.

Belsky, Y. L., & Dmitriev, A. S. (1995). The effect of Perturbing Factors on the operation of an Information Transmission System with a Chaotic Carrier. *Radiotekhnika I Elektronika*, *40*(2), 265–281.

Belykh, V. N., Belykh, I. V., Colding-Jorgensen, M., & Mosekilde, E. (2000). Homoclinic bifurcations leading to bursting oscillations in cell models. *The European Physical Journal E*, *3*(3), 205–219. doi:10.1007/s101890070012

Berger, M. (2009). *Geometry I*. Berlin: Springer-Verlag.

Bertram, J. (2000). The molecular biology of cancer. *Molecular Aspects of Medicine*, *21*(6), 167–223. doi:10.1016/S0098-2997(00)00007-8

Besançon, G. (1999). A viewpoint on observability and observer design for nonlinear systems. *Lecture Notes in Control and Information Sciences*, *244*, 3–22. doi:10.1007/BFb0109918

Beyer, H.-G. (1995). Toward a theory of evolution strategies: On the benefit of sex- the (μ/μ, λ)-strategy. *Evolutionary Computation*, *3*(1), 81–111. doi:10.1162/evco.1995.3.1.81

Beyer, H.-G. (2001). *The Theory of Evolution Strategies*. Berlin: Springer.

Beyer, H.-G., & Schwefel, H. P. (2002). Evolution Strategies: A Comprehensive Introduction. *Natural Computing*, *1*(1), 3–52. doi:10.1023/A:1015059928466

Bilotta, E., Di Blasi, G., Stranges, F., & Pantano, P. (2007). A Gallery of Chua Attractors, Part VI. *International Journal of Bifurcation and Chaos in Applied Sciences and Engineering*, *17*, 1801–1910. doi:10.1142/S0218127407018105

Blasius, B., Huppert, A., & Stone, L. (1999). Complex dynamics and phase synchronization in spatially extended ecological systems. *Nature*, *399*, 354–359. doi:10.1038/20676

Blekhman, I. I., Mosekilde, E., & Fradkov, A. L. (Eds.). (2002). *Special Issue on Chaos Synchronization and Control* (*Vol. 58*). Amsterdam: Elsevier.

Blickle, T., & Thiele, L. (1995a). A Mathematical Analysis of tournament selection. In L. Eshelman (Ed.), *Proceedings of the Sixth International Conference on Genetic Algorithms (ICGA95)*. San Fransisco, CA: Morgan Kaufmann Publishers.

Blickle, T., & Thiele, L. (1995b). A Comparison of Selection Schemes used in Genetic Algorithms (2nd Ed.). (TIK Report No. 11). Zürich, Switzerland: Swiss Federal Institute of Technology (ETH), Computer Engineering and Communication Networks Lab (TIK).

Blondel, V. D., Sontag, E. D., Vidyasagar, M., & Willems, J. C. (1999). *Open Problems in Mathematical Systems and Control Theory*. Communication and Control Engineering.

Boccaletti, S., Kurths, J., Osipov, G., Valladares, D. L., & Zhou, C. S. (2002). The synchronization of chaotic systems. *Physics Reports*, *366*, 1–101. doi:10.1016/S0370-1573(02)00137-0

Boccaletti, S., Kurths, J., Osipov, G., Valladares, D. L., & Zhou, C. S. (2002). The synchronization of chaotic systems. *Physics Reports*, *366*, 1–101. doi:10.1016/S0370-1573(02)00137-0

Boccaletti, S., Pecora, L. M., & Pelaez, A. (2001). Unifying framework for synchronization of coupled dynamical systems. *Physical Review E*, *63*(6), 066219/1-066219/4.

Bogris, A., Argyris, A., & Syvridis, D. (2007). Analysis of the optical amplifier noise effect on electro-optically generated hyperchaos. *IEEE Journal of Quantum Electronics*, *47*(7), 552–559. doi:10.1109/JQE.2007.898843

Bohme, F., Feldman, U., Schwartz, W., & Bauer, A. (1994 July). Information transmission by chaotizing. In *Proc. of Workshop NDES'94*, Krakov, Poland, (pp. 163-168).

Booker, L. B. (1982). Intelligent behavior as an adaptation to the task environment. (Doctoral dissertation). *Dissertation Abstracts International*, *43*(2), 469B.

Bottou, L., & Vapnik, V. (1992). Local learning algorithms . *Neural Computation*, *4*, 888–900. doi:10.1162/neco.1992.4.6.888

Boutat-Baddas, L., Barbot, J. P., Boutat, D., & Tauleigne, R. (2004, May). Sliding mode observers and observability singularity in chaotic synchronization. *Mathematical Problems in Engineering*, *1*, 11–31. doi:10.1155/S1024123X04309038

Boutayeb, M., Darouach, M., & Rafaralahy, H. (2002, March). Generalized state-space observers for chaotic synchronization and secure communications. *IEEE Transactions on Circuits and Systems. I, Fundamental Theory and Applications*, *49*(3), 345–349. doi:10.1109/81.989169

Bowong, S., & Kakmeni, F. M. M. (2004). Synchronization of uncertain chaotic systems via backstepping approach. *Chaos, Solitons, and Fractals*, *21*(4), 999–1011. doi:10.1016/j.chaos.2003.12.084

Bowong, S. (2004). Stability analysis for the synchronization of chaotic systems with different order: application to secure communications. *Physics Letters. [Part A]*, *326*, 102–113. doi:10.1016/j.physleta.2004.04.004

Bracikowski, C., & Roy, R. (1991). Chaos in a multimode solid-state laser system. *Chaos (Woodbury, N.Y.)*, *1*, 49–64. doi:10.1063/1.165817

Breakspear, M., & Terry, J. R. (2002). Detection and description of non-linear interdependence in normal multichannel human EEG data. *Clinical Neurophysiology*, *113*, 735–753. doi:10.1016/S1388-2457(02)00051-2

Breakspear, M., Terry, J. R. & Friston, K. J. (2003). Modulation of excitatory synaptic coupling facilitates synchronization and complex dynamics in a biophysical model of neuronal dynamics. *Network: Computation in Neural Computing, 14*, 703.

Brindle, A. (1981). *Genetic algorithms for function optimization* (Doctoral dissertation and Technical Report TR81-2). Edmonton, Canada: University of Alberta, Department of Computer Science.

Brown, R., Rulkov, N. F., & Tracy, E. R. (1994). Modeling and synchronizing chaotic systems from time-series data. *Physical Review E: Statistical Physics, Plasmas, Fluids, and Related Interdisciplinary Topics*, *49*, 3784–3800. doi:10.1103/PhysRevE.49.3784

Buchman, S. (2000). Cryogenic gyroscopes for the relativity mission. *Physica B, Condensed Matter*, *280*, 497. doi:10.1016/S0921-4526(99)01846-3

Bulmer, M. G. (1980). *The Mathematical Theory of Quantitative Genetics*. Oxford, UK: Clarendon Press.

Bünner, M. J., Ciofini, M., Giaquinta, A., Hegger, R., Kantz, H., Meucci, R., & Politi, A. (2000a). Reconstruction of systems with delayed feedback: Theory. *The European Physical Journal D*, *10*, 165–176. doi:10.1007/s100530050538

Bünner, M. J., Ciofini, M., Giaquinta, A., Hegger, R., Kantz, H., Meucci, R., & Politi, A. (2000b). Reconstruction of systems with delayed feedback: Application. *The European Physical Journal D*, *10*, 177–187. doi:10.1007/s100530050539

Bünner, M. J., Meyer, T., Kittel, A., & Parisi, J. (1997). Recovery of the time-evolution of time-delay systems from time series. *Physical Review E: Statistical Physics, Plasmas, Fluids, and Related Interdisciplinary Topics*, *56*, 5083–5089. doi:10.1103/PhysRevE.56.5083

Burckhardt, J. J. (1966). *Die Bewegungsgruppen der Kristallographie* (2nd ed.). Basel, Germany: Birkhauser Verlag.

Burrus, V., & Waldor, M. (2004). Shaping bacterial genomes with integrative and conjugative elements. *Research in Microbiology*, *155*(5), 376–378. doi:10.1016/j.resmic.2004.01.012

Butzer, P. L., & Westphal, U. (2000). *An introduction to fractional calculus*. Singapore: World Scientific.

Cai, C., & Chen, G. (2006). Synchronization of complex dynamical networks by the incremental ISS approach. *Physica A*, *371*, 754–766. doi:10.1016/j.physa.2006.03.052

Campi, M., & Kumar, P. R. (1996, December). *Learning dynamical systems in a stationary environment*. Presented at the 35th Conference on Decision and Control, Kobe, Japan.

Cao, J., Li, H. X., & Ho, D. W. C. (2005). Synchronization criteria of Lur'e systems with time-delay feedback control. *Chaos, Solitons, and Fractals*, *23*(4), 1285–1298.

Cao, J., Chen, G., & Li, P. (2008). Global synchronization in an array of delayed neural networks with hybrid coupling. *IEEE Trans. Systems, Man, and Cybernetics-Part B . Cybernetics*, *38*(2), 488–498.

Cao, J., & Liang, J. L. (2004). Boundedness and stability for Cohen-Grossberg neural network with time-varying delays. *Journal of Mathematical Analysis and Applications*, *296*(2), 51–65. doi:10.1016/j.jmaa.2004.04.039

Cao, J., & Wang, J. (2005). Global asymptotic and robust stability of recurrent neural networks with time delays. *IEEE Transactions on Circuits and Systems. I, Fundamental Theory and Applications*, *52*(2), 417–426. doi:10.1109/TCSI.2004.841574

Cao, J., Wang, Z., & Sun, Y. (2007). Synchronization in an array of linearly stochastically coupled networks with time delays. *Physica A*, *385*(2), 718–728. doi:10.1016/j.physa.2007.06.043

Cao, J., Yuan, K., & Li, H.-X. (2006). Global Asymptotical stability of generalized recurrent neural networks with multiple discrete delays and distributed delays. *IEEE Transactions on Neural Networks*, *17*(6), 1646–1651. doi:10.1109/TNN.2006.881488

Cao, J., Li, P., & Wang, W. (2006). Global synchronization in arrays of delayed neural networks with constant and delayed coupling. *Physics Letters. [Part A]*, *353*(4), 318–325. doi:10.1016/j.physleta.2005.12.092

Caponetto, R., Fortuna, L., Fazzino, S., & Xibilia, M. G. (2003). Chaotic sequences to improve the performance of evolutionary algorithms. *IEEE Transactions on Evolutionary Computation*, *7*(3), 289–304. doi:10.1109/TEVC.2003.810069

Caputo, M. (1967). Linear models of dissipation whose Q is almost frequency independent- II. *Geophysical Journal of the Royal Astronomical Society*, *13*, 529–539.

Carroll, T. (1995). A simple system for demonstrating regular and synchronized chaos. *American Journal of Physics*, *63*, 377–379. doi:10.1119/1.17923

Carroll, T. L., & Pecora, L. M. (1991). Synchronizing chaotic circuits. *IEEE Transactions on Circuits and Systems*, *38*(4), 453–456. doi:10.1109/31.75404

Carroll, T. C., Heagy, J. F., & Pecora, L. M. (1996). Transforming signals with chaotic synchronization. *Physical Review E: Statistical Physics, Plasmas, Fluids, and Related Interdisciplinary Topics*, *54*(5), 4676–4680. doi:10.1103/PhysRevE.54.4676

Carroll, T. L. (2001). Noise-resistant chaotic synchronization. *Physical Review E: Statistical, Nonlinear, and Soft Matter Physics*, *64*, 1–4. doi:10.1103/PhysRevE.64.015201

Carroll, T. L., & Pecora, L. M. (1993). Synchronizing nonautonomous chaotic circuits. *IEEE Trans. Circuits and Systems II*, 40, 646-650.

Celikovsky, S., & Chen, G. (2005). Secure synchronization of a class of chaotic systems from a nonlinear observer approach. *IEEE Transactions on Automatic Control*, *50*(1), 76–82. doi:10.1109/TAC.2004.841135

Chairez, I., Poznayk, A., & Poznayk, T. (2006). New sliding-mode learning law for dynamic neural network observer. *IEEE Transactions on Circuits and Systems, II*, 53, 1338–1342.

Chambers, W. G., & Frey, D. (1993). Comments on Chaotic digital encoding: an approach to secure communication and reply. *IEEE Trans. On Circuits and Systems II*, 46, 1445–1448.

Chang, Y. C. (2001). Adaptive Fuzzy-Based Tracking Control for Nonlinear SISO Systems via VSS and H∞ Approaches. *IEEE Transactions on Fuzzy Systems*, *9*, 278–292. doi:10.1109/91.919249

Chembo, Y. K., Larger, L., & Colet, P. (2008). Nonlinear dynamics and spectral Stability of optoelectronic microwave oscillators. *IEEE Journal of Quantum Electronics*, *44*(9), 858–866. doi:10.1109/JQE.2008.925121

Chen, S.-S., Chen, L.-F., Wu, Y.-T., Wu, Y.-Z., Lee, P.-L., Yeh, T.-C., & Hsieh, J.-C. (2007). Detection of synchronization between chaotic signals: An adaptive similarity-based approach. *Physical Review E: Statistical, Nonlinear, and Soft Matter Physics*, *76*, 066208. doi:10.1103/PhysRevE.76.066208

Chen, G. (1997). *Control and synchronization of chaos*. Houston, TX: University of Houston, Department of Electrical Engineering.

Chen, W. C. (2008). Nonlinear dynamics and chaos in a fractional-order financial system. *Chaos, Solitons, and Fractals*, *36*, 1305–1314. doi:10.1016/j.chaos.2006.07.051

Chen, H.-K. (2005). Global chaos synchronization of new chaotic systems via nonlinear control. *Chaos, Solitons, and Fractals*,*23*(4), 1245–1251.

Chen, J. Y., Wong, K. W., Cheng, L. M., & Shuai, J. W. (2003). A secure communication scheme based on the phase synchronization of chaotic systems. *Chaos (Woodbury, N.Y.)*, *13*(2), 508–514. doi:10.1063/1.1564934

Chen, M., & Han, Z. (2003). Controlling and synchronizing chaotic Genesio system via nonlinear feedback control. *Chaos, Solitons, and Fractals*,*17*(4), 709–716. doi:10.1016/S0960-0779(02)00487-3

Chen, S., & Lu, J. (2002). Parameters identification and synchronization of chaotic systems based upon adaptive control. *Physics Letters. [Part A]*, *299*(4), 353–358.

Chen, H. K. (2002). Chaos and chaos synchronization of a symmetric gyro with linear-plus-cubic damping. *Journal of Sound and Vibration*, *255*(4), 719–740. doi:10.1006/jsvi.2001.4186

Chen, H. K., & Lin, T. N. (2003). Synchronization of chaotic symmetric gyros by one-way coupling conditions. *Proceedings of the Institution of Mechanical Engineers. Part C, Journal of Mechanical Engineering Science*, *217*, 331–340. doi:10.1243/095440603762869993

Chen, G., & Dong, X. (1993). On feedback control of chaotic continuous-time systems. *IEEE Transactions on Circuits and Systems*, *40*, 591–601. doi:10.1109/81.244908

Chen, G. R., & Dong, X. (1998). *From Chaos to Order: Methodologies, Perspectives, and Applications*. Singapore: World Science Publishing Company.

Chen, G. R., Zhou, J., & Liu, Z. R. (2004). Global synchronization of coupled delayed neural networks and applications to chaotic CNN models. *International Journal of Bifurcation and Chaos in Applied Sciences and Engineering*, *14*(7), 2229–2240. doi:10.1142/S0218127404010655

Chen, B. S., Lee, C. H., & Chang, Y. C. (1996). H1 tracking design of uncertain nonlinear SISO systems: adaptive fuzzy approach. *IEEE Transactions on Fuzzy Systems*, *4*(1), 32–43. doi:10.1109/91.481843

Chen, B. S., Tseng, C. S., & Uang, H. J. (2000). Mixed H2/H1 fuzzy output feedback control design for nonlinear dynamical systems: an LMI approach. *IEEE Transactions on Fuzzy Systems*, *8*(3), 249–265. doi:10.1109/91.855915

Chen, B., & Wornell, G. W. (1996). Efficient channel coding for analog sources using chaotic systems. In *GLOBECOM '96. Communications: The Key to Global Prosperity* (pp. 131–135).

Chen, G., & Dong, X. (1995,April-May). *Identification and control of chaotic systems*. Paper presented at the IEEE International Symposium on Circuit and Systems, Seattle, WA.

Chen, M., Zhou, D., & Shang, Y. (2005). A new observer-based synchronization scheme for private communication. *ChaosSolitons and Fractals, 24*, 1025-1030.

Chen, Q., Hong, Y., Chen, G., & Hill, D. J. (2006). Intermittent phenomena in switched systems with high coupling strengths. *IEEE Trans Circuits and Systems I*, 53, 2692-2704.

Cheng, C. J., Liao, T. L., Yan, J. J., & Hwang, C. C. (2006). Exponential synchronization of a class of neural networks with time-varying delays. *IEEE Trans. Sys., Man, and Cyber. Part B: Cybernetics*, *36*(1), 209–215. doi:10.1109/TSMCB.2005.856144

Cheng, C. J., Liao, T. L., & Hwang, C. C. (2005). Exponential synchronization of a class of chaotic neural networks. *Chaos, Solitons, and Fractals*, *24*, 197–206.

Chien, T.-I., & Liao, T.-L. (2005). Design of secure digital communication systems using chaotic modulation, cryptography and chaotic synchronization. *Chaos, Solitons, and Fractals*, *24*, 241–242.

Chlouverakis, K. E., Argyris, A., Bogris, A., & Syvridis, D. (2008). Complexity and synchronization in chaotic fiber-optic systems. *Physica D. Nonlinear Phenomena*, *237*, 568–572. doi:10.1016/j.physd.2007.09.023

Chlouverakis, K. E., Mikroulis, S., Stamataki, I., & Syvridis, D. (2007). Chaotic dynamics of semiconductor microring lasers. *Optics Letters*, *32*, 2912–2914. doi:10.1364/OL.32.002912

Chlouverakis, K. E., & Sprott, J. C. (2007). Hyperlabyrinth chaos: From chaotic walks to spatiotemporal chaos. *Chaos (Woodbury, N.Y.)*, *17*, 023110. doi:10.1063/1.2721237

Choon, K. A. (2009). A new chaos synchronization method for Duffing oscillator . *IEICE Electronics Express*, *18*, 1355–1360.

Chu, Y. H., & Chang, S. (1999). Dynamic data encryption system based on synchronized chaotic systems. *Electronics Letters*, *35*, 271–273. doi:10.1049/el:19990148

Chu, Y. H., & Chang, S. (1999). Dynamical cryptography based on synchronized chaotic systems. *Electronics Letters*, *35*, 97–975.

Chua, L. O., Kocarev, L. J., Eckert, K., & Itoh, M. (1992). Experimental chaos synchronization in Chua's circuit. *International Journal of Bifurcation and Chaos in Applied Sciences and Engineering*, *2*, 705–708. doi:10.1142/S0218127492000811

Chua, L. O., & Lin, G. N. (1990). Canonical realization of Chua's circuit family. *IEEE Trans Circuits Systems I*, *37*, 885–902. doi:10.1109/31.55064

Chua, L. O., Wu, C. W., Huang, A., & Zhong, G. Q. (1993). A universal circuit for studying and generating chaos. Part II. Strange attractors. *IEEE Transactions on Circuits and Systems*, *40*, 732–761. doi:10.1109/81.246149

Chua, L. O., Komuro, M., & Matsumoto, T. (1986). The double scroll family. *IEEE Transactions on Circuits and Systems*, *33*(11), 1072–1118. doi:10.1109/TCS.1986.1085869

Chua, L. O., Yang, T., Zhong, G. Q., & Wu, C. W. (1996). Adaptive synchronization of Chua's Oscillators. *International Journal of Bifurcation and Chaos in Applied Sciences and Engineering*, *6*, 189–201. doi:10.1142/S0218127496001946

Chua, L. O., Yang, T., Zhong G. Q., & Wu, C. W. (1996). Synchronization of Chua's circuits with time-varying channels and parameters. *IEEE Trans. Circuits and Systems I*, 43, 862-868.

Colet, P., & Roy, R. (1994). Digital communication with synchronized chaotic lasers. *Optics Letters*, *19*, 2056–2058. doi:10.1364/OL.19.002056

Cuomo, K. M., & Oppenheim, V. (1993). Circuit implementation of synchronized chaos with application to communication. *Physical Review Letters*, *71*, 65–68. doi:10.1103/PhysRevLett.71.65

Cuomo, K. M., Oppenheim, A. V., & Strogatz, S. H. (1993). Synchronization of Lorenz-based chaotic circuits with applications to communications. *IEEE Transactions on Circuits and Systems II: Analog and Digital Signal Processing*, *40*(10), 626–633. doi:10.1109/82.246163

Dachselt, F., & Schwarz, W. (2001). Chaos and Cryptography. *IEEE Transactions on Circuits and Systems*, *48*(12), 1498–1509. doi:10.1109/TCSI.2001.972857

Daemen, J., & Paris, K. (2004). *The self-synchronzing stream cipher moustique*. Technical report, e-Stream Project. Retrieved from http://www.ecrypt.eu.org/stream/p3ciphers/mosquito/mosquito-p3.pdf

Dana, S. K., & Chakraborty, S. (2004). Generation of homoclinic oscillation in the Phase Synchronization Regime in Coupled Chua's oscillators. *International Journal of Bifurcation and Chaos in Applied Sciences and Engineering*, *14*(4), 1375. doi:10.1142/S0218127404009958

Dana, S. K., Chakraborty, S., & Aananthakrishna, G. (2005). Homoclinic bifurcation in Chua's circuit. *PRAMANA - . Journal of Physics*, *64*(3).

Dana, S. K., Sengupta, D. C., & Hu, C.-K. (2006). Spiking and bursting in Josepshon junction. *IEEE Transactions on Circuits and Wystems. II, Express Briefs*, *50*(10), 1031. doi:10.1109/TCSII.2006.882183

Dana, S. K., & Roy, P. K. (2007). Bursting near homoclinic bifurcation in two coupled Chua oscillators. *Int. J. Bifur. Chaos*, *17*(10).

De Jong, K. A. (1975). An analysis of the behavior of a class of genetic adaptive systems. (Doctoral dissertation, University of Michigan). *Dissertation Abstracts International*, *36*(10), 5140B.

Dedieu, H., Kennedy, M. P., & Hasler, M. (1993). Chaos Shift Keying: Modulation and Demodulation of a Chaotic Character Using Self-Synchronizing Chua's Circuits. *IEEE Transactions on Circuits and Systems*, *40*(10), 634–642. doi:10.1109/82.246164

Dedieu, H., & Ogorzalek, M. (1997). Identification of chaotic systems based on adaptive synchronization. In *Proc. ECCTD'97*, Budapest, Sept. 1997 (pp. 290-295).

Delfs, H., & Knebl, H. (2002). *Introduction to cryptography*. Berlin: Springer-Verlag.

DeShazer, D. J., Breban, R., Ott, E., & Roy, R. (2001). Detecting phase synchronization in a chaotic laser array. *Physical Review Letters*, *87*(4), 441011–441014. doi:10.1103/PhysRevLett.87.044101

Devaney, R. (2003). *An introduction to chaotic dynamical systems. Studies in nonlinearity*. Boulder, CO: Westview Press.

Devaney, R. L. (1989). *An introduction to Chaotic Dynamical Systems*. Redwood City, CA: Addison-Wesley.

Diethelm, K., & Ford, N. J. (2002). Analysis of fractional differential equations. *Journal of Mathematical Analysis and Applications*, *265*, 229–248. doi:10.1006/jmaa.2000.7194

Diethelm, K., Ford, N. J., & Freed, A. D. (2002). A predictor-corrector approach for the numerical solution of fractional differential equations. *Nonlinear Dynamics*, *29*, 3–22. doi:10.1023/A:1016592219341

Dixon, C. J., Woods, N. M., Cuthbertson, K. S. R., & Cobbold, P. H. (1990). Evidence for two Ca^{2+} -mobilizing purinoreceptors on rat hepatocytes. *The Biochemical Journal*, *269*, 499–502.

Dmitriev, A., Panas, A., Starkov, S., & Kuzmin, L. (1997). Experiments on RF band communications using chaos. *International Journal of Bifurcation and Chaos in Applied Sciences and Engineering*, *7*(11), 2511–2527. doi:10.1142/S0218127497001680

Dmitriev, A. S., & Panas, A. I. (2000). *Dynamic chaos: novel type of information carrier for communication systems*. Moscow: Fiz.-Mat. Lit.

Dmitriev, A. S., Panas, A. I., & Starkov, S. O. (1995). Experiments on speech and music signals transmission using chaos. *International Journal of Bifurcation and Chaos in Applied Sciences and Engineering*, *5*(4), 1249–1254. doi:10.1142/S0218127495000910

Dmitriev, A. S., Panas, A. I., & Starkov, S. O. (1994). Experiments on transmission of speech and musical signals using dynamical chaos. In *Preprint IRE RAS* N12(600), Moscow, Russia (pp. 1-42).

Donati, S., & Mirasso, C. R. (Eds.). (2002). Feature Section on Optical Chaos and Applications to Cryptography. *IEEE J. of Quantum Electron.*, 38.

Dooren, R. V. (2003). Comments on Chaos and chaos synchronization of a symmetric gyro with linear-plus-cubic damping. *Journal of Sound and Vibration*, *268*, 632–634. doi:10.1016/S0022-460X(03)00343-2

Dorizzi, B., Grammaticos, B., Le Berre, M., Pomeau, Y., Ressayre, E., & Taller, A. (1987). Statistics and dimension of chaos in differential delay systems. *Physical Review A.*, *35*, 328–339. doi:10.1103/PhysRevA.35.328

Dupont, G., & Goldbeter, A. (1993). A one-pool model for Ca^{+2} oscillations involving Ca^{+2} and inositol 1,4,5-trisphosphate as co-agonists for Ca^{+2} release. *Cell Calcium*, *14*, 311. doi:10.1016/0143-4160(93)90052-8

DuPont, G., Houart, G., & De Koninck, P. (2003). Sensitivity of CaM kinase II to the frequency of Ca^{2+} oscillations: a simple model. *Cell Calcium*, *34*, 485–497. doi:10.1016/S0143-4160(03)00152-0

Eckmann, P., & Ruelle, D. (1985). Ergodic theory of chaos and strange attractors. *Reviews of Modern Physics*, *57*, 617–656. doi:10.1103/RevModPhys.57.617

Elabbasy, E. M., Agiza, H. N., & El-Dessoky, M. M. (2004). Adaptive synchronization of Lu system with uncertain parameters. *Chaos, Solitons, and Fractals*, *21*(3), 657–667. doi:10.1016/j.chaos.2003.12.028

El-Sayed, A. M. A., El-Mesiry, A. E. M., & El-Saka, H. A. A. (2007). On the fractional-order logistic equation. *Applied Mathematics Letters*, *20*, 817–823. doi:10.1016/j.aml.2006.08.013

Epstein, I. R. (1990). Differential delay equations in chemical kinetics: Some simple linear model systems. *The Journal of Chemical Physics*, *92*, 1702–1712. doi:10.1063/1.458052

Erjaee, G. H., Atabakzade, M. H., & Saha, L. M. (2004). Interesting synchronization-like behavior. *International Journal of Bifurcation and Chaos in Applied Sciences and Engineering*, *14*, 1447–1453. doi:10.1142/S0218127404009934

Etemadi, S., Alasty, A., & Salarieh, H. (2006). Synchronization of chaotic systems with parameter uncertainties via variable structure control. *Physics Letters. [Part A]*, *357*(1), 17–21. doi:10.1016/j.physleta.2006.04.101

Falcke, M. (2003). Deterministic and stochastic models of intracellular Ca^{2+} waves. *New Journal of Physics*, *5*, 1–28. doi:10.1088/1367-2630/5/1/396

Farmer, J. D. (1982). Chaotic attractors of an infinite-dimensional dynamical system. *Physica D. Nonlinear Phenomena*, *4*(3), 366–393. doi:10.1016/0167-2789(82)90042-2

Feki, M. (2003). An adaptive chaos synchronization scheme applied to secure communication. *Chaos, Solitons, and Fractals,18*(1), 141–149. doi:10.1016/S0960-0779(02)00585-4

Feldmann, U., Hasler, M., & Schwarz, W. (1996). Communication by chaotic signals:the inverse system approach. *Int. J. of Circuit Theory Appl., 24*, 551–579. doi:10.1002/(SICI)1097-007X(199609/10)24:5<551::AID-CTA936>3.0.CO;2-H

Femat, R., & Alvarez-Ramírez, J. (1997). Synchronization of a class of strictly different chaotic oscillators. *Physics Letters. [Part A], 236*(12), 307–313. doi:10.1016/S0375-9601(97)00786-X

Femat, R., & Jauregui-Ortiz, R. (2001). A chaos-based communication scheme via robust asymptotic feedback. *IEEE Transactions on Circuits and Systems-I, 48*(10), 1161–1169. doi:10.1109/81.956010

Femat, R., Alvarez-Ramírez, J., & Fernández-Anaya, G. (2000). Adaptive synchronization of high-order chaotic systems: A feedback with low-order parametrization. *Physica D. Nonlinear Phenomena, 139*(3-4), 231–246. doi:10.1016/S0167-2789(99)00226-2

Femat, R., Alvarez-Ramirez, J., & Castillo-Toledo, B. (1999a). On robust chaos suppression in a class of nondriven oscillators: Application to Chua's circuit. *IEEE Transactions on Circuits and Systems, 46*(1), 1150.

Femat, R., Jauregui-Ortiz, R., & Solís-Perales, G. (2001). A Chaos-Based Communication Scheme via Robust Asymptotic Feedback. *IEEE Transactions on Circuits and Systems, 48*(10).

Femat, R., & Solís-Perales, G. (1999b). On the chaos synchronization phenomena. *Physics Letters. [Part A], 262*, 50–60. doi:10.1016/S0375-9601(99)00667-2

Fliess, M., Levine, J., Martin, P., & Rouchon, P. (1995). Flatness and defect of non-linear systems: introductory theory and examples. *International Journal of Control, 61*(6), 1327–1361. doi:10.1080/00207179508921959

Fogel, L. J., Owens, A. J., & Walsh, M. J. (1966). *Artificial Intelligence Through Simulated Evolution*. New York: John Wiley and Sons.

Fradkov, A. L., & Evans, R. J. (2005). Control of chaos: Methods and applications in engineering. *Annual Reviews in Control, 29*, 33–56. doi:10.1016/j.arcontrol.2005.01.001

Fradkov, A. L., & Markov, A. Y. (1997). Adaptive synchronization of chaotic systems based on speed gradient method and passification. *IEEE Transactions on Circuits and Systems. I, Fundamental Theory and Applications, 44*(10), 905–912. doi:10.1109/81.633879

Fradkov, A. L., Nijmeijer, H., & Markov, A. (2000). Adaptive observer-based synchronization for communications. *International Journal of Bifurcation and Chaos in Applied Sciences and Engineering, 10*(12), 2807–2813. doi:10.1142/S0218127400001869

Fradkov, A. L., & Pogromsky, A. Y. (1996). Speed gradient control of chaotic continuous-time systems. *IEEE Transactions on Circuits and Systems. I, Fundamental Theory and Applications, 43*(11), 907–913. doi:10.1109/81.542281

Freeman, M. (2000). Feedback control of intracellular signaling in developed. *Nature, 408*, 313–319. doi:10.1038/35042500

Frey, D. R. (1993). Chaotic digital encoding: an approach to secure communication. *IEEE Transactions on Circuits and Systems II: Analog and Digital Signal Processing, 40*(10), 660–666. doi:10.1109/82.246168

Fridman, E. (2006). Descriptor discretized Lyapunov functional method: Analysis and design. *IEEE Transactions on Automatic Control, 51*(5), 890–897. doi:10.1109/TAC.2006.872828

Fridrich, J. (1998, June). Symmetric ciphers based on two-dimensional chaotic maps. *International Journal of Bifurcation and Chaos in Applied Sciences and Engineering, 8*(6), 1259–1284. doi:10.1142/S021812749800098X

Friedman, A. (1976). *Stochastic Differential Equations and Applications*. New York: Academic Press.

Fujisaka, H., & Yamada, T. (1983). Stability theory of synchronized motion in coupled-oscillator systems. *Progress of Theoretical Physics*, *69*, 32–47. doi:10.1143/PTP.69.32

Gahinet, P., Nemirovsky, A., Laub, A. J., & Chilali, M. (1995). *LMI control Toolbox: For use with Matlab*. Natick, MA: The MATH Works, Inc.

Galias, Z., & Maggio, G. M. (2001). Quadrature chaos–shift keying: theory and performance analysis. *IEEE Transactions on Circuits and Systems-I*, *48*(12), 1510–1519. doi:10.1109/TCSI.2001.972858

Gallegos, J. A. (1994). Nonlinear regulation of a Lorenz system by feedback linearization techniques. *Dynamics and Control*, *4*, 277–298. doi:10.1007/BF01985075

Gao, H., Chen, T., & Lam, J. (2008). A new delay system approach to network-based control. *Automatica*, *44*(1), 39–52. doi:10.1016/j.automatica.2007.04.020

Gao, H., Lam, J., & Chen, G. (2006). New criteria for synchronization stability of general complex dynamical networks with coupling delays. *Physics Letters. [Part A]*, *360*(2), 263–273. doi:10.1016/j.physleta.2006.08.033

Garcia, P., & Jimenez, J. (2002). Communication through chaotic map systems. *Physics Letters. [Part A]*, *298*, 35–40. doi:10.1016/S0375-9601(02)00382-1

García, A., Poznayk, A., Chairez, I., & Poznyak, T. (2008). Differential Neural Networks Observers: development, stability analysis and implementation . In Husek, P. (Ed.), *System* (pp. 61–82). Structure and Control.

Garcia-Ojalio, J., & Roy, R. (2001). Spatiotemporal Communication with Synchronized Optical Chaos. *Physical Review Letters*, *86*, 5204. doi:10.1103/PhysRevLett.86.5204

García-Valdovinos, L.-G., Parra-Vega, V., & Arteaga, M. A. (2007). Observer-based sliding mode impedance control of bilateral teleoperation under constant unknown time delay. *Robotics and Autonomous Systems*, *55*(8), 609–617. doi:10.1016/j.robot.2007.05.011

Gaspard, P., & Wang, X.-J. (1987). Homoclinic orbits and mixed-mode oscillations in far from equilibrium systems. *Journal of Statistical Physics*, *48*, 151. doi:10.1007/BF01010405

Gauthier, D. J. (1998). Chaos Has Come Again. *Science*, *279*, 1156–1157. doi:10.1126/science.279.5354.1156

Gauthier, J. P., Hammouri, H., & Othman, S. (1992). A simple observer for nonlinear systems with application to bioreactors. *IEEE Transactions on Circuits and Systems*, *37*, 875–880.

Ge, Z. M., Chen, H. K., & Chen, H. H. (1996). The regular and chaotic motion of a symmetric heavy gyroscope with harmonic excitation. *Journal of Sound and Vibration*, *198*, 131–147. doi:10.1006/jsvi.1996.0561

Geddes, J. B., Short, K. M., & Black, K. (1999). Extraction of signals from chaotic laser data . *Physical Review Letters*, *83*, 5389–5392. doi:10.1103/PhysRevLett.83.5389

Germain, E. K. H. (2006). *Synchronization dynamics of nonlinear self-sustained oscillations with applications in physics, engineering and biology*. PhD thesis, Université d'Abomey-Calavi, Bénin.

Gero, J. S., & Kazakov, V. (2001). A Genetic Engineering Approach to Genetic Algorithms Evolutionary Computation. *Evolutionary Computation*, *9*(1), 71–92. doi:10.1162/10636560151075121

Ghosh, D. (2009). Generalized projective synchronization in time-delayed systems: Nonlinear observer approach. *Chaos (Woodbury, N.Y.)*, *19*, 013102. doi:10.1063/1.3054711

Ghosh, D., Banerjee, S., & Chowdhury, A. R. (2007). Synchronization between variable time-delayed systems and cryptography. *EPL*, *80*, 30006. doi:10.1209/0295-5075/80/30006

Ghosh, D., Banerjee, S., & Chowdhury, A. R. (2010). Generalized and projective synchronization in modulated time-delayed systems. *Physics Letters. [Part A]*, *374*, 2143–2149. doi:10.1016/j.physleta.2010.03.027

Ghosh, D., & Bhattacharya, S. (2010). (in press). Projective synchronization of new hyperchaotic system with fully unknown parameters. *Nonlinear Dynamics*, *60*.

Gickenheimer, J., & Holmes, P. (1983). *Nonlinear Oscillations, Dynamical Systems and Bifurcations of Vector Fields*. Berlin, Germany: Springer.

Glendinning, P., Abshagen, J., & Mullin, T. (2001). Imperfect homoclinic bifurcations. *Phy. Rev. E*, *64*, 036208. doi:10.1103/PhysRevE.64.036208

Glendinning, P., & Sparrow, C. (1984). Local and global behavior near homoclinic orbits. *Journal of Statistical Physics*, *35*, 645. doi:10.1007/BF01010828

Goedgebuer, J. P., Larger, L., & Porter, H. (1998a). Optical cryptosystem based on synchronization of hyperchaos generated by a delayed feedback tunable laser diode. *Physical Review Letters*, *80*, 2249–2252. doi:10.1103/PhysRevLett.80.2249

Goedgebuer, J. P., Larger, L., Porter, H., & Delorme, F. (1998b). Chaos in wavelength with a feedback tunable laser diode. *Physical Review E: Statistical Physics, Plasmas, Fluids, and Related Interdisciplinary Topics*, *57*, 2795–2798. doi:10.1103/PhysRevE.57.2795

Goldberg, D. E. (1989). *Genetic Algorithms in Search, optimization and machine learning*. Reading, MA: Addison-Wesley.

Goldberg, D. E. (1994). Genetic and Evolutionary Algorithms Come of Age. *Communications of the ACM*, *37*(3), 13–119. doi:10.1145/175247.175259

Goldberg, D. E., & Kalyanmoy, D. (1991). A Comparative Analysis of Selection Schemes Used in Genetic Algorithms . In Rawlins, G. J. E. (Ed.), *Foundations of Genetic Algorithms*. San Mateo, CA: Morgan Kaufmann Publishers.

Goldobin, D. S., & Pikovsky, A. (2005). Synchronization and desynchronization of self-sustained oscillators by common noise. *Physical Review E: Statistical, Nonlinear, and Soft Matter Physics*, *71*, 045201. doi:10.1103/PhysRevE.71.045201

Golea, N., Golea, A., & Benmahammed, K. (2003). Fuzzy Model Reference Adaptive Control. *Fuzzy Sets and Systems*, *137*(3), 353–366. doi:10.1016/S0165-0114(02)00279-8

Golub, G. H., & Van Loan, C. F. (1989). *Matrix Computating*. Baltimore, MD: The Johns Hopkins University Press.

Gomes, M. G. M., & King, G. P. (1992). Bistable chaos. II. Bifurcation analysis. *Physical Review A.*, *46*, 3100. doi:10.1103/PhysRevA.46.3100

Gómez-Rodriguez, A. (2000). Entanglement states and the singular value decomposition. *Revista Mexicana de Física*, *46*(5).

González, J., Femat, R., Alvarez-Ramírez, J., Aguilar, R., & Barrón, M. A. (1999). A discrete approach to the control and synchronization of a class of chaotic oscillators. *IEEE Transactions on Circuits and Systems . Part I*, *46*(9), 1139–1143.

Gonzalez Miranda, J. M. (2004). *Synchronization and control of chaos: an introduction for scientists and engineers*. London, UK: Imperial College Press. doi:10.1142/9781860945229

González-Miranda, J. M. (1996a). Chaotic systems with a null conditional Lyapunov exponent under nonlinear driving. *Physical Review E: Statistical Physics, Plasmas, Fluids, and Related Interdisciplinary Topics*, *53*, R5–R8. doi:10.1103/PhysRevE.53.R5

González-Miranda, J. M. (1996b). Synchronization of symmetric chaotic systems. *Physical Review E: Statistical Physics, Plasmas, Fluids, and Related Interdisciplinary Topics*, *53*, 5656–5669. doi:10.1103/PhysRevE.53.5656

González-Miranda, J. M. (1998a). Amplification and displacement of chaotic attractors by means of unidirectional chaotic driving. *Physical Review E: Statistical Physics, Plasmas, Fluids, and Related Interdisciplinary Topics*, *57*, 7321–7324. doi:10.1103/PhysRevE.57.7321

González-Miranda, J. M. (1998b). Using continuous control for amplification and displacement of chaotic signals. *The European Physical Journal B*, *6*, 411–418. doi:10.1007/s100510050568

González-Miranda, J. M. (2002a). Amplitude envelope synchronization in coupled chaotic oscillators. *Physical Review E: Statistical, Nonlinear, and Soft Matter Physics, 65*, 036232. doi:10.1103/PhysRevE.65.036232

González-Miranda, J. M. (2002b). Phase synchronization and chaos suppression in a set of two coupled nonlinear oscillators. *International Journal of Bifurcation and Chaos in Applied Sciences and Engineering, 12*, 2105–2112. doi:10.1142/S0218127402005716

González-Miranda, J. M. (2002c). Generalized synchronization in directionally coupled systems with identical individual dynamics. *Physical Review E: Statistical, Nonlinear, and Soft Matter Physics, 65*, 047202. doi:10.1103/PhysRevE.65.047202

Gorodilon, A., & Morozenko, V. (2008). Genetic Algorithm for finding the Key's length and Cryptanalysis of the Permutation Cipher. *International Journal Information Theories and Applications, 15.*

Goryachev, A., Strizhak, P., Kapral, R. (1997). Slow manifold structure and the emergence of mixed-mode oscillations. *J. Chem. Phys.*

Grassberger, P., & Procaccia, I. (1983). Characterization of Strange Attractors. *Physical Review Letters, 50*, 346–349. doi:10.1103/PhysRevLett.50.346

Grassberger, P., & Procaccia, I. (1983). Measuring the Strangeness of Strange Attractors. *Physica D. Nonlinear Phenomena, 9*, 189–208. doi:10.1016/0167-2789(83)90298-1

Grassi, G., & Mascolo, S. (1997). Nonlinear observer design to synchronize hyperchaotic systems via a scalar signal. *IEEE Transactions on Circuits and Systems, 44*(10), 1011–1014. doi:10.1109/81.633891

Grassi, G., & Mascolo, S. (1998). Observer Design for Cryptography based on Hyperchaotic Oscillators. *Electronics Letters, 34*, 1844–1846. doi:10.1049/el:19981285

Grebogi, C., Ott, E., Romeiras, E., & Yorke, J. A. (1987). Critical exponents for crisis-induced intermittency. *Physical Review A., 36*, 5365–5380. doi:10.1103/PhysRevA.36.5365

Green, A. K., Cobbold, P. H., & Dixon, J. C. (1995). Cytosolic free Ca^{2+} oscillations induced by diadenosine 5′,5′′′-P1,P3-triphosphate and diadenosine 5′,5′′′-P1,P4-tetraphosphate in single rat hepatocytes are indistinguishable from those induced by ADP and ATP respectively. *The Biochemical Journal, 310*, 629–635.

Grefenstette, J. J., & Baker, J. E. (1989). How genetic algorithms work: A critical look at implicit parallelism. In *Proceedings of the Third International Conference on Genetic Algorithms* (pp. 20-27). San Mateo, CA: Morgan Kaufmann Publishers.

Greub, W. (1975). *Linear Algebra* (4th ed.). New York: Springer-Verlag.

Grosu, I., Padmanaban, E., Roy, P. K., & Dana, S. K. (2008). Designing Coupling for Synchronization and Amplification of Chaos. *Physical Review Letters, 100*, 234102. doi:10.1103/PhysRevLett.100.234102

Gu, K., Kharitonov, V. L., & Chen, J. (2003). *Stability of time-delay systems*. Boston: Birkhauser.

Guan, S., Wan, X., & Lai, C.-H. (2006). Frequency locking by external force from a dynamical system with strange nonchaotic attractor. *Physics Letters. [Part A]*, 298–304. doi:10.1016/j.physleta.2006.01.067

Guanron, Ch., & Xianonig, D. (1993). Ordering chaos of Chua's circuits. *IEEE International Symposium on Circuits and Systems*, 4, 2604-2607.

Güémez, J., Martín, C., & Matías, M. A. (1997). Approach to the chaotic synchronized state of some driving methods. *Physical Review E: Statistical Physics, Plasmas, Fluids, and Related Interdisciplinary Topics, 55*, 122–134. doi:10.1103/PhysRevE.55.124

Guo, S. M., Liu, K. T., Tsai, J. S. H., & Shieh, L. S. (2008). An observer-based tracker for hybrid interval chaotic systems with saturating inputs: The chaos-evolutionary-programming approach. *Computers & Mathematics with Applications (Oxford, England), 55*, 1225–1249. doi:10.1016/j.camwa.2007.06.024

Habutsu, T., Nishio, Y., Sasane, I., & Mori, S. (1991). A secret key cryptosystem by iterating a chaotic map . In *Proc. Advances in Cryptology – EUROCRYPT '91* (pp. 127–140). Berlin, Germany: Springer-Verlag.

Hale, J., & Verduyn Lunel, S. M. (1993). *Introduction to functional differential equations*. New York: Springer Verlag.

Hale, J. K. (1977). *Theory of Functional Differential Equations*. Berlin: Springer-Verlag.

Halle, K. S., Wu, C. W., Itoh, M., & Chua, L. O. (1993). Spread spectrum communication through modulation of chaos in Chua's circuit. *International Journal of Bifurcation and Chaos in Applied Sciences and Engineering, 3*, 469–477. doi:10.1142/S0218127493000374

Hanias, M. (2008). Chaotic Behavior of an electrical analogue to the mechanical double pendulum. *Journal of Engineering Science and Technology Review*, *1*, 33–38.

Hanias, M. P., & Anagnostopoulos, J. A. N. (1993). Negative-differential- resistance effects in the $TlGaTe_2$ ternary semiconductor. *Physical Review B: Condensed Matter and Materials Physics*, *47*, 4261–4267. doi:10.1103/PhysRevB.47.4261

Hanias, M. P., Avgerinos, Z., & Tombras, G. S. (2009). Period doubling, Feigenbaum constant and time series prediction in an experimental chaotic RLD circuit. *Chaos, Solitons, and Fractals*, *40*, 1050–1059. doi:10.1016/j.chaos.2007.08.061

Hanias, M. P., Giannaris, G., Spyridakis, A., & Rigas, A. (2006). Time series analysis in chaotic diode resonator circuit. *Chaos, Solitons, and Fractals*, *27*, 569–573. doi:10.1016/j.chaos.2005.03.051

Hanias, M. P., Giannis, L. I., & Tombras, G. S. (2010). Chaotic operation by a single transistor circuit in the reverse active region. *Chaos (Woodbury, N.Y.)*, *20*, 0131051–0131057. doi:10.1063/1.3293133

Hanias, M. P., Kalomiros, J. A., Karakotsou, C., Anagnostopoulos, A. N., & Spyridelis, J. (1994). Quasiperiodic and chaotic self-excited voltage oscillations in $TlInTe_2$. *Physical Review B: Condensed Matter and Materials Physics*, *49*, 16994–16998. doi:10.1103/PhysRevB.49.16994

Hanias, M. P., & Karras, D. A. (2009). On efficient multistep non-linear time series prediction in chaotic diode resonator circuits by optimizing the combination of non-linear time series analysis and neural networks. *Engineering Applications of Artificial Intelligence*, *22*, 32–39. doi:10.1016/j.engappai.2008.04.016

Hanias, M. P., & Tombras, G. (2009). Time series cross – prediction in a single Transistor Chaotic Circuit. *Chaos, Solitons, and Fractals*, *41*, 1167–1172. doi:10.1016/j.chaos.2008.04.055

Hanias, M. P., & Tombras, G. S. (2009). Time series analysis in single transistor chaotic circuit. *Chaos, Solitons, and Fractals*, *40*, 246–256. doi:10.1016/j.chaos.2007.07.065

Hanias, M. P., & Karras, D. A. (2007). Efficient Non Linear Time Series Prediction Using Non Linear Signal Analysis and Neural Networks in Chaotic Diode Resonator Circuits. In *Advances in Data Mining. Theoretical Aspects and Applications* (LNCS 4597, pp. 329-338). Berlin: Springer.

Hanias, M. P., & Karras, D. A. (2007). Improved Multistep Nonlinear Time Series Prediction by applying Deterministic Chaos and Neural Network Techniques in Diode Resonator Circuits. In *IEEE International Symposium on Intelligent Signal Processing, WISP 2007* (pp. 1 - 6).

Hanias, M. P., Magafas, L., & Kalomoiros, J. (2008). Non- linear Analysis in RL-LED optoelectronic circuit. *Optoelectronics and advanced materials – Rapid Communications, 2*, 126 – 129.

Hansen, N., Müller, S. D., & Koumoutsakos, P. (2003). Reducing the Time Complexity of the Derandomized Evolution Strategy with Covariance Matrix Adaptation (CMA-ES). *Evolutionary Computation*, *11*(1). doi:10.1162/106365603321828970

Hansen, N., & Ostermeier, A. (2001). Completely Derandomized Self-Adaptation in Evolution Strategies. *Evolutionary Computation*, *9*(1), 159–195. doi:10.1162/106365601750190398

Harrison, R. G., & Biswas, D. J. (1986). Chaos in Light. *Nature*, *321*, 394–401. doi:10.1038/321394a0

Hartmut, D., & Hartmut, H. J. (1997). *Fundamentals of Wavelets*. New York: Wiley.

Hasler, M. (1998, April). Synchronization of chaotic systems and transmission of information. *International Journal of Bifurcation and Chaos in Applied Sciences and Engineering*, *8*(4). doi:10.1142/S0218127498000450

Hasler, M., Delgado-Restituto, M., & Rodriguez-Vasquez, A. (1996). Markov maps for communications with chaos. In *Proc. of the 1996's Nonlinear Dynamics in Electronic Systems (NDES'96)*, Sevilla (pp. 161–166).

Haykin, S. (1994). *Neural Networks. A comprehensive Foundation*. New York: IEEE Press.

He, R., & Vaidya, P. G. (1998). Implementation of chaotic cryptograph with chaotic synchronization. *Physical Review E: Statistical Physics, Plasmas, Fluids, and Related Interdisciplinary Topics*, *57*, 1532–1535. doi:10.1103/PhysRevE.57.1532

He, R., & Vaidya, P. G. (1992). Analysis and synthesis of synchronous periodic and chaotic systems. *Physical Review A.*, *46*, 7387. doi:10.1103/PhysRevA.46.7387

He, R., & Vaidya, P. G. (1999). Time delayed chaotic systems and their synchronization. *Physical Review E: Statistical Physics, Plasmas, Fluids, and Related Interdisciplinary Topics*, *59*, 4048. doi:10.1103/PhysRevE.59.4048

He, Y., Wang, Q., Wu, M., & Lin, C. (2006). Delay-dependent state estimation for delayed neural networks. *IEEE Transactions on Neural Networks*, *17*, 1077–1081. doi:10.1109/TNN.2006.875969

He, Y., Wang, Q., & Zheng, W. (2005). Global robust stability for delayed neural networks with polytopic type uncertainties. *Chaos, Solitons, and Fractals*, *26*, 1349–1354. doi:10.1016/j.chaos.2005.04.005

He, W., & Cao, J. (2008). Adaptive synchronization of a class of chaotic neural networks with known or unknown parameters. *Physics Letters. [Part A]*, *372*, 408–416. doi:10.1016/j.physleta.2007.07.050

He, R., & Vaidya, P. G. (1998). Implementation of chaotic cryptograph with chaotic synchronization. *Physical Review E: Statistical Physics, Plasmas, Fluids, and Related Interdisciplinary Topics*, *57*, 1532–1535. doi:10.1103/PhysRevE.57.1532

Heagy, J. F., Carroll, T. L., & Pecora, L. M. (1994). Synchronous chaos in coupled oscillator systems. *Physical Review E: Statistical Physics, Plasmas, Fluids, and Related Interdisciplinary Topics*, *50*(3), 1874–1885. doi:10.1103/PhysRevE.50.1874

Heagy, J. F., Platt, N., & Hammel, S. M. (1994). Characterization of on-off intermittency. *Physical Review E: Statistical Physics, Plasmas, Fluids, and Related Interdisciplinary Topics*, *49*(2), 1140–1150. doi:10.1103/PhysRevE.49.1140

Healy, J. J., Broomhead, D. S., Cliffe, K. A., Jones, R., & Mullin, T. (1991). The Origins of Chaos in a Modified Van der Pol Oscillator. *Physica D. Nonlinear Phenomena*, *48*, 322. doi:10.1016/0167-2789(91)90091-M

Heaviside, O. (1971). *Electromagnetic theory.* New York: Chelsea.

Hegazi, A. S., Agiza, H. N., & El-Dessoky, M. M. (2001). Synchronization and adaptive synchronization of nuclear spin generator system. *Chaos, Solitons, and Fractals*, *12*, 1091–1099. doi:10.1016/S0960-0779(00)00022-9

Hegazi, A. S., Agiza, H. N., & El-Dessoky, M. M. (2002). Adaptive synchronization for Rössler and Chua's circuit systems. *International Journal of Bifurcation and Chaos in Applied Sciences and Engineering*, *12*, 1579–1597. doi:10.1142/S0218127402005388

Hegger, R., Bünner, M. J., Kantz, H., & Giaquinta, A. (1998). Identifying and modeling delay feedback systems. *Physical Review Letters*, *81*, 558–561. doi:10.1103/PhysRevLett.81.558

Herrero, R., Pons, R., Farjas, J., Pi, F., & Orriols, G. (1996). Homoclinic dynamics in experimental Shil'nikov attractors. *Phy. Rev. E, 53*(6), 5627. doi:10.1103/PhysRevE.53.5627

Hilfer, R. (Ed.). (2000). *Applications of fractional calculus in physics*. Upper Saddle River, NJ: World Scientific.

Hirsch, M. W., Smale, S., & Devaney, R. L. (2004). *Differential Equations, Dynamical Systems and an Introduction to Chaos* (2nd ed.). San Diego, CA: Academic Press/Elsevier.

Ho, D. W. C., Liang, J. L., & Lam, J. (2006). Global exponential stability of impulsive high-order BAM neural networks with time-varying delays. *Neural Networks, 19*, 1581–1590. doi:10.1016/j.neunet.2006.02.006

Hojati, M., & Gazor, S. (2002). Hybrid Adaptive Fuzzy Identification and Control of Nonlinear Systems. *IEEE Transactions on Fuzzy Systems, 10*(2), 198–210. doi:10.1109/91.995121

Holland, J. H. (1975). *Adaptation in natural and artificial systems*. Ann Arbor, MI: The University of Michigan Press.

Hornby, A. S. (1984). *Oxford Advanced Learner's Dictionary*. Oxford, UK: Oxford University Press.

Hou, Y. Y., Liao, T. L., & Yan, J. J. (2007). H_∞ synchronization of chaotic systems using output feedback control design. *Physica A, 379*, 81–89. doi:10.1016/j.physa.2006.12.033

Houart, G., DuPont, G., & Goldbeter, A. (1999). Bursting, chaos and birhythmicity originating from self-modulation of the inositol 1,4,5-trisphosphate signal in a model for intracellular Ca^{+2} oscillations. *Bulletin of Mathematical Biology, 61*, 507. doi:10.1006/bulm.1999.0095

Hramov, A. E., & Koronovskii, A. A. (2005). Generalized synchronization: A modified system approach. *Physical Review E: Statistical, Nonlinear, and Soft Matter Physics, 71*, 067201. doi:10.1103/PhysRevE.71.067201

Hramov, A. E., & Koronovskii, A. A. (2005c). Intermittent generalized synchronization in unidirectionally coupled chaotic oscillators. *Europhysics Letters, 70*, 169–175. doi:10.1209/epl/i2004-10488-6

Hramov, A. E., Koronovskii, A. A., & Levin, Y. I. (2005b). Synchronization of Chaotic Oscillator Time Scales. *Journal of Experimental and Theoretical Physics, 100*, 784. doi:10.1134/1.1926439

Hsiao, M. Y., Li, T. H. S., Lee, J. Z., Chao, C. H., & Tsai, S. H. (2008). Design of interval type-2 fuzzy sliding-mode controller. *Information Science, 178*, 1696–1716. doi:10.1016/j.ins.2007.10.019

Huaguang, Z., Huanxin, G., & Zhanshan, W. (2007). Adaptive synchronization of neural networks with different attractors. *Progress in Natural Science, 17*(6), 687–695. doi:10.1080/10002007088537459

Huang, T. S. (1979). *Topics in Applied Physics: Picture Processing and Digital Filtering*. New York: Springer-Verlag.

Huang, H., Feng, G., & Cao, J. (2008). Exponential synchronization of chaotic Luré systems with delayed feedback control. *Nonlinear Dynamics*. doi:.doi:10.1007/s11071-008-9454-z

Huang, D. B. (2004). Synchronization-based estimation of all parameters of chaotic systems from time series. *Physical Review E: Statistical, Nonlinear, and Soft Matter Physics, 69*, 067201. doi:10.1103/PhysRevE.69.067201

Huang, D. B. (2005). Simple adaptive-feedback controller for identical chaos synchronization. *Physical Review E: Statistical, Nonlinear, and Soft Matter Physics, 71*, 037203. doi:10.1103/PhysRevE.71.037203

Huang, D. B. (2006). Adaptive-feedback control algorithm. *Physical Review E: Statistical, Nonlinear, and Soft Matter Physics, 73*, 066204. doi:10.1103/PhysRevE.73.066204

Huijberts, H. J. C., Nijmeijer, H., & Willems, R. (2000). System identification in communication with chaotic systems. *IEEE Transactions on Circuits and Systems. I, Fundamental Theory and Applications, 47*(6), 800–808. doi:10.1109/81.852932

Husainy, A., & Muhammed, A. F. (2006). Image Encryption Using Genetic Algorithm. *Information Technology Journal, 5*(3), 516–519. doi:10.3923/itj.2006.516.519

Huygens, C. (1986). *Horologium Oscillatorium* [The Pendulum clock]. Iowa City, IA: Iowa State University Press. (Original manuscript published in 1673).

Hwang, E. J., Hyun, C. H., Kim, E., & Park, M. (2009). Fuzzy model based adaptive synchronization of uncertain chaotic systems: Robust tracking control approach. *Physics Letters. [Part A], 373*, 1935–1939. doi:10.1016/j.physleta.2009.03.057

Ikeda, K. (1979). Multiple-valued stationary state and its instability of the transmitted light by a ring cavity system. *Optics Communications, 30*, 257–261. doi:10.1016/0030-4018(79)90090-7

Ikeda, K., Kondo, K., & Akimoto, O. (1982). Successive higher-harmonic bifurcations in systems with delayed feedback. *Physical Review Letters, 49*, 1467–1470. doi:10.1103/PhysRevLett.49.1467

Inoue, E., & Ushio, T. (2001). Chaos communication using unknown input observers. *Electronics and Communications in Japan (Part III Fundamental Electronic Science), 84*(12), 21–27. doi:10.1002/ecjc.1053

Isidori, A. (1995). *Nonlinear Control Theory*. New York: Springer-Verlag.

Itoh, M., Wu, C. W., & Chua, L. O. (1997). Communications systems via chaotic signals from a reconstruction viewpoint. *International Journal of Bifurcation and Chaos in Applied Sciences and Engineering, 7*(2), 275–286. doi:10.1142/S0218127497000194

Itoh, M. (1998). Chaos-based spread spectrum communication systems. In *Proc. Industrial Electronics ISIE 98 IEEE Int. Symp.* (Vol. 2, pp. 430–435).

Izhikevich, E. M. (2000). Neural excitability, spiking, and bursting. *International Journal of Bifurcation and Chaos in Applied Sciences and Engineering, 10*(6), 1171. doi:10.1142/S0218127400000840

Jackson, E. A. (1991). *Perspectives of Nonlinear Dynamics*. New York: Cambridge University Press.

Jakimoski, G., & Kocarev, L. (2001). Analysis of some recently proposed chaos-based encryption algorithms. *Physics Letters. [Part A], 291*, 381–384. doi:10.1016/S0375-9601(01)00771-X

Jakimoski, G., & Kocarev, L. (2001). Chaos and Cryptography: Block Encryption Cipher Based on Chaotic Maps. *IEEE Transactions on Circuits and Systems. I, Fundamental Theory and Applications, 48*(2). doi:10.1109/81.904880

Jiang, Z. (2002). A note on chaotic secure communication systems. *IEEE Transactions on Circuits and Systems-I, 49*(1), 92–96. doi:10.1109/81.974882

Jiang, G. P., Tang, K. S., & Chen, G. (2003). A simple global synchronization criterion for coupled chaotic systems. *Chaos, Solitons, and Fractals, 15*, 925–935. doi:10.1016/S0960-0779(02)00214-X

Jin, J., & Shi, J. (1999). Feature-preserving data compression of stamping tonnage information using wavelets. *Technometrics, 41*(4), 327–339. doi:10.2307/1271349

John, K. (1994). synchronization of unstable orbits using adaptive control. *Physical Review E: Statistical Physics, Plasmas, Fluids, and Related Interdisciplinary Topics, 49*(6), 4843–4848. doi:10.1103/PhysRevE.49.4843

Jovic, B., Berber, S., & Unsworth, C. P. (2006). A novel mathematical analysis for predicting master–slave synchronization for the simplest quadratic chaotic flow and Ueda chaotic system with application to communications. *Physica D. Nonlinear Phenomena, 213*, 31. doi:10.1016/j.physd.2005.10.013

Jozwiak, L., & Postula, A. (2002). Genetic engineering versus natural evolution Genetic algorithms with deterministic operators. *Journal of Systems Architecture, 48*, 99–112. doi:10.1016/S1383-7621(02)00094-2

Kantz, H., & Schreiber, T. (1997). *Nonlinear Time Series Analysis* (2nd ed.). Cambridge, UK: Cambridge University Press.

Kapitaniak, T. (1994). Synchronization of chaos using continuous control. *Physical Review E: Statistical Physics, Plasmas, Fluids, and Related Interdisciplinary Topics*, *50*(2), 1642–1644. doi:10.1103/PhysRevE.50.1642

Kaplan, J., & Yorke, J. (1979). Chaotic behavior of multidimensional difference equations . In Peitgen, H. O., & Walther, H. O. (Eds.), *Functional Differential Equations and Approximation of Fixed Points* (pp. 204–227). New York: Springer. doi:10.1007/BFb0064319

Karimi, H. R., & Maass, P. (2009). Delay-range-dependent exponential H_∞ synchronization of a class of delayed neural networks. *Chaos, Solitons, and Fractals*, *41*(3), 1125–1135. doi:10.1016/j.chaos.2008.04.051

Karnik, N. N., Mendel, J. M., & Liang, Q. (1999). Type-2 fuzzy logic systems. *IEEE Transactions on Fuzzy Systems*, *7*, 643–658. doi:10.1109/91.811231

Katok, A., & Hasselblatt, B. (1995). Introduction to the modern theory of dynamical systems . In *Encyclopedia of Mathematics and its Applications* (*Vol. 54*). Cambridge, UK: Cambridge University Press.

Kennedy, M. P. (1992). Robust op amp realization of Chua's circuit. *Frequuenz*, *46*(3-4), 66–80.

Kennedy, M. P., & Kolumbán, G. (2000). Digital communications using chaos. *Signal Processing*, *80*, 1307–1320. doi:10.1016/S0165-1684(00)00038-4

Kennedy, M. (1994). Chaos in Colpitts oscillator. *IEEE Transaction on Circuits and Systems – I, 41*, 1771-774.

Kennedy, M. P. (1993). Three steps to chaso-A Chua's circuit primer. *IEEE Trans. Cir. Systs., 40*.

Kennedy, M. P., & Ogorzalek, M. J. (Eds.). (1997). Special Issue. Chaos synchronization and control: theory and applications. *IEEE Trans. Circuits. Syst. I: Fundamental Theo. Appl., 44*(10), 853–1039.

Kenneth, S. M., & Bertram, R. (1993). *An introduction to the fractional calculus and fractional differential equations*. New York: Wiley-Interscience.

Khargonekar, P. P., Petersen, I. R., & Zhou, K. (1990). Robust stabilization of uncertain linear systems: Quadratic stabilizability and H_∞ control Theory. *IEEE Transactions on Automatic Control*, *35*, 356–361. doi:10.1109/9.50357

Kheireddine, C., Lamir, S., Mouna, G., & Hier, B. (2007). Indirect adaptive interval type-2 fuzzy control for nonlinear systems. *International Journal of Modeling, Identification and Control, 2*(2), 106-119.

Kim, D., Yang, H., & Hong, S. (2003). An Indirect Adaptive Fuzzy Sliding-Mode Control for Decoupled Nonlinear Systems. *IEEE Transactions on Fuzzy Systems*, *6*(2), 315–321.

Kipnis, A., & Shamir, A. (1999). Cryptanalysis of the hfe public key cryptosystem by relinearization. *Lecture Notes in Computer Science*, *1666*, 19–30. doi:10.1007/3-540-48405-1_2

Kiss, I. Z., & Hudson, J. L. (2001). Phase synchronization and suppression of Chaos through intermittency in forcing of an electrochemical oscillator. *Physical Review E: Statistical, Nonlinear, and Soft Matter Physics*, *64*(4II), 462151–462158. doi:10.1103/PhysRevE.64.046215

Kittel, A., Pyragas, K., & Richter, R. (1994). Pre-recorded history of a system as an experiment tool to control chaos. *Physical Review E: Statistical Physics, Plasmas, Fluids, and Related Interdisciplinary Topics*, *50*, 262–268. doi:10.1103/PhysRevE.50.262

Klein, E., Mislovaty, R., Kanter, I., & Kinzel, W. (2005). Public-channel cryptography using chaos synchronization. *Physical Review E: Statistical, Nonlinear, and Soft Matter Physics*, *72*, 016214. doi:10.1103/PhysRevE.72.016214

Knuth, D. E. (1998). *The Art of Computer Programming* (*Vol. 2*). Reading, MA: Addison-Wesley.

Kocarev, L., & Parlitz, U. (1996). Generalized synchronization, predictability, and equivalence of unidirectionally coupled dynamical systems. *Physical Review Letters*, *76*, 1816–1819. doi:10.1103/PhysRevLett.76.1816

Kocarev, L., Halle, K. S., Eckert, K., Chua, L. O., & Parlitz, U. (1992). Experimental demonstration of secure communication via chaotic synchronization. *International Journal of Bifurcation and Chaos in Applied Sciences and Engineering*, *2*(3), 709–713. doi:10.1142/S0218127492000823

Kocarev, L., Szczepanski, J., Amigó, J. M., & Tomosvski, I. (2006). Discrete chaos: part I. *IEEE Transactions on Circuits and Systems*, *I*, 53.

Kocarev, L., Jakimoski, G., & Tasev, Z. (2003). Pseudorandom bits generated by chaotic maps. *IEEE Transactions on Circuits and Systems-I*, *50*(1), 123–126. doi:10.1109/TCSI.2002.804550

Kocarev, L., & Parlitz, U. (1995). General approach for chaotic synchronization with application to communication. *Physical Review Letters*, *74*(25), 5028–5031. doi:10.1103/PhysRevLett.74.5028

Kocarev, L. (2001). Chaos-based cryptography: A brief overview. *IEEE Circuits and Systems Magazine*, *1*(3), 6–21. doi:10.1109/7384.963463

Kocarev, L., Jakimoski, G., Stojanovski, T., & Parlitz, U. (1998). From chaotic maps to encryption schemes. In *Proc. IEEE Int. Symposium Circuits and Systems, 4*, 514-517.

Koeller, R. C. (1984). Application of fractional calculus to the theory of viscoelasticity. *Journal of Applied Mechanics*, *51*, 294–298. doi:10.1115/1.3167616

Kolumban, G., Kennedy, M. P., & Chua, L. O. (1998, November). The role of synchronization in digital communications using chaos - part ii: Chaotic modulation and chaotic synchronization. *IEEE Transactions on Circuits and Systems. I, Fundamental Theory and Applications*, *45*, 1129–1140. doi:10.1109/81.735435

Kolumban, G., Kennedy, M. P., & Chua, L. O. (1997). The role of synchronization in digital communications using chaos – part I: fundamentals of digital communications, *IEEE Trans. Circuits and Systems I*, 44, 927-936.

Kolumban, G., Kennedy, M. P., & Chua, L. O. (2000). The role of synchronization in digital communications using chaos– part II: Chaotic Modulation and Chaotic Synchronization. *IEEE Trans. Circuits and Systems I*, 47, 1129-1140.

Kolumbán, G., Kis, G., Kennedy, M., et al. (1997). FM-DCSK: A new and robust solution to chaos communications. In *Proc. Int. Symposium on Nonlinear Theory and Its Applications*, Hawaii (pp. 117-120).

Kolumbán, G., Vizvári, B., Schwarz, W., & Abel, A. (1996). Differential chaos shift keying: a robust coding for chaos communication. In *Proc. Int. Workshop on Nonlinear Dynamic of Electronic Systems*, Sevilla, Spain (pp. 87-92).

Koper, M. T. M. (1995). Bifurcations of mixed-mode oscillations in a three-variable autonomous van der Pol-Duffing model with a cross-shaped phase diagram. *Physica D. Nonlinear Phenomena*, *80*, 72–94. doi:10.1016/0167-2789(95)90061-6

Koronovskii, A. A., & Hramov, A. E. (2004). Chaotic Phase Synchronization Studied by Means of Continuous Wavelet Transform. *Technical Physics Letters*, *30*(7), 587–590. doi:10.1134/1.1783411

Kosko, B. (1994). Fuzzy systems are universal approximators. *IEEE Transactions on Computers*, *43*(11), 1329–1333. doi:10.1109/12.324566

Kotulski, Z., & Szczepanski, J. (1997). Discrete chaotic cryptography. *Annals of Physics*, *6*, 381–394.

Kotulski, Z., Szczepanski, J., Grski, K., Paszkiewicz, A., & Zugaj, A. (1999). Application of discrete chaotic dynamical systems cryptography – DCC method. *International Journal of Bifurcation and Chaos in Applied Sciences and Engineering*, *9*, 1121–1135. doi:10.1142/S0218127499000778

Kovacic, Z., Balenovic, M., & Bogdan, S. (1998). Sensitivity based self learning fuzzy logic control for a servo system. *IEEE Control Systems Magazine*, *18*(3), 41–51. doi:10.1109/37.687619

Koza, J. R. (1992). *Gentic Programming: On the Programming of computers by Means of natural selection*. Cambridge, MA: The MIT Press.

Koza, J. R. (1994). *Genetic Programming II: Automatic Discovery of Reusable Programs*. Cambridge, MA: The MIT Press.

Koza, J. R., Andre, D., Bennett, F. H., & Keane, M. A. (1999). *Genetic Programming III: Darwinian Invention and Problem Solving*. San Francisco: Morgan Kaufmann.

Koza, J. R., Keane, M. A., & Streeter, M. J. (2003). Evolving inventions. *Scientific American, 288*, 52–59. doi:10.1038/scientificamerican0203-52

Koza, J. R., Keane, M. A., Streeter, M. J., Mydlowec, W., Yu, J., & Lanza, G. (2003). *Genetic Programming IV: Routine Human-Competitive Machine Intelligence*. Hingham, MA: Kluwer Academic Press.

Kramer, U., Krajnc, B., Pahle, J., Green, A. K., Dixon, C. J., & Marhl, M. (2005). Transition from stochastic to deterministic behavior in calcium oscillations. *Biophysical Journal, 89*, 1603–1611. doi:10.1529/biophysj.104.057216

Krawiecki, A., & Matyjaśkiewicz, S. (2001). Blowout bifurcation and stability of marginal synchronization of chaos. *Physical Review E: Statistical, Nonlinear, and Soft Matter Physics, 64*, 036216. doi:10.1103/PhysRevE.64.036216

Krawiecki, A., & Sukiennicki, A. (2000). Generalizations of the concept of marginal synchronization of chaos. *Chaos, Solitons, and Fractals, 11*, 1445–1458. doi:10.1016/S0960-0779(99)00062-4

Krawiecki, A., & Sukiennicki, A. (2003). Marginal synchronization of spin-wave amplitudes in a model for chaos in parallel pumping. *Physica Status Solidi. B, Basic Research, 236*, 511–514. doi:10.1002/pssb.200301716

Kuipers, L., & Niederreiter, H. (1974). Uniform distribution of sequences . In *Pure and Applied Mathematics*. New York: Wiley-Interscience.

Kung, C. C., & Chen, T. H. (2005). Observer-based indirect adaptive fuzzy sliding mode control with state variable filters for unknown nonlinear dynamical systems. *Fuzzy Sets and Systems, 155*, 292–308. doi:10.1016/j.fss.2005.04.016

Kuramoto, Y. (1984). *Chemical Oscillations, Waves and Turbulence*. Berlin: Springer-Verlag.

Kushchu, I. (2002). Genetic programming and evolutionary generalization. *IEEE Transactions on Evolutionary Computation, 6*, 431–442. doi:10.1109/TEVC.2002.805038

Kusnezov, D., Bulgac, A., & Dang, G. D. (1999). Quantum levy processes and fractional kinetics. *Physical Review Letters, 82*, 1136–1139. doi:10.1103/PhysRevLett.82.1136

Kuznetsov, Y. A. (1995). *Elements of Applied Bifurcation Theory*. New York: Springer-Verlag.

Kye, W. H., Choi, M., Kim, M. W., Lee, S. Y., Rim, S., Kim, C. M., & Park, Y. J. (2004a). Synchronization of delayed systems in the presence of delay time modulation. *Physics Letters. [Part A], 322*, 338–343. doi:10.1016/j.physleta.2004.01.046

Kye, W. H., Choi, M., Rim, S., Kurdoglyan, M. S., Kim, C. M., & Park, Y. J. (2004b). Characteristics of a delayed system with time-dependent delay time. *Phys. Rev. E*, 69, 055202(R)1-4.

Lakshmanan, M., & Murali, K. (1996). *Chaos in nonlinear oscillators: controlling and synchronization*. Singapore: World Scientific.

Larger, L., Goedgebuer, J. P., & Delorme, F. (1998a). Optical encryption system using hyperchaos generated by an optoelectronic wavelength oscillator. *Physical Review E: Statistical Physics, Plasmas, Fluids, and Related Interdisciplinary Topics, 57*, 6618–6624. doi:10.1103/PhysRevE.57.6618

Larger, L., Goedgebuer, J. P., & Merolla, J. M. (1998b). Chaotic Oscillator in Wavelength: A New Setup for Investigating Differential Difference Equations Describing Nonlinear Dynamics. *IEEE Journal of Quantum Electronics, 34*, 594–601. doi:10.1109/3.663432

Lee, H., & Tomizuka, M. (2001). Robust Adaptive Control using a Universal Approximator for SISO Nonlinear Systems. *IEEE Transactions on Fuzzy Systems*, *8*, 95–106.

Lee, M. W., Larger, L., Udaltsov, V., Génin, E., & Goedgebuer, J. P. (2004). Demonstration of chaos generator with two delays. *Optics Letters*, *29*, 325–327. doi:10.1364/OL.29.000325

Lee, M. W., Rees, P., Shore, K. A., Ortín, S., Pesquera, L., & Valle, A. (2005). Dynamical characterization of laser diode subject to double optical feedback for chaotic optical communications. *IEE Proceedings. Optoelectronics*, *152*, 97–102. doi:10.1049/ip-opt:20045025

Lee, C. C. (1990). Fuzzy logic in control system: Fuzzy logic controller- Parts I, II. *IEEE Trans. Syst., Man, Cybern 20, 20*, 404-435.

Lei, Y., Xu, W., & Zheng, H. (2005). Synchronization of two chaotic nonlinear gyros using active control. *Physics Letters. [Part A]*, *343*, 153–158. doi:10.1016/j.physleta.2005.06.020

Leipnik, R. B., & Newton, T. A. (1981). Double strange attractors in rigid body motion. *Physics Letters. [Part A]*, *86*, 63. doi:10.1016/0375-9601(81)90165-1

Leu, Y. G., Lee, T. T., & Wang, W. Y. (1999). Observer-based adaptive fuzzy-neural control for unknown nonlinear dynamical systems. *IEEE Transactions on Systems, Man, and Cybernetics*, *29*, 583–591. doi:10.1109/3477.790441

Leung, H. K. (1999, June). Dissipative effects on synchronization of nonchaotic oscillators. *The Chinese Journal of Physiology*, *37*(3).

Levant, A. (1993). Sliding order and sliding accuracy in sliding mode control. *International Journal of Control*, *58*, 1247–1263. doi:10.1080/00207179308923053

Lewis, F. L., Yesildirek, A., & Liu, K. (1996). Multilayer neural-net robot controller with guaranteed tracking performance. *IEEE Transactions on Neural Networks*, *7*(2), 1–11. doi:10.1109/72.485674

Li, C., & Yan, J. (2006). Generalized projective synchronization of chaos: The cascade synchronization approach. *Chaos, Solitons, and Fractals*, *30*, 140–146. doi:10.1016/j.chaos.2005.08.155

Li, Y., Chen, L., Cai, Z., & Zhao, X. (2003). Study on chaos synchronization in the Belousov-Zhabotinsky chemical system. *Chaos, Solitons, and Fractals*, *17*, 699–707. doi:10.1016/S0960-0779(02)00486-1

Li, Z. G., Wen, C. Y., Soh, Y. C., & Xie, W. X. (2001). The stabilization and synchronization of Chua's oscillators via impulsive control. *IEEE Transactions on Circuits and Systems. I, Fundamental Theory and Applications*, *48*(11), 1351–1355. doi:10.1109/81.964427

Li, X., Xu, W., & Xiao, Y. (2008). Adaptive tracking control of a class of uncertain chaotic systems in the presence of random perturbations. *Journal of Sound and Vibration*, *314*, 526–535. doi:10.1016/j.jsv.2008.01.035

Li, Z., Chen, G., Shi, S., & Han, C. (2003). Robust adaptive tracking control for a class of uncertain chaotic systems. *Physics Letters. [Part A]*, *310*, 40–43. doi:10.1016/S0375-9601(03)00115-4

Li, P., Cao, J. D., & Wang, Z. D. (2007). Robust impulsive synchronization of coupled delayed neural networks with uncertainties. *Physica A*, *373*, 261–272. doi:10.1016/j.physa.2006.05.029

Li, Z., & Chen, G. (2006). Global synchronization and asymptotic stability of complex dynamical networks. *IEEE Transactions on Circuits and Systems*, *53*, 28–33. doi:10.1109/TCSII.2005.854315

Li, H. X., & Tong, S. (2003). A Hybrid Adaptive Fuzzy Control for A Class of Nonlinear MIMO Systems. *IEEE Transactions on Fuzzy Systems*, *11*(1), 24–34. doi:10.1109/TFUZZ.2002.806314

Li, T. Y., & Yorke, J. A. (1975). Period three implies chaos. *The American Mathematical Monthly*, *82*, 985–992. doi:10.2307/2318254

Li, S., Chen, G., Wong, K. W., Mou, X., & Cai, Y. (2004). Baptista-type chaotic cryptosystems: Problems and countermeasures. *Physics Letters. [Part A], 332*, 368–375. doi:10.1016/j.physleta.2004.09.028

Li, S., Mou, X., & Cai, Y. (2001). Improving security of a chaotic encryption approach. *Physics Letters. [Part A], 290*, 127–133. doi:10.1016/S0375-9601(01)00612-0

Li, S., Mou, X., Ji, Z., Zhang, J., & Cai, Y. (2003). Performance analysis of Jakimoski–Kocarev attack on a class of chaotic cryptosystems. *Physics Letters. [Part A], 307*, 22. doi:10.1016/S0375-9601(02)01659-6

Li, P., Li, Z., Fetinger, S., Mao, Y., & Halang, W. A. (2007). Application of Chaos – based Pseudo – Random – Bit Generators in Internet-based Online Payments. [SCI]. *Studies in Computational Intelligence, 37*, 667–685. doi:10.1007/978-3-540-37017-8_31

Li, B. X., & Haykin, S. (1995). A new pseudo-noise generator for spread spectrum communications. In *ICASSP-95, Int. Conf. Acoustics, Speech, and Signal Process.* (Vol. 5, pp. 3603–3606).

Li, S. J., Alvarez, G., Li, Z., & Halang, W. A. (2007). Analog chaos-based secure communications and cryptanalysis: a brief survey. In *Proceedings of 3rd International IEEE Scientific Conference on Physics and Control*, Potsdam, Germany (pp. 92-98).

Lian, K. Y., & Liu, P. (2000). Synchronization with message embedded for generalized lorenz chaotic circuits and its error analysis. *IEEE Transactions on Circuits and Systems. I, Fundamental Theory and Applications, 47*(9), 1418–1424. doi:10.1109/81.883341

Liang, J., & Cao, J. (2004). Global asymptotic stability of bi-directional associative memory networks with distributed delays. *Applied Mathematics and Computation, 152*, 415–424. doi:10.1016/S0096-3003(03)00567-8

Liang, J., & Cao, J. (2006). A based-on LMI stability criterion for delayed recurrent neural networks. *Chaos, Solitons, and Fractals, 28*(1), 154–160. doi:10.1016/j.chaos.2005.04.120

Liang, J., & Cao, J. (2007). Global output convergence of recurrent neural networks with distributed delays. *Nonlinear Analysis Real World Applications, 8*(1), 187–197. doi:10.1016/j.nonrwa.2005.06.009

Liang, J., Wang, Z., & Liu, X. (2008). Exponential synchronization of stochastic delayed discrete-time complex networks. *Nonlinear Dynamics, 53*, 153–165. doi:10.1007/s11071-007-9303-5

Liang, X., Zhang, J., & Xia, X. (2008, July). Improving the Security of Chaotic Synchronization With a Δ-Modulated Cryptographic Technique. *IEEE Transactions on Circuits and Wystems. II, Express Briefs, 55*(7), 680–684. doi:10.1109/TCSII.2008.921585

Liao, T., & Tsai, S. (2000). Adaptive synchronization of chaotic systems and its application to secure communications. *Chaos, Solitons, and Fractals, 11*(9), 1387–1396. doi:10.1016/S0960-0779(99)00051-X

Liao, T. L., & Lin, S. H. (1999). Adaptive control and synchronization of Lorenz systems. *Journal of the Franklin Institute, 336*, 925–937. doi:10.1016/S0016-0032(99)00010-1

Liao, T.-L. (1998). Adaptive synchronization of two Lorenz systems. *Chaos, Solitons, and Fractals,9*(9), 1555–1561. doi:10.1016/S0960-0779(97)00161-6

Liao, T.-L., & Huang, N.-S. (1999). An observer-based approach for chaotic synchronization with applications to secure communications. *IEEE Transactions on Circuits and Systems. I, Fundamental Theory and Applications, 46*(9), 1144–1150. doi:10.1109/81.788817

Liao, T.-L., & Tsai, S.-H. (2000). Adaptive synchronization of chaotic systems and its application to secure communications. *Chaos, Solitons, and Fractals,11*(9), 2387–2396. doi:10.1016/S0960-0779(99)00051-X

Liao, X., & Chen, G. (2003). Chaos synchronization of general Luré systems via time-delay feedback control . *International Journal of Bifurcation and Chaos in Applied Sciences and Engineering, 13*(1), 207–213. doi:10.1142/S0218127403006455

Liao, T. L., & Tsai, S. H. (2008). Adaptive synchronization of chaotic systems and its application to secure communication. *Chaos, Solitons, and Fractals, 11*(9), 1387–1396. doi:10.1016/S0960-0779(99)00051-X

Lichtenberg, A. J., & Lieberman, M. A. (1992). *Regular and Chaotic Dynamics* (2nd ed.). New York: Springer-Verlag.

Lin, F. J., Hwang, W. J., & Wai, R. J. (1998). *Ultrasonic motor servo drive with on-line trained neural network model-following controller* (*Vol. 145*, pp. 105–110). Proc. Inst. Elect. Eng. Electr.

Lin, T. C., Kuo, M. J., & Hsu, C. H. (in press). Robust Adaptive Tracking Control of Multivariable Nonlinear Systems Based on Interval Type-2 Fuzzy approach. *International Journal of Innovative Computing . Information and Control.*

Lin, T. C., Liu, H. L., & Kuo, M. J. (2009). Direct adaptive interval type-2 fuzzy control of multivariable nonlinear systems. *Engineering Applications of Artificial . Intelligence, 22*, 420–430.

Lin, S. C., & Chen, Y. Y. (1994). *Design of adaptive fuzzy sliding mode for nonlinear system control.* Paper presented at the Int. Conf. Fuzzy Syst., Orlando, FL.

Lin, T. C. (2009). Analog Circuit Fault Diagnosis under Parameter Variations Based on Type-2 Fuzzy Logic Systems. *International Journal of Innovative Computing, Information and Control., accepted to be published.*

Ling, L., Xiaogang, W., & Hanping, H. (2004). Estimating system parameters of Chua's circuit from synchronizing signal. *Physics Letters. [Part A], 324*, 36–41. doi:10.1016/j.physleta.2004.02.047

Liu, Z., Lai, Y.-C., & Matías, M. A. (2003). Universal scaling of Lyapunov exponents in coupled chaotic oscillators. *Physical Review E: Statistical, Nonlinear, and Soft Matter Physics, 67*, 045203. doi:10.1103/PhysRevE.67.045203

Liu, X., & Chen, T. (2008). Boundedness and synchronization of y-coupled Lorenz systems with or without controllers. *Physica D. Nonlinear Phenomena, 237*, 630–639. doi:10.1016/j.physd.2007.10.006

Liu, F., Ren, Y., Shan, X., & Qiu, Z. (2002). A linear feedback synchronization theorem for a class of chaotic systems. *Chaos, Solitons, and Fractals,13*(4), 723–730. doi:10.1016/S0960-0779(01)00011-X

Liu, Y., & Davids, P. (2000). Dual synchronization of chaos . *Physical Review E: Statistical Physics, Plasmas, Fluids, and Related Interdisciplinary Topics, 61*, 2176–2179. doi:10.1103/PhysRevE.61.R2176

Liu, Y., Wang, Z., & Liu, X. (2007). Design of exponential state estimators for neural networks with mixed time delays. *Physics Letters. [Part A], 364*(5), 401–412. doi:10.1016/j.physleta.2006.12.018

Liu, Y., Wang, Z., & Liu, X. (2006). Global asymptotic stability of generalized bi-directional associative memory networks with discrete and distributed delays. *Chaos, Solitons, and Fractals, 28*, 793–803. doi:10.1016/j.chaos.2005.08.004

Ljung, L., & Glad, T. (1994). On global identifiability for arbitrary model parametrizations. *Automatica, 30*, 265–276. doi:10.1016/0005-1098(94)90029-9

Ljung, L. (1996 December). *Pac-learning and asymptotic system identification theory*. Presented at the 35th Conference on Decision and Control, Kobe, Japan.

Llinás, R. R. (1988). The intrinsic electrophysiological properties of mammalian neurons: insights into central nervous system function. *Science, 242*, 1654. doi:10.1126/science.3059497

Locquet, A., Ortín, S., Udaltsov, V., Larger, L., Citrin, D. S., Pesquera, L., & Valle, A. (2006). Delay-time identification in chaotic optical systems with two delays. In D. Lenstra, M. Pessa, I. H. White (Eds.) *SPIE Proceedings: Vol. 6184. Semiconductor Lasers and Laser Dynamics II.* (pp. 173-184). Bellingham, WA: SPIE.

Lonngren, K. E. (1991). Notes to accompany a student laboratory experiment on chaos. *IEEE Transactions on Education, 34*, 123–128. doi:10.1109/13.79892

Lorenz, E. N. (1963). Deterministic nonperiodic flow. *Journal of the Atmospheric Sciences, 20*, 130–141. doi:10.1175/1520-0469(1963)020<0130:DNF>2.0.CO;2

Lorenzo, M. N., Perez-Munuzuri, V., & Perez-Villar, V. (2000). Noise Performance of a synchronization scheme through compound chaotic signal. *International Journal of Bifurcation and Chaos in Applied Sciences and Engineering, 10*, 2863–2870. doi:10.1142/S0218127400001912

Lou, X. Y., & Cui, B. (2006). New LMI conditions for delay-dependent asymptotic stability of delayed Hopfield neural networks. *Neurocomputing, 69*(16–18), 2374–2378. doi:10.1016/j.neucom.2006.02.019

Lou, X., & Cui, B. (2008). Synchronization of neural networks based on parameter identification and via output or state coupling. *Journal of Computational and Applied Mathematics, 222*(2), 440–457. doi:10.1016/j.cam.2007.11.015

Lu, J., & Cao, J. (2005). Adaptive complete synchronization of two identical or different chaotic (hyperchaotic) systems with fully unknown parameters. *Chaos (Woodbury, N.Y.), 15*(4), 043901. doi:10.1063/1.2089207

Lu, H., & van Leeuwen, C. (2006). Synchronization of chaotic neural networks via output or state coupling. *Chaos, Solitons, and Fractals, 30*, 166–176. doi:10.1016/j.chaos.2005.08.175

Lu, J., & Cao, J. (2007). Synchronization-based approach for parameters identification in delayed chaotic neural networks. *Physica A, 384*, 432–443.

Lu, W., & Chen, T. (2004). Synchronization analysis of linearly coupled networks of discrete time systems. *Physica D. Nonlinear Phenomena, 198*, 148–168. doi:10.1016/j.physd.2004.08.024

Lü, J., Chen, G., Cheng, D., & Celikovsky, S. (2002). Bridge the gap between the Lorenz and the Chen system. *International Journal of Bifurcation and Chaos in Applied Sciences and Engineering, 12*, 2917–2926. doi:10.1142/S021812740200631X

Luzyanina, T., Engelborghs, K., & Roose, D. (2001). Numerical bifurcation analysis of differential equations with state-dependent delay. *International Journal of Bifurcation and Chaos in Applied Sciences and Engineering, 11*(3), 737–753. doi:10.1142/S0218127401002407

Mackey, M., & Glass, L. (1977). Oscillation and chaos in physiological control systems. *Science, 197*, 287–289. doi:10.1126/science.267326

Madan, R. (1993). *Chua's Circuit: A Paradigm for Chaos*. Singapore: World Scientific.

Maineri, R., & Rehacek, J. (1999). Projective synchronization in three-dimensional chaotic systems. *Physical Review Letters, 64*, 3042–3045. doi:10.1103/PhysRevLett.82.3042

Mané, R. (1987). *Ergodic theory and differentiable dynamics*. Berlin: Springer-Verlag.

Mao, X. (2002). A note on the LaSalle-type theorems for stochastic differential delay equations. *Journal of Mathematical Analysis and Applications, 268*, 125–142. doi:10.1006/jmaa.2001.7803

Marino, I. P., Allaria, E., Meucci, R., Boccaletti, S., & Arecchi, F. T. (2003). Information encoding in homoclinic chaotic systems. *Chaos (Woodbury, N.Y.), 13*(1), 286–290. doi:10.1063/1.1489115

Marino, I. P., Lopez, L., Miguez, J., & Sanjuan, M. A. F. (2002). A novel channel coding scheme based on continuous-time chaotic dynamics. In *14th International Conference on Digital Signal Processing* (pp. 1321–1324).

Martell, L. (2000). *Wavelet-based data reduction and de-noising procedures*. Unpublished doctoral dissertation, North Carolina State University, Raleigh, USA.

Martínez-Guerra, R., Aguilar, R., & Poznyak, A. (2004). A new robust sliding-mode observer design for monitoring in chemical reactors. *Journal of Dynamic System, Measurement and Control . ASME Journal, 126*(3), 473–478.

Martínez-Guerra, R., Cruz-Victoria, J. C., Gonzalez-Galan, R., & Aguilar-López, R. (2006). A new reduced-order Observer design for the Synchronization of Lorenz Systems. *Chaos, Solitons, and Fractals, 28*(12), 511–517. doi:10.1016/j.chaos.2005.07.011

Martínez-Guerra, R., & Wen, Y. (2008). Chaotic synchronization and secure communication via sliding-mode observer. *International Journal of Bifurcation and Chaos in Applied Sciences and Engineering, 18*(1), 235–243. doi:10.1142/S0218127408020264

Masoller, C. (2001). Anticipation in the Synchronization of Chaotic Semiconductor Lasers with Optical Feedback. *Physical Review Letters, 86*(13), 2782. doi:10.1103/PhysRevLett.86.2782

Massey, J. L. (1992). Contemporary cryptology: an introduction . In Simmons, G. J. (Ed.), *Contemporary Cryptology*. New York: IEEE Press.

Matías, M. A., Güémez, J., & Martín, C. (1997). On the behavior of coupled chaotic systems exhibiting marginal synchronization. *Physics Letters. [Part A], 226*, 264–268. doi:10.1016/S0375-9601(96)00946-2

Matouk, A. E. (2008). Dynamical analysis feedback control and synchronization of Liu dynamical system. *Nonlinear Analysis: TMA, 69*, 3213–3224. doi:10.1016/j.na.2007.09.029

Matouk, A. E. (2009a). Stability conditions, hyperchaos and control in a novel fractional order hyperchaotic system. *Physics Letters. [Part A], 373*, 2166–2173. doi:10.1016/j.physleta.2009.04.032

Matouk, A. E. (2009b). *Chaos synchronization between two different fractional systems of Lorenz family*. Mathematical Problems in Engineering.

Matouk, A. E. (in press). 2010). Chaos, feedback control and synchronization of a fractional-order modified Autonomous Van der Pol-Duffing circuit. *Communications in Nonlinear Science and Numerical Simulation*. doi:. doi:10.1016/j.cnsns.2010.04.027

Matouk, A. E., & Agiza, H. N. (2008). Bifurcations, chaos and synchronization in ADVP circuit with parallel resistor. *Journal of Mathematical Analysis and Applications, 341*, 259–269. doi:10.1016/j.jmaa.2007.09.067

Matouk, A. E. (2005). *Chaos and synchronization in some nonlinear electronic circuits*. Unpublished master thesis, University of Mansoura, Mansoura.

Matsumoto, T., Chua, L. O., & Komuro, M. (1985). The Double Scroll. *IEEE Transactions on Circuits and Systems, CAS-32*, 798–818.

Matthews, M. R., Stinner, A., & Gauld, C. F. (2005). *The Pendulum: Scientific, Historical, Philosophical and Educational Perspectives*. New York: Springer.

Matthews, R. (1989). On the derivation of a chaotic encryption algorithm. *Cryptologia, 13*, 29–41. doi:10.1080/0161-118991863745

Mei, S., Chang-Yan, Z., & Li-Xin, T. (2009). Generalized Projective Synchronization between Two Complex Networks with Time-Varying Coupling Delay. *Chinese Physics Letters, 26*, 010501. doi:10.1088/0256-307X/26/1/010501

Mendel, J. M. (2004). Computing Derivatives in Interval Type-2 Fuzzy logic Systems. *IEEE Transactions on Fuzzy Systems, 12*(1), 84–98. doi:10.1109/TFUZZ.2003.822681

Mendel, J. M. (2007). Type-2 fuzzy sets and systems: an overview. *Computational Intelligence Magazine, IEEE, 2*(1), 20–29. doi:10.1109/MCI.2007.380672

Mendel, J. M., John, R. I., & Liu, F. (2006). Interval Type-2 Fuzzy Logic Systems Made Simple. *IEEE Transactions on Fuzzy Systems, 14*(6), 808–821. doi:10.1109/TFUZZ.2006.879986

Mendel, J. M., & John, R. I. B. (2000). Type-2 fuzzy sets made simple. *IEEE Transactions on Fuzzy Systems, 10*, 117–127. doi:10.1109/91.995115

Mendoza-Camargo, J., Aguilar-Ibañez, C., Martínez-Guerra, R., & Garrido, R. (2004). On the parameters identification of the Duffing's System by means of a reduced order observer. *Physics Letters. [Part A], 331*(5), 316–324. doi:10.1016/j.physleta.2004.09.005

Menezes, A. J., Oorschot, P. V., & Vanstone, S. (1997). *Handbook of Applied Cryptography*. Boca Raton, FL: CRC Press.

Mensour, B., & Longtin, A. (1998). Synchronization of delay-differential equations with application to private communication. *Physics Letters A, 244*(1-3), 59-70.

Meyer, Y. (1992). *Wavelets and Applications*. Berlin: Springer-Verlag.

Mianovic, V., & Zaghloul, M. E. (1996). Improved masking algorithm chaotic communication systems. *Electronics Letters, 1*, 11–12. doi:10.1049/el:19960004

Midavaine, T., Dangoisse, D., & Glorieux, P. (1985). Observation of Chaos in a Frequency-Modulated CO_2 Laser. *Physical Review Letters, 55*, 1989–1992. doi:10.1103/PhysRevLett.55.1989

Millérioux, G., & Daafouz, J. (2003). An observer-based approach for input independent global chaos synchronization of discrete-time switched systems. *IEEE Transactions on Circuits and Systems. I, Fundamental Theory and Applications, 50*(10), 1270–1279. doi:10.1109/TCSI.2003.816301

Millérioux, G., & Daafouz, J. (2004, July). Input independent chaos synchronization of switched systems. *IEEE Transactions on Automatic Control, 49*(7), 1182–1187. doi:10.1109/TAC.2004.831118

Millérioux, G. (1997). Chaotic synchronization conditions based on control theory for systems described by discrete piecewise linear maps. *International Journal of Bifurcation and Chaos in Applied Sciences and Engineering, 7*(7), 1635–1649. doi:10.1142/S0218127497001266

Millérioux, G., & Daafouz, J. (2004, April). Unknown input observers for message-embedded chaos synchronization of discrete-time systems. *International Journal of Bifurcation and Chaos in Applied Sciences and Engineering, 14*(4), 1357–1368. doi:10.1142/S0218127404009831

Millérioux, G., & Daafouz, J. (2006). *Chaos in AutomatiControl-theoretical concepts in the design of symmetric cryptosystems c Control*. Boca Raton, FL: CRC Press.

Millérioux, G., & Daafouz, J. (2009). *Flatness of switched linear discrete-time systems*. IEEE Trans. on Automatic Control.

Millérioux, G., & Mira, C. (1998). Coding scheme based on chaos synchronization from noninvertible maps. *International Journal of Bifurcation and Chaos in Applied Sciences and Engineering, 8*(10), 2019–2029. doi:10.1142/S0218127498001674

Millérioux, G., Amigó, J. M., & Daafouz, J. (2008 July). A connection between chaotic and conventional cryptography. *IEEE Trans. on Circuits and Systems I: Regular Papers*, 55(6).

Millérioux, G., Bloch, G., Amigo, J. M., Bastos, A., & Anstett, F. (2003). Real-time video communication secured by a chaotic key stream cipher. In *Proc. of IEEE 16th European Conference on Circuits Theory and Design, ECCTD'03*, Krakow, Poland, September 1-4 (pp. 245–248).

Millérioux, G., Hernandez, A., & Amigo, J. M. (2005 October). Conventional cryptography and message-embedding. In *Proc. of International Symposium on Nonlinear Theory and its Applications, NOLTA'2005*, Bruges.

Mirasso, C. R., & Colet, P., & García-Fernández, P. (1996). Synchronization of chaotic semiconductor lasers: Application to encoded communications. *IEEE Photonics Technology Letters, 8*(2), 299–301. doi:10.1109/68.484273

Mitchell, T. (1996). *Machine Learning*. New York: McGraw-Hill.

Morgül, O., & Solak, E. (1996). Observed based synchronization of chaotic systems. *Physical Review E: Statistical Physics, Plasmas, Fluids, and Related Interdisciplinary Topics, 54*(5), 4803–4811. doi:10.1103/PhysRevE.54.4803

Mou, S., Gao, H., Lam, J., & Qiang, W. (2008a). A new criterion of delay-dependent asymptotic stability for Hopfield neural networks with time delay. *IEEE Transactions on Neural Networks, 19*(3), 532–535. doi:10.1109/TNN.2007.912593

Mou, S., Gao, H., Qiang, W., & Chen, K. (2008b). New delay-dependent exponential stability for neural networks with time delay. *IEEE Trans. Systems, Man and Cybernetics-Part B . Cybernetics, 38*(2), 571–576.

Moukam, K. F. M., Bowong, S., Tchawoua, C., & Kaptouom, E. (2004). Chaos control and synchronization of a Φ^6-Van der Pol oscillator. *Physics Letters. [Part A], 322*, 305–323. doi:10.1016/j.physleta.2004.01.016

Mukai, T., & Otsuka, K. (1985). New route to optical chaos: Successive-subharmonic-oscil- lation cascade in a semiconductor laser coupled to an external cavity. *Physical Review Letters, 55*, 1711–1714. doi:10.1103/PhysRevLett.55.1711

Murali, M., & Lakshmanan, M. (1994). Drive-response scenario of chaos synchronization in identical nonlinear systems. *Physical Review E: Statistical Physics, Plasmas, Fluids, and Related Interdisciplinary Topics, 49*, 4882–4885. doi:10.1103/PhysRevE.49.4882

Muraly, K. (2000). Synchronization based signal transmission with heterogeneous chaotic systems. *International Journal of Bifurcation and Chaos in Applied Sciences and Engineering, 10*(11), 2489–2497. doi:10.1142/S0218127400001729

Mycolaitis, G., Tamaševičious, A., Čenys, A., Namajunas, A., Navionis, K., & Anagnostopoulos, A. N. (1999). Globally synchronizable non-autonomous chaotic oscillator. In *Proc. of 7th International Workshop on Nonlinear Dynamics of Electronic Systems,* Denmark (pp. 277-280).

Mykolaitis, G., Tamaševičius, A., & Bumelienė, S. (2004). Experimental demonstration of chaos from the Colpitts oscillator in the VHF and the UHF ranges. *Electronics Letters, 40*, 91–92. doi:10.1049/el:20040074

Naghavi, S. V., & Safavi, A. A. (2008). Novel synchronization of discrete-time chaotic systems using neural network observer. *Chaos (Woodbury, N.Y.), 18*, 033110. doi:10.1063/1.2959140

Newcomb, R. W., & Sathyan, S. (1983). RC Op Amp chaos generator. *IEEE Transactions on Circuits and Systems, CAS-30*(1), 54–56. doi:10.1109/TCS.1983.1085277

Newcomb, R. W., & El-Leithy, N. (1986). Chaos using hysteretic circuits. In *Proceedings - IEEE International Symposium on Circuits and Systems* (pp..56-61). San Jose, CA: IEEE.

Nguang, S. K., & Shi, P. (2003). H1 fuzzy output feedback control design for nonlinear systems: an LMI approach. *IEEE Transactions on Fuzzy Systems, 11*(3), 331–340. doi:10.1109/TFUZZ.2003.812691

Nijmeijer, H. (1997). Special Issue. Control of chaos and synchronization. *Systems & Control Letters, 31*, 259–322. doi:10.1016/S0167-6911(97)00042-X

Nijmeijer, H., & Mareels, I. M. Y. (1997, October). An observer looks at synchronization. *IEEE Transactions on Circuits and Systems. I, Fundamental Theory and Applications, 44*, 882–890. doi:10.1109/81.633877

Nijmeijer, H., & Berghuis, H. (1995). On Lyapunov control of the Duffing equation. *IEEE Transactions on Circuits and Systems, 42*, 473–477. doi:10.1109/81.404059

Nijmeijer, H., & Mareels, I. M. Y. (1997b). An observer looks at synchronization. *IEEE Transactions on Circuits and Systems. I, Fundamental Theory and Applications, 44*(10), 882–890. doi:10.1109/81.633877

Nijmeijer, H., & Van der Shaft, A. (1990). *Nonlinear dynamical control systems*. New York: Springer-Verlag.

Ning, D., Lu, J., & Han, X. (2007). Dual synchronization based on two different chaotic systems: Lorenz systems and Rossler systems. *Computational & Applied Mathematics, 206*(2), 1046–1050. doi:10.1016/j.cam.2006.09.007

Noroozi, N., Roopaei, M., & Zolghadri Jahromi, M. (2009). Adaptive fuzzy sliding mode control scheme for uncertain systems. *Communications in Nonlinear Science and Numerical Simulation, 14*(11), 3978-3992.

Noroozi, N., Roopaei, M., Balas, V. E., & Lin, T. C. (2009). *Observer-based adaptive variable structure control and synchronization of unknown chaotic systems.* Paper presented at the SACI 2009 5th International Symposium on Applied Computational Intelligence and Informatics.

Noroozi, N., Roopaei, M., Karimaghaee, P., & Safavi, A. A. (2010). Simple adaptive variable structure control for unknown chaotic systems *Communications in Nonlinear Science and Numerical Simulation, 15*(3), 707-727.

Ogorzalek, M. J. (1993). Taming chaos - part I: Synchronization. *IEEE Transactions on Circuits and Systems. I, Fundamental Theory and Applications*, *40*(10), 693–699. doi:10.1109/81.246145

Ojalvo, J. G., & Roy, R. (2001). Spatiotemporal communication with synchronized optical chaos. *Physical Review Letters*, *86*(22), 5204–5207. doi:10.1103/PhysRevLett.86.5204

Omidvar, O., & Elliott, D. L. (1997). *Neural Systems for Control*. New York: Academic Publishers.

Ortín, S., Gutierrez, J. M., Pesquera, L., & Vasquez, H. (2005). Nonlinear dynamics extraction using modular neural network: synchronization and prediction. *Physica A*, *351*, 133–141. doi:10.1016/j.physa.2004.12.015

Ortín, S., Jacquot, M., Pesquera, L., Peil, M., & Larger, L. (2007b). Security analysis of chaotic communication systems based on delayed optoelectronic feedback . In Zubia, J., Aldabaldetreku, G., Durana, G., Arrue, J., & Jiménez, F. (Eds.), *Proceedings 5ª Reunión Española de Optoelectrónica* (pp. 429–434). Bilbao, Spain: Escuela Técnica Superior de Ingeniería.

Ortín, S., & Pesquera, L. (2006). Cracking Chaos-based Encryption Systems Based on Laser Diodes with Optoelectronic Feedback with Two Delays. In D. Lenstra, J. M. Dudley, J. Rarity, C. R. Mirasso (Eds.), *Proceedings ECOC: Vol. 6. CLEO Focus meeting* (Tu3_1_2, pp. 15-16). Cannes, France: European Conference on Optical Communications.

Ortín, S., Jacquot, M., Pesquera, L., Peil, M., & Larger, L. (2007a). Nonlinear dynamics reconstruction of chaotic cryptosystems based on delayed optoelectronic feedback. In CLEO-E/IQEC (Ed.), *Proceedings CLEO-E/IQEC*(JSI1-3-THU). Munich, Germany: CLEO-E/IQEC.

Ortín, S., Jacquot, M., Pesquera, L., Peil, M., & Larger, L. (2008). Time delay extraction in chaotic cryptosystems based on optoelectronic feedback with variable delay. In K. P. Panajotov, M. Sciamanna, A. A. Valle, R. Michalzik. (Eds.) *SPIE Proceedings: Vol. 6997. Semiconductor Lasers and Laser Dynamics III*. (pp. 69970E-1-12). Bellingham, WA: SPIE.

Ortín, S., Pesquera, L., Gutiérrez, J. M., Valle, A., & Cofiño, A. (2007c). Nonlinear dynamics reconstruction with neural networks of chaotic communication time-delay systems. In J. Marro, P. L. Garrido, J. J. Torres (Eds.), *AIP Proceedings: Vol. 887. Cooperative Behavior in Neural Systems.* (pp. 249-258). New York: American Institute of Physics.

Ott, E., Grebogi, C., & York, J. A. (1990). Controlling chaos. *Physical Review Letters*, *64*(11), 1196–1199. doi:10.1103/PhysRevLett.64.1196

Palacios, A., & Juarez, H. (2002). Crytography with cycling chaos. *Physics Letters. [Part A]*, *303*, 345–351. doi:10.1016/S0375-9601(02)01323-3

Palaniyandi, P., & Lakshmanan, M. (2001). Secure digital signal transmission by multistep parameter modumation and alternative driving of transmitter variables. *International Journal of Bifurcation and Chaos in Applied Sciences and Engineering*, *11*(7), 2031–2036. doi:10.1142/S021812740100319X

Palm, R. (1992). *Sliding mode fuzzy control.* Paper presented at the Int. Conf. Fuzzy Syst., San Diego, CA.

Pareek, V., Patidar, N. K., & Sud, K. K. (2003). Discrete chaotic cryptography using external key. *Physics Letters. [Part A]*, *309*, 75–82. doi:10.1016/S0375-9601(03)00122-1

Park, J. H. (2005). Adaptive synchronization of hyperchaotic Chen system with uncertain parameters. *Chaos, Solitons, and Fractals*, *26*(3), 959–964. doi:10.1016/j.chaos.2005.02.002

Park, J. H. (2006). Chaos synchronization between two different chaotic dynamical systems. *Chaos, Solitons, and Fractals*, *27*(2), 549–554. doi:10.1016/j.chaos.2005.03.049

Park, J. H. (2006). Synchronization of Genesio chaotic system via backstepping approach. *Chaos, Solitons, and Fractals*, *27*, 1369–1375. doi:10.1016/j.chaos.2005.05.001

Park, P. (1999). A delay-dependent stability criterion for systems with uncertain time-invariant delays. *IEEE Transactions on Automatic Control, 44*, 876–877. doi:10.1109/9.754838

Parker, A. T., & Short, K. M. (2001). Reconstructing the keystream from a chaotic encryption Scheme. *IEEE Trans on Circuits and Syst- I, 48*(5), 624-630.

Parlitz, U., Chua, L., Kocarev, L., Halle, K., & Shang, A. (1992). Transmission of digital signals by chaotic synchronization. *International Journal of Bifurcation and Chaos in Applied Sciences and Engineering, 2*(4), 973–977. doi:10.1142/S0218127492000562

Parmananda, P. (1998). Recursive proportional-feedback and its use to control chaos in an electrochemical system. *Physics Letters. [Part A]*, 240–255.

Peacock, T., & Mullin, T. (2001). Homoclinic bifurcations in a liquid crystal flow. *Journal of Fluid Mechanics, 432*, 369–386.

Pecora, L. M., & Carroll, T. L. (1990). Synchronization in chaotic systems. *Physical Review Letters, 64*, 821–824. doi:10.1103/PhysRevLett.64.821

Pecora, L. M., Carroll, T. L., Johnson, G., & Mar, D. (1997, November). Volume-preserving and volume-expanding synchronized chaotic systems. *Physical Review E: Statistical Physics, Plasmas, Fluids, and Related Interdisciplinary Topics, 56*(5), 5090–5100. doi:10.1103/PhysRevE.56.5090

Pecora, L. M., Carroll, T. L., Johnson, G. A., Mar, D. J., & Heagy, J. F. (1997). Fundamentals of synchronization in chaotic systems, concepts, and applications. *Chaos (Woodbury, N.Y.), 7*, 520–543. doi:10.1063/1.166278

Pecora, L. M., & Carroll, T. L. (1991). Driving systems with chaotic signals. *Physical Review A., 44*, 2374–2383. doi:10.1103/PhysRevA.44.2374

Pecora, L. M., & Caroll, T. L. (1998). Master stability functions for synchronized coupled systems. *Physical Review Letters*, 2109–2112. doi:10.1103/PhysRevLett.80.2109

Pecora, L. (1996). Hyperchaos harnessed. *Physics World, 9*(51), 17.

Peng, G. (2007). Synchronization of fractional order chaotic systems. *Physics Letters. [Part A], 363*, 426–432. doi:10.1016/j.physleta.2006.11.053

Perez, G., & Cerdeira, H. (1995). Extracting messages masked by chaos. *Physical Review Letters, 74*, 1970–1073. doi:10.1103/PhysRevLett.74.1970

Pesin Ya, B. (1977). Lyapunov characteristic exponents and smooth ergodic theory. *Russian Mathematical Surveys, 32*(4), 55–114. doi:10.1070/RM1977v032n04ABEH001639

Petrov, V., Scott, S. K., & Showalter, K. (1992). Mixed-mode oscillations in chemical systems. *The Journal of Chemical Physics, 97*(9), 6191–6198. doi:10.1063/1.463727

Pieprzyk, J., Hardjono, T., & Seberry, J. (2003). *Fundamentals of Computer Security*. Berlin: Springer Verlag.

Pikovsky, A., Rosemblum, M., & Kurths, J. (2001). *Synchronization: A Universal Concept in Nonlinear Sciences*. Cambridge, UK: Cambridge University Press.

Pikovsky, A., Rosenblum, M., & Kurths, J. (2000). Phase synchronization in regular and chaotic systems. *International Journal of Bifurcation and Chaos in Applied Sciences and Engineering, 10*, 2291–2305.

Pikovsky, A. S., Rosenblum, M. G., Osipov, G. V., & Kurths, J. (1997a). Phase synchronization of chaotic oscillators by external driving. *Physica D. Nonlinear Phenomena, 104*, 219–238. doi:10.1016/S0167-2789(96)00301-6

Pikovsky, A., Zaks, M., Rosenblum, M., Osipov, G., & Kurths, J. (1997). Phase synchronization of chaotic oscillations in terms of periodic orbits. *Chaos (Woodbury, N.Y.), 7*(4), 680–687. doi:10.1063/1.166265

Pisarchik, A. N., Meucci, R., & Arecchi, F. T. (2001). Theoretical and experimental study of discrete behavior of Shilnikov chaos in a CO2 laser. *The European Physical Journal D, 13*, 385–391. doi:10.1007/s100530170257

Platt, N., Spiegel, E. A., & Tresser, C. (1993). On-off intermittency a mechanism for bursting. *Physical Review Letters*, *70*(3), 279–282. doi:10.1103/PhysRevLett.70.279

Podlubny, I. (1999). *Fractional differential equations*. New York: Academic Press.

Poggio, T., & Girosi, F. (1990). Regularization algorithms for learning that are equivalent to multilayer networks. *Science*, *247*(27), 978–982. doi:10.1126/science.247.4945.978

Poggio, T., & Girosi, F. (2006). Networks for approximation and learning. In *Proceedings of the IEEE CDC 2006* (Vol. 78, pp. 1481—1497).

Pogromsky, A., & Nijmeijer, H. (1998). Observer-based robust synchronization of dynamical systems. *International Journal of Bifurcation and Chaos in Applied Sciences and Engineering*, *8*(11), 2243–2254. doi:10.1142/S0218127498001832

Ponomarenko, V. I., & Prokhorov, M. D. (2002). Extracting information masked by the chaotic signal of a time-delay system. *Physical Review E: Statistical, Nonlinear, and Soft Matter Physics*, *66*, 026215-1–7. doi:10.1103/PhysRevE.66.026215

Porfiri, M., & Fiorilli, F. (2010). Experiments on node-to-node pinning control of Chua's circuits. *Physica D. Nonlinear Phenomena*, *239*, 454–464. doi:10.1016/j.physd.2010.01.012

Poznyak, A. S. (2008). *Advance Mathematical tools for automatic control engineers: Deterministic technique*. London: Elsevier.

Poznyak, A. S., Sanchez, E., & Yu, W. (2001). *Differential Neural Networks for Robust Nonlinear Control (Identification, state Estimation an trajectory Tracking)*. Singapore: World Scientific. doi:10.1142/9789812811295

Poznyak, A. S., Yu, W., Ramirez, H. S., & Sanchez, E. N. (1998). Robust Identification by dynamic neural networks using sliding mode learning. *Applied Mathematics Computer Science*, *8*, 101–110.

Poznyak, A. (2004). Deterministic output noise effects in sliding mode observation . In Sabanovic, A., Fridman, L., & Spurgeon, S. (Eds.), *Variable structure systems: from principles to implementation* (pp. 45–80). London: IET Control Engineering Series.

Press, W. H., Flannery, B. P., Teukolsky, S. A., & Vetterling, W. T. (1992). *Numerical Recipes*. Cambridge, UK: Cambridge University Press.

Principe, J. C., Rathie, A., & Kuo, J. M. (1992). Prediction of chaotic time series with neural networks and the issue of dynamic modelling. *International Journal of Bifurcation and Chaos in Applied Sciences and Engineering*, *2*, 989–996. doi:10.1142/S0218127492000598

Puebla, H. (2005). Controlling intracellular calcium oscillations. *Journal of Biological System*, *13*, 173. doi:10.1142/S021833900500146X

Pyragas, K. (1998). Synchronization of coupled time-delay systems: Analytical estimations. *Physical Review E: Statistical Physics, Plasmas, Fluids, and Related Interdisciplinary Topics*, *58*, 3067–3071. doi:10.1103/PhysRevE.58.3067

Pyragas, K., & Tamasevicius, A. (1993). Experimental control of chaos by delayed self-controlling feedback. *Physics Letters. [Part A]*, *180*, 99–102. doi:10.1016/0375-9601(93)90501-P

Rajesh, S., & Ananthakrishna, G. (2000). Incomplete approach to homoclinicity in a model with bent-slow manifold geometry. *Physica D. Nonlinear Phenomena*, *140*, 193. doi:10.1016/S0167-2789(99)00241-9

Rauzy, G. (1976). *Propriétés statistiques de suites arithmétiques. Le Mathématicien, 15*. Paris: Presses Universitaires de France.

Rechenberg, I. (1973). *Evolutionsstrategie: Optimierung technischer Systeme nach Prinzipien der biologischen Evolution*. Stuttgart, Deutschland: Fromman-Holzboog.

Rechenberg, I. (1994). *Evolutionsstrategie '94*. Stuttgart, Deutschland: Frommann-Holzboog.

Robilliard, C., Huntington, E. H., & Webb, J. G. (2006). Enhancing the Security of Delayed Differential Chaotic Systems With Programmable Feedback. *IEEE Transactions on Circuits and Wystems. II, Express Briefs*, *53*, 722–726. doi:10.1109/TCSII.2006.876405

Rollins, R. W., & Hunt, E. R. (1982). Exactly Solvable Model of a Physical System Exhibiting Universal Chaotic Behavior. *Physical Review Letters*, *49*, 1295–1298. doi:10.1103/PhysRevLett.49.1295

Roopaei, M., Karimaghaee, P., & Soleimanifar, M. (2006). Control of the Chaotic Systems by the Linear State Feedback. *Nonlinear Studies Journal*, *133*, 167–173.

Roopaei, M., & Jahromi, M. Z. (2008). Synchronization of two different chaotic systems using novel adaptive fuzzy sliding mode control. *Chaos (Woodbury, N.Y.)*, *18*, 033133. doi:10.1063/1.2980046

Roopaei, M., Zolghadri Jahromi, M. (2008). Synchronization of a class of chaotic systems with fully unknown parameters using adaptive sliding mode approach. *Chaos*, 18, 043112. Retrieved from doi:10.1063/1.3013601

Roopaei, M., Zolghadri Jahromi, M., & Jafari, S. (2009). Adaptive Gain Fuzzy Sliding Mode Control for the Synchronization of Nonlinear Chaotic Gyros. *Chaos*, *19*(1), 013125-013125-9. Retrieved from doi: 10.1063/1.3072786

Roopaei, M., Zolghadri, M., & Meshksar, S. (2009). Enhanced adaptive fuzzy sliding mode control for uncertain nonlinear systems. *Communications in Nonlinear Science and Numerical Simulation*, *14*(9), 3670-3681.

Roopaei, M., Zolghadri, M., John, R., & Lin, T.-C. (in press). Unknown Nonlinear Chaotic Gyros Synchronization Using Adaptive Fuzzy Sliding Mode Control with Unknown Dead-Zone Input *Communications in Nonlinear Science and Numerical Simulation.*

Rosa, E. Jr, Ott, E., & Hess, M. H. (1998). Transition to phase synchronization of chaos. *Physical Review Letters*, *80*(8), 1642–1645. doi:10.1103/PhysRevLett.80.1642

Rosenblum, M. G., Pikovsky, A. S., & Kurths, J. (1996). Phase Synchronization of Chaotic Oscillators. *Physical Review Letters*, *76*, 1804–1807. doi:10.1103/PhysRevLett.76.1804

Rosenblum, M. G., Pikovsky, A. S., & Kurths, J. (1997). From phase to lag synchronization in coupled chaotic oscillators. *Physical Review Letters*, *78*(22), 4193–4196. doi:10.1103/PhysRevLett.78.4193

Rosier, L., Millérioux, G., & Bloch, G. (2004). Chaos synchronization on the N−torus and cryptography. *Comptes Rendus. Mécanique*, *332*(12), 969–972. doi:10.1016/j.crme.2004.09.001

Rosier, L., Millérioux, G., & Bloch, G. (2006). Chaos synchronization for a class of discrete dynamical systems on the N-dimensional torus. *Systems & Control Letters*, *55*, 223–231. doi:10.1016/j.sysconle.2005.07.003

Rosier, L. (2010). Chaotic dynamical systems associated with tilings of R^N, arXiv:1002.1125v1 (http://arxiv.org/abs/1002.1125)

Rossler, O. E. (1979). An equation for hyperchaos. *Physics Letters. [Part A]*, *71*, 155–157. doi:10.1016/0375-9601(79)90150-6

Rössler, O. E. (1976). An equation for continuous chaos. *Physics Letters. [Part A]*, *57*(5), 397–398. doi:10.1016/0375-9601(76)90101-8

Rovatti, R., & Setti, G. (1998). On the distribution of synchronization times in coupled uniform piecewise-linear Markov maps. *IEICE Transactions on Fundamentals*, *81*(9), 1769–1776.

Rovithakis, G., & Christodoulou, M. (1994). Adaptive control of unknown plants using dynamical neural networks. *IEEE Transactions on Systems, Man, and Cybernetics*, *24*, 400–412. doi:10.1109/21.278990

Roy, R. (2005). Communications technology: Chaos down the line. *Nature*, *438*, 298–299. doi:10.1038/438298b

Roy, R., & Thornburg, J. K. S. (1994). Experimental synchronization of chaotic lasers. *Physical Review Letters*, *72*(13), 2009–2012. doi:10.1103/PhysRevLett.72.2009

Roy, R., & Scott, K. (1994). Experimental synchronization of chaos. *Physical Review Letters, 72*(13), 2009–2012. doi:10.1103/PhysRevLett.72.2009

Ruan, S., & Filfil, R. (2004). Dynamics of a two-neuron system with discrete and distributed delays. *Physica D. Nonlinear Phenomena, 191*, 323–342. doi:10.1016/j.physd.2003.12.004

Ruelle, D. (1979). Ergodic theory of differentiable dynamic systems. *Publications Mathématiques de L'IHÉS, 50*, 27–58.

Rulkov, N. F., Sushchik, M. M., Tsimring, L. S., & Abarbanel, H. D. I. (1995). Generalized synchronization of chaos in directionally coupled chaotic systems. *Physical Review E: Statistical Physics, Plasmas, Fluids, and Related Interdisciplinary Topics, 51*(2), 980–994. doi:10.1103/PhysRevE.51.980

Rulkov, N. F. (1996). Images of synchronized chaos: experiments with circuits. *Chaos (Woodbury, N.Y.), 6*, 262–279. doi:10.1063/1.166174

Saha, P., Banerjee, S., & Chowdhury, A. R. (2004). Chaos, signal communication and parameter estimation. *Physics Letters. [Part A], 326*, 133–139. doi:10.1016/j.physleta.2004.04.025

Salarieh, H., & Alasty, A. (2009). Adaptive synchronization of two chaotic systems with stochastic unknown parameters. *Communications in Nonlinear Science and Numerical Simulation, 14*(2), 508–519. doi:10.1016/j.cnsns.2007.09.002

Salarieh, H., & Alasty, A.(2008-a). Chaos synchronization of nonlinear gyros in presence of stochastic excitation via sliding mode control. *Journal of Sound and Vibration, 313*(3-5), 760–771. doi:10.1016/j.jsv.2007.11.058

Salarieh, H., & Alasty, A.(2008-b). Adaptive chaos synchronization in Chua's systems with noisy parameters. *Mathematics and Computers in Simulation, 79*(3), 233–241. doi:10.1016/j.matcom.2007.11.007

Salarieh, H., & Shahrokhi, M. (2008). Adaptive synchronization of two different chaotic systems with time varying unknown parameters. *Chaos, Solitons, and Fractals,37*(1), 125–136. doi:10.1016/j.chaos.2006.08.038

Salarieh, H., & Shahrokhi, M. (2008). Dual synchronization of chaotic systems via time-varying gain proportional feedback. *Chaos, Solitons, and Fractals,38*(5), 1342–1348. doi:10.1016/j.chaos.2008.02.015

Salarieh, H., & Shahrokhi, M. (2009). Multi-synchronization of chaos via linear output feedback strategy. *Computational & Applied Mathematics, 223*(2), 842–852. doi:10.1016/j.cam.2008.03.002

Salarieh, H., & Alasty, A. (2008). Chaos synchronization of nonlinear gyros in presence of stochastic excitation via sliding mode control. *Journal of Sound and Vibration, 313*(3-5), 760–771. doi:10.1016/j.jsv.2007.11.058

Salarieh, H., & Shahrokhi, M. (2008). Adaptive synchronization of two different chaotic systems with time varying unknown parameters. *Chaos, Solitons, and Fractals, 37*, 125–136. doi:10.1016/j.chaos.2006.08.038

Sanchez, E., Matias, M. A., & Perez-Munuzuri, V. (1997). Analysis of synchronization of chaotic systems by noise: an experimental study. *Physical Review E: Statistical Physics, Plasmas, Fluids, and Related Interdisciplinary Topics, 56*, 4068–4071. doi:10.1103/PhysRevE.56.4068

Sastry, S., & Bodson, M. (1989). *Adaptive Control Stability, Convergence, and Robustness*. Englewood Cliffs, NJ: Prentice-Hall.

Schiff, S. J., So, P., Chang, T., Burke, R. E., & Sauer, T. (1996). Detecting dynamical interdependence and generalized synchrony through mutual prediction in a neural ensemble. *Physical Review E: Statistical Physics, Plasmas, Fluids, and Related Interdisciplinary Topics, 54*, 6708–6724. doi:10.1103/PhysRevE.54.6708

Schimming, T., & Hasler, M. (2003). Coded modulations based on controlled 1-D and 2-D piecewise linear chaotic maps. In *Proceedings of the 2003 International Symposium ISCAS '03* (Vol. 3, pp. 762–765).

Schmitz, R. (2001). Use of chaotic dynamical systems in cryptography. *Journal of the Franklin Institute, 338*, 429–441. doi:10.1016/S0016-0032(00)00087-9

Schneier, B. (1996). *Applied Cryptography: Protocols, Algorithm, and Source Code in C*. New York: Wiley.

Scholl, E., & Schuster, H. G. (2008). *Handbook of chaos control* (2nd ed.). Weinheim, Germany: Wiley-VCH Verlag.

Schuster, S., Marhl, M., & Hofer, T. (2002). Modeling of simple and complex calcium oscillations: from single-cell responses to intracellular signaling. *European Journal of Biochemistry, 269*, 1333. doi:10.1046/j.0014-2956.2001.02720.x

Schuster, H. G., & Just, W. (2005). *Deterministic Chaos: An Introduction*. Berlin: Wiley-VCH Verlag. doi:10.1002/3527604804

Schwefel, H. P. (1977). *Numerische Optimierung von Computermodellen mittels der Evolutionsstrategie*. Basel: Birkhaeuser.

Schwefel, H. P. (1995). *Evolution and Optimum Seeking*. New York: Wiley.

Shahverdiev, E. M.-O. (1999). Boundedness of dynamical systems and chaos synchronization. *Physical Review E: Statistical Physics, Plasmas, Fluids, and Related Interdisciplinary Topics, 60*, 3905–3909. doi:10.1103/PhysRevE.60.3905

Shahverdiev, E. M., Sivaprakasam, S., & Shore, K. A. (2003). Dual and dual-cross synchronization in chaotic systems. *Optics Communications, 216*, 179–183. doi:10.1016/S0030-4018(02)02286-1

Shahverdiev, E. M., Sivaprakasam, S., & Shore, K. A. (2002). Lag-synchronization in time-delayed systems. *Physics Letters. [Part A], 292*, 320–324. doi:10.1016/S0375-9601(01)00824-6

Shang, F., Sun, K. H., & Cai, Y. Q. (2008). A new efficient MPEG video encryption system based on chaotic maps. In *Proceedings-1st International Congress on Image and Signal Processing* (Vol. 3, pp. 12-16).

Shannon, C. E. (1949). Communication theory of secrecy systems. *The Bell System Technical Journal, 28*, 657–715.

Shannon, C. E. (1949). Communication theory of secrecy systems. *The Bell System Technical Journal, 28*, 656–715.

Sheng, L. Y., Cao, L. L., & Sun, K. H. (2005). Pseudorandom number generator based on TD-ERCS chaos and it's statistic characteristic s analysis. *ACTA PHYSICA SINICA, 54*(9), 4031–4037.

Shihua, C., Ja, H., Changping, W., & Jinhu, L. (2004). Adaptive synchronization of uncertain Rössler hyper chaotic system based on parameter identification. *Physics Letters. [Part A], 321*, 50–55. doi:10.1016/j.physleta.2003.12.011

Shilnikov, A. L., & Rulkov, N. F. (2003). Origin of chaos in a two-dimensional map modeling spiking-bursting neural activity. *International Journal of Bifurcation and Chaos in Applied Sciences and Engineering, 13*(11), 3325. doi:10.1142/S0218127403008521

Shimada, I., & Nagashima, T. (1979). A numerical approach to ergodic problem of dissipative dynamical systems. *Progress of Theoretical Physics, 61*, 1605–1616. doi:10.1143/PTP.61.1605

Short, K. M. (1994). Steps toward unmasking secure communications. *International Journal of Bifurcation and Chaos in Applied Sciences and Engineering, 4*, 959–977. doi:10.1142/S021812749400068X

Short, K. M. (1996). Unmasking a modulated chaotic communications scheme. *International Journal of Bifurcation and Chaos in Applied Sciences and Engineering, 6*, 367–375. doi:10.1142/S0218127496000114

Short, K. M., & Parker, A. T. (1998). Unmasking a hyperchaotic communication scheme. *Physical Review E: Statistical Physics, Plasmas, Fluids, and Related Interdisciplinary Topics, 58*, 1159–1162. doi:10.1103/PhysRevE.58.1159

Shtessel, Y. B., Shkolnikov, I. A., & Brown, D. J. (2003). An asymptotic second order smooth sliding mode control. *Asian Journal of Control, 5*(4), 498–504.

Shuai, J.-W., & Durand, D. M. (1999). Phase synchronization in two coupled chaotic neurons. *Physics Letters. [Part A]*, *264*(4), 289–297. doi:10.1016/S0375-9601(99)00816-6

Skogestad, S., & Postlethwaite, I. (1996). *Multivariable Feedback Control Analysis and Design*. Sussex, UK: John Wiley & Sons.

Slotine, J. J. E., & Li, W. (1991). *Applied Nonlinear Control*. Englewood Cliffs, NJ: Prentice Hall.

Sobhy, M. I., & Shehata, A.-E. R. (2001). Methods of attacking chaotic encryption and countermeasures. *Proc IEEE Int Conf Acoustics Speech Signal Process (2)*,1001–1004.

Solak, E. (2004). A reduced-order observer for the synchronization of Lorenz Systems. *Physics Letters. [Part A]*, *325*, 276–278. doi:10.1016/j.physleta.2004.04.001

Song, Q., & Wang, Z. (2008). Stability analysis of impulsive stochastic Cohen-Grossberg neural networks with mixed time delays. *Physica A: Statistical Mechanics and its Applications, 387*(13), 3314-3326.

Spillman, R. (1993, April). Use of a genetic algorithm in the cryptanalysis of simple substitution ciphers. *Cryptologia, 17*(1), 187–201. doi:10.1080/0161-119391867746

Sprott, J. C. (1997). Simplest dissipative chaotic flow. *Physics Letters. [Part A]*, *228*, 271–274. doi:10.1016/S0375-9601(97)00088-1

Sprott, J. C. (1997). Some simple chaotic jerk functions. *American Journal of Physics*, *65*, 537–543. doi:10.1119/1.18585

Sprott, J. C. (2003). *Chaos and Time-Series Analysis*. Oxford, UK: Oxford University Press.

Sprott, J. C. (2007). A simple chaotic delay differential equation. *Physics Letters. [Part A]*, *366*, 397–402. doi:10.1016/j.physleta.2007.01.083

Sprott, J. C. (1994). Some simple chaotic flows. *Physical Review E: Statistical Physics, Plasmas, Fluids, and Related Interdisciplinary Topics*, *50*, 647–650. doi:10.1103/PhysRevE.50.R647

Sprott, J. C. (2004). *Chaos and time series analysis*. Oxford, UK: Oxford University Press.

Srinivas, M., & Patnaik, L. M. (1994). Genetic Algorithms: A Survey. *IEEE Computer,* 17-26.

Stark, H. (1987). *Image Recovery: Theory and application*. New York: Academic Press.

Stavrinides, S. G., Deliolanis, N. C., Miliou, A. N., Laopoulos, T., & Anagnostopoulos, A. N. (2008). Internal Crisis in a Second-Order Non-Linear Non-Autonomous Electronic Oscillator. *J. of Chaos Solitons and Fractals*, *36*, 1055–1061. doi:10.1016/j.chaos.2006.07.025

Stavrinides, S. G., Miliou, A. N., Laopoulos, Th., & Anagnostopoulos, A. N. (2008). The Intermittency Route to Chaos of an Electronic Digital Oscillator. *International Journal of Bifurcation and Chaos in Applied Sciences and Engineering*, *18*, 1561–1566. doi:10.1142/S0218127408021178

Stavrinides, S. G. (2007). *Characterization of the behavior of a nonlinear electronic oscillator producing chaotic signals.* Unpublished doctoral dissertation, Aristotle University of Thessaloniki.

Stavrinides, S. G., Kyritsi, K. G., Deliolanis, N.C., Anagnostopoulos, A. N., Tamaševičious, A., & Čenys, A. (2004). The period doubling route to chaos of a second order non-linear non-autonomous chaotic oscillator – Part I. *J. of Chaos Solitons and Fractals*, 20, 849-854.

Stavrinides, S. G., Laopoulos, T., & Anagnostopoulos, A. N. (2005). An Automated Acquisition setup for the analysis of chaotic systems. In *Proc. IEEE IDAACS 2005*, Sofia, Bulgaria (pp. 628-632).

Stavroulakis, P. (2005). *Chaos applications in telecommunications*. Philadelphia, PA: Taylor & Francis Group. doi:10.1201/9780203025314

Stedman, G. E. (1997). Ring laser tests of fundamental physics and geophysics. *Reports on Progress in Physics*, *60*, 615–688. doi:10.1088/0034-4885/60/6/001

Stepanyan, V., & Hovakimyan, N. (2007). Robust adaptive observer design for uncertain systems with bounded disturbances. *IEEE Transactions on Neural Networks*, *18*(5), 1392–1403. doi:10.1109/TNN.2007.895837

Stone, E. F. (1992). Frequency entrainment of a phase coherent attractor. *Physics Letters. [Part A]*, *163*, 367–374. doi:10.1016/0375-9601(92)90841-9

Suetani, H., Iba, Y., & Aihara, K. (2006). Detecting generalized synchronization between chaotic signals: a kernel-based approach. *Journal of Physics. A, Mathematical and General*, *39*, 10723–10742. doi:10.1088/0305-4470/39/34/009

Sugawara, T., Tachikawa, M., Tsukamoto, T., & Shimizu, T. Y. (1994). Observation of synchronization in Laser chaos. *Physical Review Letters*, *72*(22), 3502–3505. doi:10.1103/PhysRevLett.72.3502

Sun, H. H., Abdelwahab, A. A., & Onaral, B. (1984). Linear approximation of transfer function with a pole of fractional order. *IEEE Transactions on Automatic Control*, *29*, 441–444. doi:10.1109/TAC.1984.1103551

Sun, Y., Cao, J., & Wang, Z. (2007). Exponential synchronization of stochastic perturbed chaotic delayed neural networks. *Neurocomputing*, *70*(13), 2477–2485. doi:10.1016/j.neucom.2006.09.006

Sun, Y., & Cao, J. (2007). Adaptive lag synchronization of unknown chaotic delayed neural networks with noise perturbation. *Physics Letters. [Part A]*, *364*, 277–285. doi:10.1016/j.physleta.2006.12.019

Szczepanski, J., Amigó, J. M., Michalek, T., & Kocarev, L. (2005, February). Crytographically secure substitutions based on the approximation of mixing maps. *IEEE Transactions on Circuits and Systems. I, Regular Papers*, *52*(2), 443–453. doi:10.1109/TCSI.2004.841602

Szebehely, V. (1984). Review of the concept of stability. *Celestial. Mechanical Engineering (New York, N.Y.)*, *34*, 49–64.

Taherionl, S., & Lai, Y. C. (1999). Observability of lag synchronization of coupled chaotic oscillators. *Physical Review E: Statistical Physics, Plasmas, Fluids, and Related Interdisciplinary Topics*, *59*, R6247–R6250. doi:10.1103/PhysRevE.59.R6247

Takens F. (1981). Detecting strange attractors in turbulence. *Lecture Notes in Mathematics*, 366-381.

Tamaševičious, A., Čenys, A., Mycolaitis, G., & Namajunas, A. (1998). Synchronizing hyperchaos in infinite-dimensional dynamical systems. *J. of Chaos Solitons and Fractals*, *9*, 1403–1408. doi:10.1016/S0960-0779(98)00072-1

Tan, X., Zhang, J., & Yang, Y. (2003). Synchronizing chaotic systems using backstepping design. *Chaos, Solitons, and Fractals*, *16*(1), 37–45. doi:10.1016/S0960-0779(02)00153-4

Tang, Y., Fang, J., & Miao, Q. (2009a). Synchronization of stochastic delayed neural networks with Markovian switching and its application. *International Journal of Neural Systems*, *19*, 43–56. doi:10.1142/S0129065709001823

Tang, Y., Fang, J., & Miao, Q. (2009b). On the exponential synchronization of stochastic jumping chaotic neural networks with mixed delays and sector-bounded nonlinearities. *Neurocomputing*, *72*, 1694–1701. doi:10.1016/j.neucom.2008.08.007

Tang, Y., & Fang, J. A. (2008). General methods for modified projective synchronization of hyperchaotic systems with known or unknown parameters. *Physics Letters. [Part A]*, *372*, 1816–1826. doi:10.1016/j.physleta.2007.10.043

Tang, Y., Qiu, R. H., Fang, J. A., Miao, Q., & Xia, M. (2008). Adaptive lag synchronization for unknown stochastic chaotic neural networks with discrete and distributed time-varying delays. *Physics Letters. [Part A]*, *372*, 4425–4433. doi:10.1016/j.physleta.2008.04.032

Tang, Y., Wang, Z., & Fang, J. (2009c). Pinning control of fractional-order weighted complex networks. *Chaos (Woodbury, N.Y.)*, *19*, 013112. doi:10.1063/1.3068350

Tenny, R., & Tsimring, L. S. (2004). Steps towards cryptanalysis of chaotic active/passive decomposition encryption schemes using average dynamics estimation. *International Journal of Bifurcation and Chaos in Applied Sciences and Engineering, 14*, 3949–3968. doi:10.1142/S0218127404011727

Ticos, C. M., Rosa, E. Jr, & Pardo, W. B. (2000). Walkenstein, J.A., Monti, M., Experimental real-time phase synchronization of a paced chaotic plasma discharge. *Physical Review Letters, 85*(14), 2929–2932. doi:10.1103/PhysRevLett.85.2929

Tigan, G. (2004). Analysis of a dynamical system derived from the Lorenz system. In *Mathematics in engineering and numerical physics* (pp. 265-272).

Toemeh, R., & Arumugam, S. (2007). Breaking Transposition Cipher with Genetic Algorithm. *Electronics and Electrica Engineering, 7*(79).

Tong, X., & Mrad, N. (2001). Chaotic motion of a symmetric gyro subjected to a harmonic base excitation. *Journal of Applied Mechanics— . Transactions of the American Society of Mechanical Engineers, 68*, 681–684.

Toniolo, C., Provenzale, A., & Spiegel, E. A. (2002). Signature of on-off intermittency in measured signals. *Physical Review E: Statistical, Nonlinear, and Soft Matter Physics, 66*(066209), 1–066209.

Tragha, A., Omary, F., & Kriouile, A. (2005). Genetic Algorithms Inspired Cryptography. *A.M.S.E As-sociation for the Advancement of Modeling & Simulation Techniques in Enterprises, Series D: Computer Science and Satistics.*

Tragha, A., Omary, F., & Moloudi, A. (2006). ICIGA: Improved Cryptography Inspired by Genetic Algorithms. In *Proceedings of the International Conference on Hybrid Information Technology (ICHIT'06)* (pp. 335-341).

Tseng, C. S., & Chen, B. S. (2001). H1 decentralized fuzzy model reference tracking control design for nonlinear interconnected systems. *IEEE Transactions on Fuzzy Systems, 9*(6), 795–809. doi:10.1109/91.971729

Tsimring, L. S., & Sushchik, M. M. (1996). Multiplexing chaotic signals using synchronization. *Physics Letters. [Part A], 213*, 155–166. doi:10.1016/0375-9601(96)00118-1

Tyson, J. J., Chen, K. C., & Novak, B. (2003). Sniffers, buzzers, toggles and blinkers: dynamics of regulatory and signaling pathways in the cell. *Current Opinion in Cell Biology, 15*, 221. doi:10.1016/S0955-0674(03)00017-6

Ucar, A., Lonngren, K. E., & Bai, E.-W. (2006). Synchronization of the unified chaotic systems via active control. *Chaos, Solitons, and Fractals, 27*(5), 1292–1297. doi:10.1016/j.chaos.2005.04.104

Ucar, A. (2003). On the chaotic behaviour of a prototype delayed dynamical system. *Chaos, Solitons, and Fractals, 16*(2), 187. doi:10.1016/S0960-0779(02)00160-1

Uchida, A., Kinugawa, S., & Yoshimori, S. (2003). Synchronization of chaos in two microchip lasers by using incoherent feedback method. *Chaos, Solitons, and Fractals, 17*, 363–368. doi:10.1016/S0960-0779(02)00375-2

Uchida, A., Kawano, M., & Yoshimori, S.(2003-a). Dual synchronization of chaos in Colpitts electronic oscillators and its applications for communications. *Physical Review E, 68*, 056207. Uchida, A., Kinugawa, S., Matsuura, T., & Yoshimori, S. (2003-b). Dual synchronization of chaos in microchip lasers. *Optics Letters, 28*, 19–21. doi:10.1364/OL.28.000019

Uchida, A., Rogister, F., García-Ojalvo, J., & Roy, R. (2005). Synchronization and communication with chaotic laser systems . In Wolf, E. (Ed.), *Progress in Optics* (*Vol. 48*, pp. 203–341). Amsterdam, The Netherlands: Elsevier B. V.

Udaltsov, V. S., Goedgebuer, J. P., Larger, L., Cuenot, J. B., Levy, P., & Rhodes, W. T. (2003). Cracking chaos-based encryption systems ruled by nonlinear delay differential equations. *Physics Letters. [Part A], 308*, 54–60. doi:10.1016/S0375-9601(02)01776-0

Van Wiggeren, G. D., & Roy, R. (1998). Communication with Chaotic Lasers. *Science, 279*, 1198–1200. doi:10.1126/science.279.5354.1198

Van Wyk, M. A., & Steeb, W.-H. (1997). Chaos in electronics . In *Mathematical Modelling: Theory and Applications* (*Vol. 2*). Dordrecht, the Netherlands: Kluwer Academic Publishers.

Vesentini, E. (1999). An introduction to topological dynamics in dimension one. *Sem. Mat. Univ. Politec. Torino*, *55*(4), 303–357.

Vicente, R., Dauden, J., Colet, P., & Toral, R. (2005). Analysis and characterization of the hyperchaos generated by a semiconductor laser subject to a delayed feedback loop. *IEEE Journal of Quantum Electronics*, *41*(4), 541–548. doi:10.1109/JQE.2005.843606

Vidyasagar, M. (1993). *Nonlinear Systems Analysis* (2nd ed.). Upper Saddle River, NJ: Prentice-Hall.

Vo Tan, P., Millérioux, G., & Daafouz, J. (2007 August). A comparison between the message-embedded cryptosystem and the self-synchronous stream cipher mosquito. In *Proc. of IEEE 18th European Conference on Circuits Theory and Design, ECCTD'07*, Sevilla, Spain.

Volkovskii, A. R., & Rul'kov, N. V. (1993). Synchronous chaotic response of a nonlinear oscillating system as the detection principle of chaos informational component. *Pis'ma v GTF, 19*(3), 71-75.

Voss, H. U. (2000). Anticipating chaotic synchronization. *Physical Review E: Statistical Physics, Plasmas, Fluids, and Related Interdisciplinary Topics*, *61*, 5115–5119. doi:10.1103/PhysRevE.61.5115

Walters, P. (1982). *An introduction to ergodic theory*. New York: Springer-Verlag.

Wan, L., & Sun, J. (2005). Mean square exponential stability of stochastic delayed Hopfield neural networks. *Physics Letters. [Part A]*, *343*(4), 306–318. doi:10.1016/j.physleta.2005.06.024

Wang, C., & Ge, S. S. (2001). Adaptive synchronization of uncertain chaotic systems via backstepping design. *Chaos, Solitons, and Fractals*, *12*(7), 1199–1206. doi:10.1016/S0960-0779(00)00089-8

Wang, C. P., & Su, J. P. (2004). A new adaptive variable structure control for chaotic synchronization and secure communication. *Chaos, Solitons, and Fractals*, *20*(5), 967–977. doi:10.1016/j.chaos.2003.10.026

Wang, Y., Guan, Z. H., & Wen, X. (2004). Adaptive synchronization for Chen chaotic system with fully unknown parameters. *Chaos, Solitons, and Fractals*, *19*(4), 899–903. doi:10.1016/S0960-0779(03)00256-X

Wang, L., & Cao, J. (2009). Global robust point dissipativity of interval neural networks with mixed time-varying delays. *Nonlinear Dynamics*, *55*(1-2), 169–178. doi:10.1007/s11071-008-9352-4

Wang, Y., Wang, Z., & Liang, J. (2008). A delay fractioning approach to global synchronization of delayed complex networks with stochastic disturbances. *Physics Letters. [Part A]*, *372*(39), 6066–6073. doi:10.1016/j.physleta.2008.08.008

Wang, Y. W., Wen, C., Soh, Y. C., & Xiao, J. W. (2006a). Adaptive control and synchronization for a class of nonlinear chaotic systems using partial system states. *Physics Letters. [Part A]*, *351*(1-2), 79–84. doi:10.1016/j.physleta.2005.10.055

Wang, Z., Ho, D. W. C., & Liu, X. (2005b). State estimation for delayed neural networks. *IEEE Transactions on Neural Networks*, *16*(1), 279–284. doi:10.1109/TNN.2004.841813

Wang, Z., Lauria, S., Fang, J., & Liu, X. (2007). Exponential stability of uncertain stochastic neural networks with mixed time-delays. *Chaos, Solitons, and Fractals*, *32*, 62–72. doi:10.1016/j.chaos.2005.10.061

Wang, Z., Liu, Y., Fraser, K., & Liu, X. (2006c). Stochastic stability of uncertain Hopfield neural networks with discrete and distributed delays. *Physics Letters. [Part A]*, *354*(4), 288–297. doi:10.1016/j.physleta.2006.01.061

Wang, Z., Liu, Y., & Liu, X. (2005a). On global asymptotic stability of neural networks with discrete and distributed delays. *Physics Letters. [Part A]*, *345*, 299–308. doi:10.1016/j.physleta.2005.07.025

Wang, Z., Liu, Y., Yu, L., & Liu, X. (2006d). Exponential stability of delayed recurrent neural networks with Markovian jumping parameters. *Physics Letters. [Part A], 356*(4), 346–352. doi:10.1016/j.physleta.2006.03.078

Wang, Z., Shu, H., Liu, Y., Ho, D. W. C., & Liu, X. (2006b). Robust stability analysis of generalized neural networks with discrete and distributed time delays. *Chaos, Solitons, and Fractals, 30*(4), 886–896. doi:10.1016/j.chaos.2005.08.166

Wang, K., Teng, Z., & Jiang, H. (2008). Adaptive synchronization of neural networks with time-varying delay and distributed delay. *Physica A, 387*, 631–647. doi:10.1016/j.physa.2007.09.016

Wang, Z., Fang, J. A., & Liu, X. (2008). Global stability of stochastic high-order neural networks with discrete and distributed delays. *Chaos, Solitons, and Fractals, 36*, 388–396. doi:10.1016/j.chaos.2006.06.063

Wang, Z., Lauria, S., Fang, J. A., & Liu, X. (2007). Exponential stability of uncertain stochastic neural networks with mixed time-delays. *Chaos, Solitons, and Fractals, 32*, 62–72. doi:10.1016/j.chaos.2005.10.061

Wang, Z., Liu, Y., Fraser, K., & Liu, X. (2006). Stochastic stability of uncertain Hopfield neural networks with discrete and distributed delays. *Physics Letters. [Part A], 354*, 288–297. doi:10.1016/j.physleta.2006.01.061

Wang, C. H., Cheng, C. S. C. S., & Lee, T. T. (2004). Dynamical Optimal Training for Interval Type-2 Fuzzy Neural Network (T2FNN). *IEEE Trans. Systems, Man and Cybernetics,* . *Part B, 34*(3), 1462–1477.

Wang, C. H., Lin, T. C., Lee, T. T., & Liu, H. L. (2002). Adaptive Hybrid Intelligent Control for Uncertain Nonlinear Dynamical Systems. *IEEE Trans. Syst., Man . Cybern., 32*(5), 583–597.

Wang, C. H., Liu, H. L., & Lin, T. C. (2002). Direct Adaptive Fuzzy-Neural Control with State Observer and Supervisory Control for Unknown Nonlinear Dynamical Systems. *IEEE Transactions on Fuzzy Systems, 10*(1), 39–49. doi:10.1109/91.983277

Wang, J. S., & Lee, C. S. G. (2002). Self-Adaptive Neuro-Fuzzy Inference Systems For Classification Application. *IEEE Transactions on Fuzzy Systems, 10*(6), 790–802. doi:10.1109/TFUZZ.2002.805880

Wang, L. X. (1993). Stable adaptive fuzzy control of nonlinear systems. *IEEE Transactions on Fuzzy Systems, 1*(1), 146–155. doi:10.1109/91.227383

Wang, L. X. (1994). *Adaptive Fuzzy Systems and Control: Design and Stability Analysis*. Englewood Cliffs, NJ: Prentice-Hall.

Wang, L. X., & Mendel, M. (1992). Fuzzy basis functions universal approximation, and orthogonal least squares learning. *IEEE Transactions on Neural Networks, 1*(3), 804–814.

Wang, W. Y., Chan, M. L., Hsu, C. C. J., & Lee, T. T. (2002). H∞ Tracking-Based Sliding Mode Control for Uncertain Nonlinear Systems via Adaptive Fuzzy-Neural Approach. *IEEE Transactions on Systems, Man, and Cybernetics, 32*, 483–492. doi:10.1109/TSMCB.2002.1018767

Wang, X. Y., & Yu, Q. (2009). A block encryption algorithm based on dynamic sequences of multiple chaotic systems. *Communications in Nonlinear Science and Numerical Simulation, 14*, 574–581. doi:10.1016/j.cnsns.2007.10.011

Wang, M., Hou, Z., & Xin, H. (2005). Internal noise-enhanced phase synchronization of coupled chemical chaotic oscillators. *Journal of Physics. A. Mathematical Nuclear and General, 38*, 145–152. doi:10.1088/0305-4470/38/1/010

Wang, L. X. (1997). *A Course in Fuzzy Systems and Control*. Englewood Cliffs, NJ: Prentice Hall.

Wang, W., & Slotine, J.-J. E. (2003). *On partial contraction analysis for coupled nonlinear oscillators*. NSL-030301. Retrieved from http://www.springerlink.com

Watson, J. D., Hopkins, N. H., Roberts, J. W., Steitz, J. A., & Weiner, A. M. (1987). *Molecular Biology of the Gene*. Menlo Park, CA: Benjamin Cummings.

Wen, G., Wang, Q. G., Lin, C., Han, X., & Li, G. (2006). Synthesis for robust synchronization of chaotic systems under output feedback control with multiple random delays. *Chaos, Solitons, and Fractals, 29*(5), 1142–1146. doi:10.1016/j.chaos.2005.08.078

Wenbo, M. (2004). *Modern Cryptography: Theory and Practice*. Upper Saddle River, NJ: Prentice Hall.

Weste, N., & Eshraghian, K. (1993). *Principles of CMOS VLSI Design: A systems Prospective*. AT&T.

Wiegand, T., Sullivan, G. J., Bjontegaard, G., & Luthra, A. (2003). Overview of the H.264/AVC video coding standard. *IEEE Trans. Circuits and Systems for Video Technology, 7*, 560–576. doi:10.1109/TCSVT.2003.815165

Wiggins, S. (1990). *Introduction to Applied Nonlinear Dynamical Systems and Chaos*. New York: Springer-Verlag.

Winful, H., & Rahman, L. (1990). Synchronized chaos and spatiotemporal chaos in arrays of coupled lasers. *Physical Review Letters, 65*(13), 1575–1578. doi:10.1103/PhysRevLett.65.1575

Wolf, A., Swift, J. B., Swinney, H. L., & Vastano, J. A. (1985). Determining Lyapunov exponents from a time series. *Physica D. Nonlinear Phenomena, 16*(3), 285–317. doi:10.1016/0167-2789(85)90011-9

Wolf, J. A. (1984). *Spaces of constant curvature* (5th ed.). Wilmington, DE: Publish or Perish, Inc.

Wong, K. W. (2002). A fast chaotic cryptographic scheme with dynamic look-up table. *Physics Letters. [Part A], 298*, 238–242. doi:10.1016/S0375-9601(02)00431-0

Wong, K. W. (2003). A combined chaotic cryptographic and hashing scheme. *Physics Letters. [Part A], 307*, 292–298. doi:10.1016/S0375-9601(02)01770-X

Wong, K. W., Ho, S. W., & Yung, C. K. (2003). A chaotic cryptography scheme for generating short ciphertext. *Physics Letters. [Part A], 310*, 67–73. doi:10.1016/S0375-9601(03)00259-7

Wong, W. K., Lee, L. P., & Wong, K. W. (2001). A modified chaotic cryptographic method. *Computer Physics Communications, 138*, 234–236. doi:10.1016/S0010-4655(01)00220-X

Woodman, K. F., Franks, P. W., & Richards, M. D. (1987). The nuclear magnetic resonance gyroscope: a review. *Navigation, 40*, 366–384. doi:10.1017/S037346330000062X

Wu, X., Li, J., & Chen, G. (2008). Chaos in the fractional order unified system and its synchronization. *Journal of the Franklin Institute, 345*, 392–401. doi:10.1016/j.jfranklin.2007.11.003

Wu, C., Fang, T., & Rong, H. (2007). Chaos synchronization of two stochastic Duffing oscillators by feedback control. *Chaos, Solitons, and Fractals, 32*(3), 1201–1207. doi:10.1016/j.chaos.2005.11.042

Wu, C., Lei, Y., & Fang, T. (2005). Stochastic chaos in a Duffing oscillator and its control. *Chaos, Solitons, and Fractals, 27*, 459–469. doi:10.1016/j.chaos.2005.04.035

Wu, C. W., & Chua, L. O. (1995). Synchronization in an array of linearly coupled dynamical systems. *IEEE Transactions on Circuits and Systems. I, Fundamental Theory and Applications, 42*(8), 430–447. doi:10.1109/81.404047

Wu, C. W., Yang, T., & Chua, L. O. (1996). On adaptive synchronization and control of nonlinear dynamical systems. *International Journal of Bifurcation and Chaos in Applied Sciences and Engineering, 6*, 455–471. doi:10.1142/S0218127496000187

Wu, C. W., & Chua, L. O. (1993). A simple way to synchronize chaotic systems with applications to secure communications systems. *International Journal of Bifurcation and Chaos in Applied Sciences and Engineering, 3*(6), 1619–1627. doi:10.1142/S0218127493001288

Wu, T., Cheng, Y., Tan, J., & Zhou, T. (2008). The Application of Chaos Genetic Algorithm in the PID Parameter Optimization. In *Proceedings of 3rd International Conference on Intelligent System and Knowledge Engineering.*

Xu, D., & Chee, C. Y. (2002). Controlling the ultimate state of projective synchronization in chaotic systems of arbitrary dimension. *Physical Review E: Statistical, Nonlinear, and Soft Matter Physics*, *66*, 046218. doi:10.1103/PhysRevE.66.046218

Xu, D., Chee, C. Y., & Li, C. (2004). A necessary condition of projective synchronization in discrete-time systems of arbitrary dimensions. *Chaos, Solitons, and Fractals*, *22*, 175–180. doi:10.1016/j.chaos.2004.01.012

Xu, D., Ong, W. L., & Li, Z. (2002). Criteria for the occurrence of projective synchronization in chaotic systems of arbitrary dimension. *Physics Letters. [Part A]*, *305*, 167–172. doi:10.1016/S0375-9601(02)01445-7

Xu, S., Lam, J., & Ho, D. W. C. (2006). A new LMI condition for delay-dependent asymptotic stability of delayed Hopfield neural networks. *IEEE Transactions on Circuits and Wystems. II, Express Briefs*, *53*(3), 230–234. doi:10.1109/TCSII.2005.857764

Xu, S., Lam, J., Ho, D. W. C., & Zou, Y. (2005). Novel global asymptotic stability criteria for delayed cellular neural networks. *IEEE Transactions on Circuits and Wystems. II, Express Briefs*, *52*(6), 349–353. doi:10.1109/TCSII.2005.849000

Xu, S., Chen, T., & Lam, J. (2003). Robust H Filtering for Uncertain Markovian Jump Systems With Mode-Dependent Time Delays. *IEEE Transactions on Automatic Control*, *48*, 900–907. doi:10.1109/TAC.2003.811277

Yalçınkaya, T., & Lai, Y.-C. (1997). Phase Characterization of Chaos. *Physical Review Letters*, *79*, 3885–3888. doi:10.1103/PhysRevLett.79.3885

Yamada, T., & Fujisaka, H. (1983). Stability theory of synchronized motion in coupled oscillator systems II. *Progress of Theoretical Physics*, *70*, 1240–1248. doi:10.1143/PTP.70.1240

Yan, J. J., Hung, M. L., & Liao, T. L. (2006). Adaptive sliding mode control for synchronization of chaotic gyros with fully unknown parameters. *Journal of Sound and Vibration*, *298*, 298–306. doi:10.1016/j.jsv.2006.05.017

Yan, J. J., Hung, M. L., Lin, J. S., & Liao, T. L. (2007). Controlling chaos of a chaotic nonlinear gyro using variable structure control. *Mechanical Systems and Signal Processing*, *21*(6), 2515–2522. doi:10.1016/j.ymssp.2006.07.002

Yan, J. J., Hung, M. L., Chiang, T. Y., & Yang, Y. S. (2006). Robust synchronization of chaotic systems via adaptive sliding mode control. *Physics Letters. [Part A]*, *356*(3), 220–225. doi:10.1016/j.physleta.2006.03.047

Yan, J., & Li, C. (2005, November). Generalized projective synchronization of a unified chaotic system. *Chaos, Solitons, and Fractals*, *26*(4), 1119–1124. doi:10.1016/j.chaos.2005.02.034

Yan, X. F., Chen, D. Z., & Hu, S. X. (2003). Chaos-genetic algorithms for optimizing the operating conditions based on RBF-PLS model. *Computers & Chemical Engineering*, *27*, 1393–1404. doi:10.1016/S0098-1354(03)00074-7

Yang, T. (1996). Recovery of digital signal from chaotic switching. *Int J Circuit Theory Appl.*, *23*, 611–615. doi:10.1002/cta.4490230607

Yang, T. (1996). Secure communication via chaotic parameter modulation. *IEEE Transactions on Circuits and Systems. I, Fundamental Theory and Applications*, *43*(9), 817–819. doi:10.1109/81.536758

Yang, T., & Chua, L. O. (1997). Impulsive stabilization for control and synchronization of chaotic systems: Theory and application to secure communication. *IEEE Transactions on Circuits and Systems*, *44*(10), 976–988. doi:10.1109/81.633887

Yang, Y., & Zhou, C. (2005). Adaptive fuzzy H1 stabilization for strict-feedback canonical nonlinear systems via backstepping and small-gain approach. *IEEE Transactions on Fuzzy Systems*, *13*(1), 104–114. doi:10.1109/TFUZZ.2004.839663

Yang, L. B., & Yang, T. (1997). Synchronization of Chua's circuits with parameter mismatching using adaptive model-following control. [English Version]. *Chinese Journal of Electronics*, *6*, 90–96.

Yang, T. (2004). A survey of chaotic secure communication systems. *International Journal of Bifurcation and Chaos in Applied Sciences and Engineering, 2*, 81–139.

Yang, T., & Chua, L. O. (1997). Impulsive control and synchronization of nonlinear dynamical systems and application to secure communication. *International Journal of Bifurcation and Chaos in Applied Sciences and Engineering, 7*, 645–664. doi:10.1142/S0218127497000443

Yang, T., & Sharuz, S. (1997). Channel-independent chaotic secure communication system using general chaotic synchronization. *Telecommunications Review, 7*, 240–254.

Yang, T., Yang, L. B., & Yang, C. M. (1997). Impulsive synchronization of Lorenz systems. *Physics Letters. [Part A], 226*, 349–354. doi:10.1016/S0375-9601(97)00004-2

Yang, X., Wu, T. X., & Jaggard, D. L. (2002). Synchronization recovery of chaotic wave through an imperfect channel. *IEEE Antennas and Wireless Propagation Letters, 1*, 154–156. doi:10.1109/LAWP.2002.807569

Yang, T., Wu, C. W., & Chua, L. O. (1997, May). Cryptography based on chaotic systems. *IEEE Trans. Circuits Syst. I., 44*, 69.

Yang, T., & Shao, H. H. (2002). Synchronizing chaotic dynamics with uncertainties based on a sliding mode control design. *Physical Review E, 65*(4), 046210/1-046210/7.

Yang, T., Yang, L. B., & Yang, C. M. (1998). Breaking Chaotic Switching Using Generalized Synchronization: Examples. *IEEE Trans. Circuits and Systems I*, 45, 1062-1067.

Yassen, M. T. (2005). Chaos synchronization between two different chaotic systems using active control. *Chaos, Solitons, and Fractals,23*(1), 131–140. doi:10.1016/j.chaos.2004.03.038

Yassen, M. T. (2006). Chaos control of chaotic dynamical systems using backstepping design. *Chaos, Solitons, and Fractals, 27*, 537–548. doi:10.1016/j.chaos.2005.03.046

Yassen, M. T. (2007). Controlling, synchronization and tracking chaotic Liu system using active backstepping design. *Physics Letters. [Part A], 360*, 582–587. doi:10.1016/j.physleta.2006.08.067

Yau, H. T. (2004). Design of adaptive sliding mode controller for chaos synchronization with uncertainties. *Chaos, Solitons, and Fractals,22*(2), 341–347. doi:10.1016/j.chaos.2004.02.004

Yau, H. T. (2008). Chaos synchronization of two uncertain chaotic nonlinear gyros using fuzzy sliding mode control. *Mechanical Systems and Signal Processing, 22*, 408–418. doi:10.1016/j.ymssp.2007.08.007

Yaz, E., & Azemi, A. (1994). Robust adaptive observers for systems having uncertain functions with unknown bounds. In *Proceedings of American Control Conference*, (pp. 73-74).

Yen, J. C., & Guo, J. I. (2000). A new chaotic key-based design for image encryption and decryption. In *IEEE International Symposium on ISCAS*, Geneva, IV-49-IV-52

Yin, X., Ren, Y., & Shan, X. (2002). Synchronization of discrete spatiotemporal chaos by using variable structure control. *Chaos, Solitons, and Fractals, 14*, 1077–1082. doi:10.1016/S0960-0779(02)00048-6

Yin, M., & Wang, L. W. (2008). A new study in encryption based on fractional order chaotic system. *Journal of Electronic Science and Technology of China, 6*(3), 238–241.

Yiqiang, G., Yanbin, W., Zhengshan, J., Jun, W., & Luyan, Z. (2009). *Remote sensing image classification by the Chaos Genetic Algorithm in monitoring land use changes.* Mathematical and Computer Modelling.

Yoo, B., & Ham, W. (1998). Adaptive Fuzzy Sliding Mode Control of Nonlinear System. *IEEE Transactions on Fuzzy Systems, 6*(2), 315–321. doi:10.1109/91.669032

Yu, H., Peng, J., & Liu, Y. (2006). Projective synchronization of unidentical chaotic systems based on stability criterion. *International Journal of Bifurcation and Chaos in Applied Sciences and Engineering, 16*, 1049–1056. doi:10.1142/S0218127406015301

Yu, W., & Cao, J. (2007). Adaptive synchronization and lag synchronization of uncertain dynamical system with time delay based on parameter identification. *Physica A*, *375*(2), 467–482. doi:10.1016/j.physa.2006.09.020

Yu, W., Cao, J., & Lu, J. (2008). Global synchronization of linearly hybrid coupled networks with time-varying delay. *SIAM Journal on Applied Dynamical Systems*, *7*(1), 108–133. doi:10.1137/070679090

Yu, W., & Cao, J. (2007). Synchronization control of stochastic delayed neural networks. *Physica A*, *373*, 252–260. doi:10.1016/j.physa.2006.04.105

Zadeh, L. A. (1965). Fuzzy Sets. *Information and Control*, *8*(3), 338–353. doi:10.1016/S0019-9958(65)90241-X

Zak, H., & Walcott, B. L. (1990). State observation of nonlinear control systems via the method of Lyapunov. In Zinober, A. S. I. (Ed.), *Deterministic Control of Uncertain Systems* (pp. 333–350). Washington, DC: IEEE Control Engineering Series.

Zeitz, M. (1987). The extended Luenberger observer for nonlinear systems. *Systems & Control Letters*, *9*(28), 149–156. doi:10.1016/0167-6911(87)90021-1

Zeng, Y., & Singh, S. N. (1997). Adaptive control of chaos in Lorenz system. *Dynamics and Control*, *7*, 143–154. doi:10.1023/A:1008275800168

Zhan, M., Wei, G. W., & Lai, C. H. (2002). Transition from intermittency to periodicity in lag synchronization in coupled Rössler oscillators. *Physical Review E: Statistical, Nonlinear, and Soft Matter Physics*, *65*, 036202–036205. doi:10.1103/PhysRevE.65.036202

Zhang, L., & Chen, J. L. G. (2005). Extension of Lyapunov second method by fractional calculus. *Pure and Applied Mathematics*, *21*, 291–294.

Zhang, H., Huang, W., Wang, Z., & Chai, T. (2006). Adaptive synchronization between two different chaotic systems with unknown parameters. *Physics Letters. [Part A]*, *350*(5-6), 363–366. doi:10.1016/j.physleta.2005.10.033

Zhang, H., Ma, X.-K., & Liu, W.-Z. (2004). Synchronization of chaotic systems with parametric uncertainty using active sliding mode control. *Chaos, Solitons, and Fractals*, *21*(5), 1249–1257. doi:10.1016/j.chaos.2003.12.073

Zhang, Q., Wen, X., & Xu, J. (2005). Delay-dependent exponential stability of cellular neural networks with time-varying delays. *Chaos, Solitons, and Fractals*, *23*, 1363–1369. doi:10.1016/j.chaos.2004.06.036

Zhao, H. (2004b). Existence and global attractivity of almost periodic solution for cellular neural network with distributed delays. *Applied Mathematics and Computation*, *154*, 683–695. doi:10.1016/S0096-3003(03)00743-4

Zhao, H. (2004). Global asymptotic stability of Hopfield neural network involving distributed delays. *Neural Networks*, *17*, 47–53. doi:10.1016/S0893-6080(03)00077-7

Zhao, H. X., & Gan, J. (2005). An improved algorithm of generate chaotic sequence. *Computer Applications*, *25*(B12), 78–79.

Zheng, Z., & Hu, G. (2000). Generalized synchronization versus phase synchronization. *Physical Review E: Statistical Physics, Plasmas, Fluids, and Related Interdisciplinary Topics*, *62*(6B), 7882–7885. doi:10.1103/PhysRevE.62.7882

Zheng, Z., Hu, G., & Hu, B. (1998). Phase slips and phase synchronization of coupled oscillators. *Physical Review Letters*, *81*(24), 5318–5321. doi:10.1103/PhysRevLett.81.5318

Zheng, S. H., Li, C. D., & Liao, X. F. (2003). Pseudo-random bit generator based on coupled Logistic maps and its applications in chaotic stream-cipher algorithms. *Computer Science*, *30*(12), 95–98.

Zheng, Y. Q., Liu, Y. J., Tong, S. C., & Li, T. S. (2009). *Combined Adaptive Fuzzy Control for Uncertain MIMO Nonlinear Systems.* Paper presented at the American Control Conference.

Zhou, T., & Li, C. (2005). Synchronization in fractional order differential systems. *Physica D. Nonlinear Phenomena*, *212*, 111–125. doi:10.1016/j.physd.2005.09.012

Zhou, D., Shen, T., & Tamura, K. (2006). Adaptive nonlinear synchronization control of twin gyro procession. *ASME Journal of Dynamics System Measurement and Control*, *128*, 592–599. doi:10.1115/1.2232683

Zhou, C., & Lai, C. (1999). Extracting messages masked by chaotic signals of time-delay systems. *Physical Review E: Statistical Physics, Plasmas, Fluids, and Related Interdisciplinary Topics*, *60*, 320–323. doi:10.1103/PhysRevE.60.320

Zhu, L., & Lai, Y.-C. (2001). Experimental observation of generalized time-lagged chaotic synchronization. *Physical Review E: Statistical, Nonlinear, and Soft Matter Physics*, *64*, 045205. doi:10.1103/PhysRevE.64.045205

Zhu, G. B., Cao, C. X., & Hu, Z. Y. (2003). An image scrambling and encryption algorithm based on affine transformation. *Journal of Computer-Aided Design & Computer Graphics*, *15*(6), 711–715.

About the Contributors

Santo Banerjee received his PhD degree in Physics from Jadavpur University, India. He was working as an assistant professor in Army Institute of Management, Kolkata in the department of Operations. Currently he is working as a senior research associate in department of Mathematics, Politecnico di Torino, Torino, Italy. He has another affiliation as a research scientist in Micro and Nanotechnology Unit inTechfab s.r.l., Chivasso, Italy. He has 2 books and around 40 research articles in the field of nonlinear physics. His current research area includes Chaotic systems, Synchronization, Cryptography, Genetic engineering and Neural networks.

***

Hassan Adloo was born in Shiraz, Iran in 1982. He received the BS degree in Electrical Engineering from Sadjad Institute of Higher Education, Mashhad, Iran in 2005, and MS degree in Shiraz University, Shiraz, Iran, in 2009. His research interests include Iterative learning control, Sliding mode control, Adaptive control and 2-D systems control.

José María Amigó received the PhD in Theoretical Physics from the University of Göttingen (Germany) in 1987. He was a postdoctoral fellow at the National Aerospace Laboratory in Tokyo (1989-1990) and a system analyst with the Space Division of Construcciones Aeronáuticas S.A. in Madrid (1991-1997), before joining the faculty of the Miguel Hernández University in Elche (Spain). Currently he is Professor of Applied Mathematics, Chairman of the Department of Statistics, Mathematics and Computer Science, and affiliated with the Operations Research Centre. Prof. Amigó is the author of the book "Permutation Complexity in Dynamical Systems", published by Springer Verlag in 2010.

Antonis N. Anagnostopoulos received his Diploma in Physics from the Aristotle University of Thessaloniki in 1969 and his PhD Degree in Physics from the Technical University of Karlsruhe, Germany, in 1978. From 1977 to 1978 he was a Research Associate at the Institut fur Angewandte Physik, University of Karlsruhe, Germany. Since 1980 he joined the faculty of the Department of Physics, Aristotle University of Thessaloniki. His research interests include among others analysis, design and simulation of electronic circuits and study of non-linear chaotic electronic properties. Prof. Anagnostopoulos has authored or co-authored more than 128 journal and conference publications including invited contributions and he participated in numerous national, EU-funded and NATO projects as a coordinator or a researcher. Prof. Anagnostopoulos is a member of Deutsche Physikalische Gesellschaft, European Physical Society, Deutsche Gesellschaft fór Kristallwachstum und Kristallzuechtung eV, Material Research Society (European), American Physical Society, Selenium-Tellurium Development Association, Inc.

Muhammad Rezal Kamel Ariffin holds the degrees Bac. Sc. (Hons.) –Mathematics, Masters of Science from Universiti Putra Malaysia and the PhD from Universiti Kebangsaan Malaysia. His research interest revolves around cryptosystems based on chaotic dynamical systems. He is a senior lecturer at the Department of Mathematics, Faculty of Science, Universiti Putra Malaysia and is an associate researcher with the Theoretical Laboratory, Institute for Mathematical Research, Universiti Putra Malaysia. He is a member of the International Association for Cryptologic Research (IACR), committee member of the Malaysian Society for Cryptology Research (MSCR) and a member of the Malaysia Society for Mathematical Sciences (PERSAMA). He is on the editorial board of the International Journal for Cryptology Research (ISSN 1985-5753) a journal published by the MSCR. Since joining Universiti Putra Malaysia as a lecturer in 2005, he is active in research and played a pivotal role in enhancing cryptologic research in the university.

J. Isaac Chairez is with the Professional Interdisciplinary Unit of Biotechnology. He received his master's degree (2004) and PhD (2007) from the Advanced Researching Centre form National Polytechnic Institute in Mexico City. Mexico. He is Member of System of National Investigators (SNI -1). His current interests are in the control automatic, the chemical engineering control process, biomedical engineering and image processing method in medicine.

Satyabrata Chakraborty is a student at the Indian Institute of Chemical Biology in Kolkata, India, working with Prof. Syamal Kumar Dana.

Ming-Che Chen was born in Changhua, Taiwan, in 1986. He received the B.S. degree in computer science & information engineering from the National formosa University, Yunlin, Taiwan in 2008, the M.S. in electrical engineering from the Feng Chia University in 2010. His current research interests and publications are in the areas of adaptive control, fuzzy-neural-network.

Konstantinos Chlouverakis was born in Athens, Greece, in 1978. He received the Physics degree in 2001 from the Aristotle University of Thessaloniki, Greece and the PhD degree in 2005 from the University of Essex, Colchester, Essex, UK, in the Photonics Group of the Electronic Systems Engineering Department. He is currently working as a research associate at the Optical Communications Laboratory, Department of Informatics and Telecommunications, University of Athens, Greece. He has been a research collaborator with the Physics department of the University of Madison since 2002, Wisconsin, USA and with the COSA (Complex Systems and Applications) group of the Demokritos EKEFE institute since 2007. His current research interests include non-linear dynamics in semiconductor lasers, chaotic encryption, high-dimensional chaos in delay-differential equations and quantum dot lasers. He has authored and co-authored 29 articles published in international scientific journals (20) and conferences (9) and serves as a reviewer for journals of the IEEE, Elsevier, Institute of Physics (IOP), Worldscinet and American Institute of Physics (AIP).

Syamal Kumar Dana is presently Scientist F and Head, Central Instrumentation Indian Institute of Chemical Biology, Kolkata 700032, India. His research interest is synchronization of chaotic systems and his main focus is to experiment on synchronization using electronic chaotic circuits. His current interest is extended to designing of electronic version of synthetic genetic networks to explore their collective behaviors. He is Head of Central Instrumentation at Indian Institute of Chemical Biology in Kolkata, India

Alexander S. Dmitriev is a leading researcher of the Kotelnikov Institute of Radio Engineering and Electronics of RAS (IRE RAS). He graduated from the Moscow Institute for Physics and Technology (MIPT) in 1971 (Department of General and Applied Physics). He received a PhD degree in 1974 from MIPT in the field of physics and mathematics and a Dr.Sci. degree from the same institute in 1988. In 1995 Alexander Dmitriev received his degree of Professorship in physics and mathematics. From 1971 to 1981 he worked in space agency "Energia". His research interests at that time were concentrated in the field of radio wave propagation, antennas and space communication. Since 1981 Alexander Dmitriev works in IRE RAS while he delivers an annual course of lectures "Dynamic Chaos and Applications" in MIPT.

Elena V. Efremova received her MS degree in 2000 and Ph.D. in 2003 from the Moscow Institute for Physics and Technology (MIPT), and her Dr.Sci. in 2008 from the Kotelnikov Institute of Radio Engineering and Electronics of RAS (IRE RAS). Since Feb. 2003 she works at IRE RAS, now in the position of senior researcher.

Jian-an Fang received the BS, MS, and PhD degree in electrical engineering from Donghua University (China Textile University), Shanghai, China in 1988, 1991 and 1994 respectively. Subsequently, he joined the College of Information Science and Technology, Donghua University, Shanghai, China, where he became a dean and professor in 2001. During February 1998 and May 1998, he was the visiting scholar in the University of Michigan at Ann Arbor. During May 1998 and Feb 1999, he was the visiting scholar in the University of Maryland at College Park. During June 2005 and August 2005, he was the senior visiting scholar in the University of Southern California. In 2005 and 2006, Prof. Fang was elected as council member of Shanghai Automation Association and a council member of Shanghai Microcomputer Applications, respectively. His research interests are mainly in complex system modeling and control, intelligent control systems, chaotic system control and synchronization, and digitalized technique for textile and fashion. He is also an active reviewer of some international journals, such as IEEE Trans. on Automatic Control, Physics Letters A, Applied Mathematics and Computation, Communications in Nonlinear Science and Numerical Simulation. He has published more than 40 papers in refereed international journals.

Ricardo Femat was born in Orizaba, Veracruz, México, in 1968. He received the Bc.Sc. degree from the Universidad de Guadalajara, Guadalajara, México, in 1992 and the D.Sc. degree from the Universidad Autónoma Metropolitana-Iztapalapa (UAM-I), México, in 1997. Since 2006, he has been a Senior Professor at the Applied Mathematics Department, IPICyT, San Luis Potosí, México, where he is currently the Head of the Applied Mathematics Department. His research interests include suppression and synchronization of chaotic systems, study of the regulation of the glucose level via feedback control, and dynamics and control of nonlinear systems. He is the author of 70 journal papers, about 63 contributions in international conferences, has edited one book, and authored one more book. Dr. Femat received the Medal of Academic Merit from UAM-I in 1998, the "El Potosí" Prize in 2005, is a regular member of the Mexican Academy of Sciences since 2003,and was President of the Asociacion de Mexico de COntrol Automatico(AMCA); which is the national member organization of International Federation of Automatic Control (IFAC) in México.

Alejandro García earned the Bachelor's degree on Biomedical Engineering from the Professional Interdisciplinary Unit of Biotechnology of the National Polytechnic Institute of Mexico (2001) and the M. Sc and Ph. D. Degrees on Automatic Control in 2006 and 2010, respectively, from the Research and Advanced Studies Center of the National Polytechnic Institute of Mexico (CINVESTAV - IPN). He is currently in a post-doctoral position in the project "Neuro-Control of environmental remediation processes, represented by partial differential equations", supported by the National Council of Science of Technology. His researching interests include the topics on dynamic neural networks, automatic control, hybrid systems, rehabilitation and biomechanical applications, and control processes area.

Rafael Martínez-Guerra was born in Mexico City in 1959, He earns his PhD Degree from Universidad Autonoma Metropolitana in 1996. Currently, he is a researcher at the Automatic Control Department of the CINVESTAV-IPN (Center of Research and Advanced Studies of IPN) and member of the National System of Researchers since 1992 (level II, currently). He is author and co-author of more 46 papers in International Journals and more than 90 contributions in International Conferences with Proceedings. His main research interests are in the field of Nonlinear Systems, Differential Geometric and Differential Algebraic methods, Nonlinear Observers, Fault Detection Problems and Chaos.

Michael P. Hanias was born in Athens, Greece in 1961. He received the B.Sc. degree in Physics from Aristotelian University of Thessaloniki and the Ph.D. degree from Aristotelian University of Thessaloniki, in 1984, and 1993 respectively. From 1989 to 1993 he was Teaching and Research Assistant at the Laboratory of Solid State, Physics Dept., Aristotelian University of Thessaloniki and, from 1994 to 2009, he was Scientific collaborator at TEI of Chalkis (Department of Physics and Chemistry). From 1992 to 2008 he was involved in various NATO& European Union R&D projects related to Chaotic electrical properties of semiconductors and semiconductors devices. Since 1989, he has published many papers and has been referee to many scientific journals. His research interests include electronic transport phenomena, Non linear chaotic circuits, Chaos in Electronics, Time series prediction, and Neural Networks.

Hamid Reza Karimi is a Professor in Control Systems at the Faculty of Engineering and Science of the University of Agder in Norway. His research interests are in the areas of nonlinear systems, networked control systems, robust control/filter design, time-delay systems, wavelets and vibration control of flexible structures with an emphasis on applications in engineering. He has published more than 130 papers in refereed journals and transactions, book chapters and conference proceedings. Dr. Karimi is a senior member of IEEE and serves as chairman of the IEEE chapter on control systems at IEEE Norway section. He is also serving as an editorial board member for some international journals, such as Journal of Mechatronics-Elsevier, Journal of Mechatronics and Applications, International Journal of Control Theory and Applications and International Journal of Artificial Intelligence, etc. He is a member of IEEE Technical Committee on Systems with Uncertainty, IFAC Technical Committee on Robust Control and IFAC Technical Committee on Automotive Control. He was the recipient of the Juan de la Cierva Research Award in 2008, Alexander-von-Humboldt-Stiftung Research Fellowship in 2006, German Academic Exchange Service (DAAD) Research Fellowship in 2003, National Presidency Prize for Distinguished PhD student of Electrical Engineering in 2005 and National Students Book Agency's Award for Distinguished Research Thesis in 2007, etc.

Lev V. Kuzmin received his BS and MS degrees in Applied Mathematics, and a Ph.D. degree in Radiophysics, in 1994, 1997, and 2000, respectively, all from the Moscow Institute for Physics and Technology (MIPT), Moscow. Since Feb. 1997 he works at the Kotelnikov Institute of Radio Engineering and Electronics of RAS (IRE RAS) (senior researcher).

Tsung-Chih Lin was born in Taichung, Taiwan, in 1961. He received the B.S. degree in electrical engineering from the Feng Chia University, Taichung, Taiwan in 1984, the M.S. in control engineering from the National Chiao Tung University in 1986 and the Ph.D. degree in the School of Microelectronic Engineering, Griffith University, Brisbane, Australia. He is currently Associate Professor in the Department of Electronic Engineering, Feng Chia University, Taichung, Taiwan. His current research interests and publications are in the areas of adaptive control, fuzzy-neural-network, robust control and analog circuit testing.

Ricardo Aguilar López was born in Mexico City in 1964. He ears their BSc, MSc and PhD degrees from Universidad Autónoma Metropolitana in 1989, 1993 and 1998, respectively, all of them in chemical engineering. He was a Postdoctoral Fellow at Instituto Mexicano del Petróleo (2000), besides he ears a Ph D in automatic control from CINVESTAV-IPN (2003). He works in robust observer design for non-linear systems and process control. He is author and co-author of 88 papers published in international journals. Currently he is scientific researcher at the Departamento de Biotecnología y Bioingeniería del Centro de Investigación y Estudios Avanzados del IPN

Ahmed Ezzat Mohamed Matouk (A. E. Matouk) had been born in March 19, 1976. He had received the BS in mathematics from Mathematics Department, Faculty of Science, Mansoura University, Mansoura, Egypt in 1999. He had got the M.S. in applied mathematics from the same department in 2005, thesis title "*Chaos and Synchronization in Some Nonlinear Electronic Circuits*". His PhD thesis entitled by "*On Synchronization of Fractional Order Dynamical Systems*". He has been with the Mathematics Department, Faculty of Science at Hail University, Hail, Saudi Arabia since 2007. His research interests include Fractional Calculus, Chaos Theory, Bifurcation Theory, Chaos Synchronization, Chaos Control, Complex networks and Mathematical modeling.

J. M. González-Miranda obtained his MSc in Sciences in 1977, and his PhD in Physical Sciences in 1982, both by the University of Barcelona. Since April 1987, he is an Associate Professor at the Department of Fundamental Physics of the University of Barcelona. His research inters have evolved along time. From 1979 to 1986, he worked in Liquid State Physics studying structural and diffusive properties of metals through Molecular Dynamics simulations. From 1986 to1995, he applied Monte Carlo simulation to the study of Non-Equilibrium Statistical Mechanics. Since 1995, his interest is focused on the field of Nonlinear Dynamics and Chaos, where he has contributed to several problems of synchronization and control of chaos, neuronal dynamics and bifurcation theory. He is the author of the book "Synchronization and Control of Chaos. An Introduction for Scientists and Engineers" published in 2004 by Imperial College Press.

Amalia N. Miliou received her Diploma in Physics from the Aristotle University of Thessaloniki in 1985 and her MSc and PhD Degrees in Electrical and Computer Engineering from the University of Florida, USA, in 1988 and 1991, respectively. From 1991 to 1992 she was a Research Associate at the

University of Florida. Since 1993 she joined the faculty of the Department of Informatics, Aristotle University of Thessaloniki. Her research interests include among others analysis, design and simulation of non-linear chaotic electronic circuits and optoelectronic devices. Dr. Miliou has authored or co-authored more than 50 journal and conference publications including invited contributions and she participated in numerous national, EU-funded and NATO projects as a coordinator or a researcher. Dr. Miliou is a member of IEEE/LEOS and OSA.

Gilles Millérioux received the PhD in automatic control from the Institut National des Sciences Appliquées de Toulouse (France) in 1997. He is currently Professor on control theory at Nancy University and a researcher at the Centre de Recherche en Automatique de Nancy (CRAN) in France. His research activities include complex dynamics and cryptography , observers and chaos synchronization, switched systems.

Mala Mitra received her PhD degree in Physics from Jadavpur University, India. She was an assestent professor in department of physics,JIS College of Engineering, kalyani, India from 2004-2008. Currently she is working as a professor in the department of Physics, Camelia School of Engineering, Kolkata, India. Her current research interst includes chaos and nonlinearity and there applications.

Seyyed Hossein Mousavi was born in Shiraz, Iran, on September 9, 1986. He received his B.S. degree in Electrical Engineering from Shiraz University, Shiraz, Iran; in 2008. He is currently a MS student in Shiraz University, Shiraz, Iran. His research interests are Adaptive control, nonlinear analysis, chaotic synchronization, system identification and Fuzzy systems.

Sumona Mukhopadhyay received her BTech degree in Computer Science & Engineering from West Bengal University of Technology, West Bengal, India in 2006. After this she pursued M.Tech in Software Engineering from School of Information Technology under West Bengal University of Technology, West Bengal, India in 2008. In 2007 she was working as faculty in Siliguri Institute of Technology, West Bengal India in the Dept of Computer Science & Engineering. Also in 2007 she was selected in Wipro Technologies as Project Engineer. In 2009 till date she is now Assistant Professor in the Department of Computer Science in Army Institute of Management, West Bengal, India. During her academic career she handled various subjects like Software Engineering, Artificial Intelligence, Distributed Operating System, Database management system, Management Information System and Fundamentals of Networking. Her research interest include optimizing techniques by evolutionary computation, mobile computing and secure communication.

Mohd. Salmi Md. Noorani holds the degrees Bac. Sc. (Hons.), Masters of Science and the PhD from Warwick. His research interest revolves around ergodic theory and dynamical systems. He is a Professor and the chairman of the School of Mathematical Sciences, Universiti Kebangsaan Malaysia. He is the President of the Malaysia Society for Mathematical Sciences (PERSAMA), member of the American Mathematical Society, member of the Society for Industrial and Applied Mathematics (SIAM), committee member of the Southeast Asian Mathematical Society and a council member of the Malaysian Head of Mathematics Departments & Professors of Mathematics. He is on the editorial board of the SAINS MALAYSIANA (ISSN 0126-6039) (which is indexed and abstracted in ISI Thomson Reuters (Science Citation Index Expanded/SciSearch®, Journal Citation Reports/Science Edition), SCOPUS, Chemical

Abstracts and Zentralblatt MATH. He is also on the editorial board of the International Journal of Open Problems in Computer Science and Mathematics IJOPCM (ISSN: 1998-6262). Since joining Universiti Kebangsaan Malaysia as a lecturer in 1986, he is active in research and played a pivotal role in enhancing his area of mathematical research in the university.

Jean Chabi Orou was born in the city of Porto-Novo in BENIN on the 24th of October 1955. He earned a Bachelor and a Master degrees in theoretical physics from the National university of BENIN. From 1980 to 1989 he taught physics and chemistry in high schools. He started to work on his PhD in 1990 which he earned in 1994. In 1991 he was a visiting scholar at the department of physics of the City University of New-York for a year before going to Tallahassee in Florida at the NASA/ FAMU Center for Nonlinear and Nonequilibrium Aeroscience .He spent there two years working with Prof. Joseph Andrew Johnson III on the evolution of a free shear layer of a supersonic flow. Dr. Chabi Orou is teaching physics (fluid mechanics and dynamical systems) and is the current head of the department of physics at université d' Abomey-Calavi in BENIN and the head of the theoretical physics research group at the Institut de Mathématiques et de Sciences Physiques in Porto-Novo.

Silvia Ortín was born in Bilbao, Spain, in 1977. She received the degree in electronic engineering from the University of Cantabria, Spain, in 2002. She is currently working toward the PhD degree at the Instituto de Física de Cantabria, in Santander, Spain. Her research activities deal with time series analysis, neural networks, nonlinear dynamics reconstruction and security in chaotic communication systems.

Andrey I. Panas (b. 1955) graduated from the Moscow Energy Institute in 1978, Department of Electron Devices. In 1988 he received a Ph.D. degree from the Kotelnikov Institute of Radio Engineering and Electronics of RAS (IRE RAS). His thesis was on chaotic oscillators based on two-mode oscillating systems. He received his Dr.Sci. degree from the same Institute in 2001 and his thesis was on communication systems with chaotic signals. Since 1978 he works with IRE RAS, he is now the Director of Fryazino branch of the Institute.

Luis Pesquera was born in Vega de Infanzones, León, Spain, in 1952. He received the MSc degree in physics in 1974 from the Universidad de Valladolid, Spain. He was a postgraduate fellow at the Université de Paris VI during 1977-1980. He received the PhD degree in physics in 1980 from the Universidad de Cantabria, Santander, Spain. In 1980 he joined the Departamento de Física Moderna of the Universidad de Cantabria. Since 1991 he has been Professor of Physics at the Universidad de Cantabria. In 1995 he joined the Instituto de Física de Cantabria (CSIC--UC). His research work started in the field of stochastic processes applied to Physics and he has made contributions to the foundations of quantum physics, fluctuations in nuclear reactors, disordered systems and laser physics. He is currently working on the modeling of noise and nonlinear properties of semiconductor lasers and their applications to optical communication systems.

Alexander S. Poznyak was graduated from Moscow Physical Technical Institute (MPhTI) in 1970. He earned PhD and Doctor Degrees from the Institute of Control Sciences of Russian Academy of Sciences in 1978 and 1989, respectively. From 1973 up to 1993 he served this institute as researcher and leading researcher, and in 1993 he accepted a post of full professor (3-E) at CINVESTAV of IPN in Mexico. He is the director of 33 Ph.D thesis's (25 in Mexico). He has published more than 140 papers in different international journals and 9 books. He is a Regular Member of Mexican Academy of Sciences and System of National Investigators (SNI -3). He is Associated Editor of Ibero-american Int. Journal on "Computations and Systems". He was also Associated Editor of CDC, ACC and Member of Editorial Board of IEEE CSS.

Bijan Ranjbar Sahraei, was born in Shiraz, Iran, on April 14, 1986. He received his BS degree in Electrical Engineering from Shiraz University, Shiraz, Iran; in 2008. He is currently a MS student in Shiraz University, Shiraz, Iran. His research interests are process control, mobile robot control, swarm intelligence, adaptive chaotic synchronization and system identification.

Lionel Rosier is currently working as a professor in Institut Elie Cartan, Universite Henri Poincare, Nancy 1, France. He has several research articles on chaotic systems, synchronization, Stability and control theory.

Mehdi Roopaei was born in Shiraz, Iran, on September 18, 1980. He received his BS and MS degrees in Control Engineering from Shiraz University, Shiraz, Iran; in 2002 and 2006 respectively. He is currently PhD student in Electrical and Computer Eng. of Shiraz University, Shiraz, Iran. His research interests are nonlinear control, adaptive fuzzy sliding mode control, Fuzzy systems, chaos synchronization.

Hassan Salarieh received his BSc degree in Mechanical Engineering and Pure Mathematics in 2004 from Sharif University of Technology, Tehran, IRAN. In 2004 and 2008 he received his MSc and PhD degrees both in Applied Mechanics from Sharif University of Technology. Currently he is a faculty member of school of Mechanical Engineering and also a member of Center of Excellence in Design, Robotics and Automation (CEDRA) at the Sharif University of Technology. His research interests include fields of complex and stochastic dynamics and control.

Mohammad Shahrokhi received his BSc degree in Chemical Engineering from Sharif University of Technology in 1977 and his PhD degree from Wisconsin University in the same field in 1981. From 1981 to 1988 he worked as a faculty member of Chemical Engineering Department of Isfahan University of Technology and in 1988 he joined the Sharif University of Technology where he is currently working as a professor of Chemical Engineering at the Department of Chemical and Petroleum Engineering. His research interest is process control especially adaptive control.

Sergei O. Starkov (b. 1956) graduated from the Moscow Institute for Physics and Technology in 1979. He received his PhD degree from the same Institute in 1987 and the Dr.Sci. degree from the Kotelnikov Institute of Radio Engineering and Electronics of RAS (IRE RAS) in 2003. Since 1981 he works with IRE RAS (senior researcher) whereas from 2004 he is the Head of the department of Obninsk State Technical University for Nuclear Power Engineering.

Stavros Stavrinides received his Physics Diploma, his MSc in Electronics and his PhD in Chaotic Electronics in 1996, 2003 and 2007, respectively; all from the Aristotle University of Thessaloniki. He is Research Associate at the Electronics and Computer Lab of Physics Dept., Aristotle University of Thessaloniki, since 2007. Dr. Stavrinides has taught numerous topics in physics and electronics, in various institutions, for more than 12 years. His research interests include, non-exhaustively, the design of analog and mixed-signal electronic circuits, measurement and instrumentation systems, as well as, chaotic electronics and their applications. Dr. Stavrinides has authored or co-authored more than 25 journal and conference papers. He has participated, as a researcher, in several national and NATO funded projects. Dr. Stavrinides is a Hellenic Physics Society and IEEE member.

Kehui Sun was born on September 4, 1968, Yiyang city, Hunan Province, China. He received his PhD on Control Theory and Control Engineering from Central South University, in 2005. At present he is working as a Professor, School of Physics Science and Technology, Central South University, China. His research interest includes Chaos theory and its application, Fractional order chaotic systems and its control, Design of intelligent instruments etc.

Yang Tang received the BS degree in electrical engineering from Donghua University, Shanghai, China in 2006. Now he is a Ph. D student with Department of Automation, Donghua University, Shanghai, China. Since January 2009, he has been a research associate in The Hong Kong Polytechnic University, Hung Hom Kowloon, Hong Kong. His current research interests include stochastic systems, neural networks, chaotic synchronization, image encryption, complex networks, data mining and evolutionary computation. He is an active reviewer of some international journals, such as IEEE Trans. on Neural Networks, Neural Networks, Physics Letters A, Neurocomputing, Communications in Nonlinear Science and Numerical Simulation.

George S. Tombras was born in Athens, Greece in 1956. He received the BSc degree in Physics from Aristotelian University of Thessaloniki, Greece, the MSc degree in Electronics from University of Southampton, UK, and the PhD degree from Aristotelian University of Thessaloniki, in 1979, 1981, and 1988 respectively. Since 1991, he has been with the Laboratory of Electronics, Faculty of Physics, University of Athens, where currently is an Associate Professor of Electronics, Director of the Department of Electronics, Computers, Telecommunications and Control, and Vice Chairman of the Faculty of Physics. His research interests include Free Space Optical Communications, Mobile Communications, Analog and Digital Circuits and Systems, Chaos in Electronics, as well as Instrumentation, Measurements and Audio Engineering. Professor Tombras is the author of the textbook "Introduction to Electronics" (in Greek) and has authored or co-authored more than 80 journal and conference refereed papers and many technical reports.

Behnam Zare was born in Shiraz, Iran, on December 18, 1985. He received his BS degree in Electrical Engineering from Shiraz University, Shiraz, Iran; in 2008. He is currently a MS student in Tarbiat Modarres University (T.M.U), Tehran, Iran. His research interests are adaptive control, chaotic synchronization, system identification and swarm intelligence.

Mansoor J. Zolghadri received the BSc degree in mechanical engineering from Shiraz University, Shiraz, Iran, in 1979, the MSc and PhD degrees in instrumentation and control from the University of Bradford, Bradford, U.K., in 1982 and 1988, respectively. Since 1988, he has been with the Department of Computer Science and Engineering, Shiraz University, where he is currently an Associate Professor. He is the author or coauthor of more than 40 papers published in various international journals and conference proceedings. His current research interests include pattern recognition, fuzzy systems, information retrieval, and web search.

Index

A

active control 128, 148
adaptation law 46
adaptive algorithm 46
adaptive control 128, 158, 165, 166, 167, 176, 177, 178, 183, 184, 205, 314
adaptive fuzzy sliding mode control 314, 334
adaptive fuzzy sliding mode controller 316
adaptive synchronization 289, 290, 417
Adler equation 249, 251
Advanced Encryption Standard (AES) 362, 367, 368
affine transformations 23, 34
Alvarez's attack 464, 465, 466
Amplitude Envelope Synchronization 106, 114, 115, 116
amplitude modulated (AM) 416
AMS subject classifications 19
analog circuits 464
anticipated synchronization 106, 113, 114
anti-damping 4
aperiodic long-term behavior 415
a priori 44, 45, 48
a-priori knowledge 388, 390, 408
artificial neural networks (ANN) 43, 45, 46, 57, 315
asymmetry 91, 93, 94, 95, 96, 97, 101
asymptotic stability 128
atomic polarization 107
attractor dimension 1, 2, 14
auto-correlation function 20
autonomous 5
autonomous systems 5

B

back propagation algorithm 315
backstepping design 128, 184, 208, 209
Baptista cryptosystem 465
bi-directional associative memory networks 261, 262, 285
bidirectional coupling 155, 441, 442
bifurcation diagram 92, 96, 97
bifurcation parameters 107, 118
bifurcation point 92, 93, 95
biological oscillations 212, 233
biological systems 210, 212, 233, 234
bipolar junction transistor (BJT) 70, 73, 75
birhythmicity 257
broadband chaotic signal 386
broadband spectrum 386
bursting 91, 92, 93, 95, 97, 98, 101, 102, 103

C

cellular mobile communication 416
cellular NNs 261, 262
chaos 1, 2, 5, 6, 8, 9, 13, 14, 15, 16, 17, 18
chaos-based communication systems 440
chaos-based cryptography 363, 415, 416, 435
chaos-based encryption schemes 387
chaos-based optical communication systems 387
chaos communications 210, 211, 212
chaos control 184, 315
chaos encryption 1
chaos generator 344, 345, 418, 426
chaos masking 153, 415, 419, 427, 428
chaos modulation 153, 415, 427, 428
chaos quantifiers 1, 2, 14

chaos shift keying (CSK) 153, 415, 421, 422, 423, 424, 425, 426, 427, 436, 437
chaos signals 416
chaos spreading spectrum 415
chaos synchronization 19, 39, 40, 41, 43, 105, 106, 107, 121, 122, 125, 127, 128, 137, 141, 142, 144, 148, 149, 150, 152, 154, 155, 156, 158, 159, 162, 165, 166, 167, 170, 172, 174, 176, 177, 178, 180, 182, 183, 202, 203, 207, 210, 211, 212, 229, 244, 289, 311, 312, 314, 315, 316, 332, 415, 419, 422, 436
chaos theory 1, 8, 210, 211
chaotic 19, 20, 21, 23, 25, 26, 27, 28, 29, 32, 33, 34, 37, 38, 39, 40, 41
chaotic AM communication system 343
chaotic attractor 156, 157, 168, 388, 390, 409
chaotic behavior 20, 32, 38, 42, 44, 45, 54, 68, 87, 92, 106, 154, 158, 184, 186, 203, 205, 247, 283, 362, 389
chaotic carrier 210, 212, 386, 387, 388, 389, 390, 391, 393, 394, 397, 399, 400, 402, 404, 408, 414, 440, 455, 459
chaotic carrier signal 388
chaotic ciphers 361, 363, 377
chaotic circuit 68, 69, 70, 72, 73, 80, 87, 89, 153, 154, 177, 178
chaotic communication 417, 437
chaotic communication system 69, 386, 387, 388, 389, 403, 408, 412, 440
chaotic cryptographic scheme 363, 464, 475
chaotic cryptography 363, 381, 415, 429, 431
chaotic cryptosystem 361, 363, 369, 377, 378, 381, 386, 388, 402, 409, 412, 463, 464, 469, 473, 474, 475
chaotic demodulator (CDM) 69
chaotic dynamical system 19, 23, 39, 41, 441
chaotic dynamics 2, 14, 44, 105, 106, 115, 387, 388, 402
chaotic encryption 463, 474
chaotic encryption systems 387
chaotic flow 5, 18
chaotic fluctuation 418
chaotic generator 211, 389, 390, 402
chaotic gyros 183, 184, 187, 203, 204, 205, 208
chaotic logistic equation 463, 464
chaotic maps 464
chaotic masking 418, 419, 420, 421, 422
chaotic model 43
chaotic modes 154
chaotic modulation 153
chaotic modulator (CM) 69
chaotic module 338, 342, 343, 344, 345, 353, 354, 355, 356, 357, 358
chaotic module oscillations 343
chaotic neural networks (CNNs) 289, 290, 291, 305, 309, 310, 312, 313
chaotic nodes 262
chaotic nonlinear gyros 185, 200, 202, 204, 207, 209
chaotic ODE 5
Chaotic On-Off Keying (COOK) 422, 426, 427
chaotic oscillation 211
chaotic oscillator 105, 106, 107, 108, 110, 111, 112, 113, 114, 118, 120, 121, 123, 125, 156, 157, 211, 212, 213, 218, 229, 244, 314
chaotic patterns 69
chaotic phenomena 212
chaotic properties 1, 2, 20
Chaotic Secure Receiver (CSR) 69
Chaotic Secure Transmitter (CST) 69
chaotic semiconductor 154, 179
chaotic sequence 415, 416, 429, 430, 431, 432, 433, 434, 438, 440
chaotic signal 20, 40, 128, 152, 153, 172, 173, 181, 210, 212, 219, 337, 338, 339, 342, 343, 344, 345, 348, 351, 355, 386, 390, 391, 413, 419, 439, 440, 443, 458, 459, 460, 463
chaotic slave systems 69, 87
chaotic solutions 3, 5
chaotic switching 422
chaotic symmetric gyro 185
chaotic synchronization 128, 149, 415, 424, 425, 436, 437, 441, 464

chaotic system 19, 20, 21, 32, 42, 43, 44, 45, 46, 54, 61, 62, 63, 68, 69, 77, 79, 87, 105, 106, 109, 122, 123, 124, 125, 126, 127, 128, 130, 133, 134, 142, 144, 149, 150, 152, 153, 154, 155, 156, 157, 158, 159, 160, 161, 165, 166, 167, 169, 172, 173, 174, 177, 178, 179, 180, 181, 182, 184, 188, 190, 194, 196, 207, 208, 211, 212, 213, 214, 229, 230, 232, 243, 245, 246, 248, 253, 254, 255, 259, 314, 316, 322, 324, 332, 334, 337, 360, 362, 377, 382, 383, 385, 415, 416, 420, 421, 422, 424, 425, 428, 429, 430, 431, 435, 437, 438, 442, 448, 459, 460, 461, 462, 463, 474
chaotic system synchronization 441
chaotic transformations 20
chaotic transmissions 440
chaotic waveform 387, 440
chaotic wavelength 389, 398
chaotic windows 92, 93, 99, 100
Chen chaotic oscillators 218
Chen model 227
Chen system 127, 142, 143, 144
Chua circuit 42, 69, 70, 87, 88, 91, 92, 93, 94, 95, 96, 102, 154, 171, 338, 344, 359
Chua's oscillators 229, 231
ciphertext 387, 388, 463, 464, 465, 466, 467, 469, 471, 473, 475
co-coupling synchronization 417
communication system in RF band 337, 338, 342, 358
communication system security 439
compact metric space 21, 23
complete synchronization 289, 290, 301, 311, 442
complex analog signals 340, 342
complex analog signal transmission system 337, 338
conditional Lyapunov exponents (CLEs) 127, 130, 132, 133, 134
continuous-time 1, 2
continuous variable feedback synchronization 417
control laws 210, 212, 214, 233
controlled slave system 263, 267
control theory 361, 363, 374, 380, 381, 384
counter measure 465, 467, 470
coupled oscillators 251, 255
coupling matrix 154
cryptanalysis 463, 464
cryptanalytic attacks 464, 465
cryptanalyzed a variation 465
cryptographic key 387
cryptographic scheme 363, 380, 464, 475
cryptographic techniques 362, 363
cryptography 19, 38, 40, 41, 361, 362, 363, 364, 367, 369, 381, 382, 384, 385, 463, 475
cryptology 362, 383, 416, 432
cryptosystem 68, 361, 362, 363, 368, 369, 374, 377, 378, 379, 380, 381, 382, 383, 384, 385, 416, 436, 463, 464, 474, 475
crystallographic group 20, 22

D

damping 3, 4, 5, 15
Data Encryption Standard (DES) 362
dead-bead synchronization 417
decryption algorithms 69
defuzzification 320
delay differential equations 1, 2, 5, 6
delay equation 1, 17
delay of memorization 365
demodulating system 153
demodulation 153
deterministic 2
deterministic behavior 184
Devil's staircase 251
Differential Chaos Shift Keying (DCSK) 422, 426, 427, 437
differential equation 1, 2, 5, 6, 8, 16, 17, 18, 43, 44
differential neural networks (DNN) 44, 45, 47, 51, 53, 54, 55, 56, 57, 59, 60
digital chaotic cryptography 363
digital circuits 464
digital communication 154
digital encryption 363
digital filter 154
digital information transmission 439, 440
digital networks 464

diode resonator circuits 68, 70, 88, 89
diode saturation current 72
discrete and distributed delays 289, 309, 310, 312, 313
discrete-time drive-response networks 262
discrete-time systems 2
discretized Lyapunov-Krasovskii functional (DLKF) 261, 263, 281, 283
dissipative 3, 18
dissipative systems 3
distributed time-delays 261, 262, 263, 264, 283
DNN theory 44
double scroll attractor 95
double scroll chaos 95
drive-response 107, 108, 109, 110, 113, 120
Duffing-Chen systems 214
dynamical behavior 261
dynamical chaos 70
dynamical embedding 21
dynamical network 262
dynamical system 1, 2, 4, 8, 11, 16, 17, 18, 19, 20, 21, 23, 24, 25, 27, 28, 29, 32, 37, 39, 40, 41, 91, 92, 128, 137
dynamic behavior 184, 212, 221, 419
dynamic chaos 337, 338, 353, 359
dynamic linear model 44
dynamic neural networks 42, 63
dynamic systems 44

E

electronic chaotic circuits 442
encrypted signal 152, 153
encryption 42, 60
encryption keys 69
entropy attacks 464, 465
equidistributed sequence 19
ergodic cipher 464, 465, 466
ergodicity 19, 463, 464
external synchronization 441

F

feedback control 184, 210, 212, 214, 220, 230, 233, 234, 235
feedback controller 214
feedback synchronization 128
field-effect transistor (FET) 70
fractional order 127, 128, 137, 138, 139, 140, 141, 142, 143, 144, 146, 147, 150, 151
fractional order Chen system 127, 142, 143, 144
fractional order Lü system 127, 143, 144
frequency modulated (FM) 416, 422, 426, 427, 437
Frequency Modulation Differential Chaos Shift Keying (FM=DCSK) 422, 426, 427, 437
frequency synchronization (FS) 128
fundamental tile 22, 23, 24, 33
fuzzy control 197, 198
fuzzy controller 315, 328, 329, 330, 331, 332
fuzzy logic 43
fuzzy logic system (FLS) 315, 316, 318, 319, 320, 321, 332
fuzzy neural network (FNN) 314, 315, 319, 320, 321, 326, 327, 328, 329
fuzzy reasoning 315
fuzzy set theory 315
fuzzy sliding mode control (FSMC) 183, 185, 197, 198, 204, 205, 207, 209, 314, 316, 323, 324, 325, 334
fuzzy sliding mode controller 316, 324, 332
fuzzy systems 315

G

Gaussian membership function 318
generalized anticipated synchronization (GAS) 156
generalized complete synchronization (GCS) 156
generalized lag synchronization (GLS) 156
generalized synchronization (GS) 106, 109, 110, 111, 115, 120, 128, 152, 155, 156, 157, 248, 254, 255, 387, 413, 442, 462
global synchronization 266, 286
gyro 183, 184, 185, 186, 187, 188, 194, 200, 202, 203, 204, 205, 207, 208, 209
gyroscope 184, 185, 186, 207, 208
gyro system 183, 184, 186, 187, 188, 194, 200, 203, 205

H

Hamiltonian systems 3
H∞ control 184

homoclinic bifurcation 91, 92, 93, 102
homoclinic chaos 91, 92, 93, 95, 96, 97, 98, 99, 101
homoclinic orbit 91, 92, 101
homoclinic point 92, 93
homoclinic spiking 93
homogeneous space 19, 20, 22, 23, 34
homolcinic bifurcation 95
Hopfield NNs 261, 262
Hurwitz fixed matrix 46
hyperchaotic attractor 107
hyperchaotic dynamics 388

I

identical synchronization (IS) 106, 109, 110, 128, 210, 211, 248, 254, 255, 256
impulse synchronization 417
integer order 127, 137
intercellular communication 210, 212, 233
interval type-2 FNN 314, 320, 321
intracellular messenger 212, 233

J

Jacobian matrix 3, 6, 11

K

key recovery attacks 464, 465
keystream 364, 365, 377, 378, 379
Kuramoto model 248

L

lag synchronization (LS) 106, 125, 128, 150, 289, 290, 292, 293, 294, 299, 300, 308, 311, 313, 442
Laplace transform theory 127, 142
large amplitude oscillations 92, 95
Laypunov stability theorem 314, 332
learning algorithms 45, 62
light emitting diode (LED) 68, 70, 72, 73, 87, 89
limit cycle 4, 5, 7, 91, 92, 93, 94, 96, 97, 98, 99, 101
limit cycle oscillator 249
linear damping 184
linearized stability analysis 1, 16
linear mappings 20
linear matrix inequality (LMI) 261, 262, 263, 267, 276, 279, 280, 283, 290
linear model 44
linear/nonlinear damping 184
linear-plus-cubic damping 185, 207
logistic map 463, 464, 465, 466, 468, 470, 472
Lorenz circuit 154
Lorenz model 226
Lorenz oscillators 153
Lorenz's attractor 68
Lorenz system 5, 6, 154, 168, 417, 418
Lorenz type 68, 87
low amplitude oscillations 92
low dimensional chaos 388
Lucifer encryption algorithm 362
Lü system 127, 143, 144
Lyapunov dimension 1, 7, 9, 14, 15, 16
Lyapunov exponent 1, 2, 5, 6, 7, 8, 9, 11, 16, 17, 18, 19, 20, 32, 127, 128, 130, 132, 133, 134
Lyapunov function 45
Lyapunov functional method 262, 284
Lyapunov stability theory 127, 134
Lyapunov theory 44, 316

M

Marginal Synchronization 106, 118, 119, 120
Markov maps 20, 39, 41
master and slave neural networks (MSNNs) 261, 263, 265, 266, 278, 279, 280, 281, 282, 283
master-slave 107, 210, 211, 229
master-slave configuration 441
master-slave synchronization 211, 229, 417
master-slave systems 262
mechanical energy 3
membership function (MF) 315, 318, 326
metric entropy 1, 2, 8, 9, 14, 15, 16
mixed mode oscillation (MMO) 91, 92, 93, 95, 97, 98, 99, 100, 101
modern ciphers 416
modified autonomous Duffing-Van der Pol system (MADVP) 127, 131, 134, 140, 141
modulated signal 153
modulo functions 20

multicellular response 210, 212, 233
Multistable Synchronization 106, 109
multi-synchronization 158, 173, 174, 176, 177
mutual coupling 107, 108, 110

N

neural networks (NN) 42, 43, 44, 63, 212, 213, 244, 261, 262, 263, 264, 266, 279, 283, 284, 285, 286, 287, 289, 290, 291, 299, 300, 301, 305, 306, 307, 308, 309, 310, 311, 312, 313
neurobiological networks 211
neuron-coupling degree 212
neuron-science 212
NN training algorithm 43
noisy channel conditions 440
non-autonomous system 5
non-chaotic solutions 3
non-chaotic systems 248
non-identical oscillators 251
non linear analysis 68
non-linear chaos theories 440
nonlinear circuits 439
nonlinear damping 184, 186
nonlinear dynamica system 91, 92
nonlinear dynamics 105, 184, 315, 386, 388, 389, 390, 391, 392, 394, 397, 401, 404, 405, 406, 407, 408, 409, 410
nonlinear dynamic system 416
nonlinear gyros 185, 200, 202, 204, 207, 208, 209
nonlinear oscillators 212, 213, 214, 225
nonlinear signal 338
nonlinear systems 43, 44, 64, 68, 105, 152, 165

O

one-time pad 364
one-time pad attack 463, 464, 465, 466, 467, 469, 473
one-way coupling 127, 128, 133, 135, 136, 141
one-way fashion 251
on-line training 44
on-off intermittency 337, 346, 348, 359, 442, 461
optical communication systems 387
optical network infrastructure 387
Ordinary Differential Equations (ODE) 1, 2, 5, 8, 16, 45, 416
oscillator system 154, 178
output coupling 289, 291, 304, 305, 309, 311
output signals 416

P

parameter identification 289, 310, 311
parametric uncertainties 43
PD bifurcation 94
Pecora and Carroll 127, 128, 129, 130, 131, 133, 138, 139, 140, 262, 417
Pecora-Carroll synchronization 417
pendulum clocks 211, 261
period-doubling (PD) 92, 94, 96, 97
periodic oscillations 211, 212, 233
periodic oscillator 156, 211
phase locked loops (PLL) 156
phase locking 248, 249, 250, 251
phase modulated (PM) 416
phase synchronization (PS) 106, 111, 112, 113, 125, 128, 148, 157, 158, 166, 177, 178, 180, 181, 182, 248, 252, 254, 442
plain text attack 463, 464, 465, 466
Poincarè-Bendixson theorem 3, 5
private-key ciphers 362
propagation delay 290, 292, 308
Proportional-Integral-Derivative (PID) 214, 222, 243
prototypical elegant 2, 5
public-key algorithm 362
public-key ciphers 362, 363, 364
public key cryptosystems 440

Q

quadratic maps 465, 466
Quadrature Chaos-Shift Keying (QCSK) 422
quantum cryptography 440
quasi-periodic solutions 415

R

Radial Basis Function (RBF) networks 44
regular tiling of R2 22, 24, 31, 33
regular tiling of RN 19, 20, 21, 23, 28, 32, 34, 36